T0328726

The study of cataclysmic variables – interacting binary stars containing a white dwarf accreting from an orbiting companion – is undergoing an exciting renaissance, as it embraces observations at all wavelengths. Cataclysmic variables allow, in particular, the direct and detailed study of equilibrium and non-equilibrium accretion discs; in turn this also helps in our understanding of X-ray binaries, black holes and active galactic nuclei. This timely volume provides the first comprehensive survey of cataclysmic variable stars, integrating theory and observation into a single, synthesized text.

An introductory chapter gives the historical background of studies of cataclysmic variables. The author then goes on to give an up-to-date review of both the observations (at all wavelengths, and over all time scales) and the theories and models of the structures and accretion processes believed to be involved. A very detailed bibliography is also provided to guide the reader to pertinent primary literature. Altogether this volume offers graduate students a single-volume introductory text while providing researchers with a timely reference on cataclysmic variable stars.

Cataclysmic Variable Stars

Cambridge astrophysics series

Series editors: Andrew King, Douglas Lin, Stephen Maran, Jim Pringle and Martin Ward

Titles available in this series

1 Active Galactic Nuclei
edited by C. Hazard and S. Mitton
2 Globular Clusters
edited by D. A. Hanes and B. F. Madore
5 The Solar Granulation
by R. J. Bray, R. E. Loughhead and C. J. Durrant
7 Spectroscopy of Astrophysical Plasmas
by A. Dalgarno and D. Layzer
10 Quasar Astronomy
by C. L. Sarazin
11 X-ray Emission from Clusters of Galaxies
by S. J. Kenyon
13 High Speed Astronomical Photometry
by B. Warner
14 The Physics of Solar Flares
by E. Tandberg-Hanssen and A. G. Emslie
15 X-ray Detectors in Astronomy
by G. W. Fraser
16 Pulsar Astronomy
by A. Lyne and F. Graham-Smith
17 Molecular Collisions in the Interstellar Medium
by D. Flower
18 Plasma Loops in the Solar Corona
by R. J. Bray, L. E. Cram, C. J. Durrant and R. E. Loughhead
19 Beams and Jets in Astrophysics
edited by P. A. Hughes
20 The Observation and Analysis of Stellar Photospheres
by David F. Gray
21 Accretion Power in Astrophysics 2nd Edition
by J. Frank, A. R. King and D. J. Raine
22 Gamma-ray Astronomy 2nd Edition
by P. V. Ramana Murthy and A. W. Wolfendale
23 The Solar Transition Region
by J. T. Mariska
24 Solar and Stellar Activity Cycles
by Peter R. Wilson
25 3K: The Cosmic Microwave Background
by R. B. Partridge
26 X-ray Binaries
by Walter H. G. Lewin, Jan van Paradijs & Edward P. J. van den Heuvel
27 RR Lyrae Stars
by Horace A. Smith
28 Cataclysmic Variable Stars
by Brian Warner

CATACLYSMIC VARIABLE STARS

BRIAN WARNER

University of Cape Town

PUBLISHED BY THE PRESS SYNDICATE OF THE UNIVERSITY OF CAMBRIDGE
The Pitt Building, Trumpington Street, Cambridge, United Kingdom

CAMBRIDGE UNIVERSITY PRESS
The Edinburgh Building, Cambridge CB2 2RU, UK
40 West 20th Street, New York NY 10011–4211, USA
477 Williamstown Road, Port Melbourne, VIC 3207, Australia
Ruiz de Alarcón 13, 28014 Madrid, Spain
Dock House, The Waterfront, Cape Town 8001, South Africa

http://www.cambridge.org

First published 1995
First paperback edition 2003

A catalogue record for this book is available from the British Library

Library of Congress Cataloguing-in-Publication Data

Warner, Brian.
Cataclysmic variable stars/Brian Warner.
p. cm.–(Cambridge astrophysics series ; 28)
Includes bibliographical references and index.
ISBN 0 521 41231 5 hardback
1. Cataclysmic variable stars. I. Title. II. Series.
QB837.5.W37 1995
523.8'44–dc20 94–30698 CIP

ISBN 0 521 41231 5 hardback
ISBN 0 521 54209 X paperback

The stars that have most glory, have no rest.

Samuel Daniel. *History of the Civil War.*

To Harry Ward and Harold (Sid) Slatter, inspiring teachers in formative years.

Contents

Preface

The history of cataclysmic variable star research mirrors the objects themselves: periods of relative inactivity punctuated by heightened or even explosive advances. Until about 1970 each resurgence of interest was a result of a distinct technological advance. In the past two decades the technological improvements have been almost continuous and the interest in cataclysmic variables has burgeoned from the realization that they have so much to offer. Not only are they of interest per se, exhibiting a challenging range of exotic phenomena covering the electromagnetic spectrum from radio waves to TeV gamma rays, and time scales from fractions of a second to millions of years, they are important for their relevance to other exciting areas of astrophysics.

For example, it has become evident that accretion discs are one of the most commonly occurring structures – probably all stars form from disc-like configurations, with material left over to provide planetary systems. A large fraction of binary stars form accretion discs at some stage of their evolution. Accretion discs are important in X-ray binaries – matter accreting onto neutron stars or black holes. Entire galaxies are initially gaseous discs, and most may develop central discs intermittently that fuel their active nuclei.

But it is in cataclysmic variables (CVs) that accretion discs are observed to best advantage – quasi-stable discs, unstable discs and transformations between them. In dwarf novae during outburst, or in nova-like variables in their high state, the light is dominated by emission from discs – and being almost two-dimensional their observed properties are strongly affected by the viewing angle. All are close double stars, and those with eclipses present unrivalled opportunities for determining spatially resolved physical structures. The CVs provide test beds for theories of accretion discs that may then be extended to more energetic regimes.

The instabilities in discs result in release of gravitational potential energy, observed as dwarf novae. Ultimately, the accumulation of hydrogen-rich material on the surfaces of the white dwarf primaries results in a thermonuclear runaway, observed as a classical or recurrent nova. These furnish unique opportunities to test models of non-equilibrium nuclear reactions, of hydrodynamics of expanding shells, of common envelope binaries and of radiation-driven stellar winds.

An added dimension appears in the effects of magnetic fields – either in providing viscosity in the accretion discs of nominally non-magnetic systems, or in the modification or total prevention of discs in systems whose primaries have field strengths in the range 10^5–10^8G. Among these are many of the most readily observed celestial X-ray sources, and optically they show strongly variable linear and circular polarization.

Understanding of the evolution of CVs is still in its infancy. The physical processes involved include magnetic braking by stellar winds, angular momentum loss through emission of gravitational radiation, mass loss through nova eruptions, common envelope evolution, and complexities caused by the occurrence of large magnetic fields in the primaries of some systems.

CVs provide unusual opportunities for amateur astronomers to make valuable contributions. Their very unpredictability and their apparent brightness during outburst has ensured that they have had a high priority among amateurs, with the result that almost continuous light curves over many decades are available for a few dozen systems. Understanding of the underlying physical processes is only now becoming sufficient to enable these light curves to be analysed in meaningful ways.

There have been many review articles and conference proceedings on CVs, and there are books devoted to the theory and observation of accretion discs, but this is the first book to introduce and review the topic of CVs in its entirety, seen through the eyes of one author. It is intended that the book shall do duty at a number of levels: it may serve as a graduate text, giving an in-depth overview of one area of interacting binaries; it will give mature researchers in adjunct disciplines a means of making contact with current issues in this exciting field; it will give the community of amateur variable star observers some insight into the importance of their labours from the professional viewpoint; and it should serve the CV community itself by providing a moderately comprehensive overview into which they can fit their often more specialized knowledge. These specialists will no doubt be at least as interested in what I have not found room for as in how I have represented their own contributions. Not all of the latter have stood the test of time; the selection that I have made inevitably is a personal choice, including some judgement of what is unlikely to be of lasting value. If this provokes further investigation I shall be satisfied in the investment of time given to what has become in my mind 'the bloody book' (after 'These bloody mathematics that rule our lives': Albert Camus, *The Plague*).

A number of simple conventions have been adopted in this book: discs *outburst* and novae *erupt*; the *amplitude* of a periodic signal is half the range of the modulation (incorrectly called 'semi-amplitude' by some authors); the mass donor is called the *secondary* star, even in those rare instances where it is more massive than the mass receiver; compact forms are used for multiple repeated types (NL for nova-like) but, to avoid possible confusion with other symbols, no distinction is made between singular and plural for terms with latinate plurals (DN for both dwarf nova and dwarf novae).

References are extensive but not exhaustive. However, areas of topicality are more comprehensively referenced in order to give the researcher rapid access to the current literature (e.g., accretion disc simulations, magnetic CVs, superhump phenomena).

Over the past 25 years I have benefitted greatly from personal contacts with the expertise and enthusiasm of the international CV community. For discussions and advice specifically related to the writing of this book I am especially grateful to J. Cannizzo, F. Meyer, E. Meyer-Hofmeister, D. O'Donoghue and D. Wickramasinghe. J. Faulkner, R.P. Kraft and D.N. Lin provided financial support, hospitality and amusement during a three month sabbatical at Santa Cruz in 1990, as did F. Meyer, E.

Meyer-Hofmeister and H. Ritter for three months at Garching in 1992 and D. Wickramasinghe for one month at Canberra in 1993. These short sabbaticals were also aided by travel funds from an Ernest Oppenheimer Travelling Fellowship, from the University of Cape Town and from the Foundation for Research Development. The last named has been largely responsible for funding my research into CVs over the past decade.

J. Cannizzo kindly computed Figure 3.31 and M. Harrop-Allin kindly prepared Figure 9.11. Typing of the manuscript through its many evolutionary stages has been in the care of Penny Dobbie, whose patience in dealing with recalcitrant software, inadequate hardware and a demanding author deserve mention in the annals of sainthood. Penny was guided out of some of the software problems by generous help from Jacqui King.

BRIAN WARNER
Cape Town

1

Historical Development

> History shows you prospects by starlight.
>
> Rufus Choate. *New England History.*

This first chapter is designed to give the reader an historical perspective on the subject of cataclysmic variable (CV) stars. Ground-based photometric and spectroscopic observational developments up to 1975 are treated in detail. Since that date instrumental methods in the optical region have been to some extent fixed, and to continue the historical approach would be repetitive of much of what appears in later chapters. The introduction of observational techniques in other wavelength regions is, however, followed beyond 1975.

1.1 Pre-1900 Observations of Novae

If the ancient philosophers had been correct in their assertion that the distant stars are immutable, incorruptible and eternal, astronomy would be the dullest of disciplines. Fortunately, they were wrong on all counts. The stars possess variability on all time scales and amplitudes, sufficient to satisfy all interests, from the exotic to the commonplace, from the plodding to the impatient.

Among these, the most prominent celestial discordants are the *novae stella*: *new stars*, challenging the ancients in their own times, but, such was the power of Aristotelian philosophy, passing almost entirely unacknowledged in European and Middle Eastern societies until the post-Copernican era (Clark & Stephenson 1977). In China, however, records of celestial events (kept mostly for astrological purposes) have been maintained since c. 1500 BC, and there are supporting and supplementary records in Japan from the seventh century AD and in Korea from c. 1000 AD (Clark & Stephenson 1976, 1977). Among these are numerous accounts of temporary objects, from which may be sifted comets, meteors, novae and supernovae.

Modern catalogues of ancient novae culled from Oriental records are given by Ho Peng Yoke (1962, 1970), Pskovski (1972) and Stephenson (1976, 1986). Earlier catalogues are listed in Payne-Gaposchkin (1957) and in Duerbeck (1987).

The supernovae of 1572 and 1604, comprehensively observed by Tycho Brahe and Johannes Kepler respectively, opened Western eyes to the mutability of the stars. The result, however, was hardly a flood of discoveries. Noting that there have been seven novae this century that reached magnitude 2.0 or brighter, and on the supposition that we do not live in particularly interesting times, it is a surprise, possibly even a scandal,

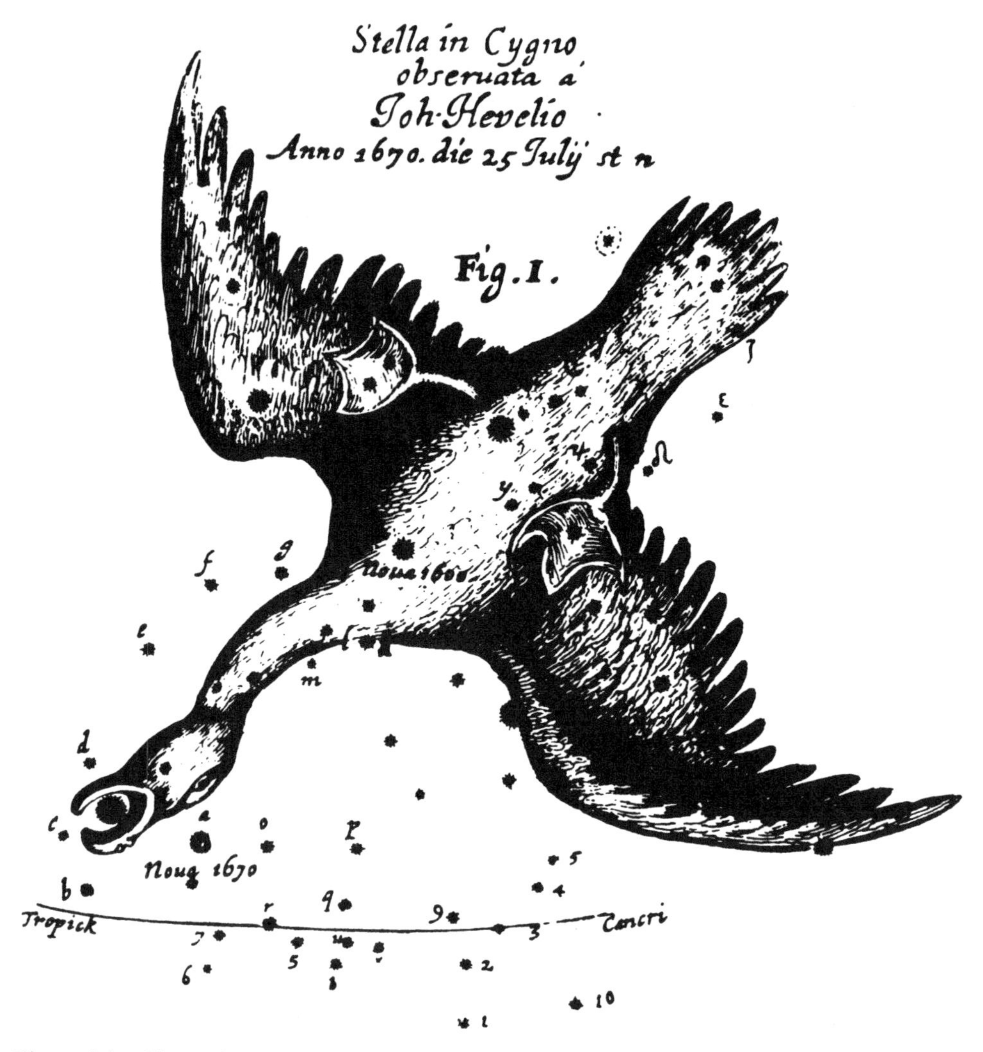

Figure 1.1 Chart of the constellation of Cygnus, published by Hevelius in 1670, showing the positions of Nova Vulpecula 1670 and 'Nova Cygni 1600' (now P Cygni). From Shara, Moffat & Webbink (1985).

to find that only one more nova was announced in the seventeenth century (the first true nova, as opposed to supernova, to be studied in Europe, discovered by the Carthusian monk Père Dom Anthelme in 1670 at second magnitude in the constellation Vulpecula: Figures 1.1 and 1.2), none in the whole of the eighteenth century and only one (Nova Ophuichi 1848) in the first half of the nineteenth century. (Some were *observed* during this time, and their positions measured, but were not *recognized* as novae at the time.) Before 1887, when novae began to be discovered on wide-field photographs of the sky, all but one of the half dozen novae found up to that time in the nineteenth century were the result of the steadily increasing activities of amateur astronomers.

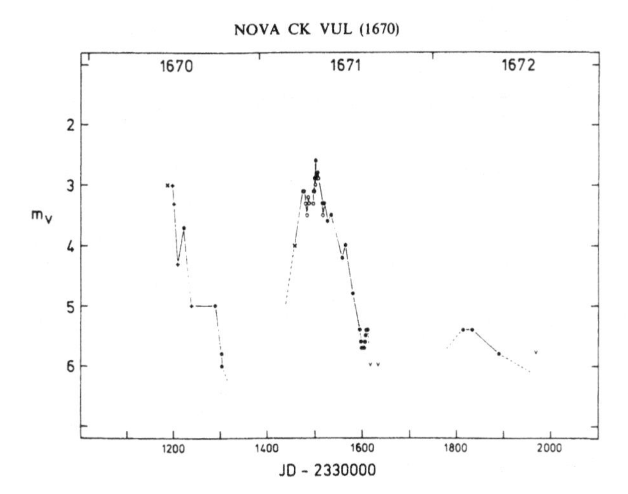

Figure 1.2 Light curve of Nova Vulpecula 1670 (CK Vul). From Shara, Moffat & Webbink (1985).

When John Russell Hind (1848) discovered Nova Oph 1848 it was the first such object to be publicly observable since 1670. Near its maximum brightness, Nova Oph was described as 'bright red' or 'scarlet' (Petersen 1848). This was during a long hiatus in the visual inspection of stellar spectra (Hearnshaw 1986), otherwise it would have been noted that the colour was due to intensely bright Hα emission. After eruption Nova Oph settled down to an easily-observable remnant of thirteenth magnitude: a unique object, at that time, as Nova Vul 1670 had faded to a level where it was unobservable.

The current status of pre-1900 novae can be found from the lists of Duerbeck (1987), Downes & Szkody (1989), and Downes & Shara (1993).

1.2 Discovery and Early Observations of Dwarf Novae

J.R. Hind, who was searching for minor planets near the ecliptic and only became a leading discoverer of variable stars *per adventure*, noted a previously unrecorded star of ninth magnitude 'shining with a very blue planetary light' on the night of 15 December 1855. After a patch of bad weather, it was observed still to be bright and in the same position nine days later, so Hind announced it as a new kind of variable star 'of a very interesting description, inasmuch as the minimum brightness appears to extend over a great part of the whole period, contrary to what happens with [the eclipsing binary] Algol and [the Mira variable] S Cancri' (Hind 1856).

Hind's confident statement about minimum light was based on the fact that he had been systematically searching this part of the sky for several years (during which in 1848 he discovered S Gem and T Gem) without previously having detected the new variable, soon designated U Geminorum. His comment on the blueness of the light of U Gem is very significant: a list of 53 variable stars known in 1856 shows that all except Nova Oph 1848 were Algol and Mira variables and hence either neutral or very red in colour (Pogson 1856). Thus U Gem was the first very hot, blue star to be studied by variable star observers; this caused some excitement over an apparent disc around the star, which was the result of optical aberrations – for a description of these and other aspects of early observations of U Gem see Warner (1986c). At minimum light, U Gem was found to be variable and just within reach of amateur telescopes at $m_V \sim 13$–14.

Three months after Hind's discovery, U Gem was found back at maximum light, this time by Pogson (1857), which proved that it was not an ordinary nova. From that time on during the nineteenth century U Gem was a favourite object among the leading amateurs, not least because of its rapid rise (less than one day) to maximum at unexpected times spaced anywhere from 60 to 250 days or more apart. Some observers followed it for over 30 years and left a mass of observations to be published posthumously (Knott 1896; Baxendell 1902; Turner 1906, 1907). The whole body of observations was discussed in a thesis by van der Bilt (1908), was the subject of one of the first harmonic analyses of light curves (Whittaker 1911; Gibb 1914), and caused Parkhurst (1897) to despair that 'Predictions with regard to it can better be made after the fact'. The early discovery of U Gem resulted in an almost continuously observed light curve (except for interference from the Sun for this ecliptic object) covering nearly 150 years.

Not until 1896 was another member of the class recognized, this time photographically, by Miss Louisa D. Wells on plates taken at Harvard College Observatory (Wells 1896). Varying from visual magnitude 7.7 to 12.4, and being far removed from the ecliptic, SS Cygni is one of the best studied of variable stars and, thanks in particular to the efforts of the American Association of Variable Star Observers (AAVSO), an almost continuous light curve from 1896 is available (Figure 1.3).

Belatedly, it was realized that T Leonis, discovered visually to be bright in 1865 by Peters (1865), and thought to be a nova, is also a member of the same class as U Gem and SS Cyg, known collectively, from their smaller amplitudes of outburst, as dwarf novae (this term was first used by Payne-Gaposchkin & Gaposchkin (1938), replacing the earlier term 'subnovae' used by Gerasimovic (1936)). Dwarf nova(e) will here be abbreviated to DN. In fact, T Leo is a member of the SU Ursa Majoris subclass, one of the three major subclasses of DN (U Gem and SS Cyg are representatives of another), which are described in Section 2.1.

The type star of the third subclass, Z Cameleopardalis, was next to be discovered (van Biesbroek 1904) and has been under continuous scrutiny since 1904 (Figure 1.4). The Z Cam stars proved stimulating as much for what they don't do as for what they do: they become stuck in an outburst state for unpredictable lengths of time.

The recognition of this new class of 'novae' resulted in distinguishing the original kind, of large amplitude, as classical novae (CN).

This is a suitable point at which to pay a tribute to the contributions made by amateur variable star organizations over the past century. Although most variable stars since the inception of wide-field photography have been discovered by professionals (a notable exception being bright novae), the responsibility of following the variations of the few hundred brightest irregular variables (Miras and other less regular red variables, novae and DN) has fallen to amateurs. Astronomy is an unusual science in that it is one in which the serious amateur can make not only useful discoveries but also long term measurements that have important statistical and interpretative value. Most

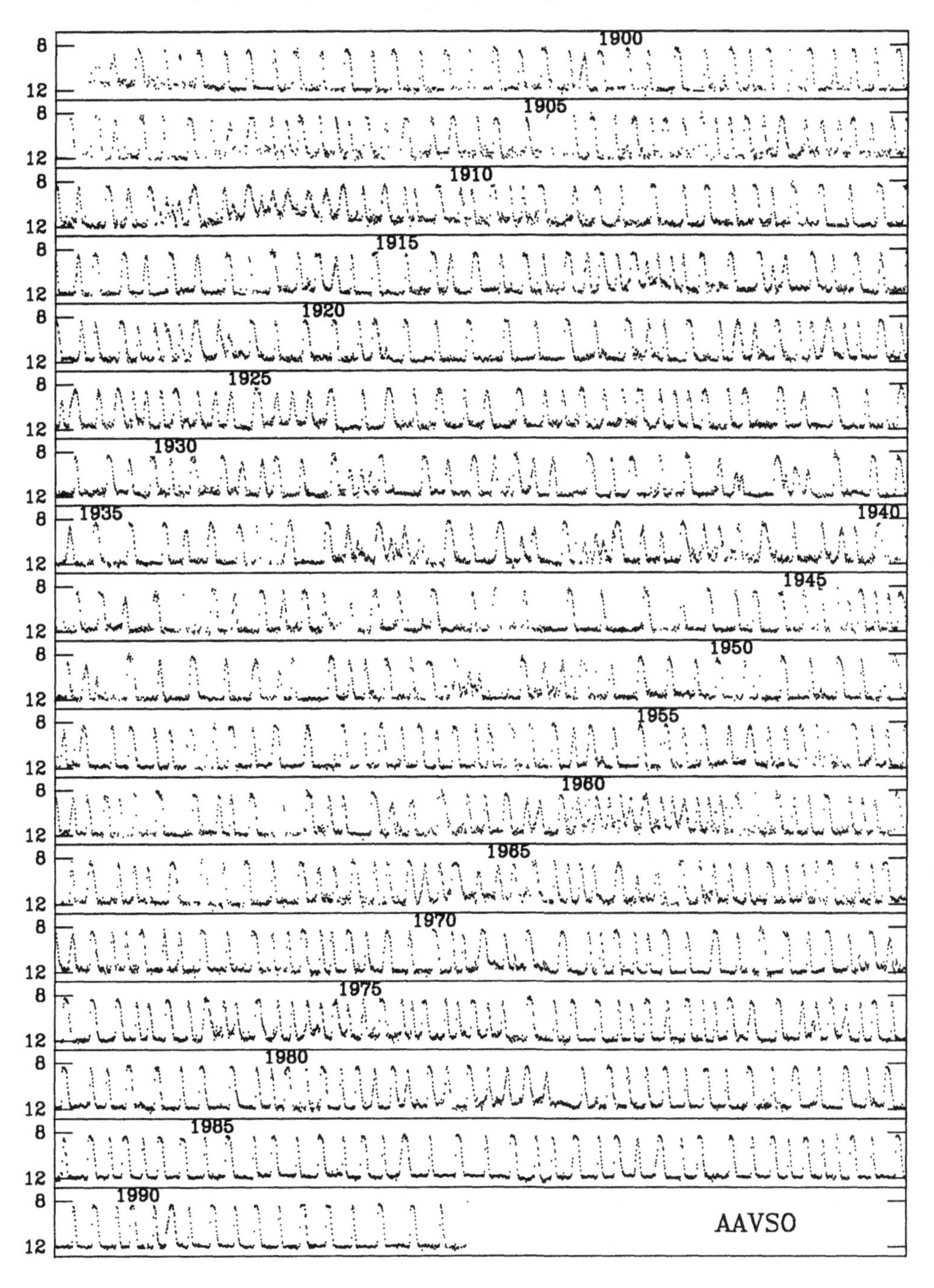

Figure 1.3 Light curve of SS Cyg, 1896–1992. The ordinate scale is visual magnitude. Constructed from observations made by the AAVSO. Courtesy J. Cannizzo.

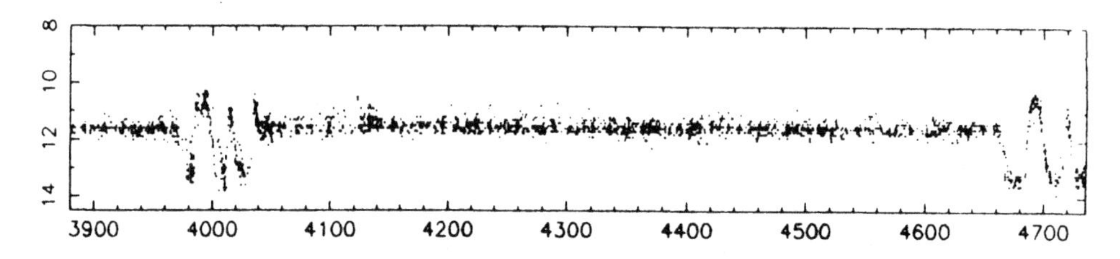

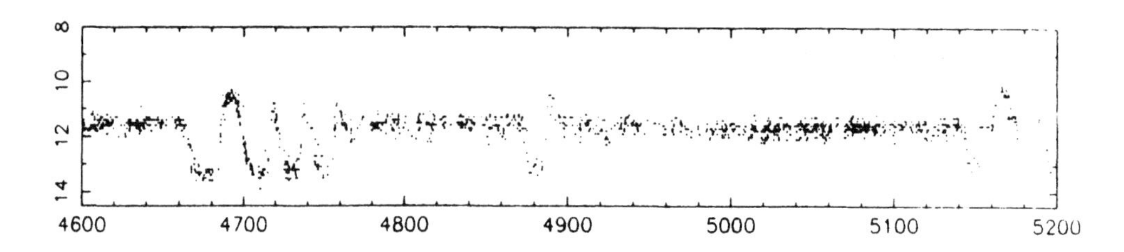

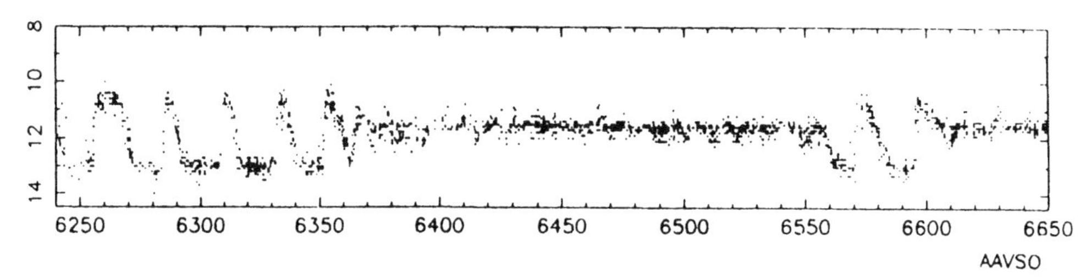

Figure 1.4 Light curves of Z Cam. The abscissa is Julian Date 2440000+; the ordinate is visual magnitude. From AAVSO observations. Courtesy J. Mattei.

of what is known about the long time scale behaviour of variable stars is derived from the concerted efforts of the amateur observers.

A crucial factor in the success of this effort has been international cooperation and central organisation, especially by the AAVSO (founded 1911) in the northern hemisphere and the Variable Star Section (founded 1927) of the Royal Astronomical Society of New Zealand in the southern hemisphere. Important long term contributions have also been made by members of the Variable Star Sections of the British Astronomical Association and the Astronomical Society of Southern Africa and also of the French Association of Observers of Variable Stars.

Thanks largely to the photographic surveys for variable stars carried out at the observatories at Harvard, Sonnenberg and Bamberg, by the time of the first edition of the General Catalogue of Variable Stars (GCVS: Kukarkin & Parenago 1948), of the 10912 variables listed, 108 were CN of various kinds, 92 were DN and 31 were nova-like (NL: see Chapter 4). By the fourth edition (1987), of 28 237 definite variables (there are thousands of suspected variables), 208 were CN, 342 were classified DN and about 35 NL (of a more restricted definition). These constitute the class of CVs that is the subject of this book.

A catalogue of 751 CVs, comprising 256 CN, 349 DN, 98 NLs (of a variety of types) and 48 unclassified, with accurate positions and finding charts, is given by Downes & Shara (1993). Charts for CN are found in Duerbeck (1987).

1.3 Photoelectric Photometry 1940–75

With the introduction of the 1P21 photomultiplier in the mid-1940s, photoelectric photometry gained sufficient sensitivity to bring many CVs within reach. The pioneering study was made by A.P. Linnell in February 1949 with the 61-in reflector of the Harvard Observatory's Oak Ridge station (Linnell 1949, 1950). The target, UX UMa, was known at that time as the shortest period eclipsing binary, with orbital period $P_{\rm orb} = 4$ h 43 min. It was not yet recognised in the GCVS as a NL.

Linnell's first observations (Figure 1.5) showed immediately the presence of intrinsic variation with amplitudes 0.01–0.2 mag on time scales from less than a minute up to several minutes, which were soon to be recognized as a characteristic of CVs (except for novae and some DN near maximum of outburst). Such *flickering*, as it was later to be termed, had been seen visually in 1946 in the recurrent nova T CrB (Petit 1946), in 1947 in the NL AE Aqr (then classified as a U Gem star) by K. Henize (1949) with the 26-in refractor at the Leander McCormick Observatory and by A.D. Thackeray and colleagues in 1949 in the NL VV Pup with the 74-in Radcliffe reflector (Thackeray, Wesselink & Oosterhoff 1950; see Section 6.3.3).

A second property discovered by Linnell, and commonly seen among other eclipsing CVs, was the variability of eclipse profile; in particular, the presence of a standstill of various lengths (or a shallower slope) on the emergent branch of the eclipse. The light curve also possessed a broad hump, lasting for about half of $P_{\rm orb}$, approximately centred on eclipse. Linnell proposed that 'a hot spot on the advancing hemisphere of the bright star' might account for the hump and eclipse profile.

UX UMa was observed further in 1952–53 by Johnson, Perkins & Hiltner (1954) and then both spectroscopically and photometrically by Walker & Herbig (1954). Their light curve (Figure 1.6) is characteristic of most CV light curves during the era of dc photometry and chart recorders over the period 1950–68. The model proposed for UX UMa by Walker & Herbig, consisting of 'a mass of hot material situated well above the surface of the primary...located asymmetrically with respect to the line joining the two stars', was an important step in the understanding of CVs.

At the same time, M.F. Walker carried out a photometric survey of CVs, stimulated by his discovery (Walker 1954a) of large amplitude flickering (up to 0.4 mag in 5 min) in the NL MacRae +43° 1 (now designated MV Lyr), in which rapid brightness variations were found in thirteen nova remnants, three NLs, four DN and a recurrent nova. During this survey, Walker (1954b) discovered that the remnant of Nova Herculis 1934, DQ Her, is an eclipsing binary with a light curve similar to that of UX UMa but with a slightly shorter orbital period of 4 h 39 min. Added to the discovery by Joy (1943, 1954a,b, 1956: Section 1.4.2) that AE Aqr and the DN SS Cyg and RU Peg are spectroscopic binaries, the evidence began to mount that all CVs are short period binaries and that flickering is in some way connected with their duplicity (Walker 1957).

An observation that was perplexing at the time but was to be of significance in the interpretation of DN was the series of light curves obtained by Grant (1955) during a 3-mag outburst of SS Cyg in which he found that, measured in intensity units, there is no increase in the amplitude of flickering during outburst: the outburst is an addition of non-flickering light to the flickering source at quiescence. Pinto & Rosino (1959) confirmed this and found that the flickering amplitude increases towards shorter wavelengths.

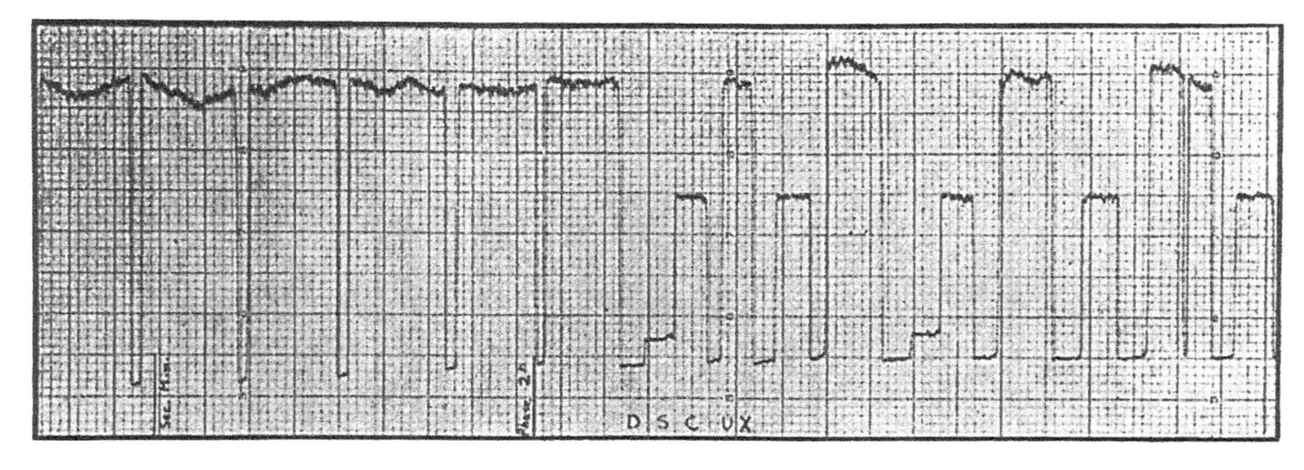

Figure 1.5 Chart recorder plot showing rapid brightness variations in UX UMa on 23–24 February 1949. Downward deflections are measurements of dark current, sky and comparison star. From Linnell (1949).

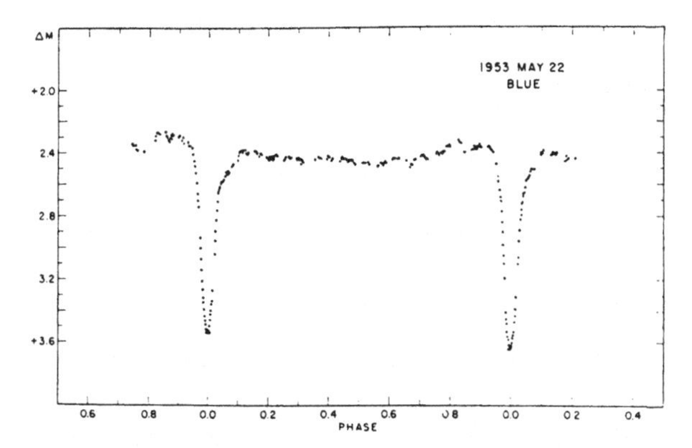

Figure 1.6 Light curve of UX UMa. B magnitudes measured every 0.5 min. The ordinate scale is the magnitude difference from a comparison star. From Walker & Herbig (1954).

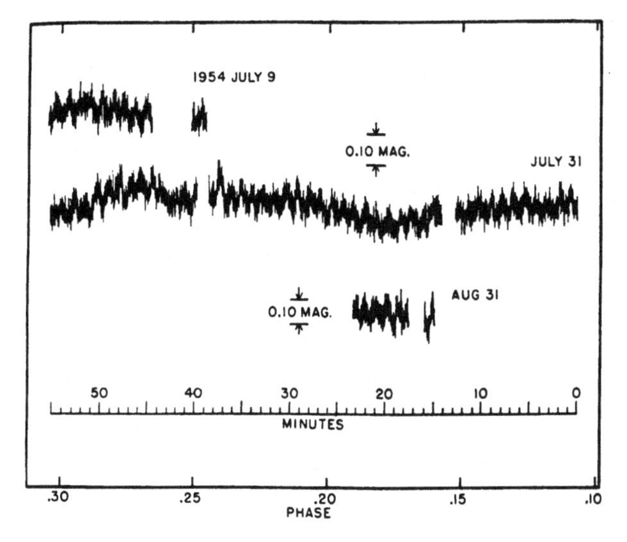

Figure 1.7 Discovery tracings of the 71 s brightness modulation in DQ Her (remnant of Nova Her 1934). The abscissa is orbital phase. From Walker (1956).

By far the most spectacular discovery to be made during Walker's survey was that DQ Her, as well as possessing flickering, shows a strictly periodic brightness modulation with a period of 71.1 s and an amplitude in the U photometric band of 0.070 mag (Walker 1956; Figure 1.7). Despite a search for similar periodicities in other CVs (Walker 1957; Mumford 1966, 1967a), it remained a unique phenomenon until technological improvements in photometry revealed further examples in the 1970s (Section 8.6).

To a large extent, the contribution of photometry during the decade following Walker's survey was simply the discovery of more eclipsing systems. By 1967, 23 CVs had known orbital periods, ranging from 82 min to 227 d, of which 19 were spectroscopic binaries and 13 were eclipsing systems (Mumford 1967a).

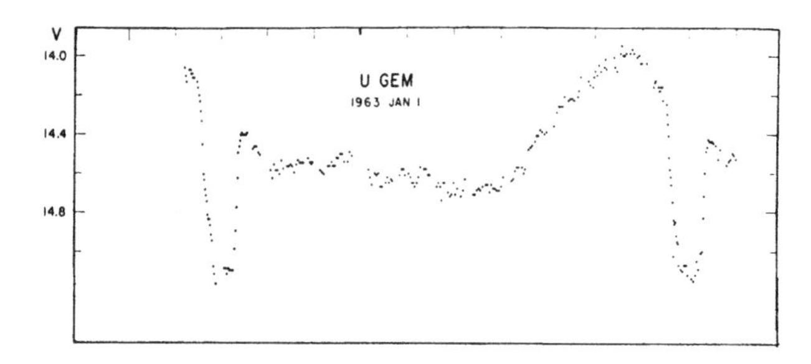

Figure 1.8 Light curve of U Gem at 0.5 min resolution. The orbital period (time between mid-eclipses) is 4 h 15 m. Adapted from Krzeminski (1965).

The most comprehensive study of a DN was that of U Gem by Krzeminski (1965). He obtained 4500 UBV measurements with a time resolution $\sim\frac{1}{2}$ min. At minimum light, U Gem showed eclipses superimposed on an orbital hump (Figure 1.8). His model for U Gem, similar to that for UX UMa, was that the hump arises from the varying aspects of a bright spot on the primary. Krzeminski's discovery that during outbursts the eclipses in U Gem become shallower and disappear at maximum, seemed to be interpretable in only one way: the seat of the outburst must be the orbital companion, not the hot primary as hitherto supposed. Further systematic effects – an increase in eclipse width during rise to maximum light and a concomitant shift in orbital phase of mid-eclipse – were interpreted as an asymmetrical expansion of the secondary star. Theoretical reasons were quickly found (Paczynski 1965a; Bath 1969) why the secondary of a close binary system could be unstable and be liable to occasional dramatic increases in luminosity.

Statistical studies of CVs by Luyten & Hughes (1965) and Kraft & Luyten (1965) showed that nova remnants have absolute visual magnitudes $M_v \approx 4$ and DN at quiescence have $M_v \approx 7.5$. They concluded that the hot primary components of CV binaries must be either white dwarfs or hot subdwarfs, confirming an earlier conclusion by Kukarkin & Parenago (1934) based on proper motions of SS Cyg and U Gem only.

The development of high speed pulse-counting photometry, reviewed in detail in Warner (1988b), revitalized the study of CVs and led rapidly to an improvement in the basic model for CVs (Warner & Nather 1971; and, independently, the same model from a different set of arguments by Smak 1971a). The realization that the hump in the light curve of U Gem lasts for at most 0.515 of P_{orb} showed that the longitudinal extent of the bright spot is no more than ~5°, which would require that its eclipse be very sudden (a few seconds) if located on the white dwarf primary. Furthermore, the high time resolution (2 s) photometry demonstrated that the amplitude of flickering is modulated in the same way as the light curve, being greatest at the maximum of the orbital hump, which implied that the flickering originates in the bright spot. The fact that flickering disappears during eclipse (Figure 1.9) showed that the hot spot is certainly eclipsed. Consequently, it was realized that *it is the accretion disc itself which brightens during a DN outburst* and that in U Gem eclipses are of a bright spot located on the outer edge of the accretion disc, where the stream of gas from the secondary impacts on the disc (Figure 1.10). The inclination in U Gem is such that most of the

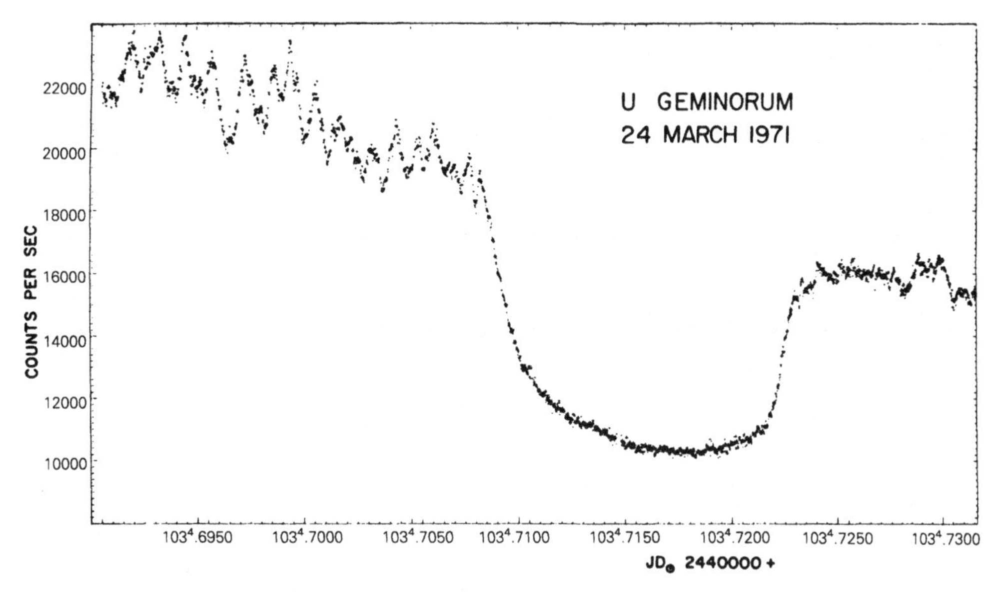

Figure 1.9 Light curve (intensity units) of U Gem at 2 s time resolution. From Warner (1976a).

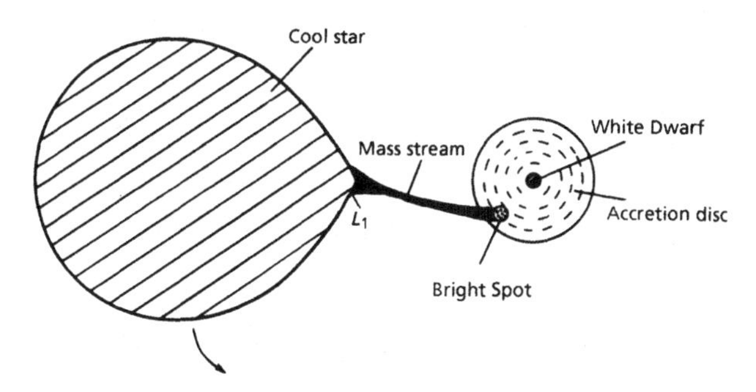

Figure 1.10 Schematic view of a cataclysmic binary system, viewed from the pole of the orbit.

disc remains uncovered at mid-eclipse. Thus, the shallow eclipses seen before or after maximum of outburst are of a nearly symmetrical disc, and mid-eclipse corresponds to inferior conjunction of the secondary; whereas at quiescence mid-eclipse occurs when the asymmetrically placed bright spot is centrally eclipsed.

The correctness of this model was demonstrated (Warner 1974a) by the continuation of deep eclipses, with the same ephemeris as at quiescence, during outbursts of the DN Z Cha, which has a high enough inclination to eclipse the entire disc. Furthermore, the shape of eclipse at maximum light shows that the principal source of light at that time is a disc viewed nearly edge-on. Similar results have been obtained for other high inclination DN (OY Car: Krzeminski & Vogt 1985; HT Cas: Patterson 1981). The model explains the asymmetry in eclipse profile observed in most CVs, first noticed in Linnell's (1949) photometry of UX UMa, as caused by the additional luminosity in the bright spot region.

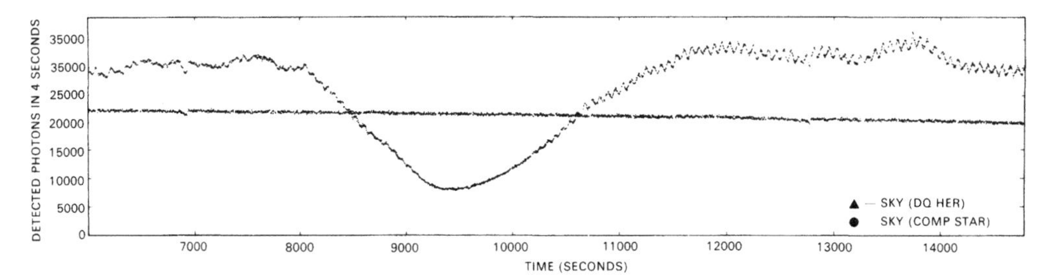

Figure 1.11 Eclipse light curve of DQ Her at 4 s resolution. The horizontal light curve through the centre of the diagram is that of a comparison star observed simultaneously with a second channel of the photometer. From Nather (1978).

A further consequence of the improved efficiency of photometry and its compatibility with computers was the successful search for rapid coherent periodicities in CVs. The light curve of DQ Her itself was better defined (Figure 1.11). A search for similar oscillations in DN during outburst, encouraged by a prediction that white dwarfs should oscillate at the start of a nova explosion (Rose & Smith 1972), was successful (Warner & Robinson 1972a: Section 8.6) even though the theory was later found to be incorrect. The oscillations were generally too low in amplitude to be seen directly in the light curve; they could only be found by power spectral analysis.

By 1975 the accretion disc, bright spot binary model for CVs was well established. This stimulated the realization (Warner 1976b), that, in NLs and in DN during outburst, stellar accretion discs are relatively bright and easy to observe; in fact CVs are the best objects in which to study the basic and exotic properties of accretion discs, whose results can have application to discs that commonly occur in protostars, X-ray binaries and possibly in active galactic nuclei.

Much of the contribution of optical astronomy since 1975 has been to refine details as improvements in detector sensitivity and instrument design have made higher signal-to-noise observations possible. There is one notable exception, however. In the two principal reviews of CVs written at that time (Robinson 1976a; Warner 1976a) there was scarcely any mention of magnetic fields. The discovery in 1976 of large and variable circular polarization in a CV (Figure 1.12) revealed a new dimension to the structure of CVs, since when the subject has burgeoned to the point where it is now necessary to devote nearly one third of this review to magnetic systems (Chapters 6, 7, 8 and part of 9).

1.4 Spectroscopic Observations

1.4.1 Early Spectroscopic Studies

The first spectrum of a CV to be studied was that of the recurrent nova (RN) T CrB when at second magnitude in May 1866. W. Huggins (1866) observed with a visual spectroscope that the spectrum contained hydrogen emission lines superimposed on a weak absorption-line spectrum in which the Na D lines were prominent. (The late-type absorption spectrum of T CrB was found later not to be a characteristic of novae in general – it is a result of the relatively high luminosity of its M giant secondary.) The

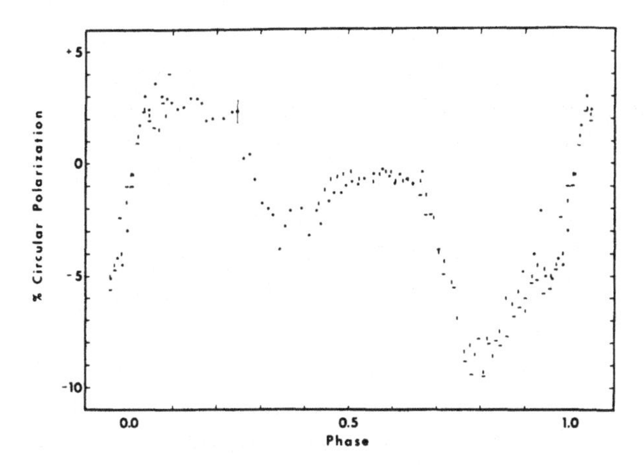

Figure 1.12 Circular polarization of AM Her observed on 17 September 1976. From Tapia (1977a).

bright line spectrum of T CrB resulted in novae being included in the same class as Wolf–Rayet stars and Mira variables in Vogel's (1874) spectral classification. Following Nova CrB 1866, the next nova studied spectroscopically was Nova Cygni 1876, in which Cornu (1876) measured eight emission lines, including 5007 Å [OIII] (then unidentified) which appeared as the nova faded. The complex development of the spectrum of Nova Aurigae 1891, photographed by Huggins & Huggins (1892a,b) and several other astronomers, and studied visually in great detail by Campbell (1892), showed that T CrB had not been a normal nova.

In the Harvard spectral classification scheme (Cannon & Pickering 1901), applied to the objective prism plates obtained in the Henry Draper Memorial programme which had commenced in 1885 and by 1911 had resulted in the spectroscopic discovery (from their emission lines) of ten novae (Fleming 1912), spectra of novae were given the designation Q. However, in her 1911 scheme, Miss Cannon used 'Pec' instead of Q for any peculiar spectra, including novae (Cannon 1912), and this has remained the practice ever since.

From Nova Aur 1891 on, the spectrographic coverage of nova outbursts became more complete (see Table 9.1 of Hearnshaw (1986)), culminating in the massive studies of Nova Pictoris 1925 (Spencer Jones 1931) and Nova Her 1934 (Stratton 1936; McLaughlin 1937; Stratton & Manning 1939). Several classification schemes were introduced in order to bring some order to the systematic changes of spectrum seen (Figure 1.13) in various kinds of novae (Lockyer 1914; Cannon 1916; Stratton 1920; Adams 1922), but these were later relegated in favour of the scheme devised by McLaughlin (1938, 1943) which was adopted by the International Astronomical Union (Stratton 1950). This is described in Section 5.3.

The first spectrum seen of a DN was that of U Gem (Copeland 1882; Knott 1882). During the Henry Draper objective prism survey, U Gem and SS Cyg were occasionally bright enough to have their spectra registered (Fleming 1912). The earliest is of U Gem at maximum in January 1891; the earliest of SS Cyg is in January 1897. Near maximum light the spectra were in absorption, described as resembling type F

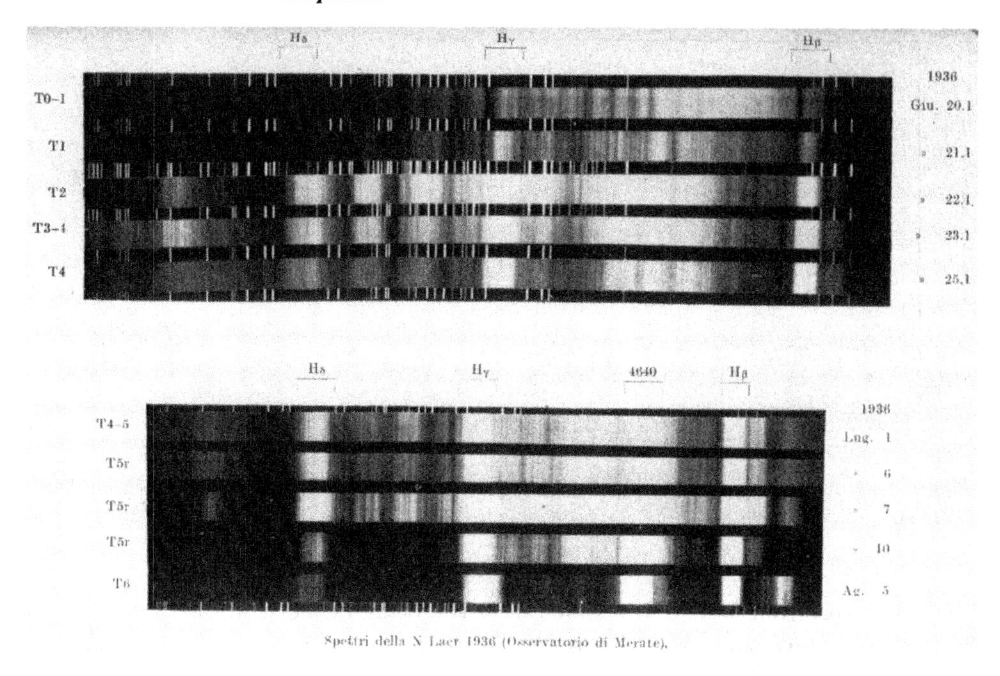

Figure 1.13 Development of the spectrum of Nova Lacertae 1936, from 20 June to 5 August. These are positive reproductions of the spectra: emission bands are bright, absorption lines are dark. From Cecchini & Gratton (1942).

and frequently called 'hazy' from the unusual width and shallowness of the lines. In some of the spectra of SS Cyg the Balmer lines were seen in emission.

The variations of spectrum from minimum to maximum light of DN were studied by Adams & Joy (1921), Wachmann (1935), Joy (1940), Elvey & Babcock (1940, 1943) and Hinderer (1949). At maximum the flux distribution matched that of spectral type A1–A2, and was either continuous or had broad absorption lines. During decline the absorption lines weakened and became filled by emission cores which eventually dominated at minimum. (As with T CrB, the secondary of SS Cyg happens to be an important contributor at minimum light, so its G-type flux distribution is not typical of the class.) Emission lines of H, HeI, HeII and CaII were recorded at minimum light, together with the Balmer continuum in emission. At maximum, broad absorptions of H and occasionally HeI, together with HeII emission, were seen.

Summaries of the early spectroscopic work on DN are given in Joy (1960) and Warner (1976a).

A survey of novae at minimum light was made by Humason (1938), who found weak emission lines on a very blue continuous spectrum. McLaughlin (1953) and Greenstein (1960) extended the survey with better spectral resolution and found broad Balmer absorptions in some objects. These characteristics were found among some already known variable stars (e.g., UX UMa, of which Struve had already obtained spectra in 1948 (Struve 1948; Linnell 1949)) and in the star MacRae +43° 1 (Greenstein 1954) which derived from a photographic survey for faint blue stars (MacRae 1952). From this *spectroscopic definition*, together with the presence of rapid light variability, arose the definition of NLs that is generally used in the context of CVs.

Summaries of the spectroscopic characteristics of old novae and NLs are given in Greenstein (1960), Warner (1976a) and Payne-Gaposchkin (1977). Comments on individual stars can be found in Payne-Gaposchkin (1957) and bibliographic information appears in Bode & Evans (1989) and Duerbeck (1987).

1.4.2 Spectroscopic observations 1940–75

The spectroscopic observations which, together with photometric results, began gradually to reveal the common underlying structure of CVs, started with Joy's (1940) observation that at minimum light the DN RU Peg has a dG3 absorption spectrum as well as the emission-line spectrum, suggesting duplicity. Later he found that SS Cyg also has a composite spectrum and was able to find an orbital period of 6 h 38 min from the radial velocity variations (Joy 1956). These results, coupled with the discovery that DQ Her is an eclipsing binary (Walker 1954b), 'led to the speculation that all cataclysmic variables might be binary systems' (Kraft 1962).

AE Aqr was also shown by Joy (1954a,b) to be a spectroscopic binary and was chosen by Crawford & Kraft (1956) for detailed study. Their realisation that the secondary (estimated to be of spectral type K5IV–V) evidently 'fills one lobe of the inner zero-velocity surface' was the first major forward step in modelling the dynamical structure of CVs. The *Roche lobe* (described in more detail in Section 2.3) is the largest volume that a star can take up in the neighbourhood of its companion. Any attempt at further expansion (or, equivalently, shrinkage of the binary orbit) results in loss of material from the point nearest the companion. Crawford & Kraft interpreted the broad emission lines in AE Aqr as arising from turbulent gas passing from the secondary to swirl around the white dwarf primary (Figure 1.14), perhaps forming a ring as already concluded by Kuiper (1941) for other close binaries. They suggested that the accretion of mass by the primary may in some way be connected with the outbursts of AE Aqr but that 'the mechanism responsible for the outbursts of AE Aqr

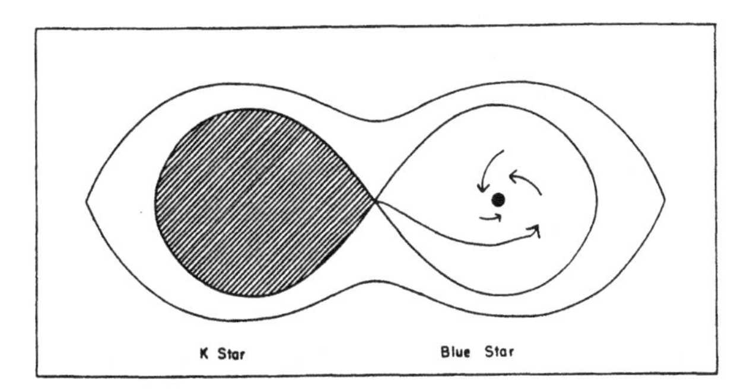

Figure 1.14 The model of AE Aqr proposed by Crawford & Kraft (1956). This is a pole-on view, and shows the inner and outer zero velocity surfaces (see Sections 2.3 and 2.4.2). The arrows represent gas streaming.

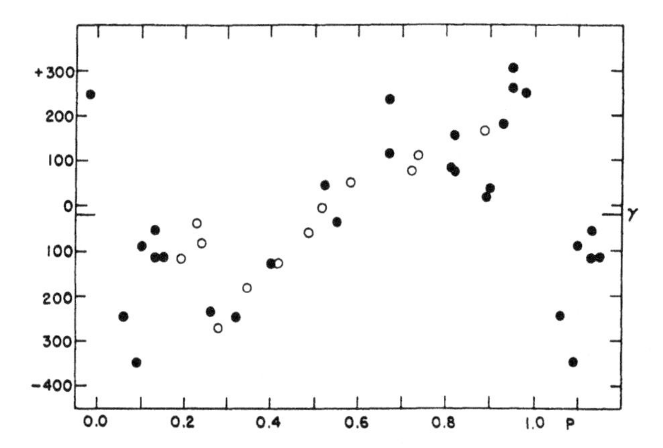

Figure 1.15 Radial velocities of HeII (λ4686) in spectra of DQ Herculis. The ordinate scale is km s^{-1}. From Greenstein & Kraft (1959).

and stars of the U Gem class need not be identical in kind with that responsible for the explosions of repeating novae'.

The same model was shown to be applicable to T CrB (Kraft 1958) and DQ Her (Kraft 1959). In the latter, Greenstein & Kraft (1959) found that the HeII emission-line radial velocity curve shows a classic rotational disturbance through eclipse (Figure 1.15), caused by progressive immersion and emersion of Doppler-shifted gas, confirming that an accretion disc with prograde rotation is present. The doubled emission lines in many CVs are also evidence of the existence of a disc, as shown by the calculations of line profiles by Smak (1969).

Emission lines are produced not only in the disc but also in the vicinity of the bright spot. As the binary rotates, the projection of the velocity of the gas passing through the bright spot region varies. On spectra in which time resolution is arranged by trailing the star slowly down the slit, this results in a characteristic S-wave from the bright spot component (Kraft 1961a; Krzeminsky & Kraft 1964). In low resolution spectra, the velocity variations of the disc (which follows those of the primary around the centre of gravity of the binary system) and the bright spot component are often not separable, with the result that the varying asymmetry of the emission-line profiles can seriously distort the orbital radial velocity curve (Smak 1970).

Kraft's extensive survey of spectra obtained with the prime focus spectrograph on the 200-in reflector (Figure 1.16), in which he found that orbital motion could be measured in almost all CVs (DN: Kraft 1962; NRs: Kraft 1964a), the few null results being attributable to low inclination, showed definitively that all CVs are close binaries transferring mass from a companion (usually of main sequence dimensions, but evolved in a few cases) to a white dwarf (Kraft 1963, 1964b).

1.5 Observations from Space: UV

The very blue continua (i.e., increasing flux towards shorter wavelengths) of CVs observed from the ground foretold that they would be rewarding objects at wavelengths shorter than 3000 Å. This was borne out by the detection of several CVs

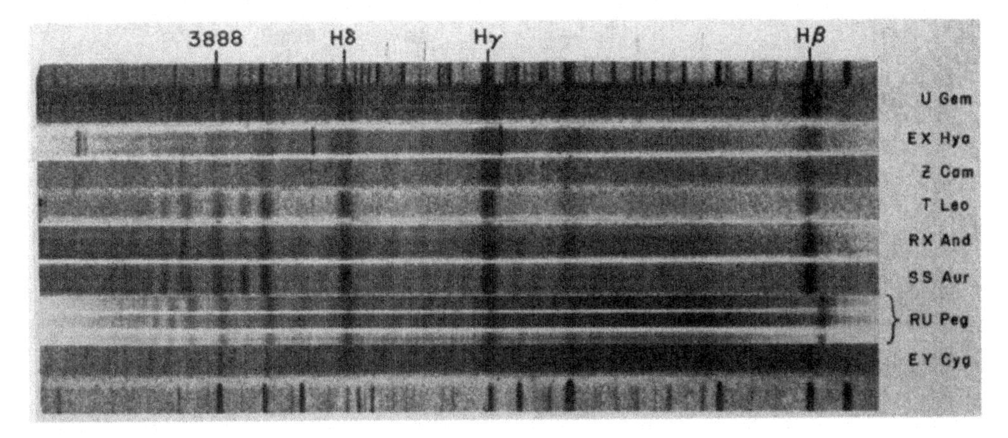

Figure 1.16 Photographic spectra (negatives) of DN at quiescence. Most of these were obtained with the prime-focus spectrograph on the 200-in telescope with exposures ~30 min. From Kraft (1962).

by the Wisconsin Experiment Package on the OAO-2 satellite. Nova Serpentis 1970 was fortuitously caught near maximum light and followed for 53 days (Gallagher & Code 1974), from which it was concluded that as the nova faded visually, the flux distribution shifted towards the ultraviolet (UV), thereby maintaining almost constant luminosity. The NRs V603 Aql and RR Pic were also observed (Gallagher & Holm 1974) and colour temperatures of 25 000 K and 30 000 K deduced respectively, together with minimum bolometric luminosities $\sim 10\ L_{\odot}$. SS Cyg was observed through an outburst and at quiescence (Holm & Gallagher 1974).

Some CVs were also observed by the Astronomical Netherlands Satellite (ANS) (e.g., Nova Cyg 1975: Wu & Kester 1977), but by far the largest body of observations has come from the International Ultraviolet Explorer (IUE) which was launched on 26 January 1978. IUE is equipped with two echelle spectrographs, each with two SEC vidicon detectors and primary and redundant cameras, designated LWP, LWR for the long wavelength (1900–3200 Å region, and SWP, SWR for the short (1150–1950 Å), with resolutions ~7 Å and ~0.2 Å in, respectively, low and high resolution modes (Boggess & Wilson 1987).

The IUE observations may be used in a number of ways. Figure 1.17 shows how individual continuum measurements are combined with broad band optical and infrared (IR) fluxes to obtain the flux distribution over the range 1150 Å–2 μm. For the star in question, the DN VW Hyi at minimum, the distribution has contributions from a 20 000 K white dwarf atmosphere and an accretion disc (Mateo & Szkody 1984). Low resolution spectra of the DN RX And in outburst and quiescence (Figure 1.18) show emission lines of ions of light elements, with prominent P Cyg shortward shifted absorption components present during outburst caused by a stellar wind (Verbunt *et al.* 1984). High resolution spectra of Nova Corona Austrinae 1981 (Figure 1.19) show a P Cyg profile with superimposed narrow absorption lines from interstellar ions (Sion *et al.* 1986).

An incidental use of IUE spectra is the determination of reddening in the 2200 Å

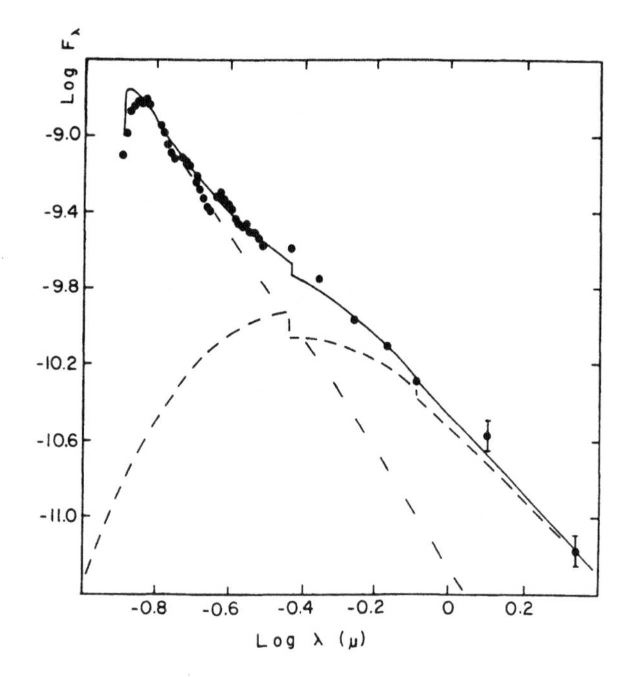

Figure 1.17 Flux distribution (F_λ erg cm^{-2} s^{-1}) in the dwarf nova VW Hyi at quiescence from 0.1 μm to 2.2 μm. The dashed lines are a 20 000 K, log $g = 9$ stellar atmosphere and a $\dot{M} = 10^{-11} M_\odot$ y^{-1} steady state accretion disc (Section 2.6.1). The solid line is the sum of these components. From Mateo & Szkody (1984).

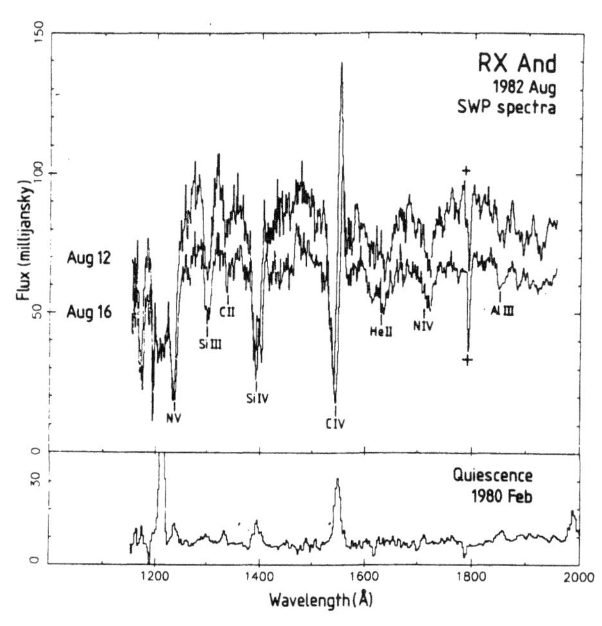

Figure 1.18 IUE spectra of the DN RX And near maximum of outburst (August 12) and on decline (August 16), compared with the quiescent spectrum. From Verbunt *et al.* (1984).

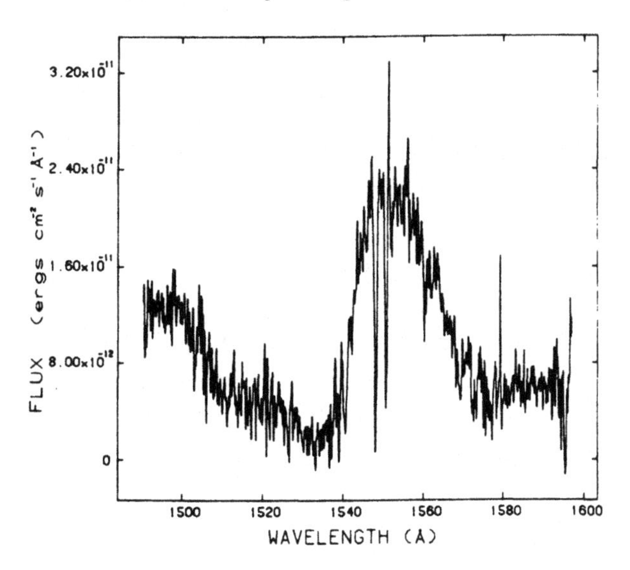

Figure 1.19 IUE spectrum of Nova Corona Austrinae 1981 in the region of the CIV lines at 1548 and 1551Å. From Sion *et al.* (1986).

interstellar absorption dip. A compilation is given by Verbunt (1987). A catalogue and display of IUE spectra of DN and NLs can be found in la Dous (1989c, 1990) and in la Dous (1991) which contains additional values of E_{B-V}.

Reviews of the contribution of IUE to the study of CVs are given by Starrfield & Snijders (1987) and Cordova & Howarth (1987). Models of the flux distributions of accretion discs (Section 2.6.1.1) predict that much of the energy is radiated shortward of 1200 Å. The far-ultraviolet (FUV: 912–1200 Å) and extreme ultraviolet (EUV: 200–912 Å) therefore acquire particular importance.

Voyagers 1 and 2 carry UV spectrometers with a resolution of 18 Å and a sensitivity over the range 500–1700 Å (Broadfoot *et al.* 1977). Because of absorption in the Lyman continuum of interstellar hydrogen, little flux is expected in the region 500 – 912 Å except from the nearest and hottest objects. Observations made with the Voyagers (Polidan & Carone 1987; Polidan & Holberg 1987; Polidan, Mauche & Wade 1990) failed to detect any EUV even during outbursts of VW Hyi, which is a nearby CV (at a distance ~65 pc). The FUV observed fluxes and the absence of any detectable EUV flux in VW Hyi are at variance with the flux distributions of model discs (Carone, Polidan & Wade 1986; Wade 1988; Polidan, Mauche & Wade 1990), which emphasizes the need for further observations at these wavelengths.

It is interesting to note (Polidan, Mauche & Wade 1990) that the amount of interstellar hydrogen absorption in the region 500–912 Å can be indirectly measured from the strengths of interstellar absorption lines observed by IUE (Mauche, Raymond & Cordova 1988), whereas shortward of 504 Å HeI is an important absorber, the estimation of which requires accurate knowledge of the state of ionization of the interstellar gas. In the region longward of 912 Å interstellar dust is the principal absorbing agent.

The detectors on the Voyagers have very small collecting areas (~0.3 cm^2). Great improvements in sensitivity have or will be given by ROSAT, EUVE, ASTRO and Lyman.

In particular, ROSAT, launched in June 1990, carries a Wide Field Camera with an angular field of 5° diameter and passbands 62–110 eV (110–200 Å) and 90–206 eV (60–140 Å). This has detected 17 CVs (Pounds *et al.* 1993). Because of interstellar absorption these sources are predominantly away from the Galactic centre and Galactic plane.

The Space Shuttle mission Astro-1 carried the Hopkins Ultraviolet Telescope (HUT: Davidsen *et al.* 1991), optimised for 900–1200 Å. Observation of Z Cam in outburst (Long *et al.* 1991) found a turn-down in flux below 1050 Å, as in VW Hyi.

The Hubble Space Telescope (HST) has three instruments with great potential for CV observations (Hall 1982; Wood 1992). The Faint Object Spectrograph can perform time-resolved spectrophotometry (down to 30 ms integration times) and spectropolarimetry over the range 1100–9000 Å and can reach $m_v \sim 23$ at moderate resolution in one hour. The High Resolution Spectrograph covers the range 1100–3200 Å at resolutions of 10^5, 2×10^4 or 2×10^3 and with time resolutions as short as 200 ms. The brightest CVs are within range of the HRS in its highest resolution mode. The High Speed Photometer (HSP) covered the range 1200–7000 Å through 22 filters. A time resolution of 10 ms was available. Linear polarization measurements could be made with a time resolution of 1 ms. The HSP was designed to give very high photometric accuracy, but this was not achieved prior to its removal from HST, to be replaced by the optics necessary to correct the primary mirror. Only a modest amount of high speed photometry was obtained before this sacrifice.

1.6 Observations from Space: X-ray Emission

The first known CV to appear in an all-sky survey of X-ray sources was EX Hya (Warner 1972a, 1973a) in the Uhuru catalogue. The Uhuru satellite's detector passband was 2.0–6.0 keV, and EX Hya, earlier thought to be an unusual DN, is now known to be a magnetic CV (see Chapter 7). Rocket flights carrying soft X-ray (< 1 keV) detectors produced a probable (later shown to be certain) detection of SS Cyg (Rappaport *et al.* 1974) and useful upper limits for the DN RX And and U Gem (Henry *et al.* 1975).

Since the advent of more sensitive X-ray satellites, the number of detections of CVs has grown steadily. The ANS, launched in 1974, observed SS Cyg to be a soft X-ray source both in outburst and at quiescence (Heise *et al.* 1978). This was confirmed by observations from SAS-3 (Bowyer *et al.* 1976; Hearn, Richardson & Li 1976). The launch of Ariel 5 in October 1974 and of HEAO-1 in August 1977 gave the first opportunities for all-sky searches for CVs, which produced detections of only a small fraction of the total known (Watson, Sherrington & Jameson 1978; Cordova *et al.* 1980b, Cordova, Jensen & Nugent 1981). However, later systematic searches through the HEAO-1 archived observations (Remillard *et al.* 1986a) are steadily producing more CV identifications (e.g., Tuohy *et al.* 1986).

The higher sensitivity of the Einstein satellite (HEAO 2) launched in November 1978, resulted in a larger number of CV detections (Cordova, Mason & Nelson 1981; Becker & Marshall 1981; Becker 1981; Cordova & Mason 1983, 1984a), amounting to about 70% of the CVs observed with the imaging detectors in the energy range 0.1–4.5

keV at a detection level $\sim 10^{29}$ (d/100 pc)^2 erg s^{-1}, and a few CVs, principally the magnetic ones and two DN in outburst, in soft X-rays (~0.1 keV).

The EXOSAT satellite operated from May 1983 until April 1986 and proved particularly useful in providing long uninterrupted X-ray light curves of CVs. These have been reviewed by van der Woerd (1987,1988) and Osborne (1986,1988). Other recent satellites that have contributed to the study of X-rays from CVs are Kvant 1 (USSR) and Ginga (Japan).

ROSAT has carried out an all-sky survey in the 0.1–2 keV band and is now engaged in detailed follow-up of selected sources. An analysis of the fainter X-ray sources, found serendipitously in Einstein pointed observations of previously known sources, showed that 23% were probably CVs which are concentrated towards the Galactic bulge (Hertz & Grindlay 1984, 1988). In consequence, it was expected that the deeper survey conducted by ROSAT, at five times the sensitivity of Einstein, would discover thousands of new CVs among its expected total of ~100 000 sources (Beatty 1990). This has proved not to be the case: of the 225 X-ray sources found in the first 200 square degrees sampled, only one, already known, CV was included (Motch *et al.* 1991).

Examples of light curves are given in Figures 1.20–1.23. Reviews of observations of X-rays from CVs can be found in Cordova & Mason (1983), Watson (1986) and Osborne (1988). A review of the technology of astronomical X-ray detection is given by Fraser (1989).

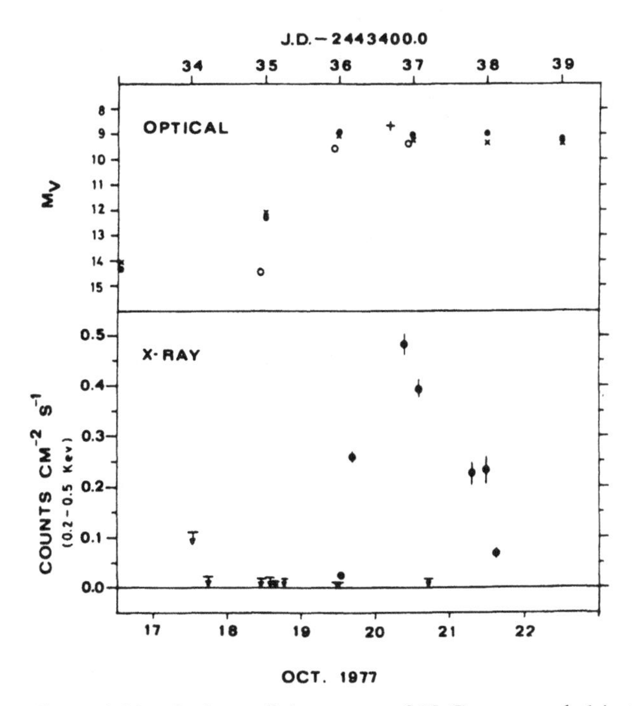

Figure 1.20 Outburst light curves of U Gem recorded in the optical and X-ray regions. The latter is a HEAO-1 observation with the low energy (0.15-2.5 keV) detector. The delay between rise in the optical and the soft X-rays is discussed in Section 3.3.6. From Mason *et al.* (1978).

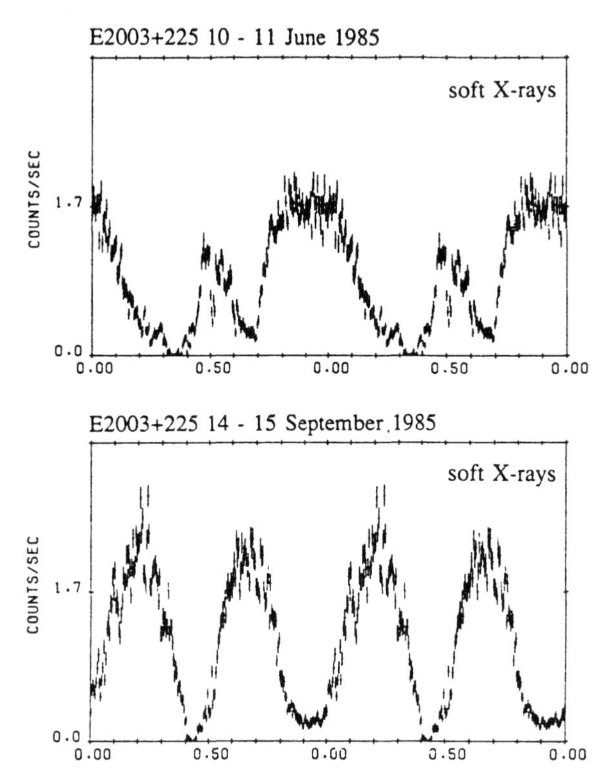

Figure 1.21 Soft X-ray fluxes of the polar QQ Vul (E2003+225) folded on its orbital period. These are discussed in Section 6.4. From Osborne (1988).

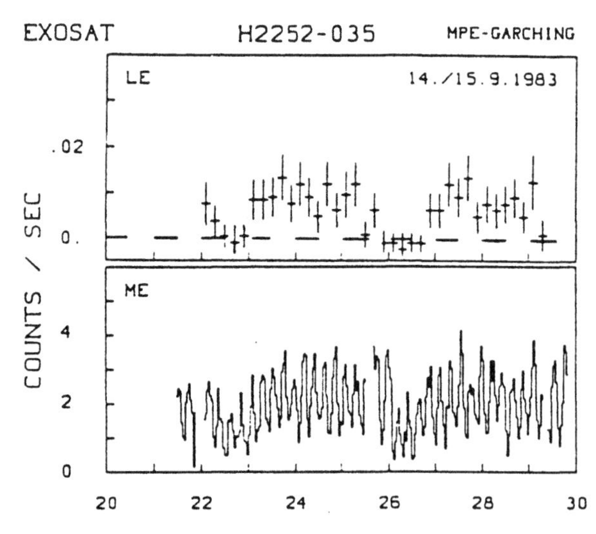

Figure 1.22 Soft (LE) and Hard (ME) X-ray fluxes in the intermediate polar AO Psc (H2252-035). The abscissa is marked in hours. The white dwarf rotation period (13.4 min) is seen in the ME (2–8 keV) light curve. Modulation at the orbital period (3.6 h) is seen in both curves. For discussion see Section 7.5.1. From Pietsch *et al.* (1987).

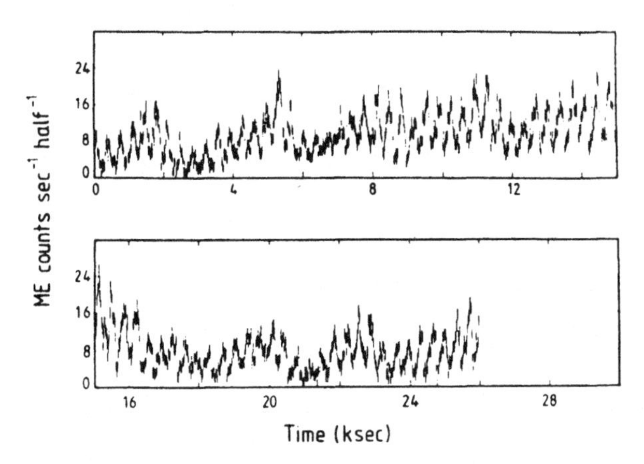

Figure 1.23 Hard X-ray (2.5–11 keV) light curve of the NR GK Per observed with EXOSAT. The 351 s modulation caused by rotation of the primary is clearly visible. This intermediate polar is discussed in Chapter 7. From Watson, King & Osborne (1985).

1.7 IR Observations

IR photometry of CVs has contributed in two distinct areas: (a) in following the development of the ejected shell from a nova eruption and (b) in unravelling the complexities caused by the multiple component structure of CVs.

The first extensive IR broad–band monitoring of a nova during eruption was that for Nova Serpentis 1970 (Geisel, Kleinman & Low 1970). As the nova started a rapid decline in the visible region, some 65 days after the eruption began, an IR excess developed which maintained the total (bolometric) luminosity approximately constant. From this it was deduced that the loss of visual flux was due to the formation of absorbing dust in the nova ejecta, an idea that was originally proposed by McLaughlin (1935) in connection with the deep minimum seen in the light curve of Nova Her 1934 and some other novae. Although many novae produce optically thick dust shells, most are optically thin or have no dust at all (Gehrz 1988).

The eruptions of over a dozen novae have been followed photometrically in the IR, generally to 28 μm from ground-based observations, but with measurements at 60 μm and 100 μm by IRAS for four novae (Harrison & Gehrz 1988). The IR emission from dust shells around several nova remnants was also detected by IRAS.

IR spectra of novae in the 2–20 μm region during their eruptive state show hydrogen emission lines from the higher series (Brackett, Pfund etc.) in early phases and coronal forbidden lines (Figure 1.24) at later stages. The 12.8 μm [NeII] line is very strong and an important source of cooling for the ejecta. The dust produces broad emission bands centred on 10 μm for SiO2 and 11 μm for SiC (Figure 1.25).

Reviews of IR observations of novae are given by Gallagher & Starrfield (1978) and Gehrz (1988).

Observations in the IR of quiescent CVs emphasize the cooler components of their structure. The two dominant contributors are the atmosphere of the secondary and the cool outer regions of the accretion disc. The latter may be optically thin or optically thick. In a binary of moderate to high inclination, the asymmetric Roche lobe causes

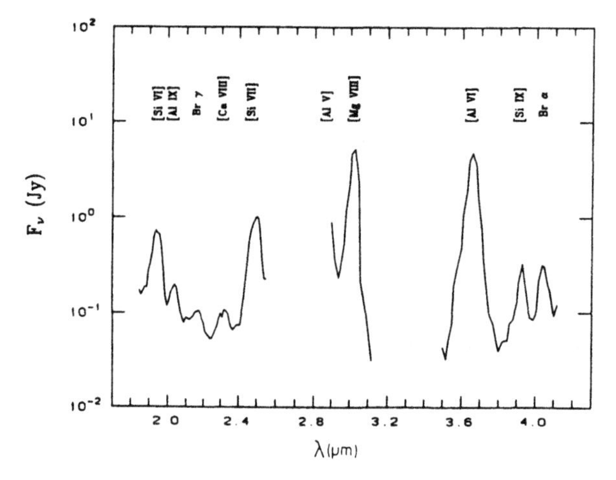

Figure 1.24 IR spectra of Nova Vulpecula 1984 No. 2 on 30 July 1986, showing coronal lines. From Greenhouse *et al.* (1988).

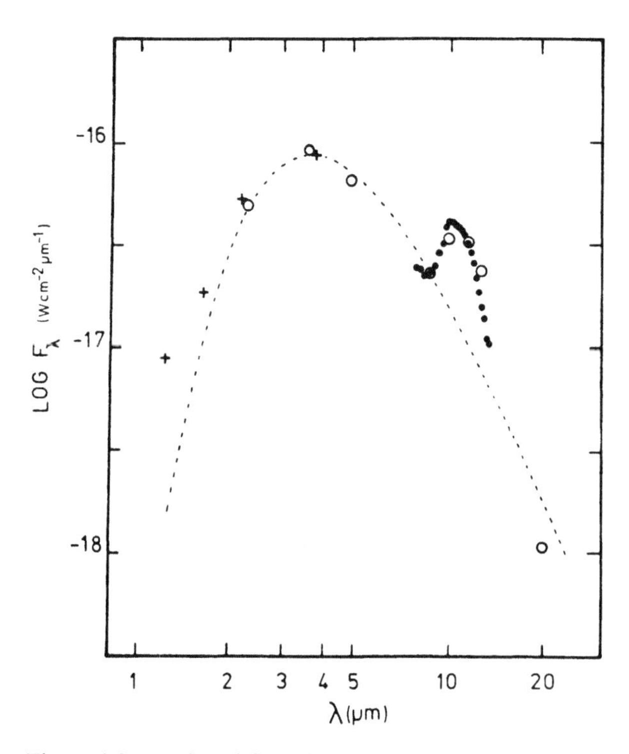

Figure 1.25 Infrared flux distribution of Nova Aquilae 1982. An 8–13 μm spectrum has been combined with photometric magnitudes at shorter wavelengths. The broken line is the flux from an 800 K blackbody. The excess flux near 10 μm is caused by silicate emission. From Bode *et al.* (1984).

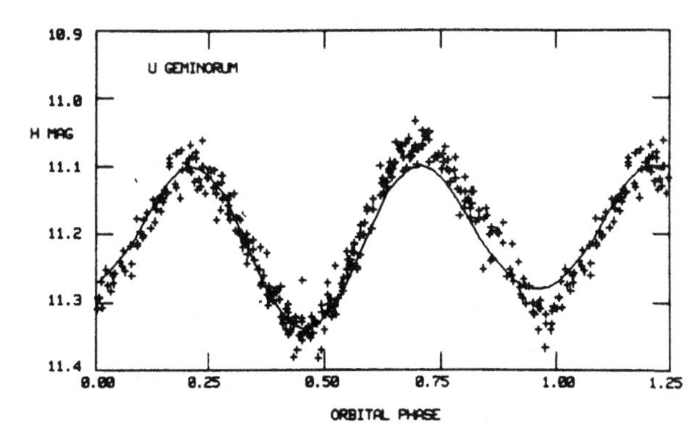

Figure 1.26 IR (H band) light curve of U Gem, folded at the orbital period. The solid line is a theoretical curve for a Roche lobe-filling secondary star. From Panek & Eaton (1982).

the flux from the secondary to be modulated twice per orbital period, with maximum brightness when the lobe is seen sideways on (Figure 1.26). Apart from directly confirming the highly distorted shape of the secondary, this modulation can aid in determining the inclination of the binary (Berriman *et al.* 1983).

The separation of contributions from the secondary and the disc is often difficult; the proportion of light supplied by optically thin and optically thick parts of the disc differs greatly from system to system (Berriman, Szkody & Capps 1985).

A large number of CVs other than novae have been detected by IRAS (Harrison & Gehrz 1992). A detection of SS Cyg in outburst is interpreted as $\sim$400 K dust with a mass $\sim 10^{-11}\mathrm{M}_\odot$ condensed from a wind (Jameson *et al.* 1987).

In strongly magnetic CVs, where no disc is present, the IR flux comes from an accretion column and may be highly polarized (Chapters 6 and 7).

1.8 Radio Observations

The first radio emissions detected from CVs were those from Nova Ser 1970 and Nova Delphini 1967 in observations by Hjellming & Wade (1970) with the NRAO three-element interferometer. The availability of the Very Large Array from 1981 increased sensitivity by up to 100 times and improved resolution, with the result that many novae and RN have been observed, including resolution of the expanding shell, and a few DN and magnetic CVs have also been detected (Sections 3.3.7, 4.6, 5.4, 6.8).

In novae, the shell is optically thick at first, but later becomes optically thin, which provides a valuable means of estimating the total mass of gas ejected (Hjellming 1990). Radio emission from DN and magnetic CVs probably originates in the magnetospheres of their secondaries.

A general review of radio emission from astrophysical sources is given by Dulk (1985) and applications to CVs are given by Chanmugam (1987). A general survey of CVs was made by Nelson & Spencer (1988).

1.9 Gamma-Ray Observations

Gamma rays from pulsars and X-ray binaries have been detected by means of satellites (0.1 MeV–30 GeV) and atmospheric Cerenkov showers (10 GeV–100 TeV).

Observational techniques in these energy ranges are reviewed by Hillier (1984), Ramana Murthy & Wolfendale (1986) and Weekes (1988).

The satellite observations use direct detection of γ rays by means of spark chambers. Flights by the Vela satellites, SAS-2 and COS-B in the 1970s (Bignami & Hermsen 1983) and more recently Gamma 1 and GRO have resulted in the discovery of many steady and transient celestial γ-ray sources. Although a possible positional correlation between γ-ray bursters and CVs was noted early on (Vidal & Wickramasinghe 1974; Vahia & Rao 1988), only recently have two magnetic CVs been tentatively identified in the MeV range (Bhat, Richardson & Wolfendale 1989: Section 6.9).

Ground-based γ-ray observatories use an array of mirrors to detect optical Cerenkov radiation from airshowers that are the result of passage of high energy γ rays through the Earth's atmosphere. The flux of optical photons from a 1 TeV γ ray is ~50 m^{-2} over a radius ~100 m. In order to discriminate against background signal, modulated γ-ray sources are desirable. Orbital period modulation for two magnetic CVs and the rotation period of another have possibly been detected (Section 6.9). The best studied signal is the 33 s rotation modulation in AE Aqr, which has been observed at TeV energies independently by two groups (Meintjes 1990; Bowden *et al.* 1991; Meintjes *et al.* 1992: Section 8.2.5).

Vahia, Rao & Singh (1991) find a positional correlation between γ-ray bursters and the highest energy cosmic rays; they propose that CVs may be the dominant source of cosmic rays above 10^{15} eV.

2

Structure of Non-Magnetic Systems

....duplicity with honour.

Saul Bellow. *Humboldt's Gift.*

Two stars keep not their motion in one sphere.

William Shakespeare. *King Henry IV, Part I.*

It is probable that magnetic fields play a rôle in all CVs – even in nominally non-magnetic systems by driving orbital evolution through magnetic braking (Section 9.1.2.1) or in providing the source of viscosity in accretion discs (Section 2.5.2.2). However, in systems where the primary has a relatively weak magnetic field ($\lesssim 10^5$ G) the movement of gas from the secondary to the primary is determined predominantly by dynamical and hydrodynamical flows.

This chapter provides an overview of how the physical properties of the various components in non-magnetic CVs can be treated theoretically and deduced observationally.

2.1 Classification of CVs

The present system of assigning individual CVs to specific types is a simplification of earlier, more detailed schemes that were developed in response to the steadily revealed diversity of CV behaviour. All early classifications were based on morphology of light curve. The distinction between novae and DN was maintained until the discovery of RN and the later realization that the recurrence times and outburst ranges of RN and DN overlap. Then appeal to their different spectroscopic characteristics became necessary.

Detailed properties of the subtypes that exist among the CN, DN, RN, NL and magnetic CV classifications are given in the respective chapters. For the purpose of this chapter, which examines the underlying binary and accretion disc structure common to all non-strongly-magnetic CVs, it is necessary only to expand a little on the types already introduced in the first chapter.

Classical novae have, by definition, only one *observed* eruption. The range from pre-nova brightness to maximum brightness is from 6 to greater than 19 magnitudes and is strongly correlated with the rate at which the nova fades after maximum. The largest amplitude eruptions, of shortest duration, are in very fast novae; the lowest amplitudes, in eruptions that may last for years, are in the slow novae. CN eruptions are satisfactorily modelled as *thermonuclear runaways* of the hydrogen-rich material that accretes on to the surface of the white dwarf primary.

Dwarf novae have outbursts of typically 2–5 mag, with some rare objects (e.g., WZ Sge) with up to 8 mag range. The interval between outbursts is from ~10 d to tens of years with a well-defined time scale for each object; the duration of normal outbursts is from 2–20 d, correlated with interval between outbursts.

There are three distinct subtypes of DN, based on the morphology of the outburst light curve:

Z Cam stars show protracted standstills about 0.7 mag below maximum brightness, during which outbursts cease for intervals of tens of days to years;

SU UMa stars have occasional superoutbursts in which the star achieves a brighter state (by ~ 0.7 mag) at maximum and remains in outburst for ~5 times the duration of an ordinary outburst;

U Gem stars include all the DN that are neither Z Cam nor SU UMa stars.

The dwarf nova outburst is reasonably well understood as a release of *gravitational* energy, caused by a temporary large increase in rate of mass transfer through the disc. A few CN also show DN outbursts.

Recurrent novae are, by definition, previously recognized CN that are found to repeat their eruptions. The distinction between RN and DN is made spectroscopically: in RN (as in CN) a substantial shell is ejected at high velocities; in DN no shell is lost (but there may be an enhanced stellar wind during outburst).

Nova-like variables include all the 'non-eruptive' CVs. This apparent oxymoron is removed by the belief that the NLs include pre-novae, post-novae and perhaps Z Cam stars effectively in permanent standstill, for all of which our observational baseline (typically ~ 1 century) is too short to reveal their cataclysms. In addition, the VY Scl stars are included, which show occasional *reductions* in brightness from an approximately constant maximal magnitude, caused by temporary lowering of the rate of mass transfer from the secondary.

Most NLs have emission-line spectra, but a subgroup show in addition broad absorption lines. The latter will here be termed UX UMa stars; this terminology has occasionally been used for all NLs.

Magnetic CVs are usually included among the NLs (except for the few that are already recognized as CN or other defined types). The magnetic fields of the primaries can disrupt the accretion disc, either partially or totally, and are accordingly treated separately (Chapters 6–8). The two populous subclasses are:

polars which have the strongest magnetic fields, and

intermediate Polars which have weaker fields; DQ Her stars are a subset of these.

2.2 Distributions of Orbital Periods

The orbital period of a CV is usually its most (and, all too frequently, its only) precisely known physical parameter. On its own, $P_{\rm orb}$ reveals something about the scale of the binary (equation (2.1b)); collectively the $P_{\rm orb}$ samples are from an equilibrium

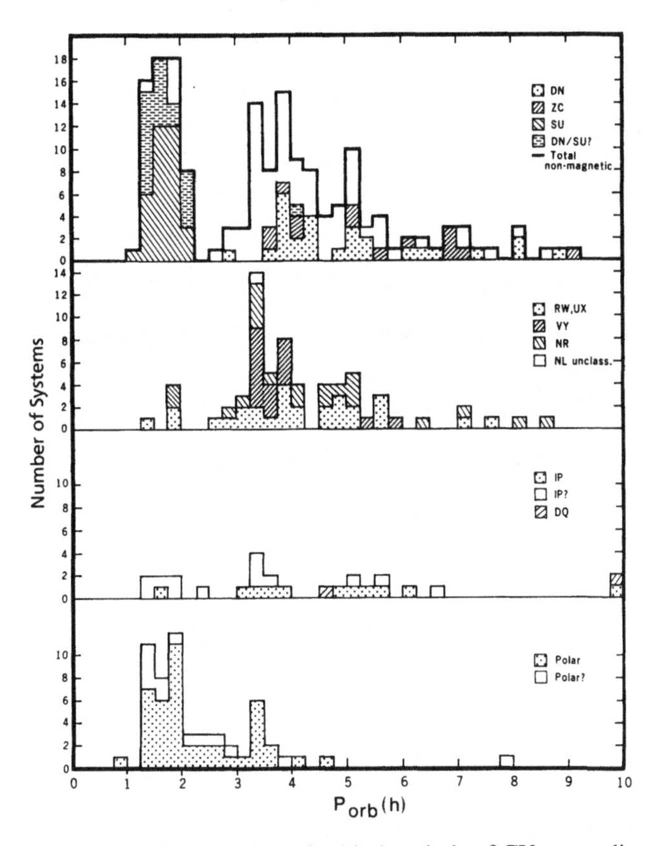

Figure 2.1 Distribution of orbital periods of CVs according to subclass.

state in which CVs appear with some initial frequency distribution of P_{orb} values, change their periods on an evolutionary time scale, and ultimately disappear from the CV classification to other states. These evolutionary matters are principally discussed in Chapter 9, but the steady state distribution of P_{orb} is of immediate interest in its correlations with CV types.

The known orbital periods of CVs are listed in the appropriate chapters. From these the P_{orb} distributions shown in Figure 2.1 are constructed. These are to a significant extent affected by different kinds of observational selection (Robinson 1983), only one of which distorts the distribution within a given subclass, viz: the relative ease of detecting orbital brightness modulations in systems with $P_{orb} \lesssim 6$ h as compared with longer periods. The other selection effects concern the relative representation of the classes. Thus CN and DN draw attention to themselves through their explosive behaviour, whereas NLs are found less easily via photometric or spectroscopic surveys. (For example, the brightest CV in the sky, the low inclination IX Vel with $m_V = 9.4$, is a NL discovered spectroscopically as recently as 1984 (Garrison *et al.* 1984).) The completeness within a class also varies between classes according to topicality – thus all magnetic CVs have known orbital periods, as do almost all SU UMa stars, but there are many relatively bright DN and NLs without known P_{orb}.

From Figure 2.1 the following conclusions are drawn:

(a) There is a highly significant relative deficiency of non-magnetic CVs with 2.2 $\leq P_{orb} \leq 2.8$ h. When this was first noted (Warner 1976a) there were *no* stars in the 2–3 h range – a situation that continued for a dozen years (with encroachment to narrow the range) and led to the terminology *orbital period gap*. In the past few years several stars (mostly novae) have been discovered with P_{orb} within the gap. In most, perhaps all, of these there is indirect evidence of magnetic fields, which moves them to (h) or (i) below. In any case, there is clearly a significantly lower space density of non-magnetic CVs with $P_{orb} \sim 2.5$ h, and the use of 'period gap' is appropriate.

(b) There is a clear minimum observed period at $P_{orb} \approx 75$ min. (There are a few CVs, discussed in Chapter 9, that have periods down to 17.5 min; these stars differ from the majority of the CVs in being hydrogen-deficient, so they are omitted from Figure 2.1.)

(c) Most non-magnetic novae have P_{orb} above the period gap. The only one below, RW UMi, has an uncertain P_{orb} (Table 4.2).

(d) All but one of the SU UMa stars lie below the period gap; the exception is TU Men which has $P_{orb} = 2.8$ h, i.e., just at the high end of the gap.

(e) All definite U Gem and Z Cam stars lie above the gap; in fact, all but AB Dra have $P_{orb} \gtrsim 3.8$ h.

(f) Almost all NLs lie above the period gap.

(g) All but two (of uncertain P_{orb}) VY Scl stars have $3.2 \leq P_{orb}(\mathrm{h}) \leq 4.0$.

(h) All polars have $P_{orb} \leq 4.6$ h, with little evidence for a period gap, although there is a minimum frequency near $P_{orb} \sim 3$ h. Polars concentrate towards shorter orbital periods.

(i) All but one of the definite intermediate polars have their P_{orb} above the gap.

The existence of the gap and the minimum P_{orb} result from the way in which the secondary responds to mass loss, as is explained in Section 9.3. The significance of some of the correlations between P_{orb} distribution and CV type is not fully understood.

A list of classifications and other properties for CVs *with known orbital periods* is updated regularly by Ritter (1984, 1987, 1990).

2.3 Roche Lobe Geometry

By definition, *close binaries* are binary systems in which some significant interaction other than simple inverse square law gravitational attraction between point masses takes place. The interaction may be radiative, as in the heating of the face of one component by a hot companion, or it may be tidal, distorting both components through the combination of gravitational and centrifugal effects. In CVs the secondary star is always greatly distorted through the gravitational influence of the white dwarf primary, but the small radius of the latter leaves it immune to tidal influence except in rare circumstances (Section 6.7).

The effects of tidal interaction on the secondary cause it to rotate synchronously with the orbital revolution and eliminate any initial eccentricity of orbit. The time scale for these conditions to be satisfied is very short compared to the lifetime of a CV (Section

9.1.3), so we may expect CVs (except perhaps the few with $P_{orb} \gg 1$ d) to have circular orbits and synchronously rotating secondaries. Then Newton's generalization of Kepler's third law gives

$$P_{orb}^2 = \frac{4\pi^2 a^3}{G[M(1) + M(2)]} \tag{2.1a}$$

or

$$a = 3.53 \times 10^{10} M_1^{1/3}(1)(1+q)^{1/3} P_{orb}^{2/3}(\mathrm{h}) \quad \mathrm{cm}, \tag{2.1b}$$

where a is the separation between the centres of mass of the binary components, $M(1)$ and $M(2)$ are the masses of the primary and secondary respectively, q is the *mass ratio*[1] $M(2)/M(1)$ and $M_1(1) = M(1)/\mathrm{M}_\odot$. From equation (2.1b), we see that with $M_1(1) \approx M_1(2) \approx 1$, orbital periods of 1–10 h imply component separations ~ 0.5–3 times the radius of the Sun.

In order to interpret the shapes of eclipses, interpret quantitatively the IR modulation of the secondary, compute the flow of gas from the secondary, etc., it is essential to know the shape of the highly distorted secondary. To find this exactly requires a model for the density distribution inside the secondary and much numerical computation (Pringle 1985). Fortunately, the secondary stars in most CVs are closely similar to single main sequence stars or to giants, in which the mass is quite centrally condensed, and therefore the Roche approximation (Kopal 1959) can be adopted, which assumes that the gravitational field of a star is that for no distortion. This is equivalent to putting all of the mass of the star at its centre.

Taking a set of Cartesian coordinates (x,y,z) *rotating with the binary*, with origin at the primary, where the x-axis lies along the line of centres, the z-axis is perpendicular to the orbital plane and the y-axis is in the direction of orbital motion of the primary, the total potential at any point, which is the sum of the gravitational potentials of the two stars and the effective potential of the fictitious centrifugal force, is (Kruszewski 1966; Pringle 1985; Frank, King & Raine 1985)

$$\Phi_R = -\frac{GM(1)}{(x^2+y^2+z^2)^{1/2}} - \frac{GM(2)}{[(x-a)^2+y^2+z^2]^{1/2}} - \frac{1}{2}\Omega_{orb}^2[(x-\mu a)^2+y^2], \tag{2.2}$$

where $\mu = M(2)/[M(1)+M(2)]$ and $\Omega_{orb} = 2\pi/P_{orb}$.

From equations (2.1a) and (2.2) we see that

$$\Phi_R = \frac{GM(1)}{a} F\left(\frac{x}{a}, \frac{y}{a}, \frac{z}{a}, q\right).$$

Therefore the *shapes* of the Roche equipotentials, Φ_R = const, are functions only of q and their scale is determined by a.

The Roche equipotential sections in the plane of the orbit ($z = 0$) for $q = 0.5$ are shown in Figure 2.2. As it is readily shown (Pringle 1985) that the surface of a

[1] Throughout this book q is the ratio of the mass of the mass-losing star to that of the mass-receiving degenerate primary. Some texts invert this definition.

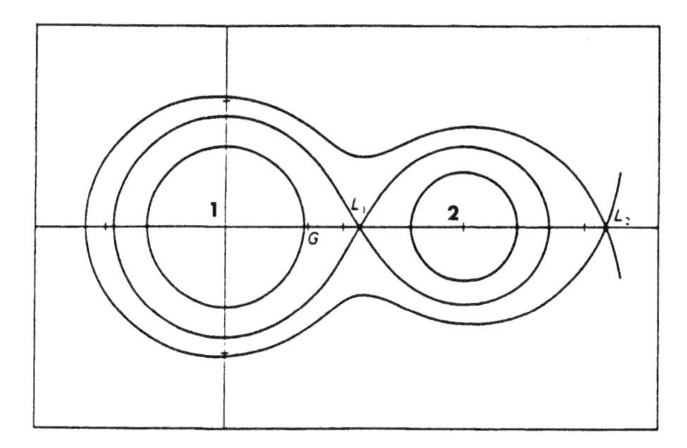

Figure 2.2 Roche equipotentials. The centres of the primary (1), secondary (2) and the centre of gravity (G) are shown for mass ratio $q = 0.5$. L_1 and L_2 are the inner and outer Lagrangian points respectively. Modified from Pringle (1985).

synchronously rotating star lies on Φ_R = const, the equipotentials define the shape of the secondary. Note that the shape is determined by two factors: rotation, which flattens the star along its rotation axis, and tidal force, which elongates the star in the direction of its companion.

If its radius is relatively small, the secondary will be almost spherical, as shown by the innermost equipotential around star 2 in Figure 2.2. If the secondary expands, its surface becomes more distorted until it fills the surface which passes through the point labelled L_1, called the *inner Lagrangian point*, which is a saddle point of Φ_R. That surface, known as the *Roche lobe* of the secondary, is the largest *closed* equipotential that can contain the mass of the secondary. Any further expansion will result in mass moving from the secondary into the Roche lobe of the primary. A binary in which both stars lie *within* their Roche lobes is called a *detached* system. A CV, in which the secondary fills its Roche lobe but the primary lies well within its, is *semi-detached.*

The point L_1 is common to the closed equipotentials of both stars and must lie at the point where the effective attractions of the stars exactly balance, i.e., $\partial\Phi_R/\partial x = 0$. A second Lagrangian point, L_2, satisfies the same condition and lies on the largest closed equipotential of a *contact* binary system (Figure 2.2). The point $(x_1, 0, 0)$ of L_1 is found from $x_1^{-1} - x_1 = q[(1 - x_1)^{-2} - (1 - x_1)]$ (Kopal 1959; Plavec & Kratochvil 1964). Solutions are approximated by the formulae given in Table 2.1. There is a third Lagrangian point, L_3, on the opposite side of the system from L_2.

In order to find an approximate spherical stellar model for a Roche lobe-filling star, appeal is often made to the *volume radius* $R_L(2)$ of the Roche lobe. Computations have been made by Kopal (1959), Plavec & Kratochvil (1964) and Eggleton (1983a). Approximate analytical formulae are given in Table 2.1.

The mean density of a lobe-filling secondary then becomes

$$\bar{\rho} = \frac{M(2)}{\frac{4}{3}\pi R_L^3(2)}, \tag{2.3a}$$

Table 2.1. *Approximate Formulae involving the Roche Lobe.*

1. Distance R_{L_1} from centre of primary to inner Lagrangian point

$$\frac{R_{L_1}}{a} = 1 - w + \frac{1}{3}w^2 + \frac{1}{9}w^3 \qquad (2.4a)$$

where

Formula	Range	Reference	Eq.
$w^3 = \frac{q}{3(1+q)}$	$q \leq 0.1$	Kopal (1959)	
$\frac{R_{L_1}}{a} = 0.500 - 0.227 \log q$	$0.1 \leq q \leq 10$	Plavec & Kratochvil (1964)	(2.4b)
$= (1.0015 + q^{0.4056})^{-1}$	$0.04 \leq q \leq 1$ error $< 1\%$	Silber (1992)	(2.4c)

2. Volume radius $R_L(2)$ of the Roche lobe of the secondary

Formula	Range	Reference	Eq.
$\frac{R_L(2)}{a} = 0.38 + 0.20 \log q$	$0.3 < q < 20$ accurate to 2%	Paczynski (1971)	(2.5a)
$\frac{R_L(2)}{a} = 0.462\left(\frac{q}{1+q}\right)^{1/3}$	$0 < q < 0.3$ accurate to 2%	Paczynski (1971)	(2.5b)
$\frac{R_L(2)}{a} = \frac{0.49q^{2/3}}{0.6q^{2/3} + \ln(1+q^{1/3})}$	$0 < q < \infty$ accurate to better than 1%	Eggleton (1983)	(2.5c)

3. Equatorial Roche lobe radius (y direction) of the secondary

Formula	Range	Reference	Eq.
$\frac{R_L(\text{eq})}{a} = 0.378q^{-0.2084}$	$0.1 < q < 1$ accurate to 1% over $0.2 \leq q \leq 1$	Plavec & Kratochvil (1964)	(2.5d)

4. Volume radius $R_L(1)$ of the Roche lobe of the primary:
Put $q \rightarrow q^{-1}$ in equations (2.5a), (2.5b) or (2.5c), or use

$$\frac{R_L(1)}{a} = 0.396q^{-1/6} \qquad 0.07 \leq q \leq 0.6 \qquad (2.6)$$

which agrees with (2.5c) to within 2%.

which, from equation (2.1) and equation (2.5c) of Table 2.1, gives

$$\bar{\rho}(2) = 107P_{\text{orb}}^{-2}(\text{h}) \quad \text{g cm}^{-3}. \qquad (2.3b)$$

Equation (2.3b) is accurate to 3%, independent of q, over the range $0.01 \leq q \leq 1$ (Faulkner, Flannery & Warner 1972; Eggleton 1983). For $1 < P_{\text{orb}} < 10$ h, equation (2.3b) shows that stars with densities of typical lower main sequence stars ($\rho \sim 1$–100 g cm^{-3}: Allen 1976) can fill their Roche lobes. Systems with $P_{\text{orb}} \gtrsim 10$ h must contain evolved secondaries.

Another useful quantity, required in comparing the radius of the accretion disc around the primary with the size of the Roche lobe of the primary, is the nearest point of the Roche lobe to the primary. This is approximately equal to the distance of the lobe along the y-axis (y_5 in Table 3-1 of Kopal (1959)). A formula for this is given in Table 2.1; it does not differ greatly from the volume radius $R_L(1)$.

2.4 Mass Transfer

It is not possible in the space available to give the derivations of all of the equations pertinent to the flow of gas from the secondary star via the stream to the impact zone on the accretion disc. A selection of the most important conclusions and the relevant physical principles will suffice for the later requirements of the text.

2.4.1 Roche Lobe Overflow

At L_1 gas can escape from the atmosphere of the secondary into the Roche lobe of the primary (Figure 2.3). The flow resembles the escape of gas through a nozzle into a vacuum; the flow velocity is approximately the thermal velocity of the atoms in the gas. Details of the stream lines in the vicinity of L_1 are given by Lubow & Shu (1975). The stream leaving L_1 has a core that is denser than the outer regions; the density profile should be approximately Gaussian.

The rate at which mass is lost from the secondary is

$$\dot{M}(2) = Q\rho_{L_1} c_s, \tag{2.7}$$

where Q is the effective cross-section of the stream, ρ_{L_1} is the density at L_1, averaged over the cross-section and c_s is the similarly averaged isothermal sound velocity. ($\dot{M}(2)$ is a negative quantity but its modulus will be used except where the sign has significance.)

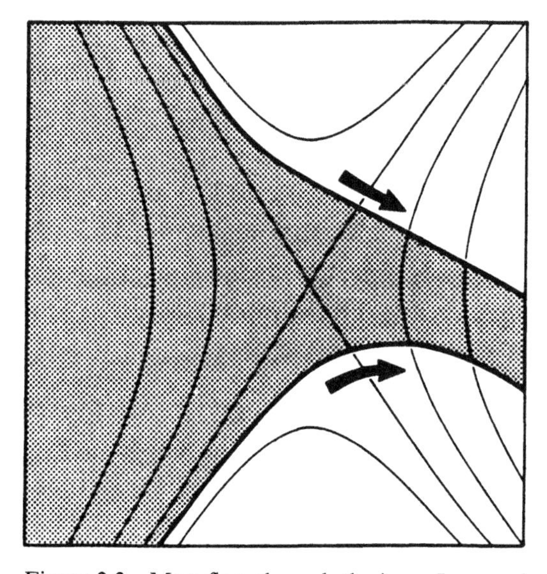

Figure 2.3 Mass flow through the inner Lagrangian point. From Pringle (1985).

An approximate estimate of the radius $W(=Q^{1/2}/\pi^{1/2})$ of the stream can be obtained by requiring that c_s exceed the escape velocity from the secondary. This gives $W \sim [H(2)R(2)]^{1/2}$, where $H(2)$ is the scale height of the unperturbed atmosphere (i.e., far from L_1) of the secondary (Pringle 1985):

$$H(2) = \frac{kT_s(2)R^2(2)}{\mu_m m_H GM(2)}, \tag{2.8}$$

where T_s is the surface temperature, μ_m is the mean molecular weight and m_H is the mass of the hydrogen atom.

A more detailed approach (Papaloizou & Bath 1975; Meyer & Meyer-Hofmeister 1983a; Hameury, King & Lasota 1986a; Ritter 1988; Kovetz, Prialnik & Shara 1988; Sarna 1990) gives

$$Q = \frac{2\pi c_s^2 a^3}{G[M(1) + M(2)]k_R}, \tag{2.9a}$$

where k_R is a factor (~7) obtained from the Roche geometry (equation (18) of Meyer & Meyer-Hofmeister (1983a)). From equation (2.1a) Q then becomes

$$Q \approx 2.4 \times 10^{17} \left(\frac{T_s}{10^4 \text{ K}}\right) P_{\text{orb}}^2(\text{h}) \quad \text{cm}^2, \tag{2.9b}$$

where use has been made of

$$c_s = \left[\frac{kT}{\mu_m m_H}\right]^{1/2} \tag{2.10a}$$

$$= 9.12 \times 10^3 T^{1/2} \quad \text{cm s}^{-1} \text{ for H atoms.} \tag{2.10b}$$

The density ρ_{L_1} can be evaluated by supposing that the atmosphere is isothermal down to a depth ΔR below the Roche surface of the secondary. Because of the rapid variation of effective gravity in the vicinity of L_1, the density does not vary exponentially with height; instead

$$\rho_{L_1} = \rho_0 \text{e}^{-(\Delta R/H')^2} \tag{2.11}$$

where ρ_0 is the density at the base of the isothermal atmosphere and $H' = P_{\text{orb}} c_s / 2\pi (A + 1/2)^{1/2}$, where A is a numerical factor ~8, dependent on q, given by Lubow & Shu (1975).

If the envelope of the secondary is a polytrope of index 1.5, appropriate for the convective state of a cool star, then it can be shown (Paczynski & Sienkiewicz 1975; Pringle 1985; Edwards & Pringle 1987a) that

$$\dot{M}(2) \approx -C \frac{M(2)}{P_{\text{orb}}} \left[\frac{\Delta R}{R(2)}\right]^3, \tag{2.12}$$

where C is a dimensionless constant ~10–20 and ΔR is now the amount by which the secondary *overfills* its Roche lobe, i.e. $\Delta R = R(2) - R_L(2)$. $\dot{M}(2)$ is therefore very sensitive to the amount of overfill, and in consequence mass can always escape at a sufficient rate to maintain ΔR small compared to $R(2)$. For example, $\Delta R/R(2) \sim 10^{-4}$ transfers all the mass of the secondary in 10^7 y (Lubow & Shu 1975; Pringle 1975).

2.4.2 *Stream Trajectory*

As the stream of gas flows away from L_1 the stream lines are deflected by the Coriolis effect and make an angle with the x-axis that is a function only of q (Lubow & Shu 1975): see Figure 2.3. After leaving L_1 the gas particles fall towards the primary, which increases their original sonic velocities to highly supersonic. The stream expands transversely at the velocity of sound, so pressure forces are soon negligible and the stream trajectory is well described by following the orbits of single particles ejected from L_1 in all directions at sonic velocities. As seen in Figure 2.4, when first initiated the stream maintains its coherence through subsequent passage past the primary.

The stream trajectory is found from integrating the equations of motion for a particle in the rotating binary frame (Flannery 1975). In conserving energy along the trajectory, a particle obeys the equation

$$\frac{1}{2}\dot{r}^2 + \Phi_R = \text{const.} \tag{2.13}$$

It follows that, if a particle starts with almost zero velocity ($\dot{r} = 0$) on the Roche lobe, it does not have sufficient energy to cross the lobe at any other point. Thus the trajectory lies entirely within the Roche lobe of the primary and whenever the particle approaches

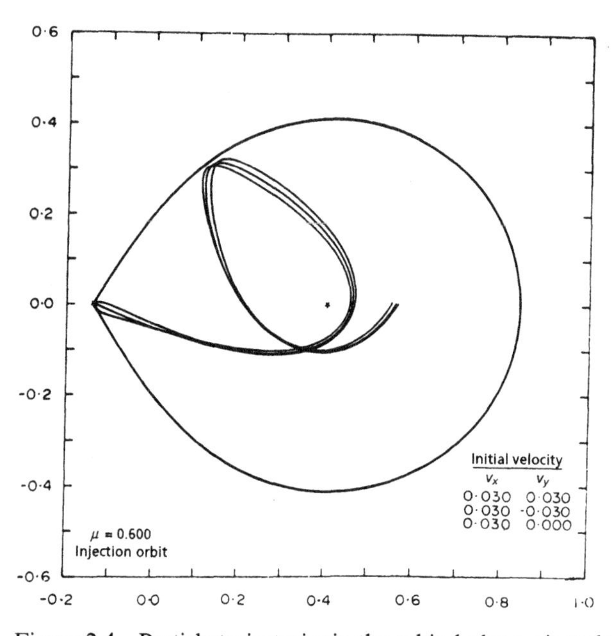

Figure 2.4 Particle trajectories in the orbital plane, ejected through the inner Lagrangian point at low velocities in a system with mass ratio $q = 0.67$. The origin of the coordinate system is the centre of gravity. Initial projection velocities are in units of $G[M(1) + M(2)]/a^{1/2}$. From Flannery (1975).

the lobe it does so with low velocity. The Roche lobes are therefore also known as *zero velocity surfaces.*

The stream has a distance of closest approach $r_{\min}$ from the centre of the primary, obtainable from trajectory computations (Lubow & Shu 1975), and approximated to 1% accuracy by

$$\frac{r_{\min}}{a} = 0.0488q^{-0.464} \qquad 0.05 < q < 1 \tag{2.14}$$

(see Ulrich & Burger (1976) for a different formulation). As a white dwarf has $R(1) \lesssim 1.0 \times 10^9$ cm, equations (2.1b) and (2.14) show that the stream will not impact on the white dwarf provided that

$$P_{\rm orb} > 0.44 \frac{q^{0.7}}{(1+q)^{1/2}} M_1^{-1/2}(1) \quad \text{h}, \tag{2.15}$$

which is satisfied by all known CVs.

As a result, when a stream first begins to flow it passes by the primary (unless there is a strong magnetic field, the consequences of which are treated in Chapter 6) and, the trajectory lying entirely in the orbital plane of the binary, collides with itself at a point well within the Roche lobe and relatively close to the primary (Figure 2.4). This collision at supersonic speed shocks the gas to high temperatures, thereby radiating away the relative kinetic energy of the impact. However, angular momentum is conserved and as a circular orbit has the least energy for a given angular momentum, the dissipation will tend to produce a ring of gas (Figure 2.5).

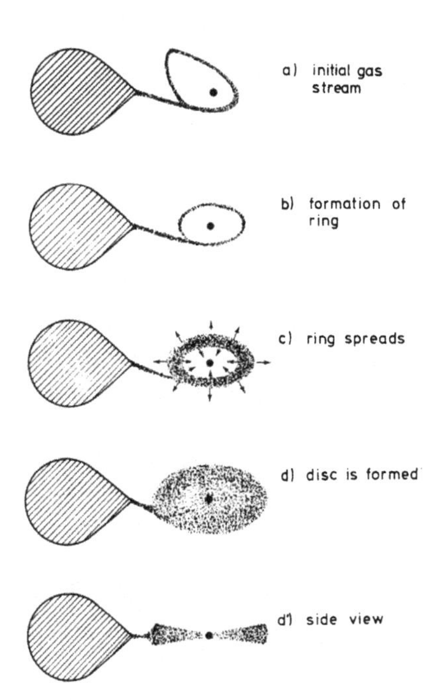

Figure 2.5 Schematic illustration of the initial formation of a ring and its evolution into a disc. From Verbunt (1982).

An instructive rough estimate of the radius r_r of this ring can be obtained by assuming that after a particle leaves L_1 its angular momentum about the primary is conserved. Noting that a particle in a circular orbit at a distance r from the primary has a Keplerian velocity $v_K(r)$ given by

$$v_K(r) = \left[\frac{GM(1)}{r}\right]^{\frac{1}{2}} \tag{2.16}$$

and conservation of angular momentum requires that

$$r_r v_K(r_r) \approx \frac{2\pi}{P_{orb}} R_{L_1}^2, \tag{2.17}$$

we have

$$\frac{r_r}{a} \approx \left(\frac{R_{L_1}}{a}\right)^4 (1+q) \tag{2.18}$$

where R_{L_1}/a (a function only of q) can be found from Table 2.1.

From single particle trajectories (Lubow & Shu 1975), in which angular momentum lost to the secondary is automatically allowed for, a more accurate value of r_r is obtained, approximated by (Hessman & Hopp 1990)

$$\frac{r_r}{a} = 0.0859 q^{-0.426} \quad 0.05 \leq q < 1 \tag{2.19}$$

which is accurate to 1% (see Verbunt & Rappaport (1988) for $0.001 \leq q \leq 0.05$). Comparison of equations (2.14) and (2.19) shows that $r_r \sim 1.75 r_{min}$.

The ring that is formed has a finite radial extent and, from equation (2.16), rotates differentially. Any viscous processes that are present in the gas will generate heat from this shearing flow. As the heat is radiated, the energy drain is met by particles moving deeper into the gravitational potential of the primary. At the same time a few particles must move outwards to conserve angular momentum. Thus the ring spreads into a disc (Figure 2.5). Clearly r_r is the *minimum* outer radius that a disc can have.

2.4.3 Bright Spot

With the disc fully established, the stream impacts onto its outer rim at supersonic speeds, creating a shock-heated area that may radiate as much or more energy at optical wavelengths as all the other components (primary, secondary, disc) combined. The location of this *bright spot*[2] is obviously determined by the intersection of the stream trajectory with the outer edge of the disc. For inviscid discs, i.e., discs in which the viscosity is so low that their radii can be assumed to be equal to r_r, the predicted geometry (Figure 2.6) is a function only of q and can be found in Flannery (1975) or Lubow & Shu (1975). However, radii of discs deduced from eclipses of CVs (Sulkanen, Brasure & Patterson 1981) are 2–3 times larger than r_r, so it is generally necessary to use a *measured* disc radius r_d together with predicted stream trajectories as a means of determining q (e.g., Cook & Warner 1984).

[2] This is often referred to as the hot spot, but is not particularly hot compared with most of the disc.

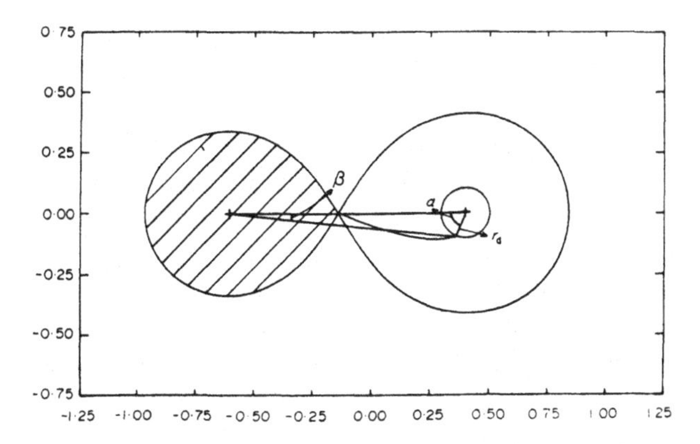

Figure 2.6 Geometry of bright spot formation at the intersection of particle trajectories and the disc. From Flannery (1975).

For a given CV, q is constant, so *variations* in position (r_d, α) of the bright spot can be derived from eclipse observations (Section 2.6) even if the absolute position may be somewhat uncertain. The results show that the spot can be relatively stable for some weeks, or can move significantly on time scales of an hour (Cook & Warner 1984; Cook 1985a; Wood *et al.* 1986).

There are, in addition, more rapid variations in the structure and position of the bright spot which are a result of the turbulence generated on impact (Shu 1976). This is almost certainly the origin of the flickering in U Gem (Figure 1.9), although it is also possible that the stream itself is inhomogeneous as a result of turbulent flow through the L_1 nozzle. The Reynolds number for neutral hydrogen is $Re \approx 2 \times 10^4 \rho L v T^{-1/2}$ (Lang 1974); with $L \sim W$ and $v \sim c_s$, equation (2.7) then gives

$$Re \sim 2 \times 10^4 \dot{M}(2) T_s^{-1/2} W^{-1} \qquad (2.20)$$

$\sim 10^{11}$ for $\dot{M}(2) = 10^{18}$ g s^{-1}, so the flow should be highly turbulent. Similarly, Re in the bright spot is $\sim 10^{11}$ (Shu 1976).

Preliminary attempts at three-dimensional pseudoparticle code and two-dimensional hydrodynamic modelling of the stream–disc impact have been carried out (Livio, Soker & Dgani 1986; Rozyczka & Schwarzenberg-Czerny 1987; Rozyczka 1988; Dgani, Livio & Soker 1989; Hirose, Osaki & Mineshige 1991; Lanzafame, Belvedere & Molteni 1992, 1993) as well as parameterized analytical studies (Bath, Edwards & Mantle 1983). These show four important complications in the interaction of stream and disc: (a) if the impact region is optically thick, so that the energy of impact is not quickly radiated, then part of the stream bounces off the disc and is sprayed into the Roche lobe; (b) the denser core of the stream can penetrate into the edge of the disc releasing its kinetic energy at optical depths greater than unity, thus locally heating the rim, increasing its scale height and causing a bulge that runs around the edge of the disc for typically half the perimeter (see also Section 2.5.3); (c) two shocks may form, with the stream material passing through one and the disc material through the other – these two shock faces form a triangular shaped region with an opening angle ~$M - 1$, where M is the Mach number of the converged downstream flow; (d) part of the stream can flow over the rim of the disc and continue approximately along the single particle trajectory over the face of the disc until it impacts the disc at a later time (see also Lubow & Shu (1975)).

The last point has been investigated by Lubow (1989), who finds that the ultimate impact point for stream material continuing over the face of the disc is in the vicinity of the point of closest approach to the primary, creating there a second bright spot at a position $(r, \alpha) \approx (r_{\rm min}, 148°)$.

If the bulk of the stream flow impacts at the rim bright spot, as from the observed orbital phase and amplitude of the maximum of the hump it evidently does in U Gem (Figure 1.8), its luminosity will be given approximately by the energy released on allowing mass to fall at a rate $\dot{M}(2)$ from infinity to a distance r_d from the primary, i.e.

$$L(\mathrm{sp}) \approx \frac{GM(1)\dot{M}(2)}{r_d} \tag{2.21a}$$

or[3]

$$L_{31}(\mathrm{sp}) \approx 2.7 M_1(1)\dot{M}_{16}(2)(r_d/5\times10^{10}\ \mathrm{cm})^{-1}. \tag{2.21b}$$

This is an upper limit because (i) the fall is from L_1 not ∞ and (ii) the stream meets the rotating disc edge obliquely so only the 'projected' difference in kinetic energy is lost on impact. By comparison, the luminosity of the accretion disc, through which mass is flowing at a rate $\dot{M}(\mathrm{d})$, will be

$$L(\mathrm{d}) \approx 1/2\frac{GM(1)\dot{M}(\mathrm{d})}{R(1)} \tag{2.22a}$$

or

$$L_{32}(\mathrm{d}) \approx 6.7 M_1(1)\dot{M}_{16}(2)R_9^{-1}(1). \tag{2.22b}$$

(The other half of the accretion luminosity may be released at the boundary between disc and star: Section 2.5.4.)

As r_d is typically ~$30R(1)$, the fact that in U Gem $L(\mathrm{sp}) > L(\mathrm{d})$, as evinced by the prominent orbital hump, is a first indication that in some discs we must be prepared to find $\dot{M}(\mathrm{d}) < \dot{M}(2)$. A more complete discussion is given in Section 2.6.5.

A review of the stream–disc interaction has been given by Livio (1993a).

2.5 Accretion Discs

CVs present the richest opportunities for the observation and interpretation of astrophysical discs. Both steady and non-equilibrium discs can be found, often in the same object at different times. The light of many CVs in their steady state, and certainly of all DN at peak of outburst (Warner 1976b), is dominated by disc emission. Such objects behave almost like two-dimensional stars, their thicknesses having only minor effects on observational properties. The theory of accretion discs is therefore central to the interpretation of many aspects of CVs.

To cover the theory fully, with all of its successes and failures, would require a great

[3] Here and throughout the book the notation X_n means that the parameter X is expressed as fraction of 10^n units. The units are cgs unless otherwise stated, the only exceptions being the use of M_1 as representing *solar* mass units (as already defined in (equation 2.1b)) and R_1 for *solar* radius units.

extension of the text of this book, much of it repetitive of what has already appeared in review or book form. Instead, the essential physics and derived results are summarized; these constitute a minimum necessary introduction for understanding the workings of accretion discs.

For fully developed explanations of the theory of accretion discs, Chapters 4–6 of Frank, King & Raine (1985) should be read in conjunction with Pringle (1985), Bath & Pringle (1985) and King (1989b). The pioneering papers of Pringle & Rees (1972), Shakura & Sunyaev (1973), Lightman (1974) and Lynden-Bell & Pringle (1974) and the later review by Pringle (1981) will then be more readily appreciated. Reviews in recent conference proceedings (Belvedere 1989; Meyer *et al.* 1989; Bertout *et al.* 1991; Pringle 1992; Vishniac 1994) bring the subject up to date. The connection between accretion discs in CVs and those around neutron stars and black holes is reviewed in an introductory way by Treves, Maraschi & Abramowicz (1988).

2.5.1 Theory of Accretion Discs

As the stream from the secondary lies in the orbital plane, and the turbulent impact at the bright spot is unlikely to raise a significant fraction of the stream material to large heights above the plane, the inflowing gas has momentum predominantly in the plane. The thickness of the disc is therefore determined by hydrostatic equilibrium (unlike the inner regions of discs around neutron stars, radiation pressure is not important in CV discs) which, as seen below, leads to a thin disc structure: to a first approximation, the disc is a two-dimensional flow.

Much of the disc is sufficiently close to the primary that the gravitational field of the secondary may be ignored – tidal interaction with the outer regions of the disc is, however, important in some circumstances. Throughout most of the disc the angular velocity $\Omega(\mathrm{r})$ of particles will differ negligibly from the Keplerian circular value

$$\Omega_{\mathrm{K}}(r) = \left[\frac{GM(1)}{r^3}\right]^{1/2}. \tag{2.23}$$

There must, however, be a radial drift of material as the viscous shear between adjacent annuli causes most particles to move in towards the primary (except near the edge of the disc where conservation of angular momentum may require a predominantly outward drift). The radial drift velocity is v_{rad}.

We define a *surface density* $\Sigma = 2\int \rho \mathrm{d}z$, which is the mass per unit area integrated through the disc in the z direction. Σ and v_{rad} are functions of r and will also be functions of time whenever non-equilibrium conditions obtain. The time-dependent equations that determine disc structure are then

$$\frac{\partial \Sigma}{\partial t} = \frac{1}{r}\frac{\partial}{\partial r}(r\Sigma v_{\mathrm{rad}}) \tag{2.24}$$

from conservation of mass, and

$$\frac{\partial}{\partial t}(\Sigma r^2 \Omega) + \frac{1}{r}\frac{\partial}{\partial r}(r\Sigma v_{\mathrm{rad}} r^2 \Omega) = \frac{1\partial}{r\partial r}\left(r^3 v_{\mathrm{k}} \Sigma \frac{\mathrm{d}\Omega}{\mathrm{d}r}\right) \tag{2.25}$$

from conservation of angular momentum. Here v_{k} is the coefficient of effective kinematic viscosity of the gas.

Eliminating $v_{\rm rad}$ from equations (2.24) and (2.25) gives

$$\frac{\partial \Sigma}{\partial t} = -\frac{1}{r}\frac{\partial}{\partial r}\left[\frac{1}{\mathrm{d}(r^2\Omega)/\mathrm{d}r}\frac{\partial}{\partial r}\left(\nu_{\rm k}\Sigma r^3\frac{\mathrm{d}\Omega}{\mathrm{d}r}\right)\right] \tag{2.26}$$

which is the equation for the surface density as a function of time. For circular Keplerian orbits, $\Omega \propto r^{-3/2}$ (equation (2.23)), and equation (2.26) simplifies to

$$\frac{\partial \Sigma}{\partial t} = \frac{3}{r}\frac{\partial}{\partial r}\left[r^{1/2}\frac{\partial}{\partial r}\left(\nu_{\rm k}\Sigma r^{1/2}\right)\right]. \tag{2.27}$$

This is a *non-linear diffusion equation*: matter diffuses inwards to the primary and angular momentum diffuses outwards to the outer edge of the disc. It results in Σ being able to change only on the *viscous* time scale

$$t_\nu(r) \sim r^2/\nu_{\rm k} \tag{2.28}$$

and implies a radial drift velocity

$$v_{\rm rad} \sim \nu_{\rm k}/r \tag{2.29}$$

which also means that $t_\nu(r)$ can be expressed as

$$t_\nu(r) \sim r/v_{\rm rad} \sim \frac{r^2}{\nu_k}, \tag{2.30}$$

i.e., the time taken for a density perturbation at r to be spread over a radial distance r.

An example of the use of equation (2.27) is given by the evolution of the ring of gas at $r_{\rm r}$ set up at the initiation of mass transfer from the secondary (Section 2.4). With the assumption $\nu_{\rm k}$ = const, equation (2.27) can be solved (Pringle 1981; Frank, King & Raine 1985) to show how the ring spreads out into a disc (Figure 2.7).

If the rate $\dot{M}(2)$ at which mass is transferred to the disc changes on a time scale that is much longer than t_ν the disc will settle into a steady state (if such is possible) with

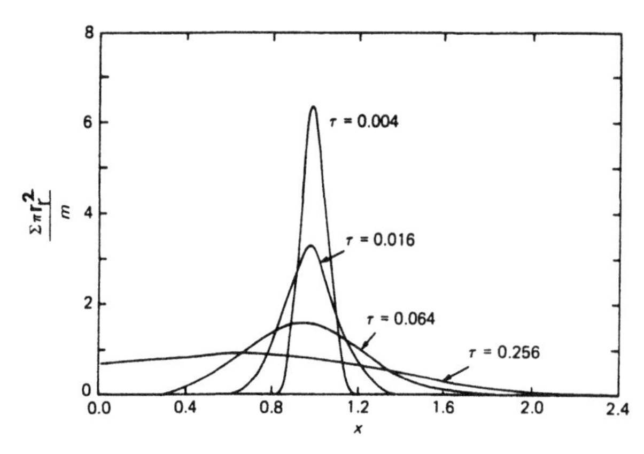

Figure 2.7 The spreading of a ring of mass m due to viscous torques. The surface density Σ is shown as a function of $x = r/r_{\rm r}$ and the dimensionless variable $\tau = 12\nu_{\rm k}tr_{\rm r}^{-2}$. From Pringle (1981).

$\partial\Sigma/\partial t = 0$. The mass conservation equation (2.24) is then integrated to give the obvious result

$$\dot{M}(d) = 2\pi r(-v_{\rm rad})\Sigma \tag{2.31}$$

(since $v_{\rm rad} < 0$). To integrate equations (2.26) or (2.27) requires adoption of boundary conditions at the inner and outer edges of the disc. In the absence of a magnetic field strong enough to affect flow near the primary, the disc extends down to the surface of the white dwarf, i.e., to $r = R(1)$. As the star rotates with an angular velocity $\Omega(1) < \Omega_{\rm K}[R(1)]$ there must exist a *boundary layer* within which the Keplerian velocity of the disc material is decelerated to match the equatorial velocity of the primary. The boundary layer is expected to be very thin in radial extent ($\ll R(1)$), so the conditions that $d\Omega/dr = 0$ (i.e., zero torque on the disc) occurs close to $r = R(1)$, and $\Omega = \Omega_{\rm K}$ there, provide an inner boundary condition (see Koen (1988) and Duschl & Tscharnuter (1991) for critical comments) which gives, from equations (2.27) and (2.31),

$$\nu_{\rm k}\Sigma = \frac{\dot{M}(\mathrm{d})}{3\pi}\left\{1 - \left[\frac{R(1)}{r}\right]^{1/2}\right\}. \tag{2.32}$$

At the outer boundary it is assumed that tidal interaction drains the outward flow of angular momentum (see Section 2.5.5). The rate of energy generation from viscous shear is

$$D(r) = \nu_{\rm k}\Sigma\left(r\frac{d\Omega}{dr}\right)^2 \tag{2.33}$$

$$= \frac{3GM(1)\dot{M}(\mathrm{d})}{4\pi r^3}\left\{1 - \left[\frac{R(1)}{r}\right]^{1/2}\right\} \tag{2.34a}$$

$$\sim \frac{3}{4\pi}\Omega_{\rm K}^2(r)\dot{M}(\mathrm{d}) \qquad \text{for } r \gg R(1) \tag{2.34b}$$

from equations (2.23) and (2.33). This energy is radiated from the two faces of the disc at a rate $2\sigma T_{\rm eff}^4(r)$, where $T_{\rm eff}$ is the effective temperature and σ is the Stephan-Boltzman constant. Hence the radial temperature structure of the disc is given by

$$T_{\rm eff}(r) = T_*\left[\frac{r}{R(1)}\right]^{-3/4}\left\{1 - \left[\frac{R(1)}{r}\right]^{1/2}\right\}^{1/4}, \tag{2.35}$$

where

$$T_* = \left[\frac{3GM(1)\dot{M}(\mathrm{d})}{8\pi\sigma R^3(1)}\right]^{1/4} \tag{2.36a}$$

$$= 4.10 \times 10^4 R_9^{-3/4}(1)M_1^{1/4}(1)\dot{M}_{16}^{1/4}(\mathrm{d}) \quad \mathrm{K}. \tag{2.36b}$$

By differentiating equation (2.35) it is found that the maximum temperature in the disc occurs at $r = (49/36)R(1)$ and has a value $0.488T_*$. For $r \gg R(1)$, equation (2.35) becomes

$$T_{\mathrm{eff}}(r) \approx T_* \left(\frac{r}{R(1)}\right)^{-3/4}. \tag{2.37}$$

Equation (2.35) predicts that a steady state accretion disc has a surface temperature that increases from its outer edge to a maximum near the primary. For typical $\dot{M}(\mathrm{d}) \approx 10^{16}$–$10^{18}$ g s^{-1} (Section 2.6.1.3), accretion discs around white dwarfs are expected to be strong UV emitters (and FUV and EUV for the highest $\dot{M}(\mathrm{d})$) in their inner regions but can be IR radiators in their outer parts. The temperature stratification is clearly very different from that of a stellar atmosphere.

Equation (2.35) is the first point in the development of the theory of steady accretion discs at which a comparison with observation becomes possible. This can be achieved either by comparing the predicted flux distribution (integrated over the surface of the disc, allowing for inclination and limb darkening) with observed distributions, or by using eclipses to derive spatial intensity maps (at a variety of wavelengths) of accretion discs, which may be compared with intensity distributions predicted from equation (2.35). These comparisons are made in Sections 2.6 and 3.5.4.4 and depend for their credibility on whether there are complicating factors in real discs that may distort the flux and spatial distributions, and whether the discs are optically thin or optically thick (or possibly of different optical thickness at various radii).

The question of optical thickness involves the vertical structure of the disc. In the z direction, for a disc of negligible mass there is a hydrostatic balance between the pressure gradient and the z component of the primary's gravitational field:

$$\frac{\partial P}{\partial z} = \rho \frac{\partial}{\partial z}\left[\frac{GM(1)}{(r^2 + z^2)^{1/2}}\right] \tag{2.38a}$$

$$= \frac{G\rho M(1)z}{r^3} \quad \text{for a thin disc } (z \ll r) \tag{2.38b}$$

$$= g\rho \frac{z}{r}. \tag{2.38c}$$

If the disc is optically thick in the z direction then, as for stellar atmospheres, equation (2.38b) must be solved simultaneously with the equation of radiative transfer (in non-convective transport: for the more general case see Meyer & Meyer-Hofmeister (1982)) for the radiative flux $F(z)$ through the face of the disc:

$$F(z) = \frac{-16\sigma T^3(z)}{3\kappa_{\mathrm{R}}\rho} \frac{\partial T}{\partial z}, \tag{2.39a}$$

where κ_{R} is the Rosseland mean absorption coefficient.

Solving equation (2.39a) in the Eddington Approximation gives the mid-plane temperature

$$T^4(\mathrm{mid}) = \frac{3}{8}\kappa_{\mathrm{R}}\Sigma T_{\mathrm{eff}}^4 \tag{2.39b}$$

or, in the optically thin case, assumed to be isothermal, the flux emitted is

$$F = \sigma T_{\rm eff}^4 = \Sigma\kappa_{\rm R}\sigma T^4({\rm mid}). \tag{2.39c}$$

For an *isothermal* z-structure equation (2.38b) is solved to give

$$\rho(r, z) = \rho_{\rm c}(r){\rm e}^{-z^2/2H^2}, \tag{2.40}$$

where $\rho_{\rm c}(r)$ is the density in the central plane of the disc and H is the scale height, which may be evaluated by noting that from equation (2.40)

$$H^2 = -z\Big/\frac{1}{\rho}\frac{{\rm d}\rho}{{\rm d}z} \tag{2.41}$$

which, from $P = \rho c_{\rm s}^2$ and equation (2.38b), gives

$$H = \left[\frac{r^3}{GM(1)}\right]^{1/2} c_{\rm s} \tag{2.42a}$$

$$= c_{\rm s}/\Omega_{\rm K}(r) \tag{2.42b}$$

or

$$\frac{H}{r} = \frac{c_{\rm s}}{v_{\rm K}(r)} \tag{2.43}$$

from equation (2.16). Thus the density falls off rapidly with height above the plane. Also, a thin disc is one in which the local Keplerian velocity is highly supersonic.

The condition for the disc to be optically thick in the z direction is then

$$\rho H\kappa_{\rm R} = \frac{1}{2}\Sigma\kappa_{\rm R} \gg 1. \tag{2.44}$$

If (2.44) is satisfied, the energy generated by viscous dissipation is radiated from an optically thick medium and is therefore not greatly different from blackbody emission. However, if (2.44) is not satisfied, although the radiation escapes more easily from the central plane region of the disc, the emissivity of the gas is lower than for a blackbody so the gas temperature must be higher than in the optically thick case.

The observed flux distribution of a disc as seen by an observer at a distance d is

$$F_\nu = \frac{2\pi\cos i}{d^2}\int_{R(1)}^{r_{\rm d}} I_\nu r{\rm d}r, \tag{2.45}$$

where i is the inclination of the disc. Approximating intensity I_ν with a blackbody distribution

$$B_\nu(T) = \frac{2\pi\nu^3}{c^2}({\rm e}^{-h\nu/kT} - 1)^{-1} \tag{2.46}$$

in the usual notation, and using the radial temperature distribution of equation (2.35) gives a first estimate of the continuous spectrum of a steady state disc (Figure 2.8).

For the region of the disc where $r \gg R(1)$ the temperature distribution of equation (2.37) obtains, and equation (2.46) simplifies in the frequency range

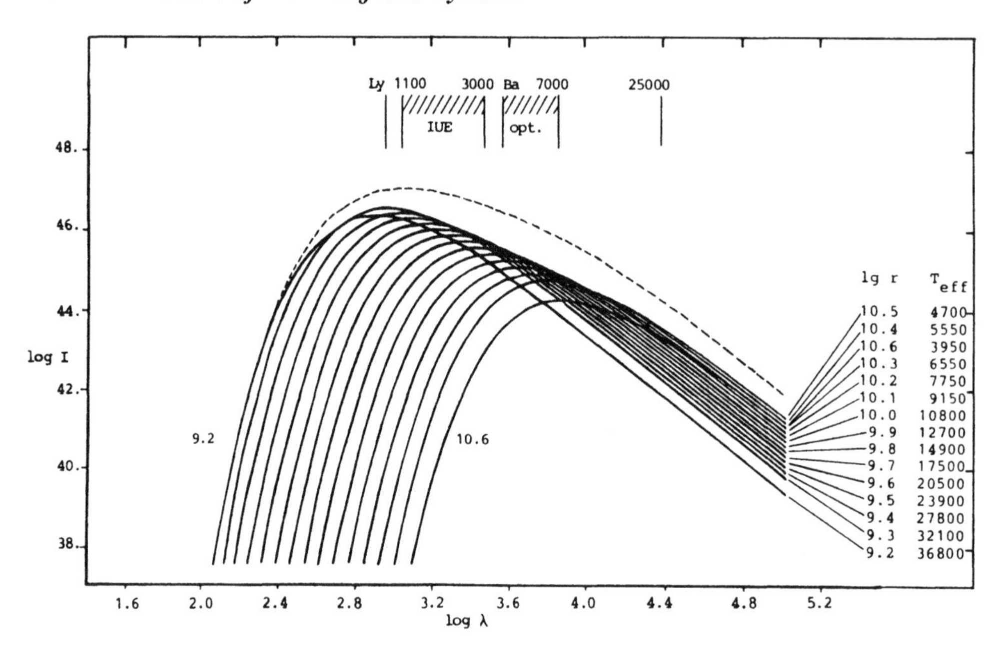

Figure 2.8 Contributions of blackbody annuli to the total intensity distribution of an accretion disc. These are for $M_1(1) = 1$ and $\dot{M}(\mathrm{d}) = 10^{-9}\ \mathrm{M}_\odot\ \mathrm{y}^{-1}$. The intensity emitted by each annulus is weighted by the area of the annulus. The effective temperature (K) and radius (in cm) of each annulus is indicated. Wavelengths and the Lyman and Balmer edges are given at the top of the diagram. From la Dous (1989a).

$kT(r_\mathrm{d})/h \ll \nu \ll kT_*/h$ to

$$F_\nu \,\tilde{\propto}\, \nu^{1/3} \int_0^\infty \frac{x^{5/3}}{\mathrm{e}^x - 1}\,\mathrm{d}x \tag{2.47}$$

where $x = h\nu/kT$. Therefore, provided $r_\mathrm{d} \gg R(2)$ (and therefore $T(r_\mathrm{d}) \ll T_*$), the spectrum of a steady state disc is characterized by $F_\nu \,\tilde{\propto}\, \nu^{1/3}$ or $F_\lambda \propto \lambda^{-7/3}$ (Lynden-Bell 1969).

equation (2.35), is independent of viscosity. This is a consequence of the assumptions of a steady state (the viscosity adjusting itself to provide the steady mass flow $\dot{M}(\mathrm{d})$) and optical thickness and is fortunate because the physical nature of viscosity in accretion discs is uncertain. Conversely, observations of steady state discs cannot be used to investigate the nature of their viscosity. The Shakura & Sunyaev (1973) approach, which is still commonly in use, is to parameterize ν_k as

$$\nu_\mathrm{k} = \alpha c_\mathrm{s} H, \tag{2.48a}$$

$$= \alpha H^2 \Omega_\mathrm{K}(r) \tag{2.48b}$$

which may be thought of as the viscosity generated by turbulent eddies of size H and turnover speed αc_s. For subsonic turbulence $\alpha < 1$. α itself may be a function of H/r (see Section 3.5.4.2).

From equation (2.29) we then have

$$v_{\text{rad}} \sim \alpha \left(\frac{H}{r} \right) c_s, \tag{2.49}$$

which shows that the radial drift velocity is highly subsonic, and, from equation (2.31),

$$\dot{M}(\text{d}) \sim 2\pi\alpha\Omega_{\text{K}}(r) H^2 \Sigma. \tag{2.50}$$

Solution of the equations of steady state disc structure with $\alpha = \text{const}$ and Kramers opacity then gives (Shakura & Sunyaev 1973; Frank, King & Raine 1985)

$$\frac{H}{r} = 1.72 \times 10^{-2} \alpha^{-1/10} \dot{M}_{16}^{3/20}(\text{d}) M_1^{3/8}(1) r_{10}^{1/8} \left\{ 1 - \left[\frac{R(1)}{r} \right]^{1/2} \right\}^{3/5} \tag{2.51a}$$

which, for $r \gg R(1)$, leads to

$$H \propto r^{9/8}. \tag{2.51b}$$

Therefore the faces of a steady state accretion disc are concave and may be illuminated by radiation from the primary and boundary layer. As already noted, the boundary layer may emit as much radiation as the entire disc, so a significant amount of heating of the disc, and consequent departure from the temperature distribution specified in equation (2.35), may occur.

Computations of vertical disc structure, including radiative and convective zones, have been made by Meyer & Meyer-Hofmeister (1982) and Smak (1992). These give the height of the photosphere ($\tau = 1$) above the central plane. In particular, the height h_{d} of the outer rim of the disc for typical CV disc radii is approximated by

$$\frac{h_{\text{d}}}{r_{\text{d}}} \cong 0.038 \dot{M}_{16}^{3/20}(\text{d}). \tag{2.52}$$

The optical thickness vertically through the disc obtained from models is a function of many parameters. For very large values of $\alpha (\geq 1)$ the densities are low and discs are mostly optically thin (Williams 1980). More realistic values of α (~0.3) lead to optically thick discs except at low temperatures ($T \lesssim 4000$ K) in the inner regions (Tylenda 1981a; Smak 1984a, 1992). With low $\dot{M}(\text{d})$ and $\alpha \sim 0.05$, as in DN at quiescence (Section 3.5.4.4), only the region $r < 1 \times 10^{10}$ cm is optically thin.

Results of more general computations are reviewed by Meyer-Hofmeister & Ritter (1993) and Smak (1992).

Another source of heating of the outer regions is the energy dissipated by tidal interaction between the secondary and the outer regions of the disc. This will only be important if $r_{\text{d}} \gtrsim 0.8 \text{R}_{\text{L}}(1)$ (which may commonly be the case: Section 2.6.2), in which case the disc is truncated by the tidal force, with transfer to the secondary of the angular momentum of the outward diffusing disc particles and a luminosity $\sim GM(1)\dot{M}(\text{d})/r_{\text{d}}$ being released in the outer disc. Further possible complications arise if the stream penetrates significantly into the disc, thereby releasing more energy than it would at the edge, but spread over a larger volume of the disc. As the time scale for establishing hydrostatic equilibrium in the z-direction $t_z \sim H/c_s = \Omega_{\text{K}}^{-1}(r_{\text{d}})$ from equations (2.23) and (2.43), any vertical bulge in the accretion disc caused by stream penetration at the bright spot will survive for most of the rotation period of the outer rim of the disc (see also Section 2.5.3).

As the properties of steady state discs are to a large extent independent of α, it is to discs adjusting to changes of $\dot{M}(\mathrm{d})$ that appeal must be made for empirical information on the order of magnitude of α. From equations (2.30), (2.42) and (2.49) we have

$$\alpha = \left(\frac{r}{H}\right)^2 \frac{1}{t_\nu \Omega_K}. \tag{2.53}$$

Hence the observed time scale t_ν required to move from one equilibrium state to another can (from observationally deduced values of r/H and Ω_K) provide α. This is discussed in Section 3.5.4.

2.5.2 *Viscosity*

The α-prescription was introduced by Shakura & Sunyaev as a measure of desperation in the absence of a theory of viscosity in differentially rotating fluid discs. From the theory outlined in the previous section, it is evident that *a priori* models of discs cannot be made without knowledge of ν_k or α, both of which are likely to be functions of many physical variables. Some assistance is obtained from observations of non-steady CV accretion discs, which show that $\alpha \sim 0.01$–1 (see, e.g., Verbunt (1982)): noticeably larger than the value for pure molecular viscosity ($\alpha \sim 10^{-11}$), but with no indication as to the cause.

Considerable effort has been expended on seeking the physical mechanism that generates $\alpha \sim 1$ in discs. The theories split broadly into three categories: turbulence, magnetic stresses, and collective hydrodynamic effects.

2.5.2.1 *Turbulence*

Although there is strong shear in a Keplerian disc, and the Reynolds number is large, the Rayleigh condition for shear instability (Hunter 1972)

$$\frac{\partial}{\partial r}(r^2\Omega) < 0$$

is far from satisfied. There is therefore no immediate reason to expect shear-driven turbulence. None appears in two-dimensional simulations (Kaisig 1989a). However, it has been established from simulations of laboratory fluid flows that turbulence is non-linear and a three-dimensional phenomenon (Orszay & Kells 1980), and it therefore remains an open question whether Keplerian discs will be spontaneously turbulent.

However, if the disc is convectively unstable, which is probable in those regions where the surface temperature $T_s < 10^4$ K so that hydrogen or helium ionization zones exist in the disc structure, then a velocity field is established which may, in a first approximation, be calculated from the mixing length theory approach as applied to stellar interiors (Section 14.8 of Cox & Giuli (1968)). The criterion for convective instability in a disc can differ, however, from the Schwarzschild one for non-rotating stars (Livio & Shaviv 1977; Tayler 1980; Lin & Papaloizou 1980; Elstner, Rüdiger & Tschäpe 1989).

Models for discs, assuming zero viscosity in radiative regions and convective viscosity elsewhere (with ν_k taken as a function of the ratio of convective speed to c_s), have been constructed (Vila 1978; Lin & Papaloizou 1980; Lin 1981; Smak 1982a). In a new approach, Cabot *et al.* (1987a,b) use the large scale turbulence model developed by Canuto & Goldman (1985) (see also Canuto, Goldman & Hubickyj (1984)) to specify the properties of convective turbulence. This theory, applied by Cannizzo & Cameron (1988), who allow for viscous dissipation also in the radiative regions, is able to generate the $\alpha \sim 0.01$–1 deduced for CV discs without any assumptions of the kind that appear in the mixing length theory. Scott (1990) has developed an alternative model for convective turbulent viscosity, which gives even larger values of α. The effective α is $\sim M_t^2$, where M_t is the Mach number of the characteristic turbulent velocity (Geertsema & Achterberg 1992).

However, as convection is caused by heating generated by viscous dissipation, which itself is a result of convection, it has been questioned whether this is a self-sustaining process (Vishniac & Diamond 1989) and answered in the negative (Ryu & Goodman 1992).

2.5.2.2 Magnetic Stresses

The traditional picture of the influence of a magnetic field in the disc is that if the disc is provided with a seed magnetic field, either by mixing in of field lines from the global field of the secondary or by field lines dragged over by the accretion stream, it will be sheared by differential rotation, the radial component being transformed into an azimuthal component on a time scale $\sim \Omega_K^{-1}(r)$. This also stretches and amplifies the field. The effective α is $\sim B^2/4\pi P$, where B is the field strength and P is the gas pressure (Geertsma & Achterberg 1992). A limiting process operates when the magnetic and gas pressures become comparable: flux tubes will be buoyed out of the disc by the Parker instability (Parker 1979), which is in essence a magnetic Rayleigh–Taylor instability[4] (Eardley & Lightman 1975; Galeev, Rosner & Vaiana 1979; Stella & Rosner 1984; Kato 1984; Kato & Horiuchi 1985, 1986). The energy used in amplifying the field is drawn from the kinetic energy of the disc, which constitutes a constant drain if there is a balance between shear amplification and reconnection (e.g., Coroniti (1981)). However, the loss of flux from buoyancy and from leakage at the outer edge of the disc (van Ballegooijen 1989) are efficient enough to remove the flux on time scales $\sim \Omega_K^{-1}(r)$ (Stellar & Rosner 1984). Only if a self-sustaining dynamo acts in the disc, which can regenerate the field, does magnetic viscosity appear feasible.

The existence of a permanent large scale magnetic field would have consequences in addition to providing a magnetic turbulent viscosity. Any stellar wind from the disc will become coupled to the field and act as a drain of angular momentum from the disc. The wind itself may be a result of centrifugal acceleration of gas attached to the field lines (Henriksen & Rayburn 1971; Blandford & Payne 1982; Koen 1986; Cannizzo & Pudritz 1988; van Ballegooijen 1989). Dynamo models for accretion discs have been discussed by Blandford (1976, 1989), Ichimura (1976), Pudritz (1981a,b), Meyer & Meyer-Hofmeister (1983b), Kato (1984), Stepinski & Levy (1988); van Ballegooijen (1989), Kônigl (1989) and Vishniac, Jin & Diamond (1990). Although it is commonly postulated that an $\alpha - \omega$ dynamo, driven by a combination of differential rotation and

[4]The Rayleigh–Taylor instability occurs when a dense fluid overlies a less dense fluid. Cumulus cloud formation is an example of such an instability.

meridional circulation, can exist in an accretion disc, questions of equilibrium field strength and stability of the process remain (Blandford 1989).

Given that a dynamo does act, turbulence in the disc can propagate Alfvén waves along the magnetic fields that arch out of the disc and hence heat a disc corona (Galeev, Rosner & Vaiana 1979; Kuperus & Ionson 1985; Burm 1985, 1986; Stepinski 1991). The shear stress in the corona can give an effective $\alpha \sim 0.01$ (Burm & Kuperus 1988).

The traditional requirement for equilibrium fields of considerable strength has been overthrown by Balbus & Hawley (1991, 1992b, 1994) (see also Hawley & Balbus (1991, 1992)), who have discovered a dynamical instability in accretion disc flow in the presence even of very weak magnetic fields. The instability requires only a field with a toroidal component and $d\Omega/dr < 0$; it is found that the Rayleigh criterion is irrelevant. The growth rate of the instability is independent of field strength and gives a growth time scale $\sim \Omega_K^{-1}$. Building on this, Tout & Pringle (1992b) show that magnetic reconnection, the Parker instability and the Balbus–Hawley instability when combined lead to a self-sustaining dynamo with an unsteady field that oscillates around an equilibrium value. The model is self-consistent in that the shear that drives the dynamo is provided by the magnetically-generated viscosity. The model gives $0.1 < \alpha < 0.7$ and has exactly the nature of a Shakura–Sunyaev disc (as opposed to the models considered in the next section, which generate an *effective* α). This is a universal value of α, independent of the disc properties, which at first sight seems in contradiction with observations of DN, which require larger values of α in outburst then in quiescence (Section 3.5.4.2). However, Tout & Pringle find that the time scale of changes during the DN outburst is too short to allow the maintenance of an equilibrium dynamo and this may result in the complete shut-down of the dynamo as the system returns to normal, leading to low values of α in quiescence.

Other discussions of these processes are given by Tagger, Pellat & Coroniti (1992), Balbus & Hawley (1992a), Knobloch (1992), Vishniac & Diamond (1992), Geertsma & Achterberg (1992) and Goodman (1993).

2.5.2.3 *Collective Hydrodynamic Effects*

In numerical simulations (see next section) Sawada, Matsuda & Hachisu (1986) found tidally induced spiral shock waves in accretion discs. Gas passing through the shocks loses energy and angular momentum even in the absence of turbulent or magnetic viscosity. By postulating that the shocks are stationary (rather than quasi-stationary as found in numerical simulations) Spruit (1987) was able to find analytical solutions for the gas flow. A minimum of two shocks, in the form of logarithmic spirals, is found. The effective α is $\alpha_{\rm eff} \sim n_s(M_s - 1)^3$, where n_s is the number of shocks and M_s is the Mach number of the shocks; an upper limit $\alpha_{\rm eff} \sim 10^{-2}$ was found (Spruit 1987; Kaisig 1989a,b), though later computations result in an effective $\alpha \sim 0.1$ (Matusuda *et al.* 1990). The computations by Sawada *et al.* are adiabatic and result in high temperature discs. For discs appropriate to CVs the spiral waves would be very tightly wound and $\alpha \sim 10^{-3}$ (Spruit 1989; Livio & Spruit 1991; Rozyczka & Spruit 1993). Such structures would be scrambled by most other viscous processes which generate $\alpha \gtrsim 10^{-3}$.

These studies are two-dimensional, and it is has been claimed that in a real disc the waves would refract away from the denser mid-plane region of the disc and dissipate in the atmosphere (Lin 1989). However, Rozyczka & Spruit (1993) point out that the waves will be reflected back from the atmosphere as happens in, e.g., the solar 5 min oscillations. The possible presence of incoherent internal waves will tend to destroy any coherent global shock waves (Vishniac & Diamond 1989).

The discovery (Papaloizou & Pringle 1984, 1985) of a global non-axisymmetric instability in differentially rotating discs, which grows on a dynamical time scale ($\sim \Omega_K^{-1}$), has opened other possibilities for transfer of angular momentum. This instability has been shown to be equivalent to an internal wave that interferes constructively after reflection from the inner or outer boundary (Goldreich & Narayan 1985). This was first studied by Drury (1980). Although the effective viscosity generated by this mechanism may be concentrated principally near the reflecting boundary (Hanawa 1988), the possibility of other, non-standing-wave transport of angular momentum is suggested. Vishniac & Diamond (1989) find that inward travelling internal waves (analogous to deep ocean waves) with dipole symmetry ($m = \pm 1$) carry significant amounts of negative angular momentum. These waves may be excited by the impact of the stream on the rim of the disc or by tidal interaction of the secondary (Vishniac, Jin & Diamond 1991). In a neutral disc the effective $\alpha \sim (H/r)^2 \sim 10^{-4}$, but the additional sources of energy loss in an ionized disc with a large scale magnetic field increase α to $\sim (H/r)^{6/5}$(Vishniac, Jin & Diamond 1991), which is of the correct order of magnitude to account for observed values of α. If excitation is via impact of the stream, the effective α will be proportional to $\dot{M}(\mathrm{d})$.

Transference of angular momentum by collective motions occurs even in an inviscid disc. It is important, therefore, to distinguish between the *effective* α generated by such processes, and any 'local' α caused by, for example, turbulence. Spiral shocks in accretions discs have been reviewed by Spruit (1989, 1991).

2.5.3 Numerical Simulation of Discs

A realistic treatment of the structure of an accretion disc would require solution of the full three-dimensional hydrodynamic equations of fluid flow (expressing conservation of mass, momentum and energy) together with the equations of radiative and convective energy transport. The outer and inner boundary conditions would have to be physically realistic – including the inner boundary layer and the effects both of tidal dissipation and stream impact at the outer edge – and the equation of state would have to be included; the mechanism of viscosity must be specified. This is not yet possible even with the largest computers.

Two approaches to approximating the solutions have been taken: numerical integration of the hydrodynamic equations under a restricted set of conditions, or multi-particle approximations with viscous interactions. Examples of the former may be found in Kaisig (1989a), Schwarzenberg-Czerny & Rozyczka (1988) and Rozyczka & Spruit (1993) based on a numerical code developed by Rozyczka (1985), the last mentioned includes local radiative cooling in a simple approximation; two-dimensional examples of the latter technique, which was introduced into the computation of CV discs by Lin & Pringle (1976), can be found in Gingold & Monaghan (1977), Hensler (1982), Sawada *et al.* (1987), Whitehurst (1988a), Hirose & Osaki (1989), Geyer, Herold & Ruder (1990) and Lubow (1991a,b).

Whitehurst's approach is an extension of that of Lin & Pringle (1976) and of Larson (1978). It allows particles to orbit in the combined gravitational potentials of the stellar

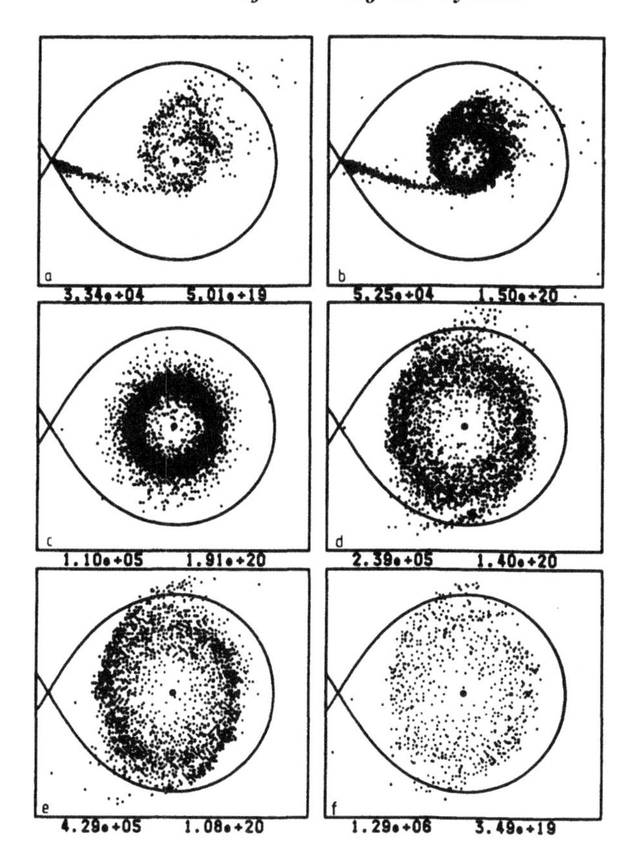

Figure 2.9 Particle simulation of a mass transfer burst in a system with mass ratio $q = 0.615$. Each instantaneous picture is labelled with time in seconds and the disc mass in kilograms. The orbital period is 2.2×10^4s. From Whitehurst (1988a).

components, interacting in a way that simulates (parameterized) viscosity. The flow is compressible; dissipation is measured from the loss of kinetic energy during particle collisions, but details of radiative transfer are not included (two-dimensional models usually assume instant cooling). Figure 2.9 shows the particle simulation of a burst of mass transfer from the secondary; it will be noted that, unlike the single particle trajectories of Figure 2.4, interactions can cause particles to pass through the Roche lobe and be lost from the primary. The disc takes up a form that maintains equilibrium between viscous and tidal forces; the elongation perpendicular to the line of stellar centres is in agreement with the inviscid (non-interacting single particle) calculations of Paczynski (1977).

Three-dimensional simulations assuming a polytropic equation of state have been made by Molteni, Belvedere & Lanzafame (1991) and Lanzafame, Belvedere & Molteni (1992, 1993); these do not include details of radiative transfer (three-dimensional models are usually computed on the adiabatic approximation, retaining heat and therefore becoming too hot relative to real discs). For a polytropic index $\gamma \approx 1$, equivalent to a nearly isothermal structure, there is agreement with the analytical thin disc results of Section 2.5.1. For $\gamma \gtrsim 1.1$ there are large differences between two-dimensional and three-dimensional results. A more complete approach is taken by Meglicki, Wickramasinghe & Bicknell (1993) who use three-dimensional particle simulation which includes pressure terms explicitly. They find turbulent flow with eddy sizes ranging up to half the thickness of the disc. In the early development of the disc, gas tends to accumulate near the radius r_r and some gas from the incoming stream runs over the face of the disc as far as this ring. The two-dimensional hydrodynamic computations of Rozyczka & Spruit (1993) for low α discs also show a strong build-up of gas in a ring of nearly constant angular momentum.

Other three-dimensional computations by Hirose, Osaki & Mineshige (1991) extend the Livio, Soker & Dgani (1986) study of the bright spot region to the whole disc. Results similar to Whitehurst's (1988a: Figure 2.9) are obtained for $q \geq 0.25$: a steady state disc with eccentricity $e \sim 0.1$ is confirmed. An important new result (see also Meglicki, Wickramasinghe & Bicknell (1993)) is the effect that stream impact (Section 2.4.3) has on the entire outer rim of the disc. With the stream in action, $H/r \sim 0.15$ at the rim, which is considerably larger than expected from hydrostatic equilibrium (equations (2.43) and (2.52)), This is a result of a vertical velocity component being generated in the stream impact. Furthermore, this causes a vertical oscillation of gas in the outer regions, leading to thickenings of the disc at orbital phases $\varphi \sim 0.2$ and 0.8 ($\varphi = 0$ at conjunction of the stars), and a less prominent effect at $\varphi \sim 0.5$ (the stream impact is at $\varphi \sim 0.9$). Such outer rim structure can cause variable obscuration of the bright inner regions of the disc in high inclination systems.

2.5.4 Boundary Layer

The boundary layer (BL) is the region over which gas moving at Keplerian velocities in the disc is decelerated to match the surface velocity of the primary. As a result, radial support of the accreting gas is given largely by the pressure gradient instead of the centrifugal effect.

From conservation of energy and angular momentum it can be shown (Kley 1991) that the energy released in the BL is

$$L(\mathrm{BL}) = L(\mathrm{d})\left\{1 - \frac{\Omega(1)}{\Omega_{\mathrm{K}}[R(1)]}\right\}^2 \tag{2.54a}$$

$$= \zeta L(\mathrm{d}) \tag{2.54b}$$

(This differs from a formula often quoted, which is derived from energy considerations alone.)

Therefore, if the primary is rotating well below break-up velocity, disc and BL rival each other in luminosity (see Duschl & Tscharnuter (1991) for quantitative estimates). Where the luminosity of the BL appears in the spectrum depends on its optical thickness.

If the BL is optically thick, the luminosity that is deposited must diffuse through a distance $\sim H$ before emerging over an area $\sim 2\pi R(1)2H$ (Figure 2.10). The effective temperature of the BL for a non-rotating primary will then be given by

$$4\pi R(1)H\sigma T_{\mathrm{BL}}^4 \approx \frac{1}{2}\frac{GM(1)\dot{M}(\mathrm{d})}{R(1)} \tag{2.55a}$$

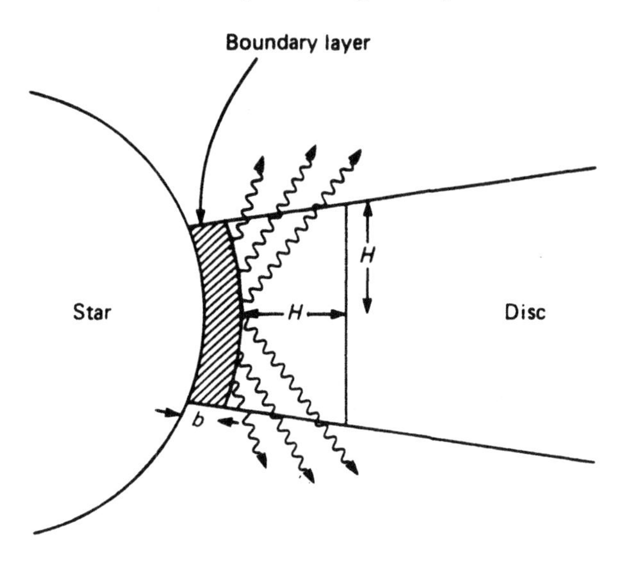

Figure 2.10 Schematic view, in a plane perpendicular to the plane of the disc, showing an optically thick boundary layer. From Frank, King & Raine (1985).

$$T_{\rm BL} \approx 2.9 \times 10^5 M_1^{1/3}(1) R_9^{-7/9} \dot{M}_{18}^{2/9}(\mathrm{d}) \quad \mathrm{K} \tag{2.55b}$$

from equations (2.10b) and (2.42a), provided that radiation pressure may be neglected when determining the scale height H. The bulk of the radiation from an optically thick BL should therefore be emitted in the soft X-ray and EUV regions.

If the BL is optically thin then radiation escapes directly from the shock front that forms as the circulating gas meets the surface of the primary. For a perfect gas the temperature of the post-shock gas is (e.g., Frank, King & Raine (1985))

$$T_{\rm sh} = \frac{3}{16} \frac{\mu_{\rm m} m_{\rm H}}{k} v_{\rm p}^2 \tag{2.56}$$

where $v_{\rm p}$ is the pre-shock velocity. Then the shock temperature for gas arriving at the primary *in a circular orbit* is

$$T_{\rm sh} = \frac{3}{16} \frac{\mu_{\rm m} m_{\rm H}}{k} \frac{GM(1)}{R(1)} \tag{2.57a}$$

$$= 1.85 \times 10^8 M_1(1) R_9^{-1}(1) \quad \mathrm{K}. \tag{2.57b}$$

Thus an optically thin BL should radiate in hard X-rays, with energies ~ 20 keV.

There is, however, an important additional consideration: at temperatures ~10^8 K cooling of an optically thin gas occurs through free–free emission, which is relatively inefficient. If the cooling time $t_{\rm cool}$ is longer than the adiabatic expansion time scale $t_{\rm ad}$ then the BL will expand to form a hot, X-ray emitting corona (Pringle & Savonije 1979; King & Shaviv 1984a; King 1986; Narayan & Popham 1993). The formation of a hot corona when $t_{\rm cool} > t_{\rm ad}$ is principally a condition on the rate of mass transfer $\dot{M}(\mathrm{d})$: the cooling time is

$$t_{\rm cool} = 2kT/n\Lambda, \tag{2.58}$$

where n is the number density of particles and Λ is the radiative cooling coefficient. The adiabatic expansion time scale is

$$t_{\rm ad} = H/c_{\rm s}. \tag{2.59}$$

The number of atoms in the BL is $\dot{M}(\mathrm{d})b/\mu_{\rm m}m_{\rm H}v_{\rm rad}$ and the volume of the BL is $\sim 2\pi R(1)2Hb$, where b is the radial extent ($\ll R(1)$) of the BL and the scale height is determined by the temperature in the BL. From equations (2.10a), (2.42a), (2.49), (2.58) and (2.59) it follows that $t_{\rm cool} > t_{\rm ad}$ when

$$\dot{M}(\mathrm{d}) \lesssim \frac{8\pi\alpha}{\Lambda}\left[\frac{R^3(1)}{GM(1)}\right]^{1/2} \frac{(kT)^{5/2}}{(\mu_{\rm m}m_{\rm H})^{1/2}} \tag{2.60a}$$

$$= 5 \times 10^{16} \alpha R_9^{3/2}(1) M_1^{1/2}(1) T_8^{5/2} \quad \mathrm{g\ s^{-1}} \tag{2.60b}$$

for $\Lambda = 3 \times 10^{-23}$ erg cm^3 s^{-1} (Osterbrock 1989; for general cooling rates for astrophysical plasmas see Raymond & Smith (1977) and Sutherland & Dopita (1993)).

Therefore, with $\alpha \sim 0.4$, we may expect the BL to expand into a hard X-ray emitting corona if $\dot{M}_{16}(\mathrm{d}) \lesssim 3$, but for higher rates of mass transfer the BL will remain geometrically thin. In that case the hard X-rays must make their way through the disc material (Figure 2.10) which becomes optically thick also at $\dot{M}_{16}(\mathrm{d}) \sim 3$ (Pringle & Savonije 1979; Narayan & Popham 1993). The condition $\dot{M}_{16}(\mathrm{d}) \sim 3$ thus separates optically thick, soft X-ray emitting BLs from coronal, hard X-ray emitting BLs.

It is likely, however, that, even when the contact between the central plane region of the disc and the primary is at large optical depth, the contact of the upper atmosphere of the disc with the primary will be able to generate low luminosity hard X-rays. A realistic model therefore must take the atmospheric density gradient into account (Patterson & Raymond 1985a,b).

Modelling of the flow of gas through the BL and onto the primary requires specification of the nature of viscosity in the layer. The radiation hydrodynamic equations are as difficult to solve as those for the accretion disc, and radiation pressure must be included (Pringle 1977). Approximate models have been produced by Tylenda (1977,1981b), Regev (1983,1989), King & Shaviv (1984a), Robertson & Frank (1986), Regev & Hougerat (1988) and reviewed by Shaviv (1987).

The most complete optically thick two-dimensional models are those of Kley (Kley & Hensler 1987; Kley 1989a,b,c, 1990, 1991), an example of which is shown in Figures 2.11 and 2.12. Most of the matter accumulates in an equatorial belt around the star, with a very small amount flowing towards higher stellar latitudes (see Robertson & Frank (1986), Kley & Hensler (1987) and Pringle (1988) for other discussions). The belt extends to $\sim \pm 5°$ either side of the equator and the BL has a thickness $b \sim 5 \times 10^{-3} R(1)$.

Flow and stability in the BL are strongly dependent on the nature of the viscosity, which is almost certainly non-isotropic. Large fluctuations in $L(\mathrm{BL})$ on a time scale

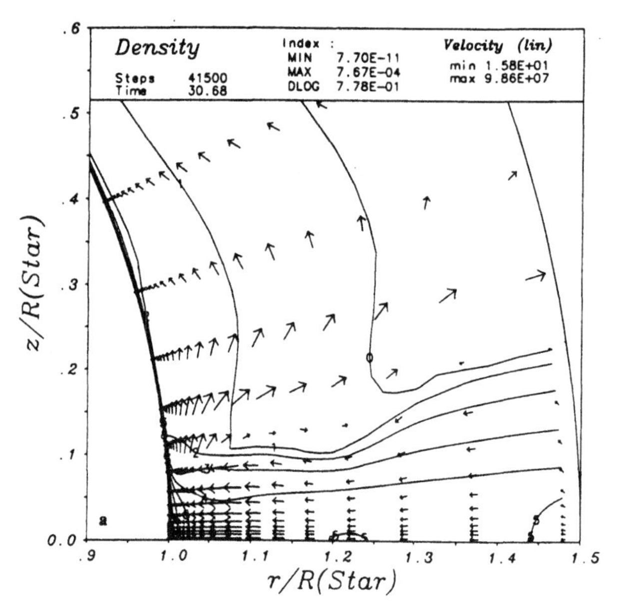

Figure 2.11 Density contours in the boundary layer for a non-rotating primary. Arrows indicate velocity flow vectors. From Kley (1989b).

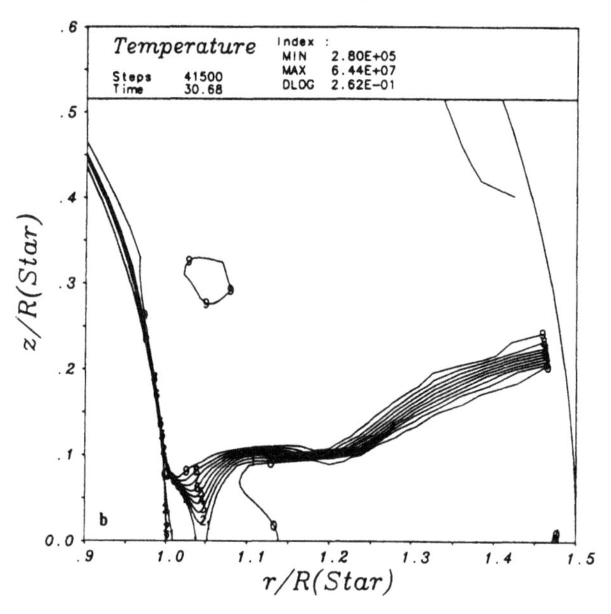

Figure 2.12 Temperature contours for the model shown in Figure 2.11. From Kley (1989b).

$\sim \Omega_K^{-1}[R(1)]$ can occur (Papaloizou & Stanley 1986; Popham & Narayan 1991; Kley 1991).

Furthermore, the radial velocity $v_{\rm rad}$ reaches a maximum in the BL, and unless careful consideration is given to realistic behaviour of viscosity, supersonic infall velocities may appear which destroy the possibility of causal contact between the star and the disc (Pringle 1977). Several existing models violate this condition, but Popham & Narayan (1992) make physically reasonable changes to the viscosity prescription to eliminate the effect. Their generalization of the α-viscosity uses $(H^{-1} + H_r^{-1})^{-1}$ instead of H in equation (2.48a), where H_r is the radial pressure scale height (which, in the BL, is much smaller than the vertical pressure scale height H), and limits turbulent velocities to $< c_s$. Glatzel (1992) has independently arrived at similar conclusions.

There is observational evidence of strong stellar winds from the inner disc or BL in systems with high $\dot{M}$(d) and that in these systems $T_{BL} \lesssim 1 \times 10^5$ K (Section 2.7.3). These results are incompatible with the model shown in Figures 2.11 and 2.12, which has $T_{BL} > 2.8 \times 10^5$ K and no pronounced wind. There is a possibility, therefore, that (a) part of the energy that would be radiated in the BL goes into kinetic energy of the wind, and (b) the wind-generating mechanism increases the surface area of the inner disc/BL region, leading to a lower effective temperature, or (c) the primary is rotating near break-up speed (equation (2.54)) (Hoare & Drew 1991).

The last of these possibilities introduces a need to consider the connection between disc and primary when $\Omega(1) \approx \Omega_K[R(1)]$. Surprisingly, the primary can continue to accrete mass without accreting angular momentum; in fact, in simplified models that treat the star and disc as a single polytropic fluid, angular momentum may flow from star to disc (Pringle 1989; Popham & Narayan 1991; Paczynski 1991; Colpi, Nannurelli & Calvani 1991).

2.5.5 *Tidal Limitation*

For discs that approach $R_L(1)$ in radius the outer parts will be significantly distorted by the gravitational influence of the secondary star. If pressure and viscosity effects can be neglected, streamlines of the disc flow are closely the same as simple periodic orbits in the restricted three-body problem. There will be a largest orbit that does not intersect any other orbit; at larger radii intersecting orbits, which are equivalent to tidal shear, will produce dissipation and prevent the disc from growing any larger.

Paczynski (1977; see also Whitehurst 1988a and Whitehurst & King 1991) computed single particle orbits and tabulated the maximum radius vector of the last non-intersecting orbit. His results are well approximated by

$$\frac{r_d(\max)}{a} = \frac{0.60}{1+q} \qquad 0.03 < q < 1. \tag{2.61}$$

The ratio $r_d(\max)/R_L(1)$ may then be obtained from the formulae in Table 2.1.

A different formalism was adopted by Papaloizou & Pringle (1977), with similar results, but they point out that the introduction of even a small amount of viscosity ensures that the streamlines do not intersect, and therefore a disc can expand a little beyond $r_d(\max)$. It is in the non-linear region just beyond $r_d(\max)$ that tidal stresses are strongly generated. The energy generation caused by the non-axisymmetric tidal perturbations is seen in two-dimensional hydrodynamic modelling (Lin & Pringle 1976; Schwarzenberg-Czerny & Rozyczka 1988; Hirose & Osaki 1990; Zhang & Chen 1992; Osaki, Hirose & Ichikawa 1993). Dissipation causes a phase lag between the perturbations and the tidal impulse of the secondary, which generates a torque on the latter with the result that angular momentum at the outer edge of the disc is fed (back) into the secondary. The two-dimensional computations show that tidal stress and viscous stress become comparable at $r_d(\max)$ and do indeed truncate the disc. Goodman (1993) has identified a local three-dimensional instability that may be very effective in limiting disc radii for low- α discs.

The actual radius of a disc is the result of competition between disc viscosity, effects of the mass transfer stream (which adds matter of low specific angular momentum to the outer edge of the disc) and tidal dissipation.

A review of tidal effects on accretion discs is given by Osaki, Hirose & Ichikawa (1993).

2.5.6 *Disc Precession*

Observations of some X-ray binaries appear to require a tilted, precessing disc for their interpretation (e.g., the 35 d cycle in the 1.70 d binary Her X-1 (Priedhorsky & Holt 1987)). Although the origin and maintenance of a tilted disc are problematical, the presence of cyclical luminosity modulations with periods ~$20P_{\rm orb}$ in the CVs (Section 2.6.6) suggests that a similar phenomenon occurs in them.

Several mechanisms have been proposed for generating a tilt to an accretion disc (Katz 1973; Roberts 1974; Boynton, Crosa & Deeter 1980; Kondo, van Flandern & Wolff 1983; Barrett, O'Donoghue & Warner 1988). Of these, only that involving a disc freely precessing under the influence of the gravitational field of the secondary is probably relevant to CVs (Barrett, O'Donoghue & Warner 1988; Schwarzenberg-Czerny 1992).

Two major theoretical problems associated with precessing discs are (a) that each annulus will precess with a different period (unless the viscosity is very high), producing a twisted disc (Petterson 1977a,b) with no single precession period, and (b) a mechanism is required to feed gas to the tilted disc in a way that maintains the tilt. One possibility is that the stream from the secondary is initially launched out of the orbital plane because of channelling by magnetic fields on the secondary (Barrett, O'Donoghue & Warner 1988). The dynamical evolution of twisted accretion discs subjected to external torques has been studied by Papaloizou & Pringle (1983) and by Pringle (1992).

Although the maintenance of a tilted disc is still fundamentally a mystery, the need to explain the observed long periods necessitates a quantitative evaluation of the model. The precession of the outer annulus of a tilted disc is retrograde with a period (Kumar 1986,1989)

$$P_{\rm pr} \approx \frac{8\pi}{3} \frac{M^{1/2}(1)a^{3/2}}{G^{1/2}M(2)} \left(\frac{r_d}{a}\right)^{-3/2} \tag{2.62a}$$

which, from equation (2.1a), gives

$$\frac{P_{\rm pr}}{P_{\rm orb}} \approx \frac{4}{3} \frac{(1+q)^{1/2}}{q} \left(\frac{r_d}{a}\right)^{-3/2}. \tag{2.62b}$$

If we allow the accretion disc to have its maximum size, given by equation (2.61), then

$$\frac{P_{pr}}{P_{orb}} \approx 2.86 \frac{(1+q)^2}{q}. \tag{2.62c}$$

Therefore $P_{pr}/P_{orb} \sim 10$–35 for the range $0.1 < q < 1$.

2.6 Comparison with Observation

The two most direct ways of confronting the predictions from accretion disc models with observations of discs in CVs are through (a) the integrated flux distributions as a function of wavelength from X-ray to IR and (b) spatial flux distributions over the surfaces of discs, as derived from eclipse simulation or deconvolution.

2.6.1 Flux Distributions

As optically thick, steady state accretion disc radial temperature distributions are largely independent of the viscosity mechanism (Section 2.5.1) it is for these that the most optimism for good accord with observation may be held. Using absorption-line spectra as diagnostic of optical thickness, the candidates for comparison are (a) the UX UMa stars, (b) certain NRs, (c) Z Cam stars at standstill and (d) DN near maximum of outburst. In anticipation of the properties of CN remnants and NLs deduced in Chapter 4, it is not unreasonable to include all members of these classes as possessors of steady-state, optically thick discs; but it would not be unexpected if the rapid changes that occur during a DN outburst (Chapter 3) result in a steady state not being approximated. DN at minimum almost certainly are not in a steady state and have solely emission-line spectra which may arise at least partly from optically thin regions.

2.6.1.1 Computed Flux Distributions

For comparison with observations, models should be made as realistic as possible. Early disc models generally adopted blackbody emissivity (Bath *et al.* 1974; Bath 1976; Schwarzenberg-Czerny & Rozyczka 1977; Tylenda 1977). More physically realistic models fit a stellar model atmosphere of the appropriate gravity and flux (i.e., effective temperature) to annuli of finite width (Koen 1976; Mayo, Whelan & Wickramasinghe 1980; Herter *et al.* 1979; Kiplinger 1979a, 1980; Pacharintanakul & Katz 1980; Jameson, King & Sherrington 1980; Williams 1980; Wade 1984, 1988; Hassall 1985a; la Dous 1989a; Hubený 1989), the later studies often making use of Kurucz atmospheres (Kurucz 1979). (However, stellar atmospheres have constant g, whereas g varies significantly in accretion discs over the region where the spectrum is formed.) A hybrid approach adopts blackbody emissivity reduced by a factor $(1 - e^{-\tau_\nu})$ in optically thin parts of the disc, which requires choice of a functional form for $\nu_{rad}(r)$ in order to calculate the local optical depth τ_ν (Tylenda 1981a; Williams & Ferguson 1982; Smak 1984a). The most recent models abandon flux-constant atmospheres to include viscous energy deposition in the optically thin upper layers of the disc atmosphere (Kriz & Hubený 1986; Shaviv & Wehrse 1986, 1991, 1993; Adam *et al.* 1988).

By incorporating the physics of line formation into the flux code, the full spectrum of a disc can be computed. In general local thermodynamic equilibrium (LTE) is

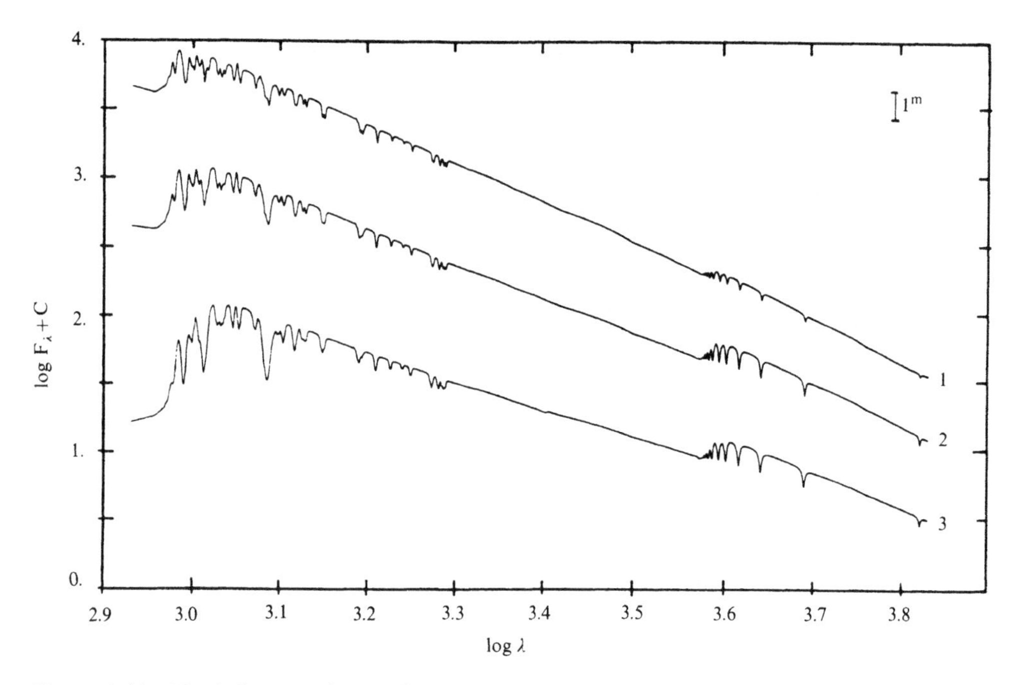

Figure 2.13 The influence of rate of mass transfer on the integrated spectrum of steady state accretion discs. The model has $M_1(1) = 1.0$, $\log r_i = 9.2$ where r_i is the inner radius of the disc, in cm. The mass transfer rates are (1) 10^{-7} M$_\odot$ y^{-1}, (2) 10^{-8} M$_\odot$ y^{-1}, (3) 10^{-9} M$_\odot$ y^{-1}. From la Dous (1989a).

assumed, which produces absorption spectra for optically thick discs, but inclusion of non-LTE conditions in the upper atmospheric regions – or the temperature inversion that occurs in models that include viscous heating throughout the atmosphere – can generate emission cores to the absorption profiles.

Figures 2.13 and 2.14 illustrate the variations of emitted spectrum with changes of the dominant disc parameters (la Dous 1989a). Variations of $\dot{M}$(d), holding all other parameters constant, produce significant changes of continuum slope, Balmer jump and line depth (Figure 2.13). In Figure 2.14(a) the effect of changing the mass of the primary (but keeping the inner radius r_0 of the disc constant) is shown; as expected, the total energy emitted is reduced at the lower masses.

In reality, as $R(1) \tilde{\propto} M^{-1/3}(1)$ (Section 2.8.1), the ratio $M(1)/R^3(1)$, which determines the maximum temperature in the disc through equation (2.33), varies as $M^2(1)$. Part of this sensitivity to $M(1)$ comes from the increase in potential energy and part from the further distance the gas has to fall and the smaller disc area near the primary it has in which to radiate its energy. Thus Figure 2.14(a) takes into account only one aspect of the $M(1)$ variation. In Figure 2.14(b) the effect of inclination of the disc is investigated: the effects of limb darkening and reduced projected area greatly reduce the total energy emitted at high inclinations; furthermore, the flux distribution becomes flatter at high inclinations as the emitted flux is weighted towards the cooler upper regions of the atmosphere.

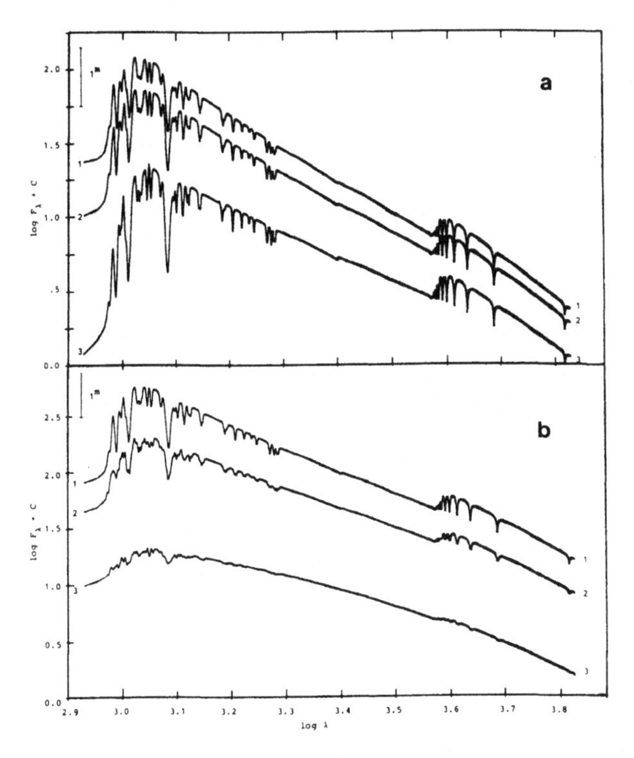

Figure 2.14 The influence of (a) primary mass and (b) inclination on the integrated spectrum of a steady state accretion disc. The white dwarf masses are (1) 1.3 $M_\odot$, (2) 1.0 $M_\odot$, (3) 0.5 $M_\odot$, and $\dot{M}(d) = 10^{-9}$ $M_\odot$ y^{-1}, log $r_0 = 9.2$. The inclinations are (1) 28°, (2) 60°, (3) 83°. Adapted from la Dous (1989a).

In Figure 2.15 the similarity in the UV between model flux distributions of two discs in which Kurucz atmospheres have been used at each annulus and that of a single star is shown (Wade 1984). In the $1000 < \lambda < 4000$ Å region the fluxes quite closely follow the $\nu^{1/3}$ law for extended discs (Section 2.5.1); only in the visible region are the slopes of the flux distributions significantly different.

These variations in spectra show that, in the absence of accurate independent determinations of $M(1)$ or i, it may often be impossible to derive an unambiguous set of system parameters from flux and spectrum line profiles alone.

In most models of disc fluxes, aimed at comparison with observations in the FUV–IR range, the contributions of the BL have been omitted. The soft X-ray and EUV emission from the BL are modelled separately (Tylenda 1981b; Kley 1989c). The concavity of the standard accretion disc implies that some radiation from the BL will be intercepted by the face of the disc, but the heating from this does not significantly affect the continuum flux from the disc (Pacharintanakul & Katz 1980; Wade 1988; Smak 1989b). Note, however, that in the case of recent novae and DN after outburst the high temperature of the primary *can* cause important heating of the disc (Friedjung 1985; Smak 1989b), and that in general a thin disc of large radius intercepts more than one quarter of the flux emitted by the primary (Adams & Shu 1986).

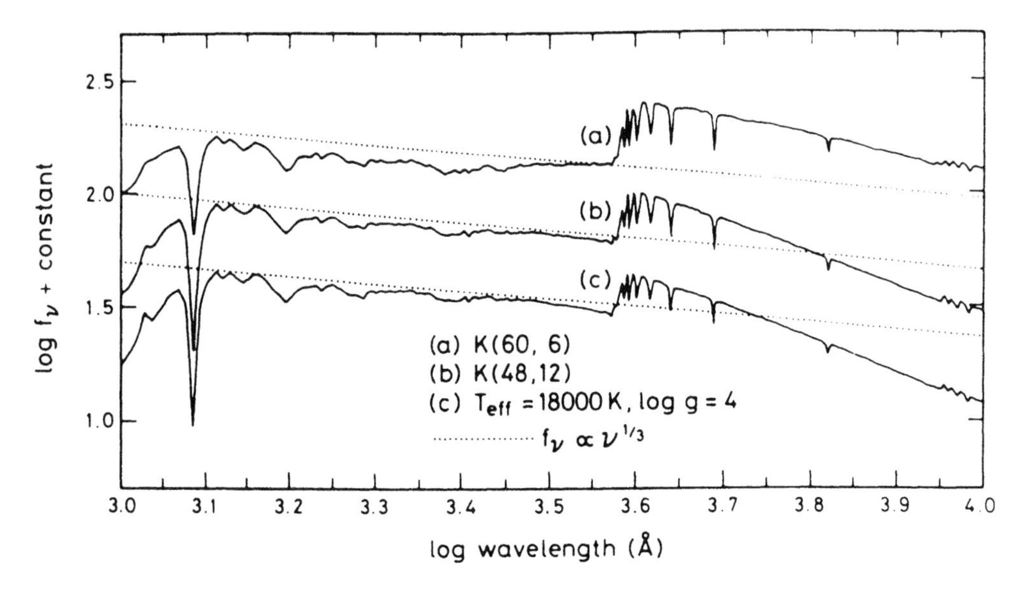

Figure 2.15 Computed spectra of two accretion discs and a stellar atmosphere in the region 1000 Å–1 μm. Spectrum (a) is for $T_* = 6.0 \times 10^4$ K, $r_d/R(1) = 22.4$, $\dot{M}(d) = 7.5 \times 10^{15}$ g s^{-1}; spectrum (b) is for $T_* = 4.8 \times 10^4$ K, $r_d/R(1) = 16.3$, $\dot{M}(d) = 3.1 \times 10^{15}$ g s^{-1}; spectrum (c) is for a stellar atmosphere with $T_{eff} = 1.8 \times 10^4$ K, log g = 4.0. From Wade (1984).

2.6.1.2 Observed Flux Distributions

A complication in the comparison of model with observed CV flux distributions is the need in the latter to subtract possible contributions from the secondary, bright spot and primary. Fortunately, at the higher mass transfer rates ($\dot{M}(d) \geq 10^{-9} M_\odot$ y^{-1}) the disc is usually the dominant contributor, so uncertainties in the removal of other contributors are relatively minor.

Observed flux distributions (e.g., Oke & Wade (1982)) in the visible region of DN near maximum of outburst or Z Cam stars at standstill are represented quite well by model discs (Kiplinger 1979a, 1980: Figures 2.16 & 2.17) but models are inadequate at minimum light (Figure 2.18). Note that in some cases the departure from a disc distribution may be due to the presence of a strong contribution from the primary, as shown for VW Hyi in Figure 1.17. Commonly it is not possible to fit the observed small Balmer jumps; one solution is to add a component of hydrogen recombination spectrum (e.g., Schwarzenberg-Czerny (1981)) whose emission Balmer jump partially fills in the absorption (Figure 2.19: Hassall 1985a). It is hoped that future models with dissipation in the atmosphere will automatically produce Balmer jumps of the correct size.

As so much of the radiation from a CV is emitted below 3000 Å, most comparisons between theoretical and observational discs have been based on IUE flux distributions. Commonly, two wavelengths are selected at which continuum fluxes may be measured uncontaminated by emission lines (e.g., 1460 Å and 2880 Å) and their ratio compared with model predictions. The results (Szkody 1985a; Verbunt 1987; Wade 1988) show moderately good, but not perfect, agreement between theory and observation for those

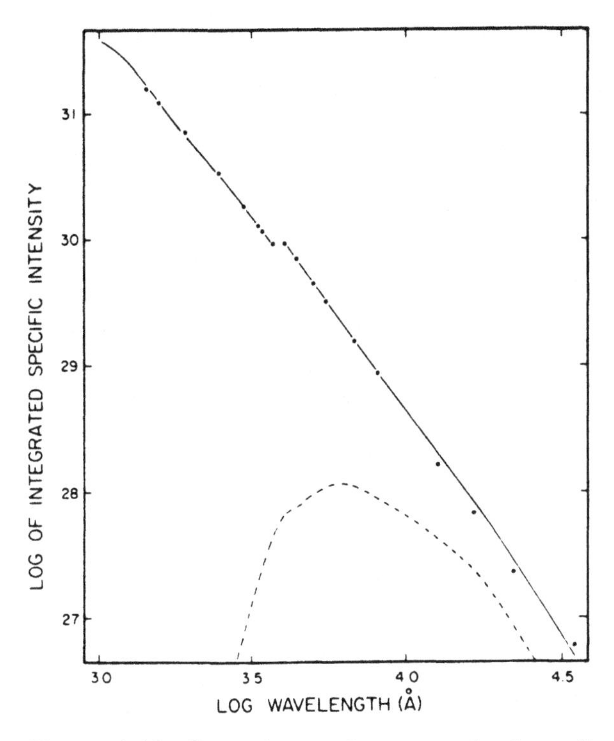

Figure 2.16 Comparison of computed flux distribution for an accretion disc ($\dot{M}(\mathrm{d}) = 8.5 \times 10^{-8}\ \mathrm{M}_\odot\ \mathrm{y}^{-1}$) and the observed flux distribution of SS Cyg at maximum light. The dashed curve is the contribution of the secondary star. From Kiplinger (1979a).

CVs expected to approximate steady discs. The great sensitivity of UV flux to the mass of the primary requires that masses more accurate than usually available (Section 2.8) be employed if $\dot{M}(\mathrm{d})$ is to be reliably estimated from the slope of the UV spectrum (Verbunt 1987).

In the FUV high $\dot{M}(\mathrm{d})$ systems show a deficiency of flux relative to standard α-disc predictions (Long *et al.* 1994). This implies major departures from the expected structure of the inner region of the disc; some suggested explanations are given in Sections 4.3.4 and 4.4.4.2.

2.6.1.3 $\dot{M}(\mathrm{d})$ *from* M_v

Similarly, use of flux gradients in the visible, or absolute visual magnitude M_v of the disc, to obtain $\dot{M}(\mathrm{d})$ is sensitive to system parameters. Smak (1989a), from blackbody disc annuli, shows that at low $\dot{M}(\mathrm{d})$, $\mathrm{d}M_\mathrm{v}/\mathrm{dlog}\dot{M}(\mathrm{d}) \approx -2$, $\mathrm{d}M_\mathrm{v}/\mathrm{dlog}\dot{M}(1) \approx -4$ and that there is little dependence of M_v on $r_\mathrm{d}/R(1)$; whereas at high $\dot{M}(\mathrm{d})$, $\mathrm{d}M_\mathrm{v}/\mathrm{dlog}\dot{M}(\mathrm{d}) \approx -1$ and $\mathrm{d}M_\mathrm{v}/\mathrm{d}[r_\mathrm{d}/R(1)] \approx -2.5$ with little dependence on $M(1)$. His computations of the $\dot{M}(\mathrm{d})$–M_v relationships offer a helpful first estimate for $\dot{M}(\mathrm{d})$ from M_v that are an improvement over use of more restrictive relationships (e.g., Tylenda (1981a)) computed for fixed $r_\mathrm{d}/R(1)$. Further computations, including radiation from the side of the disc, give M_v about 0.5 mag brighter at a given $\dot{M}(\mathrm{d})$ (Smak 1994a).

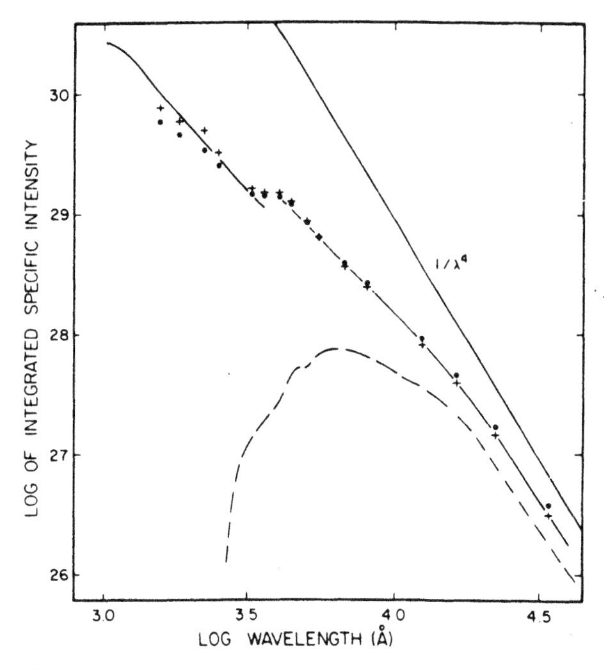

Figure 2.17 Comparison of observed (points) and computed ($\dot{M}(d) = 4.8 \times 10^{-9}$ $M_\odot$ y^{-1}) flux distributions for Z Cam at standstill. Plus signs show the effect of allowance for reddening. The dashed curve is the contribution of the secondary. The upper line is a Rayleigh–Jeans distribution for a blackbody of infinite temperature. From Kiplinger (1980).

As the apparent luminosity of a disc depends sensitively on its inclination, a convention is required to define absolute magnitude. Paczynski & Schwarzenberg-Czerny (1980), introducing a standard limb darkening formula into equation (2.45), derive a relationship which, with a limb-darkening coefficient of 0.6 which fits well the model discs of Mayo, Whelan & Wickramasinghe (1980), becomes

$$\Delta M_v(i) = -2.50 \log\left\{(1 + \frac{3}{2}\cos i) \cos i\right\} \tag{2.63}$$

where ΔM_v is the correction from *apparent absolute magnitude to absolute magnitude* (Warner 1987a). As $\Delta M_v = 0$ for $i = 56°.7$, this defines the standard inclination. It may be noted that, averaging over all inclinations, $\overline{\Delta M_v} = -0.367 = \Delta M_v(46°)$.

The effect of $\Delta M_v(i)$ is seen directly in the apparent absolute magnitudes of nova discs, which are optically thick and at approximately constant M_v (Figure 2.20). The same effect is seen in the UV luminosities (Selvelli *et al.* 1990). For large inclinations the effective inclination of the disc is $i_{\rm eff} = i + \sin^{-1}(h_d/r_d)$. Equation (2.52) shows that for $\dot{M}(d) = 10^{18}$g s^{-1}, the opening angle of the disc is > 4°. At large inclinations the side of the disc contributes significantly to the luminosity (Smak 1994a) reducing the sensitivity of ΔM_v(i) to inclination for $i \gtrsim 85°$.

From such relationships, together with values of M_v for the various subgroups of CVs (as discussed in later chapters), the following rough estimates are obtained: for nova remnants, NLs and DN during outburst, $\dot{M}(d) \sim 3 \times 10^{-9}$–$1 \times 10^{-8}$ $M_\odot$ y^{-1}

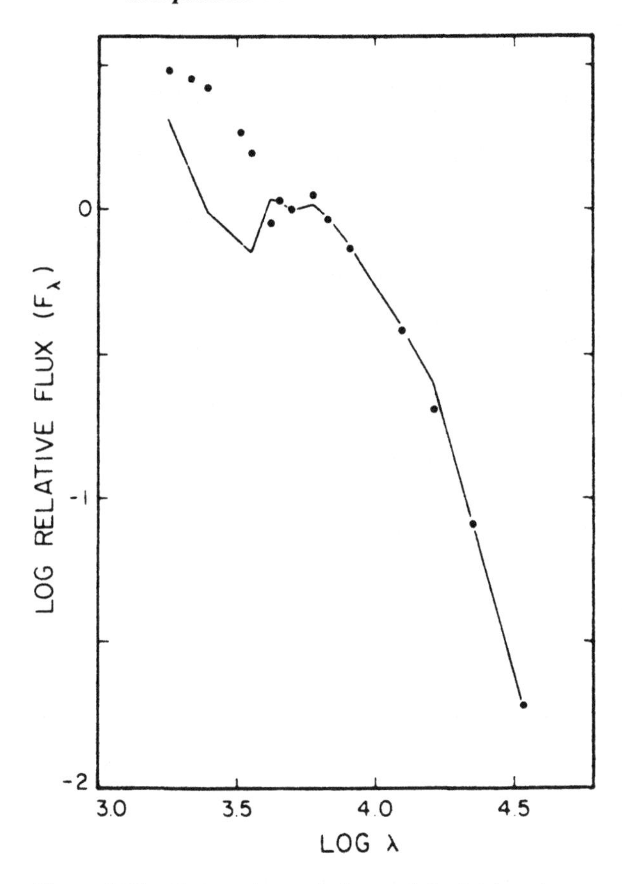

Figure 2.18 Comparison of observed (points) and computed ($\dot{M}(d) = 5 \times 10^{-11}$ $M_\odot$ y^{-1}) flux distributions of SS Cyg at quiescence. The contribution from the secondary is included. From Kiplinger (1979a).

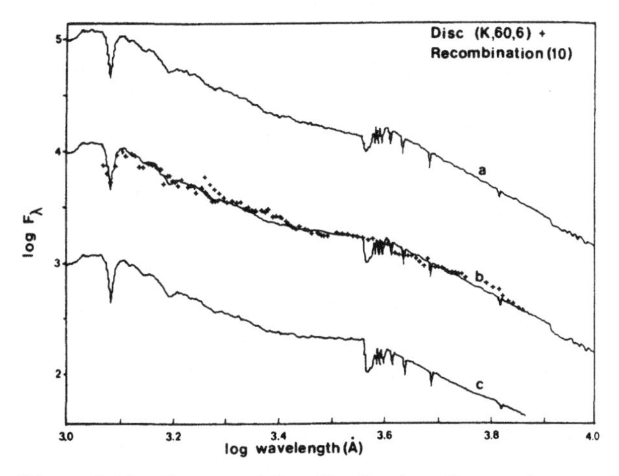

Figure 2.19 Computed flux distributions for steady state discs with Kurucz atmospheres plus a recombination spectrum. The emission measure in spectra (a), (b) and (c) increases in the ratio 1:2:3. The plus signs are the observed flux distribution of dwarf nova EK TrA during outburst.

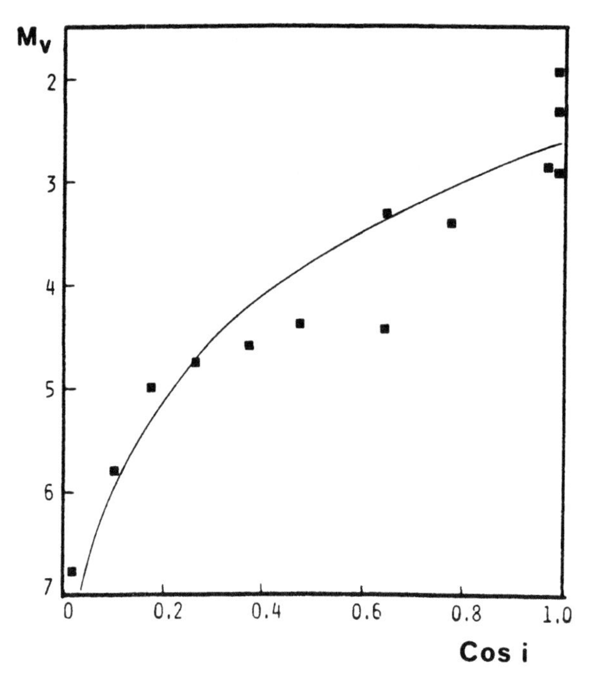

Figure 2.20 Observed relationship between apparent absolute magnitude and inclination for NRs. The curved line is equation (2.63). From Warner (1986e).

(2×10^{17}–6×10^{17} g s^{-1}) and for DN at quiescence $\dot{M}(\mathrm{d}) \sim 2 \times 10^{-11}$ –$5 \times 10^{-10}\,\mathrm{M}_\odot\,\mathrm{y}^{-1}$ (1×10^{15}–3×10^{16} g s^{-1}). DN during rise and Z Cam stars at standstill typically have $\dot{M}(\mathrm{d}) \sim 5 \times 10^{-10}$–$3 \times 10^{-9}\,\mathrm{M}_\odot\,\mathrm{y}^{-1}$

2.6.1.4 IR Observations

The early study by Sherrington *et al.* (1980) showed agreement between IR colours and predictions from steady state discs, but later observations have shown a wider range of colours in conflict with disc properties (Szkody 1981a; Sherrington & Jameson 1983). The problem is that in the IR there may be contributions from the (optically thick) atmosphere of the cool secondary and from both optically thick and optically thin parts of the outer regions of the disc.

The observed IR colours of CVs differ significantly from predicted colours of discs with blackbody emissivity (Figure 2.21). Berriman, Szkody & Capps (1985) show how the observed colours may be separated into their optically thick and thin components. The proportion of light from each component varies widely from star to star. The optically thin component has the same emission measure as is found from the Balmer emission lines and evidently arises in the disc chromosphere (Section 2.7.1); this same component can be an important continuum source in the visible region (as already noted in Figure 2.19). To be optically thin at $\lambda \sim 2\,\mu$m requires an electron density $\lesssim 10^{13}$ cm^{-3} for a temperature of 10^4 K (Berriman, Szkody & Capps 1985). Low resolution spectra in the 1–2.5 μm region (Ramseyer *et al.* 1993b) require accretion disc emission-line models for their interpretation.

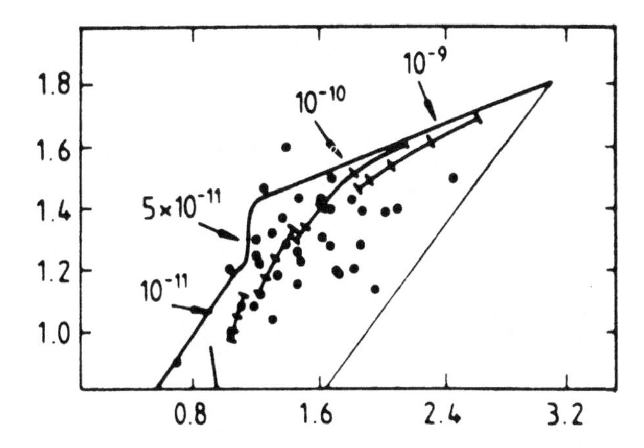

Figure 2.21 IR two-colour diagram for CVs. The ordinate is $F_\nu(H)/F_\nu(K)$ and the abscissa is $F_\nu(J)/F_\nu(K)$. Dots are the observed flux ratios. The upper boundary curve is the relationship for optically thick spectra. The short curves are computed relationships for steady state discs of various radii with the mass transfer rates as marked. From Berriman, Szkody & Capps (1985).

Harrison & Gehrz (1992) have detected ~ 80% of the DN and NLs looked for in IRAS photometry (12–100 μm). The flux distributions require line emission to be the source of the IR flux, but there is as yet no explanation for this unexpected result.

2.6.1.5 X-ray Observations

Most of the optically bright CVs have been detected in hard ($\gtrsim 1$ keV) X-rays (Beuermann & Thomas 1993). In comparing the X-ray luminosities L_x with $\dot{M}(d)$ (deduced in general from an M_v–$\dot{M}(d)$ relationship) it is found that for $\dot{M}_{16}(d) \leq 1$ there is reasonable agreement between $L_x \sim GM(1)\dot{M}(d)/2R(1)$ and observation (Patterson & Raymond 1985a); i.e., the BL is optically thin at low rates of mass transfer, as predicted in Section 2.5.4. For $\dot{M}_{16}(d) \gtrsim 1$ the observed L_x does not decrease to zero but instead remains at $L_x \sim 10^{32}$ erg s^{-1} (Figure 2.22), which is taken as evidence for hard X-ray generation from a small coronal region produced by accretion of the upper, optically thin regions of the disc onto the primary. Figure 2.22 compares the observations with (a) an optically thin BL, (b) an homogeneous BL which becomes optically thick at $\dot{M}_{16}(d) \sim 10$ (Tylenda 1981b) and (c) a model including a vertical density variation of the form given by equation (2.40) (Patterson & Raymond 1985a). The last model accounts generally for the observed relationship between L_x and $\dot{M}(d)$, and also the observed variations of the ratio of X-ray to optical flux as a function of $\dot{M}(d)$; the optical flux F_v is that generated by the flow $\dot{M}(d)$ through the disc. The most discrepant points in Figure 2.22 – the open circles that fall well below the predicted relationships – are DN during outburst which are considered in Section 3.3.6.

Figure 2.22 shows the general decrease in F_x/F_v with increasing $\dot{M}(d)$; as a result a flux-limited X-ray survey will detect low $\dot{M}$ systems to much fainter optical limits than high $\dot{M}$ systems. For example, ROSAT should detect DN down to $m_v \sim 17$, but high state NLs only to $m_v \sim 13$ (Beuermann & Thomas 1993).

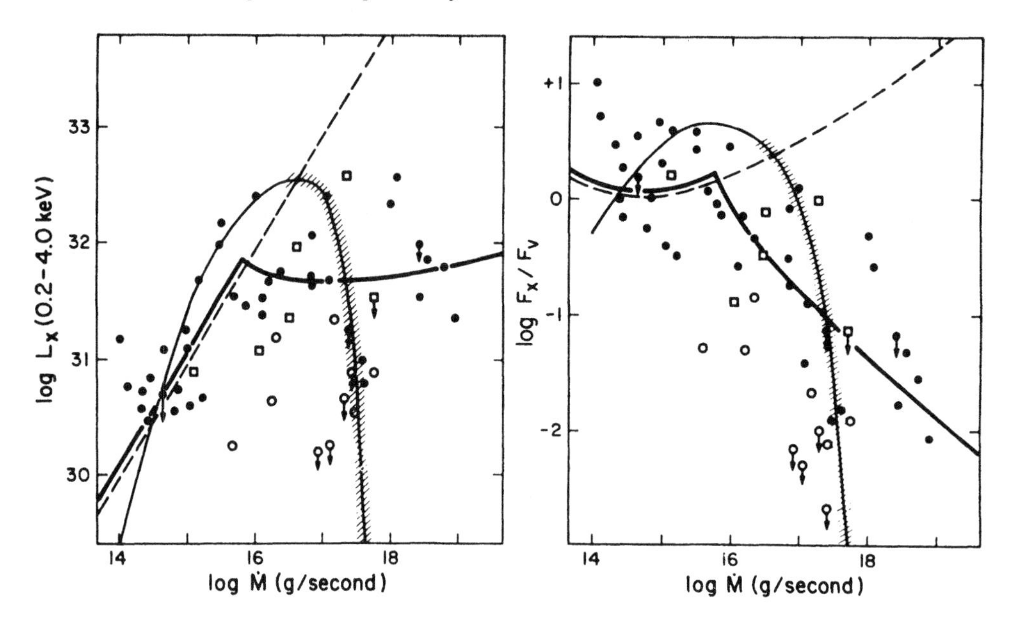

Figure 2.22 Dependences of X-ray luminosity and ratio of X-ray to optical fluxes on mass transfer rate. The dashed curve is an optically thin model; the light solid curve is Tylenda's (1981b) model and the heavy solid curve is the model including a vertical density gradient. From Patterson & Raymond (1985a).

The ratio $0.001 < F_X/F_V < 10$ is characteristic of non-magnetic CVs. In polars the ratio lies in the range 3–300 (Section 6.4). In low mass X-ray binaries, with accretion onto neutron star primaries, the ratio is 100–1000.

The derivation of L_X from observed fluxes at different energies requires integration over a flux distribution obtained from fitting an adopted parameterized energy distribution (e.g., optically thin bremsstrahlung) multiplied by a factor $e^{-N_H\kappa(E)}$, where N_H is the interstellar and/or circumstellar hydrogen column density and $\kappa(E)$ is the absorption coefficient (e.g., from Brown & Gould 1970). Often there is not a unique separation of N_H and kT.

For a group of 32 (at most weakly magnetic) CVs observed with the Einstein satellite, Eracleous, Halpern & Patterson (1991a) find a median $kT \approx 3$ keV and $N_H \sim 10^{20}$–10^{21} cm^{-2}. The latter is in agreement with interstellar N_H inferred from 21 cm emission and Lyα absorption-line strengths. A few stars show evidence for additional, probably circumstellar absorption.

The detectability of EUV and soft X-rays emerging from the optically thick BLs of high $\dot{M}$(d) systems depends crucially on the effective temperature of the thermalized BL emission. At the temperature $\sim 2 \times 10^5$ K predicted by equation (2.55b) a blackbody distribution peaks at ~ 0.05 keV, which is well below the low energy cut-off for most satellite detectors, and lies in the region of high absorption by the Lyman continuum of interstellar hydrogen. (Note, however, that to avoid detection by *both* IUE *and* ROSAT only a narrow range around $kT \sim 10$ eV is possible.) Consequently, most of the emission from optically thick, high $\dot{M}$(d) BLs (up to half of the total luminosity from such systems) could be unobservable (Pringle 1977; Cordova & Mason 1983; Patterson & Raymond 1985b). The fluxes observed by ROSAT in DN at quiescence are inadequate to explain the observed HeII emission flux, again suggesting a hidden ~10 eV photon flux (Vrtilek *et al.* 1994). An observable soft X-ray flux arises only for those nearby high $\dot{M}(\mathrm{d})$ systems where $M(1)$ is large and the interstellar column density is low; these conditions are satisfied for the DN SS Cyg, U Gem and VW Hyi, where there is good accord between observation and prediction (Section 3.3.6).

In Section 2.7.3 evidence from emission lines suggests that $T_{\mathrm{BL}} \lesssim 10^5$ K. These lower temperatures are compatible with the observed soft X-ray fluxes provided large column densities exist (Hoare and Drew 1991; see also Section 2.7.3). Large *circumstellar* column densities are consistent with the stellar wind models on which the low T_{BL} are based.

2.6.1.6 Polarization from Discs

Linear polarization is produced by electron or Rayleigh scattering in atmospheres of stars or discs. In the case of single scattering and non-symmetrical averaging over the object, the polarization may be preserved in integrated light at a level of a few tenths of a per cent. Linear polarization from the discs of Be stars is an observed example (e.g., Serkowski (1971)).

Cheng *et al.* (1988) have computed the polarization from steady state CV accretion discs with Kurucz atmospheres. For highly inclined discs they find much larger polarizations (up to 1.4%) in the UV (1360 Å) than in the visible. The degree of polarization is proportional to the ratio of electron scattering to free–free opacity at unit optical depth, which, through equation (2.51a), results in a weak positive dependence on $\dot{M}(\mathrm{d})$. As a result, there should be a variation of polarization during the outbursts of DN. The degree of polarization is a strong function of inclination (being zero for pole-on discs) and there are significant variations through eclipse.

2.6.2 Eclipse Analysis

The eclipsing CVs (amounting to ~20% of the total known) offer the greatest potential for providing details of the arrangement and relative brightness of the various components. This information may be extracted in a number of ways, the most direct being identification in the eclipse profile of the successive stages of immersion and emersion of the components, combined with simple geometric considerations.

Figure 2.23 displays the contact points identifiable in an ideal eclipse light curve. From geometry alone, assuming that the secondary fills its Roche lobe, the approximate relative sizes of the primary, the accretion disc, the bright spot and the secondary may be derived. With the addition of radial velocity (RV) observations some absolute dimensions are gained.

The duration $\Delta\varphi$ of total eclipse of the primary is a function only of q and i. For Roche geometry this function is not expressible in analytical form, but is available in tabular and graphical form (Chanan, Middleditch & Nelson 1976; Horne 1985a: Figure 2.24). Even if the eclipse of the primary is not identifiable, the width of eclipse at half depth (often called $\Delta\varphi_{1/2}$) is a reasonable measure of its duration if the disc is

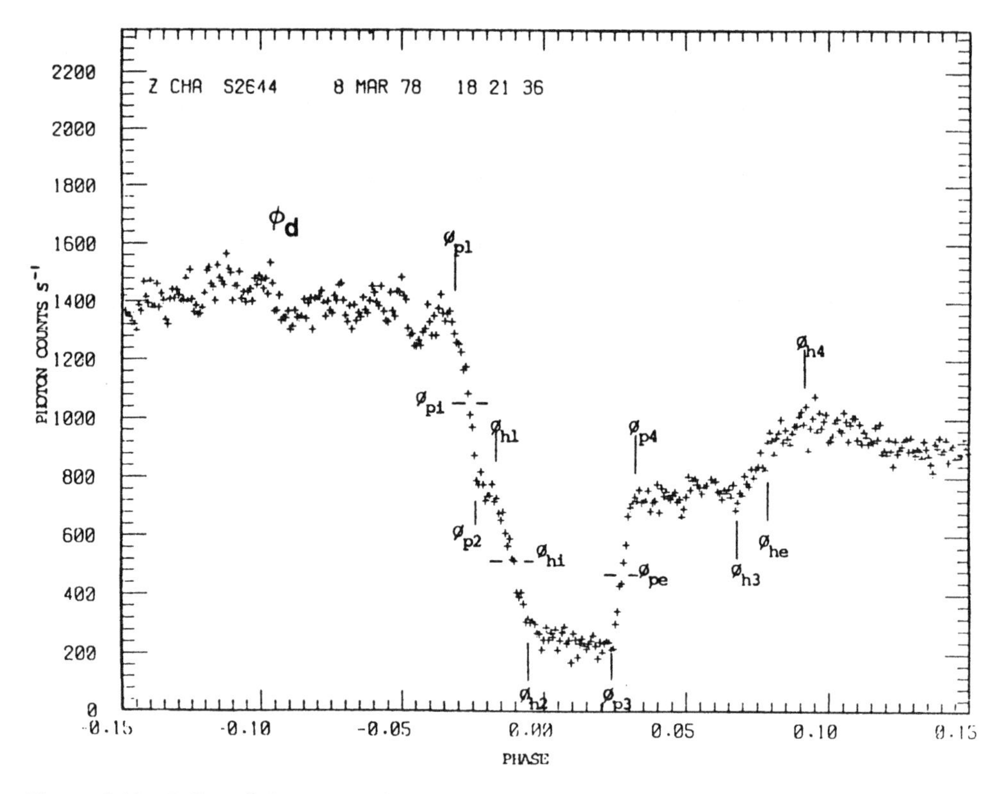

Figure 2.23 Eclipse light curve of Z Cha with contact phases marked. φ_d is the probable beginning of eclipse of the disc. The other phases are marked as φ_{pn} for the primary or φ_{hn} for the bright spot, where $n = 1$ is first contact, $n = 2$ is second contact, etc. Mid-ingress is $n =$i, and mid-egress is $n =$ e. From Cook & Warner (1984).

axially symmetric (i.e., not seriously distorted by a bright spot as in UX UMa: Figure 1.6); with a non-symmetric disc, the points of greatest slope in the eclipse profile occur at the times when the brightest and smallest object is being eclipsed, which is usually the centre of the disc (but may on occasion be the bright spot).

Approximate results are obtained by ignoring the distorted shape of the secondary. Treating it as a sphere of volume radius $R_L(2)$, a relationship between q and i is given by

$$\sin^2 i \approx \frac{1 - [R_L(2)/a]^2}{\cos^2 2\pi\varphi_p}, \tag{2.64}$$

where $R_L(2)/a$ is given as a function of q in Table 2.1 and $\pm\varphi_p$ are the phases of mid-immersion and mid-emergence of the primary. (Orbital phase zero is defined as the time of inferior conjunction of the secondary; if the system possesses a non-axially symmetric distribution of light over the disc, this may differ considerably from the time of eclipse minimum.) Although there may be some reluctance to use approximations to Roche geometry, Horne, Lanning & Gomer (1982) find that equations (2.5a) and (2.5b), together with equation (2.64), agree very well with exact computations. From

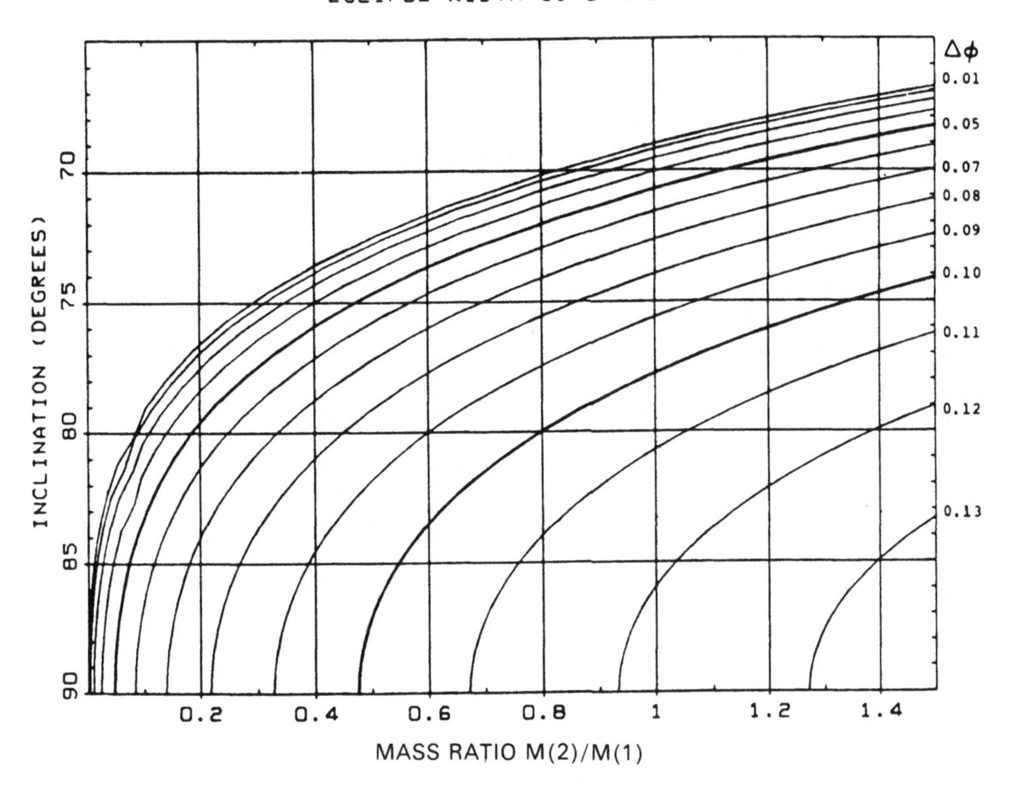

Figure 2.24 Relationship between mass ratio q and inclination i for different values of the eclipse width $\Delta\varphi$ at the centre of the disc. From Horne (1985a).

equation (2.64) it follows that the primary is eclipsed only if $i > i_{\min} = \cos^{-1}[R_L(2)/a]$; accurate values of $i_{\min}(q)$ are available in Chanan, Middleditch & Nelson (1976) and Bailey (1990).

General formulae that relate the observed eclipse contact points to the structure of the system, and also enable relative luminosities of the components to be derived, are given in Smak (1971a, 1979a), Lin (1975), Robinson, Nather & Patterson (1978) and Schwarzenberg-Czerny (1984a).

The radius of the disc may be found from its first and last contact points, occurring at phases $\pm\varphi_d$. To first order, for deep eclipses,

$$r_d/a \approx \tan 2\pi\varphi_d - \tan 2\pi\varphi_p, \tag{2.65}$$

so a rough estimate of the size of the disc is quickly obtained from the total width of eclipse and its width at half-depth. Systems with deep eclipses (e.g., Figure 1.6) typically have widths $\varphi_d \sim 0.10$ and $\varphi_p \sim 0.05$, giving $r_d/a \sim 0.4$ or $r_d/R_L(1) \sim 1$. A more detailed study shows that for high $\dot{M}$(d) systems $r_d/R_L(1) \sim 0.7$–0.8 (Ritter 1980a; Sulkanen, Brasure & Patterson 1981).[5] Comparison with r_r (equation(2.19)) then shows that discs in CVs have radii ~ 3 times those of inviscid discs.

[5] Some authors express r_d in units of R_{L_1}. From Table 2.1, $R_L(1) \approx 0.76\ R_{L_1}$ independent of q.

2.6.3 *Eclipse Simulation*

The next stage of sophistication in extracting information from eclipse light curves is to simulate eclipses of model accretion discs. This may be performed for blackbody discs (Bath 1974) or discs with atmospheres (Schwarzenberg-Czerny 1984b) but in either case the temperature distribution of equation (2.35) has usually been assumed. Furthermore, without a model for the bright spot only discs effectively free from bright spot emission can be studied in this way.

An improved procedure is to construct a system from parameterized models, allowing q, i, $R(1)$, $R(2)$, r_0, r_d, polar coordinates of the bright spot and the relative brightnesses and colour temperatures of the components all to be variables determined by a least squares (or other statistical procedural) fit to the observed multi-colour eclipse curves. Examples are given by Frank & King (1981), Frank *et al.* (1981a), Schwarzenberg-Czerny (1984b) and Matvienko, Cherepashchuk & Yagola (1988). The first two groups studied the NLs RW Tri and UX UMa, and U Gem at minimum light, in both the visual and IR regions. Good accord was found for the NLs with steady state optically thick discs, but no satisfactory model was obtained for U Gem.

The most complete parameterized models are those employed by Zhang, Robinson & Nather (1986) and Zhang & Robinson (1987). In addition to the parameters listed above, and the implementation of full Roche geometry, the size of the bright spot is introduced and limb-darkening of all luminous components and gravity darkening and reflection from the secondary is taken into account. A power law

$$T(r) = T(r_d)\left(\frac{r}{r_d}\right)^{\zeta} \tag{2.66}$$

was adopted for the analysis of eclipses of HT Cas (Zhang, Robinson & Nather 1986).

The computational procedure adopts 31 × 50 grid points each for both stars and for the disc (assumed geometrically thin). The total amount of light unobscured by the secondary is found for successive orbital phases. To test whether a given pixel on the disc is visible at a particular phase, the Roche potential Φ_R equation (2.2), is evaluated at small intervals along a line joining the pixel to the observer; if Φ_R falls below the constant value that defines the surface of the secondary, then the pixel is occulted (Horne 1985a). The computed flux distributions are folded through the response functions of the observational filters (e.g., for UBVRI: Johnson (1965)) and the differences between observed and computed eclipse profiles minimized according to the procedure detailed in Zhang, Robinson & Nather (1986).

Figure 2.25 compares observed and modelled eclipses for HT Cas for the optimal solution, in which $\zeta = -0.66$, and for the best solution when ζ is held constant at -0.40. This demonstrates that the result $\zeta = -0.66 \pm 0.05$, which is quite close to the $\zeta = -0.75$ predicted by equation (2.37), is well determined.

In their study of U Gem, Zhang & Robinson (1987) could only obtain a satisfactory fit to the eclipse profile (which is of the bright spot and nearside of the disc – Section 1.3) by setting the disc temperature to zero for $r < r_0$ and holding it constant at T_d for $r_0 < r < r_d$. Their optical solution gave $r_0 = 1.2(\pm 0.5) \times 10^{10}$ cm, $r_d = 3.2(\pm 0.4) \times 10^{10}$ cm and $T_d = 4800 \pm 300$ K. Furthermore, the bright spot could be modelled only by a circular disc tangent to the rim of the accretion disc, and

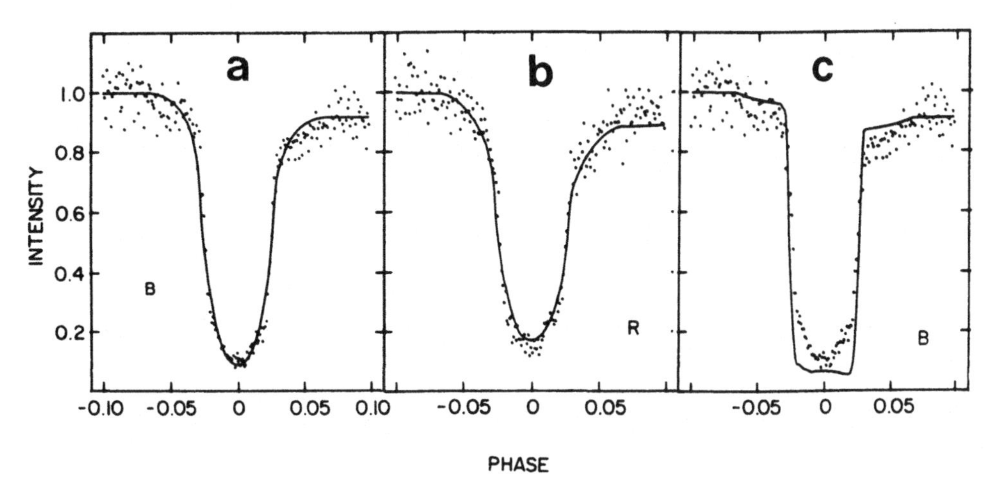

Figure 2.25 Observed average light curves (B or R band) and synthetic light curves for HT Cas in quiescence. In (a) the optimal solution is shown, for which $\zeta = -0.66$. In (b) and (c) solutions are shown for which ζ is held fixed at -0.40 and 1.50 respectively. Adapted from Zhang, Robinson & Nather (1986).

not by a spot radiating like a hemisphere. The successful and some unsuccessful model fits are shown in Figure 2.26.

An eclipse simulation technique has also been described by Berriman, Kenyon & Bailey (1986) and applied to IR light curves (see also Berriman (1987b)). In modelling the contribution from the primary, allowance must be made for obscuration by the inner edge of the disc, which can leave only one hemisphere visible, or both, according to $\dot{M}$(d) and i (Smak 1992).

A more restrictive procedure, designed to measure q and the sizes and locations of the primary and bright spot in the few systems where eclipse profiles clearly show all their phases of immersion and emersion (Figure 2.27), is given by Cook & Warner (1984; see also Cook 1985a). From the observed duration of the eclipse of the primary, a relationship between i and q is deduced. Then for a given choice of q the eclipse contact points can be mapped onto the accretion disc as projections of the limb of the secondary. As seen in Figure 2.27, the intersections of these boundaries typically indicate an elongated bright spot extended 'downstream' in the rotating disc. The *light centre*, defined as the intersection of the boundaries corresponding to mid-ingress and mid-egress of the bright spot, also typically lies near the head of the elongated spot region. Figure 2.27 also shows the stream trajectory, calculated from single particle trajectories (Section 2.4.2) and its width according to the methods of Lubow & Shu (1975). The determination of q is made by repeating the calculations shown in Figure 2.27 until the centre of the stream trajectory passes close to the light centre (see Cook & Warner (1984) for more complete explanation).

It was found (Cook & Warner 1984) that, although in general the bright spot is located at $r \sim 0.30a$, in Z Cha it extended into the inner regions of the disc to $r \sim 0.18a$, suggesting a stream thickness greater than that of the edge of the accretion disc and consequent continuation of the stream over the face of the disc. Considerable variation

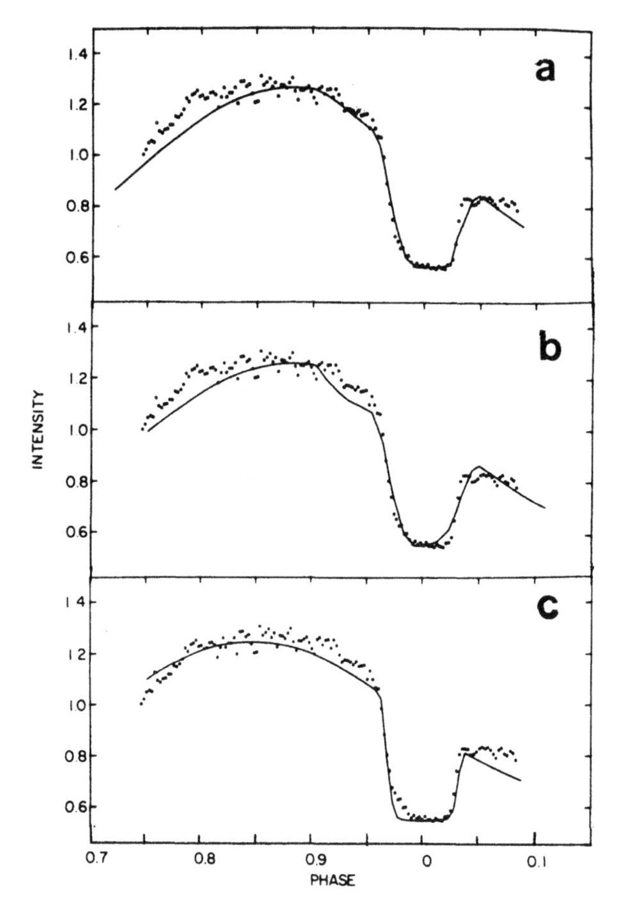

Figure 2.26 Observed average B-band light curves and synthetic light curves for U Gem in quiescence. In (a) the optimal fit using a planar bright spot is shown; in (b) the bright spot is hemispherical and does not give a good fit to the eclipse profile; in (c) ζ is fixed at 2.2. Adapted from Zhang & Robinson (1987).

in location of the bright spot from one orbit to the next ($P_{\mathrm{orb}} = 107$ min) was also occasionally present.

2.6.4 Eclipse Mapping

To generate an intensity map over the surface of the accretion disc a direct deconvolution of the eclipse profile is obviously desirable. The principle involved is seen from consideration of the eclipse process (Figure 2.28). Immergence consists of successive obscuration of strips on the disc lying along the projection of the limb of the secondary. Emergence involves successive uncovering of another set of strips, which intersect the first set because a different limb is involved and the binary has rotated. The process of deconvolution thus becomes analogous to the inversion of tomographic intensity distributions, but with the limitation that only two slices through the body are available, and the added complication that immersion of one strip and emersion of another may occur simultaneously.

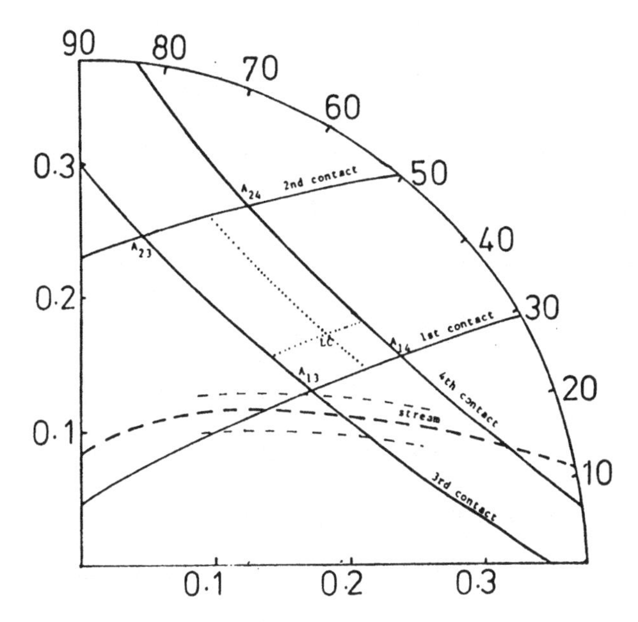

Figure 2.27 Boundaries in the orbital plane of Z Cha defined by the four contact phases of the bright spot. The origin is at the primary, the abscissa is in the direction of the secondary. The scales are fractions of the orbital separation. The bright spot must lie within the box defined by the four boundaries: the dotted lines within the box are lines of mid-ingress and mid-egress. The theoretical path and thickness of accretion stream for $q = 0.5$ are taken from Lubow & Shu (1975). From Cook & Warner (1984).

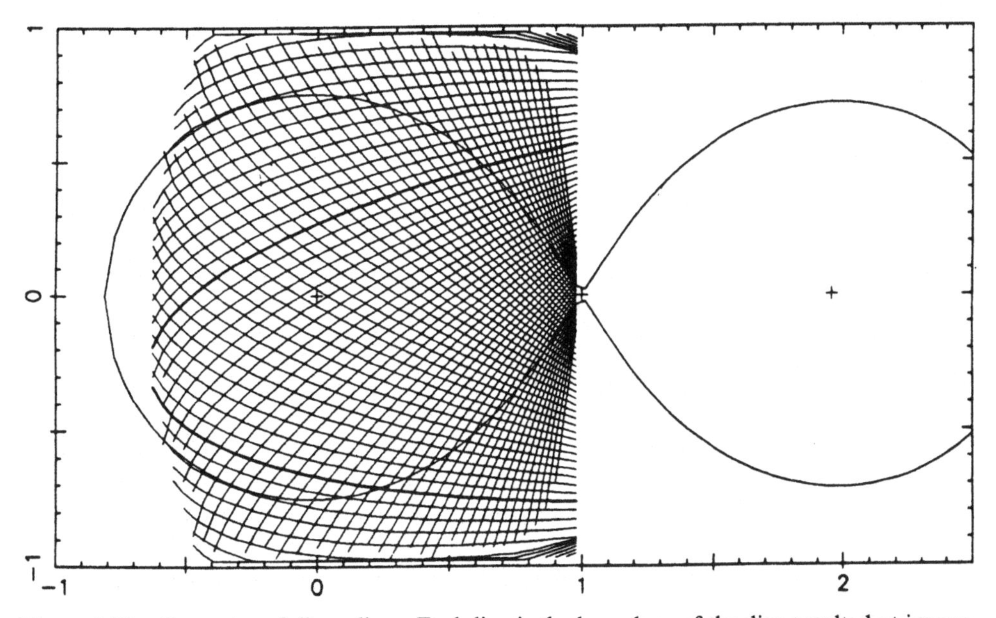

Figure 2.28 Geometry of disc eclipse. Each line is the boundary of the disc occulted at ingress or egress by the secondary at successive binary phases. The scale is R_{L_1}, and the diagram is computed for $q = 0.9, i = 75°$, which gives $\Delta\varphi = 0.081$. From Horne (1985a).

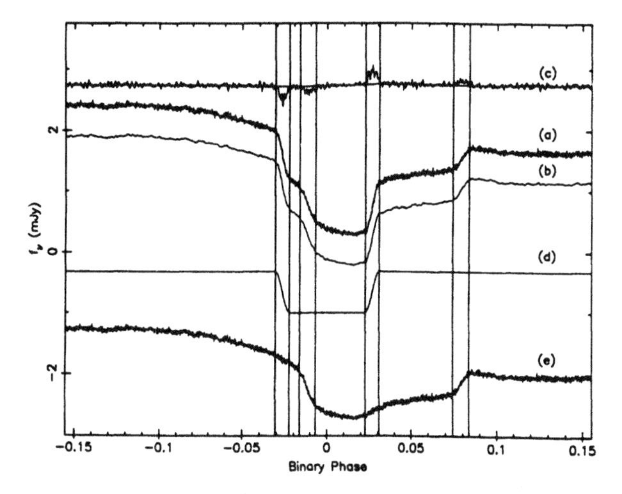

Figure 2.29 Decomposition of an average of 17 eclipse light curves of Z Cha. For explanation of the individual panels see text. From Wood *et al.* (1986).

The process of analysis is aggravated by noise from atmospheric scintillation, photon statistics and flickering intrinsic to the CV. A brief brightening of any part of the star will appear as an increased intensity of the strips undergoing emersion. In order to reduce the influence of such noise, the averaging of many eclipses may be necessary (in which case only a long-term mean intensity distribution can be produced). It can also simplify the procedure if the eclipse of the primary and the variation of intensity caused by the change of aspect of the bright spot are subtracted from the eclipse profile. A technique for decomposing a light curve into its component parts is given by Wood, Irwin & Pringle (1985; see also Smak 1994a), illustrated in Figure 2.29. The light curve (a), which may be the average of many, is first smoothed (b) with a median filter (Tukey 1971), whose width is chosen to avoid loss of information in the most rapid changes in the eclipse. The smoothed light curve is differentiated numerically (c) to enhance regions of rapid brightness variation. Appropriate smoothing then allows accurate location of the various contact points. The filtered derivative is then integrated with the assumption that the primary has constant brightness before and after eclipse (and is zero between its second and third contacts) to give a reconstituted eclipse of the primary (d). This can then be subtracted to furnish an eclipse profile freed from the rapid variations caused by the white dwarf eclipse (e). Applied to the entire light curve (Figure 2.30), with the assumption that the primary and disc have constant brightness outside of eclipse, the variations of disc, bright spot and primary become separately visible (Wood *et al.* 1986). (The 'tail' of brightness seen in Figure 2.29 from phase 0.1 to 0.5 in the bright spot component is further evidence in Z Cha for the emission downstream from the bright spot discussed in the previous Section.)

The process of deconvolution (or *image reconstruction*, as it is is usually termed) may at first sight appear a hopeless task: the eclipse profile is a one-dimensional function which cannot uniquely specify the two-dimensional intensity distribution. To select one from the infinite number of possible distributions that would lead to the same eclipse profile requires added constraints; as stated by Horne (1985b) 'the reconstruction

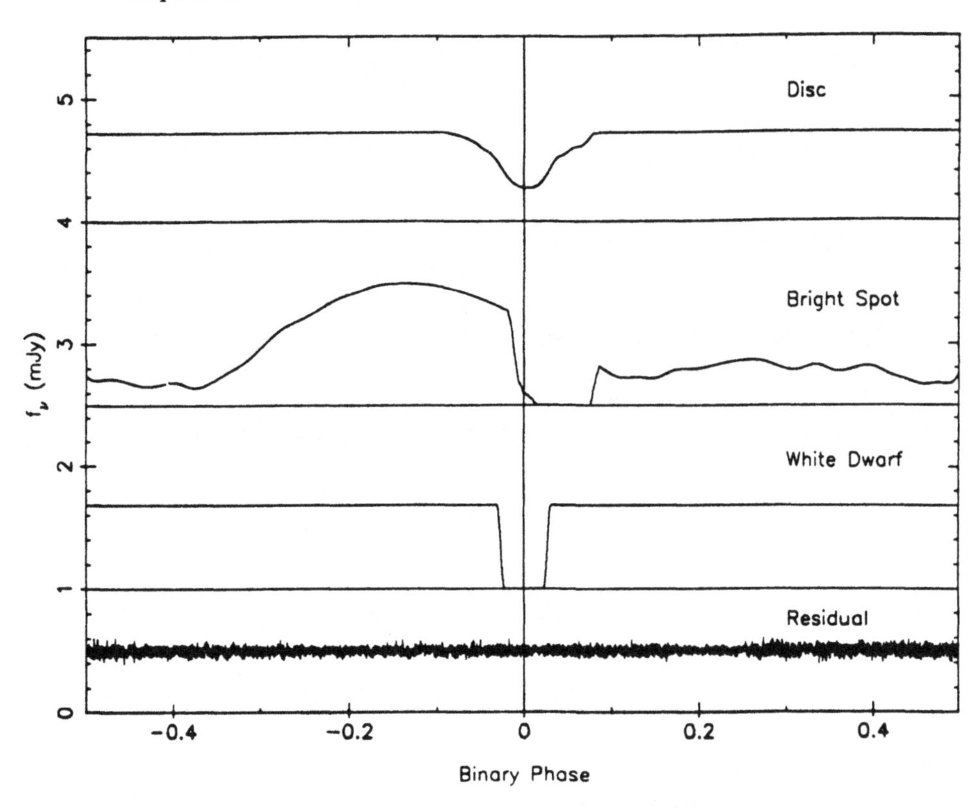

Figure 2.30 Decomposition of a complete orbital light curve of Z Cha into contribuitons from the disc, bright spot and white dwarf. The residual from the original light curve is shown in the bottom panel. From Wood *et al.* (1986).

method must be carefully chosen to reflect the aims of the investigator'. The method chosen by Horne (1985a,b) has proved of great value in determining the radial temperature distribution in CV accretion discs.

The principle of Horne's technique is based on the concept of image entropy introduced by Gull & Daniell (1978) and Bryan & Skilling (1980). Representing the accretion disc on a square array of pixels, a particular intensity distribution $I(j)$ has an entropy S relative to a *default* distribution $D(j)$ defined by

$$S = -\sum_{j=1}^{N^2} I(j)\left[\ln\frac{I(j)}{D(j)} - 1\right] \tag{2.67}$$

where N is the size of the pixel array. By differentiating equation (2.67) it is seen that, in the absence of any constraints on I(j), the entropy is maximized when I(j) and D(j) are identical. Thus, in general terms, S measures the similarity of the two intensity distributions.

For a given $I(j)$, eclipse profiles are computed as described in the previous section – recognizing that now there are N^2 free parameters to be adjusted. The difference between the observed eclipse profile (at M orbital phases k), $O(k)$, and the computed one is quantified by the χ^2 statistic

$$\chi^2 = \frac{1}{M}\sum_{k=1}^{M}\left[\frac{O(k) - C(k)}{\sigma(k)}\right]^2 \qquad (2.68)$$

where $\sigma(k)$ is an estimate of the standard deviation in $O(k)$.

The intensity distribution $I(k)$ is obtained by maximizing S subject to the constraint that χ^2 is no larger than some chosen value χ_0^2. The resulting $I(j)$ is thus the result of a balance between two opposing influences: if χ_0^2 is relatively undemanding then $I(j)$ will not depart greatly from $D(j)$; if χ_0^2 is highly demanding then $I(j)$ will differ considerably from $D(j)$ (barring accidental choice of $D(j)$ that happens to be consistent with the observed eclipse). $D(j)$ thus acts as a store of *a priori* information, and $I(j)$ will only differ from it if the need to attain χ_0^2 requires it. Consequently, $D(j)$ can be chosen according to the requirements of the investigator: a uniform $D(j)$, set at the average intensity, will result in an $I(j)$ that is the most uniform distribution consistent (by the χ_0^2 requirement) with the observed eclipse $O(k)$, whereas a $D(j)$ formed by averaging intensities around annuli centred on the primary will produce an $I(j)$ that is most nearly axisymmetric. If there is a conspicuous bright spot on the rim of the disc these two $D(j)$ produce very different $I(j)$: the maximally uniform map has arcs of intensity produced by the need to spread the cause of the sudden disappearance and reappearance of the bright spot over as great an area as possible (in this case, along the projected limb of the secondary), whereas the maximally axisymmetric map removes the arcs but spreads the bright spot to some extent around the rim of the disc (Horne 1985a,b). For purposes of measuring the radial intensity distribution of the disc, the latter choice is clearly preferable.

Most applications of the technique have assumed a flat disc. However, at large inclinations ($i > 75°$) and large $\dot{M}$ (equation (2.52)) the disc thickness is sufficient to obscure part of the central regions, and the edge of the disc can contribute significant intensity. At $\dot{M}_{17}(\mathrm{d}) \sim 1$ these are not important problems (Wood, Abbott & Shafter 1992), but at $\dot{M}_{17}(\mathrm{d}) \sim 10$ spurious results may ensue (Smak 1994b).

Results of applying the maximum entropy method (MEM) to various CV discs will be given in later chapters. A critical review has been given by Horne (1993a) and alternative approaches have been suggested by Baptista & Steiner (1991, 1993). Here two examples are chosen to illustrate the comparison of derived temperature distribution with the steady state disc prediction of equation (2.35).

Figure 2.31 illustrates in pictorial form the maximally axisymmetric intensity distributions in the U, B and V bands derived from eclipses of Z Cha during a normal outburst (Horne & Cook 1985). In Figure 2.32 the brightness temperature in the V band is shown as a function of radial distance. There is relatively good agreement with the radial temperature distribution predicted for a disc with $\dot{M}(\mathrm{d}) \sim 2 \times 10^{-9}$ $\mathrm{M}_\odot$ y^{-1}, with perhaps evidence for an increase of $\dot{M}(\mathrm{d})$ towards the centre of the disc (but see Section 3.5.4.4).

In contrast, at quiescence the temperature distribution in Z Cha is quite unlike a steady state disc (Wood *et al.* 1986) and shows almost constant temperature over the disc (Figure 2.33). The interpretation of this is discussed in Section 3.5.4.4.

In Figures 2.32 and 2.33 the abscissa is $x = r/R_{L_1}$. The temperature distribution

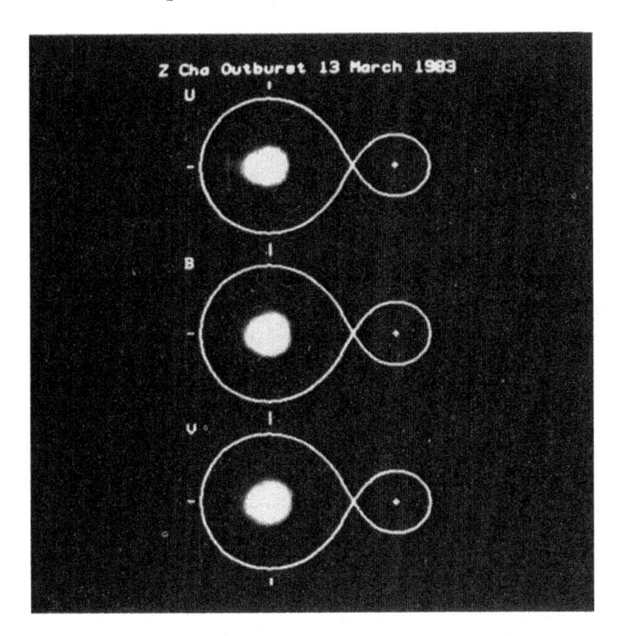

Figure 2.31 Accretion disc surface brightness distributions of Z Cha in outburst, reconstructed from observed eclipse light curves in the U,B and V bands. From Horne & Cook (1985).

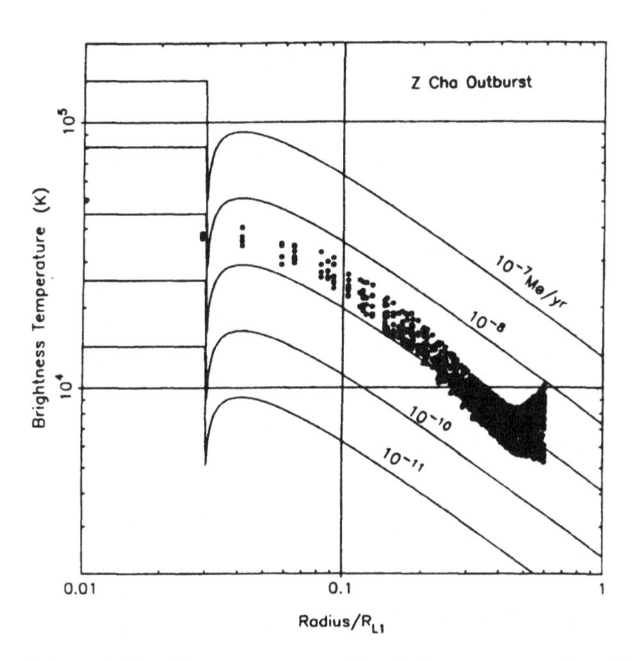

Figure 2.32 Temperature profile of the accretion disc of Z Cha in outburst compared with predicted profiles from equation (2.35). Pixels with $r < 0.03R_{L_1}$ are on the surface of the primary, which has a temperature $\sim 4 \times 10^4$ K. From Horne & Cook (1985).

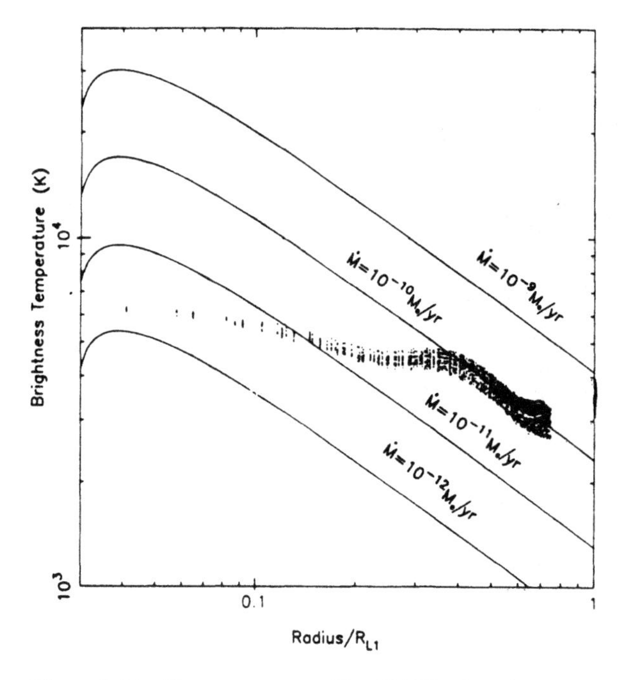

Figure 2.33 Temperature profile of Z Cha in quiescence, compared with predicted profiles from equation (2.35). From Wood *et al.* (1986).

$T_{\rm eff}(x)$ is found by combining equations (2.1a), (2.35) and (2.36a) to give

$$T_{\rm eff}^4(x) = \frac{3\pi \dot{M}(\mathrm{d})}{2\sigma P_{\rm orb}^2}\left(\frac{a}{R_{L_1}}\right)^3 (1+q)^{-1} x^{-3}\left\{1 - \left[\frac{R(1)}{R_{L_1}}\right]^{1/2} x^{-1/2}\right\} \qquad (2.69)$$

where a/R_{L_1} can be found from equations (2.4b) or (2.4c).

The MEM technique can be applied to spectra as well as broad band photometry. A study of the NL UX UMa (Rutten *et al.* 1993) derived radial flux distributions for each of 110 spectral regions, which enabled spectra to be reconstructed for selected disc annuli. The disc shows an optically thick absorption-line spectrum in its inner regions, changing to predominantly emission lines in the outer disc. (The range of derived temperatures – 17 000 K in the inner region to 6000 K in the outer – will be affected by the peculiar flux distribution observed in 4000–5000 Å; the negative slope in the Rutten *et al.* observations is quite unlike the previously determined flux distribution (Schlegel, Honeycutt & Kaitchuck 1983).)

2.6.5 The Bright Spot and Flickering

The location and dimensions of the bright spots in eclipsing DN have been found using the techniques of Section 2.6.3. In Z Cha, Cook & Warner (1984) noted that eclipse profiles can be classified in two extremes (there are intermediate examples): in Type 1 there is a long standstill between primary egress (p4 in Figure 2.23) and bright spot egress (h3), followed by quite rapid emergence of the bright spot. In Type 2 there is little or no standstill and bright spot egress is very slow.

Type 1 corresponds to a compact bright spot; Type 2 to an extended bright spot reaching into the inner parts of the disc (from $r_d/a \sim 0.35$ at the rim to $r_d/a \sim 0.18$). The extension is along the stream trajectory and is evidently caused by stream overflow (Section 2.4.3). The angular extent $\Delta\theta$ of the bright spot on the outer edge of the disc was found to vary over the range 14–40°. Wood *et al.* (1986), from a different set of observations, found $\Delta\theta \sim 10°$, a radial extent Δr_s of the bright spot $\Delta r_s/a \sim 0.05$ and a vertical thickness $\Delta z_s/a \sim 0.03$.

In OY Car, Cook (1985a) found $\Delta\theta \sim 15$–30° and an angle ~5° between first contact of the stream with the disc and the maximum of flux from the bright spot – this is an effect of cooling downstream in the disc flow. (Smak (1971a) was the first to note that downstream cooling would produce a bright spot elongated along the disc rim and used it to explain the poor definition of second contact of eclipses in U Gem.) The bright spot varies in position along the stream trajectory, which could be caused either by variable r_d or changing density in disc or stream which allows variable stream penetration. Wood *et al.* (1989a) found $\Delta\theta \sim 7$–19° and $\Delta r_s/a = 0.027$–0.077.

In IP Peg, Wood & Crawford (1986) found $\Delta\theta \sim 12°$, $\Delta r_s/a \sim 0.03$ and $\Delta z_s/a \sim 0.04$–0.11. Marsh (1988) deduced a spot diameter of 5.5×10^9 cm (or $\Delta r_s/a = 0.08$) and noted that this is approximately twice the scale height of the edge of the disc, in agreement with theory (Lubow & Shu 1975; Section 2.4.3).

Early methods of modelling (e.g., Smak (1971a), Warner & Nather (1971)) used the observed orbital phase of maximum light (i.e., the peak of the orbital hump) to locate the azimuthal position of the bright spot on the disc. With the independent method of requiring the bright spot to lie on a stream trajectory, the two methods can be compared. In U Gem the orbital phase of hump maximum is $\varphi = 0.85$; if the impact shock lay along the stream, maximum flux would occur at $\varphi = 0.78$, and if along the edge of the disc, at $\varphi = 0.95$. Therefore the shock lies between those two streaming directions (Marsh 1988). Similarly, in OY Car, if hump maximum is normal to the plane of the shock then the shock is tipped ~ 19° forward from the azimuthal direction (Wood *et al.* 1989a), and HST observations confirm that the maximum emission is roughly midway between the stream and disc flows (Horne *et al.* 1994). From Doppler imaging (Section 2.7.5) of U Gem, Marsh *et al.* (1990) find that the bright spot emission is a mixture of disc and stream material, each of which has passed through a shock: the velocity of the bright spot emission lines is an average of the disc and stream velocities. Thus the emission comes from the region between the two shock fronts described in Section 2.4.3.

The orbital humps are very closely approximated by half sinusoids, which suggests modelling the light curves as foreshortening of a bright region 'painted' onto the outer edge of the disc (Smak 1971a; Warner & Nather 1971; Wood *et al.* 1986, 1989a).

Marsh (1988) isolated the spectrum of the bright spot in IP Peg and found that it is closely matched (but with shallower Balmer jump) by that of a B8Ia star having $T_{eff} = 11\,200$ K. For Z Cha, Wood *et al.* (1986) found a blackbody brightness temperature of $12\,500 \pm 650$ K and a colour temperature of $11\,300 \pm 2000$ K for the bright spot. In OY Car, Wood *et al.* (1989b) obtained $T = 13\,800 \pm 1300$ K and 9000 K respectively, whereas Schoembs & Hartmann (1983) found $T = 15\,000$ K and Berriman (1987b) from IR photometry initially found $T < 7000$ K but agreed that $T \sim 15\,000$ K was possible if the bright spot is elongated along the rim as deduced by

Table 2.2. *Determinations of $\dot{M}(2)$ from Bright Spot Luminosities.*

Star	P_{orb} (h)	$\dot{M}(2)$ g s^{-1}	$\dot{M}(2)$ $M_\odot y^{-1}$	References
WZ Sge	1.36	2×10^{15}	3×10^{-11}	1
OY Car	1.52	4×10^{15}	6×10^{-11}	2
Z Cha	1.79	2×10^{15}	3×10^{-11}	3
IP Peg	3.80	1.4×10^{16}	2.2×10^{-10}	4
U Gem[a]	4.25	3.2×10^{16}	5.1×10^{-10}	5

[a] Scaled to a distance of 81 pc (Warner 1987a).
References: 1. Smak 1993c; 2. Wood *et al.*, 1989b; 3. Wood *et al.*, 1986; 4. Marsh 1988; 5. Zhang & Robinson 1987.

Cook (1985a). Robinson, Nather & Patterson (1978) found $T = 16\,000$ K for WZ Sge and $\Delta r_s = 2 \times 10^9$ cm, and Zhang & Robinson (1987) derived $T = 11\,600 \pm 500$ K in U Gem.

The effective temperatures and sizes of the bright spots provide luminosities which, through equation (2.21b) or a more accurate treatment, lead to estimates of the rate of mass transfer into the bright spot. This is the only direct way of determining $\dot{M}(2)$. The results are summarized in Table 2.2. There is poor relative agreement with $\dot{M}(2)$ indicated by other methods (Section 3.3.3.3).

Bright spots would be expected to be most prominent in high inclination systems where the disc is seen nearly edge-on but the spot is viewed nearly face-on. This leads to the large orbital hump characteristic of high inclination DN (e.g., Figures 1.8 and 2.30). However, eclipsing NL and CN have small or no orbital humps (but the bright spot is usually detectable through asymmetry of eclipse (e.g., Rutten, van Paradijs & Tinbergen (1992)) and almost all CVs show line emission – an S-wave – from the bright spot (Honeycutt, Kaitchuck & Schlegel 1987)), so there is a distinctive difference between high $\dot{M}$ and low $\dot{M}$ systems.

The prominence of an orbital hump, in the ideal case of an optically thick steady state accretion disc and a simple planar bright spot on the edge of the disc, is given by the ratio of bright spot visual luminosity $L_V(\mathrm{sp})$ to disc visual luminosity $L_V(\mathrm{d})$:

$$\frac{L_V(\mathrm{sp})}{L_V(\mathrm{d})} = 2\left(\frac{3-u_d}{3-u_s}\right)\left(\frac{1-u_s+u_s\sin i}{1-u_d+u_d\cos i}\right)\tan i\,\frac{\langle L_V(\mathrm{sp})\rangle}{\langle L_V(\mathrm{d})\rangle}, \tag{2.70}$$

where u_s and u_d are the limb-darkening coefficients and $\langle\rangle$ denotes averaging over all directions (see, e.g., Paczynski & Schwarzenberg-Czerny (1980)).

Assuming that the boundary layer contributes nothing to the visible flux, and that $r_d \gg R(1)$, equation (2.22a) provides the bolometric luminosity of the disc. For the bright spot, an amount of energy $GM(1)\dot{M}(2)(r_d^{-1} - R_{L_1}^{-1})$ is available, of which about half is used in matching the stream velocity with the local disc velocity, and the other half is deposited. As the core temperature of the bright spot is much greater than the local disc temperature (Pringle 1977) the outward and inward temperature gradients are comparable and only about one half of the deposited energy will be radiated locally. Noting that observations (Section 2.6.2) give $r_d \sim 0.5R_{L_1}$, we adopt for the bolometric radiated energy from the bright spot

$$L(\mathrm{sp}) = \left(\frac{f}{8}\right)\frac{GM(1)\dot{M}(2)}{r_\mathrm{d}},$$

where f is an 'efficiency' factor ~ 1. Then

$$\frac{L_\mathrm{v}(\mathrm{sp})}{L_\mathrm{v}(\mathrm{d})} = f\frac{\tan i}{1+\frac{3}{2}\cos i}\frac{\dot{M}(2)}{\dot{M}(\mathrm{d})}\frac{R(1)}{r_\mathrm{d}}10^{0.4(B_\mathrm{s}-B_\mathrm{d})}, \tag{2.71}$$

where B_s and B_d are bolometric corrections (< 0) for the spot and disc respectively and we have adopted $u_\mathrm{d} = 0.6$ (Section 2.6.1.3) and $u_\mathrm{s} = 0$ (because the orbital humps are closely sinusoidal).

Equations (2.1a), (2.5b), (2.83), (2.100) and taking $r_\mathrm{d} = 0.70\ R_L(1)$ from Section 2.6.2, give

$$\frac{R(1)}{r_\mathrm{d}} \approx \frac{0.40q^{2/3}}{P_\mathrm{orb}^{3/2}(\mathrm{h})}, \tag{2.72}$$

which, as q statistically decreases with increasing P_orb (Section 2.8.2), shows that the prominence of the bright spot should decrease rapidly with increasing P_orb (as first deduced by Lin (1975)). The principal reason is that at larger P_orb the disc is larger, so the stream material does not fall so deeply into the potential well of the primary.

Neglecting the bolometric corrections and setting $f = 1$, evaluation of equation (2.71) can be made by comparing observed spot/disc flux ratios with the function

$$B = \frac{\tan i}{1 + 3/2\cos i}\frac{R(1)}{r_\mathrm{d}}, \tag{2.73}$$

which is done in Table 2.3, where (except for V347 Pup) *observed* values of $R(1)/r_\mathrm{d}$ are employed.

The comparison shows that $\dot{M}(\mathrm{d}) = \dot{M}(2)$ gives good agreement for the NL systems but that $\dot{M}(\mathrm{d}) \ll \dot{M}(2)$ for DN. This latter inequality was first noted by Osaki (1974) and discussed by Paczynski (1978) and Paczynski & Schwarzenberg-Czerny (1980). The discrepancies are too large to be caused by the bolometric correction term in equation (2.71). In fact, for bright spot temperatures ~12 000 – 15 000 K, B_s is not greatly different from B_d for a steady state disc (Paczynski & Schwarzenberg-Czerny 1980). At the brightness temperatures actually observed for the discs of U Gem ($T \approx 4800$ K: Zhang & Robinson 1987), OY Car ($T \approx 4000$ K: Wood *et al.* 1989b) and Z Cha ($T \approx 5000$ K: Wood *et al.* 1986), B_d is also ~ B_s, but the discs are probably not steady state (Wood 1990b). The low $\dot{M}(\mathrm{d})$ in U Gem and IP Peg, by comparison with those in the NLs, will be seen in the absolute magnitudes of the discs of these stars (Figures 3.5 and 4.15).

High speed flickering in U Gem disappears during eclipse and therefore (because the central regions of the disc are not eclipsed: Section 1.3) is clearly associated with the bright spot (Warner & Nather 1971; Figures 1.8 and 1.9). However, the above analysis shows that the bright spot is particularly prominent in U Gem, and therefore its flickering activity can dominate. HST observations show the same phenomena in the

Table 2.3. *Comparison of Observed and Calculated Spot/Disc Luminosities.*

Star	Type	P_{orb}(h)	i^{o}	$R(1)/r_d$ $\times 10^{-2}$	B	$\frac{L_V(sp)}{L_V(d)}$	References
OY Car	DN(SU)	1.52	83.3	5.83	0.42	2.9	1
Z Cha	DN(SU)	1.79	82	5.86	0.34	1.4	2
SW Sex	NL	3.23	79	4.9	0.19	0.13	3,4
IP Peg	DN	3.80	81	1.77	0.61	4:	5,6
LX Ser	NL	3.80	75	5.2	0.14	0.11	3
U Gem	DN	4.25	70	1.33	0.024	2	7,8,9
UX UMa	NL	4.72	71	3.3	0.064	0.07	3
V347 Pup	NL	5.57	88	1.4:*	0.38	0.50:	10
V363 Aur	NL	7.71	73	1.2	0.027	0.03	3

* Here and elsewhere in the book a colon indicates an uncertain value

References: 1. Wood *et al.* 1989b; 2. Wood *et al.* 1986; 3. Rutten, van Paradijs & Tinbergen 1992; 4. Liebert *et al.* 1982; 5. Wood *et al.* 1989a; 6. Marsh & Horne 1990; 7. Marsh *et al.* 1990; 8. Smak 1984c; 9. Harwood 1973; 10. Mauche *et al.* 1993.

UV of IP Peg (Horne 1993b). Flickering in other systems, where the bright spot contribution is smaller, may originate elsewhere. Vogt *et al.* (1981) noted strong flickering in OY Car even when the bright spot was eclipsed, as did Patterson (1981) in HT Cas. Flares seen in OY Car by HST (Horne *et al.* 1994) are distributed uniformly around orbit and are associated with the inner disc. Horne & Stiening (1985) and O'Donoghue, Fairall & Warner (1987) deduced that flickering in the eclipsing NLs RW Tri and VZ Scl is located near the centres of their discs. It is therefore probable that in all CVs there are two sources of flickering – the bright spot and a turbulent inner region to the disc – the relative importance of which varies from system to system. Furthermore, variations in amplitude of flickering during outbursts of DN are no longer as clearly connected with changes in $\dot{M}(2)$ as was once thought (e.g., Robinson (1973a,b,d)).

Bruch (1992) has made a systematic observational study of flickering in CVs, showing that the power spectrum is non-stationary and very different from object to object. On average, amplitude $\tilde{\propto}$ (frequency)$^{-1.0\pm0.5}$. Elsworth & James (1982, 1986) found similar behaviour in a more restricted study.

The accretion disc models of Geertsema & Achterberg (1992), which include magnetohydrodynamic turbulence, generate just such a power spectrum from fluctuations in the inner disc.

2.6.6 *Evidence for Disc Precession*

There is as yet no clear demonstration of tilted, precessing discs in CVs, but the presence of otherwise inexplicable cyclical brightness modulations may be indirect evidence (see also Section 3.6.6.3 for an alternative explanation as a precessing elliptical disc and Section 4.5 for discussion). The stars with these brightness

Table 2.4. *Possible Precession Periods in CVs.*

Star	P_{orb}(d)	P_{pr}(d)	P_{pr}/P_{orb} observed	q*	P_{pr}/P_{orb} eqn (2.62c)	References
AM CVn	0.0119	0.558	46.9	0.067:	48.6:	8
V503 Cyg	0.0722	1.5	20			11
PG 0917+342	0.075:	1.7	23:	0.15:	25:	9
V795 Her	0.1083	1.53	14.1	0.20	20.6	1
MV Lyr	0.1334	3.8	28	0.17:	23:	2
TT Ari	0.1376	3.5:	25:	0.25	17.9	3
V603 Aql	0.1383	2.5	18	0.25	17.9	4
HR Del	0.2142	2.5	12	0.38	14.3	5
TV Col	0.229	4.02	17.6	0.41	13.9	6, 7, 10

* Estimated according to the methods of Section 2.8

References: 1. Shafter *et al.* 1990 , Thorstensen, Smak & Hessman 1985; 2. Borisov 1992; 3. Udalski 1988a; 4. Udalski & Schwarzenberg-Czerny 1989; 5. Bruch 1982; 6. Bonnet-Bidaud, Motch & Mouchet 1985; 7. Barrett, O'Donoghue & Warner 1988; 8. Patterson, Halpern & Shambrook 1993; 9. Skillman & Patterson 1993; 10. Hellier 1993a; 11. Szkody 1993.

modulations are listed in Table 2.4. The periods P_{pr} manifest themselves either as differences between spectroscopic and photometric periods (in which case P_{pr} is the beat period of these) or directly as a brightness modulation at P_{pr}. Examples of the last phenomenon, generally interpreted as resulting from a precessing disc, are also present in the X-ray binaries Her X-1 (Deeter *et al.* 1976) and LMC X-4 (Ilovaisky *et al.* 1984).

2.7 Lines from Discs

Optical and IR spectra of most CVs characteristically show emission lines from low ionization states: HI, HeI, HeII, FeII, CaII, OI, with a blend of CIII, OII and NIII at 4640–4660 Å enhanced by Bowen fluorescence and other processes (McClintock, Canizares & Tarter 1975). Weaker lines of other elements are difficult to detect because of the considerable line broadening from disc rotation. In the UV, lines from more highly ionized species are commonly observed (Figures 1.18 and 1.19): Lyα, HeII, CII–CIV, NIII–NV, OI, OIII, OV, MgII, AlIII, SiII–IV (Jameson, King & Sherrington 1980; Fabian *et al.* 1980).

The wide range of ionization states indicates that there is more than one site of production of the lines, and, indeed, several observational clues confirm this. For example, in U Gem and WZ Sge the S-wave shows that HeI 4471 Å emission comes almost entirely from the bright spot (this is the case also for HeII 4686 Å in U Gem (Honeycutt, Kaitchuck & Schlegel 1986); in the deeply eclipsing NL V1315 Aql there is hardly any eclipse of the Balmer lines but the CIII/NIII and HeII lines are totally eclipsed (Dhillon, Marsh & Jones 1990) and it is commonly the case that emission lines are less deeply eclipsed than the continuum (e.g., Holm, Panek & Schiffer (1982)); eclipses of the Balmer lines may be of longer duration than for HeII (Young & Schneider 1980); the UV lines often have P Cyg profiles (Figure 1.19) although none are seen in the optical; the UV lines show variations in strength uncorrelated with those of lines in the optical; whereas the *equivalent width* of HeII 4686 Å often *increases* during DN outburst (implying a very large increase in HeII flux), the Balmer lines fluxes are often almost constant throughout (e.g., Warner, O'Donoghue & Wargau (1989)).

This evidence of complexity, together with considerable uncertainties in the theory of emission-line formation, results in a strong interchange between observation and theory, but it is not yet possible to identify with certainty the dominant mechanisms of line formation in any given object. The options currently available are as follow (Sections 2.7.1, 2.7.2 and 2.7.3).

2.7.1 Optically Thin Emission

In any parts of an accretion disc where the vertical optical thickness in the continuum is <1 but the optical thickness in the lines is significant, an emission spectrum is produced. The importance of this line emission depends on the relative extent of optically thin regions and the spectral region under review. High viscosity ($\alpha \gtrsim 1$) discs (Williams 1980, 1991; Tylenda 1981; Williams & Ferguson 1982; Lin, Williams & Stover 1988; Williams & Shipman 1988) do have extensive optically thin regions (from equation (2.32), for a given $\dot{M}(\mathrm{d})$, $\Sigma \propto v_{\mathrm{k}}^{-1}$, so high α discs have low densities) especially for $\dot{M}(\mathrm{d}) < 10^{16}$ g s^{-1}, and their computed spectra have strong emission lines, although generally weaker than observed lines by a factor of a few (cf. observations of Williams (1983) and Patterson (1984)). The regions of the models which are optically thin have $T \sim 6000$ K, similar to that observed in Z Cha at quiescence (Figure 2.33). Marsh (1987) finds, however, that it is not possible to model the emission lines in Z Cha at quiescence with LTE optically thin conditions and concludes that an overlying hot chromosphere is probably present.

At sufficiently high inclinations probably all discs are optically thick in the continuum, so the emission lines would be expected to be weak or absent. Selecting DN at quiescence for which $\dot{M}(\mathrm{d}) < 10^{16}$g s^{-1} and $i \gtrsim 80°$ (e.g., WZ Sge, OY Car, Z Cha) it is nevertheless found that emission lines are very strong. It appears, therefore, that optically thin emission is not adequate to account for the lines in low $\dot{M}(\mathrm{d})$ systems. Furthermore, high $\dot{M}(\mathrm{d})$ ($\sim 10^{18}$ g s^{-1}) systems such as CN remnants and NLs also show such an opposite correlation: the emission-line equivalent widths are largest for the highest inclination discs (Warner 1987a; Figure 2.34). Therefore, in these systems at least, an alternative mechanism to produce emission lines must be sought – and that mechanism may well be expected to be important even at low $\dot{M}(\mathrm{d})$.

Nevertheless, the demonstration of optically thin continuum emission in the IR (Section 2.6.1.4) cautions that some IR emission lines may arise in optically thin outer regions of discs.

2.7.2 Chromospheric Emission

There are at least three mechanisms that can heat the optically thin upper layers of the disc atmosphere. The first, viscous dissipation, has been mentioned in Sections 2.5.2.2 and 2.6.1 but no predictions of resulting emission-line spectra have yet been made. Turbulence refracted out of the disc, or magnetic loops arching out of the disc with consequent MHD heating of a corona and downward flow of energy to heat a chromosphere, are further possibilities (Sections 2.5.2.2, 2.5.2.3; Shakura & Sunyaev

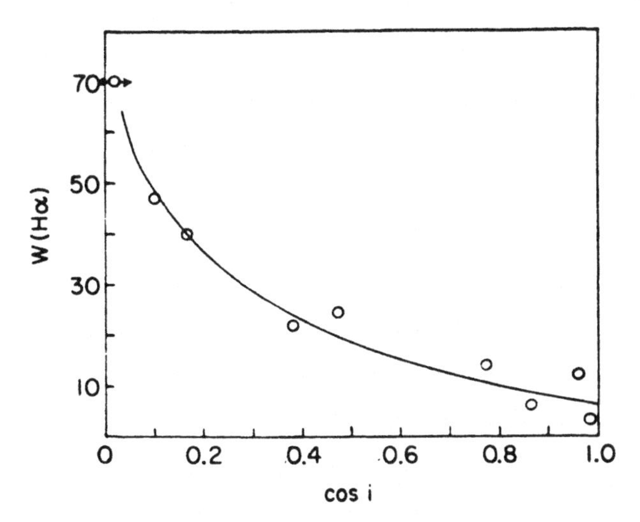

Figure 2.34 Dependence of the Hα emission-line equivalent width on orbital inclination for CN remnants. From Warner (1987a).

1973; Liang & Price 1977), and some indirect evidence for the latter is given below, but the process that is observationally demonstrated to occur, and will be of some consequence to all concave discs, is irradiation by the primary, BL or hot inner regions of the disc.

From geometrical considerations the ratio of flux absorbed from the BL to that generated locally is ~0.1 (Pacharintanakul & Katz 1980; Schwarzenberg-Czerny 1981; Smak 1989a). Because the irradiation is so oblique ($\gtrsim 80°$), unit optical depth is reached high in the atmosphere of the disc and is effective both at producing a temperature inversion (Jameson, King & Sherrington 1980; Mineshige & Wood 1990) and (from the EUV and X-ray photons) in photoionizing and exciting the gas. The continuum of the resulting recombination spectrum (added to an optically thick disc continuum) can account for the Balmer emission continuum observed in DN at quiescence and for their peculiar photometric colours (Schwarzenberg-Czerny 1981).

One observational result clearly demonstrates the importance of irradiation from the central region, although this may only apply to the special conditions in the central region of magnetic CVs: the HeII 4686 Å line in DQ Her is pulsed with the same 71 s period as the beam of high energy radiation from the rotating primary (Chanan, Nelson & Margon 1978: this is described in detail in Section 8.1.3).

More generally, the correlation of Balmer emission strength with X-ray flux (Figure 2.35: Patterson & Raymond 1985a) demonstrates a connection between chromosphere and BL; but this is not necessarily causal, because, as seen in the tight correlation between HeII (λλ 1640 and 4686) emission strengths and $\dot{M}(\mathrm{d})$ (Figure 2.36: Patterson & Raymond 1985b), there is a scaling of all emission properties with $\dot{M}(\mathrm{d})$. Patterson & Raymond were able to account quantitatively for the HeII line fluxes by soft X-ray irradiation from the BL for the higher $\dot{M}(\mathrm{d})$ ($> 10^{17}$ g s^{-1}) systems (Figure 2.36), and the inadequacy at lower $\dot{M}(\mathrm{d})$ they attributed to omission for their model of a (then unobservable) flux of X-rays with $kT \sim 0.5$ keV (Patterson & Raymond 1985b).

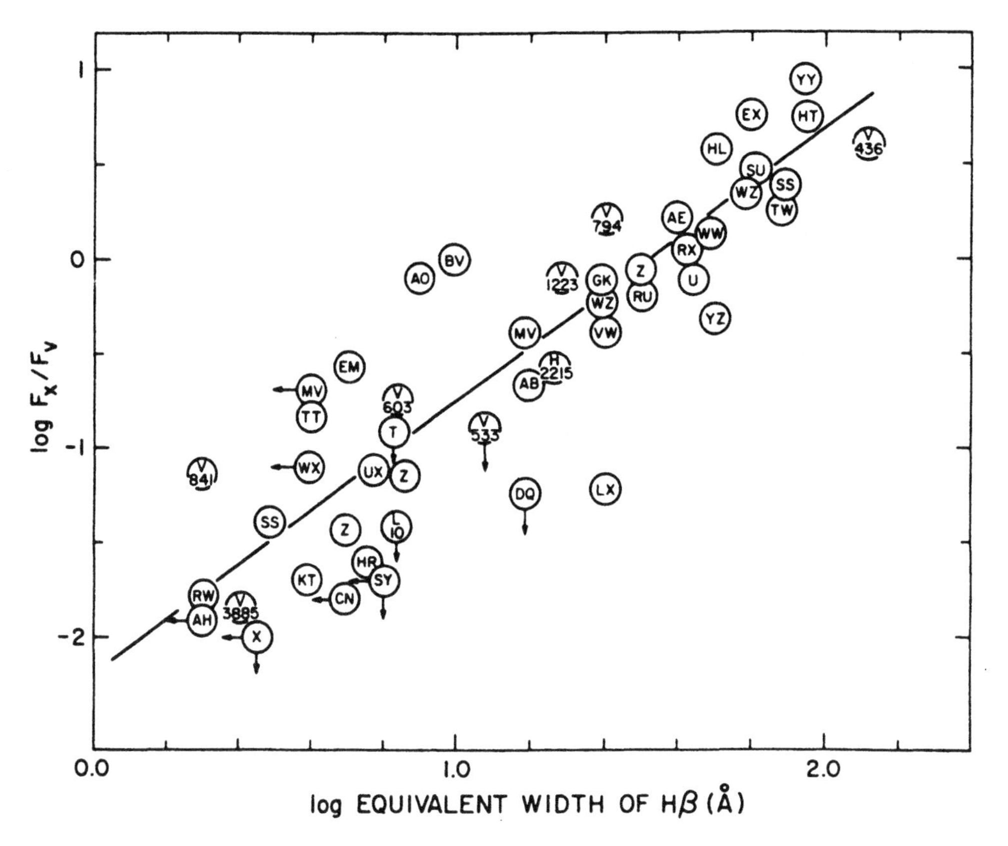

Figure 2.35 Correlation of F_X/F_V with equivalent width of Hβ emission. The straight line is a linear best fit. From Patterson & Raymond (1985a).

However, Marsh & Horne (1990) and Hoare & Drew (1991) attribute the apparent deficiency to an underestimate of the $\dot{M}$(d) used in the models, and to the omission of HeII emission from the bright spot. With these adjustments, the BL model is in good agreement with observations.

Horne & Saar (1991; see also Horne 1993b) point out that the emission-line surface brightness distribution $I(r) \tilde{\propto} r^{-3/2}$ found for DN in quiescence (see below: Sections 2.7.4 and 2.7.5) implies $I(r) \tilde{\propto} \Omega_K(r)$ (equation (2.23)), which is a similar relationship to that found between H and K CaII emission and rotation frequency among late-type stars. The similar scaling could imply a common mechanism – probably dynamo action and chromospheric heating.

The increase of emission-line equivalent width with inclination (Figure 2.34) is readily understandable if due to chromospheric emission: whereas the photospheric continuum luminosity varies as $(1 + \frac{3}{2}\cos i) \cos i$ (equation (2.63)), the chromospheric lines are of nearly constant flux until optical depth effects (in the chromospheric continuum) arise at very large inclinations. In this context, the correlation between Hβ strength and M_V(d) given by Patterson (1984) probably contains a mixture of optically thick and thin discs and does not allow for any dependence of equivalent width on i.

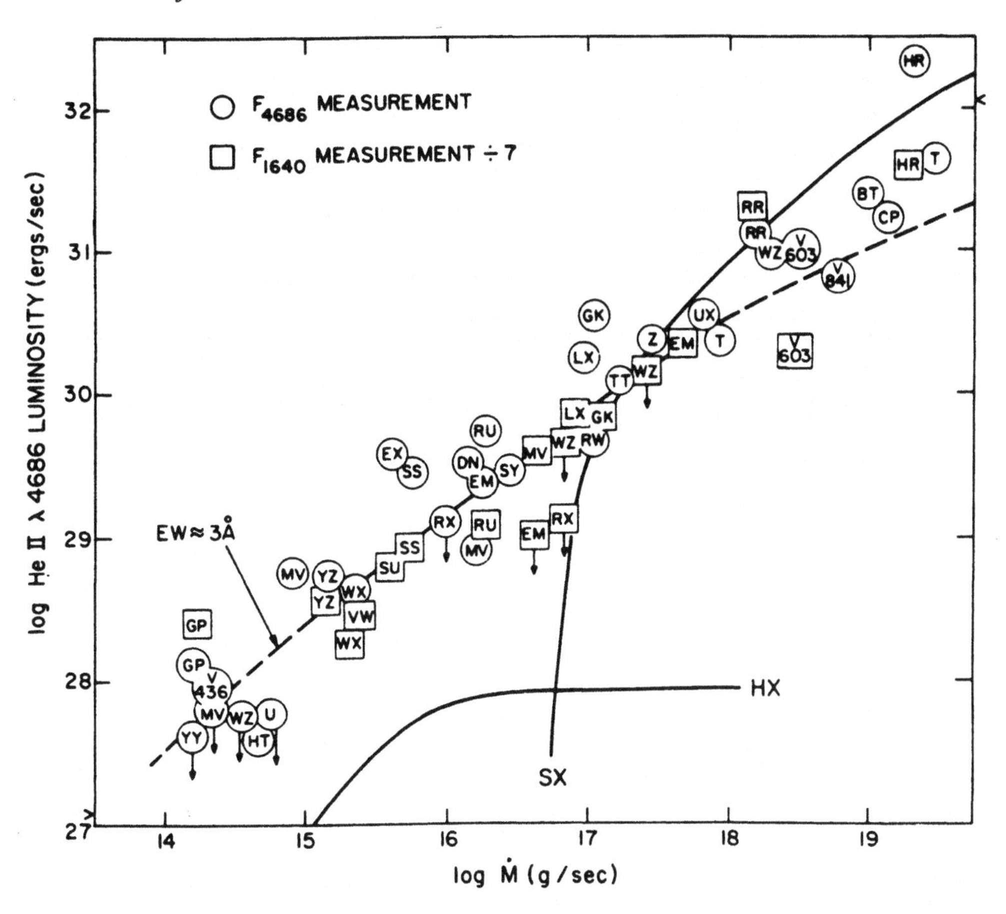

Figure 2.36 Correlation of HeII λ4686 emission luminosities with rate of mass transfer. The dashed line is the predicted relationship for accretion discs with $M_1(1) = 0.7$ and a HeII λ4686 equivalent width of 3 Å. Solid curves show the predicted luminosities from reprocessing of soft X-rays and hard X-rays. From Patterson & Raymond (1985b).

Nevertheless, using $M_V(d)$ for DN from Warner (1987a), there is little correlation between $W(\beta)$ and i, but there is a systematic increase of $W(\beta)$ with $M_V(d)$ in the range $6.5 \lesssim M_V(d) \lesssim 10$.

For the UV lines it can be shown that it is photoionization rather than collisional ionization that determines the ionization balance. For example, the observed NV1238 Å/Lyα and CIV1549 Å/NV1231 Å ratios are mutually incompatible with a collisional model (Jameson, King & Sherrington 1980). More generally, from the observed line ratios most of the ionization structure is determined by a photoionization–radiative recombination balance, and excitation of the levels is determined by recombination for H and He but by collisions for the heavier elements (Jameson, King & Sherrington 1980). The radiation required to produce the species observed has an equivalent blackbody $T \sim 10^5$ K, characteristic of the BL.

The electron density N_e in the emission region may be bracketed very easily (Warner 1976a). In nearly pole-on systems, where Doppler broadening is at a minimum, the

Balmer series can be seen to principal quantum number $n \sim 16$. The Inglis–Teller formula[6] thus provides $N_e \lesssim 10^{14}$ cm^{-3}. Treated as optically thin, the observed flux in $H\beta$ is $F(H\beta) = \alpha_4 f N_e^2 h\nu V_{ch}$, where α_4 is the recombination rate onto the fourth level, f is the fraction of transitions from that level that go through $H\beta$, and V_{ch} is the volume of the chromosphere. $F(H\beta)$ is typically $\sim 1 \times 10^{30}$ erg s^{-1} and $V_{ch} \sim \pi r_d^2 \times$ (chromospheric thickness) $\sim 3 \times 10^{30}$ cm^3, which leads to $N_e \gtrsim 10^{13}$ cm^{-3}. The shallow Balmer decrement (Kiplinger 1979a) shows that there are optical depth effects present, which first become important at $N_e \sim 10^{12}$ cm^{-3}. (The absence of the 'semi-forbidden' lines CIII] 1909 Å and NIV] 1486 Å gives a somewhat weaker limit of $N_e \gtrsim 10^{10}$ cm^{-3}.) Thus mean electron densities $\sim 10^{13}$ cm^{-3} are indicated.

Some abundance determinations made from emission-line intensities claim large enhancements of He and CNO relative to solar composition (Williams & Ferguson 1982, 1983), although others give more normal abundances (Jameson, King & Sherrington 1980; Ferland *et al.* 1984). The reason for finding anomalous abundances appears to be due to the use of optically thin discs rather than irradiated chromospheres to interpret the line intensities (Horne 1990).

Tabulations of emission-line equivalent widths, or plots of spectra, may be found in Williams (1980), Williams & Ferguson (1982,1983), Williams (1983), Szkody (1987a) and la Dous (1990).

2.7.3 *Winds from CVs*

Observations with IUE show that the UV resonance lines are commonly, but not universally, partially reversed into P Cyg profiles, implying the presence of mass loss in stellar winds (Holm, Panek & Schiffer 1982; Cordova & Mason 1982b, 1984b; Cordova, Ladd & Mason 1986). The lines most conspicuously affected are CIV(λ1549), SiIV(λ1397) and NV(λ1240) (Figure 1.19), with HeII(λ1640) and NIV(λ1719) also reversed in RW Sex (Drew 1990). The P Cyg profiles are seen only in CVs with high $\dot{M}$ and $i \lesssim 65°$, i.e., in non-eclipsing CN remnants, NLs, and DN during outburst, though not all stars in these classes necessarily show the effect.

The total widths of blueshifted absorption components lie in the range 3000–5000 km s^{-1}, which indicates a terminal velocity comparable to the velocity of escape from the primary. As there is no obvious reason why a wind from the outer regions of the accretion disc should be accelerated to just this amount, the wind almost certainly originates from the inner disc or boundary layer (Cordova & Mason 1982b). In the higher inclination systems the UV resonance lines are in emission and approximately symmetrical with total widths also in the range 3000–5000 km s^{-1}.

The line-forming region must be effectively bipolar in order to effect such a strong inclination dependence. Although the radiation field from a disc is naturally bipolar (and enhanced by limb darkening), which has been thought sufficient to produce the observed effect (Mauche & Raymond 1987), models by Drew (1987) show that some bipolarity in the wind flow itself is also required.

From observations by IUE of eclipsing systems, which show that in UX UMa and RW Tri the CIV resonance line suffers no eclipse and NV, SiIV are hardly eclipsed (Holm, Panek & Schiffer 1982; King *et al.* 1983; Drew & Verbunt 1985; Cordova &

[6] Application to the Paschen series is given by Kurochka & Maslennikova (1970).

Mason 1985; Drew 1990; and similarly in OY Car: Naylor *et al.* 1988), it was deduced that for NV, SiIV the line-forming region is larger than the secondary and for CIV may be much larger than the binary separation. However, from HST observations of UX UMa, Mason *et al.* (1995) discovered narrow absorption components in CIV emission lines which disappear during eclipse, leading to a net increase in emission during eclipse. The variations during eclipse depend, therefore, on the different eclipses of the emission and absorption components. Although coming from extended wind regions, the dimensions need not be as large as the secondary star. At the low densities implied by these dimensions, and the relatively low temperatures, collisional excitation is unimportant; the emission lines in the wind are therefore produced predominantly by resonance scattering (Drew 1986, 1990; Verbunt 1991).

Orbital phase dependence of the profiles of the resonance lines has been found in a few systems: the DN SU UMa (Drew 1990), YZ Cnc and IR Gem (Drew & Verbunt 1988; Woods *et al.* 1992), Z Cam (Szkody & Mateo 1986a) and the NLs V3885 Sgr (Woods *et al.* 1992), IX Vel (Mauche 1991) and BZ Cam (Woods, Drew & Verbunt 1990) and the CN HR Del (Friedjung, Andrillat & Puget 1982). An example is given in Figure 2.37, in which the P Cyg profile almost disappears at some orbital phases; the profile changes are not accompanied by variations in the continuum flux, and are thus different in nature from the general change from pure emission to P Cyg profiles on the rise of a DN. The orbital phase at which the absorption component is deepest varies from system to system, but the profiles of all lines vary in phase, which implies that the SiIV, CIV and NV line-forming regions are not very distant from each other (Woods *et al.* 1992). These observations appear to eliminate the proposed model of a bipolar wind with its axis tilted from normal to the disc by Coriolis effect (Drew & Verbunt 1988, Woods *et al.* 1992).

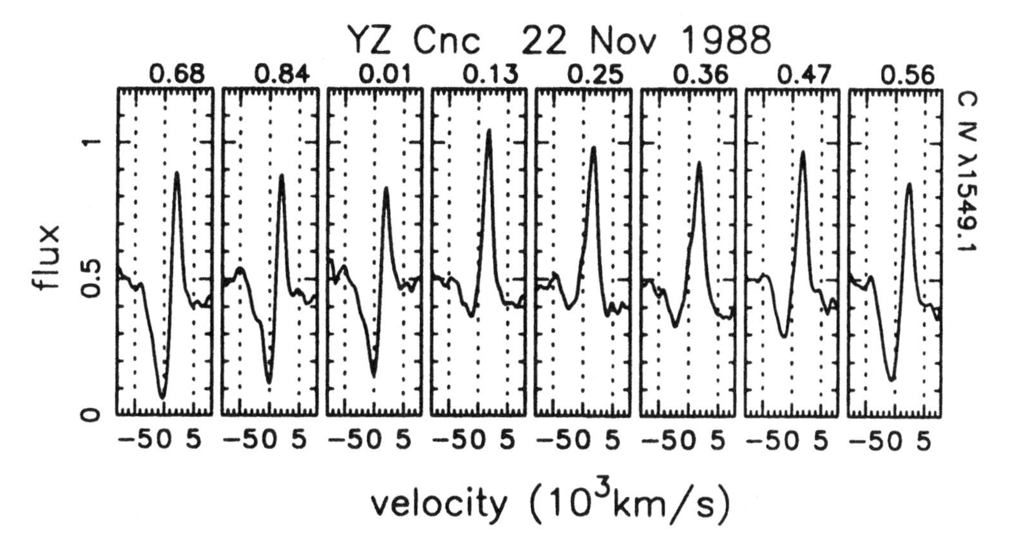

Figure 2.37 Variation of the CIV λ1549 profile with orbital phase for the DN YZ Cnc during outburst. The flux is in units of 10^{-12} erg cm^{-2} s^{-1} Å^{-1}. Orbital phases are indicated along the top of the diagram. From Wood *et al.* (1992).

On the other hand, in IX Vel (Mauche 1991) the region of SiIV absorption is demonstrably different from that of CIV: the mean velocity of the former is 400 km s^{-1} and of the latter 2000–3000 km s^{-1}. Furthermore, the SiIV absorption lines partake in the orbital motion of the primary. These results, and the deeper eclipse of SiIV than CIV in high inclination systems, require the SiIV to be concentrated near the primary. The detailed variation of velocity of Si IV absorption suggests that the wind speed is lowest in the direction of the secondary and highest in the opposite direction; alternatively this may be a difference in ionization structure in the two directions.

The ionization state of the wind is largely determined by the flux from the BL. Initial models all assumed a spherically symmetric wind (Kallman 1983; Drew & Verbunt 1985; Kallman & Jensen 1985; Mauche & Raymond 1987). Adopting the expected BL temperature $T_{BL} \sim 3 \times 10^5$ K (see equation (2.55)), Drew & Verbunt (1985) could not account for the observed strengths of the CIV and SiIV lines – the strong He^+ continuum should ionize those species. However, by using an extreme mass loss rate in the wind, such that $\dot{M}_{wind} \sim 0.3\dot{M}(d)$, Mauche & Raymond (1987) showed that the He^+ continuum can be sufficiently absorbed by the inner wind regions to remove the discordancy.

An alternative to such extraordinary values of $\dot{M}_{wind}$ has been found by turning the problem around to use the UV and visible HeII lines as a diagnostic for the BL flux (Hoare & Drew 1991). The HeII recombination lines ($\lambda\lambda$1640, 4686) are produced probably in both the disc and the wind. By applying the Zanstra (1931) method, on the assumption that the wind is optically thick in the HeII Lyman continuum (which is implied by the high abundances of CIV and SiIV noted above), a Zanstra temperature can be derived for photons with energies > 54 eV. If part of the HeII flux arises in the disc, this method sets an *upper limit* on T_{BL}.

The results (Hoare & Drew 1991) for a group of NLs and DN in outburst are 50 000 K $\lesssim T_{BL} \lesssim$100 000 K, i.e., much lower than the expected optically thick BL temperature. Alternatively, if $T_{BL} \sim 3 \times 10^5$ K, then the BL luminosity can only be a very small fraction of the disc luminosity. These conclusions remove the problem encountered with standard BL models (Mauche & Raymond 1987) which produce about 100 times the observed HeII λ1640 flux.

Following this departure from belief in the 'classical' BL, Hoare & Drew (1993) have constructed detailed photoionization models for winds from discs with cool BLs and canonical luminosity for $\dot{M}(d) = 5 \times 10^{-9}$ $M_\odot$ y^{-1}, and for discs with no BLs at all (corresponding to primaries rotating at break-up speed). For the latter, in order to have inner disc temperatures 6–9 $\times 10^4$ K, accretion rates of $\dot{M}(d) = 1\text{–}5 \times 10^{-8}$ $M_\odot$ y^{-1} were necessary. Good agreement with observed CIV and NV line strengths was obtained, and $\dot{M}_{wind}$ of only $\sim 6 \times 10^{-10}$ $M_\odot$ y^{-1} was deduced. In these circumstances, radiation pressure from the inner disc is adequate to drive the wind. Observed SiIV line strengths are considerably stronger than predicted, which may be caused by clumping into cooled masses.

Confirmation of this approach comes from models that employ biconical rotating outflow from discs instead of non-rotating radial outflows (Shlosman & Vitello 1993; Vitello & Shlosman 1993). Good agreement with observed line profiles is obtained with $T_{BL} \lesssim$ 80 000 K and $T_{eff}(1) \lesssim$ 40 000 K (most of the ionizing flux comes from the inner disc) and typical $\dot{M}_{wind} \sim 1 \times 10^{-9} M_\odot$ y^{-1}.

To the EUV and FUV observations of VW Hyi in outburst (Section 1.5) have been added low energy X-ray measurements from EXOSAT (Mauche *et al.* 1991). These give $kT_{BL} \sim 10.5$ eV and $L_{BL} \sim 6 \times 10^{32}$ erg s^{-1}, or $\zeta \approx 0.04$ in equation (2.54b), which is compatible with the Hoare & Drew deduction. In VW Hyi there are 14 s X-ray periodicities, discussed in Section 8.4, which independently suggest rapid rotation of the primary and hence a small value of ζ. Whether, however, the same reason can be given for the low values of L_{BL} found by Hoare & Drew in other CVs remains uncertain.

From models of the P Cyg profiles an interesting result emerges (Drew 1987; Mauche & Raymond 1987). In contrast with the winds from hot stars, where the initial acceleration is very high, which leads to most of the wind travelling near terminal velocity and a maximum depth of the absorption profile near its blueward edge, in CVs the maximum depth is near the line centre (Cordova & Howarth 1987 and Figure 2.37). The latter requires the terminal velocity to be reached at a distance $\sim r_d$ from the primary.

Although almost all of the evidence for CV winds comes from UV observations, there is some evidence at optical wavelengths. Honeycutt, Schlegel & Kaitchuck (1986) and Marsh & Horne (1990) deduced the presence of an extended HeII emission region from emission-line eclipses in the NL SW Sex and in the DN IP Peg in outburst. Robinson (1973c) found stationary components in the Hα emission of Z Cam; Szkody & Wade (1981) recalibrated the Hα flux and found an implied mass loss rate of 7×10^{-10} M$_\odot$ y^{-1}. A similar component is seen in Hβ in IP Peg (Piché & Szkody 1989). It is surprising that such stationary emission components have not been seen in more systems. A line at 1.9 μm, ascribed to [SiVI], has been detected (Ramseyer *et al.* 1993b) that may be combined with the SiIV lines observed in the UV to probe wind structures.

It should be noted that winds can produce pure absorption lines, without any apparent emission component (Drew 1987, Mauche & Raymond 1987) – the emission, although present, merely partially fills the absorption component. For example, Szkody & Mateo (1986a) deduced that NV and SiIV absorption lines in Z Cam are wind-formed because they are blueshifted by ~5 Å.

2.7.4 Disc Emission-Line Profiles

Disc emission lines in the optical region are seen to be broad and often double peaked (Figures 2.38 and 2.39). As $v_K(r)$ is hundreds to thousands of km s^{-1}, whereas $c_s \lesssim 10$ km s^{-1}, the profiles of the lines are largely determined by Doppler shifts from the macroscopic motions; the profiles carry information about both the velocity and the intensity distributions.[7] Adopting a Keplerian velocity distribution reduces the problem to one where the size of a disc and the radial dependence of intensity may be derived. This process has been described by Smak (1969, 1981), Huang (1972), Stover (1981a), Young, Schneider & Schectman (1981b) and generalised by Horne & Marsh (1986).

For some point (r, θ) in a Keplerian disc, where θ is the azimuth of the observer

[7] The broad lines, however, prevent measurement of Zeeman splitting to obtain magnetic field strengths in discs.

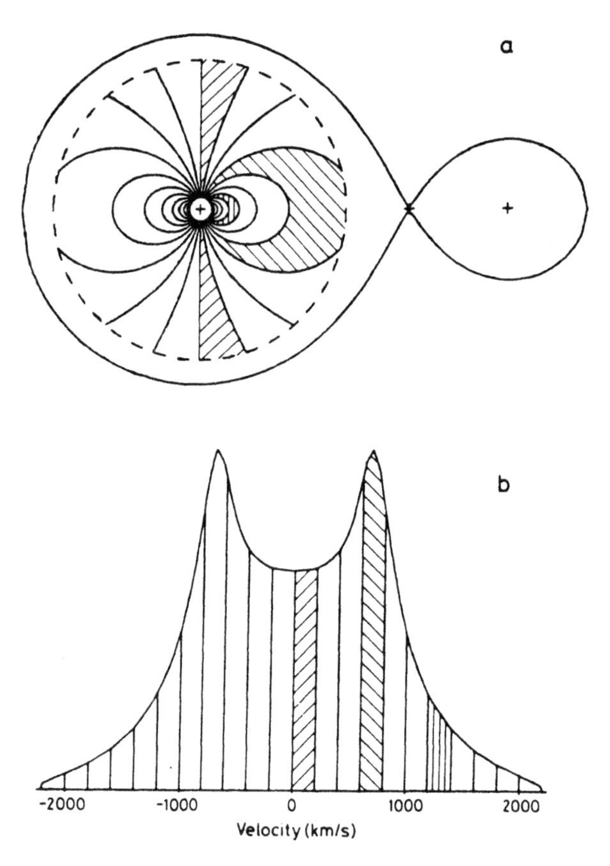

Figure 2.38 (a) Loci of constant radial velocity in a Keplerian disc in a binary of mass ratio $q = 0.15$ viewed at quadrature. (b) Velocity profile of emission lines from the disc. Emission in the shaded velocity ranges originates in corresponding shaded regions on the disc. From Horne & Marsh (1986).

relative to the radius vector, the Doppler velocity is

$$v_{\rm D} = v_{\rm K}(r) \sin\theta \sin i. \tag{2.74}$$

Lines of constant v_D lie on a dipole pattern on the surface of the disc (Figure 2.38). The intensity $I(v_D)$ observed in a given range of velocities in the profile of the line is therefore a sum over the contributions from a range of r (shaded areas in Figure 2.38), weighted by the local intensity $I(r)$ (assumed isotropic):

$$\begin{aligned} I(v_{\rm D}) &\propto \int\!\!\int I(r) r \mathrm{d}r \, \mathrm{d}\theta \\ &\propto \int_{r_0}^{r_{\rm min}} \frac{I(r) r^{3/2)} \mathrm{d}r}{(1 - v_{\rm D}^2 r)^{1/2}} \end{aligned} \tag{2.75}$$

from equations (2.16) and (2.74) and $r_{\rm min} = \min(r_{\rm d},\ v_{\rm D}^{-2})$.

For $I(r) \propto r^{-n}$, where n is integer or half-integer, equation (2.75) can be evaluated analytically (Smak 1981). Examples of profiles for such cases are given by Smak (1981)

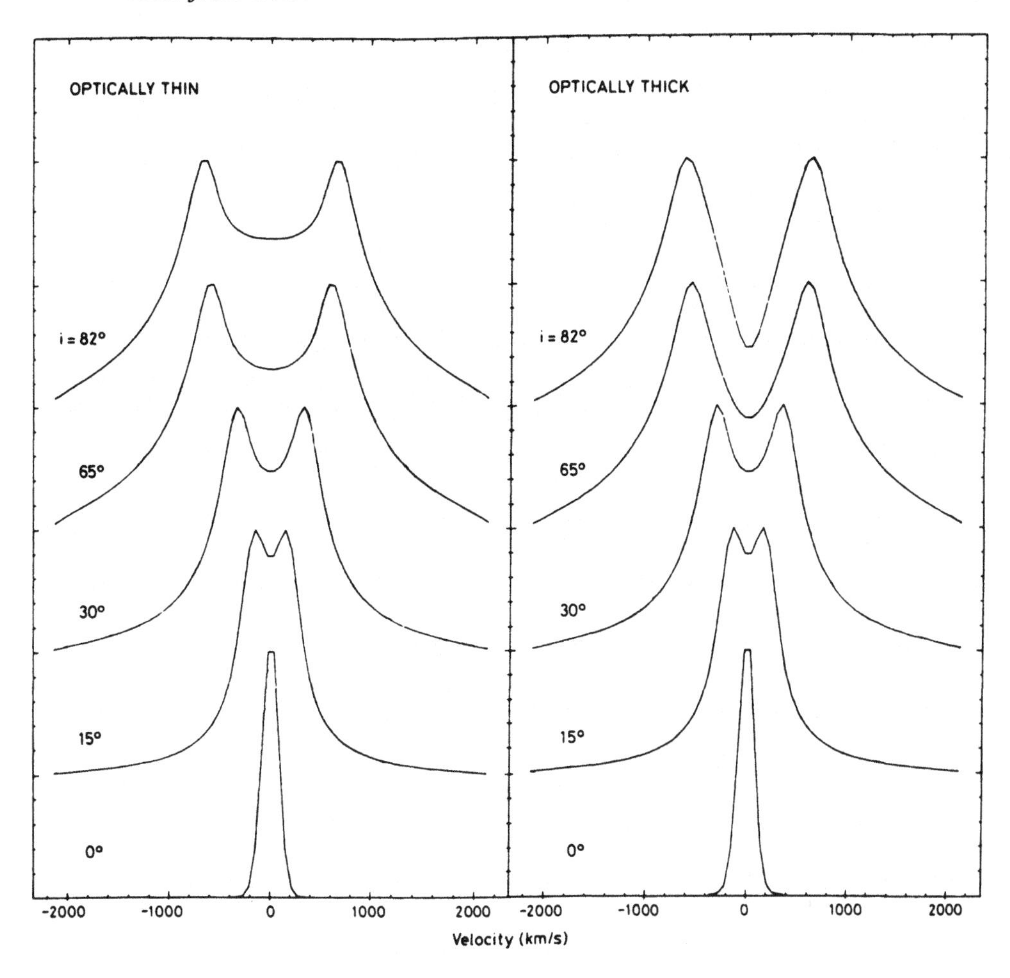

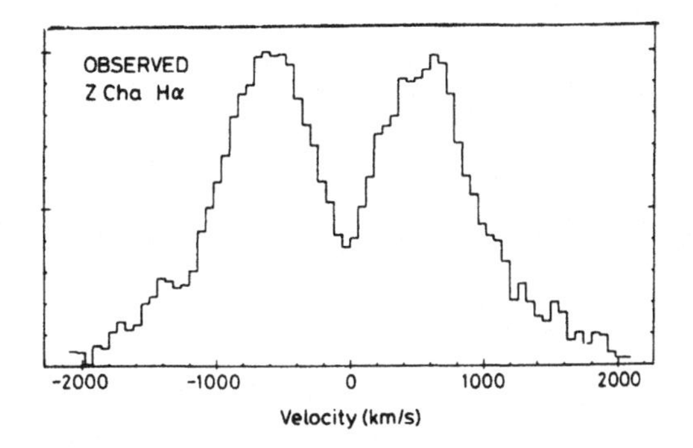

Figure 2.39 Synthetic emission-line profiles for optically thick and optically thin lines at various inclinations, compared with the observed profile of Z Cha ($i = 82°$) in quiescence. Adapted from Horne & Marsh (1986).

and in all cases, unless there is large intrinsic broadening, the profiles are double peaked. As can be seen from Figure 2.38, the largest area of disc that contributes to a specific value of v_D is that for $v_D \sim v_K(r_d)\sin i$; Smak (1981) shows that for values of n typical of observed discs, the maximum I_{max} of $I(v_D)$ occurs at

$$v_K(r_d)\sin i = (0.95 \pm 0.05)v_D(I_{max}). \tag{2.76}$$

Thus half the separation between the peaks of the emission line profile is a useful indicator of the projected velocity of the outer edge of the disc. If i and $M(1)$ are known then application of equation (2.16) gives an estimate of r_d. By comparison with other methods of finding r_d (eclipse width of the emission lines, radius vector of the bright spot determined from eclipse measurements), this usually gives larger values (Wade & Horne 1988; Marsh 1988). This has been interpreted (Marsh, Horne & Shipman 1987; Wade & Horne 1988) as implying that velocities in the disc are sub-Keplerian, but Marsh (1988) points out that there is no definite outer cut-off in the distribution of line emission, so different radii will inevitably be measured by different techniques. $v_D(I_{max})$ can be measured for different species, of which HI, HeI and HeII are commonly employed. However, Persson (1988) points out that the CaII IR triplet is formed in the coolest, outermost parts of the disc and is therefore most appropriate. It gives peak separations that are smaller than those from HI or FeII.

When line profiles are observed with high signal-to-noise, equation (2.75) can be solved by an inverse Abel transform to give n and r_0. (Note, in particular, that in the wings $I(v_D) \propto v^{2n-5}$.) Smak (1981) finds in this way that $n = 2.0 \pm 0.2$ for several CVs, including Z Cha and U Gem in quiescence in which the continuum emission rises in the outer regions of the disc (Section 2.6.3). This result stresses the independence of line and continuum emissivities.

An $n \sim 1.5$–2.0 law for the Balmer lines has been found also by the more sophisticated analyses described below, but $n \sim 0$ is more appropriate for HeII (Marsh 1988; Marsh & Horne 1990). Such laws are not in agreement with the predictions of line emission from optically thin discs (Marsh & Horne 1990).

In optically thin discs the density in the emission-line forming regions can be far greater than in chromospheres. Lin, Williams & Stover (1988; see also Thorstensen, Wade & Oke (1986)) show that in the ranges $10^{14} < N_e < 10^{17}$ cm^{-3} and $6000 < T < 12\,000$ K Stark broadening is comparable with rotational Doppler broadening in the Balmer lines. In chromosphere emission, with $N_e \sim 10^{13}$ cm^{-3}, such collision broadening would not be significant, so Marsh (1987) concludes that Stark broadening can only be important for a mechanism which produces line emission from optically thick parts of the disc and points out that the increase in line width in higher members of the Balmer series (Thorstensen, Wade & Oke 1986) is not a definitive diagnostic for the presence of the Stark effect.

The line profiles computed for isotropic $I(r)$ only apply to optically thin emission lines. The flat Balmer decrement in most CVs implies considerable optical thickness at the line centres. However, in directions in which differential rotation of the disc produces a strong velocity gradient, photons that would otherwise be absorbed in the line centre are able to escape more easily in the Doppler-shifted wings (Gorbatskii 1965; Rybicki & Hummer 1983; Horne & Marsh 1986; Marsh 1987). The resulting anisotropy of emission, which has a sin 2θ dependence, can have a substantial effect on line profiles (Figure 2.39), with the result that the rounded emission peaks, which have been interpreted as evidence for large turbulent velocities in the disc (Stover 1981a; Young, Schneider & Schectman 1981b), are in fact more probably due to Keplerian shear broadening (Horne & Marsh 1986).

As noted by Horne & Marsh (1986), the line broadening can be divided into three regimes: (a) at very low inclinations, $\sin i \lesssim H/r$, thermal broadening with velocity $\sim (H/r)v_K(r)$ dominates, (b) at small inclinations with $(H/r)^2 \lesssim \sin i \tan i \lesssim 1$, rotational broadening with velocity $v_K(r)\sin i$ dominates and produces the double-peaked profile, and (c) at moderately large inclinations, $1 \lesssim \sin i \tan i \lesssim r/H$, shear broadening with velocity $\sim v_K(r)(H/r)\sin i \tan i$ dominates.

Eclipses of the line-emitting regions can be modelled or analysed in the same manner as for the continuum sources (Sections 2.6.3 and 2.6.4), but with the added benefit that eclipse profiles are available for each RV range. Thus, as well as providing emission-line intensity maps, constraints on the dynamics of the system are obtained. The theory for this is given by Young & Schneider (1980), who applied it to DQ Her; an application to Z Cha is given by Marsh, Horne & Shipman (1987) and to the DN IP Peg by Marsh (1988). The latter finds strong evidence for Keplerian rotation of the disc, the only detected deviations being caused by the stream.

In Z Cha at quiescence, where the Balmer emission is eclipsed almost as deeply as the continuum, the vertical thickness of the emission layer appears to be no greater than the scale height of the disc (Marsh, Horne & Shipman 1987).

In contrast, in at least some high luminosity (or high $\dot{M}(d)$) CVs there is evidence that optical emission lines are formed in a region that has large vertical extent, as evinced by relatively shallow eclipses of the emission lines in the NLs V1315 Aql (Dhillon, Marsh & Jones 1990) and SW Sex (Honeycutt, Schlegel & Kaitchuck 1986). These and some other highly inclined CVs do not show the doubled emission-line profiles characteristic of a disc (BT Mon, SW Sex: Williams 1989) or are only occasionally double (VZ Scl: Williams 1989). On the other hand, some other high inclination CVs of high luminosity do show double emission peaks (DQ Her, RW Tri, AC Cnc). All low luminosity (DN at quiescence) discs show line doubling for $i \gtrsim 60°$, but there is evidence for a low velocity component partially filling the space between the emission peaks in some systems (Robinson 1973c; Smak 1981; Clarke, Capel & Bowyer 1984).

The singular appearance of the lines in some of these high $\dot{M}(d)$ systems is accompanied by an absence of rotational disturbance at eclipse, which Williams (1989) interprets as implying a lack of any true disc-like structure – possibly non-disc accretion caused by a large magnetic field on the primary (discussed in detail in Section 7.2.2). However, Honeycutt, Schlegel & Kaitchuck (1986) find that they can account for the observations of SW Sex if the lines are emitted by a wind over a height $H \sim r_d$. Such winds from high luminosity discs could account for the non-Keplerian motions deduced in RW Tri (Kaitchuck, Honeycutt & Schlegel 1983), UX UMa (Schlegel, Honeycutt & Kaitchuck 1983) and AC Cnc (Schlegel, Kaitchuck & Honeycutt 1984).

These same high luminosity systems also show an asymmetry in the strengths of the red and violet[8] components of the doubled emission in the sense that $R/V < 1$ *when*

[8] R and V are traditionally used by spectroscopists, although violet is no longer used to denote the shortward direction of wavelength.

averaged over the orbit. In some stars (e.g., AC Cnc) the blue component is *always* the strongest. Any asymmetry of disc emission, as for example that caused by emission near the bright spot, should average out over an orbit if the emission is isotropic. Honeycutt, Schlegel & Kaitchuck (1986) suggest that this is another manifestation of a wind from the disc: for large i (but not very close to 90°) the observer sees preferentially the near-side approaching wind while the far-side receding wind is obscured by the disc. It should be noted, however, that this type of behaviour is not restricted to high luminosity CVs: it is seen in many DN at quiescence (Marsh, Horne & Shipman 1987). In KT Per, for example, the red component is always the strongest (Ratering, Bruch & Diaz 1993).

There is a further departure from symmetry in the line emission of these objects (see references to Honeycutt and coworkers cited above), which remains unexplained: there is evidence for a region of enhanced emission in the outer part of the disc along the line of centres of the stars (this may, however, be located on the secondary star, near the L_1 point).

In a few objects R/V reaches a minimum at $\varphi \sim 0.3$ instead of $\varphi \sim 0.7$ as expected from the bright spot contribution. These include the DN SW UMa (Shafter, Szkody & Thorstensen 1986) and WZ Sge (Gilliland, Kemper & Suntzeff 1986) and the NR CP Pup (White, Honeycutt & Horne 1993). The reason for this emission region on the disc diametrically opposite the bright spot is not known.

2.7.5 *Doppler Imaging of Discs*

As a CV binary system rotates, the observed emission-line profiles provide a series of one-dimensional projections of the two-dimensional velocity field. Applying the same techniques of tomographic restoration as widely used in medical physics it is possible to deduce the velocity field, without any prior assumptions about its nature. The velocity coordinate system (rotating with the binary, with the y-axis pointing in the direction of motion of the secondary) adopted by Marsh & Horne (1988) is shown in Figure 2.40. In circularly symmetric motion, the outer and inner edges of the accretion disc transform into, respectively, the inner and outer edges of circles centred on the

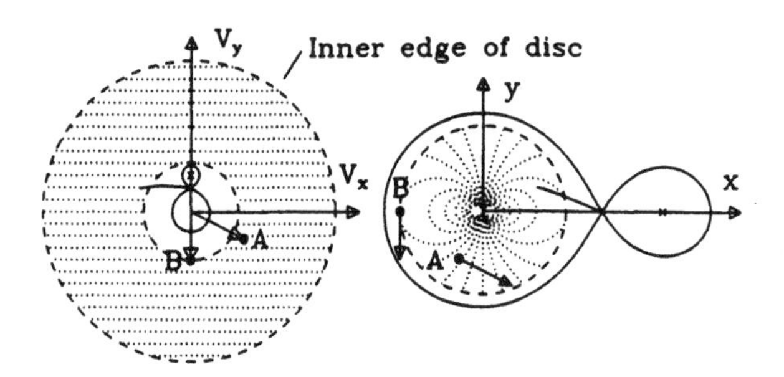

Figure 2.40 Lines of constant radial velocity in a Keplerian accretion disc (cf. Figure 2.38(a)) and their equivalents in velocity coordinates. The points marked A and B in the two diagrams are equivalent to each other. Note the Roche surfaces and the mass transfer stream in velocity space. From Marsh & Horne (1988).

primary's velocity coordinates (defined to be the origin). Lines of constant radial velocity transform in velocity space to straight lines perpendicular to the line of sight and, for a Keplerian disc, are the equivalents of the dipole pattern seen in Figures 2.38 and 2.40. As the velocities of the surface of the secondary and of any point in the stream have constant values in the rotating frame, they transform into fixed patterns (turned through 90°) in the velocity frame.

Although inversion of the observed line profiles, which are summations of the intensity distribution along the lines of constant radial velocity, can be carried out by standard tomographic processes, the line profiles are sometimes incomplete, usually noisy, and always blurred by instrumental resolution. A MEM technique has been developed by Marsh & Horne (1988) to optimize the inversion. Compared with eclipse mapping (Section 2.6.4), which in essence is based on only two slices through the disc, the inversion process for Doppler tomography is well constrained by the observations and is not sensitive to the default map. This process of Doppler tomography has been applied to IP Peg (Marsh 1988; Marsh & Horne 1990; Harlaftis *et al.* 1993d), U Gem (Marsh *et al.* 1990) and KT Per (Ratering, Bruch & Diaz 1993) in quiescence. In the first two Horne & Saar (1991) find an average radial intensity distribution $\tilde{\propto} r^{-3/2}$ for Hβ. Doppler maps at quiescence and through outburst have been constructed for the DN SU UMa and YZ Cnc (Harlaftis *et al.* 1993a,c). Doppler maps for NR CP Pup (White, Honeycutt & Horne 1993), NLs GD552 (Hessman & Hopp 1990), V795 Her (Haswell *et al.* 1994) and V1315 Aql (Dhillon, Marsh & Jones 1991) and polar VVPup (Diaz & Steiner 1994) are also available. An atlas of Doppler maps for eighteen CVs has been published by Kaitchuck *et al.* (1994).

An example of an image of a CV in the velocity plane is given in Figure 2.41. This shows that in IP Peg at quiescence a substantial amount of emission arises from stream overflow. A detailed comparison between Doppler tomography and eclipse mapping from emission lines has yet to be made. The combination of both techniques, with simultaneous application in the UV and optical, both of continuum distribution and emission-line distribution (and, for example, mapping of the Balmer discontinuity), holds great promise for detailed physical models of CVs in their various stages of quiescence and outburst.

2.7.6 *RV Curves*

One approach to determining masses in CVs requires RV measurements around orbit for both the primary and the secondary. The spectrum of the primary itself is rarely observable, so recourse may be made to the emission lines of the disc, which shares the motion of the primary in orbit.

The early RV curves showed apparent eccentricities of orbit and large departures of times of crossing of the γ-velocity (i.e., the velocity of the centre-of-gravity of the system) from times of inferior and superior conjunction (as determined from eclipse observations or from RV curves of absorption lines from the secondaries). At the low spectral resolution of these observations, part of the distortion of the RV curve was caused by blending of the bright spot S-wave with the lines from the disc. This was modelled by Smak (1970), who showed that errors of up to 50% could be made in the RV amplitude, $K(1)$, of the primary and that the anomalous phase shifts could be understood.

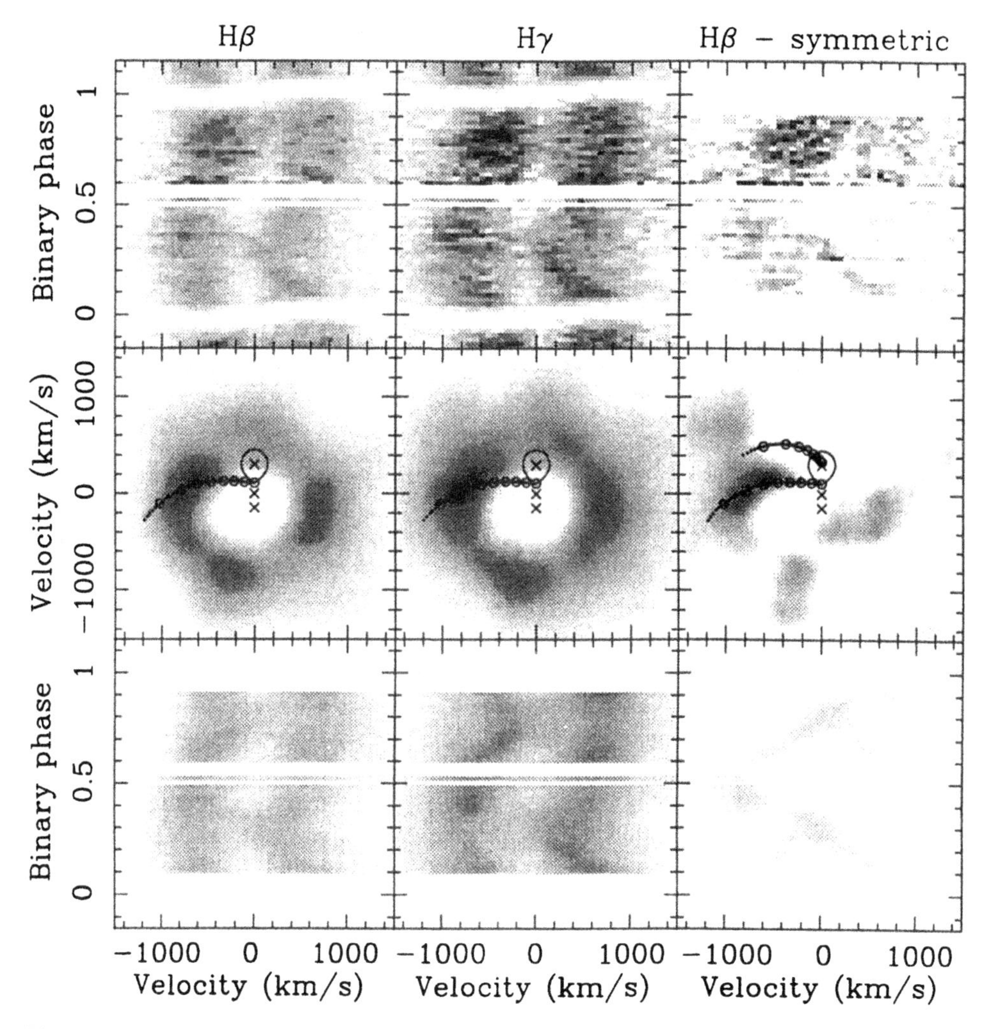

Figure 2.41 Doppler maps of the DN IP Peg in quiescence from emission lines of Hβ and Hγ. The upper panels show the observed time-resolved spectra. The central panels are the deduced Doppler maps, with superposed crosses marking the centres of mass of the primary, secondary and the binary system (zero velocity). The predicted surface of the secondary and the path of the gas stream (running leftwards from the inner Lagrangian point) are also shown. The bottom panels are time-resolved spectra reconstituted from the Doppler maps. The right hand column of panels shows the Hβ results after subtraction of all symmetric emission components. The upper arc is the locus of velocities in the disc underneath the trajectory of the stream across the face of the disc. Stream emission is seen to extend across the face, rather than round the rim of the disc. From Marsh & Horne (1990).

Modern high resolution spectroscopy is able to separate the S-wave from the disc emission lines, but there are still distortions present in the RV curve. In the hope that any asymmetric emission distribution is confined to the outer region of the disc, where stream and tidal interaction may dominate, RV curves have been obtained from the wings of emission lines using, e.g., the double-Gaussian method suggested by Schneider & Young (1980a) (see also Horne, Wade & Szkody (1986)). However, this does not

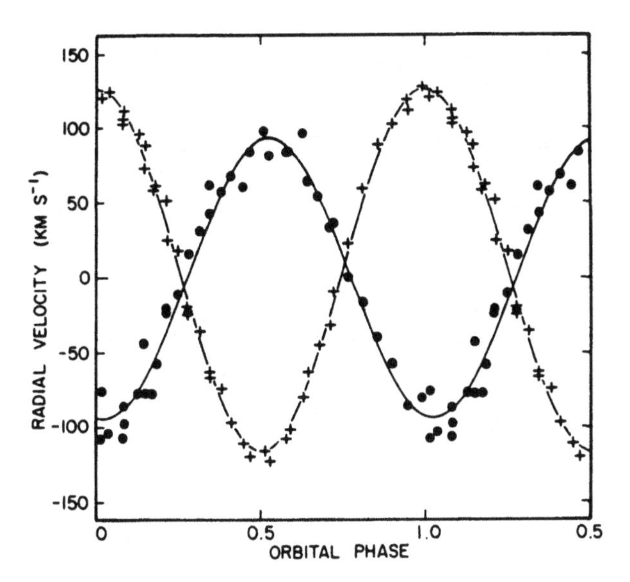

Figure 2.42 RVs from the absorption lines of the secondary (pluses) and emission lines (dots) in the DN RU Peg at quiescence. Solid lines are the best-fitting orbital solution. Orbital phase zero is defined as the time of maximum positive velocity of the secondary's velocity curve. There is a 9° phase offset between the two curves. From Stover (1981b).

remove the phase anomalies (Figure 2.42: Stover 1981b; Shafter 1984a; Watts *et al.* 1986; Hessman 1987), which can amount to as much as 58° (Hessman 1987), and the deduced value of $K(1)$ is often a function of position in the line wings. For example, Marsh, Horne & Shipman (1987) find $K(1) \sim 150$ km s^{-1} for the emission peaks and 88 km s^{-1} for the wings in Z Cha. $K(1)$ and the γ-velocity are sometimes found to differ from line to line and to vary systematically up the Balmer series (e.g., Ratering, Bruch & Diaz (1993)).

By plotting $K(1)$, fractional error $(\sigma_K/K(1))$, γ-velocity and phase lag φ_1 as a function of position in the line profile (i.e., the separation in the double-Gaussian technique) a *diagnostic diagram* is produced which aids in locating the source of perturbations of the RV curve (Shafter, Szkody & Thorstensen 1986). Alternatively, the *light centre* method of Marsh (1988) may be used. This plots $K_x = K(1) \sin \varphi_1$ against $K_y = K(1) \cos \varphi_1$ and extrapolation to the K_y-axis should give $K(1)$ free of distortion.

The distortions of the RV curves can be the result of either or both of asymmetric brightness distributions (Stover 1981b; Shafter 1984a; Watts *et al.* 1986) or non-Keplerian velocity distributions (Schoembs & Hartman 1983). Marsh, Horne & Shipman (1987) show that in Z Cha at quiescence the large distortions of the RV curves can be modelled by the velocity and intensity distribution of an elliptical disc, with eccentricity $e = 0.1$ and its major axis lying along the direction of the axis of the closed loop of stream trajectory seen in Figure 2.4. The resulting emission-line profiles have an asymmetry between red and blue components, alternating around the orbit, that may account for much of the V/R variations seen in most CVs. A Doppler tomogram of Z Cha would be very instructive.

A particularly severe problem arises in some eclipsing NLs, where single-peaked Balmer and HeI emissions, largely unobscured during eclipse of the continuum, show phase lags up to 76° (Thorstensen *et al.* 1991; Dhillon *et al.* 1992; Section 4.2.1.2).

The asymmetrical line emission distributions and velocity fields make derivations of RV amplitudes $K(1)$ from emission lines of very uncertain value. As is seen in Section 2.8.1, the mass function for a binary is proportional to $K^3(1)$, which virtually excludes this way of deriving masses. A discussion of the difficulties involved in $K(1)$ measurements is given by Wade (1985).

RVs of the secondaries are usually obtained by cross-correlating the CV spectrum with a selection of lines from a cool dwarf spectrum (Stover *et al.* 1980), a process which can detect and locate the spectrum of the secondary even when it is not obvious in the total spectrum (Hessman *et al.* 1984). Summation of the spectra, phased according to the RV of the secondary, can then reveal large numbers of weak lines from the secondary (Horne, Wade & Szkody 1986).

The RV amplitude $K(2)$ requires significant correction if flux is not uniform over the surface of the secondary, as can happen when the secondary is irradiated by the primary (Section 2.9.6). If the effective 'centre of gravity' of the absorption-line spectrum (obtained by weighting the line strength at each point on the secondary by the local intensity) is displaced by $\Delta R(2)$ from the optical centre of the secondary, the correction $\Delta K(2)$ can be written (Wade & Horne 1988 and equation (2.5b))

$$\frac{\Delta K(2)}{K(2)} \approx 0.462\, q^{1/3}(1+q)^{2/3}\frac{\Delta R(2)}{R(2)}. \tag{2.77}$$

In the extreme case where the hemi-lobe closest to the primary is devoid of absorption lines, $\Delta R(2)/R(2) = 4/3\pi$ and $\Delta K(2)/K(2) \approx 0.20 q^{1/3}(1+q)^{2/3}$, which is non-negligible even at $q = 0.1$.

Some secondary RV curves give an *apparent* eccentricity of orbit (e.g., Martin, Jones & Smith 1987; Martin *et al.* 1989; Davey & Smith 1992), which is an effect of the variation around the orbit of the light centre of the absorption lines (Wade & Horne 1988). The apparent eccentricity may be taken into account when deriving the orbital parameters by application of equations given by Friend *et al.* (1990a) or the models of Davey & Smith (1992).

2.7.7 *Absorption Lines from Discs*

Optically thick discs, particularly those of low inclination, often show broad, shallow absorption lines of HI, HeI and CaII. References to the computation of such lines have been given in Section 2.6.1.1; examples of observed spectra are given for DN in outburst in Figures 3.15–3.17 and for a UX UMa star in Figure 4.5.

The OI triplet $\lambda\lambda$7771.9, 7774.2, 7775.4 was observed in absorption in DQ Her (Young & Schneider 1981) and the NL LX Ser (Young, Schneider & Schectman 1981c). More recently it has been observed both in emission and in absorption in a variety of CVs (Friend *et al.* 1988). From RV measurements the OI absorption can be attributed to the accretion disc. During DN outbursts the OI lines change from emission at quiescence to absorption at maximum.

Smith (1990) concludes that the OI triplet provides a sensitive indicator of the optical thickness of a disc. He also points out that, as Lyβ emission can pump the upper level of the 8446 Å transition of OI, the (so far unobserved) nature of the latter line will be a useful indicator of Lyman emission. The Na D lines and an unidentified line at 6280 Å have been seen strongly in absorption in KT Per during outburst (Rattering, Bruch & Diaz 1993). The origin of these is unknown.

2.8 Masses and Mass Ratios

2.8.1 Methodologies of Measurement

The RV amplitude of the primary is

$$K(1) = \frac{2\pi aq \sin i}{P_{\rm orb}(1+q)} \tag{2.78}$$

which, on eliminating a through equation (2.1a), gives the standard relationship for the mass function

$$f[M(2)] = \frac{P_{\rm orb}K^3(1)}{2\pi G} = \frac{M^3(2)\sin^3 i}{[M(1)+M(2)]^2} = M(2)\left(\frac{q}{1+q}\right)^2 \sin^3 i. \tag{2.79}$$

A similar relationship exists for $f[M(1)]$.

Thus, given that $P_{\rm orb}$ must already be known, determination of $M(1)$ and $M(2)$ requires accurate measurement of $K(1)$, $K(2)$ (from which $q = K(1)/K(2)$) and i. In a single-line system it may be possible to determine q photometrically by making use of computed stream trajectories (Section 2.6.3). In a non-eclipsing system some estimate of i may be obtained from the ellipsoidal variation of the secondary (Section 1.8), but in general the only reliable values are for eclipsing systems. Because of the difficulty of interpreting $K(1)$ (which often differs considerably among observers), as outlined in the previous section, most masses for CVs (as listed, for example, in Ritter's catalogue (Ritter 1987)) are at best tentative. Wade (1985) has given a detailed critique of such mass determinations.

There are a number of additional relationships that make use of the existence of a disc around the primary, or of the Roche geometry and its relative insensitivity to q. From equation (2.16), if motion in the disc is Keplerian then measurement of $v_K(r)$ at any r will determine $M(1)$. In particular, if r_d can be measured from eclipse timings and the separation $2v_d(I_{\rm max})$ of the emission peaks is observed, then equations (2.16) and (2.76) give

$$M(1) = \frac{(0.95 \pm 0.05)^2 r_d v_D^2(I_{\rm max})}{G \sin^2 i}. \tag{2.80}$$

If the maximum observed velocity in the emission or absorption line profile[9] is $v_{\rm max}$, then, in the absence of other broadening mechanisms and if the disc reaches to the

[9] This is difficult to measure because it is based on where the wings of the line profile merge into the continuum. Often line widths are characterized by their FWHM (full width at half maximum intensity), but for rotationally Doppler-broadened lines it is the width at continuum that is required (often called FWZI – full width at zero intensity, even when it is really the width at the continuum intensity level).

surface of the primary,

$$M(1) = \frac{R(1)v_{max}^2}{G \sin^2 i}. \tag{2.81}$$

This usually provides only a lower limit on $M(1)$ and requires an estimate of $R(1)$, which, for non-rotating helium white dwarfs, is related to $M(1)$ through

$$R_9(1) = 0.779 \left\{ \left[\frac{M(1)}{M_{ch}}\right]^{-2/3} - \left[\frac{M(1)}{M_{ch}}\right]^{2/3} \right\}^{1/2}, \tag{2.82}$$

where M_{ch}, the Chandrasekhar mass, is $M_{ch} = 1.44\ M_\odot$ (Nauenberg 1972). For some purposes it is useful to approximate this in the forms

$$R_9(1) \approx 0.73 M_1^{-1/3}(1) \qquad 0.4 < M_1(1) < 0.7 \tag{2.83a}$$

$$R_9(1) \approx 1.12 \left[1 - \frac{M(1)}{M_{ch}}\right]^{3/5} \qquad 0.7 < M_1(1) < 1.3 \tag{2.83b}$$

$$R_9(1) \approx 0.900 \left[1 - \frac{M(1)}{M_{ch}}\right]^{1/2} \qquad M_1(1) > 1.2. \tag{2.83c}$$

For very low mass white dwarf *secondaries* it is necessary to use the relationship (Zapolsky & Salpeter 1969; Rappaport & Joss 1984)

$$R_9(2) = 0.876 M_1^{-1/3}(2)[1 - 5.14 \times 10^{-3} R_9(2) M_1^{-1/3}(2)]^5 \; 0.002 \lesssim M_1(2) < 0.2 \tag{2.83d}$$

Exact models for zero temperature pure carbon white dwarfs are given by Hamada & Salpeter (1961) and corrections (usually a few per cent increase in $R(1)$ for a given $M(1)$) for finite temperature and a hydrogen or helium envelope can be obtained from Koester & Schoenberner (1986).

The white dwarf limiting mass is reduced to 1.36 $M_\odot$ when general relativistic effects and particle interactions are taken into account (Cohen, Lapidus & Cameron 1969). This can be increased by at most a few per cent by very rapid solid body rotation (Ostriker 1971).

From equations (2.18) and (2.19), combined with (2.78) and (2.80), it is found that

$$\frac{K(1)}{v_D(I_{max})} = \text{function of } q \text{ (independent of } i\text{)}, \tag{2.84}$$

which offers the possibility of obtaining q from spectroscopic observation of just the disc emission lines. The form of the function of q in equation (2.84) depends on the actual radii of the accretion discs and is best found by calibrating observed $K(1)/v_D(I_{max})$ against q for systems where q is independently known (which may include other kinds of binaries containing accretion discs, e.g., Algol variables). This has been done by Kruszewski (1967), Warner (1973c), Piotrowski (1975), Webbink (1990a) and Jurcevic *et al.* (1994). A possible advantage of this approach is that although the true value of $K(1)$ may be impossible to determine (see previous section), provided the same problems exist in the calibration stars, the method goes some way to eliminating their effects. An alternative empirical calibration uses, instead of $v_D(I_{\max})$, the velocity width at half-intensity (Shafter 1984a; Webbink 1990a) or the FWZI (Schoembs & Vogt 1980).

A related approach (Friend *et al.* 1990a; Horne, Welsh & Wade 1993) uses the rotationally broadened absorption lines in the secondary. The observed Doppler broadening of the lines is $v_{\rm eq}(2)\sin i$; from equation (2.78)

$$\frac{K(2)}{v_{\rm eq}(2)\sin i} = \left[(1+q)\frac{R_{\rm L}(2)}{a}\right]^{-1}, \tag{2.85}$$

which is a known function of q from Table 2.1, or can be calibrated from systems with known q. With $K(2)$ and q then known, it requires only an estimate of i to obtain the masses. Horne, Wade & Szkody (1986) show how to derive $v_{\rm eq}(2)\sin i$ from the cross-correlation function of the secondary and standard star spectra. This requires modelling the Doppler-broadened line profile, including limb-darkening of the secondary, and should, when necessary, include the effect of irradiation by the primary (see also Friend *et al.* (1990a)).

It is interesting to note that, from equation (2.3b), the equatorial velocity of the secondary is

$$\begin{aligned} v_{\rm eq}(2) &= \frac{2\pi R_{\rm L}(2)}{P_{\rm orb}} \\ &= 140\left[\frac{M_1(2)}{R_{\rm L}(2)/R_\odot}\right]^{1/2} \text{km s}^{-1}, \end{aligned} \tag{2.86}$$

and therefore, if the mass–radius relationship is known for CV secondaries (see below), $v_{\rm eq}(2)$ can be predicted. From the observed $v_{\rm eq}(2)\sin i$ this leads to an estimate of i. As the secondaries in most CVs approximate lower main sequence stars, for which $M/{\rm M}_\odot \approx R/{\rm R}_\odot$, we see that secondaries in CVs should, in general, have absorption lines with total widths $\sim 280\sin i$ km s^{-1}, which, for high inclination, renders the lines much shallower than in non-rotating stars.

The general freedom from (other than hydrogen) emission lines in secondary spectra enables RV curves of high accuracy to be obtained even when the absorption lines are very weak from dilution by light of the disc. The usual technique is to choose a standard star with a spectral type similar to that of the secondary and cross-correlate its spectrum with that of the CV (e.g., Stover (1981b): Figure 2.42).

In pursuing the faintest secondaries, however, it is necessary to observe in the IR so that the flux contributions from other components are minimized. The 8190 Å doublet of NaI, which is very strong in late-type dwarfs, has been used successfully in this way (Friend *et al.* 1988, 1990a,b). For these lines, some contamination by Paschen emission must be allowed for (Martin *et al.* 1989).

A method (Robinson 1973a, 1976b) of determining $M(2)$ is to note that equation

(2.1a) can be cast in the form

$$M_1(2) = 0.358\left(\frac{1+q}{q}\right)^{1/2}\left[\frac{R_L(2)}{a}\right]^{3/2}\left[\frac{M_1(2)}{R_L(2)/R_\odot}\right]^{3/2} P_{orb}(\mathrm{h}). \tag{2.87}$$

The first two factors in brackets combined are almost independent of q and the last bracket can be evaluated from any adopted mass–radius relationship. The latter is discussed in Section 2.8.3, but, using the empirical result (Caillault & Patterson 1990) for low mass ($< 0.5\ \mathrm{M}_\odot$), main sequence stars

$$R_1 = 0.92\ M_1^{0.796} \tag{2.88}$$

in equation (2.87) gives a first estimate of $M_1(2)$ that may be combined with any available estimate of q to find $M_1(1)$ also. Combining equations (2.87) and (2.88) with equation (2.5b) leads to an approximate mass–period relationship for the secondary:

$$M_1(2) \approx 0.091\ P_{orb}^{1.44}(\mathrm{h}). \tag{2.89}$$

Direct estimates of $M(1)$ may be made in favourable circumstances. For radial free fall of gas onto a white dwarf (appropriate for some cases of magnetically controlled accretion – see Section 6.4.1), equations (2.56) and (2.82) predict a shock temperature

$$T_{sh} = 43.0 M_1(1)\left\{\left[\frac{M(1)}{M_{ch}}\right]^{-2/3} - \left[\frac{M(1)}{M_{ch}}\right]^{2/3}\right\}^{1/2} \mathrm{keV}. \tag{2.90}$$

In general, the actual shock temperature will be lower than this T_{sh}, so equation (2.90) enables a minimum value of $M(1)$ to be found from X-ray spectra (Mukai & Charles 1987).

If the RV curve of the secondary is well determined (see Section 2.9.7 for problems associated with irradiation), then the observed mass function $f[M(1)]$ can be used with an estimate of i to give

$$M(1) = f[M(1)](1+q)^2 \sin^{-3} i. \tag{2.91}$$

For non-eclipsing CVs with well-determined $K(1)$ and $K(2)$, the parameters $M(1)\sin^3 i$, $M(2)\sin^3 i$ and $a \sin i$ are determinable. Limits on i can be obtained by requiring that $M(1) \leq M_{ch}$ and $i < i_{min}(q)$. This usually still leaves possible a wide range for i. An unbiased way of presenting the resultant possible values of $M(1)$, $M(2)$ and a is to use a Monte Carlo technique, distributing $K(1)$ and $K(2)$ in Gaussian fashion according to their measuring uncertainties, and distributing i uniformly over $0 < i < 90°$, but discarding any results that violate the above conditions on i (Horne, Wade & Szkody 1986).

For eclipses where the contact phases of the primary (Figure 2.23) are identifiable, the eclipse duration $\Delta\varphi_p = \varphi_{p2} - \varphi_{p1} = \varphi_{p4} - \varphi_{p3}$ is (for a spherical secondary)

$$\Delta\varphi_p = \frac{R(1)}{\pi a}\left(1 - \frac{\cos^2 i}{\cos^2 i_{min}}\right)^{-1/2} \tag{2.92}$$

(i_{min} is a function only of q: Section 2.6.2). With equation (2.1a) this gives

$$\frac{R_9(1)}{M_1^{1/3}(1)} = 111\Delta\varphi_p P_{orb}^{2/3}(h)(1+q)^{1/3}\left(1-\frac{\cos^2 i}{\cos^2 i_{min}}\right)^{1/2} \tag{2.93}$$

where the left hand side can be made a function purely of $M(1)$ through use of equation (2.82). Provided approximate estimates of q and i are available, $M(1)$ can then be found. $\Delta\varphi_p$ may be underestimated if the white dwarf is strongly limb darkened; allowance for this introduces a dependence of $R(1)$ and hence $M(1)$ on the adopted value of the limb darkening coefficient u (Wood & Horne 1990).

Young & Schneider (1980) have suggested a method whereby q is found from the eclipse profiles observed at different velocities in the profile of emission lines. This is open to the criticism (Horne & Marsh 1986; Marsh, Horne & Shipman 1987) that it assumes Keplerian flow, is sensitive to the assumed intensity distribution and ignores the effects of shear broadening.

A statistical method of determining the average q for a group of eclipsing CVs has been developed by Bailey (1990). With random distribution of binary rotation axes, the cumulative distribution of $\Delta\varphi_{1/2}$ (Section 2.6.2) is a calculable function of q. If there are selection effects that work in favour or against discovery of high inclination systems then allowance for these would be necessary. Bailey points out that below the period gap no such effects appear to be operating – the fraction of CVs observed to be eclipsing agrees with the theoretical value ($= \cos i_{min}$) – but above the period gap only ~17% are eclipsing, whereas ~32% would be expected (see Wood, Abbott & Shafter (1992) for a possible explanation).

2.8.2 Masses of the Primaries

Derived masses for the primaries of CVs are listed in the compilation by Ritter (1990). Here a general overview will be given. Masses of the secondaries are discussed in Section 2.8.3.

The principal discussions of the masses of CV primaries are in Warner (1973c, 1976a), Robinson (1976b), Patterson (1984), Ritter (1990), Webbink (1990a, 1991), Bailey (1990) and Shafter (1992a). The Webbink and Bailey references provide the basis for comparing mean masses at different P_{orb}, and for evaluating the quality of masses determined by the standard methods described in the previous section.

Bailey's statistical method yields mean mass ratios $\bar{q} = 0.13 \pm 0.03$ for nine CVs below the gap ($P_{orb} \lesssim 2.2$ h) and $\bar{q} = 0.65 \pm 0.12$ for twelve CVs with $P_{orb} \gtrsim 3.1$ h. Because of the selection effects, the latter must be adjusted to $\bar{q} = 0.90$. Webbink's masses, derived from various combinations of the methods given in Section 2.8.1, give $\bar{q} = 0.29$ for 32 CVs below the gap and $\bar{q} = 0.64$ for 78 CVs above the gap. His mean masses are given in Table 2.5

Of the methodologies described in the previous section, two are commonly used for the determination of $M(1)$ for individual CVs: the purely photometric method, making use of $\Delta\varphi_p$ and equations (2.92) and (2.93), applicable only to eclipsing binaries where eclipse of the primary is clearly seen (though not necessarily at both ingress and egress), and the standard spectroscopic method employing $K(1)$ and $K(2)$ which, in order to secure reliable estimates of i, is also in general only applicable to eclipsing systems.

Table 2.5. *Average Masses of Primaries**, $\overline{M}(1)$.

Type	P_{orb} < 2.4 h	P_{orb} > 2.4 h	All P_{orb}
DN	0.50 ± 0.10(15)	0.91 ± 0.08(21)	0.74 ± 0.07(36)
CN		0.91 ± 0.06(8)	0.91 ± 0.06(8)
NLs		0.74 ± 0.05(22)	0.74 ± 0.05(22)
Magnetic CVs	0.70 ± 0.10(12)	0.79 ± 0.11(9)	0.73 ± 0.07(21)
All systems	0.61 ± 0.08(26)	0.82 ± 0.04(58)	0.74 ± 0.04(84)

* From Webbink (1990a). Numbers in parentheses are the total number of systems included in the average.

Table 2.6. *Primary Masses from Eclipse Durations.*

Star	Type	P_{orb} (h)	$\Delta\varphi_p$	i^o	q	$M_1(1)$	References
V2051 Oph	Polar	1.50	0.0093	80.5	0.15:	0.56	1,2
OY Car	DN(SU)	1.52	0.0078	83.3	0.102	0.84 ($u = 0$)	3
						0.76 ($u = 1$)	3
						0.68 ± 0.01	4
						0.89	5
						0.56 ± 0.03	6
HT Cas	DN(SU)	1.77	0.0098	81.0	0.15	0.61 ± 0.04	7
						0.84 ± 0.15	8
						0.66 ($u = 0$)	3
						0.58 ($u = 1$)	3
Z Cha	DN(SU)	1.79	0.0082	81.7	0.15	0.70 ($u = 0$)	3
						0.62 ($u = 1$)	3
						0.84 ± 0.09	9
						0.54	10
IP Peg	DN(UG)	3.80	0.0069	80.9	0.49	0.9	11

References: 1. Warner & Cropper 1984; 2. Warner & O'Donoghue 1987; 3. Wood & Horne 1990; 4. Wood *et al.* 1989b; 5. Schoembs, Dreier & Barwig 1987; 6. Berriman 1984a; 7. Horne, Wood & Stiening 1991; 8. Marsh 1990; 9. Wade & Horne 1988; 10. Wood *et al.* 1986; 11. Wood & Crawford 1986.

In Table 2.6 a list of primaries' masses derived from the photometric method is given. Where several determinations of mass are available, only the system parameters adopted in the most recent study are listed. The preponderance of DN and systems below the period gap in this Table is the result of the requirement for low $\dot{M}$(d) in order to see the eclipse of the primary.

When we look critically at the derivations of masses from usage of $K(1)$ and $K(2)$ (with i found from eclipse duration), which are principally those for HT Cas, Z Cha, IP Peg and U Gem, the conclusion is that so far no high quality results are available. In all systems there are effects of either (or both) of the distortions and phase shifts of the emission-line velocities described in Section 2.7.6 or of irradiation of the secondary, also mentioned in Section 2.7.6 and described quantitatively in Section 2.9.6.

One example will be illustrated here. The photometric solution for HT Cas (Horne, Wood & Stiening 1991: see Table 2.6) predicts $K(1) = 58 \pm 11$ km s^{-1} and $K(2) = 389 \pm 4$ km s^{-1}. The observed $K(1)$ is 115 ± 6 km s^{-1} but with a 30° phase discrepancy (Young, Schneider & Schectman 1981b). The raw observed $K(2)$ from the NaI lines is 450 ± 25 km s^{-1} (Marsh 1990), reduced to 430 ± 25 km s^{-1} after correction for non-uniform distribution over the secondary.

The purely photometric solutions generally predict values of $K(1)$ smaller than those observed (Shafter 1992a), agreeing with Bailey (1990b).

Despite the uncertainties, a spread in $M(1)$ at a given P_{orb} seems proven. Thus DQ Her and U Gem, both double-line eclipsing systems with $P_{orb} \sim 4.5$ h, have $M_1(1) = 0.60 \pm 0.07$ (Horne, Welsh & Wade 1993) and $M_1(1) = 1.2 \pm 0.2$ (Wade 1981; Zhang & Robinson 1987; Marsh *et al.* 1990) respectively.

For single white dwarfs the frequency distribution of masses shows a narrow peak centred on 0.56 $M_\odot$, with broad wings extending to 0.3 $M_\odot$ and 1.3 $M_\odot$ (McMahan 1989; Schmidt *et al.* 1992; Bergeron, Saffer & Liebert 1994).

2.8.3 *Mass–Radius Relationship and Masses for the Secondaries*

There are about two dozen CVs for which estimates of $M(1)$ and $M(2)$ can be made, using the techniques described in Section 2.8.1, without having to assume a mass–radius (M–R) relationship for the secondary (Webbink 1990a, 1991). For Roche lobe-filling secondaries the mean density is given by equations (2.3a) and (2.5c), which, as demonstrated by equation (2.3b), give a relationship between $\bar{\rho}(2)$ and P_{orb} which involves a very weak function of q and is therefore almost error-free.

The derived relationship between $\bar{\rho}(2)$ and $M_1(2)$ is shown in Figure 2.43; the spread in $M_1(2)$ at a given $\bar{\rho}(2)$ is a combination of uncertainties in derivation of $M_1(2)$ and possible dispersion (in the M–R relationship) because of composition and/or evolutionary effects. The range of mean densities, $1.0 \lesssim \bar{\rho}(2) \lesssim 50$ g cm^{-3} (corresponding to $1.4 \lesssim P_{orb} \lesssim 10$ h) is compatible with that of lower main sequence stars with masses $\lesssim 1$ $M_\odot$ for $\bar{\rho} \gtrsim 1$ g cm^{-3} (Allen 1976), but the few CVs with $\bar{\rho}(2) < 1.0$ g cm^{-3} (or $P_{orb} \gtrsim 9$h) would require main sequence stars both more massive and earlier in spectral type than those actually observed. These secondaries must therefore either have evolved off the main sequence or have internal structures in some other way different from normal dwarfs.

Compatability with main sequence structures does not establish identity of structure. Because of mass loss, there are theoretical reasons (Section 9.3) to expect significant departures from thermal equilibrium in at least some CV secondaries. Empirical masses and radii derived for individual secondaries have sometimes failed to agree with the M–R relationship for observed or modelled zero age main sequence stars (e.g., DQ Her: Smak 1980, Young & Schneider 1980; U Gem: Wade 1979, 1981). On the other hand, some recent work leads to conclusions that the secondaries are indistinguishable from main sequence stars (HT Cas: Marsh 1990; Z Cha: Wade & Horne 1988; IP Peg: Marsh 1988).

Part of the reason for apparent disagreements, as pointed out by Patterson (1984), resides in the poor determination of the empirical M–R relationship for very low mass

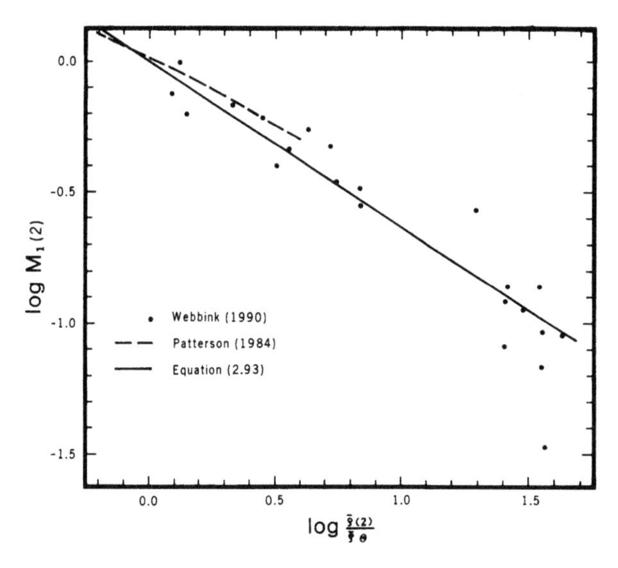

Figure 2.43 Mass–density relationship for the secondaries in CVs. The points are derived from observations (Webbink 1990a). The dashed curve is from Patterson's (1984) mass–radius relationship. The solid curve illustrates equation (2.98).

stars. The deduced radii of these are dependent on temperatures derived from blackbody fits to observed flux distributions (Veeder 1974) or other means of calibrating photometric parameters against effective temperature. Webbink (1990a) notes that most of the discrepancy between observed and theoretical low mass main sequence stars would be removed by a ~10% upward revision of T_{eff} for stars near spectral type M0.

The situation has recently shown signs of improvement. On the theoretical side, the models for 0.15–0.55 $M_\odot$ stars by Neece (1984) incorporate extensive molecular opacity computations and are an improvement on earlier models by Grossman, Hays & Graboske (1974) and Vandenberg *et al.* (1983). Neece finds

$$R_1 = -0.0869 + 2.4659M_1 - 5.0105M_1^2 + 4.8478M_1^3 \qquad (2.94a)$$

$$\approx 0.876M_1^{0.807} \qquad (0.15 < M_1 < 0.55) \qquad (2.94b)$$

For stars with $M_1 > 0.5$, the results of Copeland, Jensen & Jorgensen (1970) and Grossman, Hays & Graboske (1974) apply.

Empirical *M–R* relationships, derived from single stars or detached binaries, are the following:

$$R_1 = 0.955M_1^{0.917} \qquad 0.10 \leq M_1 \leq 1.3 \quad \text{(Lacy 1977)} \qquad (2.95)$$

$$R_1 = 0.98M_1 \qquad 0.8 \leq M_1 \leq 1.4 \qquad (2.96a)$$

$$= M_1^{0.88} \qquad 0.1 \leq M_1 \leq 0.8 \quad \text{(Patterson 1984)} \qquad (2.96b)$$

and from a reinvestigation of the lowest mass stars,

$$R_1 = 0.918 M_1^{0.796} \qquad 0.13 \leq M_1 \leq 0.6 \tag{2.97}$$

(Caillault & Patterson 1990). The last of these relationships gives radii only ~6% larger than the theoretical models of Neece (1984).

Examples of these relationships, transformed to the log M_1–log $\bar{\rho}/\bar{\rho}_\odot$ diagram, are shown in Figure 2.43 and represent the CVs well for $M_1(2) \lesssim 1$. A simple formula that fits simultaneously the CV and the detached binary measurements (Webbink 1991) in the range $0.08 \leq M_1 \leq 1.0$ is

$$\frac{\bar{\rho}(2)}{\bar{\rho}_\odot} = M_1^{-8/5}(2) \tag{2.98}$$

or

$$R_1(2) = M_1^{13/15}(2) = M_1^{0.867}(2). \tag{2.99}$$

With equations (2.5b)[10] and (2.87) this gives the mean empirical mass–period relationship

$$M_1(2) = 0.065 P_{\mathrm{orb}}^{5/4}(\mathrm{h}) \qquad 1.3 \leq P_{\mathrm{orb}}(\mathrm{h}) \leq 9 \tag{2.100}$$

and the mean empirical radius–period relationship

$$R_1(2) = 0.094 P_{\mathrm{orb}}^{13/12}(\mathrm{h}) \qquad 1.3 \leq P_{\mathrm{orb}}(\mathrm{h}) \leq 9. \tag{2.101}$$

These formulae are similar to earlier ones (Faulkner 1971; Warner 1976a; Echevarria 1983; Patterson 1984) based on different selections of observational material. As an example of the uncertainties in determination of $M(2)$, and of the use of equation (2.95) at an extreme of the mass range, Table 2.7 lists recent published values of $M(2)$ for four well-observed short-period stars, derived without assuming that the secondaries are main sequence stars. At $P_{\mathrm{orb}} = 4.65$ h, comparison is possible with results from DQ Her, a double-line eclipsing binary, for which Horne, Welsh & Wade (1993) find $M_1(2) = 0.42 \pm 0.05$ and $R_1(2) = 0.49 \pm 0.02$, compared with $M_1(2) = 0.44$ and $R_1(2) = 0.50$ from equations (2.100) and (2.101).

2.9 Primaries and Secondaries

2.9.1 Temperatures of the Primaries

Although the light from the primary is usually overwhelmed by that from the disc, bright spot or secondary, circumstances conspire to make it possible to measure its flux distribution in a number of cases. In several low $\dot{M}$(d), high inclination DN the eclipse of the white dwarf is clearly visible at optical wavelengths at quiescence (Z Cha, OY Car, HT Cas, IP Peg; Figures 2.23 and 2.29; see also calculations by Patterson & Raymond 1985a). In some systems where the primary is not eclipsed (U Gem, VW Hyi, EK TrA, ST LMi) the white dwarf is sufficiently hot that it dominates the flux

[10] This extrapolates equation (2.5b) beyond its range of decent applicability, but is compatible with the other approximations made here.

Table 2.7. *Secondary Masses in Short-Period CVs.*

Star	P_{orb}(h)	$M_1(2)$ from equation (2.100)	$M_1(2)$ observed	References
WZ Sge	1.361	0.096*	$0.06 - 0.11$	1
V4140 Sgr	1.474	0.106	0.07 ± 0.02	11
OY Car	1.515	0.11	0.09 ± 0.01	2
			0.070 ± 0.002	3
			0.10 ± 0.01	4
HT Cas	1.768	0.13	0.09 ± 0.02	5
			0.126 ± 0.02	6
			$0.11 - 0.31$	7
			0.19 ± 0.02	8
Z Cha	1.788	0.13	0.125 ± 0.014	9
			0.081 ± 0.003	10

* But see Section 3.6.3.3

References: 1. Gilliland, Kemper & Suntzeff 1988; 2. Hessman *et al.* 1989; 3. Wood *et al.* 1989b; 4. Schoembs, Dreier & Barwig 1987; 5. Horne, Wood & Stiening 1991; 6. Marsh 1990; 7. Zhang, Robinson & Nather 1986; 8. Young, Schneider & Schectman 1981; 9. Wade & Horne 1988; 10. Wood *et al.* 1986; 11. Baptista, Jablonski & Steiner 1989.

shortward of ~1500 Å. And in some higher $\dot{M}$(d) systems, temporary reduction of mass transfer can reveal the spectrum of the white dwarf (the VY Scl stars – Section 4.2.4 – and the magnetic CVs – Chapters 6 and 7).

In interpreting the observations it should be kept in mind that the primaries are not isolated white dwarfs: they may be heated by accretion or radiation in the equatorial BL (e.g., Regev (1989)); they may be observed while cooling relatively rapidly (Pringle 1988) after a DN outburst; they may be observed in the longer lived heated state following a nova eruption; or they may be observed in a slowly cooling state during the temporary cessation of an equilibrium set up during high $\dot{M}(1)$ conditions.

An example of the spectrum of the white dwarf, obtained by subtracting the mid-eclipse spectrum of Z Cha from the out-of-eclipse spectrum, is shown in Figure 2.44. The flux distribution, including Balmer line and Lyα absorption, is compared with a model white dwarf atmosphere with log $g = 8$ and $T_{eff} = 15\,000$ K (Marsh, Horne & Shipman 1987). From modelling of the primary's eclipse in U, B and R bands, Wood *et al.* (1986) find a brightness temperature of $11\,800 \pm 500$ K and a colour temperature of $10\,900 \pm 1800$ K.

Table 2.8 lists the temperatures deduced for the primaries in CVs. Because of the variable and unknown contributions of BL or accretion heating, many are upper limits to the temperatures of the undisturbed white dwarfs. The mean temperatures, from well-determined measurements, are 50 000 K for NLs, 19 200 K for DN and 13 500 K for polars, all of which are higher than the median 10 000K for field DA white dwarfs (Sion 1984). Thus, although the average CV white dwarf masses are larger than those of field white dwarfs (and hence the radii are smaller), the luminosities are high.

There is a tendency in Table 2.8 for cooler white dwarfs to occur in CVs with shorter

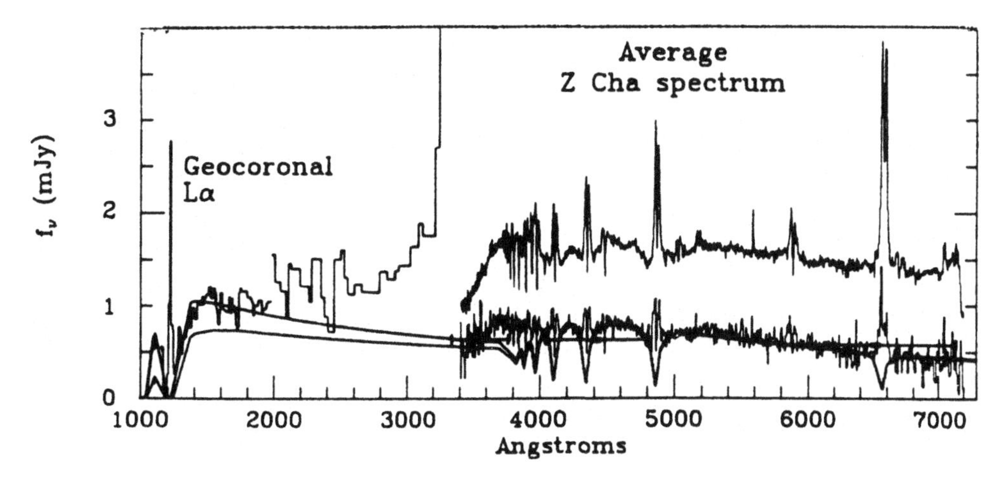

Figure 2.44 White dwarf absorption spectrum in Z Cha obtained by subtracting a spectrum taken at mid-eclipse from one taken just after egress of the primary. The optical and IUE spectra are plotted above the white dwarf spectrum. Model atmosphere spectra for log $g = 8.0$ and $T_{\text{eff}} = 1.5 \times 10^4$ K and 1.7×10^4 K are plotted over the white dwarf spectrum. Adapted from Marsh, Horne & Shipman (1987).

orbital periods (Sion 1986). This is evidently a result of the shorter P_{orb} systems having lower $\dot{M}(1)$ and hence lower accretion heating – as is seen from a strong correlation between white dwarf luminosity $L(1) = 4\pi R^2(1)\sigma T^4_{\text{eff}}$ and $\dot{M}(\text{d})$ (Smak 1984b; Shafter *et al.* 1985; Sion 1986, 1987a,b; Section 9.4.4).

The question arises (Patterson & Raymond 1985a) whether what is interpreted as flux from the primary could in fact be dominated by the BL. Optical eclipses of Z Cha, OY Car and HT Cas possess enough detail to answer this question. Wlodarczyk (1986) showed that a spherical white dwarf with a BL, used to interpret eclipses of HT Cas, leads to a range of possible white dwarf radii, and hence inferred masses. Smak (1986) analysed the primary contact points (Section 2.6.2) and deduced that the primary of Z Cha in quiescence is surrounded by a narrow equatorial BL, but Wood (1987; see also Wood *et al.* (1993)) finds in contrast that a naked white dwarf fits the observations better.

2.9.2 *Radii of the Primaries*

The most detailed modelling of the primary eclipse of Z Cha has been performed by Wood & Horne (1990) and Wood, Horne & Vennes (1992), who find best agreement with observation for a limb-darkened spherical white dwarf with no BL. The absent BL could be the result either of extremely low $\dot{M}(1)$ (much lower than $\dot{M}(2)$ – see Section 3.5) or of rapid rotation of the primary. In the latter case, it would be no longer appropriate to assume a spherical primary, but the observations are not able to distinguish between the possibilities.

The primary radii deduced from CV eclipse light curves are for HT Cas: $R(1) = 8.3(\pm 0.4) \times 10^8$ cm (Horne, Wood & Stiening 1991), for OY Car: $R(1) = 7.0(\pm 0.2) \times 10^8$ cm, and for Z Cha: $R(1) = 7.8(\pm 0.2) \times 10^8$ cm (Wood & Horne 1990; see also Horne *et al.* 1994). If there is significant polar flattening due to

Table 2.8. *White Dwarf Temperatures in CVs.*

Star	Type	P_{orb} (hr)	$T_{eff}(1)$ (10^3K)	References
WZ Sge	DN(SU)	1.36	12.5	1,2,4,8,36,57,62
SW UMa	DN (SU)	1.36	10–15	5,57
T Leo	DN (SU)	1.41	20–25	6,57
DP Leo	Polar	1.50	16	63,69
V436 Cen	DN(SU)	1.50	10–15	10,57
OY Car	DN(SU)	1.51	16.5	7–9,40,41,54,58,65
EK TrA	DN(SU)	1.53	≤20	4,57
VV Pup	Polar	1.67	9	11,43,44
V834 Cen	Polar	1.69	15	12,56
HT Cas	DN(SU)	1.77	12	8,13,35
VW Hyi	DN(SU)	1.78	18	8,14,15,36,37,62
Z Cha	DN(SU)	1.78	15.6	9,16–19,39,47,61
BL Hyi	Polar	1.89	20	20
ST LMi	Polar	1.90	11	21,45,51
MR Ser	Polar	1.90	≤8.5	22,45,57,59
AN UMa	Polar	1.92	≤20	22,46,57
YZ Cnc	DN(SU)	2.08	≤25	8,57
RE2107-05	Polar	2.08	<13	60
UZ For	Polar	2.11	11	53
AM Her	Polar	3.09	20	23–26,49,50
MV Lyr	NL	3.21	50	27,28,31,52
DW UMa*	NL	3.28	19:	29
TT Ari	NL(IP?)	3.30	50	29,30
V1315 Aql*	NL	3.35	19:	29
V1500 Cyg	N	3.35	70–120	67
UU Aql	DN	3.37	20–25	16,57
IP Peg	DN	3.80	15	29,32,42
UU Aqr	NL	3.92	34	64
YY Dra	DN	3.98	20–25	33,57
U Gem	DN	4.25	30	3,8,34,36,38,48,55,62
SS Aur	DN	4.39	≤32	8,57
RX And	DN(ZC)	5.04	35	66
V794 Aql	NL	5.5:	50	22
AH Her	DN(ZC)	6.19	≤40	8,57
SS Cyg	DN	6.60	34–40	3,8,57,66
EM Cyg	DN(ZC)	6.98	≤40	8,57
AE Aqr	NL(IP)	9.88	25	58,68
TZ Per	DN(ZC)		18	66

*See Section 4.21 for comment on temperatures of DW UMa and V1315 Aql.

References: 1. Krzeminski & Smak 1971; 2. Gilliland 1983; 3. Fabbiano *et al.* 1981; 4. Verbunt 1987; 5. Shafter 1984a; 6. Shafter & Szkody 1984; 7. Bailey & Ward 1981a; 8. Smak 1984b; 9. Schoembs & Hartmann 1983; 10. Gilliland 1982c; 11. Liebert *et al.* 1978a; 12. Maraschi *et al.* 1985; 13. Patterson 1981; 14. Hassall *et al.* 1983; 15.Mateo & Szkody 1984; 16. Patterson & Raymond 1985a; 17. Bailey 1979a; 18. Rayne & Whelan 1981; 19. Marsh, Horne & Shipman 1987; 20. Wickramasinghe, Visvanathan & Tuohy 1984; 21. Szkody, Liebert & Panek 1985 ;

rapid rotation (for which, in these stars, a case may be made according to the arguments given in Section 8.6.5.2) then it should be noted that the eclipse models give *equatorial radii*, so the *volume radii* are correspondingly smaller (and the inferred masses larger).

2.9.3 Spectra of the Primaries

Balmer absorption in optical spectra and Lyα absorption in IUE spectra are commonly observed in the spectra of CVs where emission from the primary dominates. As seen in Figure 2.44, the flux is often observed to turn down below 1400 Å (e.g., VW Hyi, WZ Sge, Z Cha, EK TrA: see Verbunt (1987)), which is characteristic of log $g \sim 8$ atmospheres with $15\,000 < T_{\mathrm{eff}} < 20\,000$ K. If the turndown were due to accretion discs with maximum temperatures $\lesssim 15\,000$ K, the predicted flux at longer wavelengths would greatly exceed that actually observed (Verbunt *et al.* 1987; Pringle *et al.* 1987). Higher members of the Lyman series are out of range of the Voyager spacecraft and HUT for CVs in low states. In the hotter primaries characteristic of NLs, absorption lines of HeII appear; for TT Ari a solar He/H ratio is derived (Shafter *et al.* 1985).

The IUE spectrum of U Gem in quiescence shows a rich collection of metallic absorption lines (Fabbiano *et al.* 1981; Panek & Holm 1984; Sion 1987a; Section 3.2.1). It is not certain that these are photospheric lines, but when analysed with the help of model atmospheres, abundances not greatly different from solar are obtained (Sion 1987a).

2.9.4 Spectral Types of the Secondaries

In many CVs with $P_{\mathrm{orb}} \gtrsim 5$ h, especially those of high inclination, the secondaries are sufficiently luminous for their absorption lines to be seen in the blue and yellow spectral region. For these, standard MK spectral types can be estimated. In shorter period systems, near-IR spectra often show lines from the secondary, notably NaI and TiO, which can furnish spectral types on the redefined MK systems for late type stars (Keenan & McNeil 1976), as described by Boeshaar (1976). Further in the IR the H_2O absorption bands at 1.4 μm and 1.9 μm are sensitive indicators of effective temperature

References to Table 2.8 (cont.)

22. Szkody, Downes & Mateo 1988; 23. Schmidt, Stockman & Margon 1981; 24. Patterson & Price 1981a; 25. Szkody, Raymond & Capps 1982; 26. Heise & Verbunt 1987; 27. Schneider, Young & Schectman 1981; 28. Szkody & Downes 1982; 29. Szkody 1987b; 30. Shafter *et al.* 1985; 31. Robinson *et al.* 1981; 32. Szkody & Mateo 1986; 33. Patterson *et al.* 1985; 34. Panek & Holm 1984; 35. Wood, Horne & Vennes 1992; 36. Szkody & Sion 1989; 37. Pringle *et al.* 1987; 38. Szkody & Kiplinger 1985; 39. Wood *et al.* 1986; 40. Wood *et al.* 1989b; 41. Berriman 1987a; 42. Martin, Jones & Smith 1987; 43. Szkody, Bailey & Hough 1983; 44. Wickramasinghe & Meggit 1982; 45. Mukai & Charles 1986; 46. Liebert *et al.* 1982b; 47. Wade & Horne 1988; 48. Wade 1979; 49. Young, Schneider & Schectman 1981b; 50. Latham, Liebert & Steiner 1981; 51. Schmidt, Stockman & Grandi 1983; 52. Chiapetti *et al.* 1982; 53. Bailey & Cropper 1991; 54. Hessman *et al.* 1989; 55. Kiplinger, Sion & Szkody 1991; 56. Beuermann *et al.* 1990b; 57. Sion 1991; 58. Horne 1993b; 59. Schwope & Beuermann 1993; 60. Glenn *et al.* 1994; 61. Wood *et al.* 1993; 62. Sion & Szkody 1990; 63. Schmidt *et al.* 1993; 64. Baptista, Steiner & Cieslinski 1994; 65. Horne *et al.* 1994; 66. Holm *et al.* 1993; 67. Schmidt, Liebert & Stockman 1995; 68. Eracleous *et al,* 1994; 69. Stockman *et al.* 1994.

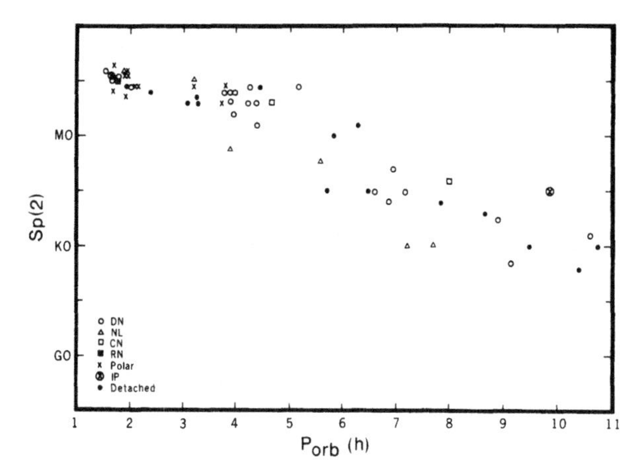

Figure 2.45 Observed spectral type of secondary stars versus orbital period for CVs and detached main sequence binaries. The latter are taken from Popper (1980).

(Bailey, Ferrario & Wickramasinghe 1991). CO absorption bands near 2.3 μm are observed in some CVs (Ramseyer *et al.* 1993b).

It is common to observe variable spectral line strengths around orbit, caused by irradiative heating of the secondary (Section 2.9.6). As a result, spectral types are likely to be earlier than what would be observed for the unirradiated hemisphere (which is only accessible near phase zero in high inclination systems).

The spectral types are shown as a function of P_{orb} in Figure 2.45.[11] Spectral types of main sequence components of detached binaries for which masses have been deduced (Popper 1980) are included in the same diagram. To convert from mass to P_{orb}, equation (2.100) has been used – i.e., it is assumed that the spectral type would not change if the isolated stars were to become lobe-filling secondaries.

There is general agreement between the spectral types of CV secondaries and isolated dwarfs of the same mass. This conclusion differs from that of Echevarria (1983), who used a misplaced linear relationship in the log$M$–spectral type relationship for detached systems.

Mateo, Szkody & Bolte (1985) give a useful calibration of radius versus spectral type for M dwarfs.

2.9.5 Photometric Properties of the Secondaries

With the masses, radii and spectral types of CV secondaries ($P_{orb} \lesssim 10$ h) apparently indistinguishable in the mean from main sequence stars, there is every expectation that the luminosities will also be similar. Taking masses and absolute visual magnitudes M_V for lower main sequence stars from the compilations of Popper (1980) and Liebert & Probst (1987), use of equation (2.100) then gives the predicted $M_V(2)$–P_{orb} relationship

[11] Kraft (1962) gives the secondary of EY Cyg a spectral type of K0V, which would be highly discrepant in Figure 2.45. However, his other Sp(2) (dG5 for SS Cyg and G8IV for RU Peg) are much earlier than those now adopted (K5V and K2-3 respectively).

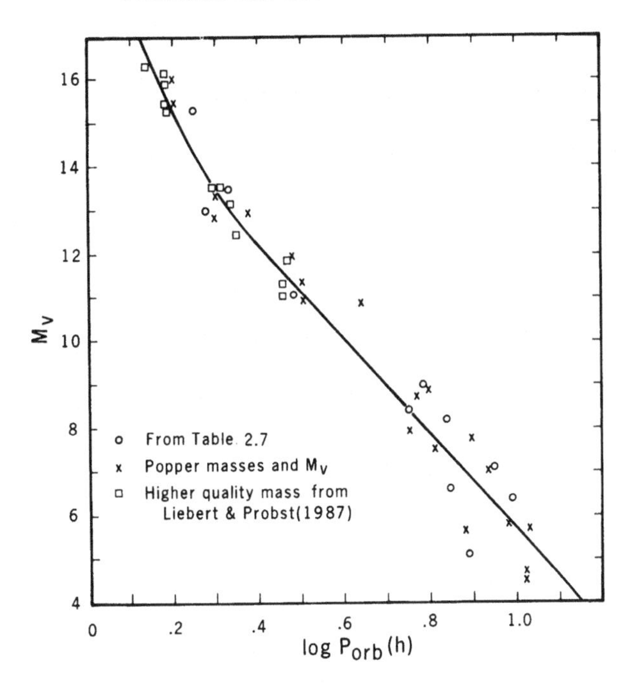

Figure 2.46 Absolute magnitude versus orbital period for CVs (Table 2.9) and for lower main sequence stars. For the latter stars the observations are from Popper (1980) and Liebert & Probst (1987), and use is made of equation (2.95) to convert from observed mass to orbital period of the lobe-filling binary.

for lobe-filling dwarfs: Figure 2.46. For $2 \lesssim P_{\rm orb}({\rm h}) \lesssim 10$ this is well represented by

$$M_{\rm v}(2) = 16.7 - 11.1 \log P_{\rm orb}({\rm h}) \tag{2.102}$$

but there is a departure from linearity at the lowest periods.

In principle, the relationship shown in Figure 2.46 can be tested with CVs of known distance (from trigonometric parallaxes, cluster, binary or common proper motion membership) and estimable contribution in the V band from the secondary. In practice, few such objects are available, so comparison must be made with indirect methods of determining distances.

Photometric methods of obtaining distances are available that make use of the Wesselink (1969) surface brightness method. In the Barnes & Evans (1976) formulation, the surface brightness in the V band is related to the unreddened V–R colour, independently of luminosity:

$$S_{\rm v} = 4.2207 - 0.1{\rm V}_0 - 0.5 \log \psi_{\rm d} = 3.841 - 0.321({\rm V} - {\rm R})_0 \tag{2.103}$$

where $\psi_{\rm d}$ is the angular diameter in milliarcseconds. Thus, given V and R magnitudes of the secondary, $\psi_{\rm d}$ is determined and the distance d may be found if $R(2)$ is taken from equation (2.101). It is rare that uncontaminated V and R magnitudes are available, but for the coolest secondaries V–R may be obtained from the calibration against TiO band strengths given by Marsh (1990).

With IR photometry (for which the contribution of the CV secondary may be

Table 2.9. *Absolute Magnitudes of Secondaries.*

Star	Type	log P_{orb}(h)	d(pc)	m_v	$\Delta(2)$	$M_v(2)$	References
HT Cas	DN(SU)	0.247				15.3	1
BL Hyi	Polar	0.277	128	18.5		13.0	2
UZ For	Polar	0.324	250	20.5		13.5	2
AM Her	Polar	0.490	75	15.5		11.1	2
RW Tri	NL	0.746	224	13.2	2.0	8.4	3,4
AH Her	DN	0.792	250	14.3	1.7	9.0	3,5
Z Cam	DN(ZC)	0.842	173	13.6	0.7	8.2	3,6
EM Cyg	DN(ZC)	0.844	350	14.2	0.1	6.6	5,7
V363 Aur	NL	0.886	900	14.2	0.7	5.1	8
RU Peg	DN	0.954	174	13.1	0.2	7.1	9
AE Aqr	NL	0.995	140	11.5	0.6	6.4	3,10

References: 1. Marsh 1990; 2. Table 6.1; 3. Warner 1976a; 4. Oke & Wade 1982; 5. Szkody 1981a; 6. Kiplinger 1979a; 7. Bailey 1981; 8. Szkody & Crosa 1981; 9. Stover 1981b; 10. Chincarini & Walker 1981.

derived using the techniques of Berriman, Szkody & Capps (1985): see Section 2.6.1.4), the surface brightness in the K band for cool stars is observed to be (Bailey 1981)

$$\begin{aligned} S_K &= 2.56 + 0.508(V - K) \qquad (V - K) < 3.5 \\ &= 4.26 + 0.058(V - K) \qquad (V - K) > 3.5. \end{aligned} \tag{2.104}$$

The weak dependence on V–K for the coolest stars (later than M3, S_J is also found to be almost independent of colour: Berriman 1987a) allows use of the approximation

$$S_K = 4.55 \qquad \text{for } (V - K) > 3.5 \tag{2.105}$$

which gives

$$\log d = \frac{K}{5} + 0.09 + \log R_1(2) \qquad P_{orb}(h) \lesssim 6, \tag{2.106}$$

where $R/R_\odot$ for CVs is available as a function of P_{orb} from equation (2.101). For longer orbital periods, a calibration of V − K versus M, and hence versus P_{orb}, leads to (Warner 1987a).

$$\log d = \frac{K}{5} - 1.06 + \log R_1(2) + 1.56 \log P_{orb}(h) \quad 6 < P_{orb}(h) < 12. \tag{2.107}$$

The following relationship connects mass to absolute K magnitude for late type dwarfs (Marcy & Moore 1989; see also Henry & McCarthy 1990):

$$\log M_1 = -0.202 M_K + 0.819, \tag{2.108}$$

Table 2.9 lists a few CVs for which $M_v(2)$ has been obtained from knowledge of their distance and the fractional contribution ($\Delta(2)$ in magnitudes) of the secondary to the V band. Three polars, for which spectra taken during low states show that the secondaries dominate, are included. $M_v(2)$ for HT Cas comes from the detailed study by Marsh (1990).

Plotted in Figure 2.46, these observed $M_v(2)$ fall along the relationship defined by the lobe-filling main sequence stars. This, and the earlier M–R and spectral type relationships, show that *on average*, as already concluded by Patterson (1984) and Ritter (1983), *the secondaries of CVs with* $P_{orb} \lesssim 10$ *h are indistinguishable from main sequence stars.*

Useful compilations of multi-colour photometry of normal K and M dwarfs have been given by Caillault & Patterson (1990) and Doyle & Butler (1990). It may also be noted that the colour at which lower main sequence stars become fully convective is estimated to be $1.27 < \mathrm{R} - \mathrm{I} < 1.38$ (Stauffer *et al.* 1991).

Separation of the contribution in the IR of the secondary from that of the disc (Section 2.6.1.4) can be aided by observation of the amplitude of the ellipsoidal variation (Section 1.7, Figure 1.26). For a Roche lobe-filling star the fractional intensity range ($\approx$ magnitude for small amplitude) is (Russell 1945; McClintock *et al.* 1983)

$$\Delta m \approx \frac{\Delta I}{I} = \frac{3}{2q}\left[\frac{R(2)}{a}\right]^3 \sin^2 i(1 + u_g)\left(\frac{15+u}{15-5u}\right), \quad (2.109a)$$

where u_g is the gravity-darkening coefficient and u is the limb-darkening coefficient. More extensive details are given by Binnendijk (1974) and applications in Sherrington *et al.* (1982). Typical values in the visible for cool atmospheres are $u_g = 0.57$, $u = 0.85$ (McClintock *et al.* 1983), so with equation (2.5b) we have

$$\Delta m \approx 0.34\frac{\sin^2 i}{1+q} \quad (2.109b)$$

which may be useful for determining i. Note, however, that u_g is very sensitive to wavelength and can be $\sim$2 at 4000 Å and $\sim$0.25 in the IR.

For examples of ellipsoidal variation see Bailey (1975b), Frank *et al.* (1981a), Jameson, King & Sherrington (1981), Bailey *et al.* (1981), Sherrington *et al.* (1982), Mateo, Szkody & Bolte (1985), Szkody & Mateo (1986b), Haug (1988), Szkody *et al.* (1992), Harlaftis *et al.* (1992), Drew, Jones & Woods (1993).

2.9.6 Irradiation of the Secondary

The atmosphere of the secondary, on the side facing the primary, is bathed in radiation from the primary, BL, disc and bright spot. The equatorial regions, however, will be shadowed by the disc from part of this radiation (Sarna 1990). Of the luminosity generated by the region on and around the primary, the fraction f_2 falling on the secondary is

$$f_2 \approx \frac{R^2(2)}{4a^2}\left[1 - \frac{4}{\pi}\frac{a\,h_d}{R(2)r_d}\right], \quad (2.110)$$

where h_d is the height of the disc at its outer edge. With $h_d/r_d \sim 0.05$ (equation (2.52)) and $R_L(2)/a$ from Table 2.1, it follows that ~ 0.2 of the flux is lost through shadowing. This, however, is a minimum value: the effective h_d/r_d will be much greater than that given by the condition that the *vertical* optical depth be unity - an integration along the light path across the face of the disc is required.

Table 2.10. *Examples of Irradiation Flux.*

$T_{eff}(1)$	$T_{eff}(2)$	$R_{10}(2)$	q	$\dot{M}_{16}(1)$	$2F_{ir}(2)/F(2)$
2×10^4	2.5×10^3	2	0.15	1	2
2×10^4	2.5×10^3	2	0.15	100	200
5×10^4	4.5×10^3	5	0.5	10	0.25
5×10^4	4.5×10^3	5	0.5	1000	18

Ignoring the contributions from the bright spot and the disc, the flux $F_{ir}(2)$ irradiating the secondary in the presence of a maximally energetic BL is

$$F_{ir}(2) \approx f_2 \left[4\pi R^2(1)\sigma T^4_{eff}(1) + \frac{GM(1)\dot{M}(1)}{2R(1)} \right], \tag{2.111}$$

where it has been assumed that the energy in the BL is radiated over the surface of the primary, rather than just in a narrow equatorial band. $F_{ir}(2)$ can be compared with the total flux $\frac{1}{2}F(2) = 2\pi R^2(2)\sigma T^4_{eff}(2)$ emitted by the hemi-lobe facing the primary.

Table 2.10 gives examples (for $M_1(1) = 1$) representing (a) a short period DN such as Z Cha at quiescence and outburst, and (b) a longer period DN such as SS Cyg at quiescence and outburst, or (in the latter case) a NL. In (a) the BL flux always dominates $F_{ir}(2)$; in (b) flux from the primary is important at quiescence, but the BL dominates at high $\dot{M}(d)$.

These examples show that irradiation is liable to have substantial observable consequences. Complete models of irradiated cool atmospheres have not yet been computed, but an important first step has been made by Brett & Smith (1993). There are a number of factors that must be considered. As much of the energy from the primary, BL and inner disc is radiated shortward of the Lyman limit, and the atmospheres of cool secondaries consist of neutral hydrogen even to subphotospheric depths, incoming soft X-rays and EUV are absorbed high in the outer layers of the secondary producing strong chromospheric emission (Hameury, King & Lasota 1986a; King 1989a). Some fraction w of the energy deposited there will be degraded and penetrate to heat the photosphere. In the standard theory of the reflection effect (e.g., Rucinski (1969)), at each point on the surface of the secondary the effective temperature becomes

$$\sigma T^4_{eff} = \sigma T^4_{eff}(2) + wF_{ir}(2), \tag{2.112}$$

where w is usually taken to be 0.5. However, King (1989a) has questioned whether w can be this large for irradiated neutral atmospheres.

Chromospheric Balmer emission is commonly observed in CVs and appears as narrow (~200 km s^{-1}) emission-line components approximately anti-phased in RV relative to the disc emission lines.[12] The secondary emission is most prominently seen in DN during outburst (U Gem: Stover 1981a, Hessman 1986; RX And: Kaitchuck,

[12] This is also seen in β Lyr (Honeycutt *et al.* 1994).

Mansperger & Hantzios 1988; IP Peg: Hessman 1989a, Marsh & Horne 1990; Harlaftis *et al.* 1994a; SS Cyg: Hessman *et al.* 1984, Hessman 1986; YZ Cnc: Harlaftis *et al.* 1994c), in NLs (RW Tri: Kaitchuck, Honeycutt & Schlegel 1983; UX UMa: Schlegel, Honeycutt & Kaitchuck 1983; IX Vel: Beuermann & Thomas 1990), and in the VY Scl and magnetic CV systems during states of low $\dot{M}(2)$ (Sections 4.2.4, 5.2.1, 6.3.1 and 7.3.1). Very weak secondary Balmer emission is seen in SS Cyg at quiescence (Hessman 1987). Usually HeI, HeII and other heavier element emissions are not seen, indicating that although hydrogen Lyman continuum radiation reaches the secondary star, helium Lyman continuum does not (Marsh & Horne 1990). Exceptions are IX Vel, in which HeI emission is present (Beuermann & Thomas 1990), and IP Peg, in which HeI and weak neutral metal emissions are observed (Hessman 1989a).

From the variation of intensity around orbit it appears that the Balmer emission is concentrated on the irradiated hemi-lobe of the secondary. Doppler tomography of IP Peg (Marsh & Horne 1990) shows that the emission appears near the poles on the irradiated side, as would result from shadowing by the disc. In U Gem (Marsh *et al.* 1990) the emission is concentrated on the leading face of the secondary, suggesting irradiation by the bright spot as the ionizing source. Rutten & Dhillon (1994) have combined Doppler tomography and line intensity measurements into 'Roche tomography', which has been used to show the concentration of line emission near the L_1 point in DW UMa.

The Balmer flux from the secondary in IP Peg is $\sim 5 \times 10^{30}$ erg s^{-1} during outburst and decreases by a factor of at least 150 at quiescence (Marsh & Horne 1990). In contrast, the most active dMe stars emit $\sim 1 \times 10^{29}$ erg s^{-1} in Balmer radiation (Kodaira 1983).

In IP Peg the observed $K_{abs}(2)$ (from the absorption lines of the secondary) is 331 km s^{-1} (Martin, Jones & Smith 1987) at a time when the irradiation correction (Section 2.7.6) amounted to -26 km s^{-1} (Marsh 1988). Hessman (1989a) measured $K_{em}(2) = 200$ km s^{-1} and finds an irradiation correction $\sim$ 130 km s^{-1} from a model of the chromosphere. The 'true' $K(2)$ for IP Peg is thus $\sim$ 300 km s^{-1}, but the value is uncertain by $\sim \pm 30$ km s^{-1} due to the effects of irradiation.

The question of photospheric heating should be answerable from observations. Martin *et al.* (1989) point out that if the temperature near L_1 is increased by ~20%, the IR ellipsoidal variation is suppressed. From the equality in amplitudes of the double sinusoidal IR light curve of U Gem at quiescence, Berriman *et al.* (1983) were able to set an upper limit $\sim 0.1 L_\odot$ for the luminosity of any central source irradiating the secondary. A positive detection is given by IX Vel, which is a high $\dot{M}(\mathrm{d})$ system ($\sim 5 \times 10^{17}$ g s^{-1}: Beuermann & Thomas 1990) in which the ellipsoidal variation at 2 μm indicates a contribution from heating of ~3% (Haug 1988). At quiescence in RU Peg and EM Cyg there is no variation of spectral type or line strengths of the secondary spectrum around orbit (Stover 1981b; Stover, Robinson & Nather 1981). In Z Cha and HT Cas at quiescence (Wade & Horne 1988; Marsh 1990) the TiO band strength ratio (an index of temperature) does not vary around orbit (Figure 2.47), but the TiO strength itself cannot be modelled correctly by a darkened secondary whose surface temperature is found from equation (2.106) plus limb-darkening: the observed TiO flux deficit at phase 0.5 is too small. Simple models of photospheric heating predict that the TiO strengths should *increase* at phase 0.5, rather than weaken as observed. Wade &

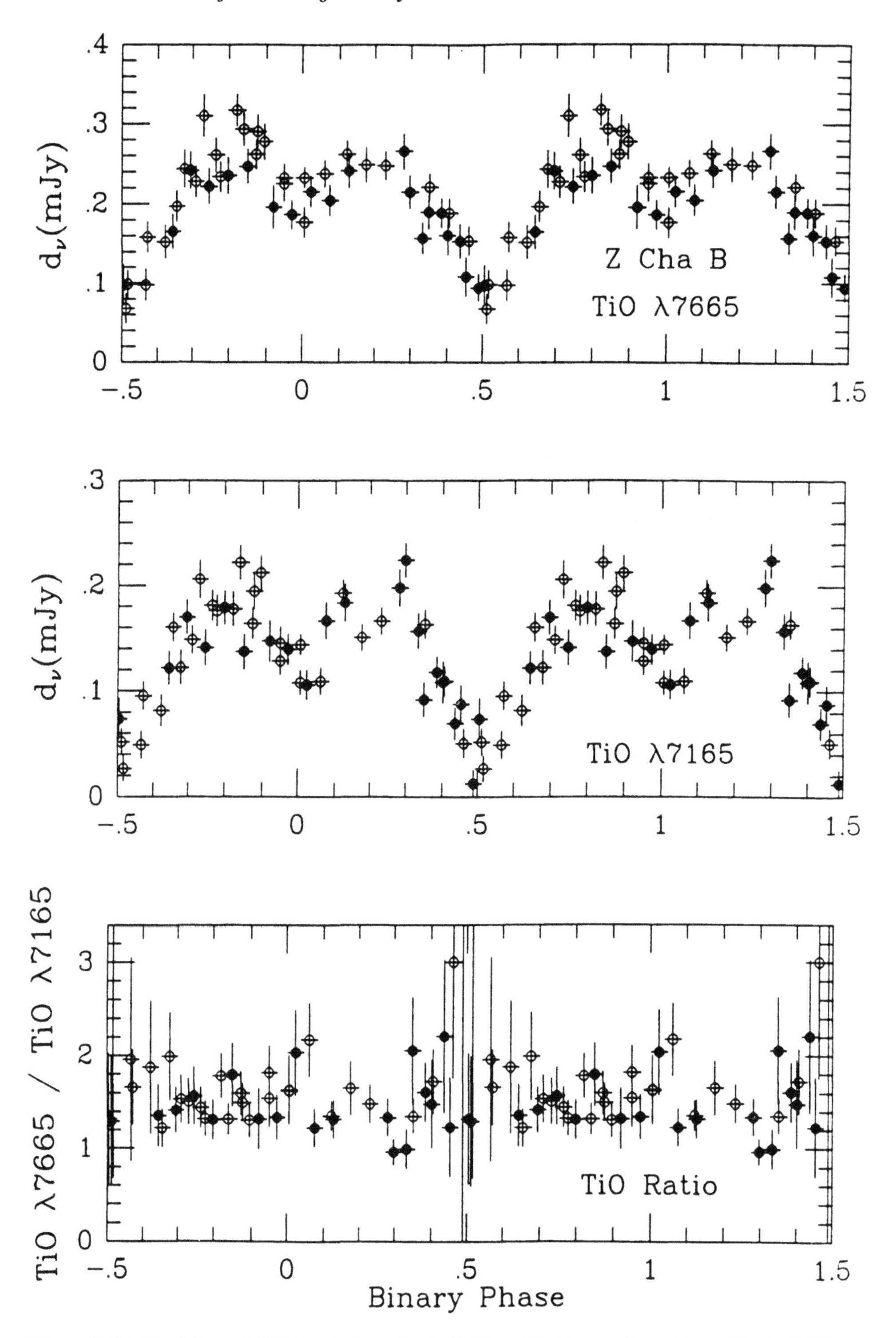

Figure 2.47 Variation of TiO band strengths in Z Cha with binary phase. From Wade & Horne (1988).

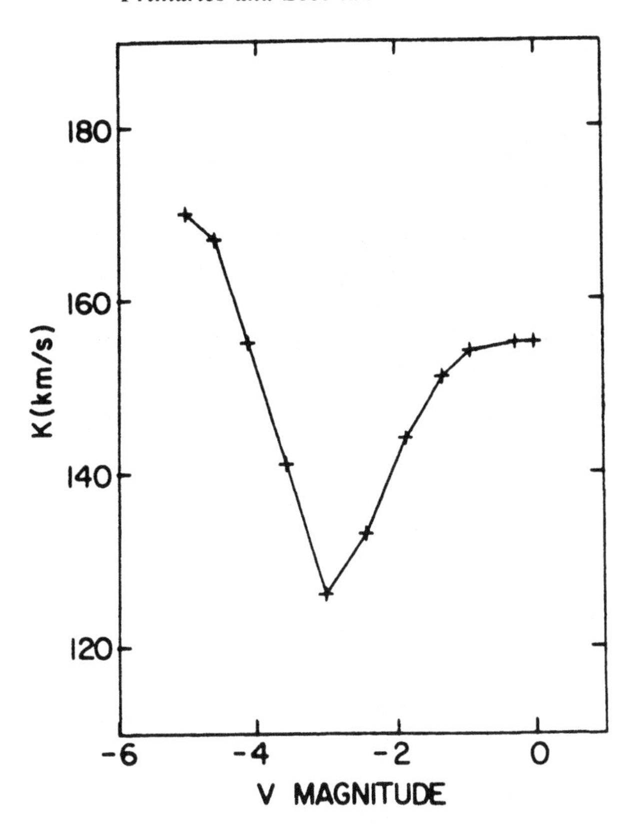

Figure 2.48 Predicted variation of radial velocity amplitude $K_{abs}(2)$ with relative visual magnitude during outburst of SS Cyg. V = 0 corresponds to quiescent magnitude and V = −4 to outburst maximum.

Horne conclude that the spectrum from the irradiated surface does not behave like that from a photosphere of higher temperature, and suggest that photospheric heating modifies the temperature gradient to the point where TiO bands are largely suppressed. (The alternative possibility – that overlying chromospheric Paschen continuum emission fills in the TiO bands – would probably require a concomitant Balmer emission continuum much greater than observed.) Brett & Smith (1993), show that more realistic modelling of irradiation is compatible with the variations of TiO (and NaI) in Z Cha.

On the other hand, the photospheric heating model has had notable success in explaining otherwise discordant $K(2)$ measurements in SS Cyg (Hessman *et al.* 1984; Robinson, Zhang & Stover 1986). At quiescence there is very little heating, so $K_{abs}(2)$ is almost equal to its true value. As the accretion disc, BL and primary become more luminous during outburst, at first the heated side of the secondary brightens but still retains its K spectral type with deep absorption lines; this moves the 'centre of gravity' of absorption line production towards the primary, decreasing $K_{abs}(2)$. At higher irradiation intensities the photospheric heating becomes so great that the spectrum moves to earlier type, with weak absorption lines, and the 'centre of gravity' moves away from the primary. The modelled behaviour is shown in Figure 2.48 and accounts

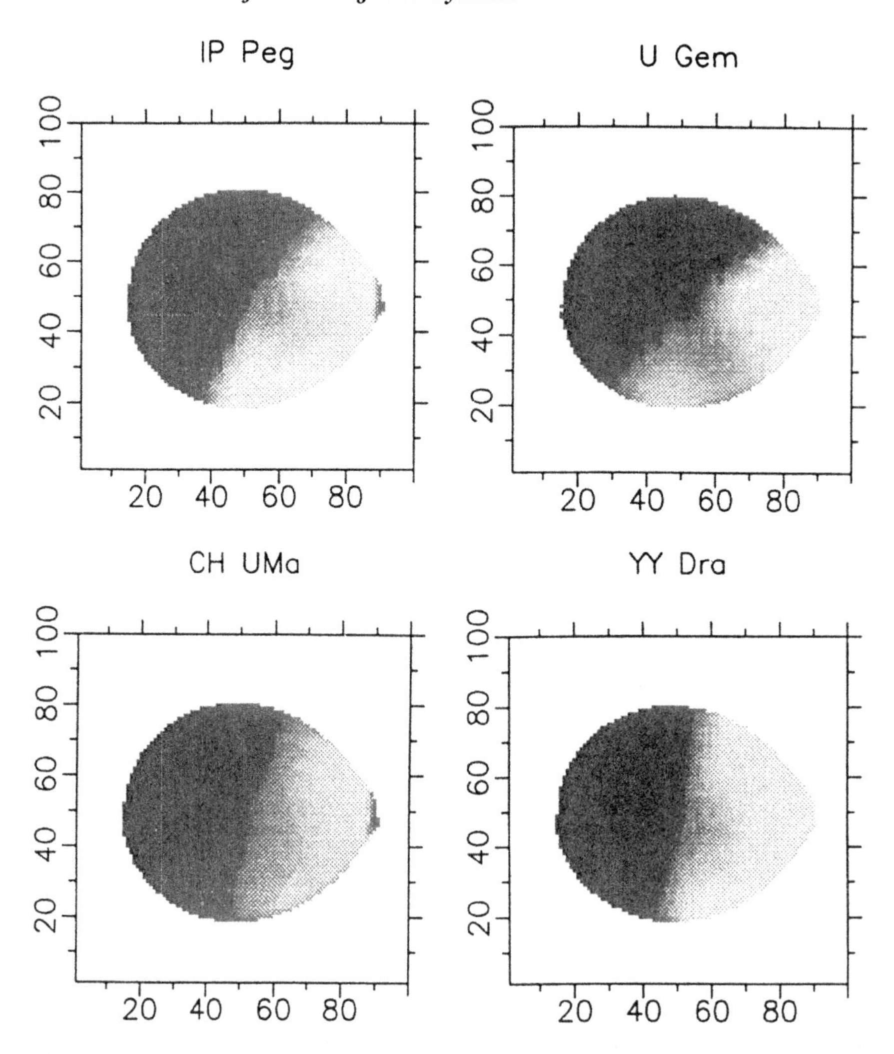

Figure 2.49 Distribution of NaI absorption line strength, and hence irradiating flux, on the surfaces of four DN at quiescence. These are viewed from above, with the primary to the right and anti-clockwise orbital motion. Adapted from Davey & Smith (1992).

well for the observed variation of $K_{abs}(2)$ as a function of time during outburst. Merely laying chromospheric continuum emission over a constant photospheric absorption spectrum would monotonically change $K_{abs}(2)$, not produce the observed bimodal distribution.

From observed departures from sinusoidality of the NaI λ8190 absorption-line RV curves of a number of CVs, Davey & Smith (1992) deduced the distribution of relative line strength over the surfaces of their secondaries. Some examples, for DN at quiescence, are shown in Figure 2.49. Weakening of the line, caused by heating of the atmosphere, is clearly present and is weighted towards the leading face of the secondary. The effect is much larger, and more asymmetric, than expected from heating by the bright spot. Davey & Smith speculate that the heating is sufficient to reduce the atmospheric temperature gradient to the point where convection is suppressed, allowing circulation currents to develop on the leading face and extending the region of heating. This is to some extent considered by Sarna (1990) and in more detail by Brett & Smith (1993) who show that the radial structure is significantly affected. Several other CVs, not in any other way discernibly different from those analysed in Figure 2.49, show no RV asymmetry and hence no evidence for significant irradiation. The reason for this is not known.

In further studies of photospheric heating, a starting point could be the pre-cataclysmic detached binaries (Section 9.2) which are simpler systems that the CVs – the absence of discs, bright spots and BLs reduces irradiation to that from a single component. Observations show (Bond & Grauer 1987) that the primaries of such stars as MT Ser and V477 Lyr have $T_{\mathrm{eff}}(1) \sim (6\text{–}10) \times 10^4$ K, which produces a tremendous reflection effect from the M dwarf secondaries. Typically, the irradiated sides of the secondaries have colour temperatures ~ 12 000 K whereas the shaded sides have $T_{\mathrm{eff}} \sim 2500$ K. The inefficiency of heat flow from the hot to the cool sides suggests the absence of significant photospheric advection or circulation currents (cf. Kirbiyik 1982). In these stars, irrespective of how the flux shortward of the Lyman limit is reprocessed by the secondaries' atmospheres, the importance of photospheric heating is demonstrated by the observed emergent flux from the secondaries *longward* of the Lyman limit – even if this were mostly chromospheric emission, approximately an equal amount of flux is being sent downward to heat the photosphere.

In the X-ray binary HZ Her, where heating of the secondary is far more intense than in CVs, the transition from high to low optical states is so rapid that the system has never been caught in between (Jones, Forman & Liller 1973; Hudec & Wenzel 1976), showing that the time scale for cooling the heated photosphere is at most a few days and therefore the heating is not deep.

The effect on $\dot{M}(2)$ of irradiation is considered in Section 9.3.1.

2.9.7 *X-Rays from Secondary Stars*

The secondaries in CVs are low mass, rapidly rotating stars with probably main sequence structures and are expected to be magnetically very active (Section 9.1.2). They could therefore be more copious emitters of coronal X-rays than the single dKe and dMe stars. Hard X-ray luminosities of single late-type dwarfs are typically $\sim 10^{27}(v_{\mathrm{eq}} \sin i)^2$ erg s^{-1} for $v_{\mathrm{eq}} \lesssim 30$ km s^{-1} (Walter & Bowyer 1981; Pallavicini *et al.* 1981). This relationship cannot, however, be extrapolated to the more rapid rotators, which appear to be limited to $L_{\mathrm{x}} \lesssim 10^{30}$ erg s^{-1} (Fleming, Gioia & Maccacaro 1989). This is more than an order of magnitude lower than typical L_{x} observed in CVs. Furthermore, the temperatures of X-ray emission from single low mass stars are factors of 3–10 lower than those observed in CVs (Eracleous, Halpern & Patterson 1991).

With the exception of CVs experiencing episodes of very low $\dot{M}$, it therefore appears unlikely that secondaries contribute a detectable amount to CV X-ray fluxes (Rucinski 1984; Eracleous, Halpern & Patterson 1991).

3

Dwarf Novae

All the noise of the dwarves lost in the night.

J.R.R. Tolkien. *The Hobbit.*

The DN, already introduced in Sections 1.2 and 1.3 with their classification scheme described in Section 2.1, are arguably the most valuable of objects for the study of accretion discs. Among them examples may be found of optically thin discs and optically thick discs, of face-on discs and edge-on discs, of non-steady discs and of nearly steady state discs and of transitions between them. Furthermore, the brightest DN at maxima reach apparent magnitudes of 8–10, at which time the entire flux is conveniently of almost pure accretion origin.

3.1 Well-Observed DN

It is inevitable that a few relatively bright DN, especially the eclipsing systems, have been preferentially observed. Although over 200 DN have been classified by their light curves, only a small fraction have been studied sufficiently to establish their orbital periods. It will be seen in this chapter that P_{orb} plays an important rôle in the systematics of DN. Among the DN in general, 12 Z Cam stars, 29 definite U Gem stars (including, slightly unconventionally , the three systems BV Cen, GK Per and V1017 Sgr with large P_{orb}), 34 SU UMa stars and 22 objects suspected of belonging in the DN class have known orbital periods. The SU UMa stars may be overrepresented because their orbital periods are easy to *estimate*, independent of inclination, from photometric observations made during superoutbursts. Orbital periods for the U Gem and Z Cam class have come predominantly from spectroscopic observations, with the addition of a few found from photometric orbital variations (eclipses, bright spot modulation, IR ellipsoidal modulation).

There is still a need to observe spectroscopically and photometrically a large number of DN, especially those that already have long term light curves from the amateur observers. From perusal of the AAVSO and RAS New Zealand circulars, the following DN all reach at least $m_v = 13.0$ at maximum, have information on light curves and frequencies of outbursts, but have no P_{orb} known: VZ Aqr, DH Aql, AT Ara, AM Cas, SV CMi, V795 Cyg, TT Ind, TU Ind, LL Lyr, HP Nor, TZ Per, FO Per, UY Pup, V1830 Sgr, FQ Sco. Finding charts for DN are given in Vogt & Bateson (1982), Bruch, Fischer & Wilmsen (1987) and Downes & Shara (1993).

Tables 3.1, 3.2 and 3.3 list the orbital periods and some of the other observational properties of the U Gem, Z Cam and SU UMa stars respectively. There remain a number of stars, Table 3.4, whose light curves strongly suggest DN type but whose categorization is in some way incomplete. All but one of these have P_{orb} below the period gap, which, if they are truly DN, means that they are probably SU UMa stars and could be transferred to Table 3.3. However, until positive evidence for superhumps (Section 3.6.4) is available, they are left in limbo. It should be noted that this has introduced some bias into Table 3.1: only stars having both classification DN and $P_{orb} \gtrsim 3$ h can appear there. If there are any true U Gem stars with $P_{orb} < 3$ h (i.e., not possessing superoutbursts with associated superhumps) then they are hidden in Table 3.4. Experience has shown, however, that every DN with $P_{orb} < 3$ h that has been sufficiently observed has been found to be an SU UMa star.

3.2 DN In Quiescence

3.2.1 Spectra of DN in Quiescence

As seen in Figure 1.16, at minimum light DN spectra are characterized by a strong Balmer emission spectrum on a blue continuum, together with other usually weaker lines of HeI and a few from heavier elements. The Balmer decrement is shallow (or even negative), indicating optically thick conditions (Szkody 1976c), and the lines converge often to give a Balmer continuum in emission. Figure 3.1 shows a typical spectral tracing. A prime characteristic of DN quiescent spectra is the large ratio Hβ/HeII λ4686. The λ4650 blend of CIII and NIII is always weak or absent.

In systems such as Z Cha, OY Car, VW Hyi, TY Psc and WZ Sge where the primary contributes substantially at optical wavelengths, its broad absorption lines can be seen as shallow troughs either side of the Balmer emission (Vogt 1976; Marsh, Horne & Shipman 1987; Bailey & Ward 1981a; Schoembs & Hartman 1983; Szkody 1985c; Smak 1979b).

In the IR a few emission lines are present with widths indicating a disc origin (Dhillon & Marsh 1993) but in the longer orbital period systems the absorption spectrum of the secondary dominates. Figure 3.2 shows the M dwarf secondary in U Gem emerging beyond 7000 Å. Equation (2.97) gives $M_V(2)$ as a function of P_{orb}; by comparing this with the observed M_V of discs for DN in quiescence, discussed in Section 3.2.3, it is seen (Figure 3.5) that signs of the secondary spectrum should appear in red spectra for $P_{orb} \gtrsim 5$ h. If the disc is highly inclined, or spectra are obtained during eclipse, it is possible to detect optical flux from the secondary in shorter period systems. These expectations are borne out in Tables 3.1, 3.2 and 3.3.

In the UV a variety of spectral behaviour occurs independent of what the optical spectrum looks like. The majority of DN show the emission lines typical of CVs with equivalent widths up to 70 Å. U Gem, however, both in the UV and FUV shows an almost entirely absorption-line spectrum (Fabbiano *et al.* 1981; Panek & Holm 1984; Kiplinger, Sion & Szkody 1991; Long *et al.* 1993) as do FO Aql, OY Car, KT Per, Z Cha, AH Her, VW Hyi, CN Ori and WZ Sge (la Dous 1991). (In all of these stars, however, the MgII λ2800 lines are in emission.) This is largely an inclination effect: emission strengths decreasing with increasing i and changing to absorption for $i \gtrsim 60°$. It is not simply an $\dot{M}$(d) dependence, for WX Hyi and YZ Cnc have very strong UV

Table 3.1. *U Gem Stars with Known Orbital Periods* ≥ 3 *h.*

Star	P_{orb} (h)	m_v (min)	m_v (max)	T_n (d)	τ_d (d)	i°	$Sp(2)$	References
GS Pav	3.7270	16.8		14				2
IP Peg	3.7969	15.8	12.3	95	1.5:	81	M4V	4-11,15,119
CN Ori	3.917	14.2	11.9	17	2.5	67	M4V	20-23,106
UU Aql	3.94:	16.1	11.0	49	1.2			1,14
AR And	3.94:	16.9	11.0	25:		47:		1,13,16,17,114
X Leo	3.946	15.8	12.4	22	1.1		M2V	16,18,19
YY Dra	3.96	16.0	10.6	1000:		42	M4V	121–126
CW Mon	4.229	16.0	12.5	122:		65:	M3V	16,19,23
U Gem	4.2458	14.6	9.4	101	1.2	70	M4.5V	24–44,112,113,130
BD Pav	4.3032	16.6	12.4		2	90:		42,45
EY Cyg	4.3495	15.5	11.4	96		0:		46–49
TW Vir	4.384	15.5	12.3	27		43:	M2–4V	13,16,23,49–53
SS Aur	4.387	14.7	10.5	43	1.4	38:	M1V	18,42, 46, 47, 51,54
HX Peg	4.819	16.6	12.9				K:	12,133,134
HS 1804+6753	5.038	14		70		86		132
CZ Ori	5.150	16.5	12.1	26	1.6			1,49,55
AR Cnc	5.150	18.7	15.3			>80	M4–5V	3,56,121
BV Pup	5.40:	15.6	13.5	19				57,58
AF Cam	5.5:	17.0	13.4	75				1,19,23,59,60
RU LMi	6.02	17.8	13.8	25:				3,56,61,62
EI UMa	6.4344	16	13.4					3,63–65
SS Cyg	6.6031	11.7	8.2	40	2.4	38:	K5V	38,42,51,66–84, 98,116–118,131
TT Crt	7.303	15.9	13.1	100:	1.8	60	K5–M0V	85,115
MU Cen	8.21	14.9	12.6	45		>45		86
CH UMa	8.23	15.9	10.7	204		21		64,86-88
RU Peg	8.990	12.7	9.0	65	3.4	33	K2–3V	51,70,71,86,89-91
DX And	10.600	15.8	11.9	300:		45:	K1V	94,99,120,129
V442 Cen	11.0:	16.5	12.3	24	2.1	27		92,93
BV Cen	14.6428	12.6	10.7	149	7	62	G5–8 IV–V	95-97,127,128
GK Per	47.9233	13.2	10.3	885	14	60:	K0IV	100-107
V1017 Sgr	137.1	13.6	10.6	6600	33		G5IIIp	108-111

References: 1 Szkody 1987a; 2 Augusteijn 1994; 3 Andronov 1986a; 4 Goranskij, Lyutyi & Shugarov 1985; 5 Wood & Crawford 1986; 6 Martin, Jones & Smith 1987; 7 Marsh 1988; 8 Hessman 1989a; 9 Martin *et al.* 1989; 10 Wood *et al.* 1989a; 11 Marsh & Horne 1990; 12 Ringwald 1992; 13 Szkody 1985a; 14 Shafter 1992b; 15 Wolf *et al.* 1993; 16 Howell & Szkody 1988; 17 Szkody, Piché & Feinswog 1990; 18 Shafter & Harkness 1986; 19 Szkody & Mateo 1986b; 20 Schoembs 1982; 21 Mantel *et al.* 1988; 22 Barrera & Vogt 1989a; 23 Howell & Szkody 1988; 24 Mumford 1964a; 25 Krzeminski 1965; 26 Smak 1971a; 27 Warner & Nather 1971; 28 Smak 1976; 29 Swank *et al.* 1978; 30 Robinson & Nather 1979; 31 Paczynski & Schwarzenberg-Czerny 1980; 32 Wade 1979; 33 Stover 1981a; 34 Wade 1981; 35 Panek & Eaton 1982; 36 Berriman *et al.* 1983; 37 Eason *et al.* 1983; 38 Polidan & Holberg 1984; 39 Smak 1984c; 40 Zhang & Robinson 1987; 41 Mason *et al.* 1988; 42 Friend *et al.* 1990a, 43 Marsh *et al.* 1990; 44 Kiplinger, Sion & Szkody 1991; 45 Barwig & Schoembs 1983, 1987; 46 Kraft 1962; 47 Kraft & Luyten 1965; 48 Hacke & Andronov 1988; 49 Szkody, Piché & Feinswog 1990; 50 Cordova &

emission lines, whereas VW Hyi, OY Car, Z Cha and WZ Sge have absorption lines – all of these are very low $\dot{M}$(d) systems (see Section 3.3.3.3). The correlation with inclination is not perfect, for HT Cas and possibly IP Peg have weak emission lines.

This behaviour contrasts with that of high $\dot{M}$(d) discs in the optical (Figure 2.34) or the UV (Section 3.3.5.1), where emission line equivalent widths rise rapidly at large inclinations. No quantitative models have been proposed, but qualitatively it appears that in DN at least the hotter inner regions of the disc are optically thin when viewed face-on, but become optically thick in most stars when viewed at high inclination.

The set of absorption lines CIII $\lambda\lambda$977,1176, NV $\lambda\lambda$1239,1243, SiII λ1265, SiIII λ1259, CII λ1335, SiIV $\lambda\lambda$1393,1402 and CIV λ1548 in U Gem can be quantitatively accounted for by a 30 000 K photospheric spectrum, but the additional lines NV λ1239 and HeII λ1640 require $T \approx 60\,000$ K (Kiplinger, Sion & Szkody 1991; Long *et al.* 1993). The location of the latter region, which persists throughout the time between outbursts, remains uncertain; the temperature is too low to be that of a boundary layer, but the emission might be from a viscously heated outer layer spun up by accretion during outburst.

HST observations of OY Car reveal a wealth of FeII absorption lines that are superimposed on the primary's 16 500 K continuum flux. The lines arise in a gas with $T = 9200$ K, which is probably the upper atmosphere of the disc (Horne 1993b; Horne *et al.* 1994).

In WZ Sge, Sion, Lekenby & Szkody (1990) find, in addition to SiII and SiIII, many

References to Table 3.1 (cont.)

Mason 1982b; 51 Shafter 1984a; 52 Mateo & Bolte 1985; 53 Mansperger & Kaitchuck 1990; 54 Efimov, Tovmasyan & Shakhovskoi 1986; 55 la Dous 1990; 56 Mukai *et al.* 1990; 57 Szkody, Howell & Kennicutt 1986; 58 Szkody & Feinswog 1988; 59 Szkody & Howell 1989; 60 Long *et al.* 1991; 61 Wagner *et al.* 1988; 62 Naylor, Smale & van Paradijs 1990; 63 Cook 1985b; 64 Thorstensen 1986; 65 Wilson *et al.* 1986; 66 Joy 1956; 67 Walker & Chincarini 1968; 68 Margon *et al.* 1978; 69 Patterson, Robinson & Kiplinger 1978; 70 Kiplinger 1979a,b; 71 Robinson & Nather 1979; 72 Cordova *et al.* 1980a, 73 Cowley, Crampton & Hutchings 1980; 74 Horne & Gomer 1980; 75 Stover *et al.* 1980; 76 Hildebrand, Spillar & Stiening 1981a, 77 Walker 1981; 78 Clarke, Capel & Bowyer 1984; 79 Hessman *et al.* 1984; 80 King, Watson & Heise 1985; 81 Hessman 1986; 82 Robinson, Zhang & Stover 1986; 83 Honey *et al.* 1989; 84 Bruch 1990; 85 Szkody *et al.* 1992; 86 Friend *et al.* 1990b; 87 Becker *et al.* 1982; 88 Green *et al.* 1982; 89 Stover 1981b; 90 Wade 1982; 91 la Dous *et al.* 1985; 92 Bateson 1977a; 93 Marino & Walker 1984; 94 Drew, Hoare & Woods 1991; 95 Vogt & Breysacher 1980; 96 Gilliland 1982c; 97 Menzies, O'Donoghue & Warner 1986; 98 Giovannelli & Martinez-Pais 1991; 99 Bruch 1989; 100 Bianchini, Hamzaoglu & Sabbadin 1981; 101 Bianchini, Sabbadin & Hamzaoglu 1982; 102 Crampton, Cowley & Hutchings 1983; 103 Bianchini & Sabbadin 1983b; 104 Bianchini *et al.* 1986; 105 Hessman 1989b; 106 Friend *et al.* 1990a; 107 Szkody, Mattei & Mateo 1985; 108 Kenyon 1986; 109 Kraft 1964a; 110 McLaughlin 1946; 111 Sekiguchi 1992a; 112 Panek & Holm 1984; 113 Fabbiano *et al.* 1981; 114 la Dous 1991; 115 Fleet 1992; 116 Voloshina & Lyutyi 1984; 117 Voloshina 1986; 118 Zuckerman 1961; 119 Harlaftis *et al.* 1994a; 120 Drew, Jones & Woods 1993; 121 Howell & Blanton 1993; 122 Patterson *et al.* 1982, 1992b; 123 Wenzel 1983a,b; 124 McNaught 1986; 125 Mateo, Szkody & Garnavich 1991; 126 Hazen 1985; 127 Hollander, Kraakman & van Paradijs 1993; 128 Williger *et al.* 1988; 129 Vrielman & Bruch 1993; 130 Smak 1993a; 131 Mansperger *et al.* 1994; 132 Fiedler 1994; 133 Honeycutt *et al.* 1994; 134 Ringwald 1994.

Table 3.2. *Z Cam Stars with Known Orbital Periods.*

Star	P_{orb} (h)	m_v (min)	m_v (stand)	m_v (max)	T_n (d)	τ_d (d)	$i°$	$Sp(2)$	References
AB Dra	3.648	15.5		11.3	13	2.5	40:		1–5
SV CMi	3.74	16.3		13	16				63
KT Per	3.903	16.0	12.3	11.7	26	1.3	57:	M3.3 ± 1	5,6,9,13,50,55,56,61
UZ Ser	4.152	15.6	13.8	12.1	21	2.5	55:		5,7,51–54
WW Cet*	4.236	15.7	13.9	9.3	31		54		1,10,11, 57–59
RX And	5.037	13.6	11.8	10.9	13	1.7	51		8,9,12–19
HL CMa	5.148	14.5	11.7	10.5	15	2.5	45:		20–23, 62
AT Cnc	5.729	15.5	12.3	11.9	14		60:		24
AH Her	6.195	14.3	12.0:	11.3	18	2.3	46	K2–M0V	4,16,25–28
V426 Oph	6.847	13.4	11.9	10.9	22		59	K2–4V	29–34,57,60
Z Cam	6.956	13.6	11.7	10.4	23	2.6	57	K7V	35–41
EM Cyg	7.178	14.2	12.9	12.0	22	3.6	63	K5V	42–47
SY Cnc	9.12	13.7	12.2	11.1	27	2.3	26	G8–9V	47–49

* It has been suggested (Warner 1987a) that WW Cet may not be a Z Cam star: $m_v = 13.9$ could be the normal quiescent state of its DN activity and $m_v = 15.7$ a 'low state' similar to that of VY Scl stars (Section 4.3.2).

References: 1 Thorstensen & Freed 1985; 2 Voloshina & Shugarov 1989; 3 Williams & Ferguson 1982; 4 Szkody 1981a; 5 la Dous 1991; 6 Robinson & Nather 1979; 7 Herbig 1944; 8 Kaitchuck 1989; 9 Clarke & Bowyer 1984; 10 Paczynski 1963; 11 Klare *et al.* 1982; 12 Kraft 1962; 13 Robinson 1973b; 14 Szkody 1976a; 15 Szkody 1981a; 16 Verbunt *et al.* 1984; 17 Kaitchuck, Mansperger & Hantzios 1988; 18 Woods, Drew & Verbunt 1990; 19 Szkody, Piché & Feinswog 1990; 20 Chlebowski, Halpern & Steiner 1981; 21 Hutchings *et al.* 1981b; 22 Bonnet-Bidaud, Mouchet & Motch 1982; 23 Mauche & Raymond 1987; 24 Goetz 1985, 1986; 25 Robinson 1973d; 26 Moffat & Shara 1984; 27 Horne, Wade & Szkody 1986; 28 Bruch 1987; 29 Szkody 1977; 30 Shugarov 1983a; 31 Hessman 1988; 32 Szkody & Mateo 1988; 33 Hellier *et al.* 1990; 34 Szkody, Kii & Osaki 1990; 35 Kraft, Krzeminski & Mumford 1969; 36 Robinson 1973a; 37 Robinson 1973c; 38 Kiplinger 1980; 39 Szkody 1981a; 40 Szkody & Wade 1981; 41 Szkody & Mateo 1986a; 42 Mumford & Krzeminski 1969; 43 Robinson 1974; 44 Mumford 1980; 45 Jameson, King & Sherrington 1981; 46 Stover, Robinson & Nather 1981; 47 Szkody 1981a; 48 Patterson 1981; 49 Shafter 1984a; 50 Ratering, Bruch & Schimpke 1991; 51 Echevarria *et al.* 1981; 52 Verbunt *et al.* 1984; 53 Echevarria 1988; 54 Dyck 1989; 55 Ratering, Bruch & Diaz 1993; 56 Sherrington & Jameson 1983; 57 Hollander, Kraakman & van Paradijs 1993; 58 Hawkins, Smith & Jones 1990; 59 Bateson & Dodson 1985; 60 Beyer 1977; 61 Thorstensen & Ringwald 1994; 62 Mansberger *et al.* 1994; 63 Augusteijn 1994.

photospheric absorption lines of CI in the range 1300–1950 Å which suggest an overabundance of C dredged from the interior of the white dwarf primary.

VW Hyi, WZ Sge, OY Car, Z Cha, EK TrA and U Gem all show a decline in UV flux below ~1400 Å (Verbunt 1987; la Dous 1991) which has been interpreted as Lyα absorption in the primary, which requires $T_{eff} < 20\,000$ K (Section 2.9.3). In modelling this feature, as with the entire UV flux distribution, it should be kept in mind that fluxes computed from steady state discs probably have no relevance to DN at quiescence. For example, the viscous time scale t_ν (equation (2.53)) in the outer disc in the quiescent state is more than an order of magnitude greater than the interval between outbursts (Cannizzo, Shafter & Wheeler 1988) and the observed temperature profiles across discs such as Z Cha (Figure 2.33) cannot be fitted by steady state discs with reasonable (i.e., <1) values of the viscosity parameter α (Wood 1990b; Wood, Horne & Vennes 1992; Section 3.5.5.4). Furthermore, as discussed in Section 8.4, discs truncated by magnetospheres are a real possibility. The evolution of quiescent fluxes during the interval between outbursts is described in Section 3.3.5.3.

To facilitate rapid identification of emission lines in DN quiescent spectra, Table 3.5 lists lines culled from a variety of publications.

3.2.2 Photometry of DN in Quiescence

From the discussion in Sections 1.3 and 2.6, and the light curves shown in Figures 1.8, 1.9, 2.23, 2.29 and 2.30, it is evident that photometry of high inclination systems provides rich opportunities for probing the physical structure of DN. Among the non-eclipsing DN some show orbital modulation of brightness attributed to varying aspect of the bright spot, others show no modulation at all and are deduced to be of low inclination. This is supported by a general correlation between hump amplitude and radial velocity amplitude $K(1)$ (e.g., Kraft 1962).

However, in some short period eclipsing systems, even when the signature of a bright spot is present in the eclipse profile, the orbital hump may be weak or absent (e.g., HT Cas: Patterson 1981; Horne, Wood & Stiening 1991), which suggests that the whole disc is optically thin at such times. In HT Cas, even when the bright spot is absent there is a 0.3 mag modulation in the U band, of maximum brightness near phase zero, which may be caused by absorption in gas above the disc at the point furthest from the L_1 point, where vertical gravity is lowest (Horne, Wood & Stiening 1991). At this same phase the absorption cores of Balmer lines are at their strongest (Young, Schneider & Schectman 1981b).

Even more extreme behaviour is shown by WZ Sge (Figure 3.3) where there are two humps in each orbit and a highly variable dip near phase 0.3. The latter is interpreted as obscuration of the bright spot by variable density fluctuations in the disc. The W UMa-like light curve can arise if the inner parts of the disc are optically thin, so the bright spot region is visible from both front and back (Robinson, Nather & Patterson 1978).

In BD Pav, with $P_{\rm orb} = 4.3$ h, a similar W UMa-like light curve has a different explanation – in that star the secondary contributes an unusually large amount to the visible region, with the result that its ellipsoidal variation is superimposed on the orbital modulation of a typical eclipsing DN (Barwig & Schoembs 1983). A characteristic of the colour-curves of eclipsing systems is that at mid-eclipse, with the primary and central disc obscured, there is strong reddening (e.g., OY Car: Cook 1985a).

Among the non-eclipsing DN of intermediate inclination, VW Hyi, which has a large orbital modulation, is the best studied (Vogt 1974, 1983a; Warner 1975b; Haefner, Schoembs & Vogt 1979; van Amerongen *et al.* 1987: Figure 3.4). The amplitude of the main hump is wavelength-dependent, but after normalization its shape and width are independent of wavelength. However, the final decay towards phase 0.7 is slower at

Table 3.3. *SU UMa Stars with Known Orbital Periods.*

Star	P_{orb} (h)	P_s (h)	P_b (d)	m_v (min)	m_v (max)	m_v (sup)	T_n (d)	T_s (d)	τ_d (d)	i°	$Sp(2)$	References
WX Cet	1.25:*	1.27:		17.5	12.5	9.5		1000:				1–7,183
LL And	1.34*	1.3681		20.0:		13.8		5000				185
WZ Sge	1.36051	1.3714	7.1±1.0	14.9		7.1		11000	0.4:	76:		10–23,179
SW UMa	1.36356	1.3999	2.2	16.5	10.8	9.4		954	1.1	45:		24–29,165
HV Vir	1.3918	1.4110	4.3	19.2		11.5		3500**	<2			8,9,174
CY UMa	1.399*	1.423		17.0		11.9	115:	297:				30,31
T Leo	1.41166	1.4436	2.7	15.5	11.0	10:		420		65:		25,32,33,176
BC UMa	1.44*	1.486		18.3		10.9						184
AQ Eri	1.4626	1.494	2.9	17.7		12.5	44	300:				34–36,173
V1159 Ori	1.497	1.543	2.1	15.5	13.6	12.5	4.1	53		40:		177,178,188
V436 Cen	1.5000	1.5308	3.0±0.5	15.5	12.4	11.3	32	335	1.1	65:		37–41
OY Car	1.51490	1.5518	2.6	15.6	12.4	11.5	160:	346		83	M6V	41,45–62, 161,163,180,181
VY Aqr	1.516	1.557	2.4	17.1	10.9	10.3	400:	800:	0.7	40:		42–44,175
ER UMa	1.522*	1.5718		15.3								184
EK TrA	1.526*	1.5581		16.6:	12.0	11.0	231	487	0.6			63,164
TV CrV	1.55*	1.6:		19:		12:						186
UV Per	1.555*	1.594		17.5	12.7	11.9		360	1.2	30:		66,170,171
TY Psc	1.638	1.68:	2.7:	15.3:	12.2	11.7	39	333	0.7:			64,65,173
SS UMi	1.641*	1.682		17.0		12.6						67–71
IR Gem	1.642	1.6982	2.1±0.6	15.5	11.7	11.2	26	174	0.8	50:		72–75,162
RZ Sge	1.646*	1.6884		16.9	12.8	12.2	77	266				76
KV And	1.663*	1.733		22.5		14.6						189
FO And	1.7186	1.75:		17.5	13.5							3,77,173
TT Boo	1.740*	1.806		19.2	12.7		45					28,184
AY Lyr	1.7616*	1.8233		18.0	13.2	12.3	24	205	1.2	60:		78–80,162
HT Cas	1.76753	1.8258	2.3	16.5	12.6	10.8	166:	400		81		25,81–89, 158,166

V1251 Cyg	1.78*	1.82		18.5		12.5						90,184
VW Hyi	1.78250	1.8509	2.0	13.3	9.5	8.5	27	179	0.7	60		38,91–113
Z Cha	1.78798	1.8576	2.0	16.0	12.4	11.9	51	218		82	M5.5V	38,60,61, 114–139,160,190
WX Hyi	1.79551	1.8564	2.2	14.7	12.5	11.4	11	195		40:		39,97,99, 140,141
AW Gem	1.829*	1.8881		18.8	13.8	13.1	98	410				28,142
SU UMa	1.8324	1.900	2.1	14.8	12.2	11.2	19	160	1.4	44:		143–147,162, 167,172
CU Vel	1.855*	1.918		15.5	11.1	10.7	165	386	1.1			169,187
BR Lup	1.903*	1.973		17.5		13.1						148
EF Peg	1.96*	2.090		18:		10.7		254:				149,150
TY PsA	2.0160	2.1036	2.0	16.5	12:	12.0				65		151–153
V344 Lyr	2.07*	2.195		>20		13.8	16:	240:				182
YZ Cnc	2.0862	2.2050	1.6±0.3	14.5	11.9	10.5	12	134	0.9	50:		25,78,144, 154–156,168,172
TU Men	2.82	3.029	1.8±0.3	16.6	12.5	12.0	37	194		65		157,159

* Estimated from the superhump period (see Section 3.6.4.3).

** Being an ecliptic object, some outbursts of HV Vir may have been missed.

References: 1 Bailey 1979b; 2 Downes & Margon 1981; 3 Szkody *et al.* 1989; 4 van Paradijs, van der Klish & Pedersen 1989; 5 Downes 1990; 6 Howell *et al.* 1991; 7 O'Donoghue *et al.* 1991; 8 Mendelson *et al.* 1992; 9 Barwig, Mantel & Ritter 1992; 10 Krzeminski & Kraft 1964; 11 Krzeminski & Smak 1971; 12 Warner & Nather 1972c; 13 Fabian *et al.* 1978; 14 Ritter 1978; 15 Robinson, Nather & Patterson 1978; 16 Fabian *et al.* 1980; 17 Gilliland & Kemper 1980; 18 Patterson 1980; 19 Walker & Bell 1980; 20 Patterson *et al.* 1981; 21 Gilliland, Kemper & Suntzeff 1986; 22 Naylor 1989; 23 Sion, Leckenby & Szkody 1990; 24 Wellmann 1952; 25 Shafter 1984a; 26 Shafter, Szkody & Thorstensen 1986; 27 Robinson *et al.* 1987; 28 Howell & Szkody 1988; 29 Szkody, Osborne & Hassall 1988; 30 Kato *et al.* 1988; 31 Watanabe *et al.* 1989; 32 Shafter & Szkody 1984; 33 Kato & Fujino 1987; 34 Szkody 1987a; 35 Kato, Fujino & Iida 1989; 36 Kato 1991a; 37 Warner 1975a; 38 Warner & Brickhill 1978; 39 Bailey 1979c; 40 Semeniuk 1980; 41 Gilliland 1982c; 42 Hendry 1983; 43 Patterson *et al.* 1993a; 44 Della Valle & Augusteijn 1990; 45 Ritter 1980c; 46 Bailey & Ward 1981a; 47 Vogt *et al.* 1981; 48 Sherrington *et al.* 1982; 49 Vogt 1983b; 50 Schoembs & Hartmann 1983; 51 Berriman 1984b; 52 Cook 1985a; 53 Schoembs 1986; 54 Berriman 1987b; 55 Naylor *et al.* 1987; 56 Schoembs, Drier & Barwig 1987; 57 Naylor *et al.* 1988; 58 Hessman *et al.* 1989; 59 Wood *et al.* 1989b; 60 Wood 1990b; 61 Wood & Horne 1990; 62 Hessman *et al.* 1992; 63 Vogt & Semeniuk 1980; 64 Szkody 1985c; 65 Szkody & Feinswog 1988; 66 Kato 1990a; 67 Mason *et al.* 1982; 68 Andronov 1986b; 69 Richter 1989; 70 Chen, Liu & Wei 1991; 71 Udalski 1990a; 72 Szkody, Shafter & Cowley 1984; 73 Feinswog, Szkody

References to Table 3.3 (cont.)

& Garnavich 1988; 74 Szkody *et al.* 1989; 75 Lazaro *et al.* 1991; 76 Bond, Kemper & Mattei 1982; 77 Bruch 1989; 78 Patterson 1979d; 79 Szkody 1982; 80 Udalski & Szymanski 1988; 81 Patterson 1981; 82 Young, Schneider & Schectman 1981b; 83 Wlodarczyk 1986; 84 Zhang, Robinson & Nather 1986; 85 Berriman, Kenyon & Boyle 1987; 86 Marsh 1990; 87 Wood & Horne 1990; 88 Horne, Wood & Stiening 1991; 89 Wood, Horne & Vennes 1992; 90 Kato 1991b; 91 Vogt 1974, 1983a; 92 Warner & Brickhill 1974; 93 Warner 1975b; 94 Haefner, Schoembs & Vogt 1977, 1979; 95 Robinson & Nather 1979; 96 Bath, Pringle & Whelan 1980; 97 Schoembs & Vogt 1980, 1981; 98 Sherrington *et al.* 1980; 99 Hassall *et al.* 1983; 100 Mateo & Szkody 1984; 101 Robinson & Warner 1984; 102 Schwarzenberg-Czerny *et al.* 1985; 103 Smak 1985b; 104 van der Woerd, Heise & Bateson 1986; 105 Polidan & Holberg 1987; 106 Pringle *et al.* 1987; 107 van Amerongen *et al.* 1987; 108 van Amerongen, Bovenschen & van Paradijs 1987; 109 van der Woerd & Heise 1987; 110 van der Woerd *et al.* 1987, 1988; 111 Verbunt *et al.* 1987; 112 Mauche *et al.* 1991; 113 Belloni *et al.* 1991; 114 Mumford 1971; 115 Bath *et al.* 1974; 116 Warner 1974a; 117 Bailey 1979a; 118 Smak 1979a; 119 Ritter 1980d; 120 Bailey *et al.* 1981; 121 Cook & Warner 1981, 1984; 122 Rayne & Whelan 1981; 123 Faulkner & Ritter 1982; 124 Vogt 1982; 125 Dmitrienko *et al.* 1983; 126 Cook 1985c; 127 Horne & Cook 1985; 128 O'Donoghue 1986; 129 Smak 1986; 130 Wood *et al.* 1986; 131 Marsh 1987; 132 Marsh, Horne & Shipman 1987; 133 Wood 1987; 134 Honey *et al.* 1988; 135 Wade & Horne 1988; 136 Warner & O'Donoghue 1988; 137 O'Donoghue 1990; 138 van Amerongen, Kuulkers & van Paradijs 1990; 139 Kuulkers *et al.* 1991a; 140 Hassall, Pringle & Verbunt 1985; 141 Kuulkers *et al.* 1991b; 142 Fujino *et al.* 1989; 143 Mumford 1966; 144 Szkody 1981a; 145 Thorstensen & Wade 1986; 146 Udalski 1990b; 147 Woods, Drew & Verbunt 1990; 148 O'Donoghue 1987; 149 Howell & Fried 1991; 150 Howell *et al.* 1993; 151 Barwig *et al.* 1982; 152 Warner, O'Donoghue & Wargau 1989; 153 O'Donoghue & Soltynski 1992; 154 Moffet & Barnes 1974; 155 Drew & Verbunt 1988; 156 Shafter & Hessman 1988; 157 Stolz & Schoembs 1984; 158 Wenzel 1987a; 159 Bateson 1979b, 1981; 160 Bateson 1990a; 161 Bateson 1990b; 162 la Dous 1991; 163 Krzeminski & Vogt 1985; 164 Hassall 1985a; 165 Granzlo 1992; 166 Wlodarczyk 1988; 167 Harlaftis *et al.* 1994c; 168 Harlaftis *et al.* 1994b; 169 Bateson 1977c; 170 Udalski & Pych 1992; 171 Howarth 1978b; 172 van Paradijs *et al.* 1994; 173 Thorstensen, Patterson & Thomas 1993; 174 Leibowitz *et al.* 1994; 175 Augusteijn 1993; 176 Lemm *et al.* 1993; 177 Jablonski & Cieslinski 1992; 178 Patterson (private communication 1994), 179 Smak 1993b; 180 Horne 1993b; 181 Horne *et al.* 1994; 182 Kato 1993; 183 Mennickent 1994; 184 Kato 1994b; 185 Howell & Hurst 1994; 186 Howell *et al.* 1995; 187 Ritter 1990; 188 Patterson *et al.* 1994; 189 Howell, Szkody & Cannizzo 1994; 190 Harlaftis *et al.* 1992a,b.

Table 3.4. *Probable DN with Known Orbital Periods*

Star	P_{orb} (h)	m_v (min)	m_v (max)	T_n (d)	$i°$	Type	References
EC23128-310	1.40	16.6				DN?	17
FS Aur	1.42:	16.2	14.4			Z Cam?	1,2
CI UMa	1.44:	18.8	13.8	34:		DN,SU?	1,3
RXJ 2353-385	1.458						36
AL Com	1.46:	20.8	12.8	325		SU?	24
V4140 Sgr	1.4743	17.5	15.5	90:	79	SU?	4,5
V2051 Oph	1.4983	15.0	13.0			DN? MCV?	6–8,29–31,33
CF Gru	1.56	19.9			> 60:	DN?	10
GO Com	1.6:	20	13.1			DN,SU?	3,9,23
BZ UMa	1.632	17.8	10.5	180		DN,SU?	1,11,12,19,25,35
V544 Her	1.66:	20:	14.5			DN?	1,3
RZ Leo	1.699	19:	11.5		< 65	(RN) SU?	1,2,19
GD552	1.712	16.5			20:	DN?	13
V503 Cyg	1.728	17.4	13.4	28	< 65	SU? MCV?	20–22
VW Vul	1.754	15.6	13.6	30	44	Z Cam SU?	1,14,15
LY Hya	1.7952	17.5			< 72	SU?	18,32,34
HS Vir	2.01	15.8	13.0			DN?	3,25,27
KK Tel	2.02	19.3	13.5		< 65	DN,SU?	1,10
DV UMa	2.0633	19.3	15.4		72	SU?	9,16,20,26
DM Dra	2.09:	20.8	15.5			DN?	1,3
CC Cnc	2.25:	17.6	13.1			DN?	1,3
ES Dra	4.24	16.3	13.9			DN?	25,28

References: 1 GCVS; 2 Howell & Szkody 1988; 3 Howell *et al.* 1990; 4 Jablonski & Steiner 1987; 5 Baptista, Jablonski & Steiner 1989, 1992; 6 Warner & Cropper 1984; 7 Berriman, Kenyon & Bailey 1986; 8 Warner & O'Donoghue 1987; 9 Mukai *et al.* 1990; 10 Howell *et al.* 1991; 11 Szkody & Feinswog 1988; 12 Ringwald & Thorstensen 1990; 13 Hessman & Hopp 1990; 14 Shafter 1985; 15 Szkody 1985c; 16 Howell *et al.* 1988; 17 Chen 1993; 18 Kubiak & Krzeminski, 1989,1992; 19 O'Donoghue *et al.* 1991; 20 Szkody & Howell 1992a; 21 Szkody 1987a; 22 Szkody *et al.* 1989; 23 Kato & Hirata 1990; 24 Abbott *et al.* 1992; 25 Ringwald 1992; 26 Howell & Blanton 1993; 27 Bruch 1989; 28 Andronov 1991; 29 Hollander, Kraakman & van Paradijs 1993; 30 Cook & Brunt 1983; 31 Watts *et al.* 1986; 32 Still, Dhillon & Marsh 1993; 33 Echevarria & Alvarez 1993; 34 Still *et al.* 1994; 35 Jurcevic *et al.* 1994; 36 Abbott, Fleming & Pasquini 1994.

longer wavelengths and an intermediate hump near phase 0.5 is sometimes present (particularly for some days after recovery from outburst) which is larger at longer wavelengths. This appears to be an intermittent manifestation of the same effect as seen in WZ Sge. It is seen also in OY Car (Schoembs & Hartmann 1983) and BV Cen (Vogt & Breysacher 1980).

SS Cyg illustrates a less easily interpreted system of lower inclination. The early photometry by Zuckermann (1961) showed no orbital modulation; nor does some recent extensive photometry (Honey *et al.* 1989). This, together with spectroscopic results, leads to the conclusion $i \sim 38°$ (Kiplinger 1979b). However, occasional orbital modulation of low amplitude (Voloshina & Lyutyi 1984; Voloshina 1986; Bruch 1990),

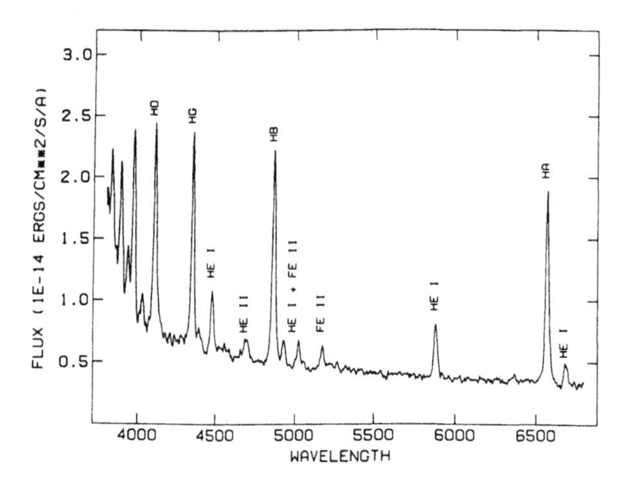

Figure 3.1 Spectrum of YZ Cnc in quiescence. From Shafter & Hessman (1988).

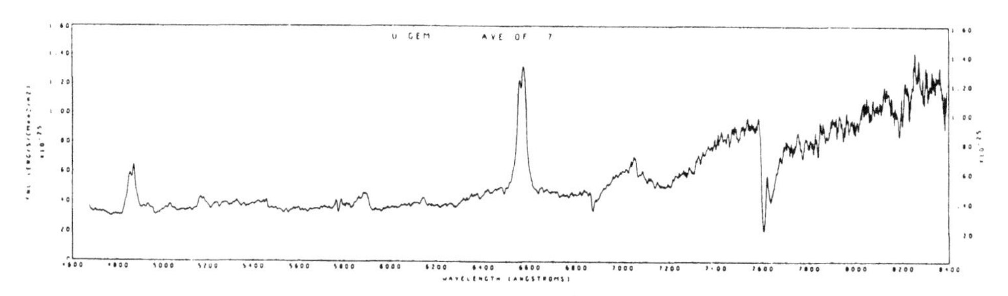

Figure 3.2 Spectrum of U Gem at quiescence, showing Hα and Hβ in emission and the rising flux in the near IR, with TiO absorption, from the secondary. From Stauffer, Spinrad & Thorstensen (1979).

also seen in the UV (Lombardi, Giovannelli & Gaudenzi 1987; Giovannelli *et al.* 1990), suggests $i \sim 50°$. This has a significant effect on deduced masses (Echevarria *et al.* 1989). Z Cam is a further example of a system with occasional orbital modulation (Kraft, Krzeminski & Mumford 1969).

Because of the faintness of DN at quiescence there was only a small amount of pre-HST time-resolved photometry in the UV. The orbital humps in U Gem and VW Hyi, prominent in the visible, are weak or absent in the UV (Wu & Panek 1982; Verbunt *et al.* 1987), showing that the bright spot has $T \lesssim 15\,000$ K (see also Section 2.6.5).

In the IR several DN are bright enough to observe at minimum light. Berriman *et al.* (1983) could detect no eclipse at phase 0.5 in U Gem, which leads to $r < 5.5 \times 10^{10}$ cm for any optically thick disc in the IR; this is compatible with the size of the optical disc at quiescence (Smak 1971a, 1976).

The longer period systems and the high inclination short period systems show ellipsoidal variations in the IR (see Section 2.9.5). In OY Car (Sherrington *et al.* 1982; Berriman 1987b) and HT Cas (Berriman, Kenyon & Boyle 1987) optically thin gas is

Table 3.5. *DN Quiescent Spectra: A Basic Emission-Line List (Wavelengths in Å).*

1026	Lyβ	3970	Hε	7065	HeI
1216	Lyα	4026	HeI	7281	HeI
1175–76	CIII	4102	Hδ	7772,74,75	OI
1239,43	NV	4121	HeI	8236	HeII
1299	SiIII	4634–42	NIII	8438	P18
1302-06	OI	4340	Hγ	8446	OI
1335	CII	4388	HeI	8467	P17
1394,03	SiIV	4471	HeI	8498	CaII
1548,51	CIV	4634–42	NIII	8502	P16
1640	HeII	4647–51	CIII	8542	CaII
1855,63	AlIII(+FeII)	4686	HeII	8545	P15
2979,03	MgII	4713	HeI	8598	P14
3722	H14	4861	Hβ	8662	CaII
3734	H13	4922	HeI	8665	P13
3750	H12	4924	FeII	8750	P12
3771	H11	5016	HeII	8863	P11
3798	H10	5018	FeI	9015	P10
3835	H9	5169	FeII	9229	P9(+FeII,MgII)
3889	H8	5317	FeII	9546	P8
3889	HeI	5876	HeI	10830	HeI
3934	CaII	6563	Hα	21660	HeI
3968	CaII	6678	HeI	Other P and Br series lines	

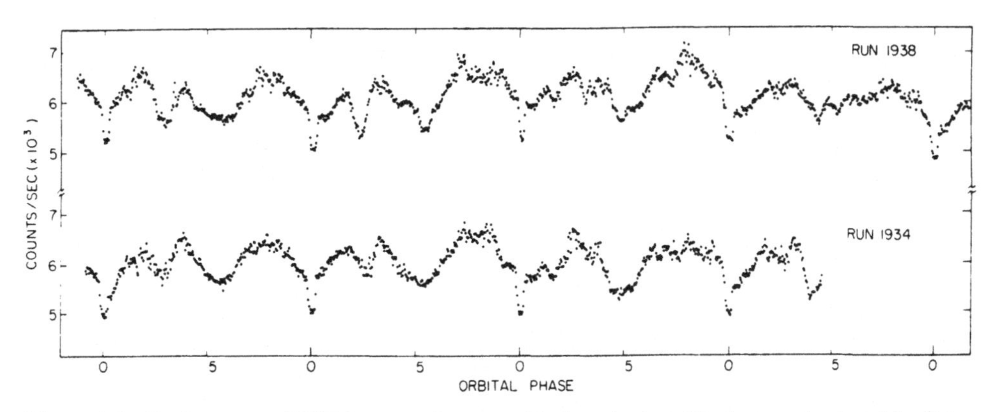

Figure 3.3 Light curve of WZ Sge at quiescence. Each point is a 20 s integration in white light. From Patterson (1980).

responsible for about half the emission in the J or H bands and is highly variable in HT Cas. This is attributed to the outer regions of the disc.

Harrison & Gehrz (1992) report the detection of a large number of DN in the 12–100 μm IR bands of IRAS. The detections appear to be unrelated to outbursts; the flux distributions suggest an emission-line origin, but there is no current explanation for this unexpected result.

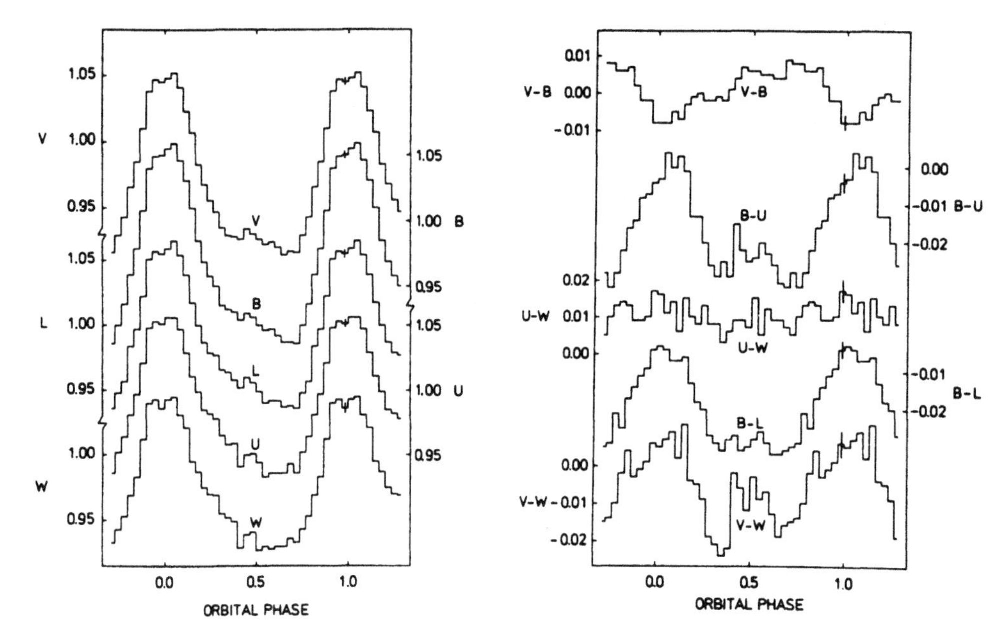

Figure 3.4 Average light and colour curves of VW Hyi at quiescence, in the passbands of Walraven photometry. From van Amerongen *et al.* (1987).

3.2.3 Absolute Magnitudes at Minimum Light: $M_V(min)$

The statistical analysis of proper motions of 25 DN and RVs of 11 DN by Kraft & Luyten (1965) returned a mean absolute magnitude $M_V = 7.5$, but with evidence for a substantial spread in luminosity. A similar result was found by Vojkhanskaya (1973). It is now clear that at least part of the spread arises from the differing disc inclinations and from differing contributions from the bright spot and secondary star.

The contribution from the secondary may be found from decomposition of the observed flux distribution, and has been assessed in Section 2.9.5. The contribution from a bright spot can be estimated from the amplitude of an orbital hump in the light curve (Section 2.6.5). For intrinsically faint systems the primary may be a substantial contributor, which can be estimated from eclipse profiles in high inclination systems (e.g., Figures 2.23, 2.30). Freed of these, the luminosity of the disc remains and can be corrected to the standard inclination of 56°.7 through equation (2.63).

Using distances determined from the K-band method described in Section 2.9.5, with the addition of the important independent distances deduced from membership of BX Pup in NGC 2482 (Moffat & Vogt 1975), the membership of SS Aur in the Hyades moving group (Eggen 1968), the common proper motion companion to RU Peg (Eggen 1968), and the location of UZ Ser in a dark cloud, which provides a reddening parallax (Schmidt-Kaler 1962), some two dozen DN become available for discussion (Warner 1987a). By employing the $M_V(\mathrm{max})$–P_{orb} relation described in Section 3.3.3.4 distances to a further dozen or so DN may be deduced and applied to give estimates of $M_V(\mathrm{min})$ for the disc.

The results from Warner (1987a), with a few adjustments for recent improvements in

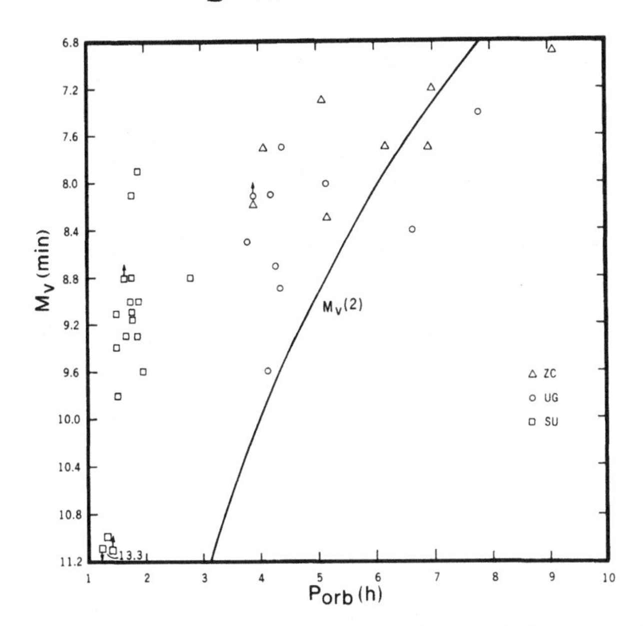

Figure 3.5 Absolute visual magnitude of DN discs at minimum light, plotted against orbital period. The subtypes (Z Cam, U Gem and SU UMa) have different symbols. The solid curve is the predicted visual brightness of the secondary (equation (2.102)).

determinations of orbital periods and inclinations[1], are shown in Figure 3.5. At a given P_{orb} there is a substantial spread of disc luminosity. For $P_{\text{orb}} < 3$ h discs are on average more than a magnitude fainter than those for $P_{\text{orb}} > 3$ h. However, there is a selection effect working against the measurement of faint discs in systems with $P_{\text{orb}} \gtrsim 6$ h, caused by the increasing dominance of the secondary star, as shown by equation (2.102) and its locus in Figure 3.5. It will become evident in Section 3.3.3.3 that much of the scatter in Figure 3.5 is caused by M_{v}(min) being a function of the interval between outbursts.

3.2.4 X-Rays in Quiescence

Hard X-rays (0.1–5 keV) have been detected from most bright (nearby) DN at quiescence, including the highly inclined systems HT Cas, Z Cha, U Gem and WZ Sge (Ricketts, King & Raine 1979; Becker 1981; Cordova, Mason & Nelson 1981; Cordova & Mason 1983, 1984a; van der Woerd, Heise & Bateson 1986; Belloni *et al.* 1991; Mukai & Shiokawa 1993; Vrtilek *et al.* 1994).

Soft X-rays ($\sim$10–30 eV) have been detected in quiescence only in VW Hyi, which happens to lie in a direction with very low interstellar gas density ($N_{\text{H}} \sim 6 \times 10^{17}\,\text{cm}^{-2}$: Polidan, Mauche & Wade 1990; Belloni *et al.* 1991).

Belloni *et al.* (1991) find that the soft *and* hard X-ray emission in VW Hyi can be fitted by a single 2.17 ± 0.15 keV optically thin thermal spectrum, which gives an X-ray flux that is $\sim 1/9$ the UV + optical flux. As half the latter is contributed by the primary, this implies that the X-ray flux is only $\sim 1/4$ that from the accretion disc, and may

[1] The M_{v}(min) for WZ Sge is from Smak (1993c), corrected for inclination.

mean (equation (2.54a)) that the primary in VW Hyi is rotating rapidly. Vrtilek *et al.* (1994) find similar ($\sim 1/2$) deficiencies in other DN. Much of this effect may be due to $\dot{M}(1)$ being very much lower than $\dot{M}(\mathrm{d})$ in the outer disc, as a result of non steady state conditions (Section 3.5.4.4).

Quiescent hard X-ray emission is variable over factors of 2–10 on a time scale of minutes; in SS Cyg there are flares with decay times ~ 100 s (Jones & Watson 1992). The spectrum has been reported to be harder as the intensity increases (Fabbiano *et al.* 1981; Watson, King & Heise 1985), but this has been attributed to incorrect background subtraction (Mukai & Shiokawa 1993). Yoshida, Inoue & Osaki (1992) also find no change of hardness with intensity. In SS Cyg there is a delay of 60 ± 15 s between > 2 keV and <1 keV X-rays (King, Watson & Heise 1985), of the same order as the cooling time scale of the primary's corona (King & Shaviv 1984b; Section 2.5.4), which is found to cover 10–100% of the surface of the primary. However, Mukai & Shiokawa, having discounted the intensity–hardness correlation which favours the coronal model, prefer the compact BL model of Patterson & Raymond (1985a). Nevertheless, there is a wide range of bremsstrahlung emission and temperature from object to object, and from day to day in a given object, the reason for which is not yet clear. Mukai & Shiokawa find 6.7 keV Fe Kα emission in most DN, and 8.0 keV Fe Kβ in SS Cyg. Vrtilek *et al.* (1994) find also ~ 0.9 keV L-shell emission from highly ionized Fe.

In VW Hyi the quiescent hard X-ray flux implies $\dot{M}(1) \sim 5 \times 10^{14}$ g s^{-1}. Approximately half of the coronal flux is radiated or conducted downwards to the primary, giving a predicted surface temperature $T(1) \sim 14\,000$ K (van der Woerd & Heise 1987). This is in good accord with that found from the Lyα profile (Mateo & Szkody 1984: Table 2.8).

Typical X-ray quiescent luminosities are $L_{\mathrm{x}} \sim 10^{31}$ erg s^{-1} (Cordova & Mason 1984a), but individual observations are from a tenth to 30 times larger (Mukai & Shiokawa 1993). For example, at its brightest, SS Cyg has $L_{\mathrm{x}}(2\text{–}10\ \mathrm{keV}) \sim 1.7 \times 10^{32}$ erg s^{-1} (Jones & Watson 1992, for a distance of 75 pc: Warner 1987a). Optical observations of SS Cyg give $\dot{M}(\mathrm{d}) \sim 3 \times 10^{15}$ g s^{-1} at quiescence (Patterson 1984; Warner 1987a); for an optically thin boundary layer (Section 2.6.1.5) the expected X-ray luminosity for a slowly rotating primary (allowing for the half of the X-ray emission which is emitted towards the primary) is $L_{\mathrm{x}} = GM(1)\dot{M}(\mathrm{d})/4R(1) = 1.8 \times 10^{32}$ erg s^{-1} for $M_1(1) = 1.0$.

Mukai & Shiokawa (1993) find that DN may be the principal contributor to the $\sim 10^{38}$ erg s^{-1} 2–10 keV emission spread around the Galactic plane at low (< 2°) latitude (e.g., Warwick *et al.* 1985).

3.3 Outbursts of DN

3.3.1 *Morphology*

Until the physical nature of CVs began to emerge in the 1960s, their study was to a large extent confined to the taxonomy of their light curves. The wide range of behaviour of DN gave opportunities for the introduction of many classes and subclasses, some of which are still worth remarking upon. The subclasses U Gem, SS Cyg, CN Ori and Z Cam listed by Prager (1934) were an extension of the SS Cyg, U

Gem division used by Müller and Hartwig (1918). The latter, based on the early light curves of SS Cyg and U Gem, which had frequent irregular maxima and infrequent more regular maxima respectively, was less convincing after SS Cyg became a U Gem type in 1907–8 (and again in 1916 and at later times: Figure 1.3) and the distinction is no longer made. CN Ori is remarkable for showing almost continually varying brightness, with rarely any interval spent at quiescence. A total of seven subclasses of DN was introduced by Brun & Petit (1952), including a redefinition of U Gem and SS Cyg types. Of these, the only new subtype to be adopted permanently was that of the SU UMa stars, which show supermaxima as well as normal maxima. A subtype of WZ Sge stars was introduced to characterize those DN with particularly large and infrequent outbursts (Bailey 1979b; Downes & Margon 1981; Patterson *et al.* 1981; Downes 1990), but these have since been assimilated into the SU UMa stars (O'Donoghue *et al.* 1991). Thus the current subtypes, already described in Sections 1.2 and 2.1, are U Gem, Z Cam and SU UMa – called generically DN.

Long term visual light curves of a few dozen DN have been steadily accumulated over the past century and appear in the publications of the AAVSO and the Variable Star Sections of the RAS of New Zealand and the British Astronomical Association.[2] A variety of morphological and statistical analyses have been made of these light curves. The most important results from these are given here as properties that any complete model of DN outbursts should aim to explain.

An obvious feature of DN outbursts is their non-reproducibility: no two outbursts in a given DN are exactly alike. There are, nevertheless, some systematic effects present. In his study of SS Cyg, Campbell (1934) drew attention to four distinct types of outburst, each having a characteristic rise time but with a variety of outburst durations (Figure 3.6). The majority (64%) fall into Class A, with the most rapid rise, taking about 2 days from quiescent brightness to maximum; the slowest rises, Class D, take about 10 days. The rise time is correlated with quiescent magnitude between outbursts, Class A arising from the faintest levels at $m_V = 11.90 \pm 0.12$ through increased quiescent brightnesses for Classes B and C to $m_V = 11.64 \pm 0.30$ at Class D (Bath & van Paradijs 1983).

Despite an initial impression of a continuum of outburst durations in Figure 3.6, one of the most distinctive properties of DN outbursts is a bimodality of duration; this is in fact readily detectable in Campbell's Class A and B outbursts in Figure 3.6. The bimodality is most dramatic for the SU UMa stars, but for the moment we consider U Gem and Z Cam stars.

Figure 3.7 shows the clear-cut division of outbursts in SS Cyg into two groups of different average duration (Bath & van Paradijs 1983). ('Duration' typically is defined as the time from a chosen magnitude on the rising branch until return through that magnitude again on the descending branch.) Similar results were obtained in other studies of SS Cyg (Kruytbosch 1928; Campbell 1934; Sterne & Campbell 1934; Martel 1961; Howarth 1978a; Cannizzo & Mattei 1992), in U Gem (van der Bilt 1908; Greep 1942; Isles 1976; Saw 1982a,b) and in over a dozen other U Gem and Z Cam type stars (Petit 1961; van Paradijs 1983, 1985; Gicgar 1987). Because of irregularities in

[2] Light curves, and conclusions drawn from them, in *The Dwarf Novae* (Glasby 1970) should be treated with reservation (Miles 1976).

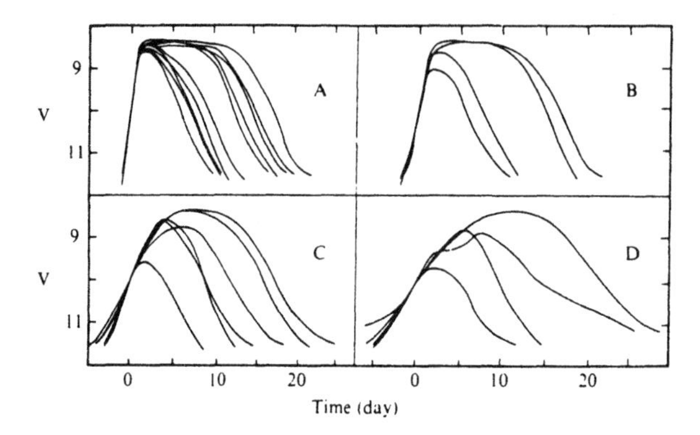

Figure 3.6 Outbursts of SS Cyg classified into four types by Campbell (1934).

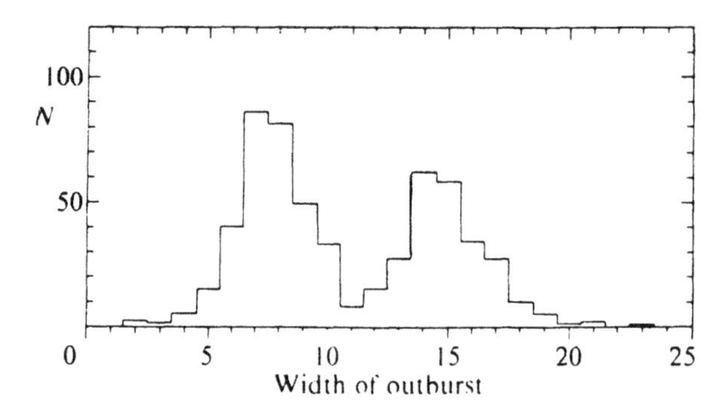

Figure 3.7 Bimodal distribution of outburst widths of SS Cyg. From Bath & van Paradijs (1983).

behaviour, any given light curve of only a few years' length may not show the bimodality very distinctly, which is probably why Szkody & Mattei (1984) could not confirm the effect in some DN AAVSO light curves of 3 years' duration.

3.3.2 Outburst Frequencies

Although not periodic, taken over a time-span of many decades the outbursts of DN show a recurrence time scale that is characteristic for each system. In a given star, the recurrence intervals may spread over a factor of 2 or 3 – e.g., 15–44 days in TW Vir and 40 – 75 days in SS Aur (Ritter 1990). Nevertheless, the mean interval T_n determined from separate long sections of light curve is a fairly stable statistic. For example, in SS Cyg the light curve for 1896–1940 gives $T_n = 51.0$ d and for 1940–85 $T_n = 47.6$ d; U Gem for 1855–1905 has $T_n = 96.5$ d and for 1905–55 $T_n = 107.6$ d, or for 1937–56 $T_n = 111$ d and 1956–77 $T_n = 112$ d (Warner 1987a).

Values of T_n are given in the following modern references: GCVS, Bateson (1979a, 1988), Szkody & Mattei (1984), van Paradijs (1985), Warner (1987a), Korth (1990), Ritter (1990). The shortest intervals between outbursts are ~ 7 d and are set by the need to distinguish between consecutive outbursts (the claim by Jablonski & Cieslinski (1992) that V1159 Ori has a 4 d interval between outbursts may change this view, but requires verification). The largest T_n are probably restricted to $\lesssim$ 80 y by the observational baseline.

Cannizzo & Mattei (1992) find in SS Cyg that $T_n = 49.5$ d and that it is normally distributed with a standard deviation σ of 15.4 d. The quiescent periods between outbursts average 38.7 d, and are also normally distributed, with $\sigma = 13.9$ d.

3.3.3 Correlations

3.3.3.1 *Statistics of Outburst Behaviour*

From early in its history, the alternation of short and long outbursts of U Gem was noted; it probably never deviated from this throughout the second half of the nineteenth century (van der Bilt 1908). SS Cyg also shows a strong tendency for short and long outbursts to alternate (Sterne & Campbell 1934), narrow following wide and vice versa at a frequency 1.45 times the random expectation (Bath & van Paradijs 1983).

In SS Cyg the amplitude of outburst is positively correlated with both the time since the previous outburst and the time until the next outburst (Sterne & Campbell 1934). It has been claimed that the 'width' of outburst is correlated only with the interval since the previous outburst, not the interval until the next, but this has been shown by Cannizzo & Mattei (1992) to be an artefact of the manner of analysis: in fact there is some correlation with both intervals. In this regard, new analyses are now warranted for other DN, for which the results so far show that in U Gem both amplitude and width are correlated with the interval since the previous outburst, not the interval until the next; in VW Hyi the same behaviour is found (Smak 1985a; van der Woerd & van Paradijs 1987); in UZ Ser and TU Men the width only, and in WW Cet the amplitude only, are correlated with the preceeding interval (Gicgar 1987). In Z Cha no such correlations are found (Kozlowska 1988). From Gicgar's analysis of 14 DN, it emerges that such strong correlations are quite rare, though it is suggestive that the most prominent correlations occur in those stars (SS Cyg, U Gem, VW Hyi) for which the observational material is most abundant.

With the exception of the few stars at the shortest known orbital periods (listed in Table 3.3) there does not appear to be any correlation between P_{orb} and T_n. Correlations connected with superoutburst behaviour are given in Section 3.6.3. SS Cyg has been tested unsuccessfully for the presence of deterministic chaos (Cannizzo & Goodings 1988; Hempelmann & Kurths 1990, 1993). Given the normal distribution of T_n (Cannizzo & Mattei 1992), the result is not surprising.

3.3.3.2 *The Kukarkin–Parenago Relation*

Kukarkin & Parenago (1934; see also Kopilov 1954; Efremov & Kholopov 1966; Kholopov & Efremov 1976) noted that a correlation exists between amplitude of outburst A_n and T_n which seemed to tie together both DN and RN. The possible application to RN will be examined in Section 5.9.4; here we consider the A_n–T_n relationship for DN.

It is obvious that if a tight correlation does exist between, for example, T_n and the average total energy emitted during a DN outburst, then that correlation will be loosened by using A_n' simply defined as $m_v(\min) - m_v(\max) = M_v(\min) - M_v(\max)$. At

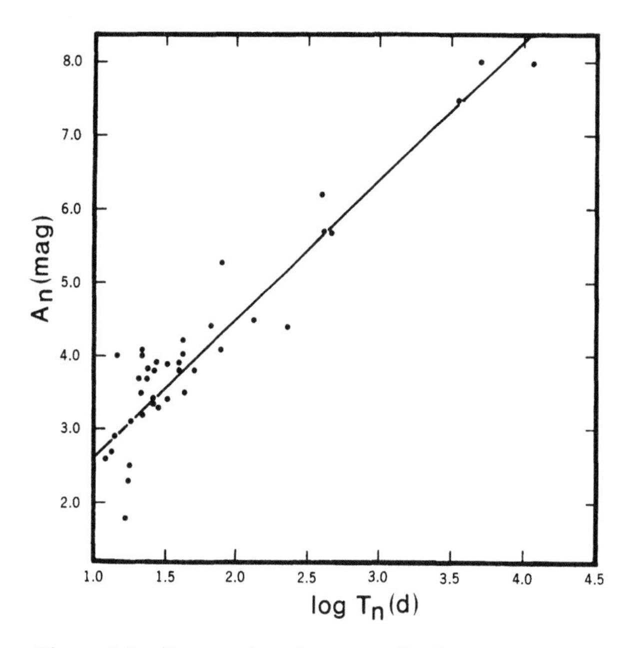

Figure 3.8 Corrected outburst amplitude versus outburst interval for DN.

maximum light M_V(max) depends on disc inclination through equation (2.63); at minimum light M_V(max) has a disc contribution similarly dependent on i but may be strongly affected by radiation from the primary, the secondary and the bright spot.

Nonetheless, a broad correlation is found between A_n', determined from GCVS magnitudes, and T_n (van Paradijs 1985). The correlation improves when m_V(min) for the disc alone is used (Warner 1987a), the relationship being

$$A_n = 0.70(\pm 0.43) + (1.90 \pm 0.22) \log T_n(\mathrm{d}) \tag{3.1}$$

largely independent of P_{orb} (see below). This is shown in Figure 3.8, with DN having $T_n < 2$ y taken from Warner (1987a) and additional points for normal outbursts of SU UMa stars from Section 3.6. Note that A_n is the true amplitude of the disc brightness, corrections having been made for contributions from other luminous sources.

Different formulations of the Kukarkin—Parenago relationship can be given. Noting (Warner 1987a) that there is no correlation between T_n and the duration $\Delta T_{0.5}$ within 0.5 mag of maximum (during which most of the outburst energy is released) equation (3.1) implies that the total energy excess radiated in the visible, relative to the luminosity of the disc at quiescence, is

$$E \mathrel{\tilde{\propto}} 10^{0.4A_n} \Delta T_{0.5} \mathrel{\tilde{\propto}} T_n^{0.76\pm 0.10}. \tag{3.2}$$

Relative to the time-averaged luminosity of the disc, $E \mathrel{\tilde{\propto}} T^{0.82\pm 0.11}$ (Warner 1987a) and for the total energy radiated $E \mathrel{\tilde{\propto}} T_n^{0.87}$ (Antipova 1987). The only significant correlation between A_n and P_{orb} is for those systems with $A_n \gtrsim 6$ mag, all of which have $P_{orb} \lesssim 1.5$ h. In particular, the stars with the shortest known P_{orb} often have very large amplitudes (Szkody 1992). However, these are also SU UMa stars in which only superoutbursts are observed. Furthermore, these stars probably have very low $\dot{M}(2)$

and low disc viscosity (Section 3.5.4.5) which can make the quiescent disc luminosity very small and lead to larger A_n.

3.3.3.3 *The Correlations of M_v(min) and M_v(mean) with T_n*

The large scatter seen in the $M_v(\mathrm{min})$–P_{orb} diagram (Figure 3.5) is caused primarily by a strong correlation between M_v(min) (at a given P_{orb}) and T_n . This will be illustrated below in a related diagram; analysis of the correlations (Warner 1987a) leads to the following equation

$$M_v(\mathrm{min}) = 7.1 + 1.64 \log T_n(\mathrm{d}\,) - 0.26 P_{orb}(\mathrm{h}). \tag{3.3}$$

In the disc instability model of DN outbursts, described in Section 3.5, most of the mass received from the secondary is stored in the disc between outbursts. The quiescent brightness, M_v(min), is therefore a measure of the mass transfer rate $\dot{M}(\mathrm{d})$ through the disc at minimum light, and not a measure of the mass transfer rate from the secondary $\dot{M}(2)$. As $\dot{M}(2)$ is the parameter that determines the overall evolutionary rate of a CV, and, in particular, is likely to be the most important factor influencing the recurrence time T_n and amplitude A_n of DN outbursts, a measure of $\dot{M}(2)$ is required. This can be approximated by the mean integrated visual luminosity of a DN over a complete outburst cycle (Patterson 1984; Warner 1987a), defining the absolute magnitude M_v(mean) as that value the system would attain if the disc actually transferred mass at the same rate that it received it. As will be seen from Section 3.4, for a Z Cam star $M_v(\mathrm{mean}) = M_v(\mathrm{standstill})$.

The $M_v(\mathrm{mean})$–P_{orb} diagram (Warner 1987a) is shown in Figure 3.9. Values of T_n have been included to show the dependence of M_v(mean) on T_n at a given P_{orb}.

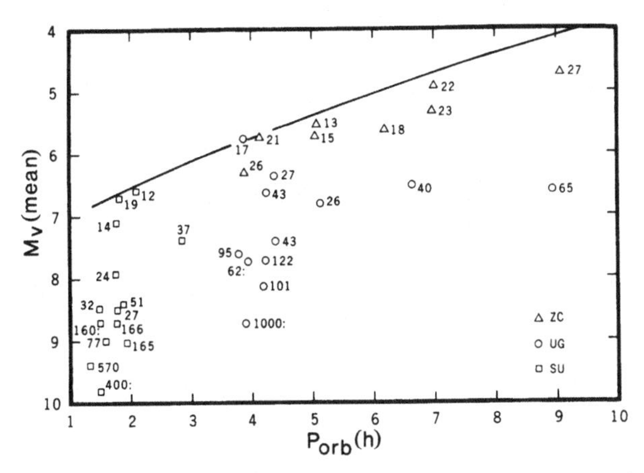

Figure 3.9 Mean (time-averaged) absolute visual magnitudes of DN accretion discs, plotted against orbital period. Subtypes (Z Cam, U Gem and SU UMa) are plotted with different symbols. Numbers adjacent to the plotted points are outburst intervals T_n(d). The solid curve is the absolute magnitude of discs with mass transfer rate $\dot{M}_{crit2}$ (Section 3.5.3.5), below which discs would be expected to be thermally unstable, leading to DN outbursts.

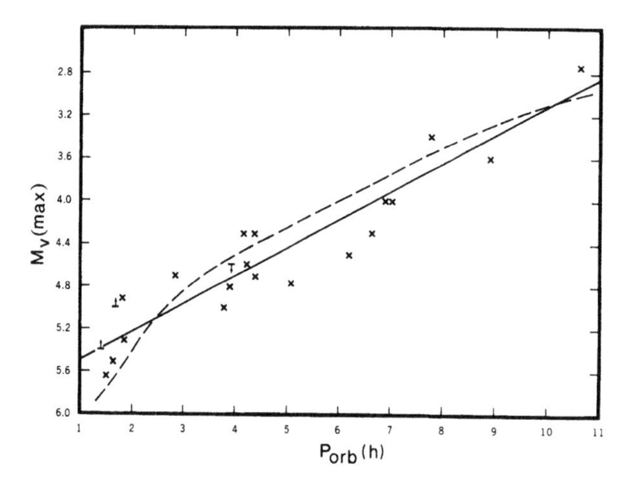

Figure 3.10 Absolute visual magnitude of DN discs at maximum of normal outbursts, plotted against orbital period. The solid line represents equation (3.4). The dashed line is the theoretical relationship, equation (3.27).

3.3.3.4 *The M_v(max)–P_{orb} Relation*

The material used for discussing M_v(min) in Section 3.2.3 can also be used for M_v(max). From the small number of DN with reliable distances available in 1981, Vogt suspected that DN have only a small range of M_v at maximum light and deduced $M_v(\max) = 4.70 \pm 0.14$ (Vogt 1981). The extended material shows (Warner 1987a) that there is indeed a restricted range of M_v(max) but there is a dependence on P_{orb}, as might be expected from the increase in area of the accretion disc with increasing P_{orb}. The observations are shown in Figure 3.10; the linear relation drawn through them has the equation

$$M_v(\max) = 5.74 - 0.259 P_{orb}(\mathrm{h}) \quad P_{orb} \lesssim 15 \text{ h}. \tag{3.4}$$

Although determined from systems with $P_{orb} < 9$ h, equation (3.4) correctly predicts M_v(max) for BV Cen, which has $P_{orb} = 14.6$ h and $M_v(\max) \approx 2.2$ (Warner 1987a). However, it fails for GK Per, which has $P_{orb} = 48$ h and $M_v(\max) \approx 1.6$ during its DN outbursts. Equation (3.4) provides a powerful means of determining distances to DN provided the inclination is known sufficiently well that equation (2.63) can be applied.

The m_v(max) listed in Tables 3.1–3.4 are mean values. Above the orbital period gap the maximal magnitudes of fully-developed outbursts do not have a large spread (e.g., Figures 1.3 and 1.4), but among the SU UMa stars there is a wide range of maximal magnitudes (e.g., Figure 3.35).

3.3.3.5 *The Decay time–P_{orb} and Outburst width –P_{orb} Relations*

As seen in Figure 3.6 for SS Cyg, the majority of outbursts in a given DN have well-defined and apparently almost invariant rates of decline, independent of the duration at maximum. Bailey (1975a) found that the decay time τ_d, expressed as a time scale d mag^{-1}, is well correlated with P_{orb}. This was confirmed for additional DN by van Paradijs (1983). The correlation is shown in Figure 3.11, with τ_d from Tables 3.1–3.3

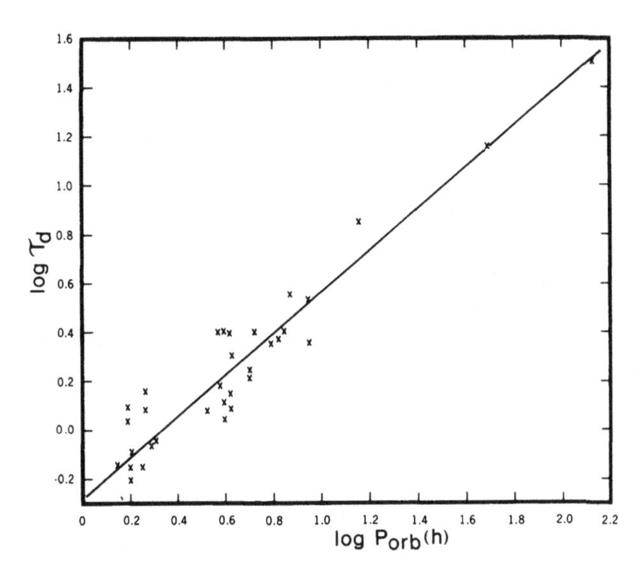

Figure 3.11 Decay time scale τ_d (d mag^{-1}) versus orbital period for DN outbursts. The solid line represents equation (3.5).

(most of these derive from van Paradijs (1983) or Szkody & Mattei (1984), with additional meaurements by the author or isolated values from the literature). The linear relation shown has the equation

$$\tau_d = 0.53 P_{orb}^{0.84}(h) \quad \text{d mag}^{-1}. \tag{3.5}$$

The time scales for outburst rise, τ_r, which are shorter and more difficult to measure accurately than τ_d, also show a correlation with P_{orb}. The measured τ_r (van Paradijs 1983; Szkody & Mattei 1984; + measurements by the author) indicate a somewhat steeper relationship with P_{orb}: for $P_{orb} < 2.1$ h the average $\overline{\tau}_r \sim \frac{1}{3}\overline{\tau}_d$, whereas for $P_{orb} >$ 3 h $\overline{\tau}_r \sim \frac{1}{2}\overline{\tau}_d$ and for the systems of longest periods (BV Cen, GK Per and V1017 Sgr; the low amplitude, $\tau_r \sim 35$ d, $T_n \sim 42$ y outbursts in V630 Cas (Honeycutt *et al.* 1993) suggest that it belongs to this category of DN, with $P_{orb} \sim 5$ d) τ_r is comparable to τ_d. Including the last systems gives

$$\tau_r = 0.14 P_{orb}^{1.15}(h) \quad \text{d mag}^{-1}. \tag{3.6}$$

Similarly, the outburst width $\Delta T_{0.5}$ is correlated with P_{orb}, and, including the long period systems, gives

$$\Delta T_{0.5} = 0.90 P_{orb}^{0.80}(h) \quad \text{d}. \tag{3.7}$$

3.3.4 *Photometric Behaviour*

During DN outbursts brightness variations are seen on a variety of time scales. Apart from the large scale changes of the outburst itself, there are modulations with various degrees of coherence, ranging from the most rapid (10–30 s) and coherent dwarf nova oscillations (DNOs: Section 8.6), through quasi-period oscillations (QPOs, 1–30 min) to orbital modulations (Section 3.3.4.2). The QPOs are probably mostly of disc origin, though some may involve effects of magnetically-controlled accretion. They are discussed collectively in Section 8.7.

3.3.4.1 Colour Variations

Comprehensive photoelectric photometry has been carried out for only a few outbursts of a few DN. Compilations of optical and IR magnitudes and colours obtained sporadically are given by Mumford (1967a), Vogt (1981, 1983c), Bruch (1984) and Echevarria (1984). Corrections for interstellar reddening can be made from values deduced from IUE flux distributions (Hassall 1985b; Verbunt 1987) or other means (Bruch 1984); in general the corrections are small for DN, but a few with $E_{B-V} > 0.1$ are known.

During an outburst cycle a DN follows a loop in the two-colour diagram (VW Hyi: Haefner, Schoembs & Vogt 1979, Bailey 1980, van Amerongen *et al.* 1987; OY Car: Vogt 1983b; SS Cyg: Bailey 1980, Hopp & Wolk 1984; Giovannelli & Martinez-Pais 1991; WX Hyi: Kuulkers *et al.* 1991b), the shape and amplitude depending partly on the contribution from the secondary. As seen in Figure 3.12, almost the entire loop is traversed during the rise to maximum, the system at first becoming very much redder in both B–V and U–B and roughly midway on the rise changing direction to become

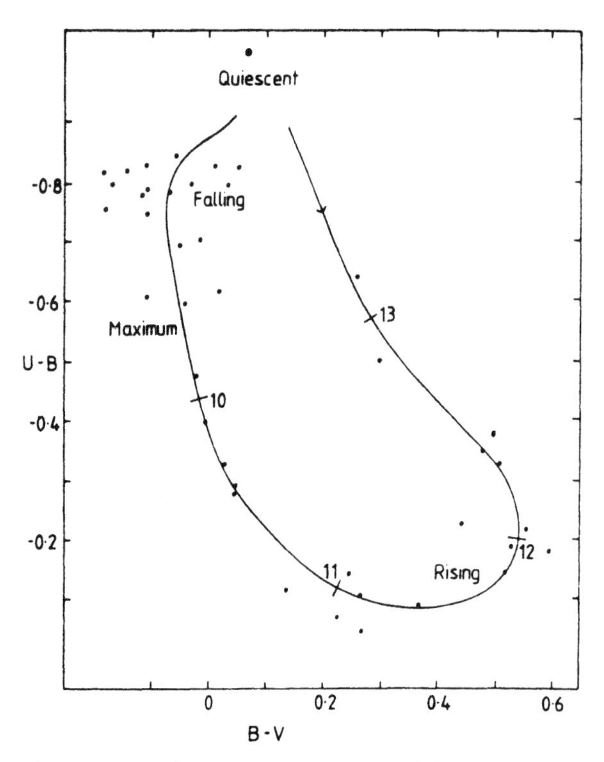

Figure 3.12 The two-colour diagram of VW Hyi during an outburst cycle. The line is the mean relationship and is marked with the V magnitude at each stage of the outburst. The loop is thus followed in a clockwise direction. From Bailey (1980).

bluer again. These two-colour wide loops only occur if the outburst rise is rapid. For slow rises (e.g., types C and D of SS Cyg: Figure 3.6) the system does not become greatly redder on the rise: U,B and V rise almost together (Bailey 1980), leading to very narrow loops (AH Her:Smak 1984a, based on photometry by Szkody (1976b) and Wargau, Rahe & Vogt (1983); RX And: Echevarria (1984)). Similarly, all declines are sufficiently slow that no large (optical) colour variations occur.

Almost all DN at maximum of outburst have colours in the range $(B-V)_0 = 0.00 \pm 0.10$, $(U-B)_0 = -0.80 \pm 0.15$ (Warner 1976b), with a tendency for the systems with longer P_{orb} to be redder in colour, expressed by the formula (Vogt 1981)

$$(B - V)_0 = 0.0293 P_{orb}(h) - 0.130 \tag{3.8}$$

which shows the effect of the increasing area of the cooler outer disc regions in the more widely separated systems. These colours match quite well the integrated colours of model discs discussed in Section 2.6.1.1, but depart significantly from observed colours of stars (because of the larger Balmer discontinuity and stronger absorption lines in the latter).

Some satellite photometry in the UV of DN during outburst was carried out (e.g., Wu and Panek 1982), but this has been superseded by observed flux distributions which are described in Section 3.3.5.

3.3.4.2 Luminosity Variations

The gradual diminution of amplitude of both the eclipse and the orbital hump during the rise to maximum in U Gem was one of the principal clues used to construct the model for that star (Smak 1971a, Warner & Nather 1971). Because of the unpredictability of outbursts, observations made during the early rise of outburst are rare. The only successes have been in U Gem (Mumford 1971), OY Car (Vogt 1983b), CN Ori (Mantel *et al.* 1988), VW Hyi (Vogt 1974; Warner 1975b; Haefner, Schoembs & Vogt 1979) and WX Hyi (Kuulkers *et al.* 1991b); in all cases the orbital hump remained of constant amplitude in intensity units even while the system brightened up a factor of 5 or 10. Observations later on the rise have been made for U Gem (Krzeminski 1965), CN Ori (Schoembs 1982; Mantel *et al.* 1988) and VW Hyi (Vogt 1983a). The orbital hump is detectable during outburst at its quiescent amplitude in VW Hyi (Haefner, Schoembs & Vogt 1979) and in CN Ori (Mantel *et al.* 1988). In other DN, there is no orbital hump visible during maxima and supermaxima, which is quantitatively consistent with it remaining at its quiescent amplitude (e.g., Cook (1985c) for Z Cha). Eclipse mapping of OY Car (Rutten *et al.* 1992) shows no increase in luminosity of the bright spot during rise to a normal maximum.

The importance of these observations is that they show no evidence for an increase in the rate of mass transfer from the secondary, $\dot{M}(2)$, during the rise to maximum. Nor is there any evidence for a pulse of increased mass transfer preceding outburst (CN Ori: Mantel *et al.* 1988; SS Cyg: Honey *et al.* 1989; Z Cha: Vogt 1983b). From emission-line mapping of IP Peg in outburst, Marsh & Horne (1990) find that emission from the stream is no stronger in outburst than at quiescence.

An exception to the rule that orbital humps are not generally seen is an observed enhancement by a factor ~20 in amplitude in a normal outburst of VW Hyi that preceded a superoutburst (Section 3.6.1) by only one day. VW Hyi at this time was ~0.6

mag brighter than its usual maximum in a normal outburst. This constitutes indirect evidence for increased $\dot{M}(2)$ during the maximal development of a superoutburst.

There is observational evidence for a short lived pause on the rise to outburst. Walker & Marino (1978) found an initial rise at -0.25 mag h^{-1}, then a $\sim$2 h pause about 1.5 mag above quiescence, with a more rapid rise at -0.5 mag h^{-1} thereafter. A similar effect is seen in WX Hyi (Kuulkers *et al.* 1991b) and possibly in TW Vir (Mansperger & Kaitchuck 1990). During maxima, DN show subdued flickering on time scales of tens of seconds, but have slow variations on longer time scales often with amplitudes of $\sim$0.15 mag or more.

In an outburst of Z Cha the H-band light curve shows the presence of ellipsoidal variations of the secondary. The disc, which contributed $\lesssim$30% of the IR luminosity at quiescence, brightens by a factor $\sim$7 and contributes $\sim$75% at maximum. The flux-ratio diagram (Figure 2.21) shows that the disc gas contributing the IR is optically thick and at (6–8) $\times$ 10^3 K (Harlaftis *et al.* 1992). However, the ratios change at mideclipse and show that the outermost parts of the disc (left uneclipsed) have considerable optically thin emission.

The IR light curve of U Gem (Figure 1.26) is considerably distorted near the end of decline from an outburst; this is caused by increased luminosity of the outer cool regions of the disc, not by heating of the secondary (Berriman *et al.* 1983).

3.3.4.3 Eclipse Profiles and Accretion Disc Radii

The rapid change in eclipse profile as the disc brightens during an outburst is shown for OY Car in Figure 3.13 (Vogt 1983b). In this case it is evident that the brightening starts at the outer edge of the disc and propagates inwards within a few hours (Smak (1971a, 1984c) found the same for U Gem). MEM deconvolution of these eclipses (Rutten *et al.* 1992) shows the brightness temperature of the outer disc rising from 7000 K to 11 000 K over the two orbital periods. At maximum light the eclipse is almost perfectly symmetrical. The small gradient as the disc enters eclipse produces large uncertainty in the eclipse contact times (Section 2.6.2) and consequently in the measurement of the radius of the luminous disc. At a maximum in Z Cha Warner & O'Donoghue (1988) found from eclipsing mapping that $r_d/a \approx 0.48$.

The variation in disc radius during an outburst cycle can be followed with precision as soon as eclipse contact times for the bright spot become identifiable on the declining branch of an outburst. Paczynski (1965b) deduced spectroscopically and from changes in eclipse width that the disc in U Gem shrinks between outbursts. From eclipse profiles, Smak (1971a, 1976, 1984c) extracted the radius variations shown in Figure 3.14.

In Z Cha very similar results are obtained (O'Donoghue 1986; Zola 1989; Smak 1991a), with r_d/a shrinking from 0.40 shortly after maximum to $\sim$0.26 just before the next outburst (i.e., r_d is a function of the fractional time to the next outburst, not simply the time since the last outburst). In HT Cas, Patterson (1981) found a 70% increase of r_d at the beginning of an outburst. Wood *et al.* (1989b) found $r_d/a = 0.313$ for OY Car in quiescence, and in IP Peg Wood *et al.* (1989a) deduced that r_d/a decreases from 0.33 a few days after maximum to 0.26 at the end of quiescence (see also Harlaftis *et al.* (1993d) and (Wolf *et al.* 1993)). In U Gem and IP Peg the disc radii follow an $\exp(t/16\ \mathrm{d})$ decay. In TY PsA the inclination ($\sim$65°) is such that eclipses of the bright

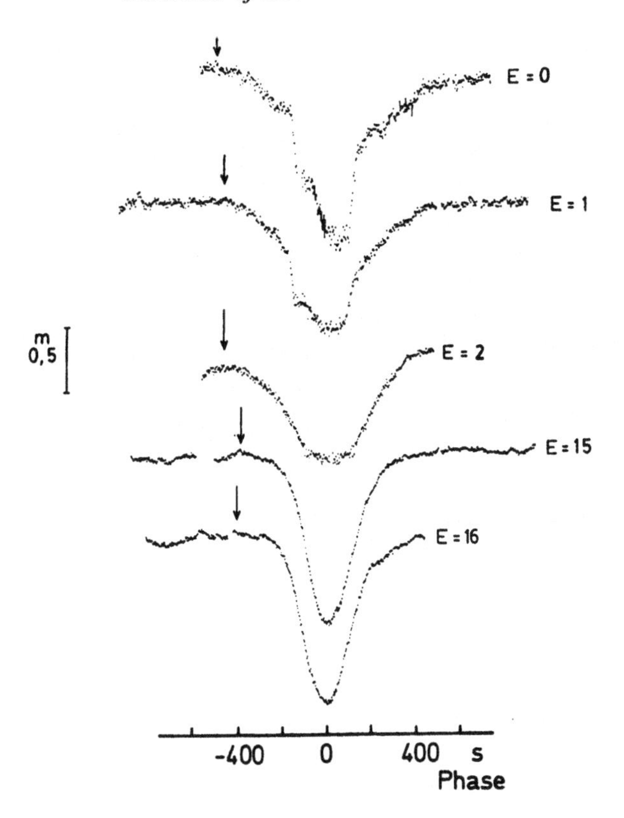

Figure 3.13 Eclipse profiles of OY Car at the start of a rise ($E = 0, 1, 2$) and at maximum ($E = 15, 16$) of outburst. Arrows mark the ingress of disc eclipse. Adapted from Vogt (1983b).

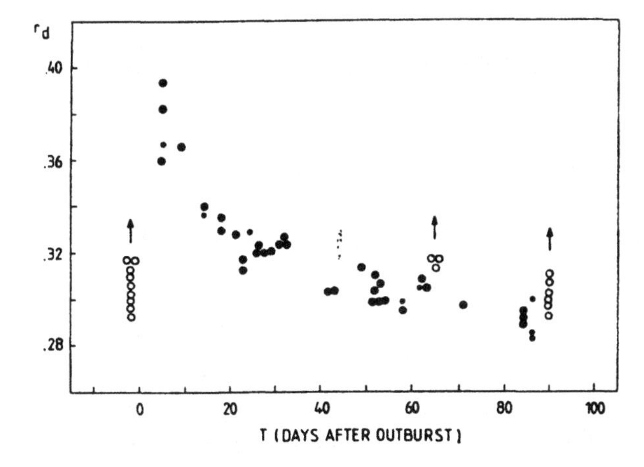

Figure 3.14 Variation of disc radius r_d in U Gem, as a function of days after outburst. r_d is in units of the orbital separation a. Open circles are radii measured early on the rise of outburst. From Smak (1984c).

spot are briefly visible during a time of expanded disc radius on decline from outbursts (Warner, O'Donoghue & Wargau 1989).

To the quiescent disc radii determinations listed above may be added $r_d/a = 0.31$ for WZ Sge (Smak 1993c) and 0.29 for U Gem (Figure 3.14). It appears, therefore, that $r_d/a \sim 0.30$ for DN over a wide range of q. This is much smaller than the tidally limited radius (equation 2.61) but is a factor of 2–3 larger than the circularization radius r_r (equation (2.19)).

Because the normal outbursts of most eclipsing DN are of short duration ($\sim$ 2 d), high speed photometry at such times is quite rare. Section 2.6.4 gives an example of eclipse mapping of Z Cha about 0.5 mag down the declining branch from maximum (Cook 1985a; Horne & Cook 1985). The intensity and temperature distributions agree well with those predicted for a steady state disc (Section 2.5.1), but the small departures are significant and of the kind expected for a disc evolving during outburst (Section 3.5.4.4).

Eclipse mapping during another outburst of Z Cha, probably caught at an earlier stage than the one analysed by Horne & Cook (because in the latter, but not in the former, a bright spot can be detected) apparently shows a bright outer rim (Warner & O'Donoghue 1988). This arises in the standard MEM deconvolution process because the eclipse is observed to be relatively shallow although quite narrow. A similar effect is seen in a normal outburst of OY Car (Rutten *et al.* 1992) and in superoutbursts of OY Car (Naylor *et al.* 1987) and Z Cha (Kuulkers 1990). The basic characteristic is that an unexpected $\sim$15% of the outburst luminosity, which develops on the rise, remains visible at mideclipse. Rutten *et al.* (1992) point out that this could be material above the plane of the disc, as in a wind, but that it is unlikely that such a large optical luminosity could be developed by a wind. An alternative explanation (Harlaftis *et al.* 1992), that the effect is caused by obscuration of the inner disc by extended vertical structure on the disc rim, is not supported by detailed modelling by Rutten *et al.* (1992) nor by the absence of any orbital modulation of the UV flux (Harlaftis *et al.* 1992).

However, a similar effect seen in an HST observation of Z Cha at maximum (Wood *et al.* 1993) can be removed if the disc has a half opening angle of 8° (which, with $i = 82°$, leaves the primary barely visible). This is much larger than predicted for a steady state disc (equation (2.52)), but the MEM-deduced brightness profile is less steep than that for an equilibrium disc. This is a first indication that DN discs thicken greatly during outburst.

There is obviously a need to follow the development of disc intensity and temperature distributions throughout an entire outburst of an eclipsing DN. This will need extraordinary cooperation in a multiwavelength, multilongitude effort to observe the unpredictable.

3.3.4.4 Optical Polarization

The predictions of relatively large polarization in the UV of high inclination DN (Section 2.6.1.6) have not yet been followed up observationally. In the visible, linear polarization measurements have been made over the outburst cycles of SS Cyg, RX And and U Gem (Krzeminski 1965; Szkody, Michalsky & Stokes 1982). No variations $\gtrsim$ 0.1% were found, but both SS Cyg and RX And possess steady linear polarization of $\sim$ 0.1–0.4% with a wavelength dependence unlike normal interstellar polarization. Less complete coverage of AH Her shows ~0.4% polarization. Only RX And showed possible variation with orbital phase. Earlier, less sensitive observations gave null results for SS Cyg, U Gem and Z Cam (Kraft 1956; Krzeminski 1965; Belakov & Shulov 1974).

Cropper (1986b) found linear polarization ~ 0.6–0.8% in V442 Cen, HL CMa, TY PsA and WZ Sge, but nothing significant in Z Cha. He did not confirm circular polarization in HL CMa previously reported by Chlebowski, Halpern & Steiner (1981). The observed large linear polarizations are greatly in excess of what is predicted for scattering in a disc and are therefore almost certainly interstellar.

Schoembs & Vogt (1980) found ~0.02% linear polarization near supermaximum in VW Hyi, increasing steadily to ~0.09% on the subsequent decline. This variation is in the correct direction but an order of magnitude greater than predicted by Cheng *et al.* (1988).

3.3.5 Spectra During Outburst

3.3.5.1 Line Spectra

The transition from emission-line spectrum at quiescence to absorption-line spectrum at maximum, noted by the early observers (Section 1.4.1) is illustrated in Figure 3.15. The emission-lines, prominent at minimum, are gradually overwhelmed by the increasing continuum and the development of broad absorption troughs that depress their bases. At maximum light in most DN emission-line cores appear, with narrower widths than the emissions at quiescence. In IP Peg, however, no absorption lines develop during outburst (Piché & Szkody 1989), which may be because its high inclination reduces the absorption-line contribution from the disc (Marsh & Horne 1990), and during a Type B outburst in SS Cyg, although the absorption lines developed on the rise, they were absent at maximum light (Horne, la Dous & Shafter 1990). The Balmer decrement is much steeper in the emission than the absorption lines,

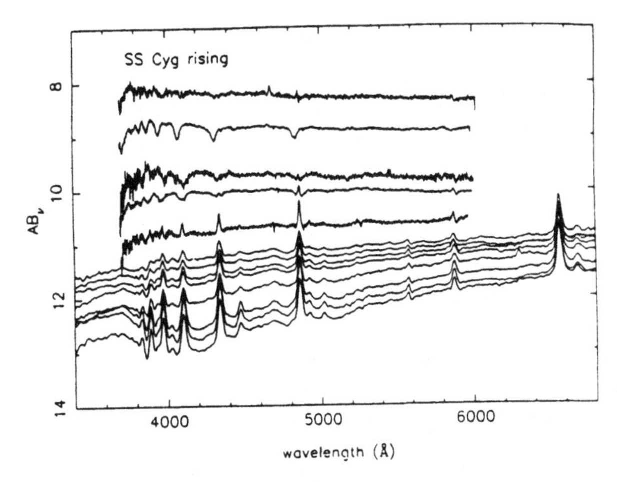

Figure 3.15 Spectral changes in SS Cyg from quiescence (lowest spectrum) to maximum of outburst (uppermost spectrum). From Horne (1991).

which can give Hα entirely in emission and high series members in absorption (e.g., KT Per: Ratering, Bruch & Diaz 1993).

The principal studies of the optical spectral development during outbursts are RX And (Clarke & Bowyer 1984; Kaitchuck, Mansperger & Hantzios 1988; Szkody, Piché & Feinswog 1990), YZ Cnc (Harlaftis *et al.* 1994b), SS Cyg (Hessman *et al.* 1984; Hessman 1986; Clark, Capel & Bowyer 1984; Horne, la Dous & Shafter 1990), IR Gem (Feinswog, Szkody & Garnavich 1988), VW Hyi (Schoembs & Vogt 1981; Schwarzenberg-Czerny *et al.* 1985), IP Peg (Piché & Szkody 1989; Marsh & Horne 1990), SU UMa (Harlaftis *et al.* 1994c), and AR And, AH Her and TW Vir (Szkody, Piché & Feinswog 1990; Mansperger & Kaitchuck 1990). These studies, and others that are less complete, show a variety of behaviour from star to star, but repeat well in different outbursts of the same object.

In SU UMa, YZ Per, VW Hyi and TY PsA the Balmer emission-line flux appears almost constant from quiescence through outburst and back to quiescence (Warner, O'Donoghue & Wargau 1989; van Paradijs *et al.* 1994). In KT Per the Hα flux increases by only a factor ~ 2 at maximum, when the system has brightened by 23 times; part of this is due to an additional emission-line source in the disc, located at $\alpha \sim 120°$ (Rattering, Bruch & Diaz 1993) and possibly caused by enhanced stream overflow.

The absorption lines (principally Balmer plus HeI λ4471) at maximum have widths 50–100 Å, similar to those of the emission lines at quiescence. Note that both H and He lines appear in absorption, unlike the DA/DB separation in white dwarfs (though a few hotter stars do show both elements together). In SS Cyg, the widths of the Balmer absorption lines remain surprisingly constant (Hessman 1986) considering the large scale changes that are taking place (e.g., a cooling transition front: Section 3.5.4.1). Clarke & Bowyer (1984) find that the absorption-line profiles are better fitted by model disc profiles than by white dwarf lines. In contrast, in the short period systems SU UMa and YZ Cnc the broad absorption lines disappear rapidly after maximum, yet in VW Hyi they persist well down the decline (Schwarzenberg-Czerny *et al.* 1985; Harlaftis *et al.* 1993c).

The location of the narrow emission cores at maximum can be studied by comparing the phases of their RV curves with those of the primary and secondary. In a number of DN (see Section 2.9.6) there is a component originating on the heated face of the secondary star. For most systems, however, the narrow emission is in phase with the accretion disc and must originate in the outer regions of the disc – probably due to complete hydrogen ionization of the chromosphere in the inner region. In SS Cyg, the CaII lines, which are in emission and double peaked with a separation of 300 km s^{-1} at quiescence, appear at maximum as doubled absorption lines with a separation of 800 km s^{-1} (Walker & Chincarini 1968). When observed, the HeII λ4686 and λ4650 emissions are generally wider (e.g., 30% wider in HL CMa: Wargau *et al.* 1983) than the Balmer emission, in accord with the more energetic conditions in the inner regions. However, the HeII emission is usually singly-peaked, probably due to a wind component (Piché & Szkody 1989).

During the decline phase, the emission components broaden back to their width at quiescence (e.g., Clarke, Capel & Bowyer 1984; Harlaftis *et al.* 1994b). In the last stages of decline, and for a few days into quiescence, there is evidence for additional heating, probably from the primary. Thus in IP Peg the absorption spectrum of the secondary is unexpectedly visible even 1 mag above quiescence and there are HeI emission lines from its chromosphere (Martin *et al.* 1989); unusually strong HeII and λ4650 emission continues into quiescence (Hessman 1989a). In VW Hyi, FeII lines are unusually strong at the end of decline (Schoembs & Vogt 1981). Echevarria *et al.* (1989) found that at the end of an SS Cyg outburst the secondary had a K2 spectral type, instead of its usual quiescent K5 type.

The appearance of the UV spectrum at maximum in DN is strongly correlated with inclination (la Dous 1991). For low inclination systems there is an almost pure absorption spectrum which grows by reversal of the strongest emission lines seen in quiescence (Table 3.5). The exception is P Cygni emission in most stars, seen in CIV λ1550, sometimes also in SiIV λ1400 and rarely in NV λ1240. The overall strength of the absorption lines diminishes at higher inclinations, until for $70 \lesssim i \lesssim 80°$ there is an almost continuous spectrum at maximum. For $i \gtrsim 80°$, however, there are strong emission lines. The same behaviour is seen in NL spectra (Section 4.2.1).

On the rising branch of outbursts of CN Ori and RX And, Cordova, Ladd & Mason (1986) noted that the flux at the bottom of the absorption lines stays almost constant during outburst and has the same distribution as the continuum at quiescence. This indicates that the rising UV continuum is associated with wind material, which is optically thick in resonance lines and is seen projected against the quiescent continuum. In Z Cha the wind origin of the UV lines is shown by the shallowness (10–30%) of their eclipse (Harlaftis *et al.* 1992). The appearances of the UV spectra in early and late decline of EK TrA are seen in Figure 3.16. In the FUV a wealth of absorption lines appears in Z Cam at maximum, including HeII, CIII, CIV, NIII–V, OIV–VI, SiIII, SVI and PV (Figure 3.17).

3.3.5.2 Flux Distributions during Outburst

Simultaneous UV and optical (plus sometimes IR and X-ray) observations of the changing flux distributions in DN as they rise to maximum have revealed two distinctively different types of behaviour (Smak 1984a,d, 1987). These correspond to the fast or slow rises seen at optical wavelengths, which, for example, separate Class A or B outbursts from Class D in SS Cyg (Figure 3.6) or lead either to wide loops or narrow loops in the two-colour diagram (Section 3.3.4.1).

In Type A, with fast optical rise and wide two-colour loops, the system brightens at longer wavelengths first, with shorter wavelengths delayed progressively. The delay from the optical region to 1000 Å is ~ 5–15 h; the delayed rise in FUV flux from Voyager observations of a rapidly rising outburst in SS Cyg is illustrated in Figure 3.18. An illustration of the evolution of the flux distribution during the rise to outburst in VW Hyi is given in Figure 3.19.

In Type B, with a slower rise and a narrow two-colour loop, the rise is almost simultaneous at all wavelengths, with at most only a small delay between optical and UV. Figure 3.15 shows spectra of SS Cyg during a slow rise (Horne 1991).

On the declining branches, in both types of outburst, fluxes at all wavelengths fall simultaneously (Pringle & Verbunt 1984), at first approximately equally, but with the UV declining fastest in the late stage as fluxes approach quiescent values (Figure 3.19; see also Figure 1 of Szkody (1977)).

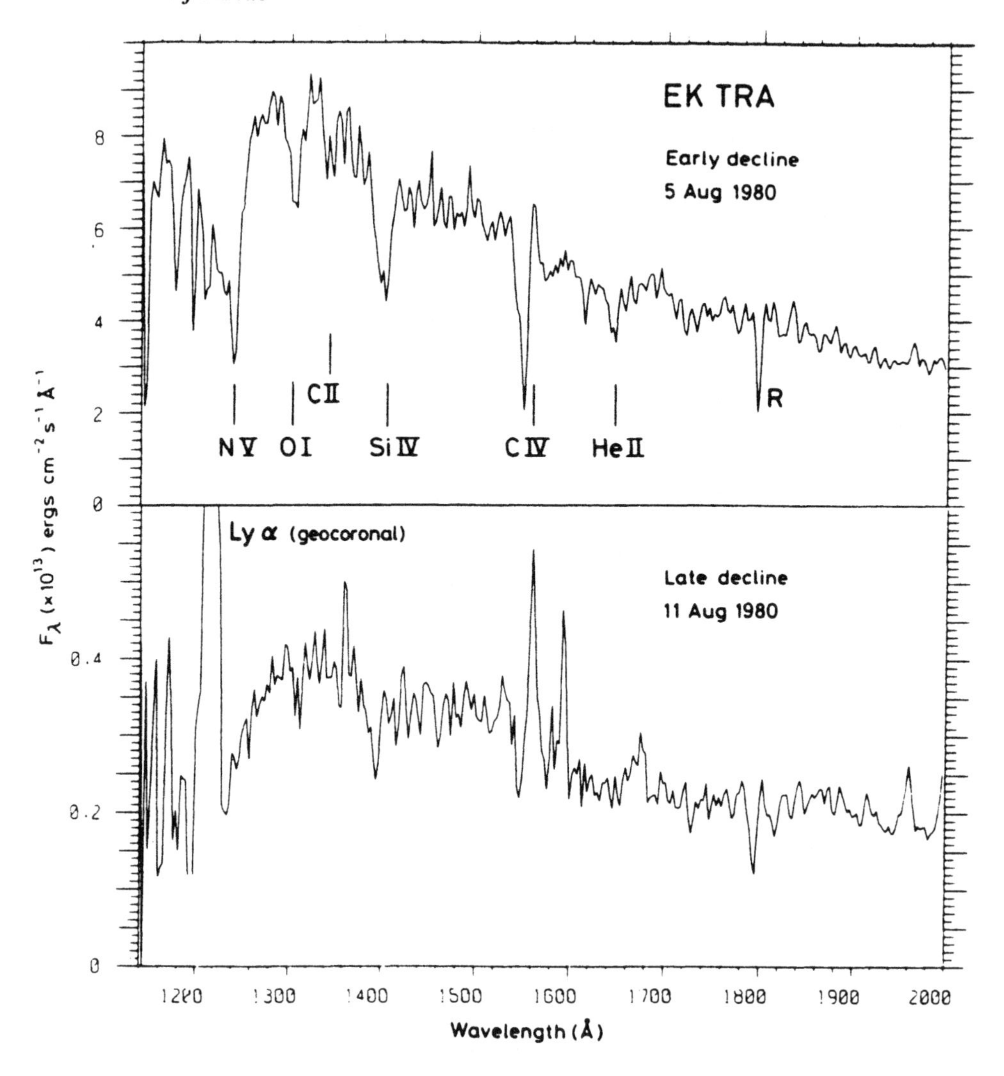

Figure 3.16 UV spectra of EK TrA on decline from a superoutburst, showing the change from an absorption-line spectrum early in decline to an emission spectrum late in decline. From Hassall (1985a).

Table 3.6 provides references to multiwavelength studies (which include the UV) of DN during outburst rise or fall. An important conclusion (e.g., Pringle & Verbunt 1984, Verbunt 1987) concerns the repeatability of development of the UV flux distribution during outbursts in a given DN, for a given kind of outburst (Type A, Type B, superoutbursts) the UV spectral distributions evolve similarly.

In Figure 2.16 an example is given of the flux distribution in a DN at maximum of outburst (SS Cyg: Kiplinger 1979a), fitted to a disc model. UV continua have commonly been fitted to a $F_\lambda \propto \lambda^\alpha$ distribution; results for 21 DN at maxima are given by Hassall (1985b) and a compilation of results for 10 DN is given by Szkody (1985a). With few exceptions the values of α lie within ± 0.3 of the value $\alpha = -7/3$ expected of

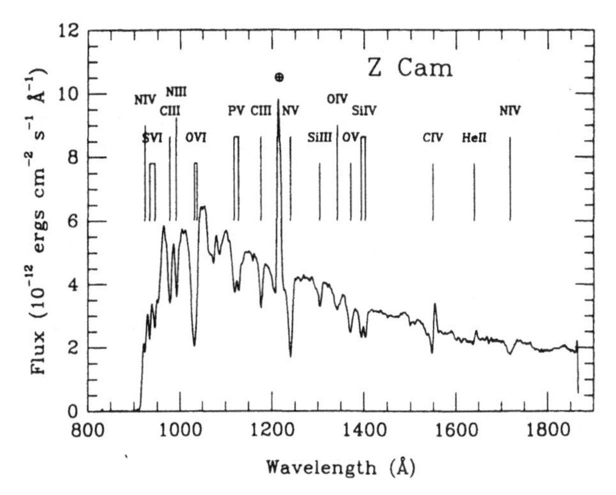

Figure 3.17 FUV spectra of Z Cam in outburst. The Lyα emission line is geocoronal. From Long *et al.* (1991).

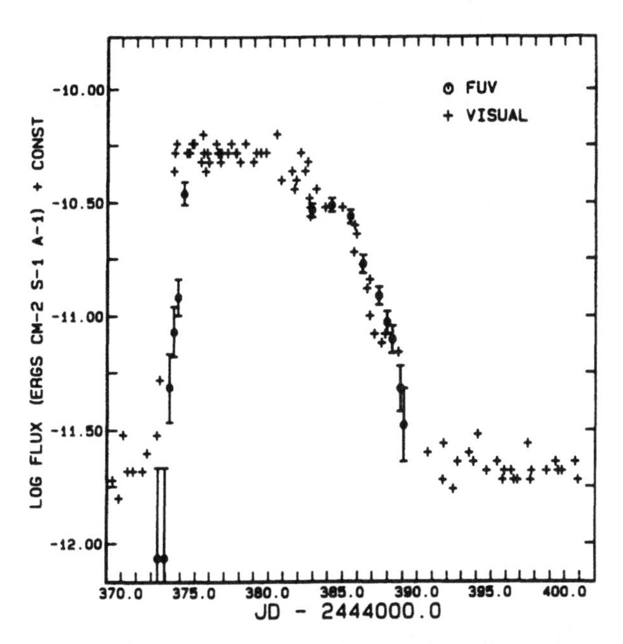

Figure 3.18 Outburst light curves of SS Cyg in the FUV from Voyager and in the visual from the AAVSO observers. Optical rise occurs $\sim$ 0.5 d before the FUV rise. From Cannizzo, Wheeler & Polidan (1986).

an infinitely large steady state disc (Section 2.5.1). Verbunt's (1987) rediscussion of all IUE spectra shows that DN at maxima closely resemble NL systems, and have $\alpha \approx -2$, equivalent to F_ν = const, but usually require slightly different values of α in the short and long wavelength regions of IUE operation.

There is strong evidence for great homogeneity among DN at maximum of outburst:

Table 3.6. *Observations of the Evolution of Flux Distribution during Outburst.*

Star	Outburst Type	References
RX And	A	1–4
DX And	B	5
SS Aur	A?	4,6
Z Cha	A	7
SS Cyg	A	8–10
	B	4,11
EM Cyg	B	4,6
U Gem	A	8
AH Her	B?	2,3
VW Hyi	A	4,12–17
WX Hyi	A	4,14,18
CN Ori	B	4,19
SU UMa	A	4,6

References: 1 Szkody 1976b, 1981a, 1982, 1985b,c; 2 Verbunt *et al.* 1984; 3 Pringle & Verbunt 1984; 4 Verbunt 1987; 5 Drew, Hoare & Woods 1991; 6 Wu & Panek 1983; 7 Harlaftis *et al.* 1992; 8. Polidan & Holberg 1984; 9 Cannizzo, Wheeler & Polidan 1986; 10 Jones & Watson 1992; 11 Horne 1991; 12 Hassall *et al.* 1983 ; 13 Schwarzenberg-Czerny *et al.* 1985; 14 Hassall, Pringle & Verbunt 1985; 15 Verbunt *et al.* 1987; 16 Polidan & Holberg 1987; 17 van Amerongen *et al.* 1987; 18 Kuulkers *et al.* 1991b; 19 Pringle, Verbunt & Wade 1986.

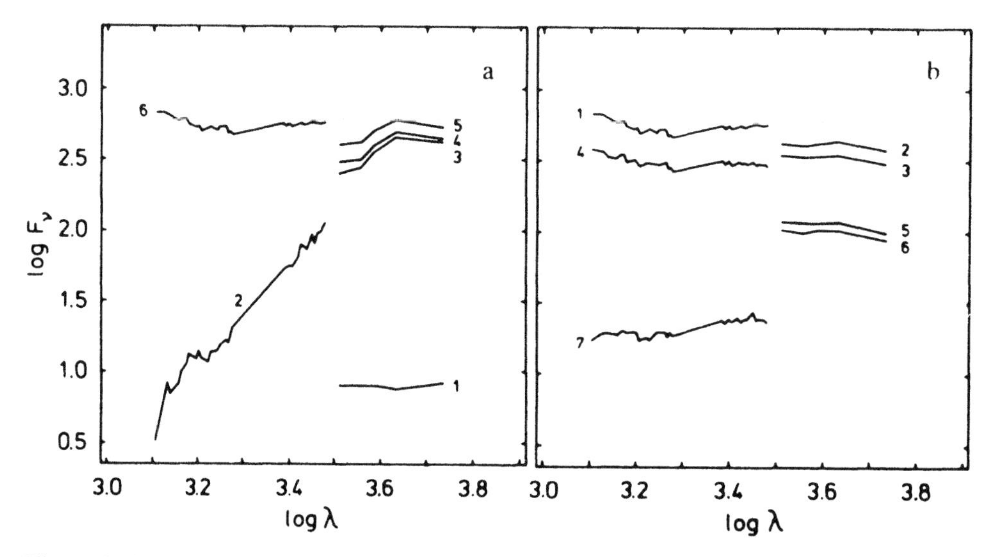

Figure 3.19 Spectral energy distributions of VW Hyi during (a) rise and (b) decline. The numbers are in (unequally spaced) chronological order. Wavelength is in Å. Adapted from van Amerongen *et al.* (1987).

la Dous (1991) finds that of 32 objects, 80% show essentially identical IUE fluxes over the range 1200–3000 Å, independent of inclination. Different outbursts of a given DN show similar repeatability. The overall flux distribution at maximum is quite well fitted, both in the UV and optical, by the continua of B2–3 V–III stars, corresponding to $T_{\rm eff} = 20\,000 \pm 2000$ K. This matches the U–B colour of DN discs at maximum, but is about 0.20 mag redder than the B–V colour (Section 3.3.4.1).

The current inability of model spectra to produce fits to observed fluxes over a wide range of wavelengths (Section 2.6.1) and the sensitivity of the UV flux ratios to $M(1)$ (Verbunt 1987) preclude confident estimation of $\dot{M}$(d) for DN at outburst. Fits to Williams & Ferguson (1982) disc models give $\dot{M}(\mathrm{d}) \sim 10^{-7} \mathrm{M}_\odot\ \mathrm{y}^{-1}$ (Szkody 1985a), but models with realistic atmospheres return values an order of magnitude lower (Hassall 1985b). However, the situation for DN outburst discs is apparently the same as for NL discs: the flux gradient and the absolute UV flux cannot be simultaneously fitted – the former gives $\dot{M}(\mathrm{d}) \sim 10^{-10} \mathrm{M}_\odot\ \mathrm{y}^{-1}$ and the latter $\sim 10^{-8} \mathrm{M}_\odot\ \mathrm{y}^{-1}$ (Wade 1988). Of these, the latter is preferable as doing the least violation to total flux requirements (note that since $\lambda F_\lambda \,\tilde{\propto}\, \lambda^{-4/3}$, most of the bolometric flux from a disc is emitted at short wavelengths).

In working towards using total fluxes, observations at the shortest wavelengths are desirable. The Voyager FUV results for U Gem, SS Cyg and VW Hyi in outburst (Polidan & Holberg 1984, 1987; Polidan & Carone 1987; Polidan, Mauche & Wade 1990) and the Astro-1 spectrum of Z Cam at maximum (Long *et al.* 1991) bring this one step closer. The flux distributions for U Gem and SS Cyg are shown in Figure 3.20. It is clear that the fluxes do not continue to rise below $\lambda \sim 1300$ Å, as extrapolated

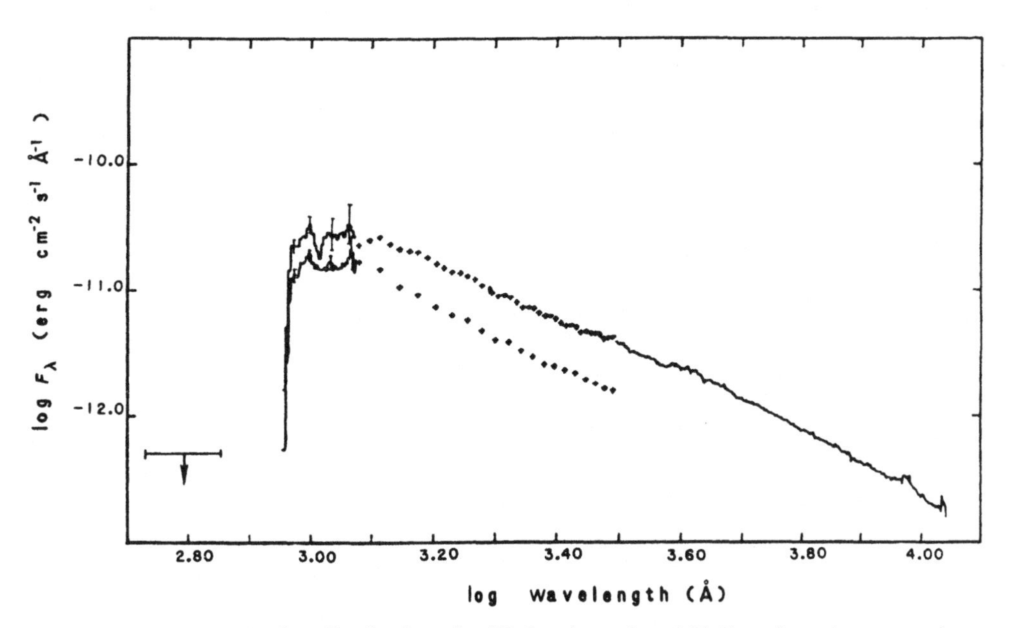

Figure 3.20 Composite flux distributions for SS Cyg (upper) and U Gem (lower) near maxima of outbursts. The EUV bar is an upper limit. From Polidan & Holberg (1984).

from IUE fluxes (Cordova & Howarth 1987) or predicted by basic models (Wade 1984, 1988). The reason for this is not certain, but line blocking is suspected – it is thought to be underestimated at $\lambda \sim 1300$ Å by factors as high as 10 in hot O subdwarf atmospheres (Bruhweiler, Kondo & McLuskey 1981) and significantly reduces fluxes in the EUV of DA white dwarfs with $T_{\rm eff} > 40\,000$ K (Barstow *et al.* 1993). Another possibility is a combination of rapid rotation of the primary and truncation of the inner disc by its magnetosphere.

It is evident that no large contribution from the boundary layer is appearing in the FUV. Therefore the integrated flux, which can be estimated with some confidence for VW Hyi from the flux distribution in Figure 3.19 and the upper limit on the EUV flux, is derived almost entirely from the disc luminosity $L = \frac{1}{2}GM(1)\dot{M}(\mathrm{d})/R(1)$, which, with $M_1(1) = 0.65$ and $L = 1.5 \times 10^{-34}$ erg s^{-1} (Polidan, Mauche & Wade 1990), gives $\dot{M}(\mathrm{d}) \approx 5 \times 10^{-9} \mathrm{M}_\odot$ y^{-1}. For SS Cyg and U Gem, the bolometric luminosities given by Polidan & Holberg (1984), scaled to distances given by Warner (1987a), and corrected for inclination effects, produce similar values of $\dot{M}(\mathrm{d})$.

These are minimum values of $\dot{M}(\mathrm{d})$ during outburst. If the inner edge of the disc is at radius r_0 determined by the magnetosphere of the primary (Section 8.4) then the deduced $\dot{M}(\mathrm{d})$ will be a factor of $r_0/R(1)$ larger.

3.3.5.3 Secular Trends Between Outbursts

It is evident from the general light curves (e.g., Figure 1.3) that there are no *strong* secular trends in optical fluxes between DN outbursts. A detailed study of Z Cha during quiescence (van Amerongen, Kuulkers & van Paradijs 1990) shows that the optical fluxes from the disc and primary are constant, but there is a ~30% increase in flux from the bright spot which is quantitatively consistent with the deeper potential well that the mass transfer stream experiences as a result of the systematic shrinkage of the disc between outbursts (Section 3.5.4.3). A similar effect is seen in YZ Cnc (van Paradijs *et al.* 1994). In WX Hyi (Kuulkers *et al.* 1991b) and VW Hyi (van Amerongen *et al.* 1987) there are no secular changes in total optical flux.

The situation in the UV is different. In WX Hyi and VW Hyi IUE fluxes continue to decline by ~20–30% during the ~20 d intervals between outbursts (Hassall, Pringle & Verbunt 1985; Verbunt *et al.* 1987), as they do also for the ~100 d interval in U Gem (Szkody & Kiplinger 1985) and even for 8 y in WZ Sge, which returned to its *optical* quiescence level 4 months after outburst (Holm 1988; Szkody & Sion 1988; Sion & Szkody 1990).

A survey of IUE results for 15 DN (Szkody *et al.* 1991) showed a variety of behaviour. Certainly SS Cyg and RX And, and probably SS Aur, Z Cam, AH Her and RU Peg, showed declining flux throughout the UV. Z Cha and T Leo declined in longer wavelength UV flux but were constant at the shorter wavelengths. YZ Cnc, TY Psc, SU UMa and possibly OY Car had increasing flux in the UV (as did the long period system BV Cen). The majority of the systems also showed decreasing line flux in the UV. The quiescent X-ray flux in VW Hyi declines by a factor 1.2–1.6 throughout an outburst interval (van der Woerd & Heise 1987), implying a decrease in $\dot{M}(1)$ and that the disc is in some manner non-steady.

It is not possible to interpret these results in a generalized way: for each star a different recipe is required, according to whether the UV is dominated by flux from the primary or the disc, and whether the bright spot contributes significantly to the longer wavelength region of the UV.

The decline in UV flux can be due to cooling of the primary, heated by accretion during outburst (Pringle 1988). This is directly observed in Z Cha (Wood *et al.* 1993), where $T_{\rm eff}(1)$ is 17 400 K immediately after outburst and cools to its quiescent value in ~16 d; in OY Car (Marsh 1993), which takes ~40 d to cool after a superoutburst; in U Gem, which is 39 400 K and 32 100 K at 13 d and 70 d after outburst respectively (Long *et al.* 1994b); and in WZ Sge, which takes ~3000 d to cool after its superoutburst (Sion & Szkody 1990). The cooling time scale can give the depth of heating of the white dwarf envelope, which, for WZ Sge, U Gem and VW Hyi, gives heated masses $(5\text{–}20) \times 10^{-10}\,{\rm M}_\odot$ (Sion & Szkody 1990). These, however, are for spherically symmetric heating, whereas in reality the mass accreting and much of the heating is concentrated at the equator, so questions of lateral heat transport arise (Pringle 1988). There is spectroscopic evidence (Section 3.2.1) for a hotter equatorial belt in U Gem. Sparks *et al.* (1993) find that they can only match the cooling curve for WZ Sge if the heated material is confined to a broad equatorial band for 10 y.

3.3.5.4 Outburst Precursors

For 2 d prior to an outburst of TW Vir, Mansperger & Kaitchuck (1990) noted narrowing and weakening of the emission lines of the kind usually seen on outburst rise. There was, however, only a slight increase in the brightness of the system. Apparently the disc was slowly heating for at least one day before the onset of the rapid rise towards maximum. This 'plateau' or precursor phase in brightness has also been seen in RX And (Kaitchuck, Mansperger & Hantzios 1988), HL CMa (Mansperger *et al.* 1994) and SS Cyg (Clarke, Capel & Bowyer 1984; Voloshina 1986). It is not certain that precursors occur in all DN – even in the same system precursors may vary (Mansperger 1990) – but they are clearly of great potential in revealing more about the triggering of DN outbursts (see Section 3.5.4.2).

Harlaftis *et al.* (1994a) noted a 30% drop in Hα emission from the disc, and an increase in chromospheric emission from the secondary, 5 d before an outburst of IP Peg.

3.3.6 X-Rays During Outburst

At quiescence the X-ray emission from DN is characteristic of the $\sim 10^8$ K corona predicted for low $\dot{M}(1)$ ($\lesssim 10^{16}$ g s^{-1}) systems (Sections 2.5.4 and 3.2.4). The rise of $\dot{M}(1)$ during an outburst should convert the BL into an optically thick $\sim 10^5$ K region, emitting predominantly soft X-rays (Section 2.5.4). This is, to some extent, verified by the observations.

On average, the hard X-ray flux is a factor of 3 lower in outburst than in quiescence (Cordova & Mason 1984a). In SS Cyg, during the progress of an outburst, the hard X-ray flux increases initially by a factor of ~ 5, reaching $L_{\rm X} \sim 8 \times 10^{32}$ erg s^{-1}, falls to below quiescent level as the visual magnitude rises through 9.0 and reappears 10 or 12 days later as the system decays through $m_{\rm v} \sim 9$ (Ricketts, King & Raine 1979; Jones & Watson 1992). Comparison with $L_{\rm X}$ in quiescence (Section 3.2.4) shows that the hard X-rays begin to be suppressed at $\dot{M}_{16}({\rm d}) \sim 1.5$, which is close to the predicted $\dot{M}_{16}({\rm d}) \sim 3$ (Section 2.5.4).

During outburst there is no additional low-energy absorption of hard X-rays (Swank 1979; Watson, King & Heise 1985; Jones & Watson 1992), showing that the disappearance of flux is not simply the result of increased optical thickeness in the high mass flow, but must be due to a change in structure.

In U Gem the hard X-ray flux rises by a factor of about 3 at the start of optical outburst and lasts for ~ 2 d, decaying before optical decline (Swank *et al.* 1978). In VW Hyi there is no observed increase, only a decrease during outburst (van der Woerd, Heise & Bateson 1986; van der Woerd & Heise 1987).

In SS Cyg, U Gem, VW Hyi and OY Car there are usually large increases in soft X-rays associated with outbursts. In U Gem the soft X-rays increase by a factor $\gtrsim 100$ about one day after start of optical outburst (Mason *et al.* 1978; Cordova *et al.* 1984; Figure 1.20), reaching $L_X \sim 3 \times 10^{32}$ erg s^{-1}, and, like the hard X-rays, decaying after 2 d. A spectral fit gives $kT \sim 25$ eV. SS Cyg has been observed just at the start of the rise in soft X-rays (Jones & Watson 1992), the time delay from the optical rise is 0.5–1.1 d – similar to that for the FUV (Polidan & Holberg 1984). The rise time is ~ 8 h, which is much shorter than the ~ 2 d optical rise time. After maximum the decay is faster than that at optical wavelengths (van der Woerd & Heise 1987; King 1986). At maximum, $kT \sim 30$ eV and $L_X \sim 10^{32}$ erg s^{-1} (Cordova *et al.* 1980a; Cordova & Mason 1983). The soft X-rays from U Gem during outburst show orbital dips due to absorption with column density $N_H \sim 3 \times 10^{20}$ cm^{-2} from material high above the orbital plane (Mason *et al.* 1988; Figure 7.9).

The rise of a normal outburst in VW Hyi has yet to be observed, but that of a superoutburst is delayed 2.5 d relative to optical rise (van der Woerd, Heise & Bateson 1986). The decline from outburst is about twice as steep in the X-ray band as in the optical, both in normal and superoutbursts (van der Woerd & Heise 1987). Surprisingly, the spectral gradient remains constant (at an implied single-component temperature (1.0–1.5) $\times 10^5$ K) even though the flux changes by a factor of 100; this almost certainly implies a multicomponent origin for the soft X-rays. During superoutbursts the average $L_X \sim 10^{34}$ erg s^{-1}. Van Teeseling, Verbunt & Heise (1993) have developed such a model with a $> 10^6$ K optically thin component and a $\sim 10^5$ K optically thick component, the latter covering ~10^{-3} of the surface of the primary. This is similar to the result of Mauche *et al.* (1991) who concluded that ζ (equation 2.54b) is ~0.04 and kT_{BL} ~ 10 eV. One normal optical outburst of VW Hyi failed to produce any soft X-ray increase (H.-C. Thomas, private communication).

The situation for OY Car, which is highly inclined, is rather different and sheds light on the VW Hyi case. In OY Car the X-rays observed during superoutburst have $kT = 0.15$ keV, or $T \sim 2 \times 10^6$ K, more characteristic of coronal gas, and $L_X \sim 10^{31}$–10^{32} erg s^{-1} (Naylor *et al.* 1988). This suggests that the very soft X-ray emitting BL is obscured in OY Car, leaving only an extended hot corona visible. The absence of X-ray eclipses supports this view. In systems of lower inclination, such as VW Hyi, both the BL and the corona would be observable. Two components are also indicated by the existence of DNOs in the softest X-rays but their absence in hard X-rays, including those from OY Car (Section 8.6.4). Similarly, soft X-rays are not observed from Z Cha (Harlaftis *et al.* 1992), although hard X-rays were detected at quiescence by Einstein, and diminished during outburst (Becker 1981).

In Section 2.6.1.5 it was noted that the hard X-ray emission of DN during outburst (and of UX UMa stars) is much less than is expected from the model of Patterson & Raymond (1985a: see Figure 2.22). In other high $\dot{M}(\mathrm{d})$ systems the hard X-rays are in satisfactory agreement with the model, in which they arise from the low density outer regions of the mass flow. The discrepancy is largely due to underestimating $\dot{M}(\mathrm{d})$ in these systems (Hoare & Drew 1991; Sections 2.7.2 and 2.7.3).

In contrast to the expected transition from predominantly hard to predominantly soft X-ray emission in outburst, SU UMa and RU Peg are found to have optically thin emission with $kT \sim 2$ keV and ~ 5 keV respectively during outburst (Silber, Vrtilek & Raymond 1994). Yet in both cases $\dot{M}(1)$, estimated from the X-ray flux (and rescaled to distances given in Warner (1987a)) is $(1\text{–}2) \times 10^{17}$ g s^{-1}, which is well above the value expected for optical thickness.

3.3.7 *Radio Observations of DN*

Cordova, Mason & Hjellming (1983) could detect no radio emission from two DN at a level of 0.1 mJy. Scaling by analogy with radio flux measured from OB supergiant winds gave $\dot{M}_{\mathrm{wind}} \lesssim 10^{-8} \mathrm{M}_{\odot}$ y^{-1} in the DN. Benz, Fürst & Kiplinger (1983, 1985) detected SU UMa at 4.75 GHz at a flux of 1.3 mJy, during two normal outbursts near the end of decline, but not in quiescence. It was not detected by later observers at 4.9 GHz (Chanmugam 1987). Turner (1985) observed 11 DN at 2.5 GHz and detected TY Psc and UZ Boo at a few mJy level. The latter is probably the result of confusion with a nearby source (Benz & Güdel 1989). EM Cyg was detected during outburst with the VLA at 49 GHz (Benz & Güdel 1989), variable on a time scale of days and strongly circularly polarized. Several other DN have been looked at but radio emission was not detected (Fürst *et al.* 1986; Torbett & Campbell 1987; Benz & Güdel 1989). The radio detections of DN seem all to have been during outbursts, but not all outbursting DN have radio emission (Benz & Güdel 1989).

Several origins for the radio emission from CVs have been considered (Benz, Fürst & Kiplinger 1983; Chanmugam 1987; Benz & Güdel 1989). Gyrosynchrotron emission from non-thermal electrons requires fields of ~ 50 G over volumes several times larger than the binary separation. This could arise from the wind/field interaction of the secondary. A more efficient process is maser emission in regions of strongly converging field lines. Benz & Güdel point out that a field $B(1) \sim 10^5$ G at the surface of the *primary*, with an outflow $\sim 10^{-11}$ $\mathrm{M}_{\odot}$ y^{-1}, can produce the highly polarized emission observed in EM Cyg. The existence of emission only during outbursts would then imply that the outflow is the wind from the disc (Sections 2.7.3 and 3.3.5.1). There is accumulating evidence for fields of $B(1) \sim 10^5$ G even in 'non-magnetic' CVs (Sections 3.5.4.2 and 8.6).

3.4 Standstills: The Z Camelopardalis Stars

In the latest GCVS 28 stars are listed as certain Z Cam stars with a further 14 possibles. Of the definites , VW Vul is almost certainly an SU UMa star, CN Ori has been lost to pure U Gem status (Mantel *et al.* 1988) and TT Ari to the NL class. Recent research (see references in Table 3.2) has added HL CMa, AT Cnc, WW Cet and UZ Ser to the list of definites, none of which are listed in the GCVS as possibles. A deduction that might be made from the latter is that 'all U Gem stars are unrecognised Z Cam stars' – it takes only one observed standstill to effect the reclassification. But comparison of

Tables 3.1 and 3.2 shows that the Z Cam stars occur only among DN with $T_n \lesssim 30$ d (and all the Z Cam stars with unknown P_{orb} but with well-determined T_n in the GCVS have $T_n \lesssim 20$ d), so a more correct deduction would be that only DN with $T_n \leq 30$ d currently classified as U Gem are potentially Z Cam stars. (In the GCVS only a small fraction of the U Gem stars have estimates of T_n so candidates are rare.) The bias in Tables 3.1, 3.2 and 3.3 towards DN with m_v(max) brighter than ~ 13, and particularly for known Z Cam stars to have $m_v(\max) \lesssim 12$, is a result of the requirement that, in general, to be well enough observed to receive a reliable subclassification, a DN must be within reach of amateurs' telescopes.

The values of M_v(min), described in Section 3.2.3 and plotted in Figure 3.5, show that the Z Cam stars are at the bright end of the range of DN quiescent magnitudes. This is even more evident when M_v(mean) is considered (Figure 3.9). As there is evidence (Zuckerman 1954; Warner 1976a) that the time-averaged flux for periods when normal outbursts occur is close to the flux at standstill, it is reasonable (and time-saving) to assume $M_v(\text{mean}) \equiv M_v(\text{stand})$. The approximation is justifiable on the disc instability model of DN outbursts (Section 3.5.3.5), where a critical value of $\dot{M}(2)$ separates stable from unstable discs: the Z Cam stars phenomenologically appear to be very close to the dividing line whereby a small increase in $\dot{M}(2)$ results in stability. Thus $\dot{M}$(stand) would be only slightly above the long-term average of $\dot{M}(2)$.

Because of their short recurrence times, many Z Cam stars spend little or no time at true quiescence. As seen in Figure 3.21, the outburst durations are often conspicuously bimodal. RX And, which has the shortest T_n, can change its behaviour from Type A

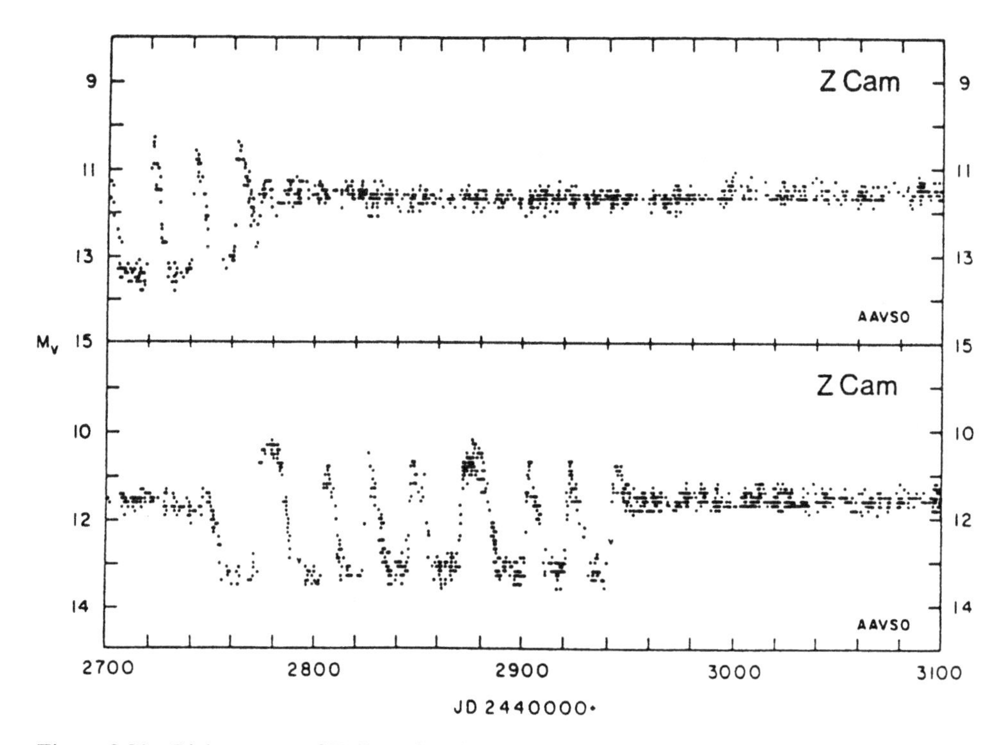

Figure 3.21 Light curves of Z Cam showing standstills. From AAVSO observations.

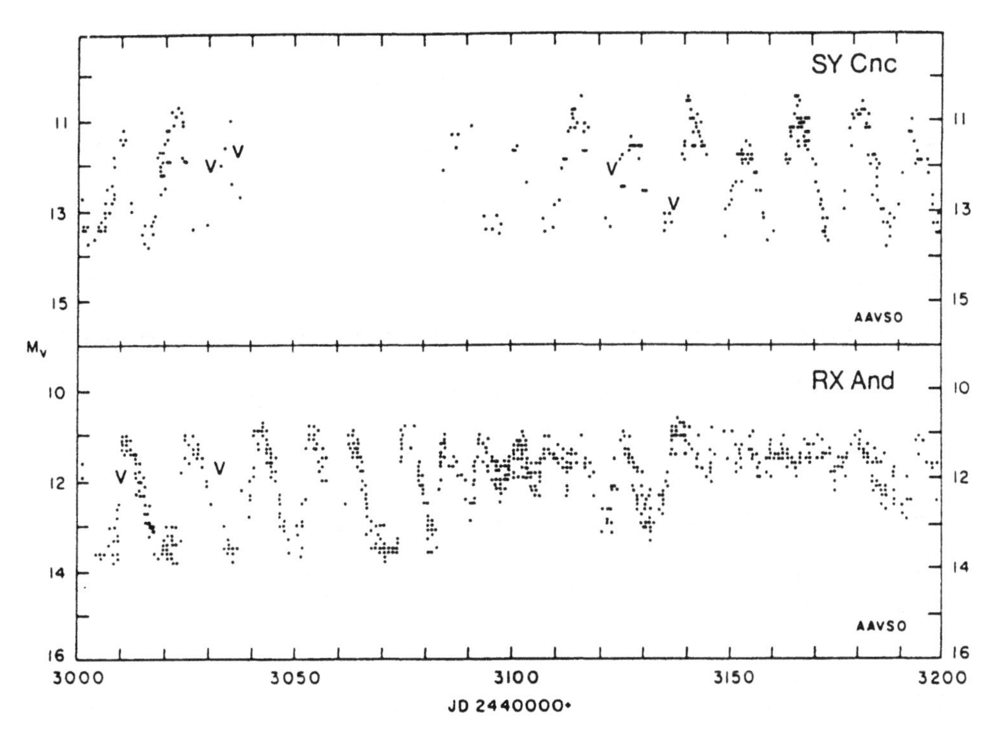

Figure 3.22 Light curves of the Z Cam stars SY Cnc and RX And, the latter showing erratic behaviour and standstills. From AAVSO observations.

outbursts with $T_n \sim 13$ d to Type B outbursts from a quiescent level just below m_V(stand) and $T_n \sim 25$ d (see Figure 3.22 and Figure 67 of Campbell & Jacchia (1941)).

An important morphological point is the way in which standstills begin and end. In the examples shown in Figure 1.4 every standstill terminates with return to m_V(min) at a slope somewhat less than that of decline of a full outburst. The decline after standstill seems to be an almost invariable rule, the only clear exception in Z Cam being the standstill at the end of 1958 which terminated in a slow rise to maximum early in 1959 (Figure 3.23). Entry into standstill is almost always from the declining phase of an outburst. The exceptions (e.g., the upper section of Figure 3.21) are those at a time of shortening of the interval between outbursts, which is accompanied by a brightening of the quiescent magnitude, in which case standstill can be reached either from below or from the decline of a very weak outburst.

In standstill itself there are often small fluctuations of brightness ~0.2 mag on time scales 10–20 d. There is, in addition, low amplitude flickering with time scales of approximately minutes typical of DN in outburst.

In respect of optical and UV light curves, UV flux gradients (Pringle & Verbunt 1984), X-ray emission and the presence of DNOs, the outbursts of Z Cam stars do not appear distinguishable from those of U Gem stars. It is necessary, therefore, only to describe observations made during standstill.

Of the one dozen Z Cam stars with known orbital periods (Table 3.2) only one, EM Cyg, is an eclipsing system. In it, eclipses are 'almost total' (Mumford & Krzeminski

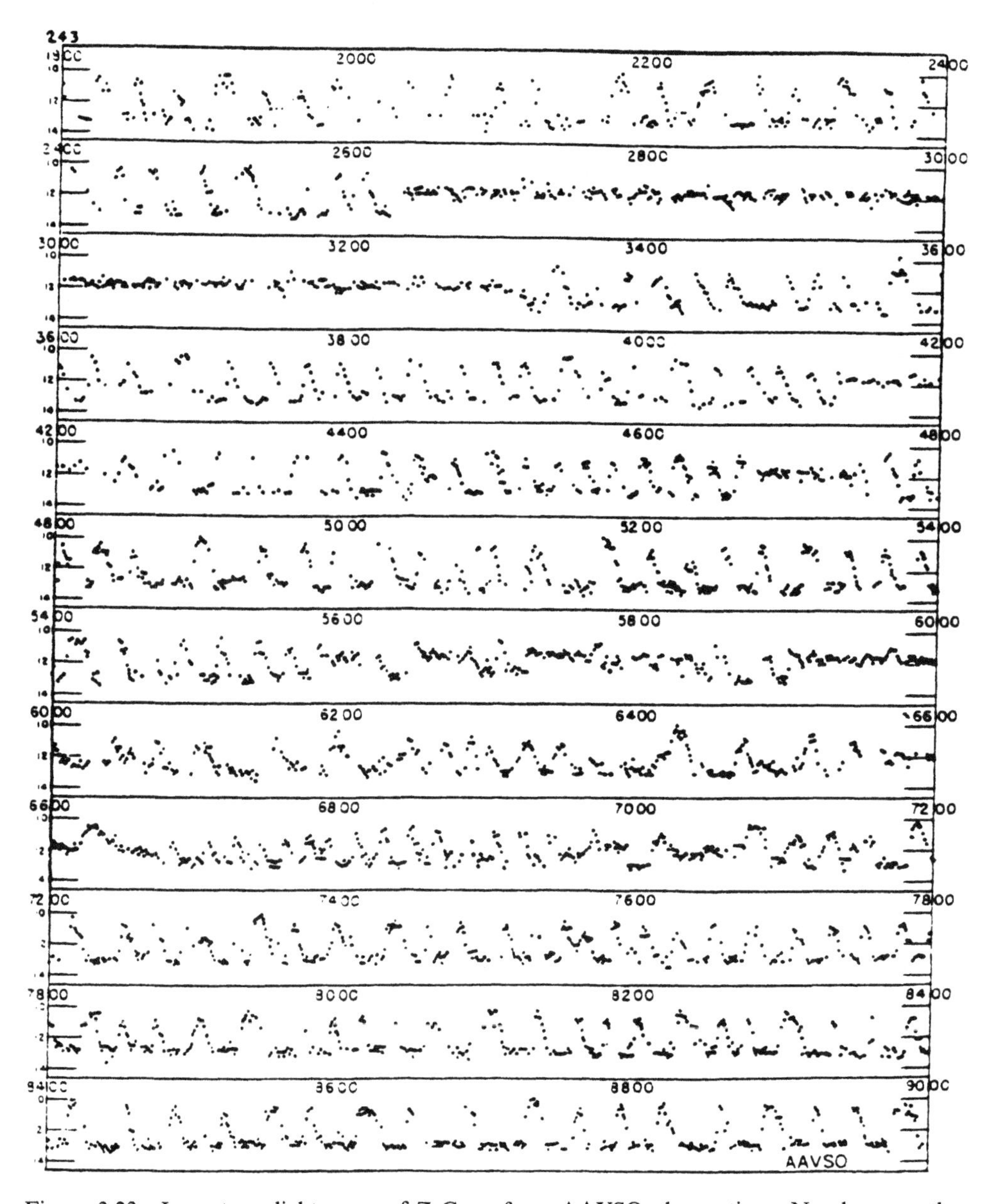

Figure 3.23 Long term light curve of Z Cam, from AAVSO observations. Numbers are the final four digits of the Julian Date. The rise from standstill at the beginning of 1959 occurs at 6620.

1969; Robinson 1974). It provides the only current opportunity to study the intensity distribution across a disc in standstill – but this has yet to be done. Mumford & Krzeminski, observing at a time when the DN nature of EM Cyg had not yet been recognised, inadvertently observed it through most phases from quiescence to outburst maximum, but not, apparently, in standstill. They comment on the unusual instability of the eclipse profiles, and Stover, Robinson & Nather (1981) similarly find unusually rapid emission-line profile changes. A more coordinated study, now that the unusual nature of EM Cyg is recognized, is warranted.

Kiplinger (1979a, 1980) found that a steady state disc flux distribution could be fitted to Z Cam in standstill, but not during its decline from outburst. His value of $\dot{M}$(d) $\sim 5 \times 10^{-9}$ $M_\odot$ y^{-1} in standstill, although compatible with the suspected higher values at the peak of outburst, suffers from the usual suspicion of results obtained from fitting to optical continua (Section 2.6.1.2). Extending the observational baseline into the UV, Szkody (1981a) found that of all the DN flux distributions, only Z Cam in standstill could be fitted convincingly by a simple $F_\lambda \propto \lambda^{-7/3}$ model.

Studies of optical spectra during standstill have been made only for Z Cam itself (Lortet-Zuckermann 1967; Szkody 1976b; Kiplinger 1980; Szkody & Wade 1981). The Balmer series and Balmer jump are in absorption, with variable strength emission cores decreasing towards higher series members. HeII λ4686 is occasionally seen in emission. The Balmer jump is stronger at standstill than at the same luminosity on decline from outburst, and is variable in strength. The FWHM of the emission components is 900 km s^{-1}, as compared with 1100 km s^{-1} at quiescence, but the total flux in Hα is the same at standstill and quiescence. The $K(1)$ and phasing of the emission components are the same at standstill as in quiescence.

Klare *et al.* (1982) observed Z Cam, TZ Per and RX And at standstill with IUE and noted strong absorption lines with some P Cyg profiles. In a more extensive study of Z Cam, Szkody & Matteo (1986a) found that the P Cyg absorption and emission components are stronger at decline from outburst than they are well into standstill, suggesting that the wind generated by the outburst is not sustainable for long periods by the conditions that obtain in standstill.

3.5 Theories and Models of DN Outbursts

3.5.1 Early Developments

The earliest theories of DN outbursts not unexpectedly associated DN with RN and CN and sought a common cause. The first serious attempt to account quantitatively for DN and RN energetics and time scales came in Schatzman's model in which a hydrogen thermonuclear runaway is detonated by He^3 burning, the whole process being initiated by vibrational instability (Schatzman 1951: see summary of Schatzman's work in Chapter 11 of Payne-Gaposchkin 1957). From Schatzman came the prediction that the ratio of energy radiated during outburst to the total energy radiated between outbursts should be approximately constant. This proved observationally to be the case (Zuckerman 1954), although some systematic difference between DN and RN was noted (Payne-Gaposchkin 1957).

Schatzman (1958, 1959) later suggested that DN, RN and CN explosions arise from undamped non-radial oscillations excited by tidal resonance in one of the binary components. This was reviewed unfavourably by Kraft (1963), who preferred a thermonuclear runaway origin based on Mestel's (1952) work on white dwarfs. Hydrodynamic models of thermonuclear runaways in white dwarfs, with $M(1) < 0.5 M_\odot$ or low CNO enhancements, can indeed produce eruptions which have expansion velocities too low to eject matter and which reach $\sim 100 L_\odot$ at maximum light. However, it requires ~ 15 y to radiate the energy injected into a white dwarf envelope during this process (Starrfield, Sparks & Truran 1974), whereas DN decline after only a few days.

The claim (Section 1.3) that in U Gem it is the secondary that brightens during outburst inadvertently started a line of thought that might never have occurred if instead it had been necessary from the outset to seek an explanation for brightening of the disc. A brief description will be given of this *theory of mass transfer instability* which, for a decade or more, was a serious competitor to the disc instability theory, but which, for reasons to be given, no longer has many adherents.

3.5.2 The Theory of Mass Transfer Instability (MTI)

Paczynski (1965a) pointed out that stars with convective envelopes increase their radii with decreasing mass (see Section 9.3.1 for a more complete discussion) and therefore a low mass secondary star filling its Roche lobe is potentially unstable. Paczynski, Ziolkowski & Zytkow (1969) speculated that the loss of mass would occur so rapidly that the convective envelope would become radiative, stabilizing the process and leading to recurrent outbursts. Bath (1969) developed the concept further, finding that the introduction of a spherical mass-losing boundary condition leads to runaway mass loss on a dynamical time scale, producing models not only of DN outburst luminosity, but also of CN luminosity. Bath's technique was criticized by Osaki (1970) who instead attributed outbursts of the secondary to variations in efficiency of convective energy transport related to the effect of mass loss on the flow over the surface of the secondary. He was able to obtain $\sim$1.5 mag outbursts with $\sim$15 d recurrence cycle. In an improved analysis, Bath (1972) continued to find DN-type outbursts in secondaries, but not ones as energetic as CN.

Following the realization that the DN outburst is not seated on the secondary, Smak (1971a,c) pointed out that the outburst must either be from the primary, inherent in the disc itself, or a result of the disc brightening because of the recurrent MTIs already under discussion. On examination of the stability of convective envelopes in a Roche potential the energetic mass loss instability found by Paczynski and Bath was converted into a less dramatic MTI having little or no effect on the luminosity of the secondary. The MTI was found to be confined to a cone, with its apex at the centre of the secondary and its base in the vicinity of the Lagrangian point (Papaloizou & Bath 1975). In a hydrodynamic treatment of the instability, Bath (1975) found the intervals between episodes of high mass transfer should be 10–200 d. In this modified form of MTI, the time scale and total energy available are controlled by the secondary star, but the outburst light curve and its spectral properties depend on the response of the disc to the mass transfer burst (and to the 'profile' of the burst).

However, the existence of the MTI itself continues to be a matter of controversy. In the original model (Bath 1975), the instability was a result of recombination in the H and HeI ionization zones. The phase of high mass transfer is terminated when these zones are advected away; the secondary then shrinks from the Roche lobe and the ionization zones are reestablished on a thermal time scale. The latter gives $\sim$10–100 d interval between outbursts. A hydrodynamical simulation by Wood (1977) found no instability in cool secondaries, only a steady mass transfer, but may be sensitive to the form of convective transport adopted (Edwards 1988). Gilliland (1985) also found stable mass transfer in a one-dimensional hydrodynamical treatment, and noted the importance of the sideways flow within the secondary needed to replace the mass lost. Edwards (1985, 1987, 1988) has explored these and other aspects of the problem and concludes that the dynamical time scale properties of flow can only satisfactorily be decided by a three-dimensional computation including at least time-dependent convection, internal flow, effects of rotation of the star, and continuation for at least a thermal time scale to be sure that even if there is an initial instability, subsequent ones are not damped out.

From an observational point of view there are several reasons to be dissatisfied with the MTI model, though it is to be admitted that there have been no attempts as yet to adjust the model with these criticisms in mind. First, there is the fact that no high $\dot{M}(2)$ systems show DN outbursts. This is made more quantitative in Section 3.5.3.5, where it is shown that in the disc instability (DI) model there exists a well-defined $\dot{M}_{crit}(d)$ (independent of α) above which discs are stable, and that observations of CV discs agree closely with this prediction. In the MTI model no $\dot{M}(2)$ above which stable mass transfer would occur has been suggested; indeed, the reverse is indicated by Bath (1976), who remarks that low $\dot{M}(2)$ would *remove* the instability. Secondly, as seen in Section 3.3.4.2, there is no evidence from bright spot luminosities for increased $\dot{M}(2)$ during or before outbursts. Thirdly, polars do not have discs, and do not show DN outbursts. Coupled with this is the fact that low mass X-ray binaries, which have mass transferring secondaries in the same mass range and evolutionary state as CVs, do not show DN outbursts. Unless the strong magnetic fields of the former, or the irradiation of the latter, can be shown to modify the MTI, this is not compatible with the MTI model. Fourthly, Type B outbursts (Section 3.3.5.2) arise from 'inside out' outbursts (Section 3.5.4.1), the triggering of which appears impossible by matter deposited at the outer edge of the disc.

There are additional problems, concerning the detailed shapes of outburst light curves, which are incompatible with the MTI model (Cannizzo 1993a; Ichikawa & Osaki 1992).

3.5.3 *The Theory of DI*

3.5.3.1 Early Development

With the observational conclusion that DN outbursts are centred on the primary and its accretion disc, Osaki (1974) turned away from the MTI and proposed a 'working model' in which the outbursts were attributed to intermittent accretion from the disk. He was influenced by the discovery of the importance of accretion luminosity in X-ray binaries, and by the earlier suggestion by Crawford & Kraft (1956) that the ~ 2 mag outbursts in AE Aqr discovered by Zinner (1938) were due to accretion onto the primary, but was apparently unaware that Smak (1971c) has already proposed a similar mechanism in broad outline.

Osaki showed that a model in which matter is stored in the disc, and then rapidly accreted via some unknown instability mechanism, agreed with the overall energetics and time scales of DN outbursts and the current estimates of $\dot{M}(2)$. The standstills of Z Cam stars and the NLs and CN remnants were interpreted as examples of stable accretion.

At the same time Bath (1973) also realised that gravitational energy released in the disc, following a MTI, could explain the nature of DN outbursts. From that time there was agreement that passage of matter through the disc is the cause of a DN outburst, but controversy, for a long time unresolved by observational evidence, over whether the MTI or a DI triggered the event.

A DI mechanism was first proposed by Hoshi (1979), who studied the consequences of the disc viscosity parameter α being small enough that $\dot{M}(\mathrm{d}) < \dot{M}(2)$. Treating the annulus in which matter accumulates as part of an accretion disc, the requirement that the surface temperature $T_{\mathrm{eff}}(r)$, given by equation (2.37), be compatible with the radiative transfer (i.e., for an optically thick annulus) equation (2.39) leads to the condition $\kappa_{\mathrm{R}}\Sigma \sim 1$. Since $\kappa_R = \kappa_R(\rho,\ T) = \kappa_{\mathrm{R}}(\Sigma/H,\ T)$ and $H = H(M(1),\ r, T)$ through equation (2.42a), for a given choice of $M(1)$ and r the condition for equilibrium becomes $\kappa_{\mathrm{R}}(\Sigma,\ T)\Sigma \sim 1$.

Hoshi noted that $\kappa_{\mathrm{R}}(\rho,\ T)$ for a typical stellar composition passes through a maximum (due to photoionization of hydrogen) near $T \sim 10^4$ K. Therefore, κ_{R} is double-valued for each choice of ρ. As a consequence *there are two values of* Σ that satisfy $\kappa_{\mathrm{R}}\Sigma \sim 1$. In turn, through equations (2.50) and (2.42b), and keeping α constant, this implies that there are two different $\dot{M}(\mathrm{d})$ that satisfy the equilibrium condition; the high value $\dot{M}_{\mathrm{h}}(\mathrm{d})$ may be identified with outburst and the low value $\dot{M}_{\mathrm{l}}(\mathrm{d})$ with quiescence.

However, Hoshi found that the equilibria are in general unstable: if $\dot{M}(2) > \dot{M}_{\mathrm{l}}(\mathrm{d})$ then Σ in the storage annulus increases until a transition to the high state occurs. If at that point $\dot{M}(2) < \dot{M}_{\mathrm{h}}(\mathrm{d})$, the disc is unstable and may eventually revert to the low state. Although he correctly identified the thermal instability, the details of Hoshi's work have been superseded by the developments described below. His pioneering analysis went largely unremarked until Pringle, at the Sixth North American Workshop on Cataclysmic Variables, held at Santa Cruz in 1981, described the general hysteresis-like curve that must exist in the Σ–T plane if DN outbursts are to be attributed to the DI mechanism. This stimulated intensive investigations and the first detailed models of the DI process (Meyer & Meyer-Hofmeister 1981; Bath & Pringle 1982; Smak 1982a; Cannizzo, Ghosh & Wheeler 1982).

A good introductory article on the DI model is that by Cannizzo & Kaitchuck (1992). Reviews can be found in Smak (1984d), Meyer (1985), Lin & Papaloizou (1988) and Cannizzo (1993a).

The connection with postulated accretion discs in active galactic nuclei is shown by the similarities in approach in such references as Lin & Shields (1986), Clarke (1988), Mineshige & Shields (1990), Cannizzo & Reiff (1992) and Cannizzo (1992). These contain much that is applicable to CVs.

3.5.3.2 *The S-Curve in the* Σ–T_{eff} *Plane*

From equations (2.23) and (2.33) the flux per unit area radiated from one side of an annulus of radius r is

$$F = \sigma T_{\mathrm{eff}}^4(r) = \frac{9}{8}\frac{GM(1)}{r^3}\Sigma\nu_{\mathrm{k}}. \tag{3.9}$$

For a chosen value of ν_{k}, equation (2.38b) and the radiative transfer equation (2.39a) or an appropriate convective transport equation, can be integrated in the z direction in the same manner as for stellar atmospheres, with equation (3.9) as a boundary value. This gives $T(z)$, $\rho(z)$ and $P(z)$ at each r. It should be noted that the low densities can result in only a small fraction of the flux being carried by convection and the usual Schwarzschild criterion for convection is modified in accretion discs (Tayler 1980).

Integrating $\rho(z)$ gives Σ and hence another $\Sigma\nu_k$–Σ relationship (for each r) to add to the diffusion equation (2.27). It also gives a Σ–T_{eff} relationship for each annulus in the disc. This has a characteristic S-shape curve, irrespective of whether the vertical structure of the disc is simply averaged (Hoshi 1979; Mineshige & Wood 1990) or integrated with an adiabatic gradient where appropriate (Faulkner, Lin & Papaloizou 1983; Pojmanski 1986) or both radiative and convective transport are included (Meyer & Meyer-Hofmeister 1981; Mineshige & Osaki 1983; Cannizzo & Wheeler 1984).

Because it turns out that $\Sigma\nu_k = f(\Sigma, T)$, a further relation between Σ and T_{eff} is required before equation (2.27) can be integrated. This is supplied by the equation of energy balance, given in Section 3.5.4.

Consider first the effect of opacity variations, ignoring convection. From equations (2.10a), (2.39b), (2.42b), (2.48a) and (3.9), the Σ–T_{eff} relationship for optically thick discs in radiative equilibrium has the form

$$T_{eff}^3 \propto \alpha\kappa_R(\rho,\ T)\Sigma^2. \tag{3.10}$$

At high temperatures, with H mostly ionized, the Kramers opacity can be used (Faulkner, Lin & Papaloizou 1983):

$$\kappa_R = 1.5 \times 10^{20}\rho T^{-2.5} \quad \text{cm}^2\ \text{g}^{-1}. \tag{3.11}$$

Noting from equations (2.10a) and (2.42a) that $\rho \mathrel{\tilde{\propto}} \Sigma T^{-1/2}$, we see that

$$T_{eff} \mathrel{\tilde{\propto}} \alpha^{1/6}\Sigma^{1/2} \tag{3.12}$$

For $5000 \lesssim T < 10\,000$ K, where H is partially ionized and H^- is an important source of opacity, κ_R is very sensitive to T (Faulkner, Lin & Papaloizou 1983):

$$\kappa_R = 1.0 \times 10^{-36}\rho^{1/3}T^{10} \quad \text{cm}^2\ \text{g}^{-1}, \tag{3.13}$$

and hence, for regions in radiative equilibrium,

$$T_{eff} \mathrel{\tilde{\propto}} \alpha^{-60/41}\Sigma^{-70/41}. \tag{3.14}$$

However, in a region where partial ionization occurs, the steep dependence of opacity on temperature and, more importantly, the reduction of the adiabatic temperature gradient (Cannizzo & Wheeler 1984) result in convective instability. From equations (2.23), (2.42a), (2.48a) and (3.9), $\alpha\Sigma T \propto T_{eff}^4$, and therefore any steep dependence of T on T_{eff} produces an inverse relationship between Σ and T_{eff}, as in the second example given above. Where convection transports energy from the midplane to the surface it lowers the temperature gradient. The connection between T and T_{eff} therefore depends sensitively on the *fraction* of flux carried by convection (Cannizzo & Wheeler 1984; Cannizzo 1993a); as a result, for $T_{eff} \lesssim 8000$ K, where convection first extends from the midplane to the surface (Cannizzo & Wheeler 1984; Pojmanski 1986), there is a range of T_{eff} for which $d\Sigma/dT_{eff} < 0$. When convection carries most of the energy, there can no longer be sensitivity to the fraction and the Σ–T_{eff} relationship

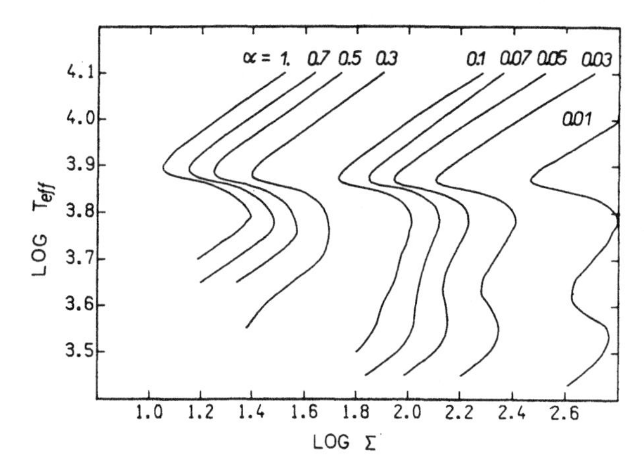

Figure 3.24 The 'S-curve' relationships between temperature and surface density for a range of viscosity parameter α. From Pojmanski (1986).

reverts to a positive form. There is thus a $\Sigma_{\max 1}$ corresponding to the condition where convection carries most of the flux.

At lower temperatures, still with T > 5000 K, convective transport decreases and a region appears in which equation (3.13) applies. At still lower temperatures, $T \lesssim 4000$ K, $\kappa_R(T)$ is dominated by molecular absorption and ceases to be sensitive to temperature, so a lower branch with $d\Sigma/dT_{\rm eff} > 0$ is formed, with an associated $\Sigma_{\max 2}$. The disc is, in general, still optically thick in the vicinity of this second density maximum, but, at lower temperatures still, becomes optically thin.

The Σ–$T_{\rm eff}$ relationships for a range of α are shown in Figure 3.24. It may be noted that, through equations (2.32) and (3.9), the ordinate $T_{\rm eff}$ is a monotonic function of $\dot{M}$ and $\Sigma\nu_k$. For $\alpha \gtrsim 0.3$ convective transport is relatively unimportant, so only one $\Sigma_{\max}$ occurs.

3.5.3.3 *Stability of an Annulus*

The S-curve is the line of thermal equilibrium, i.e., the locus of models where viscous heating balances radiation from the surface. If an annulus is given a (Σ, $T_{\rm eff}$) which places it anywhere to the right of the S-curve, viscous heating exceeds the radiation cooling so the annulus will heat until it reaches the S-curve. Similarly, an annulus starting to the left of the curve will cool until it reaches equilibrium.

Although the S-curve locates equilibrium states, not all states are stable. On a branch of the curve for which $dT_{\rm eff}/d\Sigma < 0$, given a small positive perturbation of Σ, the annulus would have to seek an equilibrium at lower $T_{\rm eff}$. But an increase of Σ causes an increase in heating (equation (2.33)) which would require an increase in $T_{\rm eff}$ to regain equilibrium. Such a region therefore is thermally unstable: discs cannot remain in equilibrium there. In contrast, regions with $dT_{\rm eff}/d\Sigma > 0$ are stable.

It should be noted that the above is a local stability analysis, it ignores the effect of adjacent annuli. The global treatment confirms that the thermal limit cycle apparent in individual annuli can affect the whole disc.

In addition to thermal instability, there is a viscous instability if $dT_{\rm eff}/d\Sigma < 0$. As this

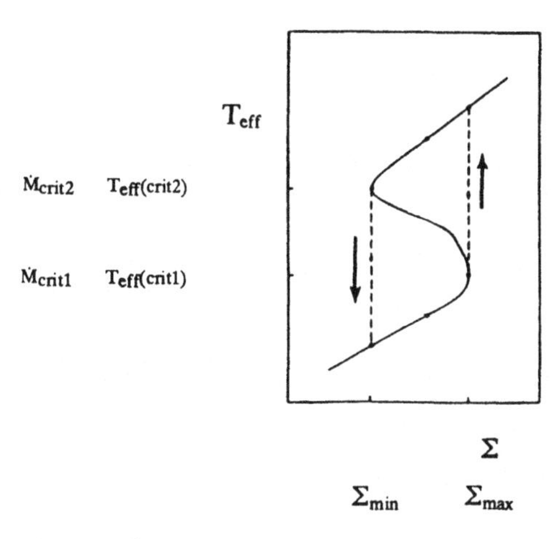

Figure 3.25 The 'S-curve' shown schematically, with heating and cooling (dashed) lines showing the thermal cycle.

is equivalent to $d\dot{M}/d\Sigma < 0$ it can be seen that any local reduction of Σ results in an increase in $\dot{M}$ which depletes the annulus more rapidly and lowers Σ further. This instability (Lightman & Eardley 1974) operates on a viscous time scale, which is much slower than the thermal time scale in thin discs (Bath & Pringle 1982). However, the actual instability in CV discs has characteristics of both types of instability (Mineshige & Osaki 1985).

3.5.3.4 The Thermal Limit Cycle

Consider a simplified S-curve (Figure 3.25) in which Σ_{min} corresponds to $\dot{M}_{crit2}$ and Σ_{max} corresponds to $\dot{M}_{crit1}$. If the rate of mass transfer $\dot{M}_{in}$ arriving at the outer edge of an annulus is less than the value $\dot{M}_{crit1}$ then the annulus can find an equilibrium on the lower branch of the S-curve. Similarly, if $\dot{M}_{in} > \dot{M}_{crit2}$ the annulus will find equilibrium on the upper branch of the S-curve.

For $\dot{M}_{crit1} < \dot{M}_{in} < \dot{M}_{crit2}$, however, there is no equilibrium available. Supposing the annulus initially to lie on the lower branch, the condition $\dot{M}_{in} > \dot{M}_{crit1}$ will result in an increase of Σ: the annulus is unable to pass on mass to inner annuli at the same rate that it receives it. Eventually Σ will increase to the maximum possible on the lower branch; this critical surface density is usually denoted Σ_{max}. At this point the annulus cannot follow the S-curve into region 2, but instead heats on a thermal time scale until it reaches equilibrium on the upper branch. It now finds itself with $\dot{M}_{in} < \dot{M}_{crit2}$ and therefore reduces Σ, running down the upper branch until it arrives at the lower limit, at surface density Σ_{min}, where it leaves the curve and cools to the lower branch.

The annulus therefore cycles between the upper and lower equilibrium states. This constitutes the foundation of the DI theory of DN outbursts. The heating phase of any thermally unstable annulus in the disc causes neighbouring annuli to become unstable, sending heating pulses inwards and outwards until most of the disc is on the upper branch. The first annulus to reduce its surface density to Σ_{min} starts the reverse process: a cooling front that runs from the outer edge of the disc (see Section 3.5.4.1) to the inner edge.

The principal critical values of Σ are given by (Cannizzo, Shafter & Wheeler 1988)

$$\Sigma_{max} = 11.4 r_{10}^{1.05} M_1^{-0.35}(1) \alpha_C^{-0.86} \quad \text{g cm}^{-2}, \tag{3.15}$$

$$\Sigma_{min} = 8.25 r_{10}^{1.05} M_1^{-0.35}(1) \alpha_H^{-0.8} \quad \text{g cm}^{-2}, \tag{3.16}$$

where, as noted in Section 3.5.4.2, Σ_{max} is determined by α_C on the lower (cold) branch and Σ_{min} by α_H on the upper (hot) branch; as these α's are not necessarily equal they are henceforth distinguished. Through equations (2.10a), (2.23), (2.42b), (2.48a), (3.9) and the relationship $T(\text{mid}) = 37700 r_{10}^{0.03}\,(\alpha/0.1)^{-0.2}$ K for the midplane temperature at Σ_{min} (Cannizzo & Wheeler 1984), equation (3.16) may be converted to

$$T_{eff}(\text{crit2}) = 7690\left(\frac{r}{3 \times 10^{10}\text{cm}}\right)^{-0.105} M_1^{0.15}(1) \quad \text{K}. \tag{3.17}$$

In a DN at quiescence the disc should be everywhere at a lower temperature than T_{eff}(crit2), as is the case in Z Cha (Figure 2.33); and in outburst (or in discs permanently in a high state) the temperature should be everywhere greater than T_{eff}(crit2), as seen for Z Cha in Figure 2.32. The observation (Figure 2.33; see also Wood, Horne & Vennes 1992) that $T_{eff}(r) \gtrsim 4000$ K in quiescence excludes the lower minima of Σ from involvement in the thermal limit cycle.

As seen from Figure 3.24, $T_{eff}(\text{crit1}) \approx 6000$ K independent of α. For a steady state disc the maximum temperature is 0.488 T_* (Section 2.5.1). From equations (2.3b) and (2.83a) it follows that an accretion disc will remain stable on the lower branch of the S-curve only if $\dot{M}_{13}(\text{d}) < 3.1\ M_1^{-2}(1)$.

3.5.3.5 *The Region of DI*

The exact value of Σ_{min}, or its equivalent $\dot{M}_{crit2}$, is model-dependent. Nevertheless, there is good agreement between the various computations – see the tabulation of Shafter, Wheeler & Cannizzo (1986), who compare the results of Meyer and Meyer-Hofmeister (1983a), Cannizzo & Wheeler (1984), Faulkner, Lin & Papaloizou (1983), Smak (1984a) and Mineshige & Osaki (1985). The Faulkner, Lin & Papaloizou formulation is

$$\dot{M}_{crit2}(r) = 1.02 \times 10^{16} \alpha_H^{3/10} r_{10}^{21/8} M_1^{-7/8}(1) \quad \text{g s}^{-1} \tag{3.18a}$$

$$\dot{M}_{crit2}(r_d) = 8.08 \times 10^{15}\left(\frac{\alpha_H}{0.3}\right)^{3/10}(1+q)^{7/8} P_{orb}^{7/4}(\text{h}) \quad \text{g s}^{-1} \tag{3.18b}$$

using the empirical result $r_d = 0.30a$ for DN (Section 3.3.4.3).

$\dot{M}_{crit2}$ is therefore largest at the outer edge of the disc. If $\dot{M}(2)$ falls below $\dot{M}_{crit2}(r_d)$ the disc will enter a thermal limit cycle. An important test of the DI theory is to examine whether unstable discs, i.e., those of DN, have $\dot{M}(2) < \dot{M}_{crit2}(r_d)$. By using equation (3.18b) $\dot{M}_{crit2}(P_{orb})$ can be converted to $M_v(P_{orb})$ through Smak's (1989a, 1994a) graphs. The resulting relationship is shown in Figure 3.9. All of the DN lie below the relationship, i.e., they have $\dot{M}(\text{mean}) = \dot{M}(2) < \dot{M}_{crit2}(P_{orb})$ and the Z Cam stars are in the vicinity of the relationship, where small changes in $\dot{M}(2)$ might be expected to carry them into and out of the instability zone, giving rise to alternating episodes of standstills and DN outbursts. It will be seen in Sections 4.3.3, 4.4.3 and 9.4.3 that the stable discs of NLs and CN remnants all lie above the $\dot{M}_{crit2}(P_{orb})$ relationship.

The agreement between observation and theory, first noted by Smak (1982b) and extended by Warner (1987a), further strengthens the case for the DI model: in the MTI theory any such relationship between a critical value in the disc and $\dot{M}(2)$ would be fortuitous.

3.5.4 *Time-Dependent Disc Models*

To proceed from the local stability analysis of an annulus to its global effect requires thermodynamic and hydrodynamic connection of adjacent annuli. The first study of this (Bath & Pringle 1982) simply switched the entire disc from the lower to the upper branch of the S-curve by increasing α and following the consequent adjustment of the mass and angular momentum distributions through solution of the diffusion equation (2.27).

If the disc is fed by gas from the secondary star, then to the right hand side of equation (2.27) must be added a source term $S_\Sigma(r, t)$ specifying where in the disc the matter arrives and how it varies with time. In the MTI model $S_\Sigma(r, t)$ defines the mass transfer pulse; in the DI model it is usually held constant.

A more detailed treatment (Faulkner, Lin and Papaloizou 1983; Lin, Papaloizou & Faulkner 1985) notes that there will be radiative flux transport F_r in a radial direction (equation (2.39a) with $\partial T/\partial r$ in place of $\partial T/\partial z$) and energy advected by mass motion. The latter must take into account work done by compression and the effects of changes in the state of ionization (or dissociation of molecules at low temperatures). After integrating vertically through the disc, the energy diffusion equation becomes

$$c_V\left[\frac{dT(\text{mid})}{dt} - (\Gamma_3 - 1)\frac{T(\text{mid})}{\Sigma}\frac{d\Sigma}{dt}\right] = \frac{9\nu_K\Omega_K^2}{4} - \frac{2F}{\Sigma} - \frac{2H}{\Sigma r}\frac{\partial(rF_r)}{\partial r} + \frac{S_T(r, t)}{\Sigma}, \tag{3.19}$$

where c_v is the specific heat and Γ_3 is the ratio of specific heats. $S_T(r, t)$ is the energy released in the region of the bright spot. This equation has the basic form $dT/dt = f(\Sigma, T, r)$ which, with equation (2.27) and the Σ, T, r relationship of the S-curves, is integrated to give $\Sigma(r, t)$ and $T_{eff}(r, t)$. The latter can be further integrated to give the total luminosity of the disc $L_d(t)$.

3.5.4.1 *Transition Waves*

For a given viscosity law, which may take the form of α varying with r and may be different for the upper and low branches of the S-curve, Σ_{min} and Σ_{max} are simply functions of r. In the case where cycling occurs, $\Sigma(r)$ increases on the lower branch until at some radius $\Sigma_{max}(r)$ is reached. The upward transition of that annulus infects its neighbours and propagates a heating front through the disc. The nature of this, and of the later cooling front, are considered in detail by Meyer (1984), Lin, Papaloizou & Faulkner (1985), Mineshige & Osaki (1985) and Mineshige (1986, 1987, 1988a).

The heating of the first unstable annulus increases its scale height H and therefore, through equation (2.48b), enhances its viscosity – *even for constant* α. The annulus spreads, as in Figure 2.7. The resultant transfer of mass to neighbouring annuli increases their Σ, and, more importantly, the steep radial temperature gradient between adjacent high and low temperature annuli diffuses heat so that the cooler annulus can be heated, triggering its own transition to the high state. In terms of the α-disc formulation, an inward moving heating front (which has a radial width $\sim H$) travels through the disc at a velocity $v_H \sim \alpha_H c_s$ (Meyer 1984; Lin, Papaloizou & Faulkner 1985; Cannizzo 1993a). This is much larger than the viscous drift velocity v_{rad} (equation (2.49)).

The inward moving front propagates with a relatively small increase of Σ because the higher viscosity region lies outside the lower viscosity region and is thus able easily to transport angular momentum outwards. The outward moving front, in contrast, cannot remove angular momentum easily, so a large Σ enhancement occurs which sweeps outwards in the disc, leaving a greatly depleted Σ behind it because $\dot{M} \tilde{\propto} \nu_k \Sigma$ is greatly increased.

From equations (2.32), (3.9) and (3.15), $\Sigma_{max}(r)$ increases with r, therefore, depending on the $\Sigma(r)$ profile that has evolved viscously since the last outburst, the inward travelling front will generally maintain $\Sigma(r) > \Sigma_{max}(r)$, but the outward moving front could reach a region where $\Sigma(r)$ is too small for it to be enhanced above $\Sigma_{max}(r)$.

The DN outburst rise time scale τ_r (Section 3.3.3.5) is related to the time t_r taken by the heating front to traverse the disc by $\tau_r \approx t_r/A_n$, where

$$t_r \approx \frac{r_d}{\alpha_H c_s} \tag{3.20}$$

for a 'complete' outburst, i.e., one where the heating front traverses the entire disc.

Taking $r_d = 0.30a$ (Section 3.3.4.3) and adopting $T = 5 \times 10^4$ K for the temperature of the transition front (Lin, Papaloizou & Faulkner 1985), gives

$$\tau_r \sim \frac{1}{3} t_r = 0.17 \left(\frac{0.1}{\alpha_H}\right) M_1^{1/3}(1)(1+q)^{1/3} P_{orb}^{2/3}(\text{h})\ \text{d} \quad (P_{orb} \lesssim 9\ \text{h}) \tag{3.21}$$

which, for $M_1(1) \approx 1$ (and noting that $(1+q) \sim P_{orb}^{0.4}$ for $1.5 \lesssim P_{orb}(\text{h}) \lesssim 8$) is reasonably close to the observed relationship, equation (3.6). This establishes that $\alpha_H \sim 0.1$ from DN time scales.

As T_{eff} increases radially inwards in a quasi-steady state disc, any heating front starting near the outer edge of the disc and travelling inwards triggers successively hotter annuli, and therefore produces an outburst that starts at longer wavelengths and moves to shorter wavelengths. This outside-in sequence is the Type A outburst of Section 3.3.5.2; an inside-out sequence produces Type B outbursts (Papaloizou, Faulkner & Lin 1983; Smak 1984d).

Further, as first shown by Smak (1984a), because an outward moving front moves more slowly than an inward moving front, the rise time of inside-out outbursts (Type B) is longer than for outside-in ones, leading to more symmetrical outburst light curves. Figure 3.26 shows the evolution of $\Sigma(r)$ during the rising phase of an outside-in outburst.

With most or all of the disc in the high viscosity state, Σ is steadily lowered everywhere until at some radius it is reduced to $\Sigma_{min}(r)$ and that annulus then falls to the lower branch of the S-curve. As the disc in the high state is in a quasi-steady state

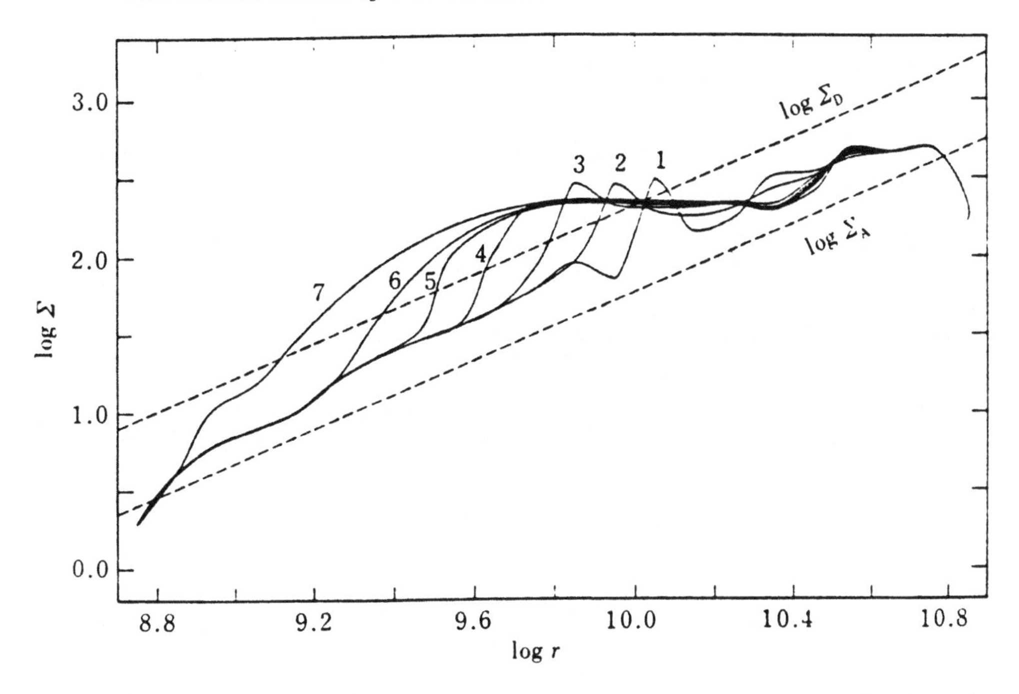

Figure 3.26 Surface density distribution Σ(r) at different stages on the rise of a Type A outburst. $\Sigma_D \equiv \Sigma_{max}$ and $\Sigma_A \equiv \Sigma_{min}$. In numerical order, the elapsed times from the initial state are 63.48, 63.82, 64.04, 64.16, 64.25 and 64.41 in units of 10^5 s. From Mineshige & Osaki (1985).

with $\Sigma(r) \mathrel{\tilde{\propto}} r^{-3/4}$ (from equations (2.50) and (2.51b)), and $\Sigma_{min}(r)$ increases monotonically with r, the outermost annulus will be the first to be depleted below the critical density; as a result the cooling wave always moves from the outer region of the disc to the inner. For this reason, both Type A and Type B outbursts of DN have similar decay light curves, with the UV falling last (Section 3.3.5.2). The effects of the cooling front on $\Sigma(r)$ are shown in Figure 3.27.

The velocity v_c of the cooling front is

$$v_c \approx \alpha_H c_s \left(\frac{H}{r}\right)\left(\frac{r}{\delta r}\right), \tag{3.22}$$

evaluated for $\Sigma = \Sigma_{min}$ in the hot state, where δ_r is the radial width of the front (which, unlike the heating front, is not narrow (Mineshige 1987; Cannizzo, Shafter and Wheeler 1988)). From the model computations, $\delta r/r \approx 0.1$, independent of r. From equation (2.43) this implies $v_H \sim$ few $\times\ v_c$.

Therefore the decay time scale t_d of a DN outburst is

$$t_d \approx \frac{r_d}{10\alpha_H c_s}\left(\frac{H}{r}\right)^{-1} \tag{3.23}$$

and

$$\tau_d \sim \frac{1}{3} t_d = 0.75\left(\frac{0.1}{\alpha_H}\right) M_1^{2/3}(1)(1+q)^{1/6} P_{orb}^{1/3}(\mathrm{h})\ \mathrm{d} \quad (P_{orb} \lesssim 9\ \mathrm{h}) \tag{3.24}$$

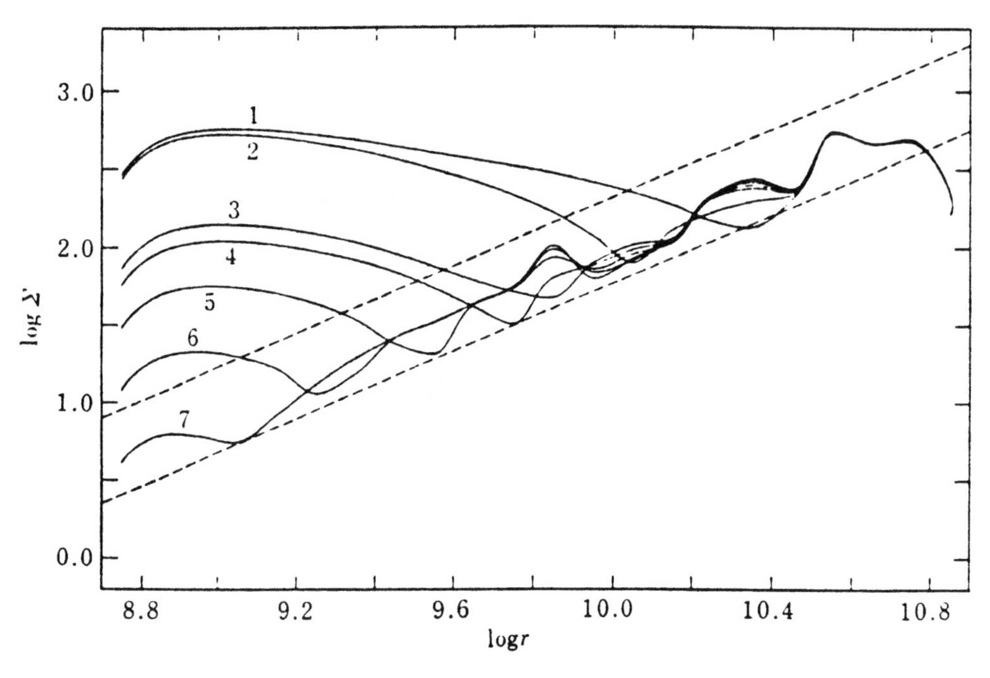

Figure 3.27 The decline branch of the outburst shown in Figure 3.26. Elapsed times are 65.94, 68.42, 73.19, 74.50, 75.58, 76.79 and 77.49 in units of 10^5 s. From Mineshige & Osaki (1985).

using equation (2.42a). Allowing for the higher average $M(1)$ above the orbital period gap (Table 2.3), this gives decay time scales in reasonable agreement with those observed (equation (3.5)), again indicating $\alpha_H \sim 0.1$ in the high state of DN discs. For comparison, the MTI model involves a purely viscous time scale and leads to $\alpha \sim 1$–3: Mantle & Bath (1983), see also equation (2.53). Smak (1984a) measured τ_d as a function of P_{orb} from model light curves and found good agreement with observation.

Note that

$$\frac{\tau_r}{\tau_d} \approx \frac{t_r}{t_d} \tilde{\propto} P_{orb}^{1/3} \tag{3.25}$$

and therefore τ_r/τ_d increases with P_{orb} as found empirically (Section 3.3.3.5).

Accretion onto the primary is enhanced only during the time $\sim \Delta T$ between the arrival of the heating and cooling fronts at the inner edge of the disc. For an inside-out outburst, the duration of the outburst depends on $\Sigma(r_d)$ when the heating front arrives at the outside of the disc. If $\Sigma(r_d) < \Sigma_{min}(r_d)$ then the cooling front immediately returns through the disc, producing a short outburst. On the other hand, if $\Sigma(r_d) > \Sigma_{min}(r_d)$, then the disc will remain in a high state until $\Sigma(r_d)$ is reduced to $\Sigma_{min}(r_d)$. This produces the flat-topped oubursts of long duration (Figure 1.3; Cannizzo 1993b). There is, consequently, usually relatively little time for matter to accrete onto the primary. If the total mass of the disc is $M(d)$ and the mass accreted during outburst is $\delta M(d)$, then outburst models show that $\delta M(d)/M(d)$ ~0.03 for short duration outbursts, increasing to ~0.30 for long outbursts (Mineshige & Osaki 1985; Cannizzo, Wheeler & Polidan 1986; Mineshige & Shields 1990; Ichikawa & Osaki 1992; Cannizzo 1993b). In the words of Cannizzo (1993a): 'the primary role for the limit cycle model is not so much the accretion of matter onto the white dwarf and the production of the outbursts which we see in DN, but rather the sloshing back and forth of the gas which constitutes the accretion disc. In going from quiescence to outburst and back to quiescence again, the bulk of the matter shifts from large radii to small radii and then back to large radii, while at the same time only a small amount of mass loss from the inner edge onto the white dwarf occurs'. This is seen graphically in Figures 3.26 and 3.27 where the initial mass distribution (line 1 in Figure 3.26) and the final mass distribution (line 7 in Figure 3.27) are not greatly different, but in the interim a density wave has run in and out again. The overall process is a relaxation oscillation – which was already suspected more than 50 years ago (Wesselink 1939).

The mass of the disc $M(\mathrm{d}) = 2\pi \int \Sigma(r)\, r\mathrm{d}r$. Therefore, from equations (3.15) and (3.16) (see also Anderson (1988)), the disc mass will be approximately bounded by $\frac{1}{3} M_{\mathrm{max}}(\mathrm{d}) \lesssim M(\mathrm{d}) \lesssim M_{\mathrm{max}}(\mathrm{d})$ where

$$M_{\mathrm{max}}(\mathrm{d}) \sim 7 \times 10^{-10} \left(\frac{\alpha_{\mathrm{C}}}{0.03}\right)^{-0.86} M_1^{-0.35}(1) \left(\frac{r_{\mathrm{d}}}{3 \times 10^{10}}\right)^3 \mathrm{M}_\odot \tag{3.26a}$$

and the minimum mass for discs permanently in the high state (i.e., NLs) is

$$M_{\mathrm{min}}(\mathrm{d}) > 2.3 \times 10^{-10} M_1^{-0.35}(1) \left(\frac{\alpha_H}{0.1}\right)^{-0.8} \left(\frac{r_{\mathrm{d}}}{3 \times 10^{10}}\right)^{3.05} \mathrm{M}_\odot \tag{3.26b}$$

The disc mass in a model for SS Cyg, its change during an outburst and the different $\dot{M}(1)$ profiles for Type A and Type B outbursts are shown in Figure 3.28.

3.5.4.2 Model Light Curves

The time-dependent equations for mass and energy transfer through a disc have been integrated for a wide range of prescriptions for α, convective transport and boundary conditions. Smak (1984a) found only low amplitude outbursts if $\alpha = $ const. Although the pioneering papers by the Faulkner, Lin, Papaloizou group obtained large amplitude outbursts even with constant α, this was because their approach automatically gave $\Sigma_{\mathrm{max}}/\Sigma_{\mathrm{min}} > 3$, which is the contrast required for such outbursts (see discussion by Cannizzo (1993a)). Furthermore, their outbursts were all of the inside-out type.

In general, in order to achieve realistic outbursts it has been found necessary to adopt a larger α in the hot state than in the cool, typically $\alpha_{\mathrm{H}} \sim 0.1$ and $\alpha_{\mathrm{C}} \sim 0.03$ (Smak 1984a; Meyer & Meyer-Hofmeister 1984; Mineshige & Osaki 1983, 1985; Cannizzo, Wheeler & Polidan 1986; Pringle, Verbunt & Wade 1986; Cannizzo & Kenyon 1987; Mineshige 1988a).

One way of introducing a smooth rather than abrupt change of α in the two states is to postulate that $\alpha = \alpha_0 (H/r)^n$. A particular model for magnetic stresses leads to $n = 3/2$ (Meyer & Meyer-Hofmeister 1983b). It may be desirable to use this relationship in conjunction with a saturation limit $\alpha \lesssim \alpha_{\mathrm{max}} \sim 0.4$ (Duschl 1986). Typically $\alpha_0 \sim$50–80. Through this and equation (2.48b), the difference in α in the high and low state implies that the *viscosity* $\nu_{\mathrm{k}} \propto \alpha^{(n+2)/n}$ is two orders of magnitude different in the two states.

Following Cannizzo (1993b), assuming that the quiescent disc fills to an amount of

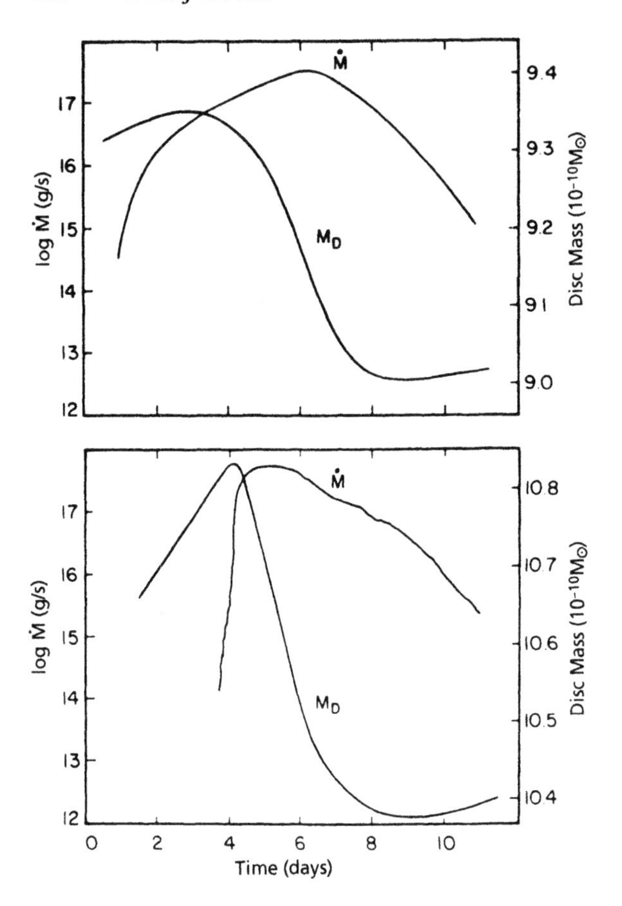

Figure 3.28 Accretion rate $\dot{M}$ at the inner edge of the disc, and disc mass M_D ($\equiv M(\mathrm{d})$) as functions of time. The upper curves are for a Type B (inside-out) outburst and the lower curves are for a Type A (outside-in) outburst. Adapted from Cannizzo, Wheeler & Polidan (1986).

$M_{\max}(\mathrm{d})$ before triggering an outburst, adopting $\Sigma(r)$ for an equilibrium disc on the upper branch of the S-curve (e.g., equation A3 of Cannizzo & Reiff (1992)), and taking $r_\mathrm{d} = 0.8R_\mathrm{L}(1)$, gives

$$\dot{M}(\mathrm{d}) = 6.36 \times 10^{16}\left(\frac{f}{0.4}\right)^{1.43}\left(\frac{\alpha_\mathrm{C}}{0.02}\right)^{-1.23}\left(\frac{\alpha_\mathrm{H}}{0.1}\right)^{1.14} M_1^{0.43}(1)(1+q)^{0.86} P_{\mathrm{orb}}^{1.18}(\mathrm{h}) \;\; \mathrm{g\,s^{-1}}. \tag{3.27}$$

From Smak's (1989a) calibration (modified as recommended in Sections 3.5.4.4 and 4.3.3), this can be converted into the $M_\mathrm{v}(\max)$–P_{orb} relationship shown in Figure 3.10, which is in good agreement with observation.

A comparison of computed outbursts of Types A and B is given in Figure 3.29 and the computed spectrum during the rise of a Type A outburst is given in Figure 3.30. The latter may be compared with the observations given in Figure 3.19. Despite the apparent agreement between observed and calculated UV delay (Smak 1984a, Cannizzo, Wheeler & Polidan 1986) when using Planckian flux distributions, discs with

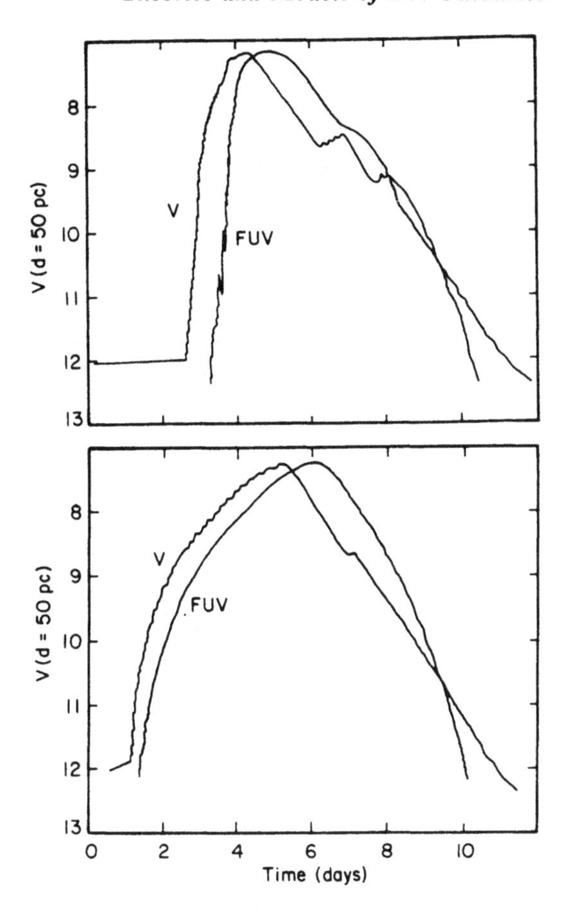

Figure 3.29 The outburst light curves for (upper) the Type B and (lower) the Type A outbursts shown in Figure 3.28. Note the more symmetrical profile and smaller FUV delay of the Type B outburst. Adapted from Cannizo, Wheeler & Polidan (1986).

detailed atmospheres give a much smaller delay (Pringle, Verbunt & Wade 1986; Cannizzo & Kenyon 1987). Modifications to the DI model have been suggested by Mineshige (1988a), who includes the large specific heat of partially ionized hydrogen in the computation of the heating front, which results in a 'stagnation' phase of a disc warmed initially to ~6000 K and only elevated to the upper branch of the S-curve after a delay $\sim$1 d (cf. Section 3.3.5.4). Further work on this model is given by Mansperger & Kaitchuck (1990). Alternatively the disc may spend time at a lower bend in the S-curve (Meyer-Hofmeister 1987; Meyer-Hofmeister & Meyer 1988). However, if viscosity is caused at least partly by the existence of convection, the disappearance of convection at $\sim$6000 K will lower viscosity and cause the disc to spend more time in the vicinity of $T \sim 6000$ K (Duschl 1989). These two mechanisms are not favoured by observations (e.g., Figure 3.19) which show that the visible flux reaches maximum (and therefore that much of the disc is already on the upper branch of the S-curve) before the UV flux has begun to increase.

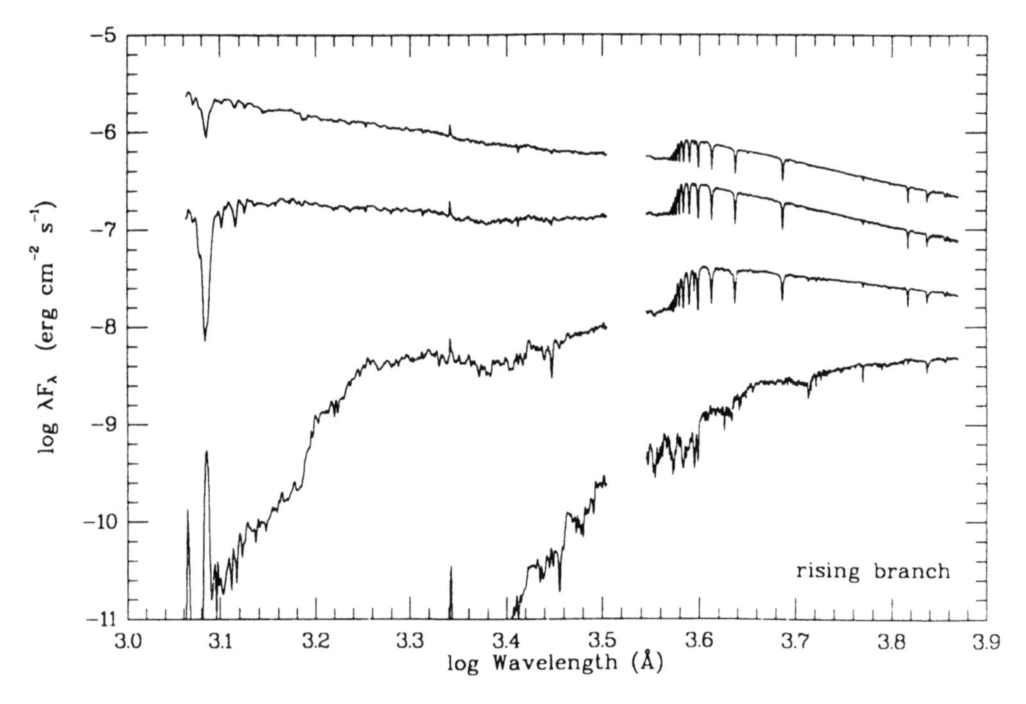

Figure 3.30 Evolution of the spectrum of the rising branch of a Type A (outside-in) outburst, from a model computation. The spectra are (from the bottom) at 0.27, 2.75, 3.02 and 5.33 d respectively. From Cannizzo & Kenyon (1987).

Another way of achieving a UV delay is to eliminate the inner regions of the disc, either by depletion during the quiescent phase via a coronal siphon flow (Meyer & Meyer-Hofmeister 1989, 1994) or through the effect of the magnetosphere of the primary. It has long been postulated (Paczynski 1978; Warner 1983a) that the properties of DNOs (Section 8.6) may be explained by the existence of weak ($\sim 10^{4-5}$ G) magnetic fields on the primaries of DN. Such a field will drain matter from the inner regions of the disc during the quiescent phase. Consequently, at the beginning of an outburst there is a delay while the hot inner regions of the disc are filled on a viscous time scale. Models show that this can produce the $\sim$ 0.5 d UV delay observed in short period DN (Livio and Pringle 1992; Section 7.2.4). HST observations of Z Cha show no sign of any boundary layer throughout quiescence, and only a narrow one immediately after an outburst (Wood *et al.* 1993).

The alternating long–short outbursts, of the kind seen in SS Cyg and several other DN (Section 3.3.1), appeared in Smak's (1984a) early models. There, and in Cannizzo's (1993b) work, they are of Type B and result from an alternation between outbursts where the heating front does not propagate all the way to the outer parts of the disc and ones where it does. The $\Sigma(r)$ remaining after one kind of outburst is that appropriate eventually to trigger the other.

In a given system (i.e., for fixed P_{orb}, $M(1)$, $M(2)$, α_C, α_H) whether outbursts of Type A or Type B occur is a race between build-up of matter at the outer edge of the disc (where the viscous time scale is long) and inward movement of matter in the inner

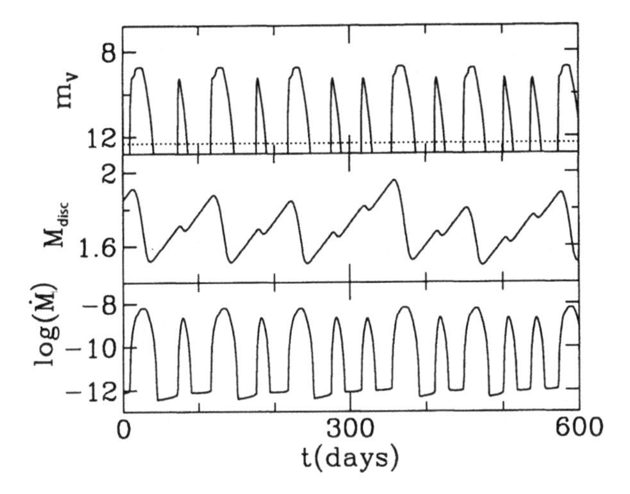

Figure 3.31 Light curve simulation of SS Cyg computed by John Cannizzo with the techniques described in Cannizzo (1993b). The top panel is the Johnson V magnitude, assuming a face-on disc at 100 pc; the middle panel is the disc mass in units of 10^{24} g the lowest panel is log $\dot{M}(1)$ in $M_\odot$ y^{-1}.

regions where the viscous time scale is much shorter. There is a narrow range of $\dot{M}(2)$ where both types could be triggered almost simultaneously (Smak 1984a). Mixed long and short outbursts can occur from both inside-out and outside-in outbursts (Cannizzo 1993b; Ichikawa & Osaki 1994). An example of a DN light curve, calculated with parameters appropriate for SS Cyg, is given in Figure 3.31. Colour variations during an outburst as given by models for Types A and B outbursts (Smak 1984a) agree well with those observed (Sections 3.3.4.1 and 3.3.5.2).

3.5.4.3 Variations in Disc Radii

The observed variations of r_d during and between outbursts (Section 3.3.4.3) provide a check on time-dependent disc models, and in particular act as a diagnostic to discriminate between the MTI and DI theories. The implications of the $r_d(t)$ observations have been discussed from model simulations (Smak 1984a, 1989c; Livio & Verbunt 1988; Wood *et al.* 1989a; Osaki 1989b; Ichikawa & Osaki 1992; Osaki, Hirose & Ichikawa 1993; Lanzafame, Belvedere & Molteni 1993) and analytically (Anderson 1988).

In the MTI model, the addition of a large amount of mass with relatively low specific angular momentum causes a short lived factor of 2 decrease in r_d, which is not present in the DI model (nor in the observations: Smak 1989c). In both models, the subsequent accretion of mass (and angular momentum) onto the primary requires a compensating viscous transfer of mass outward in the disc, with consequent increase of r_d to a maximum radius at maximum light. After maximum the disc shrinks, in the MTI model reaching a minimum as soon as quiescence is reached, that being the point at which a steady state is rapidly achieved in an $\alpha \sim 1$ disc. In the DI model the disc shrinks throughout quiescence because most of the low specific angular momentum material continually accreted into a (low) α_C disc is stored.

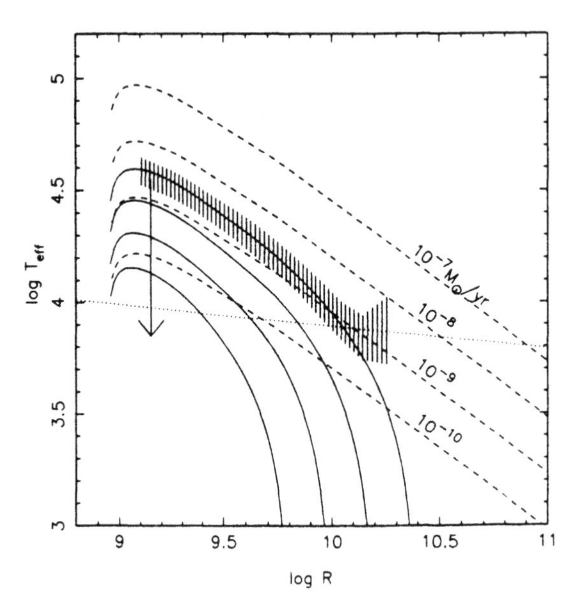

Figure 3.32 Radial temperature distributions for non-steady theoretical discs of Z Cha during outburst (solid curves) at times 0.68, 1.7, 3.2 and 5.5 d from maximum, compared with the observed distribution (hatched area). Dashed lines are steady disc temperature distributions for various mass transfer rates. The dotted line is T_e(crit1) (Section 3.5.3.3). From Mineshige (1991).

The computed variations in r_d are in good agreement with observation; for example Ichikawa & Osaki (1992) find a range $0.32 \lesssim r_d/a \lesssim 0.42$ for U Gem, which is similar to that in Figure 3.14.

3.5.4.4 Disc Radial Temperature Distributions

Just as observed disc radius variations provide independent checks of time-dependent models, so the observed radial temperature distributions (e.g., Figures 2.32 and 2.33) furnish additional opportunities for comparisons.

The temperature distribution in Z Cha in outburst is close to that predicted (equation (2.35)) for a steady state, but would not be expected to fit exactly. The departures are significant and in the direction expected for an evolving disc. Mineshige (1991) has shown that an analytical self-similar solution can be used to fit the observed $T_{\text{eff}}(r)$; the result shown in Figure 3.32 includes a prediction for $T_{\text{eff}}(r)$ later in the outburst.

The departure from steady state implies that $\dot{M}$ is not constant through the disc. Fitting steady state $T_{\text{eff}}(r)$ to the only two DN so far studied in outburst gives, for the inner regions of the discs, $\dot{M}(\text{d}) \approx 2.5 \times 10^{-9} \text{M}_\odot \text{ y}^{-1} = 1.6 \times 10^{17} \text{ g s}^{-1}$ for Z Cha (Horne & Cook 1985) and $\dot{M}(\text{d}) \approx 1.6 \times 10^{-9} \text{M}_\odot \text{ y}^{-1} = 1.0 \times 10^{17} \text{ g s}^{-1}$ for OY Car (Rutten *et al.* 1992).

In quiescence the situation is complicated by uncertainties in the optical thickness of the disc, but the need to treat the problem with non-steady discs arises from the realization that the time required to achieve a steady state will be at least the viscous time scale t_ν which, with $\alpha_C \sim 0.01$ in equation (2.53), is ~ 10 y, which is $\gg T_n$.

Viscously evolving quiescent discs for Z Cha and OY Car have been computed by Mineshige & Wood (1989), taking into account the diffusion of mass both from the $\Sigma(r)$ remaining from a previous outburst and from the mass added by the secondary star. A good fit with the observed flat temperature distributions (Figure 2.33) and the recurrence times T_n is obtained with $\alpha_C \sim 0.01$–0.03 and $\dot{M}(2) \approx 2 \times 10^{15}$ g s^{-1} for both Z Cha and OY Car; these are in good agreement with the $\dot{M}(2)$ independently estimated from the bright spots in these stars (Table 2.2). Smak's (1989a, 1994a) calibrations applied to the M_v(mean) of Figure 3.9 give $\dot{M}$(d) in agreement with these.

Wood, Horne & Vennes (1992) point out that in HT Cas, and by analogy Z Cha and OY Car, the ~6500–9500 K quiescent gas will cool so efficiently that it could only be balanced by viscous heating with $\alpha \sim 10$–200, which is both physically improbable and incompatible with the observed large T_n, and therefore a steady state is not reached.

3.5.4.5 *Theoretical Outburst Intervals*

Following an outburst the surface density in the outer regions of the disc increases by addition of mass from the secondary. At the same time viscous drift is moving matter inwards, and this happens more rapidly at small radii (equation (2.30)). If the former process wins this race then an outside-in outburst is triggered, which, in a simplified model (Ichikawa & Osaki 1994) occurs after an accumulation time

$$T_n \approx 2\pi r_r \Delta r_r \Sigma_{max}(r) \dot{M}^{-1}(2), \tag{3.28a}$$

where Δr_r is the radial width of the accumulation ring at r_r, which after a time T_n is $\sim (\nu_k T_n)^{1/2}$ (equation (2.30)). Then

$$T_n \propto r_r^2 \nu_k \dot{M}^{-2}(2). \tag{3.28b}$$

A more complete discussion (Cannizzo, Shafter & Wheeler 1988) gives

$$T_n = 43.5 \left(\frac{r_d}{10^{10}\ \mathrm{cm}}\right)^{0.42} M_1^{0.53}(1) \left(\frac{\alpha_H}{0.1}\right)^{0.34} \left(\frac{\alpha_C}{0.01}\right)^{-1.23} \xi^k\ \mathrm{d}, \tag{3.29a}$$

where $\xi = \dot{M}(2)/\dot{M}_{crit2}(\mathrm{d})$ and $\delta r/r = 0.1$ has been assumed. (Note that, through equation (3.18), T_n is independent of α_H.) Taking $r_d \propto a$ results in

$$T_n \propto M^{0.67}(1) P_{orb}^{0.28} \alpha_C^{-1.23} \xi^k \tag{3.29b}$$

which predicts that the T_n–ξ relationship should be nearly independent of P_{orb}, For Z Cam stars, at $\xi \approx 1$, equation (3.29a) requires $\alpha_C = 0.03$ (for $\alpha_H = 0.3$) to give $T_n \sim 10$ d as observed. On the other hand, if viscous drift dominates, an inside-out outburst is triggered and $T_n \sim r^2/\nu_k$ (equation (2.30)) is independent of $\dot{M}(2)$ but is inversely proportional to α.

Outburst models, varying $\dot{M}(2)$ to cover the range of possible consequences, have been computed by Duschl & Livio (1989) and Ichikawa & Osaki (1994). Their results, plotted in the T_n–ξ diagram, are given in Figure 3.33, and show the transition from a $k \sim -1$ to -2 relationship at large ξ to T_n independent of ξ for low $\dot{M}(2)$.

Observational comparison is made by transforming M_v(mean) in Figure 3.9 to $\dot{M}(2)$ via Smak's (1989a, 1994a) relationships, evaluating $\dot{M}_{crit2}$(d) from equation (3.18b) and taking T_n from Tables 3.1–3.3. The results are also shown in Figure 3.33 (the point for

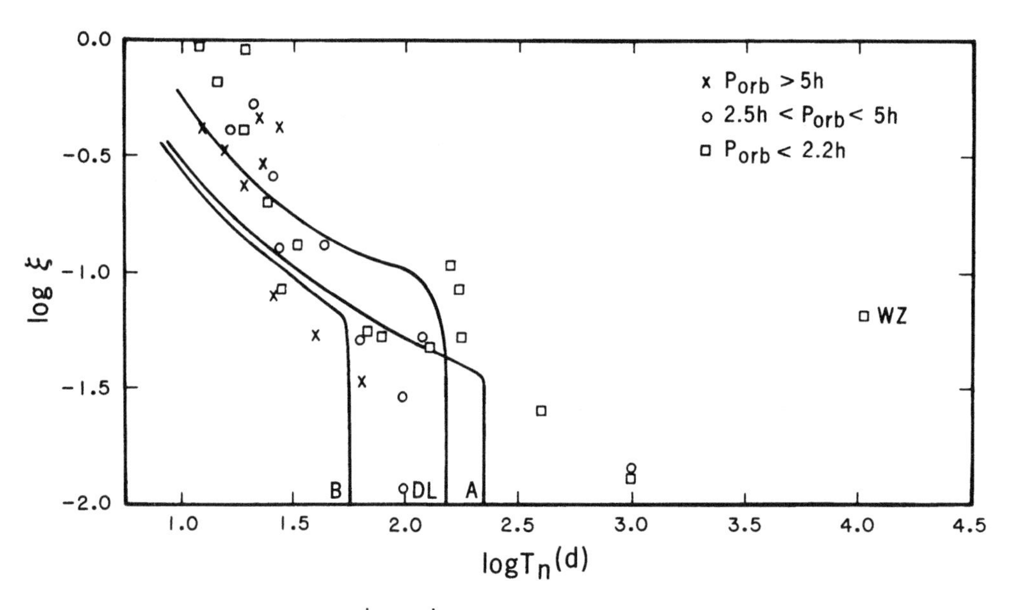

Figure 3.33 Comparison of $\xi = \dot{M}(\mathrm{d})/\dot{M}_{crit2}(\mathrm{d})$ (equation (3.29a)) with T_n, for DN segregated according to orbital period. The curves labelled A and B are from the models of Ichikawa & Osaki (1994). The curve labelled DL is from Duschl & Livio (1989); in this case the precise value of $\dot{M}_{crit2}(\mathrm{d})$ is difficult to assess, so a vertical displacement of the curve is possible. The point for WZ Sge is indicated.

WZ Sge adopts $\dot{M}(2) = 1 \times 10^{15}$ g s^{-1} from Patterson (1984), which agrees with the total energy in outburst, averaged over the 33 y outburst cycle (Sparks *et al.* 1993) with an equal contribution from quiescent mass transfer (Patterson 1984)).

As expected, there does not appear to be a dependence of T_n on P_orb. However, it is clear that some low $\dot{M}(\mathrm{d})$ systems reach parts of the diagram that the existing models do not allow. From equation (3.28) the only unknown parameter that can account for this is α_C, which requires to be lowered by a factor of 10 or more in systems with $T_\mathrm{n}(\equiv T_\mathrm{s}) \gtrsim 1000$ d.

The difficulty of accounting for the largest observed T_n without reducing α_C was noticed by Cannizo, Wheeler & Shafter (1988). Smak (1993b) and Osaki (1994) have deduced $\alpha_\mathrm{C} \leq 0.003$ in WZ Sge. As T_n for these extreme systems, which are all SU UMa stars, is determined by the viscous drift time scale in the inner disc (probably near $r = r_\mathrm{r}$) then

$$T_\mathrm{n} \approx \frac{r_\mathrm{r}^2}{\nu}$$

$$= 27.5 M_1^{1/2}(1) \left(\frac{T_\mathrm{eff}}{4000\ \mathrm{K}} \right)^{-1} r_{10}^{1/2} \alpha_\mathrm{C}^{-1} \qquad (3.30a)$$

from equations (2.10a), (2.23) and (2.48b). The fiducial value $T_\mathrm{eff}(\mathrm{d}) = 4000$ K is adopted from the almost isothermal $T_\mathrm{eff}(r)$ in Z Cha and OY Car (Section 3.5.4.4). For

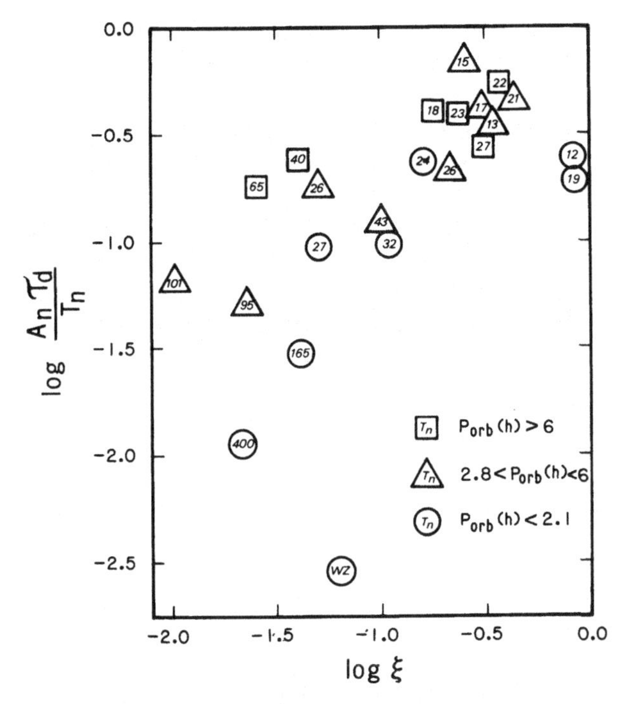

Figure 3.34 Comparison of outburst time scales with the normalized rate of mass transfer ξ (equation (3.29)). Systems are grouped into ranges of orbital period. Outburst intervals $T_{\rm n}$ are indicated.

these ultralow $\dot{M}(2)$ systems, then,

$$\alpha_{\rm C} \approx \frac{25}{T_{\rm n}({\rm d})} M_1^{5/6}(1) \left(\frac{4000\ {\rm K}}{T_{\rm eff}({\rm d})}\right) \left(\frac{0.10}{q}\right)^{0.21} P_{\rm orb}^{1/3}({\rm h}). \tag{3.30b}$$

Another comparison between observation and theory is offered by the ratio $t_{\rm d}/T_{\rm n} \approx A_{\rm n}\tau_{\rm d}/T_{\rm n}$. According to Cannizzo, Shafter & Wheeler (1988)

$$\frac{t_{\rm d}}{T_{\rm n}} \approx 0.12 \left(\frac{\alpha_{\rm C}}{0.01}\right)^{-1.1} \left(\frac{\alpha_{\rm H}}{0.3}\right)^{-1.1} \xi^{-k}. \tag{3.31}$$

The observed quantities, shown in Figure 3.34, exhibit the expected correlation and again require lower $\alpha_{\rm C}$ for systems with $T_{\rm n} \gtrsim 100$d.

3.6 Superoutbursts: The SU UMa Stars

To the traditional definition of SU UMa stars, as DN having superoutbursts, has been added the requirement that they show the superhump phenomena described in Section 3.6.4.1 This is currently the operational definition of an SU UMa star; if DN are ever found with superoutbursts lacking superhumps they will define a class of their own.

At quiescence the SU UMa stars appear to form a natural extension of U Gem stars to shorter orbital periods. With the exception of TU Men, all of the SU UMa stars (Table 3.3) have $P_{\rm orb} < 2.1$ h, i.e., they fall below the orbital gap. It is probable that, on further investigation, most of the stars in Table 3.4 will be found to be SU UMa stars;

the alternative, but not exclusive, classification is that of a magnetic CV. (SW UMa is possibly an intermediate polar and V2051 Oph is possibly a low field polar: Sections 8.1 and 6.1.)

In addition to the stars in Table 3.4, which are SU UMa candidates on the basis of large amplitude outbursts and $P_{orb} \lesssim 2.1$ h, there are a number (without known P_{orb}) judged solely on outburst behaviour. These are (Vogt & Bateson 1982; Kuulkers 1990; O'Donoghue *et al.* 1991; Richter 1992; Howell & Szkody 1995): FV Ara, UZ Boo, AK Cnc, V1504 Cyg, IO Del, DV Dra, AH Eri, V592 Her, GW Lib, V358 Lyr, AO Oct, QY Per, V336 Per, V551 Sgr, FQ Sco, HW Tau, V421 Tau, YY Tel. Reviews of the SU UMa stars are given by Warner (1976a, 1985b), Patterson (1979d), Vogt (1980) and Kuulkers (1990).

3.6.1 *Morphology of Superoutbursts*

The distinctiveness of superoutbursts, relative to normal outbursts, is seen in the light curve of VW Hyi, Figure 3.35, which is characteristic of SU UMa stars. The superoutbursts are brighter than normal ones by about 0.7 mag, have an extended 'plateau' of brightness (which is more a cuesta than a mesa) and hence are longer in duration by factors of 5–10 than normal outbursts in the same system. An obvious exception is those systems showing *only* superoutbursts, in some of which rare and as yet unobserved normal outbursts might exist. However, as it will be seen in Section 3.6.4 that superoutbursts appear to be 'triggered' by normal outbursts, the latter must be enumerated among the former.

In VW Hyi, which is the best documented case (Bateson 1977b), the majority of the superoutbursts have almost identical profiles. However, as seen in Figure 3.35, in many cases a normal outburst interferes with the beginning of a superoutburst. When the final rises to superoutburst are superimposed it is evident then that all superoutbursts are essentially identical (Marino & Walker 1979: Figure 3.36).

The sloping plateau is a universal feature of superoutbursts. This is not a common, and certainly not a recurrent, feature in the long outbursts of U Gem stars (e.g., Figures 1.3 and 1.4). The plateau region is often preceded by a somewhat brighter phase lasting one or two days.

There is no recorded instance of a normal outburst occurring on, or immediately at the end of, the declining part of a superoutburst, as would be the case if normal and superoutbursts coexisted independently of each other. There is an effect, however, that can lead to the impression of a normal outburst on superoutburst decline. WZ Sge (Patterson *et al.* 1981), AL Com (Bertola 1964) and SU UMa (Udalski 1990b) all show short lived minima of more than 2 mag depth on their declining branches. In WZ Sge (Figure 3.37) this occurs only in the 1978 outburst, and at a phase when, in the 1946 outburst, WZ Sge fell finally and irrevocably from its plateau. The appearance is not of a short lived normal outburst, but rather of a rejuvenation of a superoutburst, enabling the decline to continue slowly for a month or more, eliminating the usual plunge from the plateau.

3.6.2 *Superoutburst Frequencies*

The mean recurrence times T_s of superoutbursts are given by Bateson (1979a, 1988), van Paradijs (1985), Ritter (1990) and O'Donoghue *et al.* (1991). As can be seen from

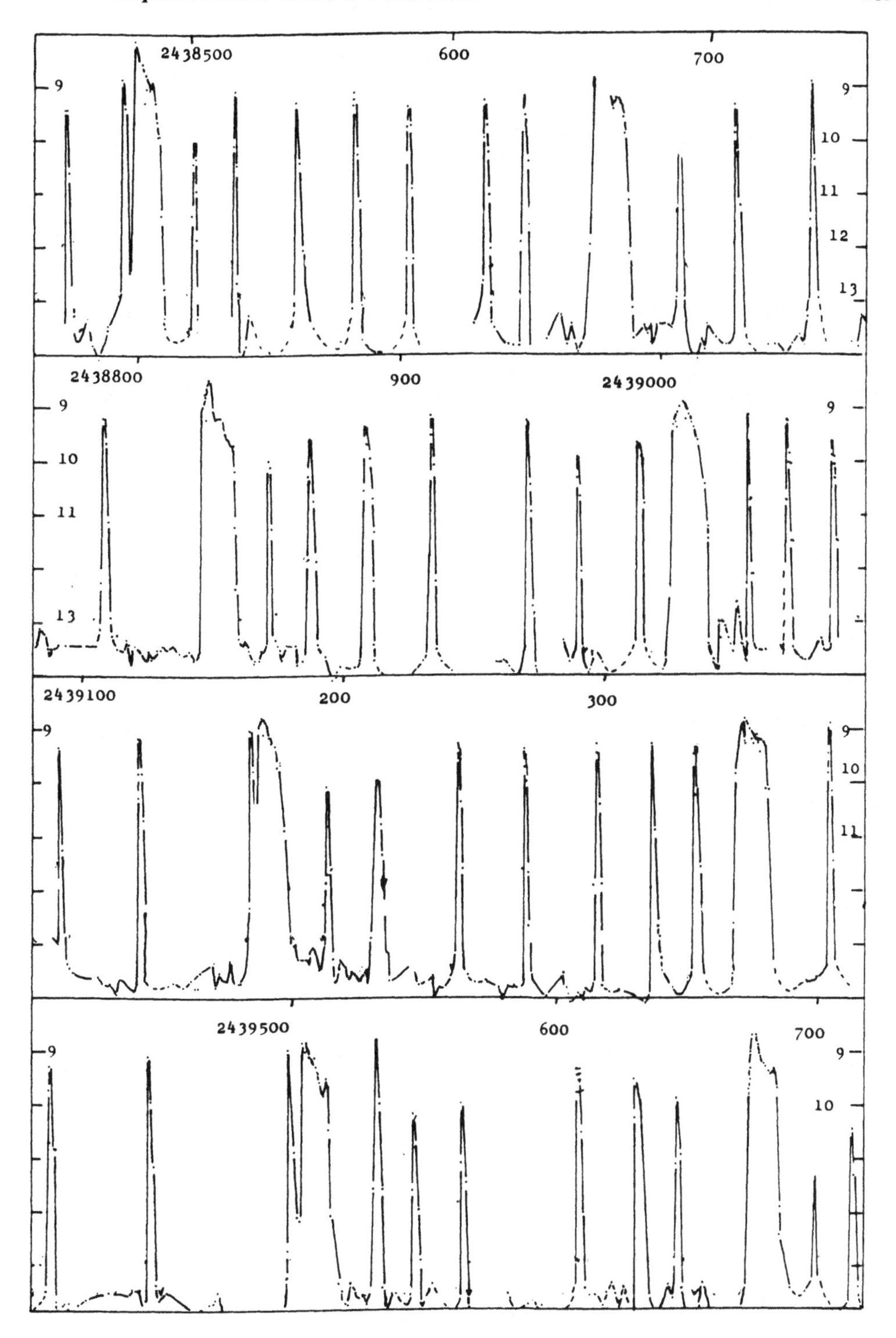

Figure 3.35 Light curve of VW Hyi, from observations made by the Variable Star Section of the Royal Astronomical Society of New Zealand. Adapted from Bateson (1977b).

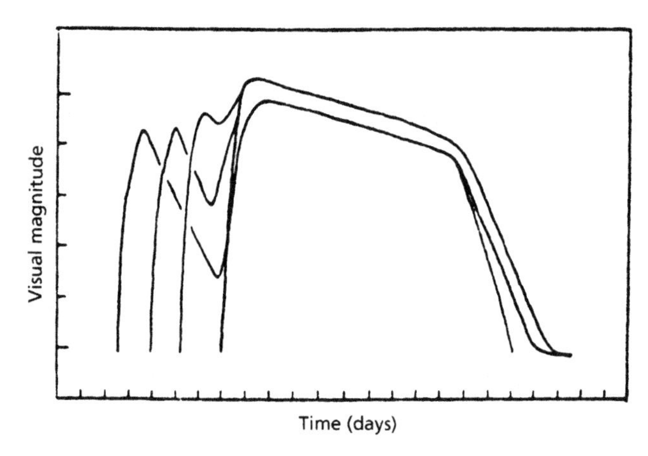

Figure 3.36 Superposition of superoutburst light curves of VW Hyi, aligned according to the final rise to supermaximum. Adapted from Marino & Walker (1979).

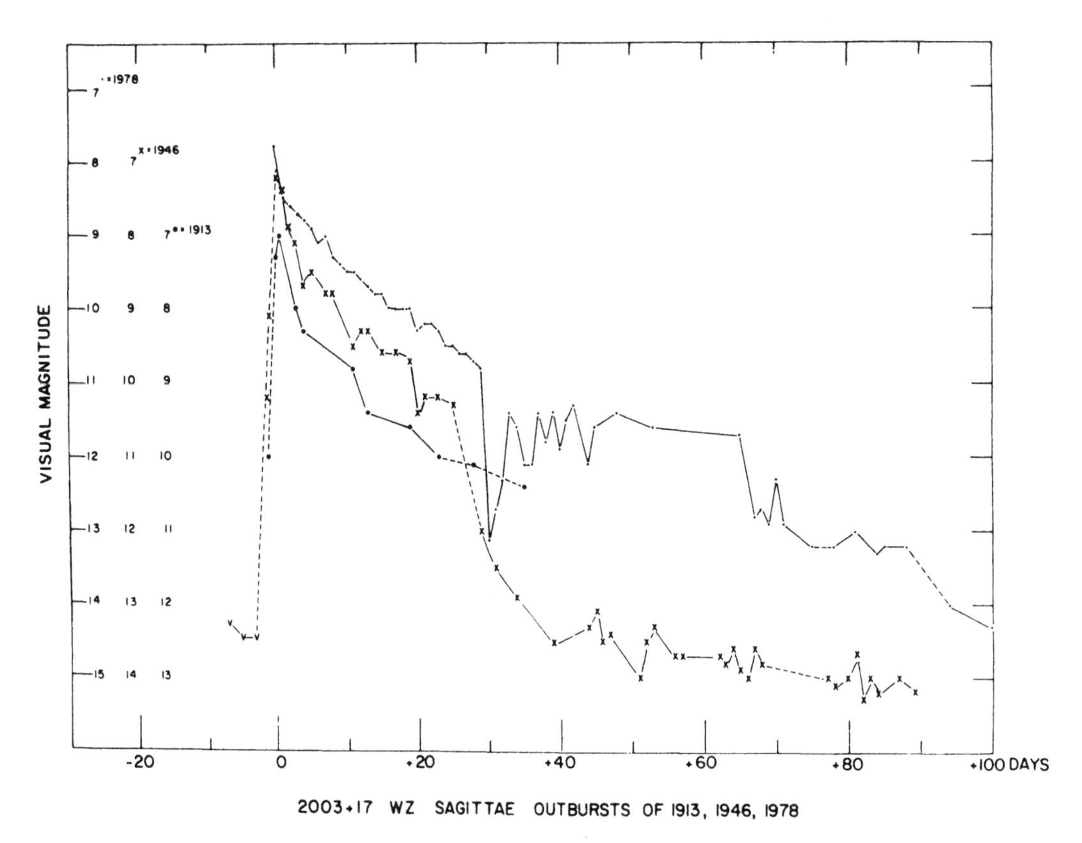

Figure 3.37 Light curves of superoutbursts of WZ Sge in 1913, 1946 and 1978, aligned according to supermaxima. From observations by AAVSO. From Patterson *et al.* (1981).

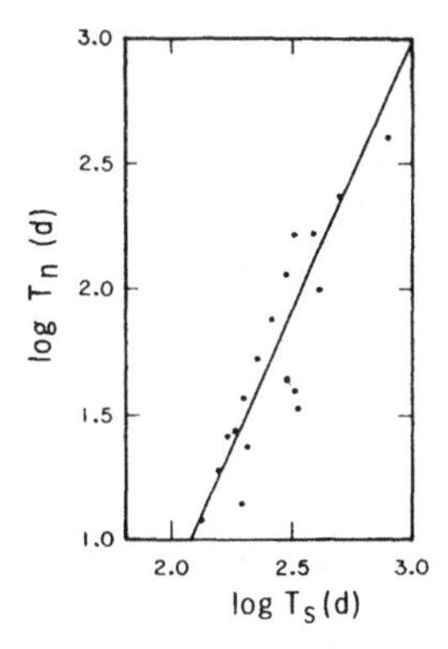

Figure 3.38 Comparison of normal and superoutburst recurrence times.

Table 3.3, the observed frequency of superoutbursts relative to normal ones lies in the range $1 \leq T_s/T_n \lesssim 14$. In a given DN the number of normal outbursts occurring between successive superoutbursts can vary widely: for VW Hyi, which has very regularly recurring superoutbursts, the number of normal outbursts ranges from 3 to 8.

3.6.3 Correlations

There is a minimum observed value $T_s \approx 130$ d and a significant correlation between T_s and T_n (Vogt 1981; Warner 1987a), shown in Figure 3.38, which has the mean relation

$$\log T_s = (0.46 \pm 0.05)\log T_n + (1.62 \pm 0.09). \tag{3.32}$$

Thus $T_s = T_n$ when $\log T_s = 2.99 \pm 0.17$, i.e., when $660 \leq T_s = T_n \leq 1450$ d, which is compatible with the apparent absence of normal outbursts (i.e., distinct from superoutbursts) for $T_n \gtrsim 400$ (Table 3.3). An implication of equation (3.32) is that the number of normal outbursts between superoutbursts, T_s/T_n, is approximately *inversely proportional* to T_s.

In VW Hyi, Smak (1985b) classified supercycles (i.e., the behaviour from one superoutburst to the next) into two kinds: Type L in which the average length of the last two normal cycles is greater than 30 d, and Type S in which it is shorter than 23 d. However, van der Woerd and van Paradijs (1987) doubt the statistical significance of the separation, and the fact that Types S and L are equally common, have the same average length and are distributed at random supports this. Nevertheless, the Type S supercycles have a spread in length (standard deviation $\sigma = 10$ d) much less than that of type L ($\sigma = 20$ d). Even for the latter the spread ($\sigma \sim 10\%$) is much less than the normal outburst intervals (which range over 8 to 76 d). This appears to be a characteristic of superoutbursts in SU UMa stars with $T_s \lesssim 300$ d (it may also be the case for at least some with longer T_s but the evidence is not conclusive).

From their studies of VW Hyi, Bateson (1977b), Smak (1985b) and van der Woerd and van Paradijs (1987) find that (a) the first normal outburst after a superoutburst occurs within a smaller range (9–24 d) than outbursts in general, (b) there is a strong dependence of mean length of outburst cycle after superoutburst, increasing from 17 d to 30 d midway through the supercycle, after which there is an increase in the spread, and (c) the last two cycles before a superoutburst are closely similar in length. This last effect is also seen in Z Cha (Kozlowska 1988).

Van der Woerd and van Paradijs (1987) give evidence for the existence of a distinct 'relaxation time' of $T_r \approx 170$ d in VW Hyi. All normal outbursts trigger a superoutburst if they occur at a time at least T_r after the previous superoutburst; if they occur before this period they only trigger a superoutburst if they are sufficiently energetic. In other SU UMa stars it is found (Vogt 1980) that the spread in supercycle lengths is roughly equal to $T_n/2$ for each system, in agreement with the proposal that superoutbursts are triggered by normal outbursts.

Vogt (1980) plotted O–C diagrams, relative to a period equal to the mean T_s, for individual superoutbursts in a number of SU UMa systems. These diagrams show a sequence of more or less linear sections, suggesting a frequent switching from one T_s to another. However, this conclusion needs to be reassessed in the light of the similar O–C diagrams produced by Koen (1992) for stochastic variation.

Van Paradijs (1983) suggested that the bimodality of outburst durations seen in the U Gem stars (Section 3.3.1) may be an extension of the super/normal outburst phenomenon to longer orbital periods. However, as seen by comparison with Section 3.3.1, the statistical behaviour of the wide outbursts in U Gem stars is quite different to that of superoutbursts. Furthermore, the phenomenon of 'superhumps' which is unique to superoutbursts (Section 3.6.1), has been found to be absent in both short and long outbursts of SS Cyg in a campaign (Honey *et al.* 1989) specially organized to look for them (nor have they been seen in any outbursts of other U Gem systems). It is interesting to note (Bateson 1979b, 1981; Warner 1985b) that the SU UMa star with the longest known orbital period, TU Men, has a trimodal distribution of outburst widths, with narrow outbursts (~ 1 d), wide (~ 8 d) and superoutbursts (~ 20 d). The last are verified to be superoutbursts by the presence of superhumps (Stoltz & Schoembs 1984). YZ Cnc, the SU UMa star with the next shortest P_{orb} after TU Men, may also be trimodal (Patterson 1979d).

It is noticeable that the four stars WX Cet, WZ Sge, SW UMa and HV Vir, which are those with exceptionally long outburst intervals (temporarily grouped into a WZ Sge subclass – see Section 3.3.1), have the shortest known orbital periods. Apart from this there is no obvious correlation between T_s and P_{orb} .

3.6.4 Photometry of Superoutbursts

The brightness and colour changes on the rise to a superoutburst are in general indistinguishable from those of normal outbursts in the same system. For example, the delay in rise at short wavelengths is the same in VW Hyi for both kinds of outburst (Verbunt *et al.* 1987) and the rates of rise in the optical are identical (Vogt 1974; Warner 1975b; Bateson 1977b; Walker & Marino 1978; Pringle *et al.* 1987). This is compatible with the idea that superoutbursts start as normal outbursts, many of which are seen as distinct precursors (Section 3.6.1). In fact, in the October 1984 superoutburst of VW Hyi, a precursor resembling an ordinary outburst was observed in the UV, more especially in the FUV fluxes, which was not detectable at optical wavelengths (Polidan & Holberg 1987).

The compilations of UBV photometry referenced in Section 3.3.4.1 include observations for superoutbursts. The supermaximum brightness, m_V(s) (see Table 3.3),

Table 3.7. *Superoutburst Time Scales.*

Star	P_{orb}(h)	T_n(d)	T_s(d)	ΔT_{pl}(d)	τ_{pl}	ΔT_{sh}(d)	References to ΔT_{sh}
WX Cet	1.25:		1000:	18:	8:	$>5,<10$	1
WZ Sge	1.36		12 000	26	10–12	13	3
SW UMa	1.36		954	20	9.5	$>3,<7$	4,13
HV Vir	1.39		3500:	23	11:	$3\frac{1}{2}$	2
AQ Eri	1.46	44	300:	7	8:		
V436 Cen	1.50	32	335	10	9	$\lesssim 1$	12
OY Car	1.51	160:	346	15	10:	2–3	5,6
VY Aqr	1.52	400:	800:	11	10		
EK TrA	1.53	231	487	13	9		
RZ Sge	1.58	77	266	11	8:		
AY Lyr	1.76	24	205	11	8:		
VW Hyi	1.78	27	179	9	7	$\lesssim\frac{1}{2}$	7
Z Cha	1.79	51	218	11	8:	~ 2	8,9
WX Hyi	1.80	14	195	10	10		
SU UMa	1.83	19	160	10	9	~ 2	10
EF Peg	1.96		254	16	9.1	$2\frac{1}{2}$	14
V344 Lyr	2.07:	16:	240:		10.6		
YZ Cnc	2.08	12	134	11	8:	<1	11
TU Men	2.82	37	194	24	12		

References: 1. O'Donoghue *et al.* 1991; 2. Leibowitz *et al.* 1994; 3. Patterson *et al.* 1981; 4. Robinson *et al.* 1987; 5. Krzeminski & Vogt 1985; 6. Naylor *et al.* 1987; 7. Haefner, Schoembs & Vogt 1979; 8. Warner & O'Donoghue 1988; 9. Kuulkers *et al.* 1991a; 10. Udalski 1990b; 11. Patterson 1979d; 12. Semeniuk 1980; 13. Granzlo 1992; 14. Howell *et al* 1993.

is 0.5 – 1.0 mag brighter than the mean normal maximum m_v(max). (Occasionally, however, normal can rival supermaxima – see Figure 3.35.) The colours at supermaxima are ~0.10 bluer in both U–B and B–V than normal maxima. The excess brightness declines through the duration of the plateau, and the colours redden, so the final rapid decline usually occurs from approximately m_v(max) and is very similar to that of a normal outburst. The rates of rise and final decline for most superoutbursts fit the relationships given in Section 3.3.3.5. In VW Hyi (Verbunt *et al.* 1987) and AY Lyr (Szkody 1982) the UV flux distributions are very similar on the declines of normal and superoutbursts.

The distinctive sloping plateau has a duration ΔT_{pl}, measured from supermaximum to the start of the rapid decline, and ranges over a factor of 4 but is correlated with T_s or P_{orb} (Table 3.7). On the other hand, the slope of the plateau, measured avoiding the short-lived peak often present at supermaximum, appears almost invariant at $\tau_{pl} \sim 9 \pm 1$ d mag^{-1} (Table 3.7; ΔT_{pl} and τ_{pl} are measurements made by the author on published light curves, mostly deriving from the AAVSO and the Royal Astronomical Society of New Zealand).

The overall impression is that a superoutburst starts as a normal outburst, but is sustained by a process that adds to its luminosity and keeps it in an outburst condition beyond its normal means.

Although the extra brightness $\Delta m = m_V(\text{max}) - m_V(\text{s})$ is ~ 1.0 for most SU UMa stars, for those of exceptionally large T_n and T_s the difference is larger. Whereas in VY Aqr, with $T_n \sim 1$ y, $\Delta m \sim 0.5$ mag (Della Valle & Augusteijn 1990), in WX Cet with $T_n \sim 3$ y, $\Delta m \sim 2$–3 mag and the same may be the case in RZ Leo (O'Donoghue *et al.* 1991). In some stars there is evidence that the brightness at supermaximum varies widely from outburst to outburst – covering the range 8.0–11.0 in VY Aqr and 9.5–10.5 in WX Cet (Bailey 1979b).

In most cases, m_V at the end of the plateau corresponds approximately to $m_V(\text{max})$. That this may be true even for the systems with large T_s is seen for WZ Sge: taking $m_V(\text{max}) = m_V(\text{end plateau}) = 10.2$ from Figure 3.37, and using a correction $\Delta M_V = -1.2$ from equation (2.63), and a distance of 45 pc (Smak 1993b) gives $m_V(\text{max}) = 5.7$, which is close to the relationship shown in Figure 3.10.

3.6.4.1 *Photometry of Superhumps*

The December 1972 superoutburst of VW Hyi was observed independently by Vogt (1974) and Warner (1975b) who found that, although the orbital hump ceased to be detectable during the bright parts of the outburst, a prominent periodic hump appeared at about the time of maximum light and was visible almost to the end of the superoutburst. Unexpectedly, the 'superhump' period was about 3% longer than the orbital period. Superhumps were also seen in the VW Hyi superoutburst one year later, but not then recognized as being periodic (Marino & Walker 1974).

Since that time, superhumps with periods a few per cent longer than P_{orb} have been observed in every SU UMa star for which high speed photometry during a superoutburst has been obtained. Figure 3.39 shows the development of superhumps during the final rise to supermaximum in an outburst of V436 Cen (Semeniuk 1980). At their full development the superhumps have a range of 0.3–0.4 mag and are equally prominent in all SU UMa stars *independent of inclination* (even stars like V436 Cen, WX Hyi, and SU UMa, which have no detectable orbital humps during quiescence, have full amplitude superhumps).

When at their maximum amplitude, soon after they have formed, the superhumps in all systems have an almost triangular profile (Figure 3.39). Their amplitude decreases faster than the system brightness, causing them to disappear at roughly the end of the long 'plateau' of superoutburst, and they broaden in profile or add further components between the principal humps (Figure 3.40; Haefner, Schoembs & Vogt 1979; Patterson 1979d; Warner 1985b; Warner & O'Donoghue 1988; Udalski 1990b). In the high inclination systems the amplitude of the superhumps is modulated at the beat period P_b (Section 3.6.4.3: Warner 1985b; Schoembs 1986), as also is the background brightness between the humps (Schoembs 1986).

The superhumps (and the late superhumps: Section 3.6.4.4) frequently show detail in their profiles that repeats, albeit in a slightly distorted form, over several cycles or even from night to night (Figure 3.41; see also Figures 2a and 2b of Schoembs & Vogt (1980)).

Multicolour photometry of superhumps shows that the superhump light is bluest at minimum and reddest at superhump maximum. This is more pronounced towards

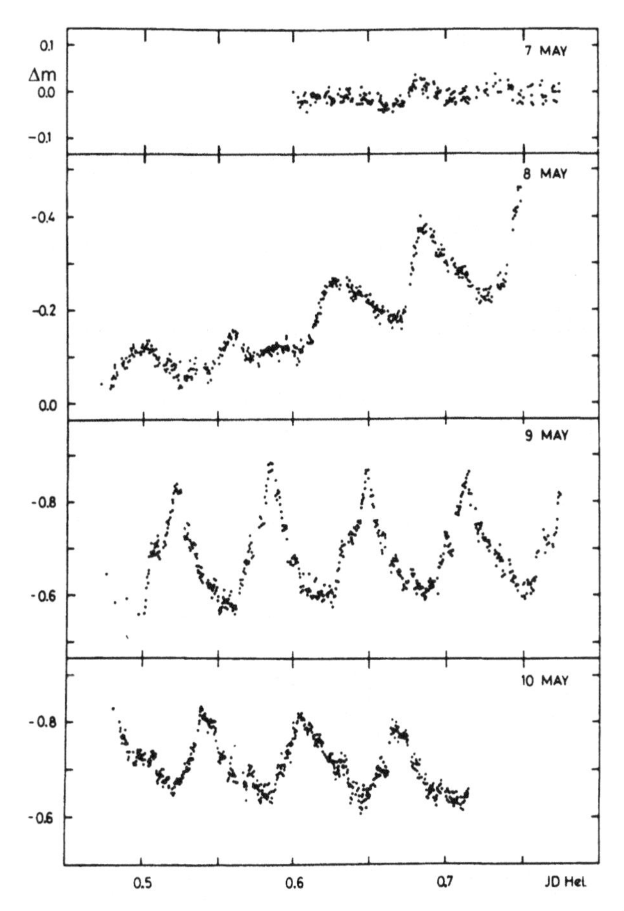

Figure 3.39 Development of superhumps in V436 Cen. The ordinate is magnitude difference from a comparison star. From Semeniuk (1980).

shorter wavelengths: the flux distribution is almost flat at B and V wavelengths (Schoembs & Vogt 1980; Stolz & Schoembs 1984; Hassall 1985a; Naylor *et al.* 1987; van Amerongen, Bovenschen & van Paradijs 1987). Thus there appears to be an inverse correlation between colour temperature and brightness, implying substantial changes in area of the superhump light source (SLS).

Colour and brightness temperatures for the SLS have been estimated as $\sim$5700 K in EK TrA (Hassall 1985a). From the observed fluxes in the superhumps, these lead to areas for the SLS of $\sim$ few $\times\ 10^{20}$ cm^2, or a substantial fraction of the area of the disc or the secondary.

3.6.4.2 Onset of Superhumps

Figure 3.39 constitutes a rare example of the direct observation of the growth of superhumps (for OY Car see Krzeminski & Vogt (1985)) – in most cases the superhumps appear at full amplitude between one night and the next. In VW Hyi and Z Cha coverage is sufficient to show that the superhumps appear about one day after

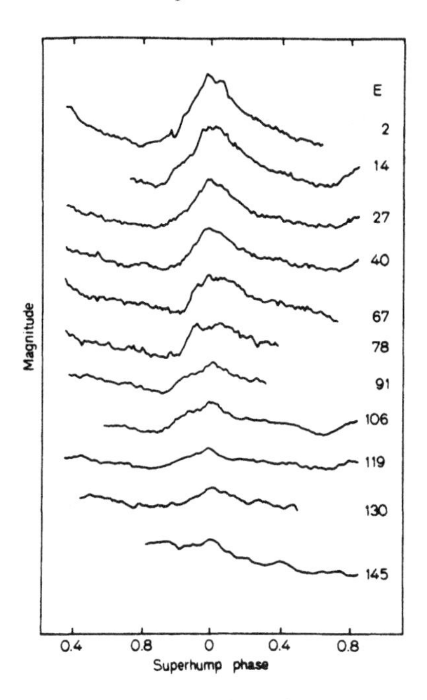

Figure 3.40 Evolution of superhump profiles during a superoutburst of VW Hyi. The values of E are the number of orbital cycles lapsed since the start of observations (VW Hyi has 13.5 cycles d^{-1}). From Haefner, Schoembs & Vogt (1979).

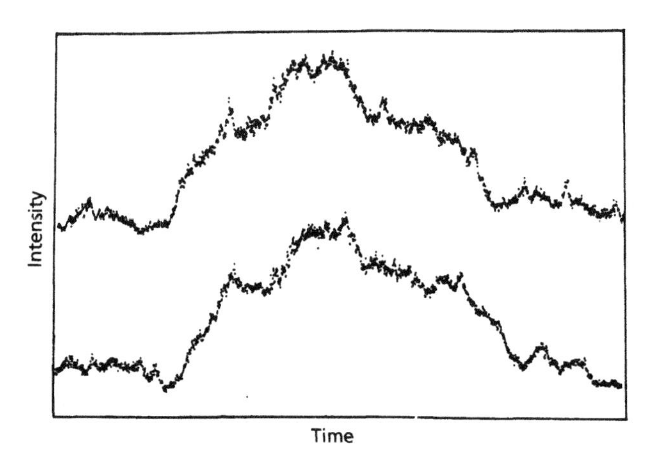

Figure 3.41 Comparison of consecutive superhumps in VW Hyi. From Warner (1986b).

supermaximum (e.g., V436 Cen: Semeniuk 1980). However, there is accumulating evidence that the delay ΔT_{sh} between time of supermaximum and first appearance of superhumps is a function of T_n, T_s or P_{orb}: see Table 3.7, which suggests $\Delta T_{sh} \tilde{\propto} T_n^{0.3}$, or a very surprising sensitivity to P_{orb} .

3.6.4.3 The Interregnum

The light curves obtained during the interregnum between attainment of maximum brightness and the onset of superhumps show a variety of behaviours, certainly dependent on i and perhaps on T_s. The implications of these have not been fully worked out.

In the non-eclipsing VW Hyi, apart from the instance of an enhanced orbital hump in a precursor normal outburst (Section 3.3.4.2), there are no modulations detectable at either P_{orb} or P_s (Marino & Walker 1979; Haefner, Schoembs & Vogt 1979; Marino *et al.* 1984). In the eclipsing systems of OY Car and Z Cha the interregnum light curve has recurrent dips at phases 0.25 and 0.75, with an associated minor maximum near phase 0.5 (Krzeminski & Vogt 1985; Naylor *et al.* 1987; Warner & O'Donoghue 1988; Kuulkers *et al.* 1991a); the same features appear in V2051 Oph during its high state (Warner & O'Donoghue 1987). Similar dips are seen in X-ray light curves of low mass X-ray binaries (Mason 1986) and are there ascribed to increases in vertical thickness of the disc at the corresponding phases. The absence of these dips in ordinary outbursts of OY Car and Z Cha could suggest that immediately before the growth of superhumps an enhanced $\dot{M}(2)$ is adding vertical structure to the disc. However, there is no prominent orbital hump, eclipses do not show the asymmetry caused by a bright spot and (in consequence) eclipse mapping does not reveal a spot (Warner & O'Donoghue 1987); therefore any enhancement of $\dot{M}(2)$ is not as great as in the singular observation of VW Hyi.

On the other hand, WZ Sge during its 1978 superoutburst showed a prominent orbital hump for the first 12 d, which, if due to enhanced mass transfer, would require an $\dot{M}(2)$ increase of $\sim$250 initially, decreasing with time, following which superhumps develop but the enhanced orbital modulation continues with an implied $\dot{M}(2)$ enhancement of $\sim$60 (Patterson *et al.* 1981). However, Smak (1993b) finds that the orbital hump can be quantitatively understood as arising from the irradiated surface of the secondary.

In IUE spectra, WZ Sge possessed an absorption dip at phase 0.7 fifteen days after maximum, which had disappeared ten days later. This is interpreted as increased vertical extent in the region of the bright spot (Naylor 1989).

3.6.4.4 Superhump Periods: The Beat Cycle

Observed over a few days the mean period P_s of the superhumps can be derived with an accuracy $\sim$0.00005 d. However, this gives a false impression of the stability of the process: the systems with greatest coverage show that in many cases there is a steady reduction in P_s during the outburst. Expressing times of superhump maxima as

$$T_{max} = T_0 + P_s E + C E^2, \tag{3.33}$$

where E is the number of elapsed cycles, VW Hyi has $C = -2.85(\pm 0.13) \times 10^{-6}$, repeating in different outbursts (Haefner, Schoembs & Vogt 1979; van Amerongen, Bovenschen & van Paradijs 1987) and other systems also show statistically significant coefficients, all negative[3] (see Table 2 of Warner (1985)b, Udalski (1990b), Kuulkers *et al.* (1991), Lemm *et al.* (1993) and Leibowitz *et al.* (1994)).

There are two ways of describing this change (Warner 1985b), either as a linear decrease in P_s amounting to about 1.25% during superoutburst, in which case the accumulated phase shift relative to the initial period is $\sim 1\frac{1}{2}$ cycles by the end of outburst, or as a quadratic phase shift relative to a constant P_s, in which case the maximum phase shifts are $\sim 90°$. It is not yet clear whether on the latter interpretation the phase shift could arise merely from a variation in the profile of the superhumps (Patterson 1979d), but if so it is not accompanied by any noticeable asymmetries (see, e.g., Figure 9 of Haefner, Schoembs & Vogt (1979)).

The values of P_s quoted in Table 3.3 are mostly taken from Ritter (1990, 1992) and are representative of the inhomogeneity of values in the literature: some are averages over several days and some are specifically of a P_s observed close to the time of supermaximum.

It can be detected in Table 3.3 that the difference $\Delta P_s = P_s - P_{orb}$ decreases as P_{orb} decreases. There are indeed good correlations in the ΔP_s or P_s/P_{orb} versus P_{orb} diagrams (Stolz & Shoembs 1984; Robinson *et al.* 1987; Molnar & Kobulnicky 1992), which may be used to predict P_{orb} from photometrically determined P_s and have been used to obtain the asterisked values of P_{orb} in Table 3.3. However, in view of the success of one particular model of superhump phenomena (Section 3.6.6), it has more physical meaning to investigate a different combination of P_s and P_{orb}, namely the *beat period*

$$P_b = \frac{P_s P_{orb}}{P_s - P_{orb}}. \tag{3.34}$$

Values of P_b are given in Table 3.3. Note that because of the inhomogeneity of observation there is an inherent *uncertainty* in P_b: at $P_{orb} = 2\,\mathrm{h}$ a range of 1.25% in P_s translates into a range of $\sim 1.0\,\mathrm{d}$ in P_b. For example, in OY Car P_b changes from 2.7 d to 3.6 d (Hessman *et al.* 1992). A further example will illustrate the problem: in SW UMa Robinson *et al.* (1987) first detected superhumps 5 d after an observing hiatus. The value of P_s when superhumps first started could therefore have been $\sim 0.5\%$ larger than their observed $P_s = 1.3999\,\mathrm{h}$, in which case P_b in Table 3.3 would become 1.8 d instead of 2.2 d. Therefore, until the value of P_s is standardized, e.g., that at the time at which superhumps first appear, any P_b may have an uncertainty (as opposed to an *error*) of $\pm 0.5\,\mathrm{d}$.

Observations of different superoutbursts of a given DN usually show the same P_s at the same stage of development, (e.g., VW Hyi: Vogt 1983a), but small differences may exist which add to the range of P_b. In some cases, e.g., TY Psc, P_{orb} or P_s is poorly determined, which leads to a considerable error in P_b. In Table 3.3 the errors due to these causes are generally $<0.2\,\mathrm{d}$ if not quoted.

Accepting the above caveats, and with the exceptions of WZ Sge and HV Vir, the P_b listed in Table 3.3 are consistent with $P_b \approx \langle P_b \rangle = 2.2\,\mathrm{d}$, independent of P_{orb}. There is perhaps a weak tendency for P_b to increase with decreasing P_{orb}.

[3] The positive C value obtained by Krzeminski & Vogt in OY Car for the first few days of superoutburst is negated by Schoembs' (1986) observation of a smaller P_s in the late stages of superoutburst.

3.6.4.5 Late Superhumps

Vogt (1983a) discovered that in VW Hyi, after the rapid brightness decrease at the end of the bright 'plateau', during the several days in which a slow decrease takes place, orbital humps are visible at approximately their quiescent amplitude, but so also is a modulation at the superhump period (or its extrapolation) shifted in phase by $\sim$180° with respect to the supermaximum humps. These are known as the 'late superhumps' and have also been seen in OY Car (Schoembs 1986; Hessman *et al.* 1992), SU UMa (Udalski 1990b), V1159 Ori (Patterson, private communication) and perhaps in V436 Cen (Warner 1983b). In VW Hyi the late superhump may develop out of a feature seen midway between the humps at maximum (Schoembs & Vogt 1980). The simultaneous presence of the late superhump and orbital hump causes a pronounced modulation with period P_b, which in both VW Hyi and OY Car is detectable well into quiescence (Haefner, Schoembs & Vogt 1979; Schoembs 1986; van der Woerd *et al.* 1988; Hessman *et al.* 1992).

The most detailed study of late superhumps is that by van der Woerd *et al.* (1988), who obtained simultaneous optical, IR and X-ray observations two days after the termination of a superoutburst of VW Hyi. By subtracting the mean quiescent orbital light curve from their observations they isolated the light curve of the late superhumps. From the wavelength independence of the amplitudes, supplemented with optical spectra, they concluded that the late superhump light source (LSLS) is optically thin in the continuum but optically thick in the Balmer emission lines. The LSLS covers an area which is a significant fraction of the area of the accretion disc.

Hessman *et al.* (1992) find that the Balmer emission-line flux closely follows that of the late superhump continuum. This is significant not only for models of the superhumps themselves: it may eventually assist in deciding between the various possible emission line mechanisms for CVs in general (Section 2.7).

3.6.4.6 Location of the SLS: Eclipses and Eclipse Mapping

In the absence of observational pointers, there are several potential sites for the luminosity that generates the superhumps – these are the secondary, the bright spot, the disc, and the primary. Models that locate the SLS in each of these places have been proposed and are briefly discussed in Section 3.6.6. Only relatively recently has it been possible, from analysis of eclipse profiles, to state definitely where the SLS is located.

That the SLS is not situated on the secondary is shown by the fact that when the superhump in Z Cha coincides with an eclipse (i.e., when the SLS is seen at its best, which would be on the side of the secondary facing the observer) then at least part of the additional light is eclipsed (Smak 1985a; Warner & O'Donoghue 1988). In general there are no rapid changes of intensity during ingress or egress, which means that the SLS cannot be as compact as the bright spot at quiescence (Horne 1984), nor be a luminous region on the primary. By process of elimination, the SLS is an extended region of the disc – involving $\sim 1/4$ of its area (Section 3.6.4.1). An exception is a unique eclipse observed early in a superoutburst of Z Cha which shows the emergence of a relatively compact source, which is found to lie very close to the quiescent position of the bright spot (Warner & O'Donoghue 1988; O'Donoghue 1990). This is evidence for enhanced $\dot{M}(2)$, but requires modulation of the bright spot region at period P_s.

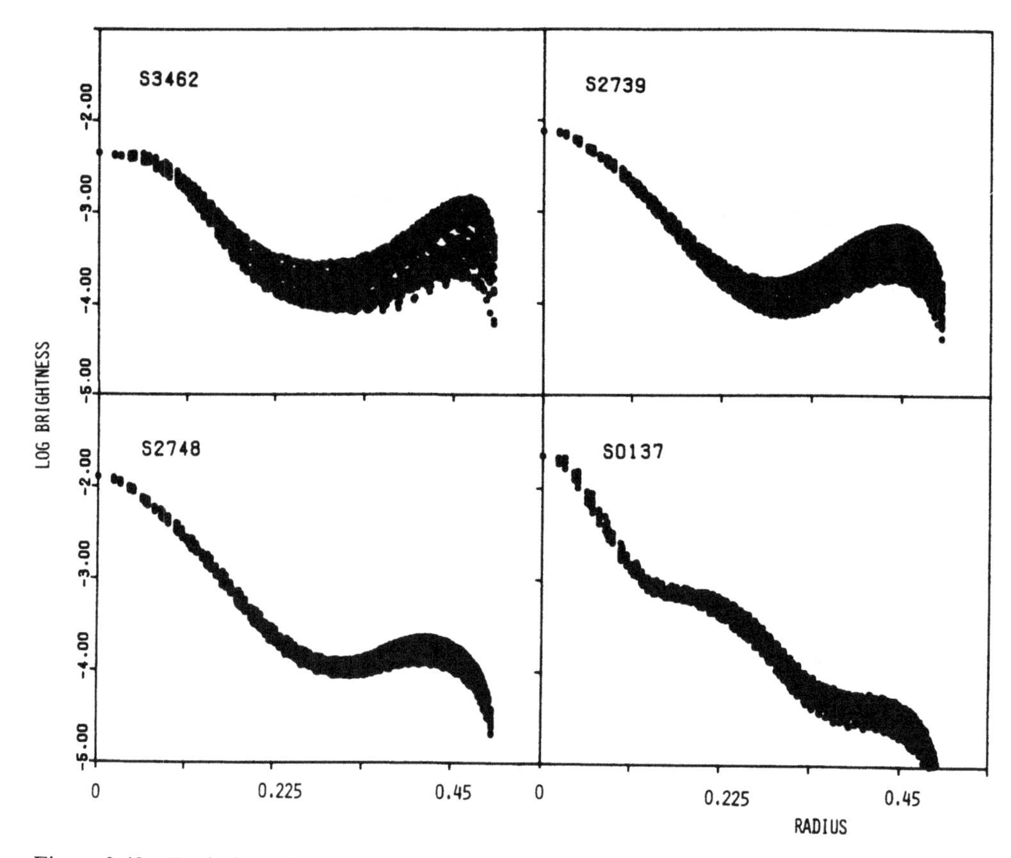

Figure 3.42 Evolution of accretion disc radial brightness distribution during a superoutburst of Z Cha. S3462 was obtained in the interregnum, S3739 soon after superhumps had developed near supermaximum, S2748 near the end of the plateau, and S0137 at the end of superoutburst, near quiescence. The abscissae are in units of the orbital separation. The ordinate scale is logarithmic with out-of-eclipse brightness defined to be unity. From Warner & O'Donoghue (1988).

Eclipses of Z Cha in superoutburst have been analysed by eclipse simulation (Section 2.6.3) and eclipse mapping (Section 2.6.4). Simulation confirmed the initial conclusions stated above and showed in addition that the SLS cannot be an extended source along the stream's trajectory into or across the disc. Instead, it was found that the SLS could be modelled as a bright rim to the disc, modulated in intensity at period P_s (Warner & O'Donoghue 1988).

The assumptions of the basic eclipse mapping technique prevent its application when there is a source of luminosity which varies on the time scale of the eclipse. However, eclipses can be mapped when the superhump is at orbital phase $\varphi \sim 0.5$. The results (Warner & O'Donoghue 1988, Figure 3.42) show a non-uniformly bright rim to the disc, which becomes less of a contributor as the superoutburst progresses and the radiation from the disc becomes more concentrated towards the centre. At least in the early stages of superoutburst, much of the SLS luminosity is concentrated on the disc near L_1.

A generalization of the eclipse mapping technique was introduced by O'Donoghue

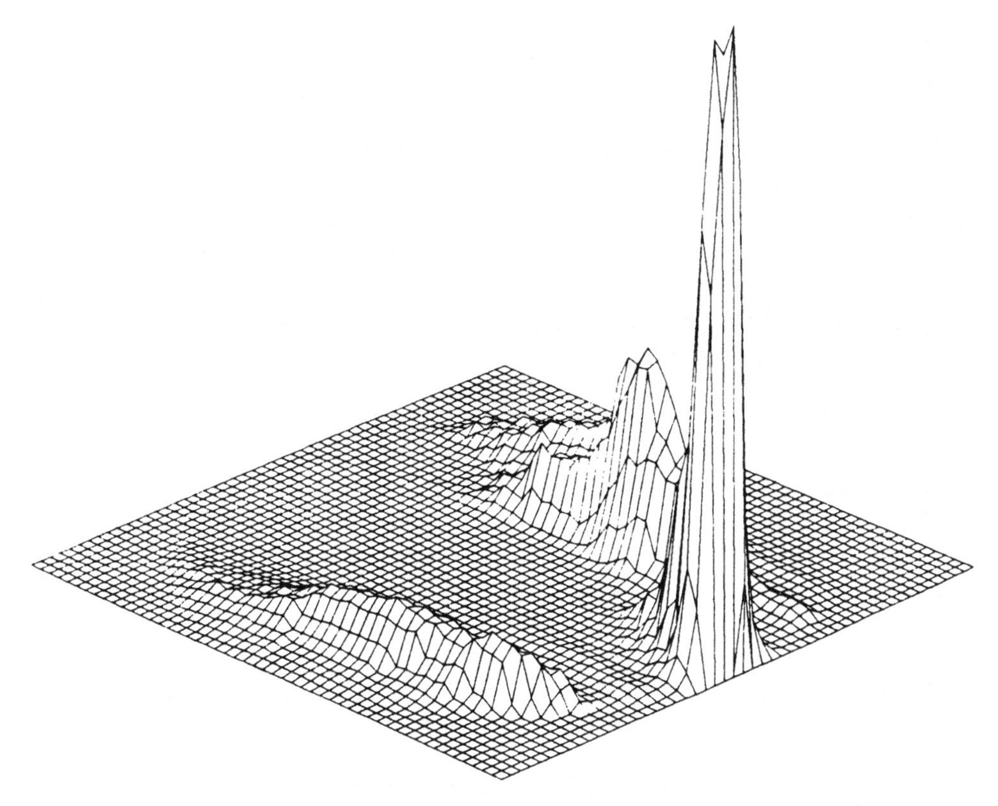

Figure 3.43 MEM deconvolved image of the SLS. The secondary star's position is towards the lower right. The primary is at the centre of the pixel array. From O'Donoghue (1990).

(1990). On the assumption that over a time $\sim P_b$ the SLS has fairly stable behaviour, it is possible to use the superhump profile observed at $\varphi = 0.5$ to predict how the SLS varied in intensity *during eclipse* when the superhump was at $\varphi = 0$. Thus, by differencing light curves obtained $\frac{1}{2}P_b$ apart it is possible to map the intensity distribution of the SLS itself.

A sample result is shown in Figure 3.43. The SLS consists of three regions, one on the disc near the L_1 point and two symmetrically placed in the disc towards the leading and trailing edges. The bright rim found in the earlier work was largely a result of maximally azimuthal symmetry acting on such regions. (The intensity map in Figure 3.43 also has maximum azimuthal symmetry, so the SLS regions may be less extended than shown.) It should be remembered that the total intensity of the three regions is modulated with period P_s; Figure 3.43 presents a frozen view close to the maximum of the superhump.

A major step forward in understanding the nature of superhumps has been made by Harlaftis *et al.* (1992a) and Billington *et al.* (1995). From IUE observations of Z Cha and HST observations of OY Car respectively, these authors find dips in the UV flux at times of maxima of superhumps. It is proposed that increased vertical height near the outer edge of the disc obscures the centre of the disc, causing the reduced UV flux, but reprocessing of the intercepted UV enhances the optical flux.

3.6.5 *Spectroscopy During Superoutbursts*

During superoutbursts the development of optical spectra is very similar to that of normal outbursts (Section 3.3.5). Some objects have only broad, shallow absorption lines at supermaximum (e.g., VW Hyi: Vogt 1976, Whelan, Rayne & Brunt 1979, Schoembs & Vogt 1980; TU Men: Stolz & Schoembs 1984), or have absorption lines with weak emission reversals (e.g., V436 Cen: Whelan, Rayne & Brunt 1979; TY PsA: Warner, O'Donoghue & Wargau 1989; YZ Cnc: Harlaftis *et al.* 1993c) or are high inclination systems having emission lines with narrow absorption cores (e.g., Z Cha: Vogt 1982, Honey *et al.* 1988; OY Car: Hessman *et al.* 1992). In Z Cha and in WZ Sge (Nather & Stover 1978) the lower Balmer series is in emission but the upper is in absorption, and after supermaximum in WZ Sge an inverse P Cyg structure develops in Hβ (Gilliland & Kemper 1980). One of the few ways in which supermaxima and normal maxima optical spectra differ, at least in Z Cha, is in a strong enhancement of HeII λ4686 and CIII/NIII λ4650 (Honey *et al.* 1988).

UV spectra and flux distributions are very similar in normal and superoutbursts (WX Hyi: Hassall *et al.* 1983; AY Lyr: Szkody 1982; EK TrA: Hassall 1985a; OY Car: Naylor *et al.* 1987; VW Hyi: Verbunt *et al.* 1987). However, a closer study of Z Cha (Harlaftis *et al.* 1990b, 1992) shows that the UV continuum has $T_{eff} \sim 12\,000$ K in superoutburst, as opposed to $T_{eff} \sim 19\,000$ K at normal outburst, and that the same probably occurs in OY Car. Furthermore, both OY Car and Z Cha have 'cooler' UV distributions than low inclination SU UMa systems. In addition, flux minima occur during superoutburts in both systems at orbital phases of $\sim$0.2 and 0.8 (as in the interregnum visible light curves: Section 3.6.4.3). Two interpretations are offered: the 'cooler' continuum and the dips may indicate obscuration of the central part of the disc by vertical disc structure caused (i) by increased mass transfer, or (ii) by increased disc thickness in the regions of tidal dissipation (Section 3.6.6.2). The dip at $\varphi \sim 0.8$ (seen also at short wavelengths in the optical: Kuulkers *et al.* 1991a) becomes shallower as the superoutburst progresses, as though higher mass transfer early in the superoutburst causes an enlarged disc impact region which diminishes later.

The brightness of WZ Sge at supermaximum enabled high resolution IUE spectra to be obtained (Fabian *et al.* 1980; Friedjung 1981). Near supermaximum the CII and SiIV lines are in absorption with FWHM $\sim$2 Å, the higher ionization potential species (CIV,NV) have deep absorption lines of width $\sim$1 Å superimposed on emission lines of total width $\sim$5000 km s^{-1}. By the time of the appearance of superhumps, when WZ Sge had dropped $2\frac{1}{2}$ mag, the shortwavelength spectrum was at times entirely in emission but highly variable on a time scale of minutes, with frequent appearance of strong absorption lines. Ten days later the spectrum was stable and predominantly in absorption.

3.6.5.1 *RVs During Superoutbursts*

During superoutbursts of Z Cha the narrow Balmer absorption lines show modulations in strength, probably with period P_{orb}: they have maximum strength near orbital phase 0.7 and are otherwise quite weak and are absent through eclipse (Vogt 1982; Honey *et al.* 1988); although having the same RV period (P_{orb}) and phase as at quiescence, their amplitude is greater and the γ-velocity itself varies with a period P_b and amplitude K_γ measured variously at $\sim$250 km s^{-1} (Vogt 1982) or $\sim$80 km s^{-1} (Honey *et al.* 1988).

Similar RV results have been found for the *broad* absorption lines in superoutbursts of TU Men (Stolz & Schoembs 1984), VW Hyi (Vogt: quoted by Stolz & Schoembs 1984) and TY PsA (Warner, O'Donoghue & Wargau 1989). In each of these, γ-velocities alternating between $\sim +200$ and -200 km s^{-1} on successive days (corresponding to changes $\sim$180° in the $\sim$2 d beat cycle) were measured. The large changes in γ-velocity seen in WZ Sge during outburst (Walker & Bell 1980; Gilliland & Kemper 1980) are also probably associated with its $\sim$7 d beat cycle.

Narrow absorption lines of considerable depth are observed in TY PsA, but, unlike Z Cha, only over a very restricted orbital phase and redshifted by 350 km s^{-1}. As TY PsA has a lower inclination than Z Cha this can be understood if the line of sight to the bright central region of the disc only just manages to intersect gas splashed up above the bright spot in TY PsA, whereas it passes through gas around much of the rim of the disc in Z Cha, being most optically thick in the region of the bright spot.

It is evident from a study of the profiles of the broad absorption lines that the apparent γ-velocity variations are caused by *changing asymmetries* rather than by bodily line shifts (Warner, O'Donoghue & Wargau 1989). Asymmetric profiles were recorded in WZ Sge even before superhumps had appeared (Walker & Bell 1980). These asymmetries show that during superoutburst a substantial fraction of at least the outer parts of the accretion disc is in non-circular motion. This non-circular component rotates with period P_s, not P_{orb}, so it is intimately linked with the superhump process.

Similar line asymmetries are seen in CN Ori during outburst, and they vary on a beat cycle of 3.7 ± 0.3 d seen in photometry during *quiescence* (Mantel *et al.* 1988). At $P_{orb} = 3.92$ h CN Ori is outside of the range of normal SU UMa stars; curiously it possesses features of normal and superhumps at quiescence but not during outburst!

3.6.6 *Theories and Models of Superoutbursts*

The superhump phenomenon is so intimately related with superoutbursts that it is not possible to consider the nature of a superoutburst itself without at the same time ensuring that it generates superhumps. However, the reverse has not been the case – many models for superhumps were proposed *ad hoc*, in the hope that they might shed light on the cause of the superoutburst. Almost all models for superhumps have now been superseded, but, for the record, a list of them is given, together with the observations that disabled them.

3.6.6.1 Historical Overview of Superhump Models

The first model, proposed by Vogt (1974) and independently (but dismissively) by Warner (1975b), and later discussed by Haefner, Schoembs & Vogt (1979), sited the SLS on a non-synchronously rotating secondary. The model was resuscitated by Whitehurst, Bath & Charles (1984) but eliminated for reasons explained in Section 3.6.4.5. Papaloizou & Pringle (1978a) proposed an intermediate polar model for VW Hyi, in which the primary's rotation period $P_{rot}(1)$ was identified with the quiescent hump period and the orbital period with P_s. This became untenable when the orbital period was determined spectroscopically and found to be equal to the quiescent hump period.

Other intermediate polar models were suggested by Vogt (1979) and Warner (1985a), who identified $P_{rot}(1) \equiv P_s$, and by Patterson (1979c). Potentially the strongest argument against these is the observation that SW UMa is possibly an intermediate polar with $P_{rot}(1) = 459$ s (Robinson *et al.* 1987). If this is confirmed, then it obviously eliminates either P_s or P_b from being equal to $P_{rot}(1)$. The discussion (Section 8.6.5.2) of DNOs strongly suggests that in most DN the primary rotates rapidly ($P_{rot}(1) \sim$ tens of seconds).

Papaloizou & Pringle (1979) suggested that the binary orbits of SU UMa stars are slightly eccentric ($e \sim 10^{-4}$), and that the eccentricity is generated and maintained through tidal resonance with the disc. (It would be lessened, and possibly eliminated, by the effect of tidal circularization discussed in Section 9.1.3, depending on the relative time scales.) The periodic variation of the separation between primary and secondary modulates the radius of the secondary's Roche lobe, which, through equation (2.12), modulates $\dot{M}(2)$. The frequency of modulation is $\Omega_m = \Omega_{orb} - \Omega_{ap}$, where Ω_{ap} is the apsidal precession frequency of the orbit. The latter is given by (Kopal 1959):

$$\frac{\Omega_{ap}}{\Omega_{orb}} = k_2 \left(1 + \frac{16}{q}\right) \left[\frac{R(2)}{a}\right]^5, \tag{3.35}$$

where k_2 is the apsidal motion constant for the secondary. For an $n = 1.5$ polytrope, which approximates the structure of fully convective low mass main sequence stars, $k_2 = 0.143$ (Cisneros-Parra 1970).

From equations (2.5b) and (3.35), and noting that $q \ll 1$ for SU UMa stars, we have

$$\frac{\Omega_{orb}}{\Omega_m} \approx \frac{\Omega_{orb}}{\Omega_{ap}} = 20.8 q^{-\frac{2}{3}} (1+q)^{\frac{5}{3}}. \tag{3.36}$$

Therefore, for $q = 0.15$, $P_s \approx P_{orb}(1 + P_{orb}/P_b) = 1.01\ P_{orb}$, if P_b is identified with $2\pi/\Omega_m$. Thus brightness modulations with period $\sim 1\%$ greater than P_{orb} can be generated in this model. The modulation of $\dot{M}(2)$ would show as a modulation of the luminosity of the bright spot and probably the whole outer rim of the disc (Clarke, Mantle & Bath 1985). To obtain a modulation visible during superoutburst requires a very large increase in $\dot{M}(2)$ over that at quiescence.

The model fails, however, on a number of counts: (a) to get the observed $P_b \sim 2$ d (Table 3.3) requires $k_2 \sim 0.5$; (b) there is no explanation for the variation of P_s during a superoutburst; (c) there is no observed modulation of the bright spot brightness at a period P_s at quiescence, yet from equation (2.12) the fractional variation in rate of mass transfer $\delta\dot{M}/\dot{M} \propto \dot{M}^{-1/3}$, so the bright spot modulation should be largest at quiescence (Edwards & Pringle 1987a); (d) the RV behaviour (Section 3.6.5.1) is difficult to understand (Warner 1985a).

Another method proposed for modulating $\dot{M}(2)$ is alignment of g-mode oscillations on an asynchronously rotating secondary (Vogt 1980). This fails, *inter alia*, to account for the RV behaviour. Gilliland & Kemper (1980) proposed that an 'excretion disc' formed during the 1978 superoutburst of WZ Sge, and that the SLS was a bright spot formed by impact on this disc of a stream from one of the outer Lagrangian points. This model cannot explain the large range of γ-velocities, and certainly does not apply to SU UMa stars in general.

From his RV observations, Vogt (1982) was inspired to suggest that the accretion disc takes up an elliptical shape during superoutburst. The line of apsides of the disc rotates with period P_b. Superhumps result from the variation of kinetic energy acquired by the mass transfer stream as its length from the secondary to the bright spot changes during an orbital cycle.[4] An eccentricity $e \sim 0.6$ is required for the latter; the range of γ-velocities of the narrow absorption lines is a result of the varying radial component of the edge velocity of the disc, seen projected against the bright centre of the disc.

Although incorporating the essence of what is currently thought to be the cause of superhumps – namely an eccentric precessing disc – Vogt's proposal has met with a number of difficulties: (a) although analysis of eclipses during superoutbursts of OY Car (Krzeminski & Vogt 1985) and Z Cha (Warner & O'Donoghue 1988) show periodic asymmetries that could be caused by an elliptical disc, they can equally well be caused by an asymmetric light distribution on a circular disc, and in any case, any eccentricity near supermaximum is < 0.2 (Warner & O'Donoghue 1988); (b) from measurements of the position of the bright spot at the end of a superoutburst of OY Car, Hessman *et al.* (1992) map out the radius vector of the disc and find $e = 0.38$, which gives far too small a variation in disc radius ($0.33 \lesssim r_d/a \lesssim 0.45$) to generate the range of superhump brightness actually observed.

Osaki (1985) and Mineshige (1988b) proposed that a slowly precessing elliptical disc develops during superoutburst, and that the irradiation of the secondary, which is partially shadowed by the disc, will vary at the synodic period, driving an $\dot{M}(2)$ modulated at period P_s. The model fails to explain why the disc becomes eccentric in the first place, and otherwise is unsatisfactory for some of the same reasons as for the eccentric orbit model of Papaloizou & Pringle described above – but it does lead to predicted values of P_s in better agreement with observation (Warner 1985c), suggesting that a precessing elliptical disc is an important element of a superhump model.

3.6.6.2 The Tidal Resonance Model

In particle simulation of a disc for SS Cyg, which has $q = 0.6$ (Figure 2.9), Whitehurst (1988a) found an elongated disc which kept constant orientation with respect to the stellar components. However, in simulating a disc for Z Cha, with $q = 0.15$, an initially circular disc transformed itself after ~ 1 d into a progradely precessing elliptical disc (Whitehurst 1988b; see also Hirose, Osaki & Mineshige 1991). In a later more detailed simulation (Whitehurst & King 1991) it became evident that it is principally a ring of higher density material, initially the outer rim of the disc, that is converted into a precessing elliptical ring (Figure 3.44). This currently is thought to be the essence of the process producing superhumps in SU UMa systems, all of which have $P_{orb} \lesssim 3$ h and, therefore, through equation (2.100), are expected to have $q \lesssim 0.3$.

The development of the elliptical rim to the disc is a result of tidal resonance. Paczynski (1977) noted that, even before reaching the tidal limit set by intersecting particle orbits (Section 2.5.5), within some discs there are unstable stream lines. Stability may be tested by giving a small perturbation Δx_i to an orbit and measuring

[4] It might be thought that maximum kinetic energy would occur when the stream hits the periastron point of the disc, but calculations show that the maximum occurs when the stream strikes the disc near the latus rectum (Warner & O'Donoghue 1988).

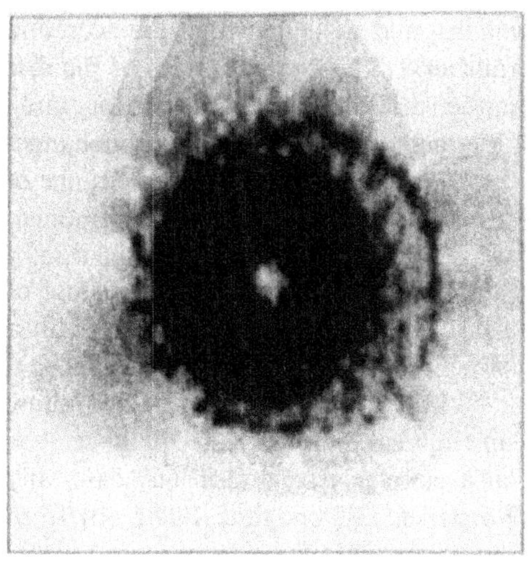
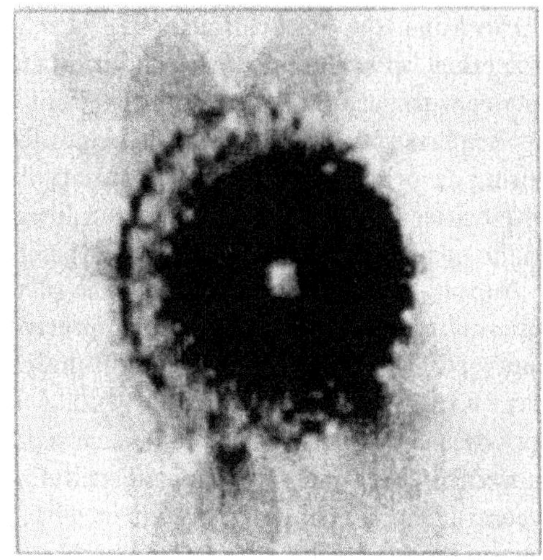

Figure 3.44 Accretion disc simulations for a system with $q = 0.12$ showing effects of tidal resonance. The images chosen are at opposite extremes of the doubly periodic orbit. From Whitehurst & King (1991).

Δx_f after computing one complete orbit; if $|\, \Delta x_f / \Delta x_i \,| > 1$ the orbit is unstable. These instabilities are in the vicinity of commensurabilities between periods of particles in the accretion disc and the orbital period of the system, with the result that tidal perturbations accumulate: the radial oscillations in orbit resonate with a multiple of the orbital frequency.

Paczynski found that the maximum radius of the last stable stream line occurs at $r_{st}(d)/a = 0.480 \pm 0.003$, independent of q, for $q \lesssim 0.25$. This intersects the tidal stability radius, $r_d(\max)$, equation (2.61), at $q = 0.25$. A more accurate computation gives $q = 0.22$ (Molnar & Kobulnicky 1992). Therefore, accumulative perturbations may be expected in discs that have radii $\gtrsim r_{st}(d)$ in systems with $q \lesssim 0.22$

Whitehurst & King (1991) point out that a range of resonances could exist. The tidally resonating particle orbit becomes non-circular and therefore precesses progradely with apsidal precession frequency ω_{pr}. It is the radial excursion of the particle that resonates with the position of the secondary. The radial, or epicyclic, frequency is $\omega - \omega_{pr}$, where ω is the mean angular frequency of the particle (in an inertial frame). Resonance therefore requires

$$k(\omega - \omega_{pr}) = j(\omega - \Omega_{orb}) \tag{3.37}$$

where j and k are positive integers. (If $\omega_{pr} = 0$ this equation reduces to $(j - k)\omega = j\Omega_{orb}$. Particles with angular frequency ω are often described as being in a $(j - k) : j$ resonance (e.g., Lubow (1991a,b)). Here the more general notation of Whitehurst & King is adopted: a j:k resonance occurs near a $(j - k) : j$ *commensurability.*)

From equations (2.1a), (2.1b) and (3.37) it follows that, for an almost circular orbit and $\omega_{pr} \ll \omega$, the resonance radius r_{jk} will be close to

$$r_{jk}/a = \frac{(j-k)^{2/3}}{j^{2/3}(1+q)^{1/3}}. \tag{3.38}$$

The condition $r_{jk} < r_d(\max)$ requires $k > 1$. The lowest resonance is $(j, k) = (3, 2)$, equivalent to a 3:1 commensurability in particle and secondary star frequencies (see also Osaki (1989b), Lubow (1991a,b, 1992b); the latter shows that the resonance operates also in a fluid disc). From equation (3.38), the mean radius of the elliptical orbit may be assumed to be near

$$\frac{r_{32}}{a} = (3)^{-2/3}(1+q)^{-1/3}. \tag{3.39}$$

This is very similar to the *average* radius of the last stable stream line in Paczynski's calculations. Although other resonances at greater radii occur and may affect the disc even if its outer radius is smaller than theirs, the 3:1 commensurability is considered to be dominant (Lubow 1991a, 1992b).

The instability of the orbits in this single particle (i.e., inviscid disc) calculation corresponds to the beginning of chaos. In the instability zone just beyond r_{32} orbits close on themselves only after two passages around the primary, i.e., the stability radius corresponds to the position of period-doubling. The doubled orbits intersect, so, in the presence of particle interactions, the whole orbit cannot be populated uniformly, hence the arcs that fill only part of the doubly-periodic orbit in Figure 3.44. As can be seen, these orbits are quite eccentric.

Collisions between the precessing arc of gas and the bulk of the disc generate additional luminosity which is one possible source of the superhump light. Hirose & Osaki (1990, 1993) have performed simulations which follow the generation of luminosity over several orbital periods. The results, Figure 3.45, have some

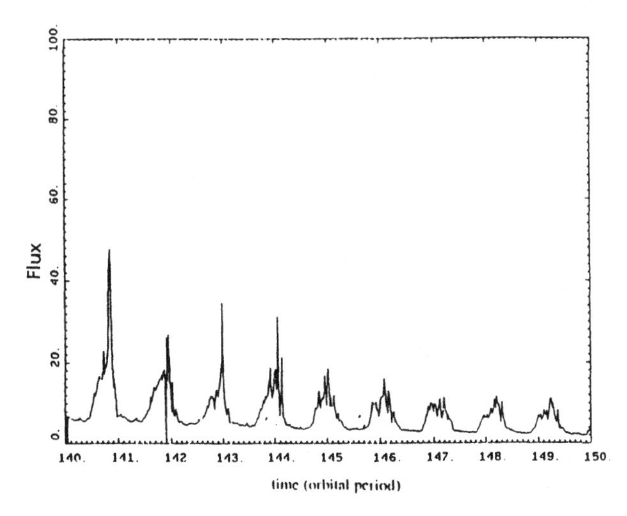

Figure 3.45 Computed flux variations in the outer part of the disc for a system with $q = 0.15$ and an eccentric disc. From Hirose & Osaki (1990).

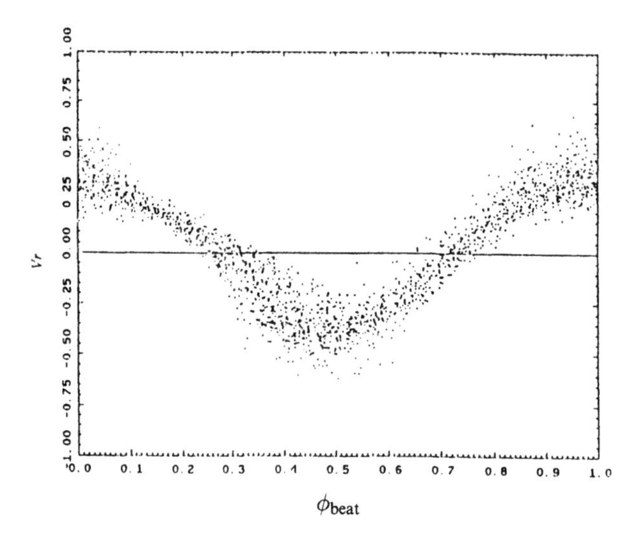

Figure 3.46 Computed γ-velocity variation over a superhump beat period. The ordinate scale is velocity in units of $a\Omega_{orb}$. From Hirose & Osaki (1990).

resemblance to the superhumps in SU UMa stars. Furthermore, by summing over the velocities of the individual particles in the disc, a RV curve at each phase in the superhump cycle can be generated. The γ-velocity variations of these curves (Figure 3.46) are similar in phase and amplitude to those observed (Section 3.6.5.1) and asymmetrical line profiles are naturally produced.

The three SLS regions found by O'Donoghue (Figure 3.43) are in the directions (from the primary) of the three regions of intersecting orbits (and therefore tidal stress) of the doubly-periodic orbits. However, the observed radii for the two, orthogonal to the line of centres, are $\sim 0.35a$, whereas the orbital intersections (for nearly circular orbits) are at $\sim 0.45a$. Note, however, that this result was obtained without allowance for increased thickness of the outer region of the disc. In the UV reprocessing model (Section 3.6.4.6) thickening of the disc probably is the result of local heating by tidal stresses, which varies with a period P_s.

There is more to be done before a complete understanding of superhump production is reached. The Whitehurst & King simulations omit the effect of material arriving in the bright spot, which is expected to influence the population of the doubly-periodic orbits and their growth rate. In the simulations, the profile of the superhump becomes double humped; the stream probably removes the symmetry between loops of the orbits and suppresses the second hump – although, in the late part of a superoutburst, the observations suggest that a second hump is able to develop and therefore the stream may not be so successful at that stage. The 3:2 resonance for $q \lesssim 0.22$ also causes a tilt to grow in the disc, but the growth rate of the disc inclination is only a few per cent of that of the eccentricity (Lubow 1992a).

There is as yet no satisfactory model of the late superhumps (Section 3.6.4.4). Osaki (1985) and Whitehurst (1988b) proposed that the eccentric disc survives for several days after the end of a superoutburst, and that modulation of the bright spot brightness, in the manner of Vogt's (1982) early model for ordinary superhumps, produces late superhumps. There are several objections to this model (van der Woerd *et al.* 1988), the most important of which is that the eccentricity determined in OY Car is too small to produce sufficient modulation (Hessman *et al.* 1992; Section 3.6.4.1). Hessman *et al.* (1992) point out that there should be considerable variation in surface density and scale height around the rim of an eccentric disc, and that this should cause variation in the appearance of the bright spot – for example, in the amount of overshooting onto the face of the disc (Section 2.4.3). Further understanding of the late superhump phenomenon could come from eclipse mapping in Z Cha and OY Car.

3.6.6.3 Superhump Chronology

The Hirose & Osaki (1990, 1993) simulations produce superhumps with periods in close agreement with observation (for a given q). More generally, the resonant orbit model can be combined with the theory of precessing orbits to arrive at an analytical expression that fairly approximates P_s.

The resonant orbits are doubly-periodic in the rotating frame of the binary, and in an inertial frame are rosette orbits which only close after three loops of the primary. Therefore $P_s = 3 \times 2\pi/\omega$, where $\omega = 3\Omega_{orb} - 2\omega_{pr}$ for a 3 : 2 resonance (equation 3.37). Therefore

$$\begin{aligned}\frac{P_{orb}}{P_b} &= 1 - \frac{P_{orb}}{P_s} \\ &= \frac{2\omega_{pr}}{3\Omega_{orb}}.\end{aligned} \tag{3.40}$$

The precession frequency is found from (Hirose & Osaki 1990)

$$\frac{\omega_{pr}}{\Omega_{orb}} = \frac{q}{(1+q)^{1/2}}\left[\frac{1}{2}\frac{1}{\rho^{1/2}}\frac{d}{d\rho}\left(\rho^2\frac{dB_0}{d\rho}\right)\right] \tag{3.41a}$$

where $\rho = r/a$ and B_0 is the Laplace coefficient (Smart 1953). Using $\rho_{32} = r_{32}/a$ from equation (3.39), together with equations (3.40) and (3.41a), gives P_b/P_{orb} as a function of q:

$$\frac{\omega_{pr}}{\Omega_{orb}} = 1/4\frac{q}{1+q}\left[1 + \frac{0.433}{(1+q)^{2/3}} + \frac{0.146}{(1+q)^{4/3}} + \frac{0.044}{(1+q)^2} + \frac{0.013}{(1+q)^{8/3}} + \cdots\right] \tag{3.41b}$$

Accurate values of q are available only for SU UMa stars of high inclination (Table 2.4). A comparison of observed and predicted values of P_b is given in Table 3.8. The result suggests that the observed P_b could provide a means of estimating q.

From equations (3.40) and (3.41b)

$$\frac{P_b}{P_{orb}} \approx \frac{3.85(1+q)}{q} \qquad 0.1 \lesssim q \lesssim 0.22 \tag{3.42}$$

or

$$P_b(\mathrm{d}) \approx \frac{2.5(1+q)M_1(1)}{P_{orb}^{1/4}(\mathrm{h})} \tag{3.43}$$

Table 3.8. *Comparison of Observed and Predicted Beat Periods.*

Star	P_{orb}(h)	q observed	P_b(d) observed	P_b(d) calculated
OY Car	1.515	0.102	2.6	2.59
HT Cas	1.768	0.15	2.3	2.17
Z Cha	1.788	0.15	2.0	2.19

from equation (2.100). As $< M_1(1) > \approx 0.7$ from Table 2.4, this accounts for the observed magnitude of P_b and shows that a range of $M(1)$ will cause a spread in P_b at a given P_{orb}.

If it is required that $q \leq 0.22$ for resonant orbits to lie within the disc (previous section), then equation (3.42) shows that there is a lower limit on P_b/P_{orb}, which on evaluation gives $P_b(\mathrm{d}) \geq 0.91 P_{orb}(\mathrm{h})$. All of the SU UMa stars in Table 3.3 satisfy this condition except for the two of longest orbital period, YZ Cnc and TU Men, which marginally disobey it. This could be the result of viscosity allowing the disc radius to increase somewhat beyond $r_d(\mathrm{max})$, which would give discs access to the 3:2 resonance for systems with q slightly larger than 0.22.

Applying the same approach (i.e., adopting the main sequence mass–period relationship) to WZ Sge and HV Vir leads to $M_1(1) > M_{ch}$. However, for very short P_{orb} CVs there is the possibility that they have passed through a minimum P_{orb} and are on an increasing P_{orb} branch with degenerate secondaries (Section 9.3.2). In this case, the mass–period relationship equation (9.42) is appropriate and equations (3.40) and (3.41b) give

$$\frac{P_b}{P_{orb}} = \frac{3.73(1+q)}{q} \qquad q \lesssim 0.05 \tag{3.44}$$

and

$$M_1(1) = 0.31 \frac{P_b(\mathrm{d})}{P_{orb}^2(\mathrm{h})}$$

$$\approx \frac{1.31 \times 10^{-2}}{P_s(\mathrm{h}) - P_{orb}(\mathrm{h})}. \tag{3.45}$$

This gives masses $M_1(1) = 1.20$ and 0.68 for WZ Sge and HV Vir respectively. The result for WZ Sge is much larger than derived by Smak (1993c), but the latter makes use of an observed $K(1)$, the interpretation of which is very uncertain.

A possible explanation of the variation of P_s during superoutburst, described in Section 3.6.4.1, is that the mean radius of the resonant orbits decreases as the outburst progresses. From equations (3.33) and (3.34)

$$C = \frac{1}{2} P_s \frac{dP_s}{dt}$$
$$= \frac{1}{2} P_s \left(\frac{P_s}{P_b}\right)^2 \frac{dP_b}{dt}$$
$$\sim 2 \times 10^{-4} \frac{d\rho}{dt} \quad d$$

from equation (3.41), evaluated for $P_{orb} = 1.75\,h$, $P_s = 1.810$ h (i.e., $P_b = 2.2\,d$) and $q = 0.15$. Therefore, a reduction of ρ of 0.10 over 10 d can give the $C \approx -2 \times 10^{-6}$ d observed (see also Lubow (1992b); Patterson *et al.* 1993a; Whitehurst (1994)).

It will be noted that equation (3.42) is numerically almost identical to equation (2.62c). It is therefore not possible to distinguish between retrogradely precessing tilted discs and progradely precessing eccentric discs on the basis of time scale alone (see also Section 4.5).

3.6.6.4 The Nature of the Superoutburst

As the rise and the fall of a superoutburst appear in essence to be those of a normal outburst (Section 3.6.4), and normal outbursts in SU UMa stars have durations of maximum brightness less than ΔT_s (so superhumps *empirically* would not be expected to develop) the central question for understanding superoutbursts is 'what sustains the outburst for ten or more days?'

Osaki (1985) suggested that an irradiation-induced mass-overflow instability could exist, in which heating of the secondary's atmosphere during a normal outburst induces a high $\dot{M}(2)$ This is shown to be possible only for low mass secondaries with small quiescent $\dot{M}(2)$ – conditions which the SU UMa stars satisfy. The mass-overflow instability has the nature of a relaxation oscillation, the time scale of which is at least as large as T_n, i.e., a second instability cannot occur before a time $T_s \geq T_n$, has passed. The model has been criticized by Hameury, King & Lasota (1986a) on the grounds (see also Section 2.9.6) that the intercepted flux cannot heat the envelope sufficiently to produce the required instability.

As an alternative, Osaki (1989a) proposed that, at the start of a superoutburst cycle, the accretion disc has small mass and a radius much less than the radius r_{32} at which instability sets in. At the first normal outburst expansion of the disc usually fails to reach r_{32}. Because of its small radius, little angular momentum is removed by tidal interaction, so in subsequent outbursts the disc is left in an increasingly massive and expanded state until finally an outburst is able to extend the disc beyond r_{32} (Figure 3.47). Strong tidal dissipation at the edge of the expanded disc maintains it in the high viscosity state which enables the accumulated angular momentum to be lost, allowing most of the gas to move to smaller radii, accrete onto the primary and prolong the outburst into a superoutburst. Typically ~10% of the disc mass is accreted during normal outbursts and ~50% in a superoutburst (Ichikawa, Hirose & Osaki 1993). Stars with $q > 0.22$ (i.e., most stars above the period gap) are in a quasi-periodic state wherein, not having access to the 3:2 resonance, tidal dissipation is just sufficient to satisfy the requirements of each normal outburst.

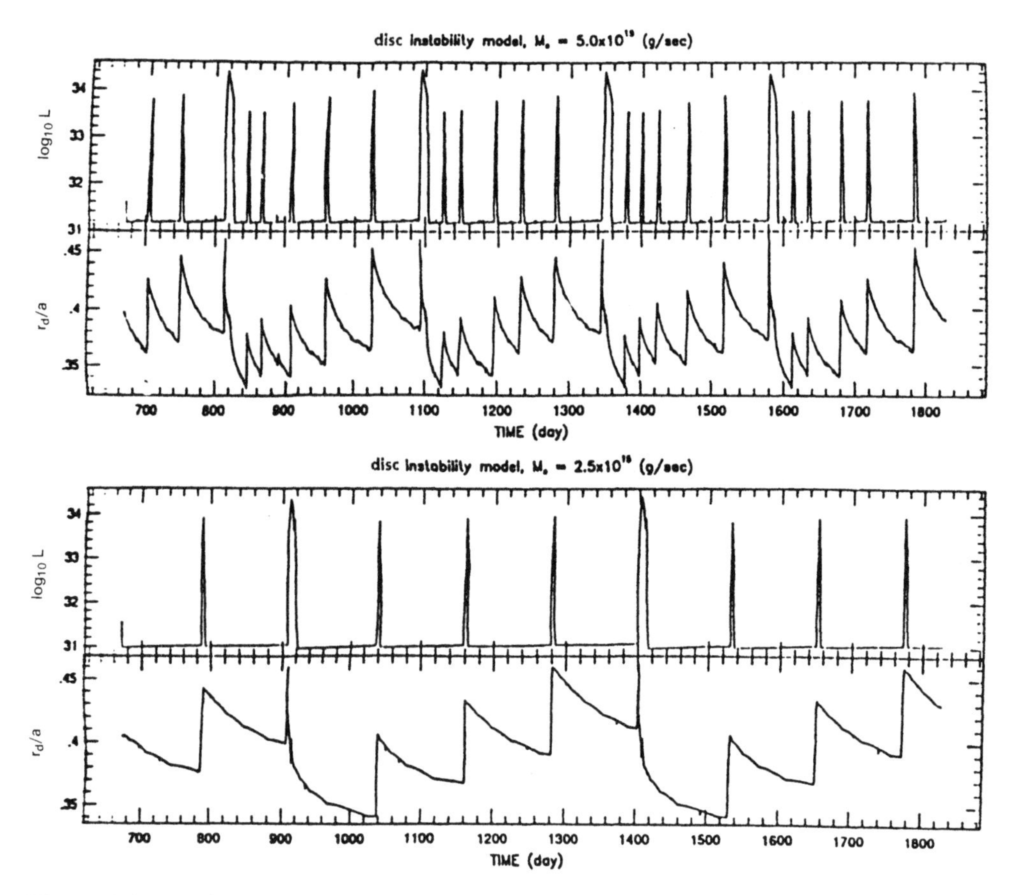

Figure 3.47 Variations of bolometric luminosity and disc radius through a number of supercycles. The upper panel has $\dot{M}_{16}(2) = 0.5$ and the lower panel has $\dot{M}_{16}(2) = 0.25$. From Osaki (1994).

In this model the mass M(d) and angular momentum J(d) stored in the disc increase linearly with time until a critical limit J_{crit}(d) is reached, corresponding to a disc with $r_{\mathrm{d}} = r_{32}$. Since $J(\mathrm{d}) \propto M(\mathrm{d}) \propto \dot{M}(2)t$, the time taken to trigger a superoutburst is $T_{\mathrm{s}} \propto J_{\mathrm{crit}}(\mathrm{d})/\dot{M}(2)$. For the SU UMa stars, with a small range of P_{orb}, J_{crit} (d) is almost constant. On the other hand, the viscosity in the low state of SU UMa stars is very low, so the time between normal outbursts $T_{\mathrm{n}} \propto \dot{M}^{-2}(2)$ (Section 3.5.4.5). (For simulation of an $\alpha \sim 0.001$ disc, with no thermal instability mechanism included, which develops an apparent superoutburst, see Rozyczka & Spruit (1993).) Hence $T_{\mathrm{s}}/T_{\mathrm{n}} \propto \dot{M}(2) \propto T_{\mathrm{s}}^{-1}$, which is the explanation of the observed T_{s}–T_{n} relationship (Section 3.6.3). A perfect correlation is not expected because a more complete analysis shows that there is an $M(1)$ dependence. Figure 3.47 demonstrates the effect of varying $\dot{M}(2)$ in the theoretical model (Osaki 1994): lowering $\dot{M}(2)$ by a factor of 2 doubles the time between superoutbursts and approximately halves the number of normal outbursts per supercycle. Observations are not yet adequate to test the predicted variation of r_{d} through the supercyle (Smak 1991a).

Whitehurst & King (1991) expressed reservations about the model, claiming that the superhumps require increased $\dot{M}(2)$ during superoutburst, and it is this that also sustains the outburst (see also Whitehurst (1988b)). They also argue that in the absence of the strong perturbations of the outer disc that enhanced $\dot{M}(2)$ provides, the growth time of orbital eccentricity is $\sim 1000\ P_{orb}$, whereas the observed ΔT_{sh} (Table 3.7) are $\sim 10P_{orb}$–$100P_{orb}$. However, Ichikawa, Hirose & Osaki (1993) find that enhanced $\dot{M}(2)$ diminishes the development of eccentricity because the disc shrinks on addition of gas of low specific angular momentum, and their numerical simulations give growth times of 1–2 d. We may also note from Section 3.6.4.2 that $\dot{M}(2)$ enhancement seems to *increase* ΔT_{sh} rather than decrease it.

If the observed ΔT_{sh} is identified with the growth time of eccentric orbits, then we must have $\Delta T_{sh} > \Delta T_{0.5}$ (Section 3.3.3.5) or else normal outbursts could develop superhumps. But the values of ΔT_{sh} (Table 3.7) are only *slightly* larger than the $\Delta T_{0.5}$ for the same stars, which would imply that, if disc radii *always* expand beyond r_{32} in normal outbursts, then it must be a cosmic accident that ΔT_{sh} is just a little smaller than $\Delta T_{0.5}$. The Osaki model does not require such fine tuning.

Duschl & Livio (1989) propose a hybrid scheme in which an MTI provides high $\dot{M}(2)$ on the superoutburst recurrence time, and in between DI provides normal outbursts. The leading edge of the high $\dot{M}(2)$ profile can trigger a precursor normal outburst if the $\Sigma(r)$ distribution in the disc happens to be of the appropriate form.

However, as seen in Figure 3.47, one attractive feature of the pure DI model is that it is able to produce the observed relatively periodic occurrence of superoutbursts while allowing irregular intervals between normal outbursts. Furthermore, following Osaki (1989a), the correct slope of the plateau region is obtained from the model, as seen in the following.

During superoutburst the disc is in a quasi-equilibrium state in which $\dot{M}(\mathrm{d}) \gg \dot{M}(2)$. Then equations (2.32) and (3.9) give, at the outer edge of the disc,

$$\dot{M}(\mathrm{d}) = \frac{8\pi\sigma}{2GM(1)} T_{\mathrm{eff}}^4(r_\mathrm{d}) r_\mathrm{d}^3 \tag{3.46}$$

and, since $\Sigma(r) \propto r^{-0.75}$, and $\Sigma(r_\mathrm{d}) \simeq \Sigma_{\min}(r_\mathrm{d})$,

$$M(\mathrm{d}) \approx \frac{2\pi}{1.25} r_\mathrm{d}^2 \Sigma_{\min}(r_\mathrm{d}). \tag{3.47}$$

Hence

$$\frac{\mathrm{d}M(\mathrm{d})}{M(\mathrm{d})} \approx -\frac{5 r_\mathrm{d} \sigma T_{\mathrm{eff}}^4(r_\mathrm{d})}{3GM(1)\Sigma_{\min}(r_\mathrm{d})} \mathrm{d}t, \tag{3.48}$$

which shows that $M(\mathrm{d})$ and $\dot{M}(\mathrm{d})$ vary exponentially with time. As the bolometric luminosity $L_{bol} \propto \dot{M}(\mathrm{d})$, equations (2.1b), (3.16), (3.17) and (3.48), with $r_\mathrm{d} = r_{32}$, give

$$\tau_{pl} \approx 7.1 M_1^{0.17}(1) \left(\frac{\alpha_H}{0.3}\right)^{-0.8} P_{orb}^{0.25}(\mathrm{h}) \quad \mathrm{d}$$

which is in good agreement with observation (Table 3.7).

3.7 Energetics of DN Outbursts

To complete this chapter we look at the total energy radiated during outbursts.

3.7.1 *Normal Outbursts*

The systematics of DN outbursts shown in Figure 3.10 suggest that it is necessary to observe only a few objects in order to calibrate the majority. Fortunately there are four objects, two above and two below the period gap, which have been observed extensively over a large range of energies:

VW Hyi Pringle *et al.* (1987) measured 2.0×10^{-3} erg cm^{-2} for the total energy received over the range 912–8000 Å. At a distance of 65 pc (Warner 1987a) this gives, for the energy emitted, $E_{\rm n} = 1.0 \times 10^{39}$ erg. As the inclination of VW Hyi is close to the standard 56°.7 (Section 2.6.1.3) there is no correction required.

Z Cha Harlaftis *et al.* (1992) measured 3.6×10^{-10} erg cm^{-2} s^{-1} over the range 1200 Å – 2 μm. If the flux distribution is similar to that of VW Hyi, this would give $\sim 5.0 \times 10^{-10}$ erg cm^{-2} s^{-1} for the range 912 Å–2 μm. For a distance of 130 pc (Warner 1987a) and an effective inclination of 85° (Section 2.6.1.3 and using $\dot{M}({\rm d}) = 1 \times 10^{-9}{\rm M}_\odot$ y^{-1} in equation (2.52)) this gives a luminosity $L_{\rm n} = 1.0 \times 10^{34}$ erg s^{-1}. The total energy radiated is $E_{\rm n} \approx L_{\rm n}\Delta T_{0.5} = 1.0 \times 10^{39}$ erg.

SS Cyg Polidan & Holberg (1984) estimate $L_{\rm n} = 1.3(\pm 0.3) \times 10^{34}$ erg s^{-1} at maximum, for a distance of 50 pc. Correcting to 75 pc (Warner 1987a) and for an inclination of 38° gives $L_{\rm n} = 1.7 \times 10^{34}$ erg s^{-1}, which, with $\Delta T_{0.5} = 7$ d, leads to $E_{\rm n} = 1.0 \times 10^{40}$ erg.

U Gem Using Polidan & Holberg's measured $L_{\rm n} = 1.5 \pm (0.5) \times 10^{34}$ erg s^{-1}, correcting to a distance of 81 pc, for an inclination of 70°, and adopting $\Delta T_{0.5} = 5$ d gives $E_{\rm n} = 1.5 \times 10^{40}$ erg.

The order of magnitude higher radiated energy for SS Cyg and U Gem, compared to VW Hyi and Z Cha, is expected from the relationships given in equations (3.4) and (3.7) and the dependence on $T_{\rm n}$ given in Section 3.3.3.2. Thus the function $P_{\rm orb}^{0.8} 10^{0.10 P_{\rm orb}} T_n^{0.87}$ takes the values 57, 74, 512 and 470 for VW Hyi, Z Cha, SS Cyg and U Gem respectively.

In physical terms the explanation is that, as the mass accreted, $\delta M({\rm d})$ is $\tilde{\propto}\ M({\rm d})$ (Section 3.5.4.1) and $M({\rm d}) \propto M^{-0.35}({\rm d}) r_{\rm d}^3$; (equation (3.26a)), the energy radiated $E_{\rm n} \propto M(1)\delta M({\rm d})/R(1)\ \tilde{\propto}\ M(1)P_{\rm orb}^2$ from equations (2.6) and (2.83). Higher primary masses above the period gap (Tables 2.5 and 2.6) enhance the $P_{\rm orb}$ dependence of $E_{\rm n}$.

3.7.2 *Superoutbursts*

VW Hyi Pringle *et al.* (1987) measured 1.61×10^{-2} erg cm^{-2} received in 912–8000 Å during a superoutburst, which produces $E_{\rm s} = 8.4 \times 10^{39}$ erg for the total radiated energy. Therefore $E_{\rm s}/E_{\rm n} = 8.4$ and the ratio of time-averaged luminosities over super and normal outbursts is $R(s/n) = E_{\rm s}T_{\rm n}/E_{\rm n}T_{\rm s} = 1.27$.

Z Cha Harlaftis *et al.* (1992) measured a bolometric luminosity at supermaximum 30% larger than at normal maximum. The duration of a superoutburst can be taken as $\Delta T_{\rm pl} + \Delta T_{0.5}$ (Table 3.7), during which the average luminosity is that at supermaximum reduced by a factor f which can be obtained by assuming that the 'plateau' region is a linear decay from $m_{\rm v}({\rm s})$ to $m_{\rm v}({\rm max})$ over a time $\Delta T_{\rm pl}$ (Section

Table 3.9. *Ratio of Time-Averaged Superoutburst and Outburst Visual Fluxes.*

Star	T_n(d)	T_s(d)	$\Delta T_{0.5}$(d)	ΔT_{pl}(d)	Δm	f	$R(s/n)$
OY Car	160:	346	2	15	0.9	0.70	6.3:
V436 Cen	32	335	1.8	10	1.1	0.63	1.1
Z Cha	51	218	1.2	11	0.5	0.85	3.2
VW Hyi	27	179	1.4	9	1.0	0.66	1.9
WX Hyi	14	195	1	10	0.9	0.70	1.3:
AY Lyr	24	205	1	11	0.9	0.70	2.2:
TU Men	37	194	4*	24	0.5	0.85	1.8:
RZ Sge	77	266	2.5	11	0.6	0.81	2.2
SU UMa	19	160	1	10	1.0	0.66	2.2:

* A weighted average of the two types of short outburst (Section 3.6.3)

3.6.4). This gives $f = \Delta m(10^{0.4\Delta m} - 1)^{-1}$, where $\Delta m = m_v(\max) - m_v(\mathrm{s})$. With the same corrections as in Section 3.7.1, this leads to $E_s \approx 1.2 \times 10^{40}$ erg. Therefore $E_s/E_n = 12$ and $R(s/n) = 2.8$.

The ratio $R(s/n)$ can be investigated more generally using purely visual measurements (Warner 1976a). Table 3.9 lists available observations and the derived

$$R(s/n) = f \times 10^{0.4\Delta m}\left(1 + \frac{\Delta T_{pl}}{\Delta T_{0.5}}\right)\frac{T_n}{T_s}.$$

The agreement between $R(s/n) = 2.8$ from bolometric measurements of VW Hyi and Z Cha and those from visual observations gives weight to the method, but some of the results are uncertain because of imprecise $\Delta T_{0.5}$. The result, $R(s/n) > 1$, is compatible with both the enhanced mass transfer model and the incomplete disc drainage model described in Section 3.6.6.4.

4

Nova-like Variables and Nova Remnants

A remnant of uneasy light.

W. Wordsworth. *Memorials of a Tour in Scotland.*

A variety of stars was early recognised to have eruptions that bear some resemblance to those of novae. Consequently, a very heterogeneous class of nova-like variables (NLs) was introduced (see, e.g., Campbell & Jacchia 1941; Kukarkin *et al.* 1958; Petit 1987), which, with hindsight, is seen to include many types of object that are totally unrelated structurally to the true novae. For example, η Car, γ Cas and P Cyg stars do not have the duplicity of CVs. Symbiotic stars, on the other hand, may all be close binaries and some may contain degenerate components as in the CVs (Kenyon 1986).

With the removal of these eruptive objects to their own classes, paradoxically only the non-eruptive residue remained as NLs (many of which, however, may have 'low states'). It is from their short time scale spectroscopic and photometric behaviour (including the evidence of binary structure) that such stars are recognized as resembling novae between eruptions (they could more appropriately have been termed 'NR-like').

4.1 Classifications

From the incomplete discovery of novae in earlier centuries (Section 1.1) it is clear that among the NLs there should be many unrecognized NRs. Similarly, the fact that many novae discovered this century can be found as blue objects on archival sky survey plates shows that among the currently known NLs must be a number of pre-novae. So far, no *previously known* variable has become a nova; yet at least four of the novae this century erupted from bright NLs: GK Per (1901), $m_V = 12.3$; V603 Aql (1918), $m_V = 11.4$; RR Pic (1925), $m_V = 12.0$; HR Del (1967), $m_V = 12.1$. There must be a high probability that within a decade or two an *already known* NL will become a nova (see Warner (1986f) for further discussion).

Optical spectra of NRs show a range of appearances, from wide, shallow Balmer and helium absorption lines with superimposed weak emission (DI Lac, V841 Oph) through weak pure emissions of H, HeI, HeII and λ4650 (V603 Aql, DN Gem) to strong emission of H, HeII, λ4650 (DQ Her, RR Pic). A variety of conditions, such as inclination, temperature of the primary (cooling from the last eruption), current $\dot{M}(2)$, and the magnetic field of the primary, must account for this range of spectral types. A similar range is found in NLs: the result is that there is no spectral (or photometric) diagnostic that can as yet distinguish pre- or post-novae among the NLs. Nor is there a clear spectroscopic signature for NLs that are systems with $\dot{M}(2)$ only slightly greater than $\dot{M}_{crit2}(d)$ (Section 3.5.3.5) and therefore closely related to the Z Cam stars.

It is possible, however, on spectroscopic and photometric grounds to define some subclasses within the NLs. Although the category UX UMa star has often been used synonymously with NL (Warner 1976b; Wade & Ward 1985), here it will denote those NLs that, like UX UMa itself, have *persistent broad Balmer absorption-line spectra*. The remainder, with pure emission line spectra (albeit occasionally with sharp absorption cores) will be termed RW Tri stars, after a venerable example. As pointed out in Section 4.2.1, the difference between RW Tri and UX UMa stars is to some extent a matter of orbital inclination.

Another distinct subclass, the SW Sex stars, has been introduced by Thorstensen *et al.* (1991) for highly inclined systems with significant RV phase shifts with respect to photometric conjunction, and transient absorption lines at certain orbital phases. Although these may merely be the most highly inclined UX UMa or RW Tri systems, and therefore not necessarily worthy of a separate name (Dhillon *et al.* 1992), they do present important problems and are therefore usefully grouped together for discussion (Section 4.2.2). They have also been called 'quiescent novae' (Vogt 1989), a term used prematurely as it begs the question.

A photometrically defined category among the NLs is the 'anti-dwarf novae' (Wade and Ward 1985), more succinctly and less misleadingly called VY Scl stars after the type star. These spend most of their time varying little about a mean magnitude, but occasionally fall in brightness by one or more magnitudes for extended periods (weeks to years) at which times they may appear spectroscopically like DN at quiescence.

As all of the NLs have long term variations about an average magnitude, the VY Scls may be no more than extreme examples. Here they are defined as showing reductions in brightness of greater than 1 mag from the mean. There is, however, usually a relatively rapid drop in the VY Scl stars, not like the slow waves that run through other NL light curves, which resembles the unpredictable drops in R CrB light curves (in which class some VY Scl stars were at first erroneously placed). It has long been supposed (Kraft 1964b; Warner and van Citters 1974) that these may be in effect Z Cam stars in extended standstills. In terms of the DI model of DN outbursts (Section 3.5.3.5) they would be interpreted as systems where $\dot{M}(2)$ is generally greater than $\dot{M}_{crit2}(d)$, with occasional excursions below. The classes UX UMa and RW Tri can include members of the SW Sex and VY Scl classes.

A list of NLs with determined orbital periods is given in Table 4.1, with subclasses where known. A list of classical and RN remnants with known orbital periods is given in Table 4.2. As with other classes of CVs, there are still many relatively bright objects without known P_{orb}; these include the NLs CL Sco, V592 Cas, KQ Mon, RX LMi and the NRs RS Car, Q Cyg, V446 Her, CP Lac, DK Lac, GI Mon, V441 Sgr, FH Ser, XX Tau and LV Vul. Others may be found in Table 12.1 of Vogt (1989) – which, however, includes many of the magnetic CVs considered in Chapters 6 and 7. The Palomar–Green survey in the northern hemisphere revealed a number of fainter NLs (Green *et al.* 1982).

There is no discussion of polarization of NLs or NRs in this chapter because any circular or non-interstellar linear polarization found in these objects makes them *de facto* magnetic CVs and they make their appearance in Chapters 6–8. The results of

Table 4.1. *Nova-like Variables with known Orbital Periods.*

Star	Alias	Subtype	P_{orb}(h)	m_v	$i°$	$Sp(2)$	References
SX LMi	CBS 31	RW	1.5:	16.0	65:		1,166,179
BK Lyn	PG 0917+342	UX	1.80:	14.1–15.1	~32		2,3,16,178,182,188
H1752+081		RW	1.8828	16.5	77	M6	181,203
V795 Her*	PG 1711+336	RW	2.5984	12.5–13.2	<65		169–177,184,201
PG 2133+115		UX	2.90:	14.3			182
EC 05114–7955		RW	3.02	15.1			191
MV Lyr	MacRae+43° 1	UX,VY	3.201	12.1–17.7	20:	M5V	4–8,163–165,180
SW Sex	PG 1012-029	RW,SW	3.2385	14.8	79		9–13,148,155
HL Aqr	PHL 227	UX	3.254	13.5	<50		14,15,189
DW UMa	PG 1030+590	RW,SW,VY	3.2786	14.9–17	80:		13,16–19,70,148,158,193
LY Hya	1329-294	RW,VY	3.2868	14.4–18.4	35:		194–196
H1933+510			3.30:	17.8	<65		181
TT Ari	BD +14° 341	UX,VY	3.3012	9.5–16.5	23:		20–34,148,202
BZ Cam	0623+71	UX,VY	3.34:	12.5–14.0	<65		35–39
WX Ari	PG 0244+104	RW,SW,VY?	3.3442	14.7–15.8	<65		16,40,167,182
V1315 Aql	KPD 1911+1212	RW,SW	3.3526	14.4	82		13,41–43,148,155,192
V442 Oph		RW,VY	3.374	12.6–15.5	60:		44,45
EC 05565-5935		UX	3.4:	14.3	<65		191
VZ Scl	Ton 120	RW,VY	3.4709	15.6–~20	85:		46–49
PX And	PG 0027+260	RW,SW,VY	3.5125	14.8–17.2:	74:		50,51,182,204,208
V425 Cas		RW,VY	3.590	14.5–18	25		52,53
PG 0859+415		UX,SW	3.6675	14.2	>65		182
BH Lyn	PG 0818+513	RW,SW,VY	3.7409	15.5–17.2	80		54,55,157
LX Ser		RW,VY	3.8024	14.5–16.5	75:		52,56–60,148,155
V380 Oph		RW	3.8	14.5–>16.1	42:		61,152
CM Del		RW,VY	3.88	13.4–15.3	30:		52,53,61
KR Aur		RW,VY	3.9072	11.3–18	38:		52,62—64,207
UU Aqr		RW	3.9259	13.3–14.0	78	K7–M0	197–200
V1776 Cyg	Lanning 90	RW	3.9537	16.2–17.2	75		65,66
V1193 Ori		UX	3.96	14.1			67–69
VY Scl	PS 141	UX,VY	3.99	12.9–18.5	30:		71–73,168,189

PG 1000 + 667		RW	4.06	15.1			182
EC 04224-2014		UX	4.2	11.5	<65		191
IX Vel	CPD −48° 1577	UX	4.6543	9.0–10.0	60		74–82,161,205
UX UMa		UX	4.7201	12.7	71	K8–M6V	83–96, 148,155,168,183,206
V345 Pav**	EC 19314–5915	RW	4.7543	13.4	70:		97
ER UMa	PG 0943 + 521	UX	4.79:	14.2			182
V825 Her	PG 1717 + 413	UX	4.94	14.1–14.4	<65		181,182
V3885 Sgr	CD −42° 14462	UX	5.191	9.6–10.5	<50		98–105,189,190
AY Psc	PG 0134 + 070	RW	5.2157	15.8	74		106,107,156,186
V794 Aql		RW,VY	5.5:	13.7–20.2	65:		52,108-111,185
RW Tri		RW	5.5652	12.5–13.4	75	K7	52,112–124,148,155,168,207
V347 Pup	LB1800	RW	5.566	13.4	87		160,187
DO Leo	PG 1038 + 155	RW	5.6284	16.0–17.0	80:		125,126
RW Sex	BD −7° 2007	UX	5.8817	10.5–10.8	34		127–133
V751 Cyg		RW, VY	6.0:	13.2–16.0	<40		134,135
AC Cnc		RW	7.2115	13.8–14.4	72	G8–K2V	136–142,148
V363 Aur	Lanning 10	RW	7.7098	14.2	73	K0V	143–145,148,155
BI Lyn	PG 0900 + 401	RW	8.12	15.1			159
RZ Gru		UX	10.01:	12.3–13.4	<20		146,147,162
H 0927 + 501			10.04	16.8	>70		181
QU Car	HDE 310376	UX	10.90	11.1–11.5	<60		149–153
WY CMa			27.464	14.5			154

* The orbital period may be 4.86h: see Section 4.2.1.

** Triple system with G companion.

References: 1 Howell & Szkody 1990; 2 Howell *et al.* 1991; 3 Dobrzycka & Howell 1992; 4 Robinson *et al.* 1981; 5 Schneider, Young & Schectman 1981; 6 Chiapetti *et al.* 1982; 7 Szkody & Downes 1982; 8 Andronov, Fuhrmann & Wenzel 1988; 9 Williams & Ferguson 1982; 10 Penning *et al.* 1984; 11 Honeycutt, Schlegel & Kaitchuck 1986; 12 Williams 1989; 13 Szkody & Piché 1990; 14 Hunger, Heber & Koester 1985; 15 Haefner & Schoembs 1987; 16 Green *et al.* 1982; 17 Kopylov *et al.* 1988; 18 Shafter, Hessman & Zhang 1988; 19 Kopylov, Somov & Somova 1991; 20 Smak & Stepien 1975; 21 Cowley *et al.* 1975; 22 Mardirossian *et al.* 1980; 23 Krautter *et al.* 1981a,b; 24 Jameson, King & Sherrington 1982; 25 Jameson *et al.* 1982; 26 Jensen *et al.* 1983; 27 Hutchings & Cote 1985; 28 Shafter *et al.* 1985; 29 Thorstensen, Smak & Hessman 1985; 30 Hutchings, Thomas & Link 1986; 31 Schwarzenberg-Czerny 1987; 32 Semeniuk *et al.* 1987; 33 Udalski 1988; 34 Volpi, Natali & D'Antona 1988; 35 Ellis, Grayson & Bond 1984; 36 Lu & Hutchings 1985; 37 Krautter, Klaas & Radons 1987; 38 Woods, Drew & Verbunt 1990; 39 Pajdos & Zola 1992; 40 Beuermann *et al.* 1992; 41 Downes *et al.* 1986; 42 Szkody 1987b; 43 Dhillon, Marsh & Jones 1991; 44 Szkody & Wade 1980; 45 Szkody & Shafter 1983; 46 Warner & Thackeray 1975;

47 Sherrington, Bailey & Jameson 1984; 48 O'Donoghue, Fairall & Warner 1987; 49 Williams 1989; 50 Li *et al.* 1990; 51 Thorstensen *et al.* 1991; 52 Shafter 1984a; 53 Szkody 1985c; 54 Andronov *et al.* 1989; 55 Thorstensen *et al.* 1991; 56 Horne 1980; 57 Szkody 1981b; 58 Young, Schneider & Schectman 1981c; 59 Eason *et al.* 1984; 60 Schwarzenberg-Czerny 1984b; 61 Shafter 1985; 62 Popova & Vitrichenko 1978; 63 Hutchings, Link & Crampton 1983; 64 Shafter 1983; 65 Shafter, Lanning & Ulrich 1983; 66 Garnavich *et al.* 1990; 67 Hamuy & Maza 1986; 68 Bond *et al.* 1987b; 69 Warner & Nather 1988; 70 Zhang 1989; 71 Burrell & Mould 1973; 72 Warner & Van Citters 1974; 73 Hutchings & Cowley 1984; 74 Wargau *et al.* 1983, 1984; 75 Eggen & Niemala 1984; 76 Garrison *et al.* 1984; 77 Williams & Hiltner 1984; 78 Sion 1985a; 79 Warner, O'Donoghue & Allen 1985; 80 Haug 1988; 81 Beuermann & Thomas 1990; 82 Mauche 1991; 83 Linnell 1949, 1950; 84 Johnson, Perkins & Hiltner 1954; 85 Walker & Herbig 1954; 86 Warner & Nather 1972b; 87 Nather & Robinson 1974; 88 Africano & Wilson 1976; 89 Kukarkin 1977; 90 Quigley & Africano 1978; 91 Frank *et al.* 1981a; 92 Holm, Panek & Schiffer 1982; 93 King *et al.* 1983; 94 Schlegel, Honeycutt & Kaitchuck 1983; 95 Shafter 1984a,b; 96 Rubenstein, Patterson & Africano 1991; 97 Buckley *et al.* 1992; 98 Bond & Landolt 1971; 99 Wegner 1972; 100 Warner 1973b; 101 Hesser, Lasker & Osmer 1974; 102 Cowley, Crampton & Hesser 1977a; 103 Bond 1978; 104 Guinan & Sion 1982a; 105 Haug & Drechsel 1985; 106 Szkody *et al.* 1989; 107 Diaz & Steiner 1990b; 108 Szkody *et al.* 1981b; 109 Honeycutt & Schlegel 1985; 110 Szkody, Downes & Mateo 1988; 111 Mukai *et al.* 1990b; 112 Walker 1963a; 113 Winkler 1977; 114 Africano *et al.* 1978; 115 Smak 1979c; 116 Frank & King 1981; 117 Longmore *et al.* 1981; 118 Young & Schneider 1981; 119 Kaitchuck, Honeycutt & Schlegel 1983; 120 Vojkhanskaya 1984; 121 Cordova & Mason 1985; 122 Horne & Stiening 1985; 123 Robinson, Shetrone & Africano 1991; 124 Rutten & Dhillon 1992; 125 Green *et al.* 1982; 126 Abbott *et al.* 1990; 127 Cowley & MacConnell 1972; 128 Hesser, Lasker & Osmer 1972; 129 Cowley, Crampton & Hesser 1977b; 130 Greenstein & Oke 1982; 131 Bolick *et al.* 1987; 132 Shylaya 1987; 133 Beuermann, Stasiewski & Schwope 1992; 134 Robinson, Nather & Kiplinger 1974; 135 Bell & Walker 1980; 136 Kurochkin & Shugarov 1980; 137 Shugarov 1982; 138 Downes 1982; 139 Okazaki, Kitamura & Yamasaki 1982; 140 Yamasaki, Okazaki & Kitamura 1983; 141 Schlegel, Kaitchuck & Honeycutt 1984; 142 Baidak & Shugarov 1986; 143 Szkody & Crosa 1981; 144 Horne, Lanning & Gomer 1982; 145 Schlegel, Honeycutt & Kaitchuck 1986; 146 Kelly, Kilkenny & Cooke 1981; 147 Stickland *et al.* 1984; 148 Honeycutt, Kaitchuck & Schlegel 1987; 149 Stephenson, Sanduleak & Schild 1968; 150 Schild 1969; 151 Hiltner & Gordon 1971; 152 Gilliland & Phillips 1982; 153 Kern & Bookmyer 1986; 154 Hacke & Richert 1990; 155 Rutten, van Paradijs & Tinbergen 1992; 156 Szkody & Howell 1992a; 157 Dhillon *et al.* 1992; 158 Hessman 1990a; 159 Lipunova & Shugarov 1990; 160 Buckley *et al.* 1990b; 161 Hessman 1990b; 162 Schaefer & Patterson 1987; 163 Greenstein 1954; 164 Walker 1954a; 165 Vojkhanskaya 1980; 166 Sion *et al.* 1991; 167 Warner 1983c; 168 Drew & Verbunt 1985; 169 Baidak *et al.* 1985; 170 Thorstensen 1986; 171 Kaluzny 1989; 172 Rosen *et al.* 1989; 173 Shafter *et al.* 1990; 174 Prinja, Rosen & Supelli 1991; 175 Zhang *et al.* 1991; 176 Prinja, Drew & Rosen 1992; 177 Wenzel, Banny & Andronov 1988; 178 Szkody & Howell 1992a,b; 179 Wagner *et al.* 1988 ; 180 Rosino, Romano & Marziani 1993; 181 Silber 1992; 182 Ringwald 1992; 183 Panek & Howell 1980; 184 Prinja & Rosen 1993; 185 Honeycutt, Cannizzo & Robertson 1994; 186 Howell & Blanton 1993; 187 Mauche et al. 1993; 188 Skillman & Patterson 1993; 189 Hollander, Kraakman & van Paradijs 1993; 190 Woods *et al.* 1992; 191 Chen 1994; 192 Smith *et al.* 1993; 193 Honeycutt, Livio & Robertson 1993; 194 Kubiak & Krzeminski 1989, 1992; 195 Haefner, Barwig & Mantel 1993; 196 Echevarria *et al.* 1983; 197 Volkov, Shugarov & Serigena 1986; 198 Diaz & Steiner 1991a; 199 Baptista *et al.* 1994; 200 Baptista, Steiner & Cieslinski 1994; 201 Haswell *et al.* 1994; 202 Robinson & Cordova 1994; 203 Silber *et al* 1994; 204 Still, Dhillon & Jones 1995; 205 Long *et al.* 1994a; 206 Smak 1994a; 207 Honeycutt *et al.* 1994; 208 Hellier & Robinson 1994.

searches for polarization among CVs can be found in Szkody, Michalsky & Stokes (1982), Cropper (1986), Buckley *et al.* (1990b), Rutten & Dhillon 1992, and Stockman *et al.* (1992).

4.2 Spectroscopic Observations

The spectra of NLs and NRs in general show signs of higher excitation conditions than those of DN. The HeII λ4686 and CIII/NIII λ4650 features are relatively stronger, though in RW Tri itself the latter is barely detectable (Kaitchuck, Honeycutt & Schlegel 1983); the ratio λ4686/Hβ may exceed unity – in QU Car in particular the Balmer lines are quite weak, suggesting a disc of unusually high temperature and $\dot{M}$(d) (Gilliland & Phillips 1982). Most of the lower excitation lines in DN, listed in Table 3.5, appear in NLs, but a few additional weak lines are often seen – Table 4.3. Unlike DN, the spectra of NLs rarely have FeII emission (e.g., DW UMa: Shafter, Hessman & Zhang 1988).

It is evident from Table 4.1 that, with one exception, no eclipsing NLs are UX UMa stars. The exception is UX UMa itself! However, at low inclinations both UX UMa and RW Tri stars occur. A simple interpretation of this partial segregation, following Section 2.7, is that all NL discs are optically thick, resulting in a diminishing contribution from the absorption-line spectrum at increasing inclination; but there must be a considerable range of total flux from the optically thin emission-line region, sufficient in some low inclination systems to fill completely the underlying absorption lines. Even in individual objects emission can vary sufficiently to fill temporarily the absorptions if they are weak (TT Ari: Vojkhanskaya 1983; UX UMa: Schlegel, Honeycutt & Kaitchuck 1983; BZ Cam: Lu & Hutchings 1985).

In most NLs there is little or no evidence for the presence of an emission-line S-wave from the bright spot, even where there is clearly a bright spot continuum affecting eclipse profiles. Thus the principal detailed studies of the following stars explicitly exclude any bright spot emission: V363 Aur, AC Cnc, MV Lyr, RW Sex, SW Sex and IX Vel. In KR Aur (Hutchings, Link & Crampton 1983), V795 Her (Shafter *et al.* 1990) and V3885 Sgr (Haug & Drechsel 1985) there is an extended emission-line region apparently on the *following* side of the disc. In some systems, although there is no S-wave there is orbital modulation of the strength of the emission lines with maximum near $\varphi = 0.75$ (SW Sex: Penning *et al.* 1984; V1776 Cyg: Garnavich *et al.* 1990). UX UMa and QU Car do have S-waves in HeII emission, although there is none in the Balmer lines (Schlegel, Honeycutt & Kaitchuck 1983; Gilliland & Phillips 1982); in addition, in UX UMa there are short lived absorption lines apparently formed from a vertically extended region above the bright spot area. These results emphasize the generally lesser contribution of bright spot emission (compared to DN at quiescence) in these high $\dot{M}$(d) systems.

The high luminosities of the discs in the NLs prevent detection of spectra of the secondaries, even for high inclination systems, in all but a few of the longest period systems or VY Scl stars (Table 4.1). Chromospheric Balmer emission from the secondary is detected in some systems (e.g., V363 Aur: Horne, Lanning & Gomer 1982; AC Cnc: Schlegel, Kaitchuck & Honeycutt 1984; RW Tri: Kaitchuck, Honeycutt & Schlegel 1983; UX UMa: Schlegel, Honeycutt & Kaitchuck 1983).

Table 4.2. *Nova Remnants with known Orbital Periods.*

Star	Nova	P_{orb}(h)	m_V	i°	$Sp(2)$	References
GQ Mus	1983	1.426	17.5			1,175
CP Pup	1942	1.471	15.0	35:		5,7,136–142,153
T Pyx	RN	1.75:	15.3	27:	M5	2–7,119,159,173
RW UMi	1956	1.9*:	18.8			8–10,98
V1974 Cyg	1992	1.9503	17:			181
V Per	1887	2.5709	18.5	80:		11,12
V2214 Oph	1988	2.8204	20.5			167
V603 Aql	1918	3.3157	11.4	17:		13–24,144,149–151,158,169
V1668 Cyg	1978	3.32	19.9			25–27
V1500 Cyg	1975	3.3507	17.2	50:		28–38,178
RR Pic	1925	3.4806	12.0	65:		14,39–44,133,146
WY Sge	1783	3.6872	19.0	70		45–50,154
DO Aql	1925	4.0263	17.9	> 0:		172
V849 Oph	1919	4.1461	17.9	80:		172
DQ Her	1934	4.6469	14.2	89	M3V	20,51–77,144,152
CT Ser	1948	4.680	17.4			165
T Aur	1891	4.9051	14.9	70:		5,78,79
V533 Her	1963	5.04	14.3	< 60		80-82
PW Vul	1984	5.13	>16.5			83
HR Del	1967	5.1400	11.9	40		84–89,133
U Leo	1855	6.42:	17.3			90
V838 Her	1991	7.1432	20.6	80:		91,92,171,176
BT Mon	1939	8.0115	15.8	84:	K5–7	93–97
QZ Aur	1964	8.580	17.2	75	K0	98,171,180
DI Lac	1910	13.0506	14.3	30:		13,99
V841 Oph	1848	14.50	13.9	30:		50,100,155
V394 CrA	RN	18.18:	20:	< 65	K	6,159
U Sco	RN	29.628	19.2	80:	F8	6,101,102,119,156,164
LMC-RN	RN	31:	20:			174
GK Per	1901	47.9233	10.2	< 73	K2–3	103–114,133–135,168
V1017 Sgr	1919	137.1	13.6		G5IIIp	115,119,159
T CrB	RN	227.59d	10.0	60:	M3III	116–126,145,147,148, 157,159–162,166
RS Oph	RN	460d	9.5–12.5	$\lesssim$30	M0–2III	119,127–132,143, 159,163,170,177,179

* It is not certain that the photometric period is actually the orbital period.

References: 1 Diaz & Steiner 1989, 1990a; 2 Joy 1954c; 3 Bruch, Duerbeck & Seitter 1982; 4 Williams 1983; 5 Szkody & Feinswog 1988; 6 Schaefer 1990; 7 Vogt *et al.* 1990; 8 Kaluzny & Chlebowski 1989; 9 Szkody *et al.* 1989; 10 Howell *et al.* 1991; 11 Shafter & Abbott 1989; 12 Wood, Abbot & Shafter 1992; 13 Kraft 1964a; 14 Gallagher & Holm 1974; 15 Cook 1981; 16 Drechsel *et al.* 1981; 17 Slovak 1981; 18 Metz 1982; 19 Drechsel *et al.* 1983; 20 Shafter 1984a; 21 Haefner & Metz 1985; 22 Haefner, Pietsch & Metz 1988; 23 Udalski & Schwarzenberg-Czerny 1989; 24 Patterson & Richman 1991; 25 Campolonghi *et al.* 1980; 26 Piccioni *et al.* 1984; 27 Kaluzny 1990; 28 Hutchings & McCall 1977; 29 Patterson 1978, 1979a; 30 Lanning & Semeniuk 1981; 31 Kruszewski, Semeniuk & Duerbeck 1983; 32 Kaluzny & Semeniuk 1987; 33 Chlebowski & Kaluzny 1988; 34 Kaluzny & Chlebowski 1988; 35 Lance, McCall & Uomoto

4.2.1 *The RW Tri Stars*

Optical spectra of DW UMa, which is representative of the higher excitation NLs, are shown in Figure 4.1. In eclipse HeII λ4686 and CII λ4267 are reduced more than the other lines, and therefore must be concentrated towards the centre of the disc. A similar stratification effect is seen during eclipses of other NLs, e.g., SW Sex (Honeycutt, Schlegel & Kaitchuck 1986).

References to Table 4.2 (cont.)
1988; 36 Stockman, Schmidt & Lamb 1988; 37 Horne & Schneider 1989; 38 Schmidt & Stockman 1991; 39 Vogt 1975; 40 Wyckoff & Wehinger 1977; 41 Krautter *et al.* 1981b; 42 Warner 1981; 43 Kubiak 1984a; 44 Warner 1986d; 45 Weaver 1951; 46 Warner 1971; 47 Shara & Moffat 1983; 48 Shara *et al.* 1984; 49 Kenyon & Berriman 1988; 50 Shara, Potter & Shara, 1989; 51 Walker 1954b; 52 Walker 1956; 53 Greenstein & Kraft 1959; 54 Kraft 1961b, 1964b; 55 Nather & Warner 1969; 56 Beer 1974; 57 Bath, Evans & Pringle 1974; 58 Katz 1975; 59 Nelson 1976; 60 Nelson & Olson 1976; 61 Liebert 1976; 62 Chanan, Nelson & Margon 1978; 63 Patterson, Robinson & Nather 1978; 64 Hutchings, Cowley & Crampton 1979; 65 Patterson 1979f; 66 Schneider & Greenstein 1979; 67 Dmitrienko & Cherapashchuk 1980; 68 Petterson 1980; 69 Smak 1980; 70 Young & Schneider 1980, 1981; 71 Balachandran, Robinson & Kepler 1983; 72 Cordova & Mason 1985; 73 Honeycutt, Kaitchuck & Schlegel 1987; 74 Dmitrienko 1988; 75 Schoembs & Rebhan 1989; 76 Horne, Welsh & Wade 1993; 77 Herbig & Smak 1992; 78 Walker 1963b; 79 Bianchini 1980; 80 Patterson 1979c; 81 Robinson & Nather 1983; 82 Hutchings 1987; 83 Hacke 1987a,b; 84 Hutchings 1979, 1980; 85 Kahoutek & Pauls 1980; 86 Krautter *et al.* 1981b; 87 Bruch 1982; 88 Friedjung, Andrillat & Puget 1982; 89 Kuerster & Barwig 1988; 90 Downes & Szkody 1989; 91 Ingram *et al.* 1992; 92 Leibowitz *et al.* 1992; 93 Robinson, Nather & Kepler 1982; 94 Schaefer & Patterson 1983; 95 Marsh, Wade & Oke, 1983; 96 Seitter 1984; 97 Williams 1989; 98 Campbell & Shafter 1992; 99 Ritter 1990; 100 Friedjung, Bianchini & Sabbadin 1988; 101 Webbink *et al.* 1987; 102 Johnston & Kulkarni 1992; 103 Gallagher & Oinas 1974; 104 Bianchini, Hamzaoglu & Sabbadin 1981; 105 Bianchini, Sabbadin & Hamzaoglu 1982; 106 Bianchini & Sabbadin 1983b; 107 Watson, King & Osborne 1985; 108 Mazeh *et al.* 1985; 109 Bianchini *et al.* 1986; 110 Crampton, Cowley & Fisher 1987; 111 Hutchings & Cote 1986; 112 Norton, Watson & King 1988; 113 Hessman 1989b; 114 Ishida *et al.* 1992; 115 Sekiguchi 1992b; 116 Kenyon & Garcia 1986; 117 Kraft 1958; 118 Lines, Lines & McFaul 1988; 119 Webbink *et al.* 1987; 120 Peel 1985; 121 Casatella, Gilmozzi & Selvelli 1985; 122 Sanford 1949; 123 Bailey 1975b; 124 Walker 1957; 125 Walker 1977; 126 Bode 1987; 127 Garcia 1986; 128 Jameson & Sherrington 1983; 129 Bruch 1986; 130 Wallerstein 1963; 131 Wallerstein & Cassinelli 1968; 132 Kenyon & Gallagher 1983; 133 Rosino, Bianchini & Rafarelli 1982; 134 Reinsch 1994; 135 Selvelli & Hack 1983; 136 Bianchini, Friedjung & Sabbadin 1985a,b, 1989; 137 Warner 1985d; 138 Duerbeck, Seitter & Duemmler 1987; 139 Barrera & Vogt 1989b; 140 O'Donoghue *et al.* 1989; 141 Vogt *et al.* 1990; 142 Diaz & Steiner 1991b; 143 Tempesti 1975; 144 Lambert & Slovak 1981; 145 Selvelli, Cassatella & Gilmozzi 1992; 146 Selvelli 1982; 147 Duerbeck *et al.* 1980; 148 Kenyon & Webbink 1984; 149 Rahe *et al.* 1980; 150 Lambert *et al.* 1981; 151 Ferland *et al.* 1982; 152 Ferland *et al.* 1984; 153 White, Honeycutt & Horne 1993; 154 Naylor *et al.* 1992; 155 Della Valle & Rosino 1987; 156 Hanes 1985; 157 Iijima 1990; 158 Patterson *et al.* 1993b; 159 Duerbeck & Seitter 1990, 160 Paczynski 1965c; 161 Bianchini & Middleditch 1976; 162 Oskanyan 1983; 163 Evans 1986; 164 Duerbeck *et al.* 1993; 165 Ringwald 1992; 166 Yudin & Munari 1993; 167 Baptista *et al.* 1993; 168 Anupama & Prabhu 1993; 169 Hollander, Kraakman & van Paradijs 1993; 170 Orio 1993; 171 Szkody & Ingram 1994; 172 Shafter, Misselt & Veal 1993; 173 Schaefer *et al.* 1992; 174 Sekiguchi 1992b; 175 Diaz & Steiner 1990a; 176 Hoard & Szkody 1994; 177 Walker 1979; 178 Schmidt, Liebert & Stockman 1995; 179 Dobrzycka & Kenyon 1994; 180 Campbell & Shafter 1995; 181 De Young & Schmidt 1994.

Table 4.3. *Additional Emission Lines in NLs (Wavelengths in Å).*

1206	SiIII	3819	HeI	4267	CII
1371	OV	4070–76	OII	4415–17	OII
1718	NIV	4089	SiII	4542	HeII
3705	HeI	4128–30	SiII	5805	CIV
3712	OIII	4143	HeI	7234	CII

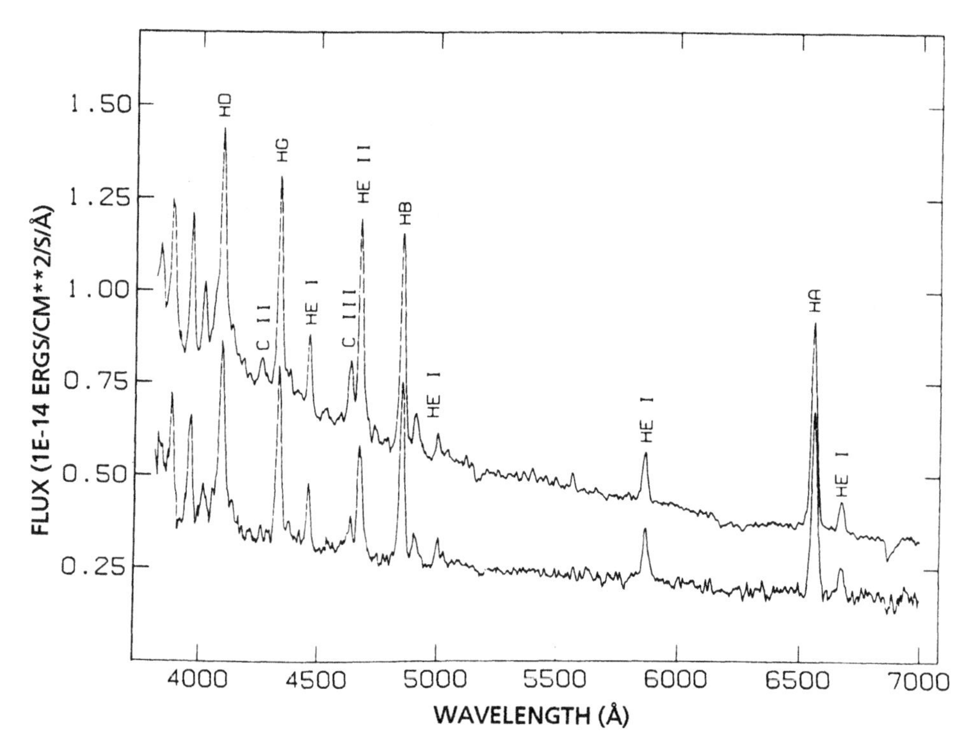

Figure 4.1. Optical spectra of DW UMa. The upper spectrum was obtained out of eclipse and the lower during eclipse. From Shafter, Hessman & Zhang (1988).

The relative strengths of the optical emission lines in RW Tri stars increase with inclination. Table 4.4 lists observed equivalent widths *W* (for systems that are also VY Scl stars, these are 'high state' spectra – the equivalent widths increase greatly as the system luminosity decreases, so there may be overestimates of *W* for any star not properly at maximum). Plotted in Figure 4.2, these show similar behaviour to the lines in NRs (Figure 2.34), discussed in Section 2.7. Substantial changes of line strength on time scales of minutes occur in many systems

In the UV the same lines appear as in DN (Table 3.5) with the addition of those in Table 4.3. Figure 4.3 illustrates the spectra of SW Sex and DW UMa in the region

Table 4.4. *Emission-Line Equivalent Widths in NLs in the High State.*

Star	$i°$	$H\alpha$(Å)	$H\beta$(Å)	HeIIλ4686(Å)	Reference
V794 Aql	65:	30	15	3	1
V1315 Aql	82		27.3	15.2	2
KR Aur	38:	12.1	5.3		19
V363 Aur	73		3.5	5.5	3
BZ Cam	<65		2.6		4
AC Cnc	72		11	4	5
V1776 Cyg	75	64			6
CM Del	30:	11	1	2	7
EC 19314	70:		1.8	1.6	8
V795 Her	<65	12	4.5	1	21
DO Leo	80:	38	21	6	9
BH Lyn	80	52.0	27.8	11.8	10
V380 Oph	42:	28	9		7
V442 Oph	40:	12.5	4.8	3.9	19
PX And	74:	45	16	9	11
AY Psc	74		17		12
LX Ser	75		25.5	16.9	13
SW Sex*	79	46	21	16	14,20
VZ Scl	85:	65	18	7	15
RW Tri	75	32	8.5	3	16,17
DW UMa	80:	45	26	22	18

* Line strengths in SW Sex vary by a factor of two or more (cf. Penning *et al.* 1984).

References: 1 Szkody *et al.* 1981b; 2 Dhillon, Marsh & Jones 1991; 3 Schlegel, Honeycutt & Kaitchuck 1986; 4 Lu & Hutchings 1985; 5 Schlegel, Kaitchuck & Honeycutt 1984; 6 Garnavich *et al.* 1990; 7 Shafter 1985; 8 Buckley *et al.* 1992; 9 Abbott *et al.* 1990; 10 Dhillon *et al.* 1992; 11 Thorstensen *et al.* 1991; 12 Szkody & Howell 1992a; 13 Young, Schneider & Schectman 1981c; 14 Szkody 1987c; 15 O'Donoghue, Fairall & Warner 1987; 16 Kaitchuck, Honeycutt & Schlegel 1983; 17 Rutten & Dhillon 1992; 18 Shafter, Hessman & Zhang 1988; 19 Williams 1983; 20 Honeycutt, Schlegel & Kaitchuck 1986; 21 Thorstensen 1986.

1000–2000 Å, which are entirely in emission. Other RW Tri stars show a transition from emisson lines at $\lambda \gtrsim 2000$ Å to absorption (or P Cyg profile) at $\lambda \lesssim 2000$ Å (KR Aur, BZ Cam, CM Del: see la Dous 1991); these may all be low inclination systems.

There is a very sensitive dependence of UV line equivalent width on inclination (la Dous 1991). This is illustrated in Figure 4.4 for the CIV λ1550 line; for $i \lesssim 60°$ the line generally has a P Cyg profile, and the emission and absorption contributions counteract each other. The absorption component equivalent widths are therefore much larger than appear in Figure 4.4. Both line strengths (la Dous 1991; Prinja, Rosen & Supelli 1991) and continuum fluxes (Verbunt 1987) in the UV vary rapidly but generally not due simply to orbital modulation.

Detailed studies of RW Tri (Cordova & Mason 1985; Drew & Verbunt 1985) and of UX UMa (Holm, Panek & Schiffer 1982; King *et al.* 1983; Drew & Verbunt 1985)

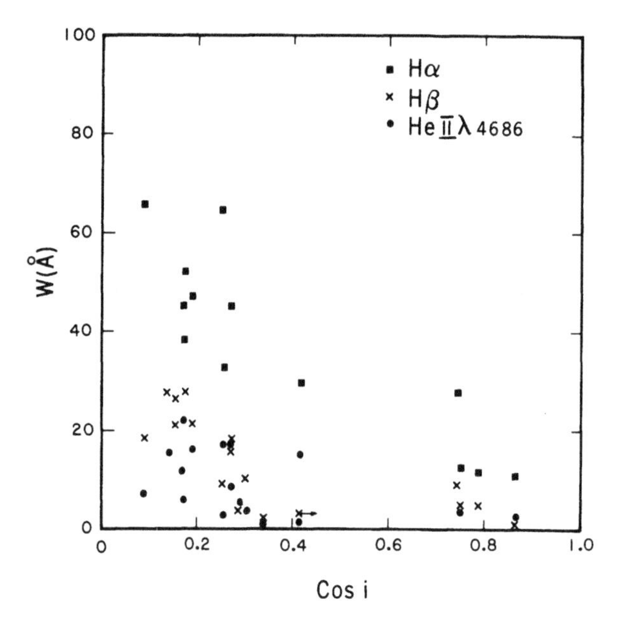

Figure 4.2. Emission–line equivalent widths of Hα, Hβ and λ4686 in NLs, plotted as a function of cosine of inclination. From Table 4.4.

show almost identical behaviour; the UV continuum is deeply eclipsed but the lines are only mildly eclipsed. The emission-line profiles are strongly asymmetric, with peaks redshifted by ~4 Å. The interpretation of the UV lines as generated by winds is discussed in Section 2.7.3.

The SiIV and CIV UV line strengths in V795 Her show a 4.86 h periodicity that is coherent over several years (Prinja & Rosen 1993). This may be the true orbital period rather than that obtained from the optical spectroscopic period (Zhang *et al.* 1991).

Unlike DN at maxima (Section 3.3.5.2), superposition of the UV spectra does not produce a concentration irrespective of inclination; instead there appears to be reduced short wavelength flux at larger inclinations (Figures 1b and 5f of la Dous (1991)). As a result, most RW Tri stars have fluxes that increase over the range 1000–3000 Å, whereas most UX UMa stars have relatively flat spectra. In general the average 1000–2000 Å spectrum of NLs is redder than those of DN in outburst.

The relatively low temperatures claimed for the primaries in DW UMa and V1315 Aql (Table 2.8) may be incorrect: the primaries may be obscured by the disc (Szkody 1987b).

4.2.2 *The UX UMa Stars*

From the implications of their absorption line spectra, UX UMa stars have been called *thick-disc* CVs by Ferguson, Green & Liebert (1984). In their study of stars found in the Palomar–Green survey, these authors identified only two previously unknown members of this class, PG1717 + 413 and PG 2133 + 115, for neither of which is P_{orb} yet known. KQ Mon and LSI +55° 08 are also UX UMa stars without known P_{orb} (Sion & Guinan 1982; Guinan & Sion 1982b).

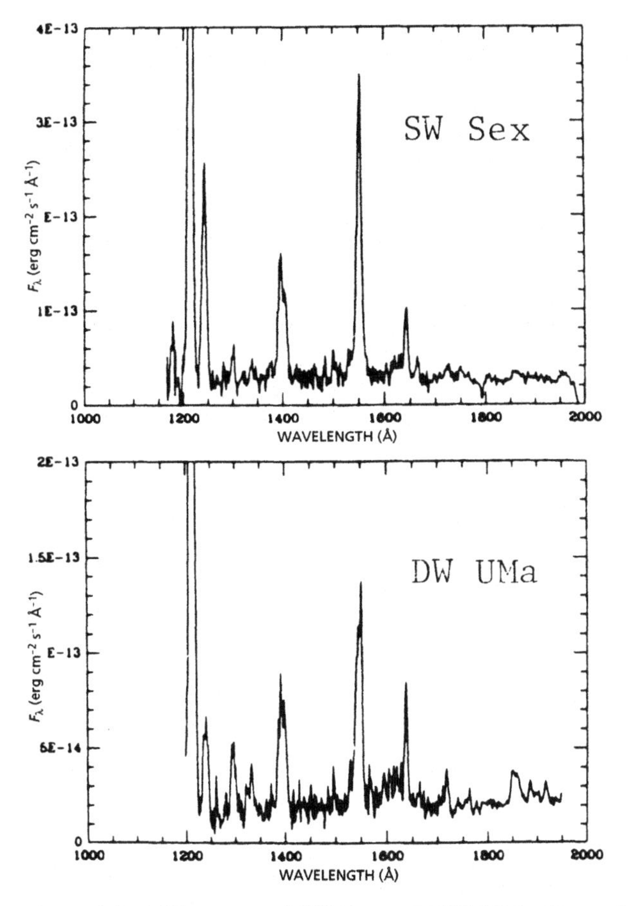

Figure 4.3 UV spectra of SW Sex and DW UMa. From Szkody (1987d).

Optical spectra of UX UMa stars exhibit a wide range of strengths of the emission-line components. Figure 4.5 shows an extremum: a pure absorption-line spectrum containing HI, HeI and CaII lines, with no sign of HeII, CIII or NIII emission. On subsequent nights, however, in this star those emissions (and weak central emission components in the Balmer lines) were present (Hessman 1990b) and at other times the Balmer emission components have appeared strongly (Beuermann & Thomas 1990). Wargau *et al.* (1983) also noted variations in strength by a factor of 2 in the HeII emission of IX Vel. (It would be desirable to obtain accurate photometry when such spectroscopic observations are made.) A less extreme example, RZ Gru, is shown in Figure 4.6, and HL Aqr and RW Sex have HeII purely in emission but the HI and HeI lines are in absorption with narrow emission components (Hunger, Heber & Koester 1985; Beuermann, Stasiewski & Schwope 1992). Large variations in emission-line strength on time scales <1 h are also seen in V3885 Sgr (Haug & Drechsel 1985).

The Balmer decrement in the emission components is actually steeper than in the absorption lines, resulting sometimes in Hα being purely in emission with higher members of the series showing progressively stronger absorption troughs.

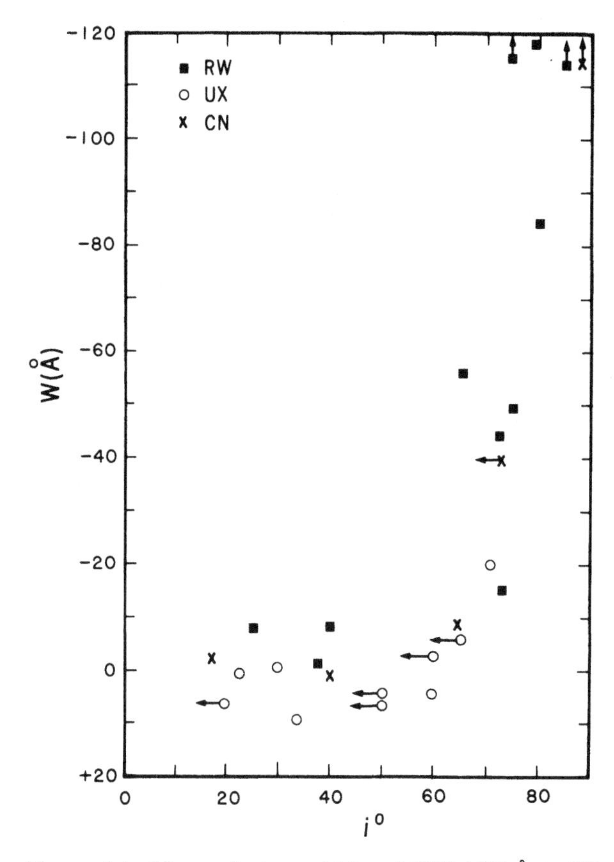

Figure 4.4 Net equivalent widths of CIV 1550 Å in NLs and NRs, plotted as a function of inclination. A distinction is made between UX UMa and RW Tri stars. The equivalent widths for NLs are from la Dous (1991), those for CN remnants are from various sources.

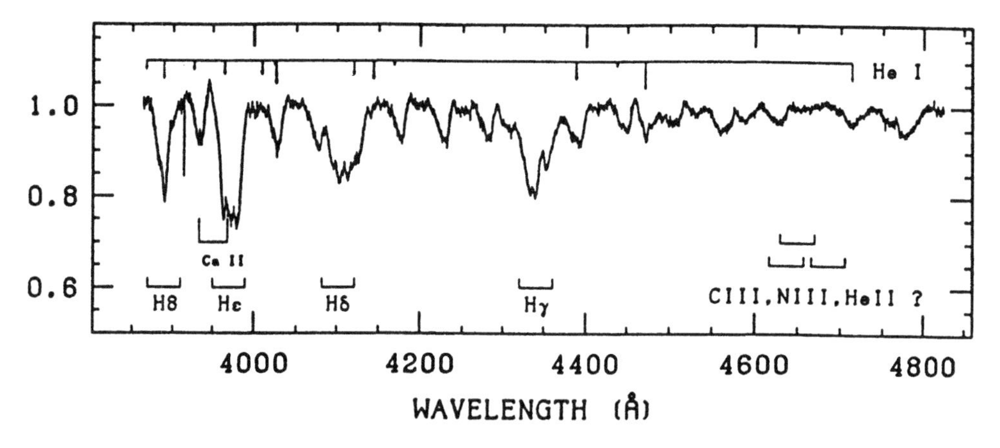

Figure 4.5 Spectrum of IX Vel. The narrow feature at 3914 Å is an instrumental artefact. The ordinate scale is relative intensity. From Hessman (1990b).

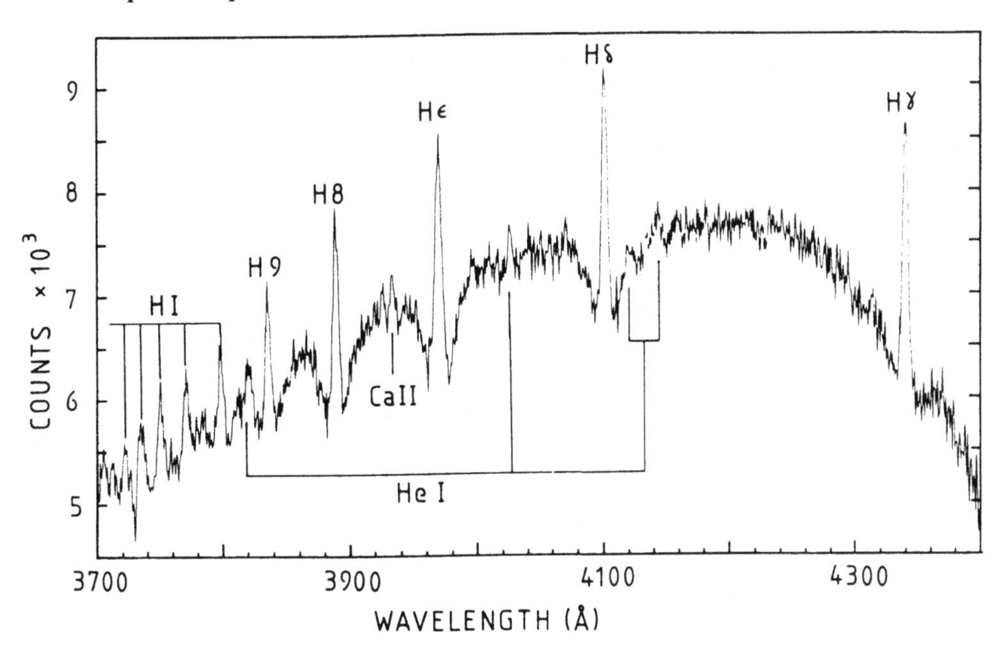

Figure 4.6 Spectrum of RZ Gru. From Stickland *et al.* (1984).

When first observed, the broad, shallow absorption lines were thought to arise from a hot subdwarf (UX UMa: Walker & Herbig 1954) or a white dwarf (V3885 Sgr: Wegner 1972). Their reinterpretation as lines from optically thick discs (Warner 1976a) has been confirmed by all later studies, in particular by the agreement between computed and observed line profiles (Mayo, Wickramasinghe & Whelan 1980; Haug 1987).

The full potential of the absorption-line profiles has yet to be realized. Computed profiles (Cheng & Lin 1989) show that the centres of the lines are Doppler broadened and hence sensitive to inclination; the wings are collision broadened and less sensitive to i, though the FWZI does decrease with increasing i as unit optical depth rises to lower densities higher in the atmosphere. At large i, as with limb spectra of stars, the effective depths of line and continuum formation are similar, causing lines to become shallow. There is thus a general decrease of absorption-line equivalent width with increasing i. This appears qualitatively in agreement with the observations: e.g., UX UMa has very shallow lines, but those in IX Vel (Figure 4.5) are noticeably deeper. The situation is confused, however, by the presence of the emission lines. Eventually, $T(r)$ and $P_e(r)$ at $\tau \sim 1$ should be derivable from the line profiles, as should also the abundances of H, He and Ca.

The UX UMa systems with well-defined absorption lines provide valuable opportunities for the measurement of RV amplitude $K(1)$ free from the complicating distortions that affect emission-line profiles (Section 2.7.6). Unfortunately, these systems are generally of low inclination, which makes an estimate of i difficult. Although no *absorption* lines from the secondary have been found in UX UMa stars in the high state (e.g., Friend *et al.* 1988), narrow Balmer *emission* is present from the

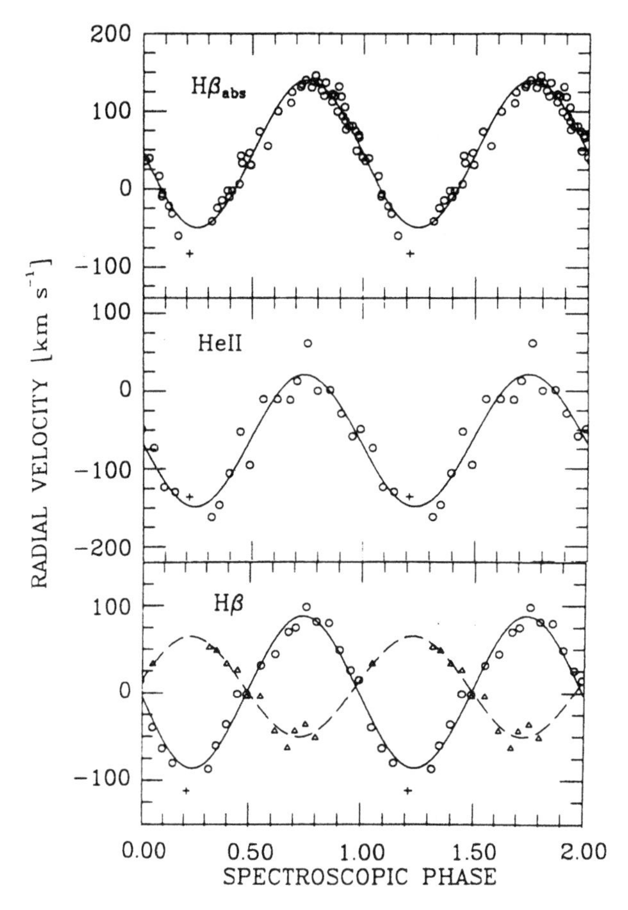

Figure 4.7 RV curves for RW Sex measured from (upper) Hβ absorption lines, (middle) HeII 4686 Å emission line and (lower) Hβ emission. Adapted from Beuermann, Stasiewski & Schwope (1992).

irradiated secondary (UX UMa: Schlegel, Honeycutt & Kaitchuck 1983; IX Vel: Beuermann & Thomas 1990; RW Sex: Beuermann, Stasiewski & Schwope 1992). Figure 4.7 illustrates the excellent agreement between the RV curves of Hβ absorption, HeII and Hβ emission from the disc, and the 180° phase-shifted Hβ emission from the secondary in RW Sex. There are no significant phase delays between the first three of these curves (cf. next Section). Similar results are obtained for IX Vel (Beuermann & Thomas 1990), where the phasing of the secondary emission agrees with the ephemeris of the secondary deduced from ellipsoidal variation in the K-band (Haug 1988). Modelling of the irradiation of the secondary (Section 2.9.6) for IX Vel leads to $i = 60 \pm 5°$, $M_1(1) = 0.80(+0.16, -0.11)$ and $M_1(2) = 0.52(+0.10, -0.07)$ (Beuermann & Thomas 1990).

UV spectra of HL Aqr, BZ Cam, RZ Gru and RW Sex show very deep absorptions of NV, SiIV and CIV, usually with P Cyg emission components. In TT Ari, KR Aur, KQ Mon, V3885 Sgr and IX Vel the absorptions and associated emissions are less

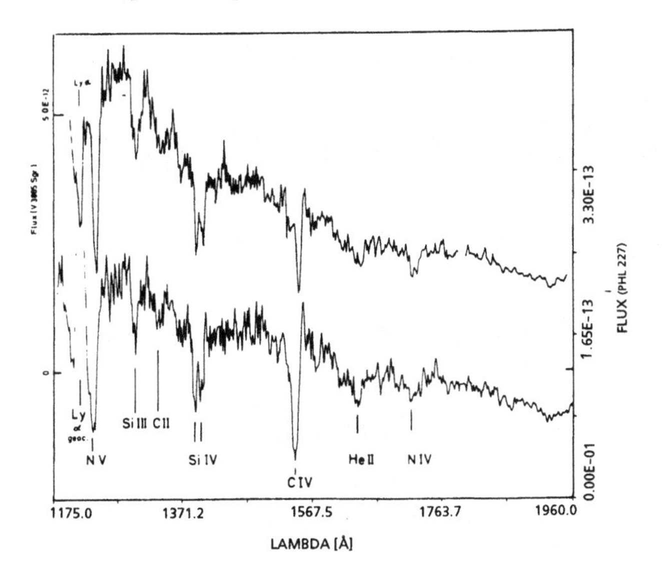

Figure 4.8 UV spectra of V3885 Sgr (upper) and HL Aqr (lower). From Hunger, Heber & Koester (1985).

strong. A comparison of HL Aqr and V3885 Sgr is given in Figure 4.8. In QU Car and VY Scl absorptions and emissions are all relatively weak, but in UX UMa only SiIII/OI is clearly in absorption, with NV, SiIV and CIV purely in emission (see Figure 1b of la Dous (1991) for illustrations of these spectra). This distinctive sequence is almost certainly one of increasing inclination, but independent estimates of i are not sufficiently accurate to provide the necessary resolution to confirm this. The UV spectrum of KR Aur is that of a UX UMa star of intermediate inclination; although currently classified as an RW Tri star, Hutchings, Link & Crampton (1983) note the occasional presence of HeI absorption, which suggests that existing optical spectra of KR Aur have not been obtained in its full high state. The FUV spectrum of IX Vel (Long *et al.* 1994a) closely resembles that of Z Cam in outburst (Figure 3.17).

Time-resolved UV spectra have been obtained of UX UMa (King *et al.* 1983, Drew & Verbunt 1985), BZ Cam (Woods, Drew & Verbunt 1990), V3885 Sgr (Woods *et al.* 1992) and IX Vel (Sion 1985a; Mauche 1991). These have contributed significantly to the picture of winds from CVs, described in Section 2.7.3. There are no comparable time-resolved observations of UV lines from low inclination RW Tri stars.

Over the range 1000–3000 Å the fluxes of UX UMa stars are relatively flat, but are systematically brighter than those of DN at maximum in the 2000–3000 Å range (compare Figures 1b, 2 and 5f of la Dous (1991)). As well as this, there are differences in line behaviour between UX UMa stars and DN at maxima. None of the latter develop the very deep NV, SiIV and CIV absorptions shown by the lowest inclination NLs. At intermediate inclinations, however, the line spectra are weak-lined and can appear very similar (e.g., the UV spectra of IX Vel and AH Her at maximum are almost identical (la Dous 1991)). Emission lines begin to appear quite strongly in NLs for $i \gtrsim 65°$, whereas they only become strong in DN at maxima for $i \gtrsim 75°$.

4.2.3 *The SW Sex Stars*

The SW Sex stars were so named by Thorstensen *et al.* (1991) to call attention to a group of eclipsing NLs with similar spectroscopic peculiarities. The five so far identified are all of RW Tri type: SW Sex (Penning *et al.* 1984; Honeycutt, Schlegel & Kaitchuck 1986; Szkody & Piché 1990), DW UMa (Shafter, Hessman & Zhang 1988; Szkody & Piché 1990), V1315 Aql (Downes *et al.* 1986; Szkody & Piché 1990; Dhillon, Marsh & Jones 1991); PX And (Thorstensen *et al.* 1991; Hellier & Robinson 1994) and BH Lyn (Thorstensen, Davis & Ringwald 1991; Dhillon *et al.* 1992). They have orbital periods in the narrow range $3.24 < P_{orb} < 3.74$ h.

The distinctive spectroscopic characteristics of the SW Sex stars are:

(i) Single-peaked Balmer and HeI emission lines which have relatively narrow FWHM ($\sim$1000 km s^{-1}), but large FWZI typical of high inclination systems: V1315 Aql 3500 km s^{-1}, BH Lyn 4000 km s^{-1} (2500 km s^{-1} for HeI), PX And 2000 km s^{-1}, SW Sex 3000 km s^{-1}, DW UMa 3000 km s^{-1}.

(ii) Narrow central absorption components in HI and HeI, increasing in strength and narrowing up the Balmer series and occasionally becoming entirely absorption in HeI, which appear only around orbital phase $\varphi = 0.5$ (Figure 4.9). These absorption components appear quite suddenly (within 0.025 phase) in the range $0.39 \leq \varphi \leq 0.42$ and disappear in $0.57 \leq \varphi \leq 0.67$ (Szkody & Piché 1990).

(iii) The low excitation lines (HI, HeI) suffer little or no eclipse at $\varphi = 0$ (Figure 4.9); the higher excitation lins (HeII, CIII/NIII, CII) are single peaked at all phases and are almost totally eclipsed at $\varphi = 0$.

(iv) RV curves of both high and low excitation lines show large ($\sim 70°$) phase lags relative to the photometric ephemeris.

Detailed studies have been made of V1315 Aql by Dhillon, Marsh & Jones (1991) and of BH Lyn by Dhillon *et al.* (1992). Both show the same basic structures. Doppler tomography (Figure 4.10) reveals that there is no emission from the secondary or the stream, and, in contrast with most CVs (e.g., IP Peg: Figure 2.41), *there is no ring of emission in velocity space characteristic of a Keplerian accretion disc.* Instead, Balmer emission is concentrated near (−200, −200) km s^{-1} (smeared out in the tomogram because of the effect of the absorption near $\varphi = 0.5$). This is a different manifestation of the phase shift φ_1.

Two early models competed to give partial explanations of the peculiar properties of the SW Sex stars:

(i) In a study of SW Sex, Honeycutt, Schlegel & Kaitchuck (1986) proposed that the single-peaked lines could be formed in an accretion disc wind. The behaviour of the low excitation optical lines is the same as for the UV resonance lines, which remain unobscured at $\varphi = 0$ in high inclination systems (Section 2.7.3). The high excitation lines are produced near the primary, but above the disc, in the outward flow near the boundary layer. They do, however, show a rotational disturbance during eclipse.

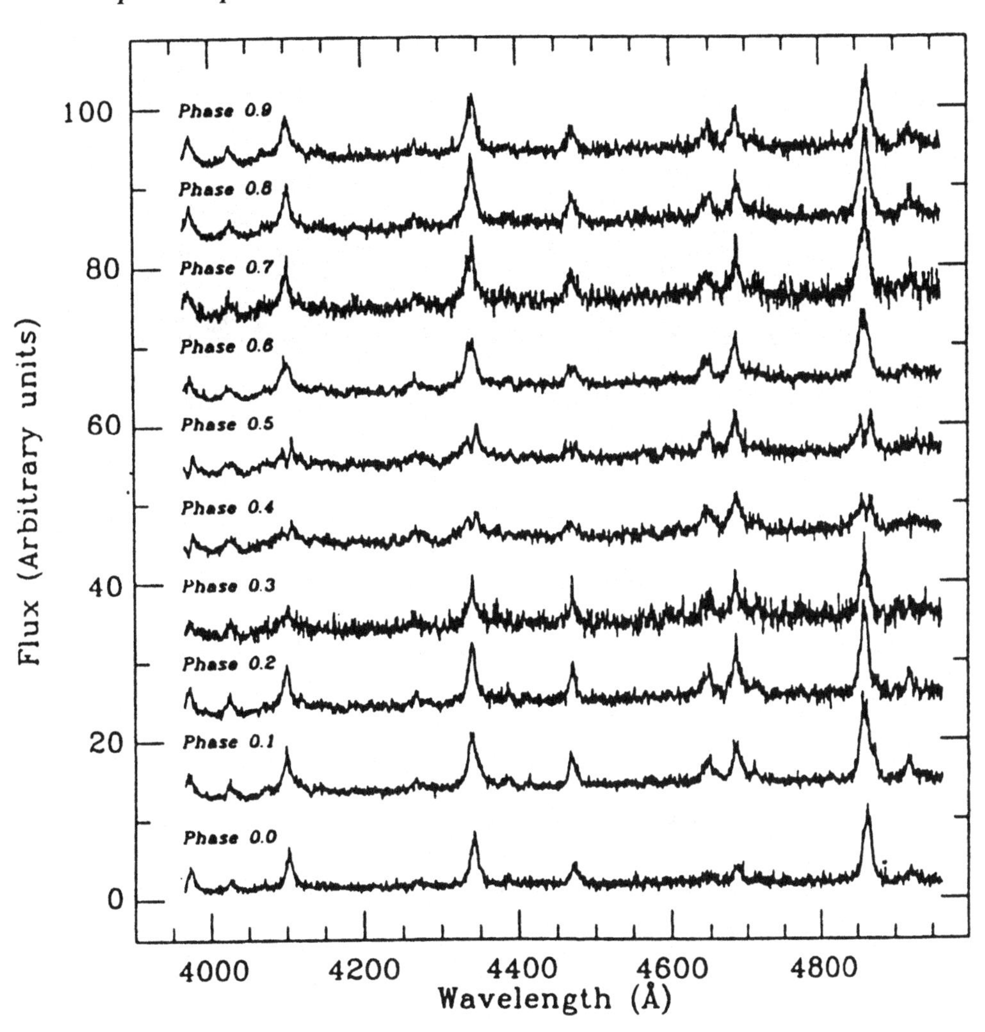

Figure 4.9 Time-resolved spectra of V1315 Aql. Note the occurrence of central absorption components to the Balmer and HeI lines near phase 0.5. From Dhillon, Marsh & Jones (1991).

Appealing though this picture is, Dhillon, Marsh & Jones (1991) point out that the high ionization state of the wind indicated by the UV lines precludes there being any significant contribution to the neutral optical transitions – unless there are high density clumps carried by the wind (Marsh & Horne 1990).

The absorption components near $\varphi = 0.5$ were attributed by Honeycutt *et al.* to gas expelled through the L_3 point (Section 2.3). Dhillon, Marsh & Jones prefer self-absorption or scattering in the wind, as any gas passing through L_3 should form a ring around the system which could not account for absorption over such a limited range of φ.

(ii) Magnetically-controlled accretion, first proposed by Williams (1989), of the kind considered in detail in Chapters 6 and 7. Absence of any circular

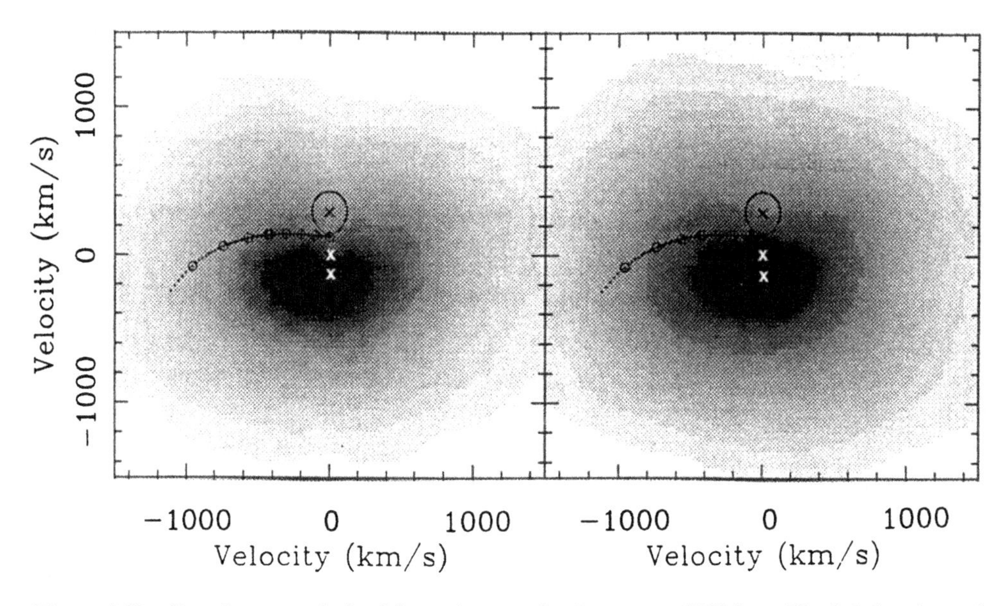

Figure 4.10 Doppler maps derived from time-resolved spectra of BH Lyn. The left hand panel is from Hβ emission and the right hand one from Hγ emission. The calculated positions of the secondary, the stream trajectory, the centre of mass of the system and the white dwarf are shown. Adapted from Dhillon *et al.* (1992).

polarization excludes a strong magnetic field, but a lower, intermediate polar structure (IP) cannot be excluded. In particular, there are no deeply eclipsing IPs yet known, so their characteristic multi-periodic light and X-ray modulations are apparently hidden at high inclinations. They are probably waiting unrecognised among the NLs – the SW Sex stars are possible candidates. See Section 6.4 for further discussion.

A third, and more successful, model has been proposed by Hellier & Robinson (1994):

(iii) Accretion stream overflow is seen projected against the bright inner regions of the disc, producing the absorption lines. Enhanced line emission originates from the region where the stream impacts the disc near the primary (Section 2.4.3); the combination of this with disc emission produces the single-peaked lines seen at most orbital phases.

V1315 Aql and SW Sex have been studied by the eclipse mapping technique (Rutten, van Paradijs & Tinbergen 1992) and, unlike RW Tri, UX UMa and V363 Aur, do not show the temperature profile of a steady state disc (equation (2.35)). Instead, $T(r)$ is much flatter in the inner region of the disc. This is compatible either with loss of energy to a wind from the inner disc, or to truncation of the inner disc by a magnetic field (see Sections 4.4.4.2 and 7.6.6).

Other high inclination NLs with orbital periods near those of the SW Sex stars, namely VZ Scl, LX Ser, CM Del and V1776 Cyg, have not been studied spectroscopically in sufficient detail to determine if they are also SW Sex stars. Of the remaining eclipsing NLs, UX UMa, RW Tri, V347 Pup, AC Cnc and V363 Aur are well studied. They show a variety of behaviour, with the Balmer lines often having multiple components and being only partially eclipsed, whereas HeII is single-peaked and deeply eclipsed. Absorption components, when present, are strongest around $\varphi \sim 0.9$–1.0. Phase lags, φ_1, are typically <0.05. In V347 Pup, Buckley *et al.* (1990b) find the behaviour compatible with absorption by the stream.

Although RW Tri has small φ_1, there is no rotational disturbance in HI and only a weak disturbance in HeII (Kaitchuck, Honeycutt & Schlegel 1993). HeII shows a rotational velocity $\sim$100 km s^{-1}, proving strongly non-Keplerian motion. The possibility that single-peaked Balmer emission is the result of single scattering of disc emission from a wind is eliminated by the absence of phase-dependent linear polarization of Hα in RW Tri (Rutten & Dhillon 1992). Absorption-line behaviour similar to that in the SW Sex stars has been seen in WX Ari, although it is not an eclipsing system (Beuermann *et al.* 1992; Hellier, Ringwald & Robinson 1994).

4.2.4 The VY Scl Stars

At maximum light the VY Scl stars are RW Tri, UX UMa or SW Sex stars, the spectra of which have been described in the previous three sections. Here the spectra at low or intermediate states are considered.

Published analyses of low state spectra are available for V794 Aql (Honeycutt & Schlegel 1985; Szkody, Downes & Mateo 1988), TT Ari (Krautter *et al.* 1981a; Vojkhanskaya 1983; Shafter *et al.* 1985; Thorstensen, Smak & Hessman 1985), KR Aur (Shafter 1983), CM Del (Szkody 1985c), BH Lyn (Dhillon *et al.* 1992), MV Lyr (Robinson *et al.* 1981; Schneider, Young & Schectman 1981; Szkody & Downes 1982; Chiapetti *et al.* 1982; Vojkhanskaya 1988; Rosino, Romano & Marziani 1993), VZ Scl (Sherrington, Bailey & Jameson 1984) and DW UMa (Dhillon, Jones & Marsh 1994).

Of these, three systems (V794 Aql, TT Ari and MV Lyr) have been observed on rare excursions to very deep minima >5 mag fainter than the normal bright state. At such times the usual broad accretion disc optical lines are replaced by very narrow (FWHM $\sim$ 150 km s^{-1}) strong Balmer emission and weaker HeI, HeII emission on a blue continuum (Figure 4.11). Their narrowness, their Balmer decrement and the observation that in TT Ari they are 180° out of phase with velocities measured in the high state (Shafter *et al.* 1985; Hutchings, Thomas & Link 1986) clearly point to an origin in the chromosphere of the secondary. At the same time, red spectra of MV Lyr (Schneider, Young & Schectman 1981, Szkody & Downes 1982) reveal the absorption spectrum of the secondary, and optical spectra of MV Lyr and TT Ari show broad absorption lines of HI and HeII which are attributable to the photospheric spectrum of the primary. (Although HeII and HI are rarely seen together in isolated white dwarfs, because He sinks out of the photosphere due to the high gravity, in a CV a fresh supply of He is continually available through accretion (Shafter *et al.* 1985), which may also induce deep mixing.) High temperatures are deduced (Table 2.6).

A $T \gtrsim 5 \times 10^4$ K blackbody appears in the UV continuum of TT Ari as observed by IUE in the lowest state (Shafter *et al.* 1985). The relative strengths of the emission lines of CIV, NV, SiIV and MgII were not characteristic of those from M dwarfs (Linsky *et al.* 1982), which suggests that some low amount of accretion was taking place, as is

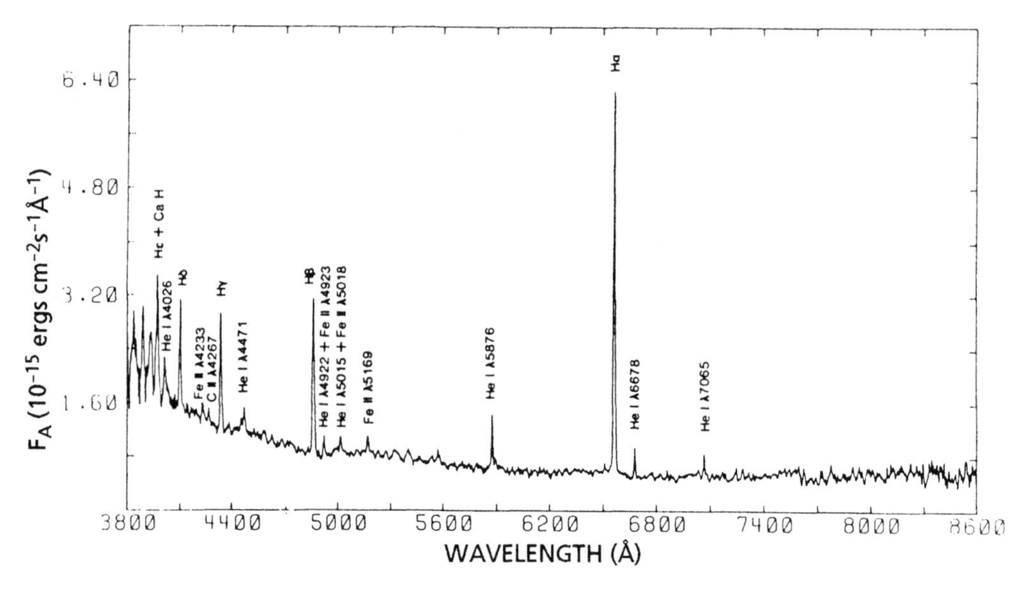

Figure 4.11 Spectrum of TT Ari in the low state ($m_V \sim 16.5$). From Shafter *et al.* (1985).

confirmed by photometric flickering activity at the time. A similarly hot primary is seen in UV spectra of MV Lyr in its lowest state (Szkody & Downes 1982).

The effect of these hot primaries is to contribute an $F_\lambda \mathrel{\tilde{\propto}} \lambda^{-4}$ distribution below ~3000 Å, with the result that although the visible flux may decrease by a factor ~100 in the low state, the UV flux decreases by factors of only 2–3.

With such hot primaries, VY Scl stars in their lowest states (with no shadowing from a disc) should resemble detached binaries such as NN Ser (Wood & Marsh 1991; Section 9.2.2.2) in which a very strong reflection effect is seen from the secondary. Unfortunately TT Ari and MV Lyr are at too low an inclination for this to produce an obvious orbital modulation.

Several systems have been observed spectroscopically while between high and low states. These are TT Ari (Thorstensen, Smak & Hessman 1985; Hutchings & Cote 1986), WX Ari (Beuermann *et al.* 1992), V425 Cas (Szkody 1985c), MV Lyr (Szkody & Downes 1982; Rosino, Romano & Marziani 1993), V442 Oph (Szkody & Shafter 1983) and VY Scl (Hutchings & Cowley 1984). In most respects the spectra are intermediate between those at high and low states; in TT Ari this even applies to the orbital phase at which maximum RV occurs – apparently the dominant location of the emission-line production changes from the secondary at low state, through the bright spot at intermediate state, to the disc as a whole at high state. (The high and low state behaviours are the opposite of what happens in DN, where chromospheric heating enhances the contribution from the secondary at maximum of outburst – Section 2.9.6). In both the optical and the UV the line spectra of VY Scl stars 2–3 mag below the high state closely resemble those of DN in quiescence.

In low state ($\Delta m \sim 3$ mag) spectra of DW UMa, Dhillon, Jones & Marsh (1994) detected both chromospheric emission from the secondary and broad H emission from a faint disc. The high excitation lines present in the high state had disappeared.

The UV flux distribution just before the end of a protracted high state is redder than immediately after recovery from a low state. This could be an indication that the inner disc takes a considerable time to settle into a steady state, and that the steady state reached involves energy radiated via mechanisms other than thermal emission (Tout, Pringle & la Dous 1993). Further work should include a detailed study of line emission, and a comparison could be made with the similar indications in Z Cam stars (Section 3.4).

4.2.5 NRs

4.2.5.1 Classical NRs

The early spectroscopic surveys of novae at minimum light (Humason 1938, Greenstein 1960, Kraft 1964a) showed a variety of types and led to the conclusion that there is less homogeneity than among the DN. In comparison with NLs, however, the NRs have a smaller range of behaviour and a different distribution of types. Only two (DI Lac and V841 Oph) have UX UMa type spectra (and have very narrow emission-line components, indicative of low inclination); the others are of RW Tri type. Although many are eclipsing (Table 4.2), no SW Sex phenomena have been recorded.

Figure 4.12 illustrates the optical spectra of CP Pup and RR Pic, which may be compared with DW UMa in Figure 4.1. The great strength of the high excitation lines is notable. RR Pic has unusually weak Balmer lines compared with spectra of other old

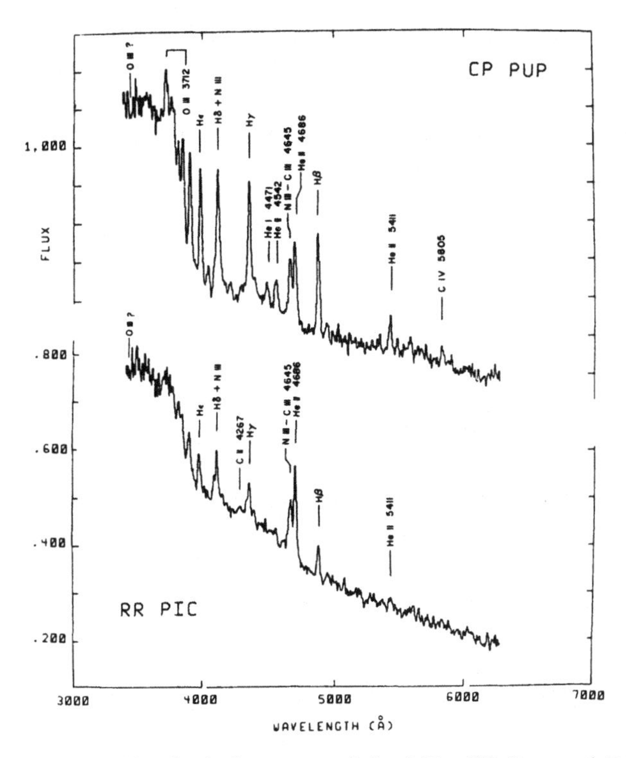

Figure 4.12 Optical spectra of the NRs CP Pup and RR Pic. Adapted from Williams & Ferguson (1983).

novae (Wyckoff & Wehinger 1977; Williams 1983; Williams & Ferguson 1982, 1983). In more recent novae, emission lines from the ejected shell (Section 5.6.3) are often superimposed on the lines from the central binary; the contributions of the former may be recognized as they do not partake in the orbital motion (e.g., HR Del: Bruch 1982).

Emission lines identified in NRs but not seen in DN (Table 3.5) or NLs (Table 4.3) are OIII λ3429; HeII $\lambda\lambda$4199, 5411, MgII λ4481; FeII $\lambda\lambda$4550, 4731 (Ferland *et al.* 1982, Shara & Moffat 1983; Williams & Ferguson 1983).

Three novae have pre-eruption low resolution spectra available on objective prism plates. These are V603 Aql, HR Del and V533 Her, taken 19, 7 and 2 years before eruption respectively. There are no discernible differences from the post-nova spectra (Seitter 1990).

Novae that erupted within the past few decades could be expected to possess very hot primaries. This is not verifiable observationally because the high $\dot{M}$(d) values preclude separation of the contributions of disc and primary (and no CN have dropped to low $\dot{M}$(2) states within a century of outburst – hence there are no $T_{\text{eff}}(1)$ values in Table 2.8). Although the hot primaries appear to contribute to ionization in the ejecta (Section 5.6.3) there is no correlation between spectral excitation level and time since eruption among the NRs. Nor is there any correlation between spectral properties and speed class of nova. Any such correlations may be diluted by the suspected presence of quite strong magnetic fields in some systems (V603 Aql: Schwarzenberg-Czerny, Udalski & Monier 1992; V Per: Wood, Abbott & Shafter 1992; GQ Mus: Diaz & Steiner 1989; V2214 Oph: Baptista *et al.* 1993; GK Per: Section 7.5.1; T Pyx: Schaefer *et al.* 1992) and the confirmed magnetic field in V1500 Cyg (Schmidt & Stockman 1991). In magnetic systems the HeII and λ4650 lines are strongly enhanced (Section 6.5.1).

The general survey of optical spectra of NRs, currently being undertaken (Bianchini *et al.* 1991, 1992), will assist in detecting correlations. Already it has been confirmed that HeII line strength increases with soft X-ray flux (measured by ROSAT), as expected from Figure 2.36. The one strong correlation that is clear in optical spectra is the increase of emission-line equivalent width with inclination (Warner 1987a: Figure 2.34). From a low dispersion survey of NRs this could be used to increase the efficiency of discovery of eclipsing systems.

UV spectra have been reviewed by Starrfield & Snijders (1987) and Friedjung (1989). Representative spectra are shown in Figure 4.13; the emission lines are similar to those in NLs. HR Del shows strong P Cyg line profiles; V603 Aql may show weak P Cyg structures at some orbital phases (Drechsel *et al.* 1981; Krautter *et al.* 1981b); RR Pic had been claimed to have P Cyg profiles but a more detailed study proves the opposite (Selvelli 1982). Equivalent widths of CIV 1550 Å, taken from the literature for V603 Aql, HR Del, DQ Her, GK Per and RR Pic, agree with the general trend with inclination seen in NLs (Figure 4.4), although an absorption spectrum would be expected for V603 Aql if its inclination is really as small as 17°.

Only DQ Her has been studied in the UV during eclipse (Cordova & Mason 1985). The UV continuum and HeII λ1640 line are deeply eclipsed, the CIV, NV and SiV lines are ~40% eclipsed; a similar effect is seen in optical spectra, where HeII is almost totally eclipsed but the Balmer lines only partially (Young & Schneider 1980). However, unlike RW Tri and UX UMa (Section 4.2.1) the out-of-eclipse line profiles in

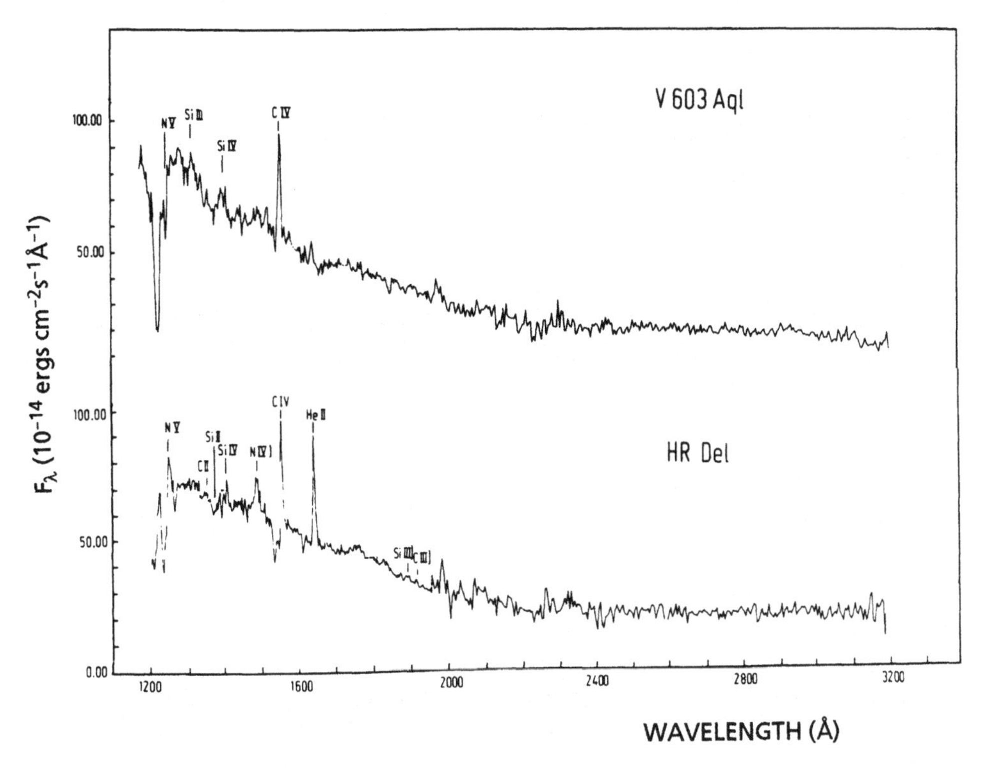

Figure 4.13 UV spectra of V603 Aql and HR Del. Adapted from Krautter *et al.* (1981b).

DQ Her are relatively narrow, symmetric and not redshifted, which suggests that the lines are generated in a symmetrical 'corona' rather than a bipolar wind. The relative unimportance of the wind in DQ Her may be the result of disruption of the inner disc by the primary's magnetosphere (Section 8.1).

From an analysis of the UV and optical spectra of V603 Aql, Ferland *et al.* (1982) concluded that the emission lines are formed in a 'corona' (density $\sim 10^{10\pm1}$ cm^{-3}) with dimensions $\sim a$ and that abundances are approximately solar. The latter is an important result because it shows that (a) the secondary has a normal composition and therefore (b) the CNO enhancements observed in the ejecta (Section 5.3.9) originate in the primary.

Ferland *et al.* (1984) used the observed ionization state of the ejecta of DQ Her to set limits on the EUV flux from the central binary.This was one of the first studies to note the absence of a hot boundary layer (Section 2.7.3). Models of DQ Her must take into account the fact that although the ejecta are affected by the overall flux from the primary and disc, the hot central regions are obscured by the disc in observations made from Earth (Section 5.6.3).

Cassatella *et al.* (1990) find, with the exception of T Aur, that the spectral index α ($F_\lambda \propto \lambda^{-\alpha}$) in the UV ranges from 0.9 for high inclinations to 2.8 for low inclinations. They also find that all NRs have a total luminosity ~ 20 L$_\odot$, implying $\dot{M}(\mathrm{d}) \sim 3 \times 10^{17}$ g s^{-1} (cf. NLs, Section 4.3.3).

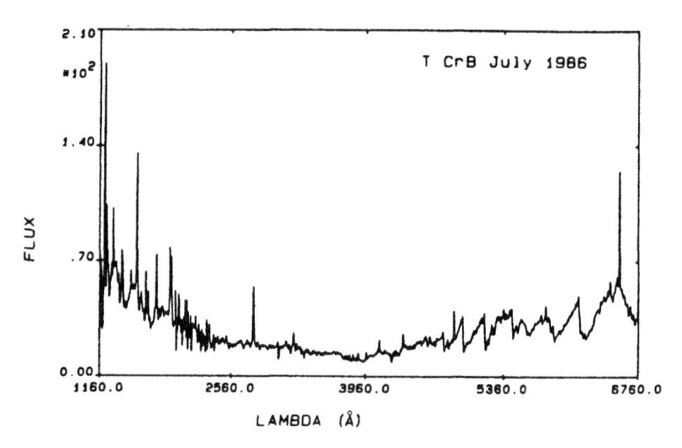

Figure 4.14 Spectrum of T CrB (corrected for reddening). Fluxes are in units of 10^{-14} erg cm^{-2} s^{-1} Å^{-1}. From Selvelli, Cassatella & Gilmozzi (1992).

4.2.5.2 RN in Quiescence

The RN T Pyx is in the orbital period gap, whereas LMC-RN, V394 CrA and U Sco form a group with relatively long $P_{\rm orb}$, implying evolved secondaries. The spectrum of T Pyx (Williams 1983; Duerbeck & Seitter 1990; Duerbeck *et al.* 1993) is of RW Tri type, quite similar to most CN remnants, but with HeII λ4686 considerably stronger than Hβ, and all lines very strong (W(Hα) = 75 Å). V394 CrA and U Sco have anomalous spectra (Duerbeck & Seitter 1990; Duerbeck *et al.* 1993). Emission lines in U Sco, once thought to be of HI (Hanes 1985), have been shown to be the Pickering series of HeII (Johnston & Kulkarni 1992); the same is the case in V394 CrA. Thus, in these stars the disc lines are entirely HeII; yet the secondary's absorption spectrum is normal and contains H absorption. Furthermore, the HeI triplet series, which characterizes regions with $N_e \gtrsim 4 \times 10^3$ cm^{-3}, occurs in absorption phased with the secondary lines.

A possible explanation is that the primaries, heated by the RN eruptions, are exceptionally hot and irradiate the disc to a level where H is kept fully ionized and only a HeII recombination spectrum is possible.

T CrB and RS Oph are RN with very long orbital periods and, consequently, giant secondaries (Table 4.2). Other RN with giant secondaries, as determined spectroscopically or from IR colours, but without known $P_{\rm orb}$, are V3890 Sgr and V745 Sco (M5III and M4III respectively: Harrison, Johnson & Spyromilio 1993). In addition, V4074 Sgr may be a RN (Webbink *et al.* 1987) but is more probably a symbiotic nova (Section 5.10).

Spectra of these systems show the dominance of the secondary in the visible and red, with TiO absorption and superimposed Balmer and disc emission (Figure 4.14). RS Oph also shows FeII emission (Kraft 1958; Williams 1983; Bruch 1986; Duerbeck & Seitter 1990).

An extended study of UV spectra of T CrB by Selvelli, Cassatella & Gilmozzi (1992) has radically changed understanding of its structure. Previously, based on measured $K(1)$ and $K(2)$, which gave $M(1) \sim 1.9\,{\rm M}_\odot$ and $M(2) \sim 2.6\,{\rm M}_\odot$ (Kraft 1958; Paczynski 1965c; Kenyon & Garcia 1986), it was believed that the primary (i.e., the mass receiving component) must be a main sequence star (because $M(1) > M_{ch}$). The optical emission lines, however, are ~330 km s^{-1} broad and lacerated by absorption lines from the giant secondary, which makes the resulting $K(1) \approx 31$ km s^{-1} more uncertain than formal errors would suggest. In any case, the usual objections (Section 2.7.6) to determining the motion of the primary from disc emission lines apply.

The UV lines in T CrB consist of those normally seen in NRs (HeII, CIV, NV) with the addition of strong semi-forbidden (intercombination) lines of SiIII], NIV], CII] and CIII]. The lines in the former group have FWZI $\geq$ 2600 km s^{-1} – much larger than for the optical lines and (with $i \sim 60°$) indicative of velocities ~1500 km s^{-1} which are too large to be associated with flow around a main sequence star. Other evidence for a white dwarf primary includes a Zanstra temperature ~10^5 K from the emission lines (typical of white dwarf boundary layers: Section 2.7.3), a UV continuum spectral index $\alpha \sim 1.26$ (typical for $i \sim 60°$ nova discs: Section 4.2.5.1) and a total UV luminosity $\geq 130 L_\odot$ (which would require $\dot{M}(2) \sim 8 \times 10^{-6} M_\odot$ y^{-1} onto a white dwarf). Furthermore, accretion onto a main sequence star should give emission mostly in the optical region, not in the UV as observed in T CrB.

There are large variations in appearance of the UV spectrum of T CrB, with at times almost complete disappearance of the blue continuum and emission lines. These are not correlated with orbital phase but may be connected with the 55 d suspected pulsation period of the secondary (Iijima 1990). This would be interesting to follow in detail: it could reveal $\dot{M}(2)$ as a function of $\Delta R(2)$ (Section 2.4.1) and the response of the disc to large changes in $\dot{M}(2)$. The variations in UV spectrum would be interesting to correlate with the large range of flickering amplitude (including almost complete absence) seen in the visible (Walker 1957, 1977; Bianchini & Middleditch 1976; Oskanyan 1983). There have also been times when an overlying blue continuum has appeared in the optical region (Joy 1938; Hachenberg & Wellman 1939; Minkowsi 1939).

As giants have strong winds ($\dot{M}_{wind} \sim 10^{-7} - 10^{-8} M_\odot$ y^{-1}: Reimers 1981; Section 9.3.4) a significant amount of mass may accrete onto the primary and the disc of long period RNs from this as well as from Roche lobe overflow. An estimate (Selvelli, Cassatella & Gilmozzi 1992), based on the theory of Livio & Warner (1984), gives $\dot{M}(1) \sim 5 \times 10^{-9} M_\odot$ y^{-1} from the wind, which generates a substantial $L_{acc} \sim 7.5 L_\odot$. The wind will fill the Roche lobe of the primary with a low density gas which can account for the semi-forbidden emission lines, which are narrower (unresolved cores + wings similar to the optical lines) than the UV permitted lines.

The accretion discs in these giant binaries are two orders of magnitude larger than in typical CVs with $P_{orb} \sim 1$–12 h. Models of such discs have been computed by Kenyon & Webbink (1984) and Duschl (1986). More contribution from the disc in the visible than is actually observed is predicted by such models, which may be a result of the disc becoming optically thin beyond $r_d \sim 4 \times 10^{10}$ cm (Selvelli, Cassatella & Gilmozzi 1992).

The situation for RS Oph is a little different, reviewed by Bruch (1986) and Garcia (1986). The spectrum is very variable on a time scale of years, often showing an ~A7 type shell spectrum in the form of absorption cores to the Balmer emissions and with absorption lines of neutral and ionized metals. The blue and visible regions are often overlaid with a blue continuum entirely obscuring the spectrum of the secondary. There is no direct measurement of $T_{eff}(1)$; although the general weakness of HeII emission argues for low temperature, the strong OIII $\lambda 3133$ Bowen fluorescence line shows that somewhere in the system there is strong HeII $\lambda 304$ (Snijders 1987).

A peculiarity of the spectrum of RS Oph is the occurrence of FeII emission lines with double peaks separated by only $\sim$ 50–80 km s^{-1}. These follow the motion of the secondary and have a strong V/R variation (Garcia 1986). Garcia proposes an origin in gas in a ring around the secondary; an alternative could be non-synchronous rotation of the secondary.

IR spectra (Harrison, Johnson & Spyromilio 1993; Ramseyer *et al.* 1993b) of T CrB and RS Oph show HI and HeI in emission, but whereas T CrB has strong CO absorption at 2.3 μm, RS Oph does not. This may be because CO emission from the disc fills in the absorptions.

4.3 Photometry of NLs

4.3.1 Long Term Brightness Variations

Detailed long term light curves derived from archival plates are available in general only for those systems which were first discovered by their large amplitudes – the VY Scl stars. Although many of the ordinary NLs are quite bright, their light curves have not been well studied (hence the absence of some magnitude ranges in Table 4.1). Exceptions are V3885 Sgr (Bond 1978), IX Vel (Wargau *et al.* 1984) and V795 Her (Wenzel, Banny & Andronov 1988). Similarly, the few NLs monitored by amateur astronomers favour the more active VY Scl stars. Nevertheless, the monthly reports of the brightness of IX Vel, since it was put on the observing programme in 1985 by the Variable Star Section of the RASNZ, represent well the general behaviour of the normal NLs: the monthly averages range over $9.2 \leq m_v \leq 9.8$, with monthly variability descriptions such as 'steady' or 'slight variations', but with occasional large irregular variations (typically $m_v \sim 9.0$–10.0) or reported steady increases or decreases over a similar range.

As the NLs are evidently in a state of high rate of mass transfer, with $\dot{M}(\mathrm{d}) > \dot{M}_{\mathrm{crit2}}$ and the resultant viscous time scale $t_\nu \sim 1$ d from equations (2.52) and (2.53), the brightness of the disc in a NL would be expected to follow $\dot{M}(2)$ with only $\sim$1 d averaging. From Smak's (1989a) calibrations for steady state discs, a range of 1 mag in m_v translates to an order of magnitude change in $\dot{M}(\mathrm{d}) \equiv \dot{M}(2)$. Such large variations are not seen in the DN, so, if real, may be connected with the greater irradiative heating of the secondary in high $\dot{M}$ systems.

The brightnesses of some NLs appear to vary quasi-periodically on times scales of approximately years. These are discussed in Section 9.4.2. Eclipse mapping of UU Aqr (Baptista *et al.* 1994; Baptista, Steiner & Cieslinski 1994) shows that its $\sim$ 0.5 mag variations are caused by changes in luminosity of the bright spot, and hence originate in variations of $\dot{M}(2)$. The increases in $\dot{M}(2)$ generate more luminosity in the outer disc, and an increase in r_{d}.

Although there are extensive measurements of variations of the orbital periods for the eclipsing systems UX UMa and RW Tri (Section 9.4.2), there are unfortunately no long term light curves to correlate with them. Johnson, Perkins & Hiltner (1954) noted variations up to 0.4 mag in out-of-eclipse brightness in UX UMa and Protich (1958) found a 1.4 mag photographic range in RW Tri, which is supported by the 1.0 mag range found photoelectrically by Walker (1963a). The latter noted a marked correlation between depth of eclipse and brightness out of eclipse: at maximum brightness eclipses are ~1.0 mag deep, at minimum ~2.5 mag deep. This must result from differing disc radii or radial intensity distributions as a function of $\dot{M}$(d). Eclipse mapping in these different states would be very instructive; the two available studies of RW Tri (Horne & Stiening 1985; Rutten, van Paradijs & Tinbergen 1992) were made when the eclipse was very deep. The eclipse depth in RW Tri is particularly sensitive to changes in r_d because r_d and $R(2)$ are closely similar (Longmore *et al.* 1981).

V1776 Cyg varies over a 1 mag range on time scales as short as a day, with no associated change of colour (Garnavich *et al.* 1990).

4.3.2 VY Scl Stars

The defining property of the VY Scl subclass of NLs is the existence of low states in the long term light curve. Only those NLs that are well above the limits of archival plates, or are serendipitously found faint at the time of a detailed study, can be assigned to this group.

Long term light curves are available for V794 Aql (Meinunger 1979; Petrochenko & Shugarov 1982), TT Ari (Hudec, Huth & Fuhrmann 1984; Wenzel *et al.* 1992), KR Aur (Liller 1980b), V425 Cas (Wenzel 1987b), V751 Cyg (Wenzel 1963), MV Lyr (Wenzel & Fuhrmann 1983; Andronov, Fuhrmann & Wenzel 1988, Wenzel & Fuhrmann 1989; Kraicheva & Genkov 1992), VY Scl (Pismis 1972; Rupprecht & Bues 1983) and LX Ser (Wenzel 1979; Liller 1980a). A schematic light curve for MV Lyr is shown in Figure 4.15.

The light curves for VY Scl stars are *distinctly different* from those of Z Cam stars (e.g., Figure 1.4). A Z Cam star that descends from standstill soon commences outbursts with maxima $\frac{1}{2}$ – 1 mag brighter than standstill; a VY Scl star that descends from high state may have occasional *apparent* outbursts but these (a) *at most* reach the high state brightness and (b) usually have rise or fall times greatly in excess of typical DN values. For example, the recent return from low state of MV Lyr took ~200 d (Fuhrmann & Wenzel 1990) and rises in DW UMa lasting 400 d have been seen (Honeycutt, Livio & Robertson 1993). Furthermore, in the high state the 10–20 d low amplitude waves that occur in Z Cam stars at standstill (Section 3.4) apparently do not occur. These observations, together with the evidence from absolute magnitudes (Figure 4.16), show that NLs are not simply Z Cam stars in permanent standstill.

There are a few instances of rapid rises, perhaps typical of DN outbursts (the rises are usually not well monitored). Some have occurred in MV Lyr (Figure 4.15 – see Wenzel & Fuhrmann (1983) for a detailed light curve) and reached $m_v \sim 14.0$. This is 1.5 mag fainter than the $m_v = 12.5$ adopted for computing M_v in Figure 4.16, and gives M_v in the vicinity of M_v(max) for DN.

With these rare exceptions there is a remarkable absence of DN outbursts in VY Scl stars. As seen in Figure 4.16, a decrease of $\dot{M}(2)$ that increases m_v by $\gtrsim 2$ mag certainly moves the accretion discs of NLs into the region of DN instability. In particular, slow falls or returns by >2 mag should result in the system remaining for substantial times with $\dot{M}(2)$ just less than $\dot{M}_{crit2}$, where DN outbursts with recurrence times ~12–20 d would be expected; yet none are observed.

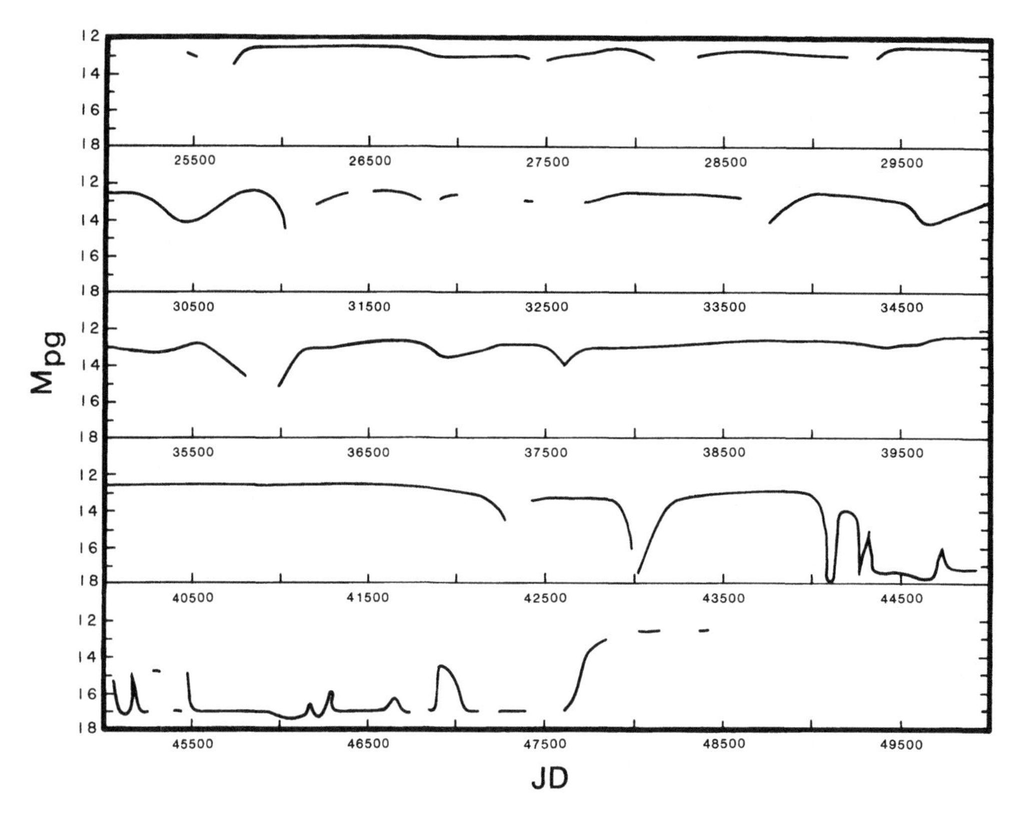

Figure 4.15 Mean light curve 1928–92 of MV Lyr, adapted from Wenzel & Fuhrmann (1983, 1989), Andronov, Fuhrmann & Wenzel (1988), Kraicheva & Genkov (1992) and Rosino, Romano & Marziani (1993). Dates are JD2400000+. The gaps in the early record occur mostly because MV Lyr went below the plate limit.

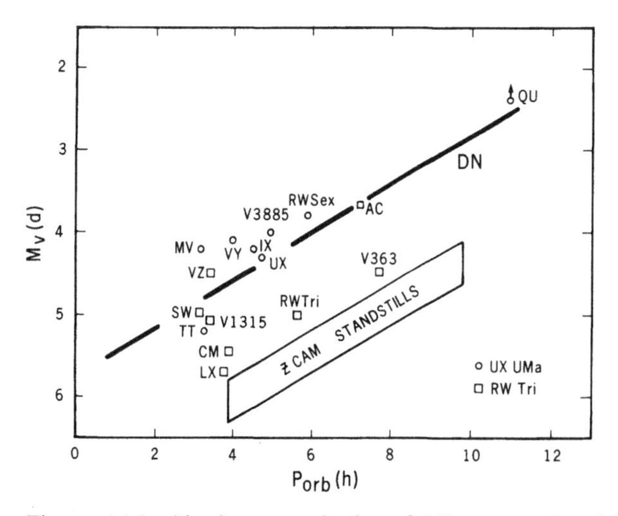

Figure 4.16 Absolute magnitudes of NLs, as a function of orbital period. The absolute magnitudes of DN at maximum light and of Z Cam stars at standstill are shown.

Rapid falls in brightness also occur, though none has been monitored in detail. In November 1980 V794 Aql fell by 3 mag within 3 d but was back in the high state one night later (Szkody *et al.* 1981b). Slow falls with rapid returns to high state have also been observed (Honeycutt, Cannizzo & Robertson 1994).

Almost all of the VY Scl stars have $3 \leq P_{\rm orb}({\rm h}) \leq 4$; the few stars in this range that are not known to be of VY Scl type have no available long term light curves. Only V794 Aql and V751 Cyg (whose $P_{\rm orb}$ is very uncertain) are VY Scl stars with $P_{\rm orb} > 4$ h. No NRs are VY Scl stars.

4.3.3 Absolute Magnitudes of NLs

Individual estimates of distances to NLs, based on reddening, proper motion or parallax measurements, indicated $M_{\rm v} \lesssim 5$ (see Warner (1976a) for details). More systematic investigations (Patterson 1984; Warner 1987a) confirm that NLs have luminosities similar to DN at maximum.

From the results of Warner (1987a), corrected for improved values given in Table 4.1, improved reddenings (la Dous 1991), and with additional distances given by Rutten, van Paradijs & Tinbergen (1992), the absolute magnitudes plotted in Figure 4.16 ensue. The NLs are all brighter than quiescent DN (Figure 3.5) and Z Cam stars at standstill (Figure 3.9). This places them in the region where $\dot{M}(2) > \dot{M}_{\rm crit2}$, in agreement with the absence of DN outbursts (Section 3.5.3.5). There is some indication that the UX UMa stars are on average slightly brighter than the RW Tri stars, implying higher $\dot{M}$(d), but the accuracy of $M_{\rm v}$ determinations does not warrant a definite conclusion. (For example, the value $i = 75°$ given for LX Ser by Young, Schneider & Schectman (1981c) seems low when considering the great depth of eclipse (Figure 4.17) and is probably based on an incorrect interpretation of the RV variations (Williams 1989) – thus LX Ser should probably be moved higher by ~1 mag.)

The absolute magnitudes can be used with Smaks's (1989a,1994a) calibrations to give $\dot{M}$(d), which may be compared with results from the eclipse mapping technique. The four NLs which show approximately steady state radial temperature distributions are RW Tri, UX UMa, V363 Aur and UU Aqr, which have $\dot{M}({\rm d}) = 2 \times 10^{17}$, 5×10^{17}, 2×10^{17} and 1.7×10^{17} g s^{-1} respectively from eclipse mapping (Rutten, van Paradijs & Tinbergen 1992; Baptista, Steiner & Cieslinski 1994). Smak's detailed calibration (Smak 1989a) gives 7×10^{17}, 1.5×10^{18}, 5×10^{17} and 6×10^{17} g s^{-1} respectively for these stars. His later work (Smak 1994a) reduces these by the factor ~3 appropriate for high $\dot{M}$(d) systems, bringing the two techniques into reasonable accord.

Compared with equation (3.18), the $\dot{M}$(d) found from eclipse mapping for these NLs are all factors 2–3 greater than $\dot{M}_{\rm crit2}$(d), in agreement with the steady state nature of their discs.

4.3.4 Orbital Brightness Variations

Unlike many DN, the NLs never have prominent orbital humps. The lessened relative importance of their bright spot luminosities is explained in Section 2.6.5. However, there is a variety of orbital modulation behaviour among the NLs, some of only short lived existence.

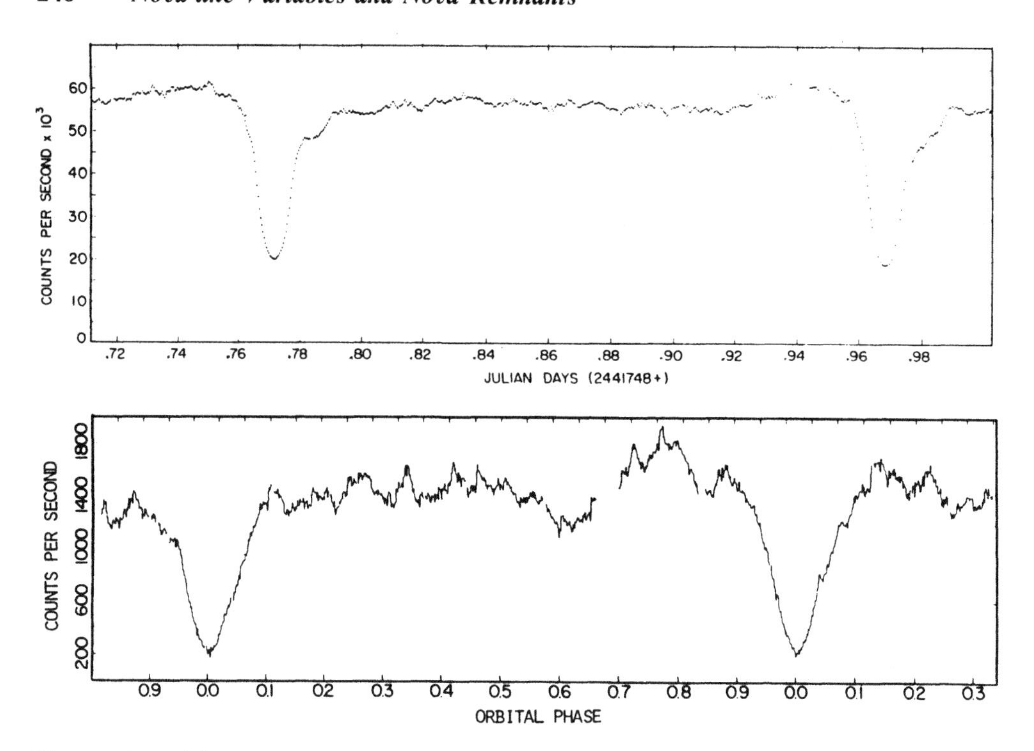

Figure 4.17 Light curves of (upper) UX UMa at 33 s time resolution and (lower) LX Ser at 15 s time resolution. Adapted from Nather & Robinson (1974) and Horne (1980).

RW Tri at times possesses a low amplitude hump, starting at $\varphi \sim 0.75$ and peaking at $\varphi \sim 0.90$; at other times it can have a substantial post-eclipse hump peaking at $\varphi \sim 0.12$, but in general there is no hump at all (Walker 1963a; Horne & Stiening 1985; Rutten, van Paradijs & Tinbergen 1992). The range is largest at short wavelengths and is attributed to variable Balmer continuum absorption by gas above the plane of the disc (Horne & Stiening 1985).

In UX UMa a low amplitude orbital hump caused by the bright spot is commonly seen, as is the delayed egress of the spot (Figures 1.6, 4.17). In 1982, when UX UMa was 0.25 mag brighter than usual, there was no orbital hump and eclipse was deeper with little asymmetry in egress (Schlegel, Honeycutt & Kaitchuck 1983). This sensitivity of structure to small changes in M_V (caused, presumably, by variations of $\dot{M}(2)$) is remarkable and requires more detailed study by eclipse mapping.

Most NLs show even greater variability of eclipse profiles than RW Tri and UX UMa. This variability is associated with the general level of short time scale activity, discussed in the next section. An example of irregular eclipse profiles in LX Ser is given in Figure 4.17.

O'Donoghue, Fairall & Warner (1987) found a smaller disc radius in VZ Scl at a time when it was recovering from a low state. Dhillon, Jones & Marsh (1994) find a smaller disc in a low state of DW UMa.

Eclipse maps of accretion discs have been made for V1315 Aql, V363 Aur, LX Ser, SW Sex, RW Tri and UX UMa (Rutten, van Paradijs & Tinbergen 1992) with a separate study of RW Tri by Horne & Stiening (1985), and for UU Aqr (Baptista *et al.* 1993; Baptista, Steiner & Cieslinski 1994). The temperature distributions are slightly flatter than steady state (equation (2.35)) for V363 Aur, RW Tri, UX UMa and UU Aqr, all of which have $P_{orb} > 3.9$ h, and very much flatter in the others, which have $3 < P_{orb} < 4$ h. The latter are all systems with single peaked emission lines showing no rotational disturbance in eclipse (Williams 1989; Dhillon, Marsh & Jones 1991), indicative probably of strong winds or magnetically-controlled accretion (Section 4.2.3). RW Tri and UX UMa are more normal (though neither show very strong rotational disturbances) but V363 Aur is more complex (Schlegel, Honeycutt & Kaitchuck 1986).

It should be noted, however, that it is only the emission lines that often do not show the classic signatures of differentially rotating discs – the large widths and depths of eclipses and the deduced light concentration towards the primary show that the continuum arises in a disc-like source.

A possible cause for the *apparent* non-steady state temperature distributions is discussed in Section 4.4.4. All of the NLs, however, have $T \gtrsim 1.6 \times 10^4$ K in the inner disc, and $T \gtrsim 7 \times 10^3$ K in the outer regions. They therefore meet the steady state requirement $T < T_e(\text{crit2})$ (equation (3.17)).

IR observations of the brightest NLs have resulted in detection of ellipsoidal variations in IX Vel (Haug 1988) and of the secondaries in VZ Scl (Sherrington, Bailey & Jameson 1984) and RW Tri (Frank & King 1981). However, most of the prominent (0.2 mag deep) secondary eclipse seen in J and K band light curves of RW Tri (Longmore *et al.* 1981) is due to eclipse of the outer cool regions of the disc (Frank & King 1981). In contrast, no secondary eclipse is seen in J band observations of UX UMa (Frank *et al.* 1981a); this is evidently the result of the lower inclination of UX UMa – in RW Tri Frank & King found that a disc thickness $h_d/r_d \sim 0.05$–0.07 (or half opening angle of $2\frac{1}{2}$–$3\frac{1}{2}°$) is required to produce the ~30% coverage of the secondary demanded by the eclipse depth. With $\dot{M}(d) \sim 2 \times 10^{17}$ g s^{-1} (Section 4.3.3), equation (2.52) predicts $h_d/r_d = 0.06$.

4.3.5 *Flickering in NLs*

There is wide variety in the flickering activity in NLs (e.g., Bruch 1992). Some systems have relatively low amplitude activity: UX UMa (Figure 4.17), V363 Aur, IX Vel, RW Sex; others have low power at time scales of approximately tens of seconds, but show considerable power on time scales of 5–20 mins, e.g., DW UMa, V1315 Aql, RW Tri, AC Cnc, V1776 Cyg; others are very active on all time scales, e.g., LX Ser (Figure 4.17), V795 Her, QU Car, V751 Aql, MV Lyr, AY Psc, VY Scl, SW Sex. There is no correlation with subtype, inclination or excitation levels of spectrum.

Flickering continues during the low states of VY Scl stars, with the exception of MV Lyr observed at an extreme minimum when there was no flickering at all (Robinson *et al.* 1981). Therefore, although the low states are caused by reductions of $\dot{M}(2)$, complete turn-off of $\dot{M}(2)$ is rare. Rapid fluctuations of large amplitude (1.4 mag in 16 min) have been observed in MV Lyr on a rise from low state (Wenzel & Fuhrmann 1989).

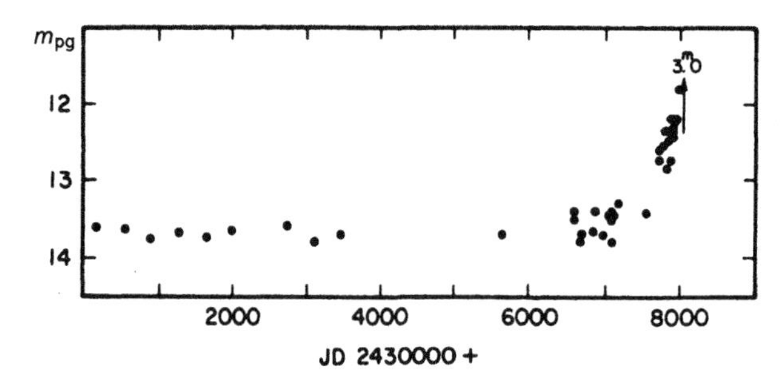

Figure 4.18 Pre-eruption light curve of V533 Her. From Robinson (1975).

4.4 Photometric Observations of Novae at Minimum Light

The eruption light curves of classical novae are described in Section 5.2.1. Archival plates enable the long term light curves of some pre-novae to be constructed. Monitoring of the brighter post-novae (i.e., NRs) is carried out both by amateur astronomers and via photographic wide-field surveys.

4.4.1 Pre-Eruption Light Curves

McLaughlin (1939, 1941, 1960a) concluded from contemporary evidence that pre- and post-nova brightnesses were essentially identical and that a nova eruption occurred without warning. Later discussion of a much larger body of observations modified these conclusions (Robinson 1975) and more recent novae have added more exceptions.

Robinson (1975) found that the magnitudes many years before eruption and the magnitudes after returning to a stable post-eruption state are the same within errors (a possible exception, BT Mon, has since been found to conform (Robinson, Nather & Kepler 1982; Schaefer 1983)). Five out of eleven novae with good pre-eruption coverage showed slow rises of 0.25–1.5 mag in the 1–15 y prior to eruption. The most dramatic of these is V533 Her: Figure 4.18.

Three very fast novae rose from exceptionally faint states but have settled down to brighter post-eruption magnitudes: Table 4.5. In all of these there is positive or suggestive evidence for strong magnetic fields. Low $\dot{M}(2)$ states are common among magnetic CVs (Sections 6.3.1 and 7.3.1). V1500 Cyg was at m_V ~13.5 for about a week before eruption (Kukarkin & Kholopov 1975). These stars are discussed further in Section 6.7.

Not all very fast novae are necessarily strongly magnetic. The fastest, Nova Her 1991, established an accretion disc only three weeks after eruption (Leibowitz *et al.* 1992) and so has at most the field strength of an intermediate polar (Chapter 7). Similarly, not all nova eruptions of large amplitude necessarily imply low $\dot{M}(2)$ states at quiescence – these may merely be high inclination disc systems (cf. Diaz & Steiner 1991b).

Three pre-novae showed probable or definite DN outbursts. V446 Her had outbursts of amplitude ~ $2\frac{1}{2}$ mag from m_V(min) ~ 17.5. Their rise time was ~10 d and recurrences after only ~60 d were observed (Stienon 1971; Robinson 1975). These are unusual properties for DN outbursts and, by analogy with BV Cen, GK Per and V1017

Table 4.5. *Pre- and Post-Eruption Magnitudes of Three Very Fast Novae.*

Star	P_{orb}	m_v(max)	m_v(pre)	m_v(post)	References
GQ Mus	1.43	7	>21	17.5	1,2
CP Pup	1.47	0.2	>17	15.0	3,4
V1500 Cyg	3.35	1.8	>21	17.2	4,5

References: 1 Krautter *et al.* 1984; 2 Diaz & Steiner 1989; 3 Gaposchkin 1946; 4 Warner 1985d; 5 Beardsley *et al.* 1975.

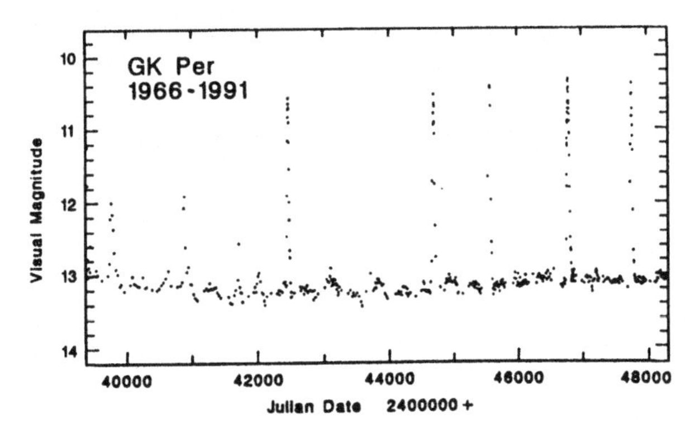

Figure 4.19 Light curve of GK Per showing DN outbursts and slow variations at minimum light. From Richman (1990).

Sgr (Section 3.3.3.5) suggest P_{orb} > 1 d. V1017 Sgr had a DN outburst in 1901, 18 y before its CN eruption (Sekiguchi 1992b). RN V3890 Sgr showed brightenings of ~1.5 mag 22 y before its 1962 eruption (Robinson 1975), but this star has an M giant secondary which places it in a different category (Section 5.9.3).

4.4.2 Post-Eruption Light Curves

Five NRs have been observed to commence DN outbursts many years after their nova eruptions. V1017 Sgr, which erupted in 1919, has had outbursts in 1973 and 1991. GK Per (1901) from 1919 to 1950 had continual irregular fluctuations over the range $12.5 < m_v < 14.0$; although irregular, many brightenings had durations ~20–50 d and amplitudes 1 mag – there is resemblance to the unstable behaviour of some Z Cam stars when near standstill (e.g., RX And: Figure 3.22). Since 1950, GK Per has had small fluctuations of brightness around $m_v = 13.1$, with DN outbursts to $m_v \sim 10.5$ and a recurrence time $T_n \sim 2$ y (Figure 4.19: Sabbadin & Bianchini 1983; Richman 1991). Further discussion of the DN properties of V1017 Sgr and GK Per is included in Chapter 3.

Q Cyg (1876) has possible DN outbursts of duration ~7 d with $T_n \sim 60$ d and amplitudes 0.6–1.0 mag (Shugarov 1983a; Bianchini 1990b). WY Sge (1783) was observed to have a DN outburst in 1982 (Shara *et al.* 1984); it is too faint (m_v(max) ~

17.4) to monitor easily for other outbursts. V446 Her has recommenced DN outbursts, with amplitude 2.0 mag and $T_n \sim 28$ d (Honeycutt *et al.* 1994). These observations strongly suggest that $\dot{M}(2)$ diminishes many decades after a nova eruption, carrying the disc from a state where $\dot{M}(d) > \dot{M}_{crit2}(d)$, through a regime where $\dot{M}(d) \sim \dot{M}_{crit2}(d)$ in which a complicated light curve ensues from interactions between incomplete DN outbursts and fluctuations in $\dot{M}(2)$, to a normal DN state where fluctuations in $\dot{M}(2)$ are rendered less conspicuous by the longer time scale of the disc.

The brightness of NRs many decades or centuries after eruption is of particular interest to investigations of CV evolution (Section 9.4.3). Irradiation from the primary, heated by the nova eruption, is expected to heat the secondary and maintain a high $\dot{M}(2)$ for a century or more after eruption. Calculations show that $\dot{M}(2) \propto t^{-0.45}$, giving a decrease by factors $\sim 10^2$ over centuries, leading to a fall in brightness $\approx 0.012\ M_{bol}\ y^{-1}$, or, through Smak's calibrations (Section 2.6.1.3), $dm_v/d \log t(y) = 0.70$ (Kovetz, Prialnik & Shara 1988; Duerbeck 1992).

A statistical analysis by Vogt (1990) of eruption amplitudes relative to m_v(min), of 97 novae at various times before and after eruption, produced a post-novae decline rate of 0.021 ± 0.006 mag y^{-1}. This analysis, however, is weakened by not having allowed for the $\Delta M_v(i)$ correction in the remnants.

Duerbeck (1992) has analysed the long series of visual observations of 11 NRs made by the amateur astronomers W.H. Steavenson and B.M. Peek between 1921 and 1956 (for detailed references see Duerbeck (1992) and Bianchini (1990b)) and of four others observed by members of the Variable Star Section of the RASNZ, supplemented with photoelectric measurements. The mean rate of decay is 0.010 mag y^{-1} or $dm_v/d \log t(y) = 1.0 \pm 0.3$, with indications of more rapid decline in systems with small P_{orb}, as expected from theory.

An extension of the time base would become possible if some of the naked eye novae recorded in early and ancient records (Section 1.1) could be relocated. CK Vul (1670), which reached $m_v(\max) \approx 2.8$, would be expected at $m_v(\min) \lesssim 17$, yet no remnant is found to $m_v \sim 21$ (Shara, Moffat & Webbink 1985; Naylor *et al.* 1992). Similarly, searches for emission lines from stars in the vicinity of T Boo (1860), U Leo (1855) and Nova Pup (1673) have been unsuccessful to limits $m_v \sim 19$ (Shara, Moffat & Potter 1990b) and searches for blue objects near the positions of Nova Gem (1892), AB Boo (1877), SY Gem (1856), VZ Gem (1856), Nova Leo (1612) and Nova Sco (1437) have also been fruitless (Downes & Szkody 1989; Shara *et al.* 1990). None of the ancient Oriental novae has been definitely recovered (Shara 1989); however, the DN HT Cas is suggested as possibly the remnant of a nova recorded in Japan in AD 722 (Duerbeck 1993). All of these observations point to a steady decline in brightness after a nova eruption with, in at least some cases, a reduction in $\dot{M}(2)$ to very low levels centuries after eruption.

As with NLs, superimposed on the light curves of NRs are variations with amplitudes up to ~ 1 mag and quasi-periods of years (Figure 4.19). References to these are given in Section 9.4. However, there are also much shorter quasi-periods present. V841 Oph has variations $12.15 \lesssim m_v \lesssim 12.57$ with mean period 51.5 d (Della Valle & Rosino 1987; Shara, Potter & Shara 1989; Della Valle & Calvani 1990; Honeycutt *et al.* 1994) and Q Cyg possibly has a low amplitude variation with period near 60 d (Shugarov 1983b; Della Valle & Calvani 1990) and RW Tri has shown 10 cycles of a

25 d period (Honeycutt *et al.* 1993b). The physical origin of these time scales has not been identified.

A 55 d quasi-period has also been seen in T CrB (Lines, Lines & McFaul 1988), which is probably the pulsation period of the M giant secondary. It has a variable amplitude (0.08–0.18 mag) in the V band, and up to 0.4 mag in U, but is not seen in IUE observations (Selvelli, Cassatella & Gilmozzi 1992). A 35 d period is suspected in RS Oph (Evans 1986).

4.4.3 *Absolute Magnitudes of NRs*

The M_V(max)–rate of decay relationship for novae (Section 5.2.3) provides a useful means of estimating M_V(min) for discs in NRs. The range $A'_{CN} = m_V(\text{max}) - m_V(\text{min}) = M_V(\text{max}) - M'_V(\text{min})$ is independent of reddening (M'_V is *apparent* absolute magnitude: Section 2.6.1.3). The $\Delta M_V(i)$ correction is required at minimum, but not at maximum, so M_V(min) can only be obtained for systems with known inclination. The technique gives results in good agreement with other methods for the few novae for which independent distance estimates are available (Warner 1987a).

Figure 4.20 shows the absolute magnitudes of the discs of CN remnants for the few systems with known P_{orb} (<15 h) and i. The observations are taken from Warner (1987a) with supplementary information for QZ Aur, V841 Oph and CP Pup. The stars lie in the same region of the diagram as the NLs (Figure 4.15), confirming that they are all high $\dot{M}$(d) systems.

The amplitude A'_{CN} is available for many CN for which no inclinations are known. An average over a large number should reduce the effect of the spread in $\Delta M_V(i)$, leaving only $\overline{\Delta M_V(i)} = -0.37$ to be applied. Table 4.6 gives the resultant $\overline{M}_V$(min) for various speed classes (Section 5.2.1) of CN. (The different values derived by Della Valle & Duerbeck (1993) are a result of their adopting the Capacciolo *et al.* (1989) M_V(max)–rate of decline relationship, criticised in Section 5.2.3, and making no allowance for $\Delta M_V(i)$.) The average over all classes is $\overline{M}_V(\text{min}) = 3.8$ which is in good agreement with the positions of individual stars in Figure 4.20.

The large P_{orb} systems GK Per and V1017 Sgr are not shown in Figure 4.20. Because both stars currently show DN outbursts (Bianchini, Sabbadin & Hamzaoglu 1982; Sekiguchi 1992b) the absolute magnitudes of their discs are of particular interest.

The 1919 brightening of V1017 Sgr was quite clearly a nova eruption (Webbink *et al.* 1987). From the rate of decline, $M_V(\text{max}) = -6.4$ and hence $M'_V(\text{min}) \sim 1.0$; the latter includes the (dominant) contribution of the G5III secondary. Kraft's (1964a) spectra show only weak emission lines, with no obvious disc continuum superimposed on the secondary absorption spectrum. The orbital inclination is unknown, but from the RV amplitude (Sekiguchi 1992b) is probably $\sim 50°$. Therefore $M_V(\text{min}) \gtrsim 3.0$ for the disc in V1017 Sgr.

GK Per has $M'_V(\text{min}) = 4.0$ (Warner 1987a), $\Delta M_V(i) \approx 0.0$ and a correction for the brightness of the companion that is at least 0.7 mag (Smak 1982b), giving $M_V(\text{min}) \gtrsim 4.7$ for the disc. Anupama & Prabhu (1993) estimate $M_V = 4.9$ for the disc.

The discs in these stars are therefore relatively faint for their large size and could well lie below the value of $M_V(\dot{M}_{crit2})$ (Section 3.5.3.5). Modelling of the disc of GK Per shows that it is indeed thermally unstable (Cannizzo & Kenyon 1986; Kim, Wheeler &

Table 4.6. *Mean Range and Mean Absolute Magnitude of NRs.*

Speed class	t_2(d)	Number	A'_{CN}	$\overline{M}_v$(min)
VF	0–10	18	13.6	3.4
F	11–25	48	11.4	3.7
MF	26–80	26	10.6	4.4
S	81–150	7	9.8	4.2
VS	151–250	4	8.1	4.2

VF = very fast, S = slow, &c (see Section 5.2.1).

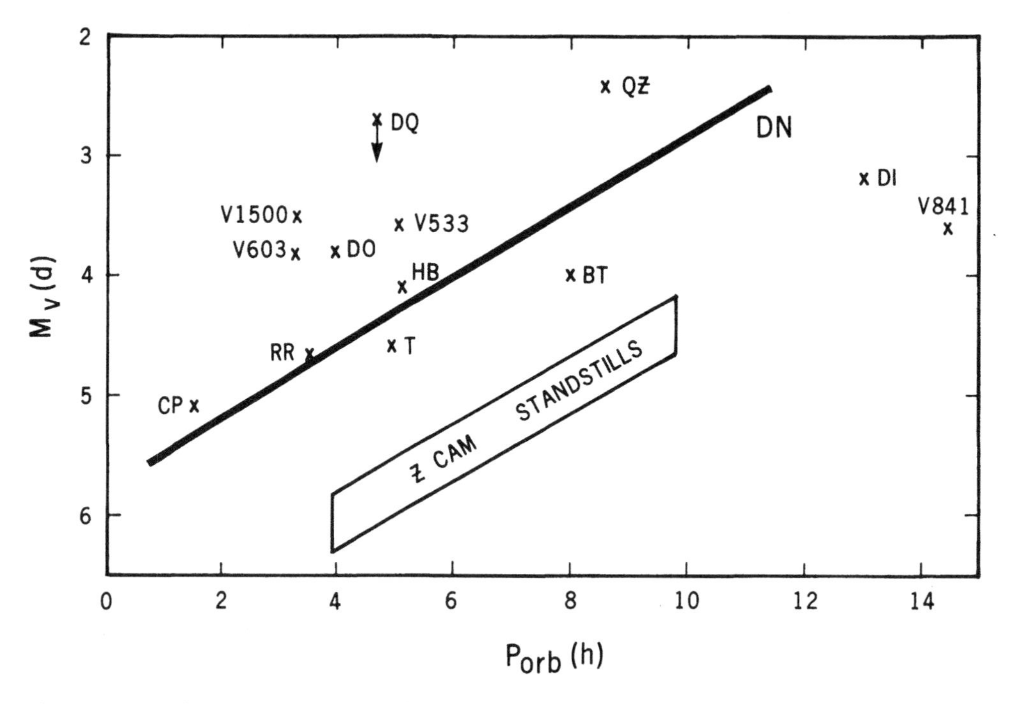

Figure 4.20 Absolute magnitudes of NRs, as a function of orbital period. The absolute magnitudes of DN at maximum and Z Cam stars at standstill are shown. The results are from Warner (1987a) with revisions from Table 4.2. Abbreviations are given of individual star designations.

Mineshige 1992). Similar models are required for V1017 Sgr and BV Cen (Menzies, O'Donoghue & Warner 1986).

WY Sge is an eclipsing NR that also shows DN outbursts (Shara *et al.* 1984). Its distance is not well determined, but lies in the range $0.15 < d < 1.5$ kpc with an extinction $1 < A_v < 3$ mag (Kenyon & Berriman 1988). These give $M_v(\text{min}) > 4.5$, which does not clearly place WY Sge in the disc instability region of Figure 3.9; however, the existence of DN outbursts requires $M_v(\text{min}) \geq 6.0$ and implies a disc fainter than those of most NRs (cf. Naylor *et al.* 1992).

Among RN, only T CrB and RS Oph have relatively reliable distances, obtained from the IR properties of their secondaries. With a distance $d = 1300$ pc and $A_V = 0.48$ (Selvelli, Cassatella & Gilmozzi 1992), $M_V(d) \sim 1.0$ in T CrB if it contributes 10% of the radiation in the visible region. In RS Oph, $d = 1800$ pc, $A_V = 2.3$, $i \sim 30°$ and dominance in the visible (Bruch 1986), $M_V(d) \sim -2.0$; however, part or most of this luminosity may arise from the wind-filled Roche lobe of the primary.

4.4.4 Brightness Modulations on Orbital Time Scales

The ease with which novae are discovered near maximum light contrasts with the difficulty of observing their remnants, some 10–14 mag fainter. This, combined with the fact that the remnants are often in crowded low Galactic latitude fields, has resulted in a strong underrepresentation of novae among the CVs with known orbital periods.

4.4.4.1 Orbital Modulation

Of the known high inclination systems, QZ Aur, V1668 Cyg, DQ Her, BT Mon, V Per and U Sco all have deep eclipses, with at most a low amplitude orbital hump at $\varphi \sim 0.9$ (BT Mon), but occasionally a hump centred on eclipse (QZ Aur, DQ Her). T Aur has eclipses only ~0.2 mag deep. Nova Her 1991 had shallow (~0.4 mag) eclipses three weeks after eruption. WY Sge has deep eclipses at quiescence which disappear during DN outburst, as in U Gem (Section 1.3). As with the NLs, eclipse profiles are often variable from orbit to orbit.

Among the lower inclination (and non-magnetic) systems almost all have had P_{orb} determined from spectroscopic observation. RR Pic, however, had a pronounced orbital brightness modulation in the 1970s which diminished and was replaced by an irregular, probably grazing, shallow eclipse in the 1980s (Warner 1986d), despite the overall brightness of the system having remained unchanged. This is probably connected with alterations in disc structure resulting from cooling of the primary, but no quantitative explanation has been given. The light curves of V1500 Cyg, GQ Mus and CP Pup appear characteristic of magnetic CVs and are considered in Section 6.7 (see also Section 4.5). V1974 Cyg shows strong orbital modulation due to heating of the secondary by the primary which was heated in the 1992 eruption (De Young & Schmidt 1994).

The dominance of the giant secondary in T CrB at visible wavelengths results in an ellipsoidal modulation with an amplitude ~0.15 mag (Isles 1974; Bailey 1975b; Peel 1985; Lines, Lines & McFaul 1988; Selvelli, Cassatella & Gilmozzi 1992; Luthardt 1992a). In the J band the amplitude is 0.09 mag (Yudin & Munari 1993). RS Oph also shows orbital modulation (Oppenheimer 1994).

4.4.4.2 Eclipse Mapping

Eclipse mapping has been carried out for V Per (Wood, Abbott & Shafter 1992) and DQ Her (unpublished: Horne 1983). V Per is a system with P_{orb} in the period gap of non-magnetic CVs. Standard eclipse deconvolution (Section 2.6.4) results in a very flat distribution for $T_e(r)$. Whereas in DN at quiescence similar flat distributions may be assigned to a non-steady state (Section 3.5.4.4), no such explanation is plausible for the high $\dot{M}(d)$ disc in V Per (which has a spectrum characteristic of NLs or NRs, not of DN (Shafter & Abbott 1989), and has no orbital hump). Instead, Wood, Abbott &

Shafter find that a steady state temperature distribution (equation (2.35)) can be deduced if the inner radius of the disc is set to $r_0 \sim (0.15\text{–}0.25)R_{L_1}$ (the range is a result of uncertainty in distance to V Per). This is indirect evidence that V Per has a strongly magnetic primary. Truncated discs in this context are discussed in Section 7.6.6.

The flat temperature distributions found for NLs (Section 4.3.4) could indicate that the central regions of their discs are also commonly missing. Smak (1994b), however, points out that at least part of this effect may be due to partial obscuration of the inner parts of the disc by the outer rim of the disc. This cannot be invoked for the low inclination IX Vel, where there is a deficiency of flux in IX Vel below 1200 Å which can be understood if the inner $r \leq 10R(1) = 0.12R_{L_1}$ is absent (Long 1993; Long *et al.* 1994).

4.5 Multiple Periodicities

All of the objects with putative precessing discs listed in Table 2.4 are NLs or NRs. The spectroscopic period is assumed $\equiv P_{\rm orb}$. In general the non-$P_{\rm orb}$ periods in these systems are imperfect clocks and therefore unlikely to be related to slow asynchronous rotation of the primary. The proposed models are (Section 2.5.6) a tilted retrogradely precessing disc and (Section 3.6.6.3) a progradely precessing elliptical disc.

In the first model, any variation in the visibility of, for example, the central region of the disc, caused by the disc precession will lead to a photometric variation with period $P_{\rm ph}$ given by $P_{\rm ph}^{-1} = P_{\rm orb}^{-1} + P_{\rm pr}^{-1}$, i.e., at a shorter period than $P_{\rm orb}$. A photometric period at $P_{\rm pr}$ would also be expected, generated by the varying aspects of the disc itself. Two stars satisfy these criteria: TT Ari (Cowley *et al.* 1975; Udalski 1988a) and TV Col (which is an intermediate polar, discussed in Chapter 7: Hutchings *et al.* 1981a; Hellier, Mason & Mittaz 1991).

In the second model, $P_{\rm orb} < P_{\rm s}$ and the beat period $P_{\rm b}$ (Section 3.6.6.3) will not necessarily appear as a photometric modulation. The stars V795 Her (Zhang *et al.* 1991), MV Lyr (Borisov 1992), V603 Aql (Patterson *et al.* 1992b) and HR Del (Bruch 1982) satisfy these criteria. V795 Her has a particularly stable photometric period ($|\dot{P}| < 2 \times 10^{-9}$: Zhang *et al.* 1991) which suggests that it may be an intermediate polar, and indeed it does lie within the period gap of non-magnetic CVs. However, the absence of X-ray emission (Rosen *et al.* 1989) and the fact that the disc in Her X-1 is a good clock (with phase noise: Ögelman 1987) require keeping V795 Her as a precessing disc candidate.

The theoretical expectations that $q \leq 0.22$ and $P_{\rm b}({\rm d}) \lesssim 0.90\, P_{\rm orb}({\rm h})$ (Section 3.6.6.3) are not satisfied for these stars. In fact, the stars of Table 2.4 form an extension of the SU UMa stars (Table 3.3) to larger $P_{\rm orb}$ and larger q. If this is a correct physical description, then the high viscosity discs in these NLs must have radii $r_{\rm d} \sim r_{32} \sim 0.90\, R_L(1)$ for $q = 0.4$.

The concept of *permanent superhumps* in NLs and NRs was introduced by Udalski (1988a) in a study of TT Ari, which ironically is one of the $P_{\rm ph} < P_{\rm orb}$ stars listed above. However, before rejecting TT Ari (and TV Col) from superhump candidature it should be noted that Lubow (1992b) shows that elliptical disc precession is driven by three competing mechanisms: (a) a prograde precession caused by the axisymmetric component of the tidal field (described in Section 3.6.6.2), (b) stresses generated by an acoustic wave originating at the 3:2 resonance radius – these can cause either prograde or retrograde precession, and (c) retrograde precession caused by pressure waves. The relative influences of these mechanisms, as a function of q and viscosity, have still to be explored. Perhaps there is a region, not accessible to systems with $q \lesssim 0.25$, in which net retrograde precession can occur. Similarly, the absence of multiple periods in most NLs and NRs may be due to a near balancing of the mechanisms, or to smaller disc radii – though observational selection may have resulted in many non-$P_{\rm orb}$ modulations being overlooked (Patterson *et al.* 1993a): note in particular that no high inclination multiple period systems are yet known (in this regard, the large and rapid variations of brightness and eclipse depth in RW Tri (Walker 1963a) require more complete study).

A sequitur is that all NLs and NRs with $P_{\rm orb} \lesssim 2$ h might be expected to be superhump systems. There is a paucity of such short period objects, principally because generally $\dot{M} < \dot{M}_{\rm crit2}$(d) below the period gap. The NL BK Lyn, which is possibly the remnant of Nova Lyn 101AD (Hertzog 1986), is a UX UMa star with a spectroscopically determined $P_{\rm orb}$ (Dobrzycka & Howell 1992) and observed brightness modulation at 0.0785 d (Skillman & Patterson 1993), variable over a range ~0.02%, giving a theoretical beat period $P_{\rm b} = 1.7$ d (no photometric $P_{\rm orb}$ modulation is detected). There seems little doubt that this is, in effect, an SU UMa star permanently in superoutburst. The low amplitude of modulation (0.027 mag) and greater stability of $P_{\rm s}$ must result from a more stable $\dot{M}$(d) than obtains at supermaxima of SU UMa stars. The decay in modulation amplitude seen on the plateaux of superoutbursts (Section 3.6.4) may be an asymptotic approach to equilibrium which is not achieved before the end of outburst. It will be interesting to see whether the modulation in BK Lyn is a normal superhump or a late superhump.

Among the NRs, however, some evidence for multiple periodicity exists. GQ Mus

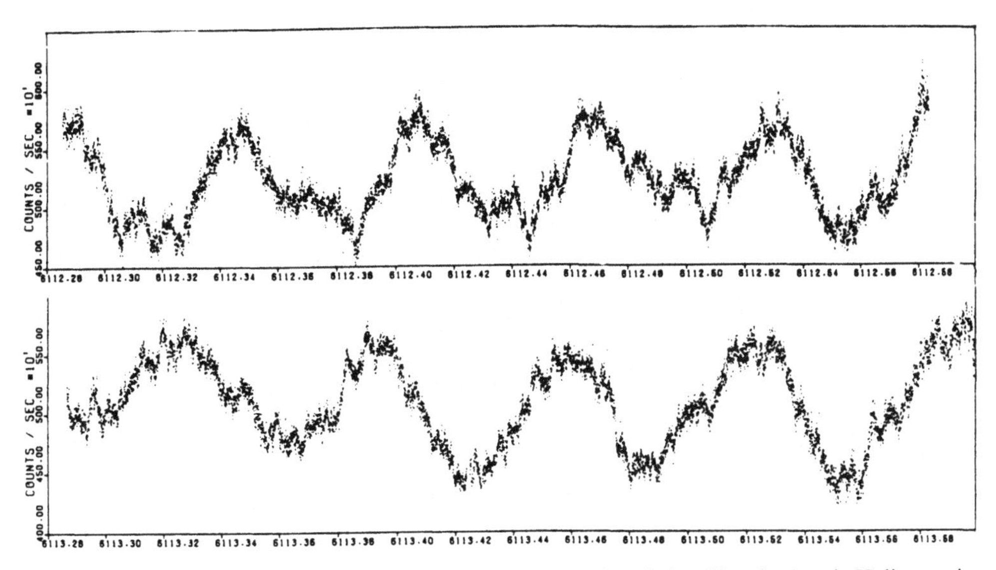

Figure 4.21 Light curve of CP Pup at 4 s resolution in white light. The abscissa is Heliocentric Julian Date 2440000+. Adapted from Warner (1985d).

has the light curve and spectrum of a polar (Diaz & Steiner 1990a) and is almost certainly strongly magnetic, but CP Pup and T Pyx both show double periods. CP Pup has a (probably variable) photometric period ~3% larger than its spectroscopic period (O'Donoghue *et al.* 1989; White, Honeycutt & Horne 1993; Patterson, private communication). The light curve of CP Pup (Figure 4.21) resembles that of SU UMa stars (Figures 3.39 and 3.41); CP Pup can also be interpreted as a weakly magnetic system (Section 6.7). T Pyx has a variable photometric period averaging 1.83 h (Schaefer *et al.* 1992); the spectroscopic period is not known.

Other, shorter quasi-periods that coexist in NLs and NRs are discussed in Section 8.7.

4.6 Radio Emission

Radio emission from AC Cnc has been detected at 6 cm (Torbett & Campbell 1987). This is compatible with the detections of DN during outburst (Section 3.3.7). There are many other bright NLs that should be looked at.

5

Novae in Eruption

All that's bright must fade, – The brightest still the fleetest.
Thomas Moore. *All That's Bright Must Fade.*

The topic of nova explosions introduces areas of physics not touched upon in previous chapters: white dwarf structure, nuclear reactions, hydrodynamics of explosive mass loss, common envelope structures, non-LTE conditions in low density ejecta, dust formation. At the same time these are areas that have been comprehensively reviewed and referenced in recent books and articles. This chapter therefore concentrates on the basics of eruption physics, expanding only in those parts that relate strongly to the properties and evolution of the classes of CVs discussed in other chapters. Novae between eruptions are discussed in Chapter 4.

Two books provide detailed reviews of observation and theory of nova eruptions: *The Galactic Novae* (Payne-Gaposchkin 1957) and *Classical Novae* (Bode & Evans 1989). The latter contains a comprehensive list of references to observational papers on novae published to the beginning of 1987. Modern review articles, and a few early ones still of value, are Stratton (1928), McLaughlin (1960a), Gallagher & Starrfield (1978), Truran (1982), Bode & Evans (1983), Starrfield (1986, 1988, 1990, 1992), Starrfield & Snijders (1987), Gehrz (1988), Shara (1989), Seitter (1990). Specialized conference proceedings appear in Friedjung (1977) and Cassatella & Viotti (1990).

Lists of novae, finding charts and references are given in Duerbeck (1987), Bode & Evans (1989) and Downes & Shara (1993).

5.1 Nova Discovery

Prior to the introduction of wide-field sky photography in the late nineteenth century novae were found generally as naked eye objects. In this century the brighter novae have almost all been found by amateurs (especially, latterly, the Japanese) specializing in nova searches, and the fainter ones by photography. After compensation for temporal non-uniformity of search coverage, Duerbeck (1990) derives the Galactic nova discovery rates given in Table 5.1. On average there is about one nova with $m_V(\max) \leq 3.0$ per decade. In the range $4 \leq m_V(\max) \leq 6$ the rate does not grow in the way expected for a Galactic disc distribution, which suggests that many bright novae are overlooked (Allen 1954; Schmidt-Kaler 1957; Duerbeck 1990). Fainter than $m_V(\max) \sim 6$ the rate increases rapidly – this is because novae in the Galactic Bulge contribute at these magnitudes. The apparent Galactic distribution of novae shows

Table 5.1. *Average Rates of Nova Discovery.*

m_V(max)	Discovery rate (y^{-1})
< 1	0.04
1–2	0.02
2–3	0.04
3–4	0.03
4–5	0.05
5–6	0.14
6–7	0.47
7–8	0.58

strong concentration towards the plane and the centre (Figure 9.11(b)). At higher spatial resolution novae are found preferentially in regions of low interstellar extinction (Plaut 1965).

The observed mean nova rate is ~ 3 y^{-1}. Correction for incompleteness and interstellar extinction is a very uncertain process; values ranging from 73 to 260 novae y^{-1} have been deduced (Allen 1954; Sharov 1972; Liller & Mayer 1987). A rate ~ 19 y^{-1} is derived indirectly by scaling from other galaxies (Della Valle 1992; Della Valle & Livio 1994); this is fraught with uncertainty because there is evidence of large optical thicknesses in galaxies (Disney, Davies & Philipps 1989; Valentijn 1990; James & Puxley 1993) which would result in a large fraction of novae being missed even in face-on spirals. On the other hand, elliptical galaxies should be free of this problem – but no nova rates are known in them.

By definition, any CN showing a second eruption joins the RN class. Several purported RN are now realized to be DN of long outburst interval (e.g., WZ Sge) or CN with DN outbursts (e.g. V1017 Sgr). Probably all CN are RN of long recurrence time T_R. There are no definite recurrences of novae from ancient records (Duerbeck 1992), suggesting T_R > 1000 y in general.

Table 5.2 lists a selection of well-observed CN. RN are listed in Table 5.3, obtained from Sekguchi (1992). Orbital periods for CN and RN are given in Table 4.2.

5.2 Photometry of Nova Eruptions

The light curves of novae show individual characteristics, the most wide ranging of which is the rate of decline from maximum light – and hence the total time taken to go from eruption to quiescence. By judicious compression of their time scales, most nova light curves can be made to resemble each other, and their spectral evolutions (Section 5.3) are brought into better accord.

5.2.1 Morphology of Nova Visual Light Curves

From a study of seven novae McLaughlin (1939, 1943) extracted the common behaviour depicted in Figure 5.1. During the 'transition' phase three alternate behaviours are shown.

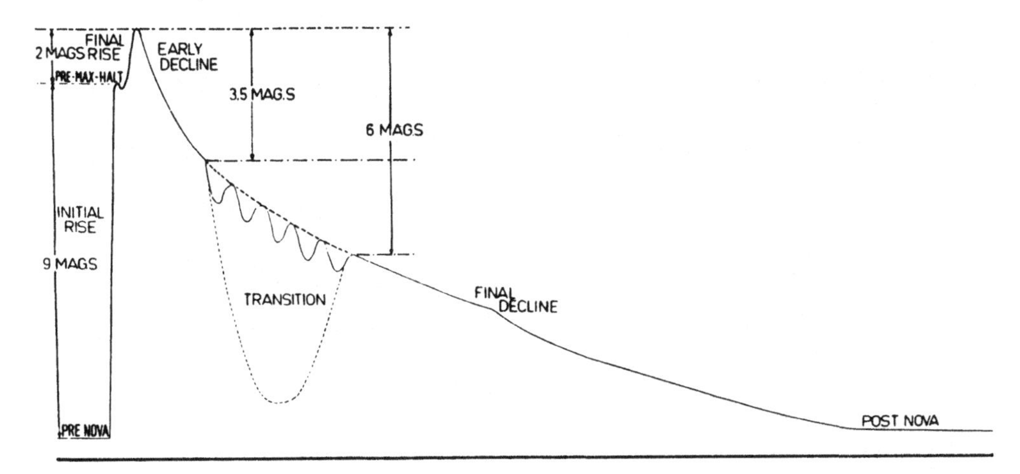

Figure 5.1 Schematic general light curve for CN eruptions. From McLaughlin (1960a).

With exceptions like V1500 Cyg and Nova LMC 1991 (Della Valle 1991), observations on the initial rise are fragmentary, but all indicate that the duration is less than 3 d. In the case of V1500 Cyg brightness measurements cover most of the rise and show that it took less than a day (Liller *et al.* 1975). In many novae there is a plateau, lasting from a few hours in fast novae to a few days in slow novae (or possibly even months in HR Del and a few others approximately (Seitter 1969, 1990)), followed by the final rise of ~2 mag which takes ~2 d for fast novae and approximately weeks for slow novae.

The decline from maximum visual luminosity is generally smooth for all except slow novae, which show variations of up to 2 mag on times scales of 1–20 d. The time taken to fall a given amount below maximum brightness provides a quantitative measure of the speed of a nova. Payne-Gaposchkin (1957) classified novae according to the speed classes given in Table 5.4, using t_2, the time (in days) taken to fall 2 mag below maximum. Others (e.g., Duerbeck 1981) use t_3; a comparison of the two parameters for the CN in Table 5.2 shows that $t_3 \approx 2.75 t_2^{0.88}$. Novae get bluer as they decline, so they decay more slowly in B than in V; van den Bergh & Younger (1987) find

$$\log t_2(\mathrm{V}) = 0.953(\pm 0.013) \log t_2(\mathrm{B}). \tag{5.1}$$

Novae show their greatest diversity of behaviour starting at 3–4 mag below visual maximum. A few (about one third of fast or very fast) novae continue to decline without interruption (e.g., CP Pup, V1500 Cyg, V1668 Cyg); others pass into a minimum 7–10 mag deep, lasting 2–3 months, after which the nova brightens and follows the extrapolated early decline (e.g., DQ Her, T Aur, LW Ser); yet others start large scale quasi-periodic brightness oscillations with amplitudes $1\text{–}1\frac{1}{2}$ mag (e.g., V603 Aql with quasi-period 12 d, GK Per with quasi-period ~5 d and DK Lac with quasi-period ~25 d: Figure 5.2).

The final decline after emergence from the transition phase is made with only small fluctuations in brightness – other than any orbital modulations and other characteristics of the post-nova phase (Section 4.4) which grow in prominence during the decline.

Table 5.2. *A Selection of Well-Observed CN.*

Star	Eruption date	m(max)	m(min)	t_2(d)	t_3(d)	v_{exp} (km s^{-1})	A_v
OS And	1986	6.3v	17.8:p	10	22	900	0.3
DO Aql	1925	8.7v	16.5p	430:	900		
V356 Aql	1936	7.7p	17.7p	145	200	500	2.0
V368 Aql	1936	5.0p	17.3p	15	42	1050	
V500 Aql	1943	6.6p	17.8p	15	42	1380	3.0
V528 Aql	1945	7.0p	18.1p	17	37	1135	2.6
V603 Aql	1918	−1.1v	12.0v	4	8	1700	0.5
V606 Aql	1899	6.7p	17.3p	25	65	750	
V1229 Aql	1970	6.7v	19.4p	20	37	750	1.55
V1370 Aql	1982	6:	19.5p	9	13	2800	0.0
T Aur	1891	4.2p	15.2p	80	100	655	1.15
QZ Aur	1964	6.0p	18.0p	13:	27:		
IV Cep	1971	7.5B	17.1B	11	37	1000	2.0
V693 CrA	1981	7.0v	23j	5.8	12	2210	0.45
Q Cyg	1876	3.0v	15.6v	5	11		0.80
V450 Cyg	1942	7.8p	16.3:p	91	108	512	1.4
V465 Cyg	1948	8.0p	17.0p	<40	140	550	
V476 Cyg	1920	2.0p	17.2B	7	16.5	790	0.85
V1500 Cyg	1975	2.2B	16.3B	2.9	3.6	1180	1.15
V1668 Cyg	1978	6.7p	20.0p	14	23	1300	1.10
V1974 Cyg	1992	4.2v	17r	16	42		0.34
HR Del	1967	3.5v	12.0v	152	230	550	0.56
DM Gem	1903	4.8v	16.7p	6	22	1100	
DN Gem	1912	3.5p	15.8p	17	37	770	0.27
DQ Her	1934	1.3v	14.5v	67	94	350	0.16
V446 Her	1960	3.0p	15.0p	5	16	1380	0.8
V533 Her	1963	3.0p	15.0p	16	44	580	0.10
V827 Her	1987	7.5v	18.0v	25	60	1000	0.0
V838 Her	1991	5.4v	20.v	1.2	3.2		6.5:
CP Lac	1936	2.1v	16.6p	5	10	2400	1.5
DI Lac	1910	4.6v	14.9p	20	43		0.48
DK Lac	1950	5.0p	15.5p	19	32	1075	1.2
HR Lyr	1919	6.5p	15.8p	45	74	725	0.45
BT Mon	1939	4.5:	15.8p	140	F?	800	0.63
GQ Mus	1983	7.2v	17.5	15:	40	535	1.4
V840 Oph	1917	6.5p	20:j	20	36		0.72
V841 Oph	1848	4.2v	13.5v	56	130		0.95
V2214 Oph	1988	8.5p	20.5j	60	100	500	
GK Per	1901	0.2v	13.0V	6	13	1300	0.7
V400 Per	1974	7.8p	20p	20:	43	630	0.2:
RR Pic	1925	1.0v	11.9p	80	150	475	0.04
CP Pup	1942	0.5v	15.0v	5	8	710	0.86
WY Sge	1783	5.4v	20.7B				
V630 Sgr	1936	1.6p	19j	4	11	2120	1.6
V928 Sgr	1947	8.9p	20.5j		150	1050	
V3645 Sgr	1970	12.6p	18p		300		

Table 5.2. *(Continued)*

Star	Eruption date	m(max)	m(min)	t_2(d)	t_3(d)	v_{exp} (km s^{-1})	A_v
V4077 Sgr	1982	8.0v	22j	20	100:	900	1.0
EU Sct	1949	8.4p	18p	20	42	310	2.6
V373 Sct	1975	7.1v	18.5p	40:	85	840	1.0:
FH Ser	1970	4.5v	16.2p	41	62	560	1.9
LW Ser	1978	8.3v	21p	40	50	1250	1.0
MU Ser	1983	7.7v	>21p	2	5	4050	
RW UMi	1956	6p	18.7V		140	950:	0.1
CK Vul	1670	2.6v	20.7v	40	100	59	2.2
LU Vul	1968	9.5p	>21p		21		1.8
LV Vul	1968	5.2v	16.9p		37	780	1.7
NQ Vul	1976	6.0v	18.5p	38	65	750	3.2
PW Vul	1984	6.4v	17:v	83	147	285	1.2
QU Vul	1984	5.6v	19p	25	40		1.0
QV Vul	1987	7.0v	19v	50	60	700	1.0

Principal sources: McLaughlin 1960a; Duerbeck 1987; Warner 1987a; van den Bergh & Younger 1987; Harrison & Gehrz 1988, 1991; Payne-Gaposchkin 1957; Downes & Shara 1993.

As seen from Table 5.3, the RN are an inhomogeneous group. T Pyx has eruption light curves typical of moderately fast novae, the other RN are very fast, and the group U Sco, V394 CrA and LMC-RN, which are a spectroscopically distinct subset (Sections 4.2.5.2 and 5.9) include the fastest novae ever observed. T CrB is unique – having faded quickly to its pre-outburst brightness it brightened again by 2 mag 106 d later (Webbink 1976; Webbink *et al.* 1987). In 1945, 260 d before its second eruption, it faded dramatically. As the system brightness at minimum is dominated by the M giant secondary, this must have originated in that component.

5.2.2 *Colours of Novae*

Van den Bergh & Younger (1987) have compiled and analysed UBV photometry of 14 CN in eruption. Discordances between observers arise because of difficulties of transforming an emission-line flux distribution to the standard system (particularly from the effect of strong Hα emission at the edge of the V passband). Dereddening is effected by measuring the colour excesses of normal stars as a function of distance in directions close to novae (Duerbeck 1981).

Novae generally redden during their rise to maximum; those with smooth light curves show a sharp 'reddening pulse' in both B–V and U–B, centred within a day of maximum. The 'half-width' of the pulse (which is strongest in U–B) ranges from $2\frac{1}{2}$ d for fast novae to ~7 d for slow novae. Two-colour plots are given by Seitter (1990).

After correction for reddening, CN at maximum have $(B–V)_0 = 0.23 \pm 0.06$ with a dispersion $\sigma = 0.16$ mag, but $(U–B)_0$ ranges from 0.26 to −0.77. Two magnitudes down from maximum (i.e., at V(max) + 2) the dispersion (which includes effects of

Table 5.3. *Recurrent Novae*

Star	Dates	m(max)	m(min)	t_2(d)	t_3(d)	P_{orb}(d)	Sp(2)	d(kpc)	A_v	References
T CrB	1866,1946	2.0p	11.3p	3.8	6.8	227.5	M3III	1.1	0.35	1,2
V394 CrA	1949,1987	7.0v	18.8B	3.3	10,5.5	0.76	K	(5)	1.0:	3,4
RS Oph	1898,1933,	4.3v	12.5v	4	9.5	230	M0-2 III	1.6	2.26	1,5,17
	1958,1967,									
	1985									
T Pyx	1890,1902,	6.5v	15.4B	62	88	0.099			1.1	1,6,7
	1920,1944,									
	1966			62	88	0..099				
V 3890 Sgr	1962,1990	8.4p	17.2p	9	17		M5III	5.2	1.5	8,9
U Sco	1863,1906,	8.8v	18.5B	2.0	7	1.23	F8	(14)	0.6	1,10,11
	1936,1979,									
	1987									
V745 Sco	1937,1989	9.7v	21j	6.6	14.9		M4III	8	3.5	9,12
LMC-RN	1968,1990	10.9v	20:	2–3	1.31			49.4	0.25	13,14,16
Candidate										
V723 Sco	1952	9.8p	19:	8	17		M0III	6.05	2.71	15

References: 1. Payne-Gaposchkin 1957; 2. Webbink 1976; 3. Sekiguchi *et al.* 1989; 4. Duerbeck 1988; 5. Bode 1987; 6. Catchpole 1969; 7. Shara *et al.* 1989; 8. González-Riestra 1992; 9. Harrison, Johnson & Spyromilio 1993; 10. Williams *et al.* 1981; 11. Sekiguchi *et al.* 1988; 12. Sekiguchi *et al.* 1990a; 13. Sekiguchi *et al.* 1990b; 14. Shore *et al.* 1991; 15. Harrison 1992; 16. Sekiguchi 1992b; 17. Snijders 1987

Table 5.4. *Speed Classes of Novae.*

Speed class	t_2(d)	dm_v/dt (mag d^{-1})
Very fast	< 10	> 0.20
Fast	11–25	0.18–0.08
Moderately fast	26–80	0.07–0.025
Slow	81–150	0.024–0.013
Very slow	151–250	0.013–0.008

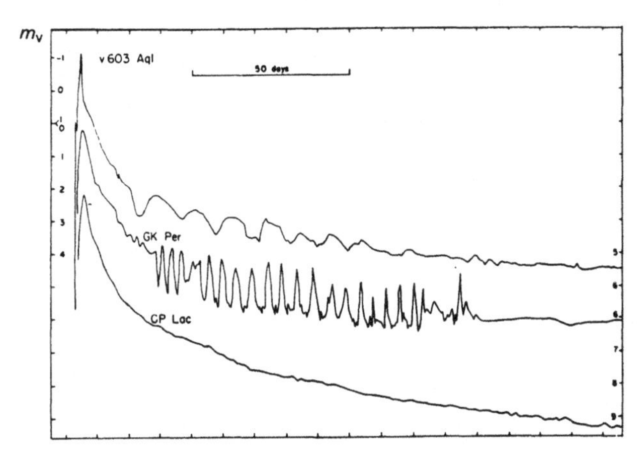

Figure 5.2 Light curves of three fast novae. From McLaughlin (1960a).

errors in the estimation of E_{B-V}) is smaller: $(B–V)_0 = -0.02 \pm 0.04$ and $\sigma = 0.12$ mag. This provides a useful method of estimating E_{B-V} and hence A_v.

5.2.3 Absolute Magnitude versus Rate of Decline

The absolute magnitudes, M(max), of CN at the maximum of eruption have long been of interest as a means of determining extragalactic distances (Schmidt 1957). Early investigations (Payne-Gaposchkin & Gaposchkin 1938; Payne-Gaposchkin 1957), using distances derived from the strengths of interstellar lines, galactic rotation, trigonometric parallaxes and expansion parallaxes (Section 5.7.1), showed that $M_v(\max) \lesssim -7$. From these studies McLaughlin (1942a) noticed that slow novae were on average ~2 mag fainter at maximum than fast novae. Hubble (1929) had already noted that bright novae in M31 fade faster than fainter novae. McLaughlin (1945) later discovered that there is a general $M(\max)$–t_2 relationship. As t_2 is relatively easy to measure, this correlation has proven a valuable means of determining distances to CN, both Galactic and extragalactic.

The relationship has usually been written in the form

$$M(\max) = a_n \log t_n + b_n. \tag{5.2}$$

Table 5.5. *Coefficients of Equation (5.2).*

M	n	a_n	b_n	References
v	3	2.5	−11.5	1
pg	3	2.4	−11.3	2
B	3	1.80 ± 0.20	−10.67 ± 0.30	3
V	2	2.41 ± 0.23	−10.70 ± 0.30	4
pg	2	3.35 ± 0.16	−12.21 ± 0.16	5

References: 1. Schmidt 1957, McLaughlin 1960a; 2. De Vaucouleurs 1978; 3. Pfau 1976; 4. Cohen 1988; 5. Capaccioli *et al.*, 1989.

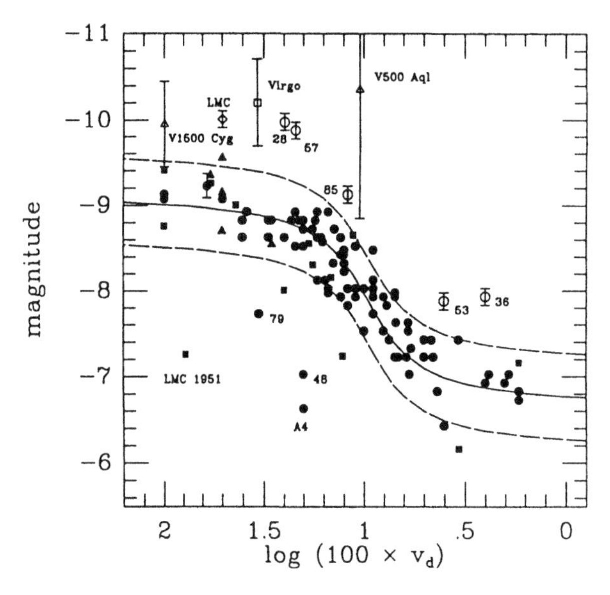

Figure 5.3 Maximum magnitude versus rate of decline for CN in M31 (filled circles) and the LMC (filled squares). Open circles are 'superbright' novae in M31. The parameter $v_d = 2/t_2$. From Della Valle (1991).

Examples of derived coefficients are given in Table 5.5. The result of Capaccioli *et al.* (1989) is based on novae in M31 with $10 < t_2(\mathrm{d}) < 50$ and is in good agreement with the relationship derived by van den Bergh & Pritchet (1986).

When the very extensive results available for novae in M31 and the Large Magellanic Cloud (LMC) are superimposed it is found that for $2 < t_2(\mathrm{d}) < 100$ the great majority fall within the ±0.5 mag S-shaped curve seen in Figure 5.3 (Capaccioli *et al.* 1989, 1990; Della Valle 1991). (The few low points are heavily reddened novae or (points labelled 48 and 79) possibly two separate eruptions of a RN (Rosino 1973).) However, the representation of CN with $t_2 \gtrsim 50$ d in Figure 5.3 ($v_d = 2/t_2$ in that diagram) is sparse, slow novae in M31 are not represented. If $M_v(\max)$ really flattened out at ~ -6.8 for $t_2 > 50$ d, the ranges of CN with large t_2 should not be less than 10

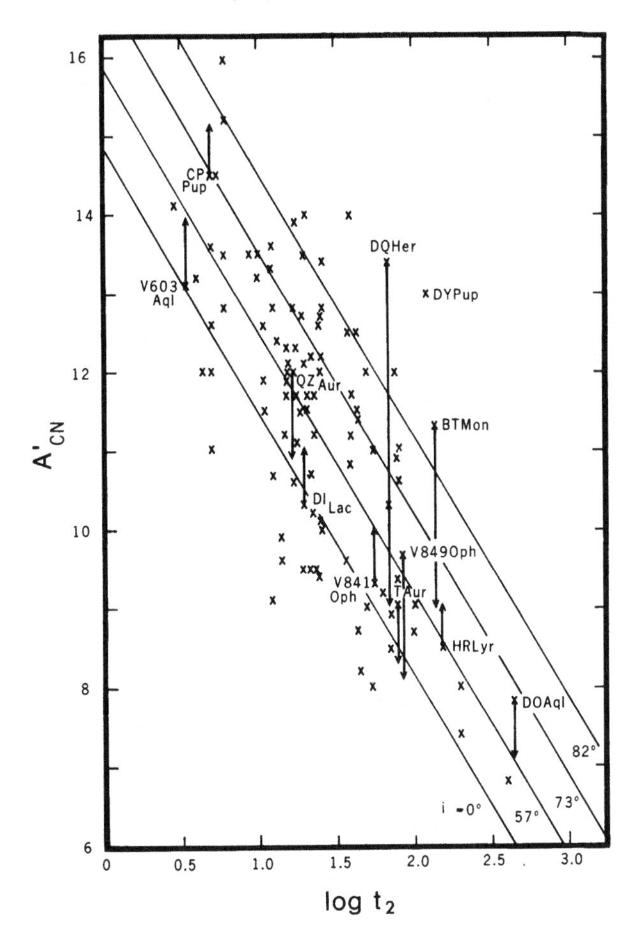

Figure 5.4 Observed range A'_{CN} versus t_2 for CN.

mag (for $\overline{M}_{\mathrm{V}}(\mathrm{min}) \sim 3.8$: Section 4.4.3), whereas they reach down to 7 mag (as will be seen in Figure 5.4). In Section 5.2.5 it is shown that the final entry in Table 5.5, rather than the S-shaped relationship, gives a good representation of the observed amplitude – t_2 relationship.

Figure 5.3 shows clear evidence for a population of novae which are brighter by ~1 mag at maximum than the great majority. This has also been noted for novae in the Virgo Cluster (van den Bergh & Pritchet 1986). Della Valle (1991) considers that the Galactic novae V1500 Cyg and V500 Aql are members of this 'super-bright' class, which amounts to ~7% of novae in each system (Galaxy, LMC, M31 and Virgo). Both V1500 Cyg and N LMC No. 1, which are overluminous in Figure 5.3, are 'neon novae' (Section 5.3.6).

Schmidt (1957) finds that the final rate of rise to maximum is correlated with the rate of decline after maximum: the time taken to rise the final two magnitudes is given by

$$\log t_{r,2} = -0.3 + 0.7 \log t_2 = -0.5 \log t_3. \tag{5.3}$$

Table 5.6. *Absolute Magnitudes 15 d after Maximum.*

Passband	$M(15)$	Reference
V	-5.2 ± 0.1	1
pg	-5.86	2
B	-5.74 ± 0.60	3
pg	-5.5 ± 0.18	4
V	-5.60 ± 0.43	5
V	-5.23 ± 0.16	6
V	-5.38	7
V	-5.69 ± 0.14	8

References: 1. Buscombe & de Vaucouleurs 1955; 2. Schmidt-Kaler 1957; 3. Pfau 1976; 4. de Vaucouleurs 1978; 5. Cohen 1985; 6. Van den Bergh & Younger 1987; 7. van den Bergh 1988; 8. Capaccioli *et al.* 1989.

5.2.4 Range versus Rate of Decline

Figure 5.4 shows the relationship between range $A'_{CN} = m(\min) - m(\max)$ and t_2 for all CN with available observations. As discussed in Section 4.4.3, the apparent range A'_{CN} is a function of inclination; corrections $\Delta M_V(i)$ are shown where known (Table 4.2) and act to reduce the vertical dispersion. In particular, note DQ Her and T Aur, which were suspected by Payne-Gaposchkin (1957) to have ranges so large that they may belong to a distinct class that does not obey the usual $M_V(\max)$–t_2 relationship. This is seen to be due merely to the high inclination of these systems. From its position in Figure 5.4, DY Pup probably also is of high inclination.

The diagonal lines in Figure 5.4 show the expected relationship for $M(\max)$ given by equation (5.2) and the Capaccioli *et al.* (1989) coefficients (Table 5.5), together with $\overline{M}_V(\min) = 3.8$ (Section 4.4.3) and $\Delta M_V(i) = -1, 0, 1$ and 2 respectively. The vertical scatter in Figure 5.4 is reasonably accounted for by random distribution of inclinations convolved with a 1 mag Gaussian dispersion caused mostly by errors in $m(\max)$ and $m(\min)$ (Warner 1986e).

5.2.5 Absolute Magnitude at $t = 15$ d

Buscombe & de Vaucouleurs (1955) noted that the absolute magnitude $M(15)$ fifteen days after maximum light appeared to be independent of speed class. There is evidence, however, that $M(15)$ does have some slight dependence on t_2 (van den Bergh & Younger 1987; Capaccioli *et al.* 1990). Deduced values of $M(15)$ are given in Table 5.6; since $(B{-}V)_0 \sim 0.0$ at $t = 15$ d, $M(15)$ should be approximately the same in B and V.

5.2.6 Flux Distributions and Luminosities

On the rise to maximum the spectrum is that of an optically thick expanding photosphere (Section 5.3.1) which falls in effective temperature to a minimum $T_{eff} \sim (4\text{–}7) \times 10^3$ K (Gallagher & Starrfield 1978).

UV observations of FH Ser (Nova Ser 1970) with the OAO-2 satellite showed that the peak of its energy distribution moved steadily from ~ 4000 Å 8 d after eruption to ~ 1500 Å 50 d later (Gallagher & Code 1974). The integrated flux between 1550 and 5500 Å was found to be almost constant until 50 d after eruption. The early decline in visual luminosity is due almost entirely to the redistribution of flux which takes place at constant bolometric luminosity. This has since been observed in a number of novae, e.g., V1668 Cyg (Stickland *et al.* 1981), V1370 Aql (Snijders *et al.* 1987), GQ Mus (Krautter & Williams 1989), and is probably universal, being the result of radiation escaping from ever deeper, hotter regions of the expanding envelope as the temperature of the outermost layers falls below 10^4 K, hydrogen recombines and the opacity falls (Gallagher & Starrfield 1976, 1978). A corollary of this behaviour is that light curves pass through their maxima later at shorter wavelengths: Cassatella & Gonzalez-Riestra (1990) find that the delay Δt between optical and λ1445 Å maxima is a function of t_3. This can be approximated by $\Delta t \approx 0.7t_3$.

During its phase of constant bolometric luminosity, FH Ser had a luminosity $L \approx 1.5 \times 10^4$ $L_\odot$ (for $d = 650$ pc). Similar results are obtained for other novae (V1370 Aql: 2.5×10^4 $L_\odot$ for $d = 5$ kpc, Snijders *et al.* 1987; V1668 Cyg: 1.7×10^4 $L_\odot$ for $d = 2.2$ kpc, Stickland *et al.* 1981; GQ Mus: $\sim 4 \times 10^4$ $L_\odot$ for $d = 4.8$ kpc, Krautter *et al.* 1984; V1500 Cyg: 2.8×10^4 $L_\odot$ for $d = 1.55$ kpc, Wu & Kester 1977, Ferland, Lambert & Woodman 1986). The peak luminosities, however, are usually several times these 'plateau' values, and in the case of very fast novae can be very much higher (V1500 Cyg: 4.7×10^5 $L_\odot$, Wu & Kester 1977; FH Ser: 1.3×10^5 $L_\odot$, Friedjung 1987a; QU Vul: 1.1×10^5 $L_\odot$, Gehrz, Grasdalen & Hackwell 1986; V1668 Cyg: 1.2×10^5 $L_\odot$, MacDonald 1983).

These luminosities approach or exceed the *Eddington luminosity* for a 1 $M_\odot$ object. This limit is found from the luminosity at which radiation pressure balances gravitational attraction:

$$L_{\rm Ed} = \frac{4\pi GcM}{\sigma_e N_e} \qquad (5.4a)$$

$$= \frac{6.41 \times 10^4}{(1+X)} M_1 \quad L_\odot \qquad (5.4b)$$

(for fully ionized gas) where σ_e is the electron scattering cross-section and X is the hydrogen fraction by mass (typically 0.3–0.6 in nova envelopes: Section 5.7.3). It is clear, therefore, that radiation pressure is a motivating force underlying ejection of nova envelopes.

5.2.7 IR Photometry

The bright, moderate speed nova FH Ser provided the first opportunity for detailed study at IR wavelengths. Concomitant with descent into the deep minimum of the transition phase a growing IR excess was observed (Geisel, Kleinmann & Low 1970; Hyland & Neugebauer 1970) which reached the same ~ $1.5 \times 10^4 L_\odot$ as the earlier UV and visual plateau luminosity. This showed that the visual minimum was due to the formation of absorbing dust grains, which reradiate the intercepted energy in the IR, and extended the duration of approximately constant bolometric luminosity. The IR excess had a thermal distribution with $T \sim 1000$ K.

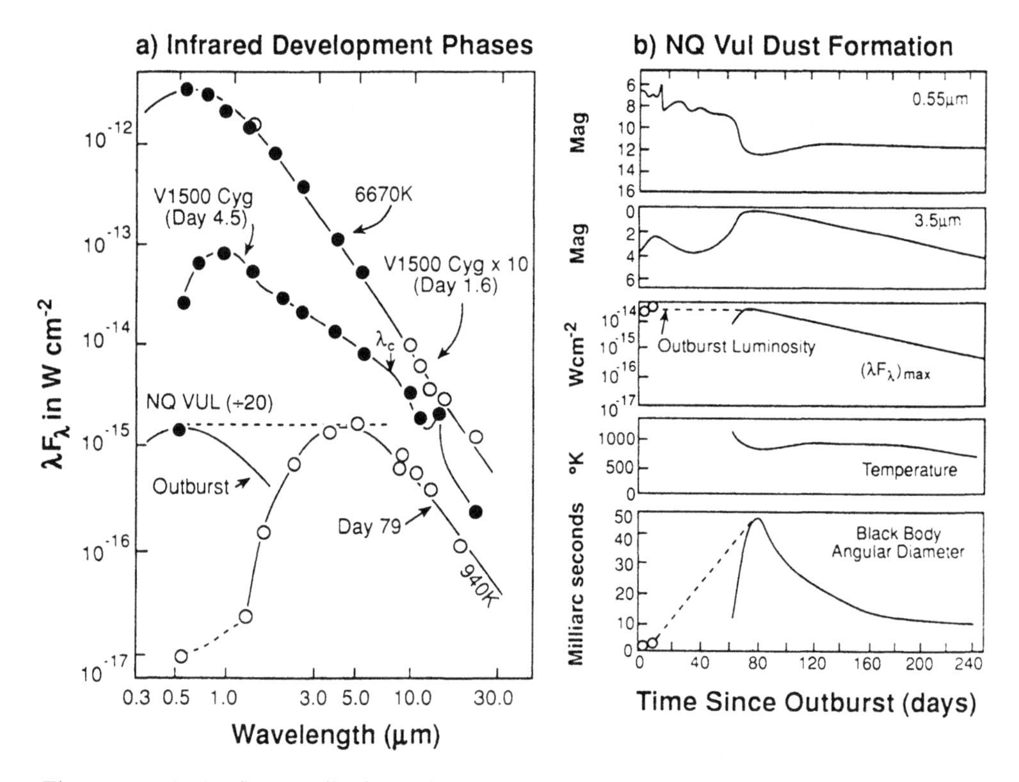

Figure 5.5 (a) IR flux distributions of V1500 Cyg and NQ Vul at two stages of development and (b) evolution of NQ Vul over its first 240 d. From Gehrz (1990).

Since 1970, extended IR observations have been made of over a dozen CN. These are comprehensively reviewed by Gehrz (1988, 1990). Examples of the IR flux distributions at two different stages of the eruptions of V1500 Cyg and NQ Vul, and the temporal development of observed and deduced parameters for NQ Vul, are shown in Figure 5.5 (Gehrz 1990). The following is a brief summary of the physical interpretation of these developments.

Prior to maximum light the flux distributions closely approximate blackbodies (Gallagher & Ney 1976). Each such observation provides a temperature T_{BB} and flux $(\lambda F_\lambda)_{max}$ at the peak of the distribution, which (for a blackbody) are related to the angular radius θ_r in arcseconds by

$$\theta_r'' = 3.38 \times 10^7 (\lambda F_\lambda)_{max}^{1/2} \, T_{BB}^{-2} \tag{5.5}$$

with λF_λ in erg cm^{-2}s^{-1}. If the distance is known, the envelope radius R_{env} is deduced. The early expansion of the optically thick envelope is measurable in this way.

When the envelope becomes optically thin at IR wavelengths the energy distribution is no longer blackbody (Figure 5.5) and is dominated by free–free emission (thermal bremsstrahlung) of ionized hydrogen. At wavelength λ_c the envelope has unit optical thickness, which causes the flux distribution to change to the Rayleigh—Jeans tail of a blackbody distribution (Gehrz, Hackwell & Jones 1974). At the time that the envelope

Table 5.7. *Envelope and Shell Masses of CN.*

Nova	$M_{env}(10^{-5}\ M_{\odot})$			$M_{shell}(10^{-5}\ M_{\odot})$		
	IR	Abs line	Emiss line	Radio	Resolved shell	References
V1370 Aql			0.4–2			14
V842 Cen			3			15
IV Cep			8			7,9
V1500 Cyg	0.5–80	1	5–15	24*		1,4,13,14,18,19
V1668 Cyg	2		20			1,8
HR Del			10	8.6*		4,12,16
DQ Her			11		4.8	10,23
V838 Her	6.4–9					2
CP Lac			3			6,7
GQ Mus	$\lesssim$0.26		8.4			1,5
RR Pic			30			11
FH Ser			1.0	4.5*		4,16
LW Ser	2					1
LV Vul		3	8			17
NQ Vul	10					1
PW Vul	0.32		>30			1,20
QU Vul			2–150	36		21,22
QV Vul	2.3		12			3

* For $T_e = 1.0 \times 10^4$ K.

References: 1. Gehrz 1988; 2. Woodward *et al.* 1992; 3. Gehrz *et al.* 1992; 4. Hjellming 1990; 5. Hassall *et al.* 1990; 6. Pottasch 1959; 7. Ferland 1979; 8. Stickland *et al.* 1981; 9. de Freitas Pacheco 1977; 10. Martin 1989c; 11. Williams & Gallagher 1979; 12. Tylenda 1978; 13. Lance, McCall & Uomoto 1988; 14. Snijders *et al.* 1987; 15. de Freitas Pacheco, da Costa & Codina 1989; 16. Hartwick & Hutchings 1978; 17. Raikova 1990; 18. Wolf 1977; 19. Ennis *et al.* 1977; 20. Saizar *et al.* 1991; 21. Taylor *et al.* 1988; 22. Saizar *et al.* 1992; 23. Ferland *et al.* 1984.

is just becoming optically thin it is so hot that electron scattering dominates; then the envelope mass is (Gehrz 1988)

$$M_{env} \approx \pi R_{env}^2 \sigma_e^{-1}, \tag{5.6}$$

where σ_e for hydrogen is 0.36 cm^2 g^{-1} and R_{env} can be obtained from equation (5.5) or from $v_{exp}t$, where v_{exp} is the expansion velocity (Section 5.3.1) and t is the time since the start of eruption.

Later, when the envelope has cooled, the relationship $N_e^2 \lambda_c^2(\text{cm}) l_{env}(\text{cm}) \sim 0.3$ (Kaplan & Pikel'ner 1970) for an optically thin ionized gas can be used with $M_{env} \approx 4\pi R_{env}^2 l_{env} N_e m_H$ to give an independent estimate. The two methods agree tolerably well and provide the values given in the IR column of Table 5.7.

IR emission from dust grains develops in most novae (Figure 1.25), and, when optically thick, enables equation (5.5) to be used for estimating the diameter of the radiating region (Figure 5.5). The physical properties of the ejected dust are not central to the interests of this review – they do, however, provide a useful means of monitoring the total luminosity of the eruption. The extended period of almost constant bolometric luminosity is a common feature of novae.

5.3 Spectroscopy of Nova Eruptions

The early studies of nova spectra, summarized by Payne-Gaposchkin (1957) and McLaughlin (1960a), were taxonomic, describing the complex systems of emission lines that develop in the post-maximum phases. The systematics of nova spectra were first organized by McLaughlin (1942b, 1943), who found that spectral development is closely correlated with phase in the light curve (Figure 5.1), independent of speed class. This development is examined here, with added physical interpretation (e.g., Gallagher & Starrfield 1978; Starrfield 1988, 1990). It is an extraordinary fact that the four successive systems of absorption lines, and the overlapping five systems of emission lines, with a variety of substructures, are followed by almost all novae. Only an outline of these can be given.

5.3.1 Pre-Maximum Spectra

On the rising branch of the optical light curve spectra are those of an optically thick, expanding, cooling shell: blueshifted broad absorption lines with occasionally a P Cygni emission component. The spectral type moves later as the rise progresses (the earliest recorded is B0 in V1500 Cyg), with similar behaviour in different novae when normalized by speed class (Figure 5.6). There is only one system of absorption lines visible at this stage; the broad lines imply a large velocity dispersion.

Although similar in appearance to supergiant spectra, the strengths of the CNO lines are anomalously strong (McLaughlin 1960a; Boyarchuk *et al.* 1977) which, from

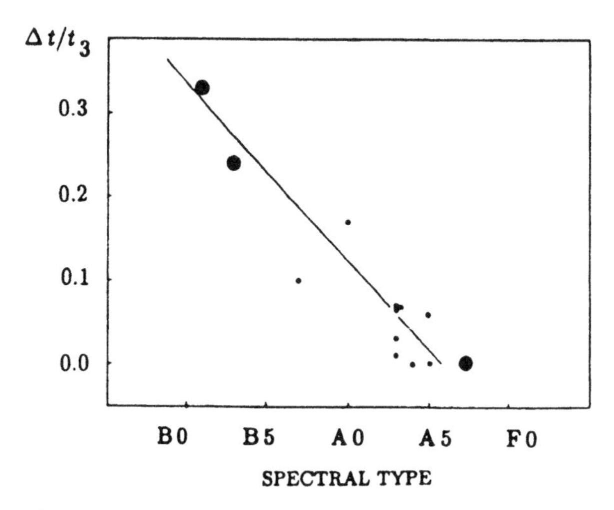

Figure 5.6 Development of pre-maximum spectra as a function of fractional time before optical maximum. From Seitter (1990).

abundance analyses of spectra later in outburst, is caused by enhanced abundances of these elements.

The absorption-line RVs remain almost constant, or decrease slightly, up to maximum; the widths of the lines are approximately equal to the expansion velocity. The velocities (Table 11 of McLaughlin (1960a)) are correlated with speed class and are approximated for $t_2 < 100$ d by $v_{exp} \approx -4750/t_2(\mathrm{d})$ km s^{-1}. v_{exp} can be combined with the rate of angular expansion derived from equation (5.5) to give an *envelope expansion parallax*, or distance d_{ee}, from

$$d_{ee} = 5.68 \times 10^{-7} \frac{v_{exp}(\mathrm{km\ s^{-1}})}{\frac{\mathrm{d}\theta_r}{\mathrm{d}t}\ ('' \text{ per day})} \text{ kpc.} \tag{5.7}$$

Barnes (1976) has used an application of the Wesselink (1969) surface brightness method, which is insensitive to reddening, to derive $\mathrm{d}\theta_r/\mathrm{d}t$ from V,R photometry.

5.3.2 The Principal Spectrum

About 0.5 mag down from maximum visual luminosity a new absorption-line system develops, known as the *principal spectrum*. This is displaced (by an amount correlated with t_2) to negative velocities relative to the pre-maximum spectrum, which itself rapidly fades from view. In strengths, the lines resemble an A or F supergiant spectrum, but with enchanced CNO. The continuum distribution agrees with the spectral class (e.g., V1668 Cyg: Cassatella *et al.* 1979). The velocities v_{prin} are approximated by (McLaughlin 1960a)

$$\log v_{prin}(\mathrm{km\ s^{-1}}) = 3.70 - 0.5 \log t_3(\mathrm{d}) = 3.57 = 0.5 \log t_2(\mathrm{d}). \tag{5.8}$$

The parameters listed in Table 5.2 agree with these relationships, but there is a range of ± 0.2 dex at a given t_2 or t_3.

The absorption lines often show multiple substructure, the evolutions of which vary among the components and differ from nova to nova. At or immediately after maximum brightness P Cyg profiles develop, with their emission components symmetrically about the systemic velocity. A large number of emission lines appear at this stage, the strongest being H, CaII, NaI and FeII in the visible (Figure 5.7). In the UV, available spectra taken close to maximum are rare; in V1668 Cyg there was P Cyg structure in MgII 2880 Å at maximum, but all other lines were simply in emission (Cassatella *et al.* 1979). Most of the mass ejected by the nova contributes to the principal spectrum (Friedjung 1987b).

As the nova declines from maximum the absorption components of the principal spectrum weaken, but it is the last of the absorption systems to disappear, leaving a pure emisson spectrum by the end of the early decline. During this phase [OI] and [NII] emission lines appear, followed by [OII]. The emission lines have complex profiles, indicative of non-uniform ejection of gas. In the UV emission arises from the resonance and the lowest-lying intercombination lines connected to the ground states of abundant species: HeII; MgII; AlII,III; SiII,III; NIII,IV,V; OIII,IV,V; CII,III,IV. In the IR, lines of HI, HeI, CI, OI and NaI are strong.

Line lists for the optical region can be found in Meinel, Aveni & Stockton (1975) and

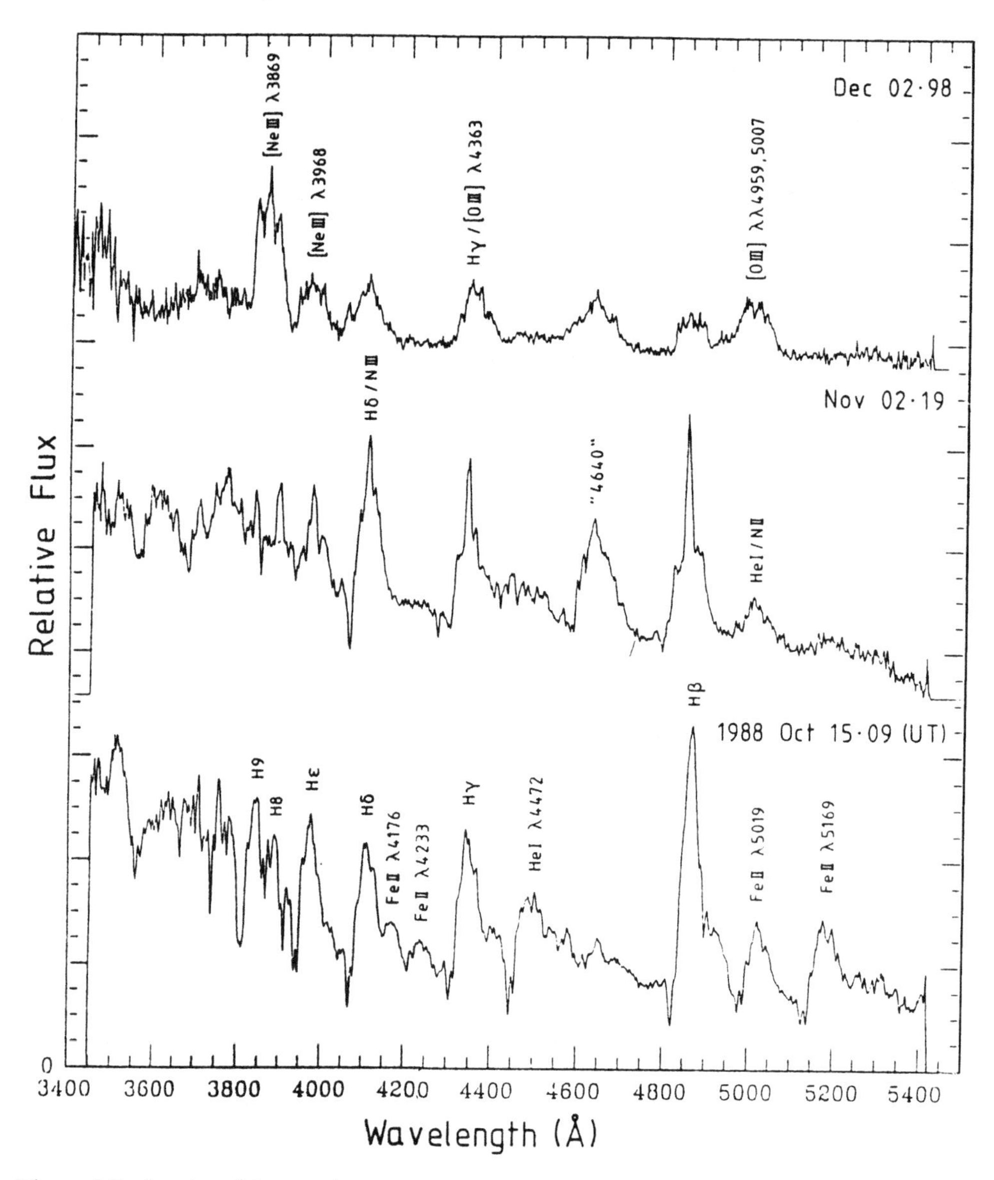

Figure 5.7 **Spectra of the very fast nova** LMC 1988 No. 2. From the lowest spectrum upwards these were obtained 1.34 d, 2.31 d and 5.29 d after optical maximum. These show respectively typical principal plus diffuse-enhanced, Orion, and early nebular spectra. From Sekiguchi *et al.* (1989).

Rosino, Ciatti & Della Valle (1986), for the UV in Williams *et al.* (1985) and Snijders *et al.* (1987), and for the IR in Ferland *et al.* (1979) and Gehrz (1988). Williams (1992) provides a list of post-1980 references to multi-epoch spectral observations of novae.

5.3.3 The Diffuse Enhanced Spectrum

A third distinctive suite of absorption lines appears soon after the principal spectrum has developed. These lines are very broad, from which they are known as the *diffuse enhanced spectrum* (Figure 5.7) and are blueshifted by about twice those of the principal spectrum. McLaughlin (1960a) gives the following relationship for the mean velocity:

$$\log v_{\mathrm{de}}(\mathrm{km\ s^{-1}}) = 3.81 - 0.41 \log t_3(\mathrm{d}) = 3.71 - 0.4 \log t_2(\mathrm{d}). \tag{5.9}$$

Again P Cyg profiles are common, their broad emissions underlying those of the principal spectrum. The duration of these lines ranges from ~10 d for very fast novae to >100 d for slow novae. In their late development they often split into multiple narrow components.

5.3.4 The Orion Spectrum

Yet a fourth coexistant absorption system, usually with single components, appears 1–2 mag down from maximum. This is displaced blueward by at least as much as the diffuse enhanced system, consists at first predominantly of HeI, CII, NII and OII (not always with attendant HI lines), later with NIII and NV emissions, and reaches its greatest strength at about the time that the diffuse enhanced spectrum disappears. The name *Orion spectrum* (Figure 5.7) derives from the similarity with the well-defined absorption lines produced by stellar winds from luminous OB stars.

Unlike the other spectral systems, the Orion spectrum shows large variations in blueshift, superimposed on a systematic increase during the decline, and remains diffuse until it disappears (about 4 mag below maximum for fast novae and 2 mag down for slow novae).

In the UV the emission-line spectrum remains similar throughout the principal, diffuse enhanced and Orion stages – the predominant change being an increase in equivalent widths as the continuum decays (Williams *et al.* 1985).

The overall development of the absorption systems has yet to be modelled in detail. The process is made complicated by the large variations in optical depth that occur in an expanding envelope[1], and the evidence for multiple and sustained mass ejection (Gallagher & Starrfield 1978; Martin 1989a; Williams 1990).

Initially the envelope is optically thick and therefore ionization-bounded, allowing an outer region of neutral gas. Neutral and low ionization lines are formed at this time. The OI $\lambda\lambda$1302,8446 emission lines, generated by resonance fluorescence of Lyβ photons trapped in the atomic hydrogen layer, are a clear diagnostic of optical thickness (Strittmater *et al.* 1977).

The drop in opacity as the outer regions expand and cool results in the photosphere moving *inwards* in radius and mass. If there is steady mass loss the photosphere will reach a steady state radius, otherwise the envelope rapidly becomes optically thin allowing the energetic radiation from the hot primary to ionize the ejecta. These extremes of behaviour characterize slow and fast novae respectively.

As the envelope expands, decreasing density and increasing hardness of the ionizing radiation account for the steady increase in species ionization seen in the absorption and emission spectra. The envelope is optically thick for $N_e \gtrsim 10^9$ cm^{-3}, forbidden lines appear at $N_e \lesssim 10^7$ cm^{-3} (Williams 1990). The persistence of the absorption-line

[1] The large optical depths that occur in the Balmer lines result in, e.g., the Hα 'photosphere' continuing to expand long after the continuum photosphere has contracted. As a result, the emission peaks days or weeks after visual maximum (Ciardullo *et al.* 1990)

systems shows the continuing existence of detached shells, or the stability of continuing mass loss. The apparent acceleration of a shell (as in the Orion spectrum) may be a material effect, or may arise from an ionization front expanding through a region with a velocity gradient (Martin 1989a). The coexistence of several velocity systems can arise from successive ejections or from gas ejected non-isotropically. The observed velocity of an absorbing system is determined by its projection against the extended continuum source and is therefore not necessarily the true radial expansion velocity. The most detailed model of nova envelope and shell ionization structure is that for DQ Her by Martin (1989b,c).

In novae that form an optically thick dust shell, an estimate of distance can be obtained in the same manner as for the optically thick expanding envelope. Equation (5.5), applied to the IR colours, gives $\theta_r(t)$ and equation (5.7) provides the expanding dust shell parallax, or distance, d_{ds}. The main uncertainty is knowing which expansion velocity to use for the dust shell. Observations of PW Vul suggest that v_{prin} should be used (Gehrz *et al.* 1988).

5.3.5 The Nebular Spectrum

The *nebular spectrum* develops at first from the emission components of the principal spectrum. [OI] and [NII] are strong throughout the early decline, [OIII] and [NeIII] appear (Figure 5.6) and strengthen relative to the permitted HI, HeI,II, NII,III emission lines. The evolution is towards the spectrum of a planetary nebula. Williams *et al.* (1991) propose a system of classification of novae spectra that denotes as 'auroral' the stage when [NII]λ5755, [OIII]λ4363 and [OII] are strong, and 'nebular' the stage when [OII]λ5007, [NII]λ6584, [NeIII] and [FeVII] are strong. The multiple structure of nebular spectra is seen in Figure 5.8, where 22 emission components are seen at high resolution in GQ Mus (Krautter *et al.* 1986).

Forbidden lines of highly ionized lines – 'coronal lines' – appear in the spectra of novae if the ionizing radiation reaches temperatures > 10^6 K. RR Pic and DQ Her

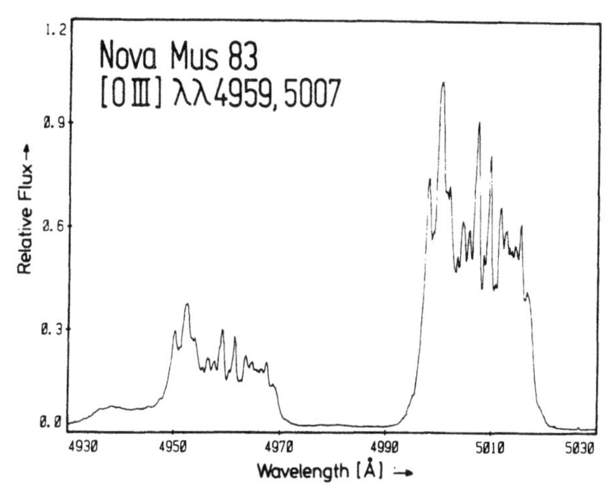

Figure 5.8 Line profiles of [OIII] lines in GQ Mus. From Krautter *et al.* (1986).

Table 5.8. *Recent Neon Novae.*

Nova	Date	Characteristic	t_2(d)	References
V1500 Cyg	1975	Optical,IR	2.9	1
V693 CrA	1981	UV	6	2
V1370 Aql	1982	UV,optical	6	3
QU Vul	1984	IR([NeII] and [NeVI])	27	4,5
V2214 Oph	1988	Optical	145	6,7
V838 Her	1991	Optical	1.2	7
V2264 Oph	1991	Optical	18:	7
N LMC No. 1	1991	Optical	4	6,7
V4160 Sgr	1991	Optical	2	7
V444 Sct	1991	Optical	4.5	7
V351 Pup	1991	Optical	13	7
V1974 Cyg	1992	IR	17	8

References: 1. Ferland & Shields 1978a,b; 2. Williams *et al.* 1985; 3. Snijders *et al.* 1987; 4. Gehrz, Grasdalen & Hackwell 1986; 5. Greenhouse *et al.* 1988; 6. Williams *et al.* 1991; 7. Williams, Phillips & Hamuy 1994; 8. Hayward *et al.* 1992.

showed successively lines of [FeII] through [FeVII] (Payne-Gaposchkin 1957; Swings & Jose 1949,1952). [FeX] and [FeXIV] have been detected (McLaughlin 1953). In V1500 Cyg lines from [FeXI], [FeX], [SVIII] and [FeVII] were observed (Ferland, Lambert & Woodman 1977) and, in the IR, lines from [MgVIII], [SiIX] and [AlIX] (Grasdalen & Joyce 1976). In the UV, V693 CrA produced lines of [NeIII,IV,V], [NaV,VI], [MgV,VII], [AlVI,VIII] and [SiVII,IX] (Williams *et al.* 1985). GQ Mus also showed strong coronal lines (Krautter & Williams 1989). Williams (1990) shows that the production mechanism for these lines is photoionization, rather than collisional ionization in the high velocity envelope. The [SiVII] 2.478 μm/[SiVI]1.960 μm (Figure 1.24) emission-line ratio is a sensitive thermometer and gives temperatures $\sim (4\text{–}5) \times 10^5$ K (Benjamin & Dinerstein 1990).

A discussion of the origin of the coronal lines, and a line list for the IR, are given by Greenhouse *et al.* (1990). The IR forbidden lines, particularly [NeII]λ12.8 μm when present, are an important cooling agent for nova shells. Ferland *et al.* (1984) and Dinerstein (1986) suggest that [OIII]52,88 μm and [NIII]57 μm emission can be significant contributors to the IR flux from old nova shells detected by IRAS.

5.3.6 Neon Novae

McLaughlin (1960b) discovered strong NeI absorption lines in the Orion spectrum of V528 Aql. This remains a unique observation, but many novae have shown exceptional strengths of forbidden Ne emission lines in UV ([NeIII,IV,V], optical ([NeIII]λ3869,3968 and [NeV]λ3426) and IR nebular spectral [NeV]λ12.8 μm) regions, attributed to large overabundances of neon and defining a class of *neon novae* (Table 5.8). An example of the spectrum of a neon nova is given in Figure 5.9.

Greenhouse *et al.* (1990) note that novae with strong IR coronal emission may all have neon overabundances. Williams *et al.* (1991), on the other hand, state that all

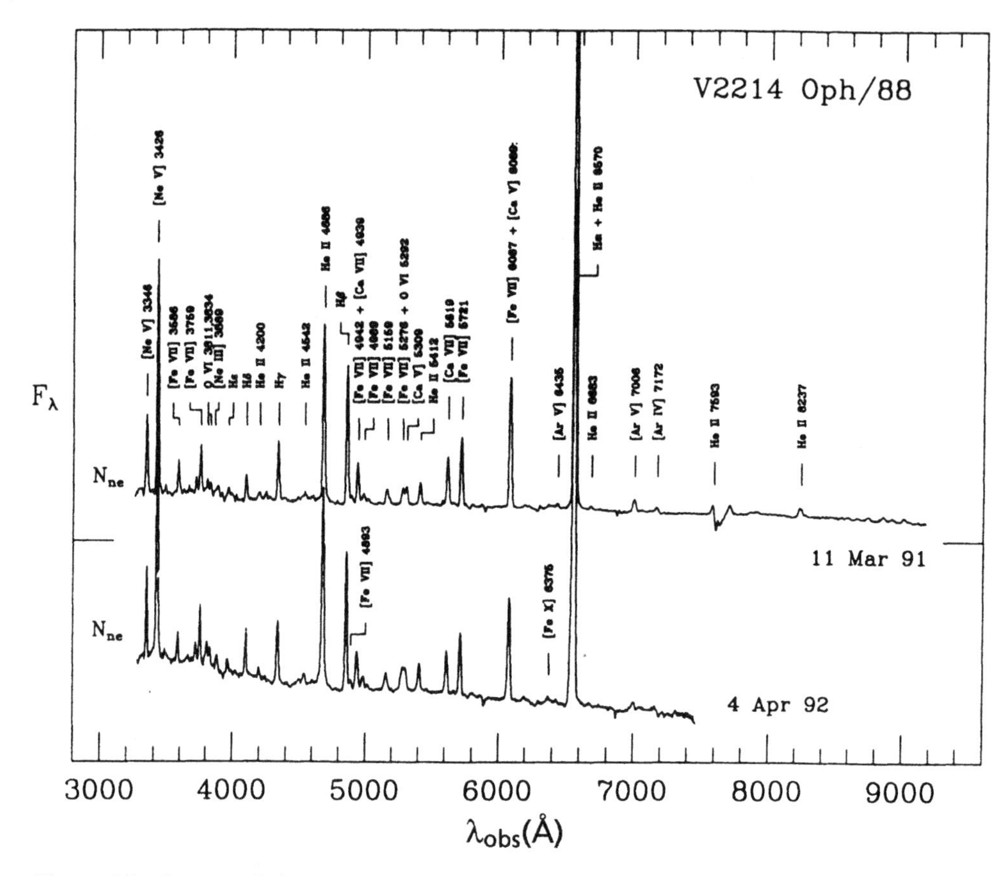

Figure 5.9 Spectra of the neon nova V2214 Oph during the nebular phase. From Williams, Phillips & Hamuy (1993).

neon novae evolve directly from the permitted line stage to the nebular phase without going through the intermediate *or coronal* phases. This apparent contradiction is merely a matter of definition: the IR observers denote as coronal all transitions in ions with ionization potentials > 100 eV, whereas the optical observers require [FeX]λ6375 to be stronger than [FeVII]λ6097. As seen from Table 5.8, all recent neon novae except V2214 Oph have been fast or very fast.

McLaughlin (1944) summarized the behaviour of [NeIII] emission in a large number of early novae and concluded 'Moderately strong [NeIII] emission is evidently a normal feature of nova spectra in the nebular stage, but very great intensity, as in Nova Persei 1901 and Nova Sagitarii 1936.7, or absence, as in Nova Cygni 1920, should be ranked as a peculiarity. These extremes are attributed to differences of abundance, since they are found in spectra otherwise very similar in composition and at similar levels of excitation'. McLaughlin's novae with exceptionally strong Ne lines are V450 Cyg, DQ Her, V849 Oph, V Per, GK Per, RR Pic, V630 Sgr, V732 Sgr and V909 Sgr, to which may be added V500 Aql (Sanford 1943), CP Pup (Gratton 1953) and V726 Sgr (Payne-Gaposchkin 1957). A few of these stars were only moderately fast and one (V450 Cyg) was slow.

Of the novae known or suspected to have strongly magnetic primaries (Sections 4.4.4.1; 4.5 and 6.7), V1500 Cyg, V2214 Oph and CP Pup are neon novae, but GQ Mus is not.

5.3.7 *Modern Spectral Classification*

The taxonomic classification devised by McLaughlin has been used extensively to describe the development of nova spectra but is not rooted in the physical processes occurring in nova shells. On the basis of the extended spectral range accessible to modern instruments, R.E. Williams has proposed a classification scheme for the emission-line spectra which takes more cognizance of the changing photoionization caused by expansion of the shell and hardening of the central radiation source (Williams 1990, 1992; Williams *et al.* 1991; Williams, Phillips & Hamuy 1994).

The high envelope densities in the early stages of eruption restrict the emission lines to permitted (P) transitions. As the density falls first 'auroral' (A) transitions ([NII]λ5755, [OIII]λ4363, [OII]λλ7319,7330) appear and then 'nebular' (N) transitions ([OIII]λ5007, [NII]λ6584, [NeIII]λ3869, [FeVII]λ6087). Finally, coronal (C) transitions ([FeX]λ6375) may appear. Each class has a number of subclasses, assigned from the strongest non-Balmer lines in the spectrum – for example, if [NeV]λ3426 is the strongest line in the nebular phase, the classification would be N_{ne}, denoting a neon nova.

Most novae fall into one of two classes: the 'FeII' and the 'He/N' classes according as the spectra near maximum have strong permitted lines of FeII or of HeI,II or NII,III. The 'FeII' spectra are formed in an optically thick wind from the white dwarf and show expansion velocities of from 1000 to 3500 km s^{-1}. The 'He/N' spectra originate in a discrete shell of gas expelled at the time of outburst and usually have expansion velocities exceeding 2500 km s^{-1}.

The novae with ejection velocities $\gtrsim$ 2500 km s^{-1} tend to evolve to Ne and/or coronal types; the lower velocities usually develop auroral spectra with N and O dominant and, constituting the bulk of novae, are designated 'standard'. A few novae evolve from initial 'FeII' type through 'He/N' type, which are designated as 'hybrids'. The relative frequencies of evolution are shown in Figure 5.10.

Continued classification of further novae, and reclassification of old novae, on this more physically-based scheme should assist quantitative interpretation of the evolution of nova spectra.

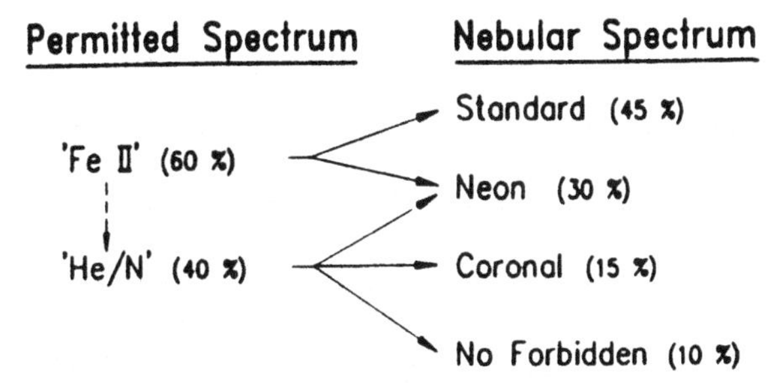

Figure 5.10 Evolution and relative frequencies of nova spectra. From Williams (1992).

Table 5.9. *Element Abundances (by Mass) in CN.*

Star	t_2	$X(H)$	$Y(He)$	Z	C	N	O	Ne	Na	Mg	Al	Si	S	Ar	Fe	References
V1370 Aql	6	0.053	0.088	0.859	0.035	0.14	0.051	0.52		6.7×10^{-3}		1.8×10^{-3}	0.10		4.5×10^{-3}	1
T Aur	80	0.47	0.40	0.13		0.079	0.051									2
V842 Cen	24	0.41	0.23	0.36	0.12	0.21	0.030	9.0×10^{-4}						9×10^{-4}		18,19
IV Cep	16	0.40	0.21	0.39												3,4
V693 CrA	6	0.29	0.32	0.39	0.0046	0.080	0.12	0.17	1.6×10^{-3}	7.6×10^{-3}	4.3×10^{-3}	2.2×10^{-3}				5,11
V1500 Cyg	2.9	0.57	0.27	0.16	0.058	0.041	0.050	0.010						4.5×10^{-4}		6
V1668 Cyg	12.1	0.45	0.23	0.32	0.047	0.14	0.13	0.0068								7
HR Del	152	0.45	0.48	0.077		0.027	0.047	0.0030								8
DQ Her	67	0.31	0.31	0.38	0.056	0.13	0.20	0.0037		7.4×10^{-4}	8.2×10^{-4}	4.6×10^{-4}	1.1×10^{-4}	1.5×10^{-3}		9
V827 Her	25	0.36	0.29	0.35	0.087	0.24	0.016	6.6×10^{-4}								19
CP Lac	5	0.60	0.26	0.14												3
DK Lac	19	0.47	0.06													10
GQ Mus	23	0.27	0.32	0.41	0.016	0.19	0.0034	1.4×10^{-3}	5.6×10^{-4}	2.8×10^{-3}	1.6×10^{-3}		4.7×10^{-4}			11
V2214 Oph	145	0.34	0.26	0.40		0.31	0.060	0.017								19
RR Pic	80	0.53	0.43	0.0039	0.022	0.0058	0.011									12
CP Pup	5	0.35:	0.17:	0.48:		0.17:										13,16
V977 Sco	40	0.51	0.39	0.10		0.042	0.030	0.026								19
V443 Sct	25	0.49	0.45	0.062		0.053	0.0070	1.4×10^{-4}								19
PW Vul	83	0.54	0.28	0.18	0.032	0.11	0.038	8×10^{-4}								11,14
QU Vul	27	0.44	0.46	0.10	0.0014	0.015	0.038	0.040		1.8×10^{-3}	2.4×10^{-3}	4.7×10^{-4}		3.0×10^{-5}		15
QV Vul	50	0.68	0.27	0.053		0.010	0.041	9.9×10^{-4}								19
Sun		0.67	0.31	0.027	0.0037	8.9×10^{-4}	8.7×10^{-3}	1.5×10^{-3}	3.1×10^{-5}	6.8×10^{-4}	5.8×10^{-5}	8.2×10^{-4}	3.5×10^{-4}	9.9×10^{-5}	1.4×10^{-3}	17

References: 1. Snijders *et al.* 1987; 2. Gallagher *et al.* 1980; 3. Ferland 1979; 4. de Freitas Pacheco 1977; 5. Williams *et al.* 1985; 6. Ferland & Shields 1978a, Lance, McCall & Uomoto 1988; 7. Stickland *et al.* 1981; 8. Tylenda 1978; 9. Martin 1989b; 10. Collin-Souffrin 1977; 11. Saizar *et al.* 1991; 17. Aller 1987; 18. de Freitas Pacheco, da Costa & Codina 1989; 19 Andrea 1992.

5.3.8 Geometry of the Ejected Envelope

The multiple absorption and emission systems seen on the early decline can be interpreted either as independently ejected spherical shells, or as non-spherical ejection. Interaction with the secondary star will lead to different behaviours in the equatorial and polar directions (e.g., Section 5.8.5), removing any initial spherical symmetry.

Hutchings (1972; see also Malakpur (1973) and Soderblom (1976)) modelled the line profiles of HR Del, FH Ser and LV Vul. HR Del was found to have two rings, inclined $\pm 15°$ to the equatorial plane, expanding at 500 km s^{-1}, and polar blobs moving outwards at 200 km s^{-1} with a viewing angle (inclination) of 62°. In contrast, FH Ser had dominating polar blobs ejected at 600 km s^{-1}, seen at an inclination of 40°. Mustel & Boyarchuk (1970) and Weaver (1974) similarly found equatorial rings and polar blobs in V603 Aql with $i \sim 20°$; they pointed out that DQ Her also had an equatorial band and polar blobs.

Further discussion is given in Section 5.6.1.

5.3.9 Ejecta Abundances and Masses

Abundances in novae can be derived independently, and with different applicable physics, at the various stages of outburst. The absorption lines seen at maximum light can be subjected to a curve of growth analysis, as in DQ Her (Mustel & Baranova 1965), but the necessity of using saturated lines formed in a non-uniform macrovelocity distribution and uncertainties in excitation and ionization temperatures make such analyses very uncertain (Williams 1977; cf. Boyarchuk & Antipova 1990).

Most nova compositions have been derived from emission lines. The superficial similarity with planetary nebulae is misleading – nova spectra show multiple components which must be resolved at high resolution, which then show the presence of differing physical conditions and compositions for the individual blobs of gas. Effects of self-absorption in the strongest permitted lines are often present. The techniques of abundance analysis have been reviewed by Collin-Souffrin (1977), Williams (1977, 1985), Peimbert & Sarmiento (1984), Martin (1989a,b,c), Boyarchuk & Antipova (1990), Snijders (1990) and Andreä (1992). Resultant abundances are given in these same references and in Truran (1985), Truran & Livio (1986) and Livio & Truran (1992).

Table 5.9 summarizes the abundances obtained for the best-observed novae[2]. He/H ratios are available for some other novae (e.g., Collin-Souffrin 1977). Solar abundances have been deduced in a very slow nova in M31 (Tomaney & Shafter 1993). The 'heavy element' abundance Z is simply $1 - X - Y$ and is dominated by CNO Ne, which, together with the helium abundance Y, differ greatly from solar and vary from nova to nova. The origin of the anomalous abundances is discussed in Section 5.8.2 and the correlation between Z and t_2 is considered in Section 5.8.3.

Abundances can also be obtained from spatially resolved nova shells (Section 5.6.3). Boyarchuk & Antipova (1990) conclude that these are the most reliable of the methods. Snijders *et al.* (1987) point out that abundances derived after dust has formed in the nova shell will show selective depletions in the gas phase, especially of O, Mg and Si.

[2] There are often differences of factors of 2 or more between observers (Livio & Truran 1994). The latest or most comprehensive analyses have been selected here.

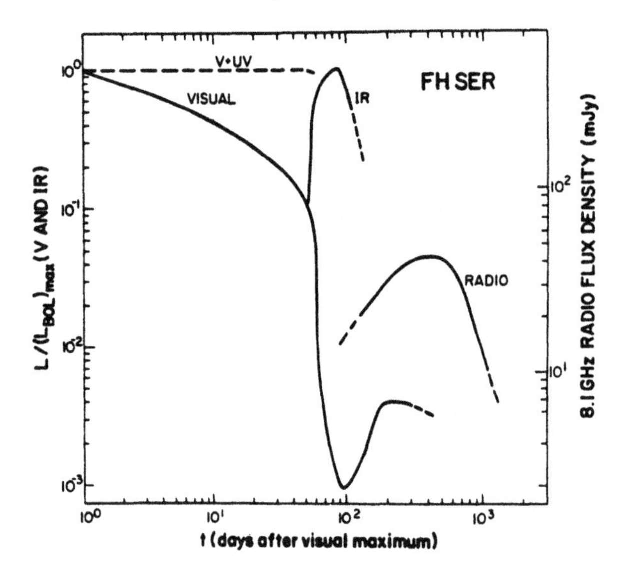

Figure 5.11 Evolution of radio flux in FH Ser, compared with other wavelengths. From Bode (1982).

Ejecta masses can be obtained by deriving N_e and T_e from the integrated luminosities of permitted and/or forbidden lines, deducing N_e/N_H from the ionization conditions, and the total mass from the relative abundances (e.g., Martin 1989a,b,c). Of course, in the early stages of eruption the derived mass refers only to the gas above unit optical depth; only when the envelope has become transparent are total masses ejected (to that time) measured. Final ejected masses (after correction for possible depletions) are in principle only found from radio fluxes (Section 5.4) or resolved shells (Section 5.6.3).

Table 5.7 summarizes measured envelope and shell masses. The results are too varied to detect any correlation with t_2. It appears that all novae eject masses $\sim 1 \times 10^{-4}$ $M_\odot$ with a possible range of a factor of 3 either side.

5.4 Radio Observations of Novae

Radio emission from novae was first observed from HR Del and FH Ser (Hjellming & Wade 1970); the fact that HR Del was still emitting three years after its eruption showed that at radio wavelengths novae continue to be detectable for much longer times. Figure 5.11 illustrates the relative luminosities at different wavelengths and their evolution. Even at their brightest, radio novae have low flux densities and therefore require the largest telescopes for their study.

Radio observations of novae have been reviewed by Hjellming (1974, 1990) and Seaquist (1989).

5.4.1 Radio Light Curves

Radio light curves have been produced for V1370 Aql, Nova Aql 1993, V1500 Cyg, V1819 Cyg, V1974 Cyg, HR Del, V827 Her, V351 Pup, FH Ser, PW Vul and QU Vul[3].

[3] In addition, HR Del, GK Per, V4077 Sgr, V368 Sct and NQ Vul have been detected many years after eruption (Seaquist 1989)

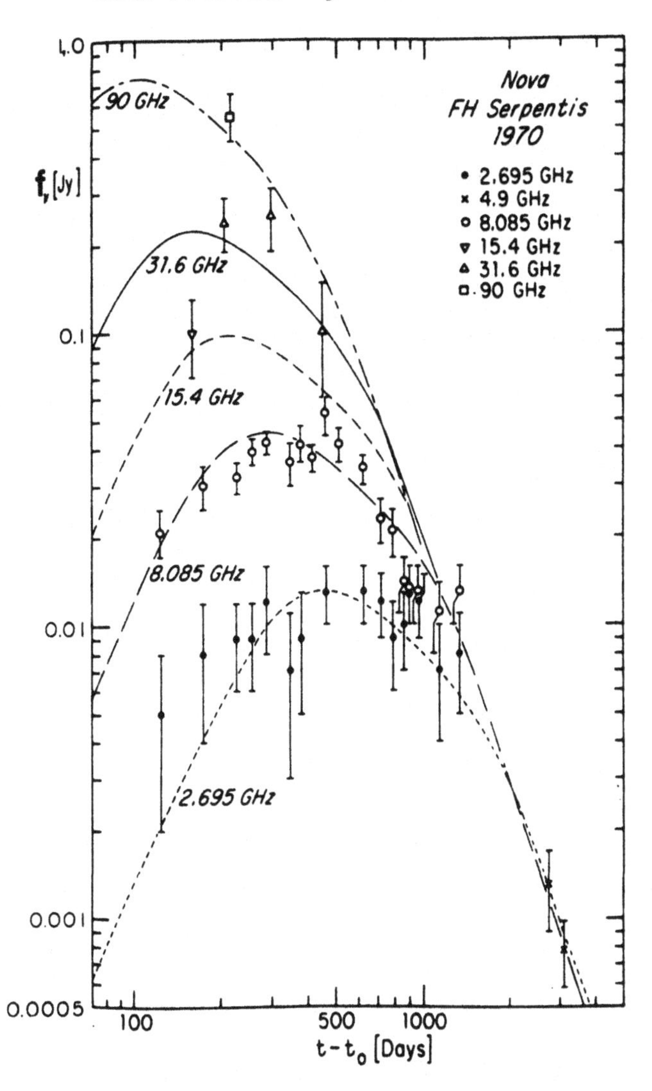

Figure 5.12 Radio light curves at different frequencies for FH Ser. Theoretical curves are for spherical shell models with inverse square density profile. From Hjellming *et al.* (1979).

All show, at different speeds, the temporal development seen in Figure 5.12. Flux maximum occurs at progressively later times at lower frequencies, and occurs much later than optical maximum. This is simply a result of the much higher opacities at radio wavelengths. The brightness temperatures on the rising part of the light curve are $\sim 10^4$ K, which are those expected of thermal (free–free) emission from an ionized shell.

As in the optical, at first the expanding envelope is optically thick; this is superseded by a transition phase in which the apparent photosphere recedes, leaving an optically thin shell. For a simple uniformly expanding spherical shell, in the three regimes the flux is proportional to $\nu^2 t^2$, $\nu^{0.6} t^{-4/3}$ and $\nu^{-0.1} t^{-4/3}$ respectively (Hjellming 1990). Observed departures from these simple relationships can occur because of deceleration of the expanding envelope or from the effects of radial gradients in density through the envelope. The fact that all nova envelopes eventually become optically thin shows that sustained mass loss from the primary only lasts for a finite time. Equation (5.7) can be applied to the observed angular expansion rate to provide a radio distance d_{rad}; this agrees well with optical determinations (Seaquist 1990).

Radio light curves can be successfully modelled in various ways, the extremes being instantaneous ejection of a shell with uniform velocity gradient and optically thick wind with variable $\dot{M}_{wind}$ (Seaquist & Palimaka 1977; Kwok 1983; Hjellming *et al.* 1979). Hjellming (1990) shows that a combination of the two can fit all gross characteristics of both radio and optical light curves.

An exception was QU Vul, which for the first 300 d had radio emission with a brightness temperature $\sim 10^5$ K, which is interpreted as bremsstrahlung from a shock propagating through the initial ejected shell, the shock having been generated by impact of a later, higher velocity ejection (Taylor *et al.* 1987).

The evolution from optically thick envelope to optically thin shell has been followed in 0.45–2.0 mm observations of V1974 Cyg (Ivison *et al.* 1993). It is pointed out that light curves at these wavelengths are particularly sensitive to the way the photosphere recedes through the outflowing gas, which is of potential value in modelling.

A major advantage of obtaining the radio flux in late stages of the radio optically thin regime is that this is generated by the *entire ejected mass*. However, the interpretation is uncertain because the electron temperature of the gas is not directly determinable from the radio observations (the derived masses scale as $T_e^{0.58}$). Typical masses for $T_e = 10^4$ K range from 5×10^{-5} to 2.4×10^{-4} $M_\odot$, which are larger by up to a factor of 10 than found from optical and IR studies (Table 5.7). The problem of unknown T_e is ameliorated when the nova shell can be resolved in the radio region.

5.4.2 Radio Images

With the Very Large Array, the expanding shell of the neon nova QU Vul was resolved after 497 d (Taylor *et al.* 1987) and its development followed until 1017 d after maximum (Taylor *et al.* 1988: Figure 5.13). The early elongated (bipolar) structure turned into a hollow spherical shell at later stages. For spherical symmetry the brightness temperature is related to the electron temperature through $T_B(\theta) = T_e(1 - e^{-\tau(\theta)})$, where θ is the angular radius (r/d). The observed $T_B(\theta)$ gives the optical depth $\tau(\theta)$, which is simply related to the *total* mass of the shell M_T. For QU Vul the result is $T_e = 1.7(\pm 0.1) \times 10^4$ K and $M_T = 3.6(\pm 0.5) \times 10^{-4}$ $M_\odot$. The deduced expansion velocities (obtained from $\theta_r(t)$ and $d = 3.6$ kpc) of the inner and outer radii of the shell are 880 km s^{-1} and 1010 km s^{-1} respectively. This is in good agreement with the 1030 km s^{-1} expansion velocity deduced from OI emission by Andrillat (1985).

5.5 X-Ray Observations

Prior to the launch of EXOSAT only nova *remnants* had been detected by X-ray satellites (Becker 1989). Observations near optical maximum of V1500 Cyg, NQ Vul and PW Vul gave upper limits $\ll L(\max)$ for any X-ray production, but GQ Mus, PW Vul and QU Vul were detected as soft X-ray ($kT \sim 1$ keV) sources from ~100 to 1000 d after eruption (Ögelman, Krautter & Beuerman 1987; Ögelman 1990). Although

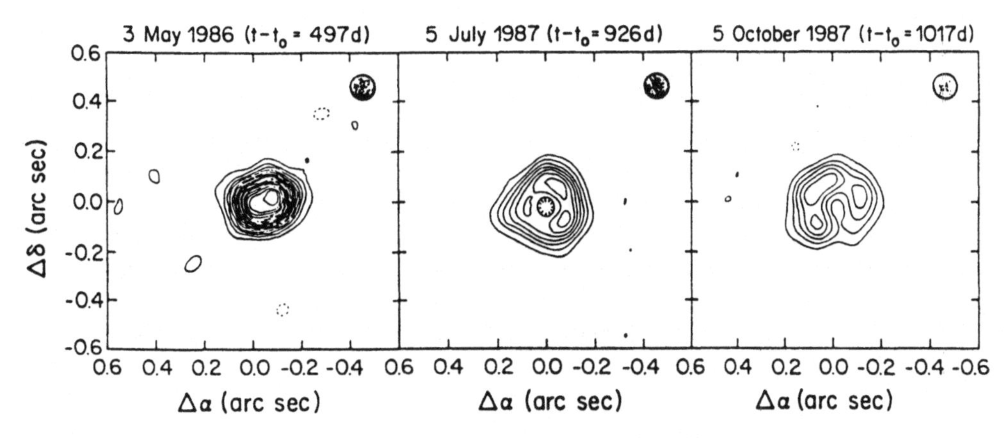

Figure 5.13 Radio images of QU Vul at a wavelength of 2 cm (15 GHz); from left to right 497 d, 925 d and 1017 d after maximum. Shaded circles are half-power beam resolutions. From Taylor *et al.* (1988).

initially interpreted as originating from the heated surface of the white dwarf (and therefore simply a continuation of the shortward movement of flux maximum (Section 5.2.5), the predicted flux curves are far more peaked than the observed ones, which opens the possibility that the X-rays are caused by the passage of a shock wave similar to what is inferred from the radio observations (Ögelman 1990).

The observation of hard X-rays 5 d after the discovery of V838 Her (Lloyd *et al.* 1992) supports the latter view. Hα emission was observed with expansion velocity $\sim$4500 km s^{-1}, which would generate $T_{sh} \sim 20$ keV (equation (2.56)) for a relative velocity $\sim$4000 km s^{-1}. This is compatible with the observed X-ray spectrum.

GQ Mus is currently an ultrasoft ($kT = 29$ ev) X-ray source: see Section 5.8.6.

5.6 Nova Shells

From equation (5.7) the angular radius of the expanding nova shell is

$$\theta_r(t) = 0''.207\left(\frac{v_{exp}}{10^3 \text{ km s}^{-1}}\right) d(\text{kpc})^{-1}(t - t_0) \qquad (5.10)$$

where $(t - t_0)$ is the time elapsed in years. A typical shell is therefore rapidly within the resolution range of optical very long baseline interferometry ($\Delta\theta \sim 0''.001$) or speckle interferometry ($\Delta\theta \sim 0''.01$), soon within range of the HST ($\Delta\theta \sim 0''.06$) and eventually ($\sim$ 10–50 y) within range of ground-based direct imaging. Cohen (1988) points out, however, that the surface brightness of the shell is $\propto [v_{exp}(t - t_0)]^{-5}$, which leads to the criterion that nova shells will only be both resolvable and detectable by terrestrial direct imaging if $m_v - A_v \lesssim 7.5$ and $t \gtrsim 15$ y. Adaptive optics technique will ameliorate these requirements (Wade 1990).

Optical interferometry with a 19.6 m baseline successfully resolved V1794 Cyg only 10 d after eruption, with $\theta_r = 0''.0019$ in the continuum and $\theta_r = 0''.0025$ in Hα (Quirrenbach *et al.* 1993). The Big Optical Array (under construction) will resolve novae with $m_v(\max) \lesssim 7$ even earlier in eruption.

Speckle interferometry was successfully applied to V1500 Cyg (Blazit *et al.* 1977) but no details have been published. Speckle images of QU Vul were obtained in the 3.02 μm [Mg VIII] emission line 925 d after eruption, giving a diameter $0''.6 \pm 0''.1$ FWHM and showing that the coronal spectrum arises from the expanded region that produced the principal spectrum (Greenhouse *et al.* 1990). The HST successfully imaged Nova Cyg 1992 467 d after eruption, with a shell radius of $0''.13$ (Paresce 1993).

A list and bibliography of 26 novae for which shell images have been obtained is given by Wade (1990). An atlas of nova shells is in preparation (Seitter & Duerbeck 1987). Ellis, Grayson & Bond (1984) note a possible nova shell around the NL BZ Cam, and Takalo & Nousek (1985) suggest that a nebulosity coincident with the X-ray source E2000 + 223 is a NR.

5.6.1 Expansion Parallaxes

Equation (5.10) provides an estimate d_{se} of the shell expansion distance for resolvable nova shells. The expansion velocity may be obtained from v_{prin} or from velocities measured in the resolved shell itself (Section 5.6.2). Figure 5.14 shows the shell around DQ Her, which illustrates the departure from spherical symmetry. This highlights the difficulty of knowing what v_{exp} to associate with a particular θ_r: e.g., for an oblate spheroid the maximum observed θ_r is always the semi-major axis, whereas the observed v_{exp} is in general less than the expansion velocity in the equatorial plane. Clearly, in order to deduce an accurate d_{se} the geometry of the shell must be known so that the tangential velocity may be inferred from the observed RV. In general such niceties have not been attended to (Cohen & Rosenthal 1983; Cohen 1985; Ford & Ciardullo 1988).

Another process that might be thought to affect shell sizes is deceleration of the outer boundary of a shell as it runs into the interstellar medium. This was treated theoretically by Oort (1946), who concluded that significant decrease in v_{exp} should occur after $\sim$50 y. Duerbeck (1986) finds confirmatory evidence, with greater decelerations for higher expansion velocities. However, Ferland (1980) points out that such deceleration would be accompanied by a $\sim 10^6$ K shock with associated coronal lines, which are not observed, and Herbig & Smak (1992) consider that, in the case of DQ Her, Deurbeck's result can be attributed to omission of a seeing correction in measuring nebula dimensions.

5.6.2 Geometry of Nova Shells

Spatially resolved spectra of nova shells provide projected expansion velocities for the individual parts of the shell, which, together with the measured θ_r of the parts, lead to a spatio-kinematic model. The distribution of emission from different atomic species can be obtained by recording images in the light of individual emission lines. Two-dimensional imaging spectroscopy has been achieved for some nova shells (Barden & Wade 1988).

The structures of only a few nova shells have been determined. These include V603 Aql (Mustel & Boyarchuk 1970; Weaver 1974), HR Del (Solf 1983), DQ Her (Mustel & Boyarchuk 1970; Cohen 1988; Ford & Ciardullo 1988), and GK Per (Duerbeck & Seitter 1987). Solf's work confirms the model produced by Hutchings from eruption spectra (Section 5.3.8). The deduced inclinations are similar, within errors, to those obtained from spectroscopic and photometric observations of the central binaries

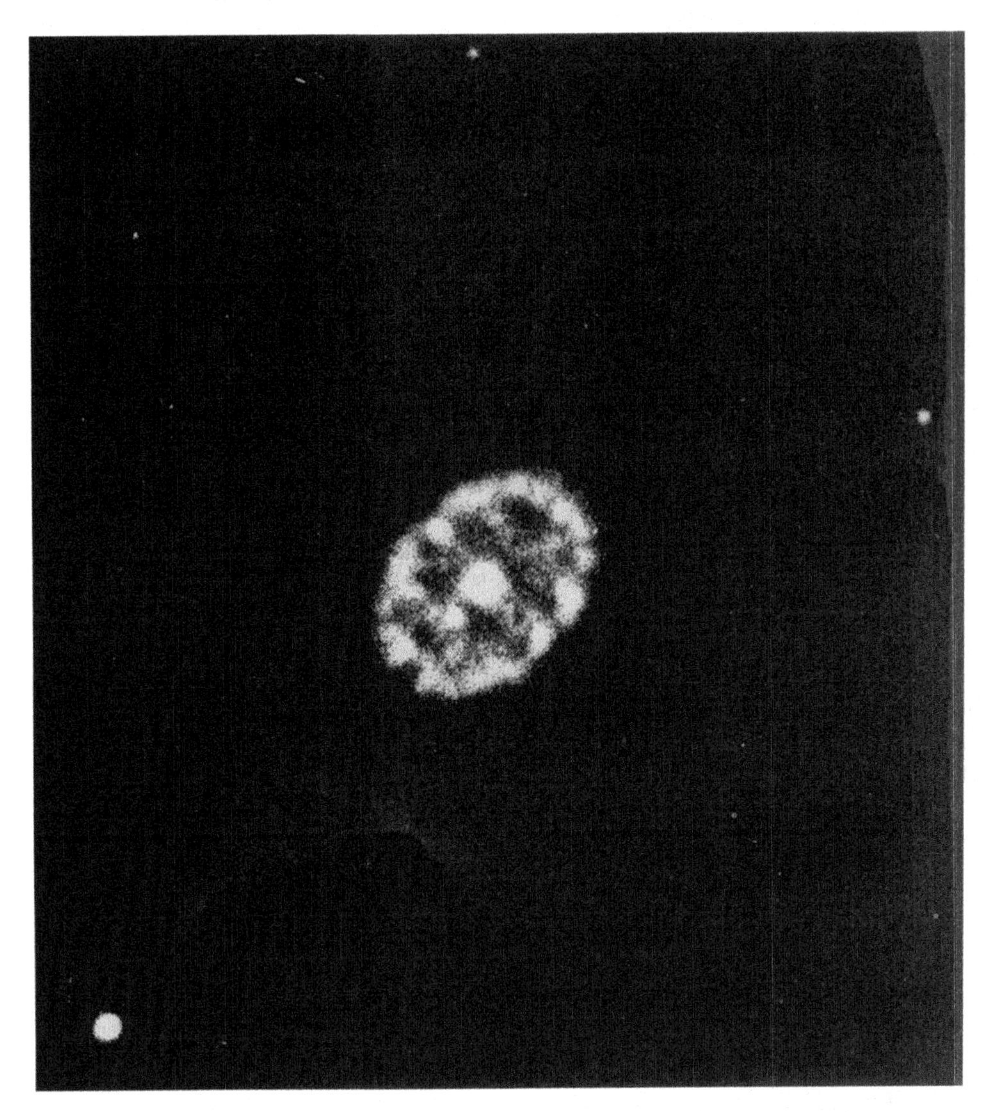

Figure 5.14 The shell around DQ Her, photographed by R. E. Williams in 1977.

(Table 4.2). With the exception of HR Del, which is oblate, nova shells are found to have prolate geometry – i.e. the polar axis is longer than the equatorial axis.

5.6.3 *Abundances in Nova Shells*

Spectra of individual pieces of spatially resolved nova shells have been used to deduce abundances and physical conditions in T Aur (Gallagher *et al.* 1980), DQ Her (Williams *et al.* 1978; Ferland *et al.* 1984), GK Per (Evans *et al.* 1992), RR Pic (Williams & Gallagher 1979; Evans *et al.* 1992), CP Pup (Williams 1982) and the RN T Pyx (Williams 1982). RR Pic and T Pyx have nebular emission spectra typical of $T_e \sim 10^4$ K gas, but T Aur, DQ Her and CP Pup have $T_e \sim$ 500–800 K. These low temperatures are the result of the exceptionally strong cooling produced by collisionally excited forbidden line emission from the abundance-enhanced CNO ions (Smits 1991a,b). The overabundances deduced by Smits from detailed modelling of the nova shells are lower than found in the earlier studies, but are nevertheless substantial. A total ejected mass for DQ Her has been deduced (Table 5.7).

In DQ Her Williams *et al.* (1978) found that forbidden line emission is strong on both axes of the nova shell, but permitted lines of CNO are prominent only along the major axis, indicating different abundances in different ejecta. Ferland *et al.* (1984) combined UV and optical spectra to demonstrate that the shell is photoionized by the central source (i.e., the DQ Her binary) and deduced its EUV flux distribution, finding no resemblance to computed disc spectra.

These and other discordant results have been largely removed by the detailed photoionization model of Petitjean, Boisson & Péquignot (1990). They find that the ionizing source has $T_{\rm eff} = 1.0 \times 10^5$ K with a total luminosity of 1.7×10^{34} erg s^{-1}. The illumination of the minor axis of the prolate ellipsoidal shell is ~0.25 that of the major axis, which is attributed to the shadow cast across the nebula by the accretion disc. In this model the different spectra along the major and minor axes of the shell are the result of the different photoionization conditions; the ejecta have the same composition everywhere.

A disc angular width of $\pm 8°$ is required to produce the shadow. This is larger than standard high $\dot{M}$(d) discs (Section 2.6.1.3). Petijtean *et al.* suggest a thickening of the disc near the radius r_r caused by the stream overflowing the disc (Section 2.4.3); however, this is incompatible with observations of the 71 s oscillations in DQ Her which show that the *whole* of the disc is illuminated by the central source (Section 8.1.4). It is pointed out in Section 2.9.6 that the effective disc thickness for transverse radiation is much greater than for vertical radiation.

If the ionizing source is an accretion disc boundary layer, then the measured luminosity, together with equation (2.54a) and $\Omega(1) = 0.0885$ s^{-1} (Section 8.1.1), gives $\dot{M}(1) = 1.0 \times 10^{-8} \mathrm{M}_\odot$ y^{-1}. A similar result is obtained by treating the ionizing source as accretion from the corotation radius to the white dwarf via an accretion column (Sections 7.2 and 8.5).

5.7 Light Echoes

For eight months after the eruption of GK Per nebulous patches were seen, apparently receding from the nova at $\sim 20'$ y^{-1} (see Campbell & Jacchia (1941) and Hessman (1989b) for illustrations). At the distance of GK Per this angular velocity corresponds linearly to the velocity of light, so it soon became evident (Kapteyn 1901) that illumination of surrounding dust clouds by the pulse of maximal light was the cause. Couderc (1939) analysed the geometry of the light echo and deduced that the dust layer was a sheet lying between the nova and the observer. Apparent superluminal expansions are possible. Only one other nova has been observed to have a light echo – V732 Sgr (Swope 1940) – although many have been looked for (van den Bergh 1977; Schaefer 1988).

The echoed image represents all reflections which have the same total path length from nova to observer. The locus of reflections is an ellipsoid with the nova at one focus and the observer at the other – the geometry is given by Schaefer (1987). In consequence, it is not possible to derive a distance from the observed angular radius of the echo.

IR echoes have been postulated as an alternative explanation of the IR emission from novae (Bode & Evans 1983; Dwek 1983), but the absence of optical echoes for the majority of novae argues against this (Gehrz 1988).

5.8 Theory of Nova Eruptions

The continued accumulation of hydrogen-rich material on the surface of a white dwarf leads inevitably to a thermonuclear runaway at the boundary of the degenerate core. This was first emphasized by Kraft (1962), based on Mestel's (1952) work on the structure and stability of white dwarfs. Although Kraft (1964a) later suspected that energy produced in this manner would be conducted away into the degenerate core, the extension of Mestel's theory by Giannone & Weigert (1967) and Saslaw (1968) showed that the runaway takes place in a region of partial degeneracy which does not have sufficiently high conductivity.

Earlier theories of nova eruptions are reviewed by Schatzman (1965). More recent reviews of the theory of the nova process are given by Gallagher & Starrfield (1978), Truran (1981, 1982), Starrfield (1986,1988,1989,1990,1992), Shara (1989) and Livio (1993b).

5.8.1 The Nova Process

A crucial part of the physics of the thermonuclear runaway is contained in the equation of state for degenerate matter, $P \propto \rho^{\gamma}$, which is independent of temperature. If the temperature and density are sufficiently high for nuclear reactions to occur, any small increase in temperature, leading to enhanced energy generation but no increase in pressure, is amplified in an exponential runaway. This is only terminated when the Fermi temperature T_F is reached, whereupon the equation of state becomes that of a perfect gas and is wildly out of balance, so expansion occurs to reduce the overpressure.

The non-degenerate envelope of a white dwarf is very thin, so the pressure at the base of the envelope is simply

$$P_b = \frac{GM(1)M_{env}}{R^2(1)} \frac{1}{4\pi R^2(1)}, \tag{5.11}$$

where M_{env} is the mass of the envelope. Fujimoto (1982a,b) and MacDonald (1983) show that the expansion rate depends mainly on P_b: if $P_b \gtrsim P_b(crit) = 1 \times 10^{20}$ dyn cm^{-2} for solar abundances of CNO then the expansion reaches escape velocity and the envelope is ejected. From equation (2.83a) $P_b \,\tilde{\propto}\, M^{7/3}(1)M_{env}$, whereas the binding energy $GM(1)M_{env}/R \propto M^{4/3}(1)M_{env}$, so a nova eruption occurs at lower M_{env} as $M(1)$ increases. The envelope mass M_{env}(crit) required to produce P_b(crit) can be obtained from equations (2.82) and (5.11) and decreases from $4.7 \times 10^{-3} M_\odot$ at $M_1(1) = 0.6$, to $3.1 \times 10^{-5} M_\odot$ at $M_1(1) = 1.3$ (Starrfield 1989); an analytic approximation to numerical models (Livio 1993b) is (for solar abundances)

$$M_{env}(\text{crit}) = 1.7 \times 10^{-4} R_9^{2.8}(1) M_1^{-0.7}(1) \quad M_\odot. \tag{5.12}$$

Two conclusions can be drawn from this. Following a nova eruption, a recurrence would be expected after a time

$$T_R = M_{env}(\mathrm{crit})/ < \dot{M}(1) > \qquad (5.13)$$

where $< \dot{M}(1) >$ is a long term average; and the number of eruptions that a CV can have is large, $\sim M_i(2)/M_{env}(\mathrm{crit})$ where $M_i(2)$ is the mass of the secondary at commencement of mass transfer (and $M_{env}(\mathrm{crit})$ must be a suitable time-average if $\dot{M}(1)$ is not constant over time). $M_{env}(\mathrm{crit})$ is similar in order of magnitude to the observed ejected masses from novae (Table 5.7).

There are a number of additional factors that complicate the nova process. If the CV system were young enough for the primary still to have substantial intrinsic luminosity the temperature gradient would be relatively steep, so the temperature at the base of the envelope would be high and runaway would occur at a lower P_b, ejecting smaller M_{env}, and would occur in less degenerate layers and be therefore less violent (e.g., Kovetz & Prialnik 1985). $\dot{M}(1)$ similarly influences the nova process by (a) compressional heating at the base of the envelope and (b) heating of the outer part of the envelope by the boundary layer (Prialnik *et al.* 1982; Shaviv & Starrfield 1987). The lower observed BL temperatures (Section 2.7.3) will reduce the effect (Regev & Shara 1989) of the latter, from which it was deduced that nova eruptions should not occur if $\dot{M}(1) \gtrsim 1 \times 10^{-9} \mathrm{M}_\odot\ \mathrm{y}^{-1}$ (Prialnik, Kovetz & Shara 1989). More recent computations, including compressional heating, give *no envelope ejection* for $\dot{M}(1) \gtrsim 7 \times 10^{-9}[2.54 - M_{ch}/M(1)]\ \mathrm{M}_\odot\ \mathrm{y}^{-1}$ (Livio 1992a; Livio & Truran 1992). A comparison of the absolute magnitudes of nova and NL discs (Figures 4.16 and 4.20) with the $\dot{M}(\mathrm{d})$ deduced for NLs (Section 4.33) gives $\dot{M}(\mathrm{d}) \lesssim 1 \times 10^{-8} \mathrm{M}_\odot\ \mathrm{y}^{-1}$ for pre-novae, in agreement with the above condition.

It may be noted that the hibernation scheme (Section 9.4.3) in which CVs remain at very low $\dot{M}(1)$ for millenia or more, can produce strong thermonuclear runaways even if $\dot{M}(1)$ is $\gtrsim 10^{-8} \mathrm{M}_\odot\ \mathrm{y}^{-1}$ after emergence from hibernation (Prialnik & Shara 1986; Livio, Shankar & Truran 1988).

5.8.2 Nuclear Reactions

After a short lived phase where the proton–proton chain generates energy, the temperature at the base of the envelope rises to $\sim 2 \times 10^7$ K and CNO reactions become important. As in the cores of high mass stars, the temperature sensitivity ($\tilde{\propto}\ T^{18}$) of the energy generation rate causes convection. The convective turn-over time scale (obtained from time-dependent mixing-length theory: Sparks, Starrfield & Truran 1978) is $\sim$100 s and at $T \sim 8 \times 10^7$ K the envelope is fully convective, so a significant number of β^+-decay nuclei (see below) can reach the surface before the decay. The nuclear reactions are far from their equilibrium rates, requiring a detailed reaction network (Figure 5.15) to follow their progress. The longest lived β^+ unstable nuclei in this network are ^{13}N, ^{14}O, ^{15}O and ^{17}F, with half-lives of 863 s, 102 s, 176 s and 92 s respectively. These become the most abundant nuclei in the envelope when $T \gtrsim 10^8$ K. Convection both transports fresh CNO into the high temperature zone and takes the β^+ unstable nuclei near to the surface, where their decay deposits large amounts of energy. This is the principal cause of expansion of the envelope and of much of the total luminosity of the eruption.

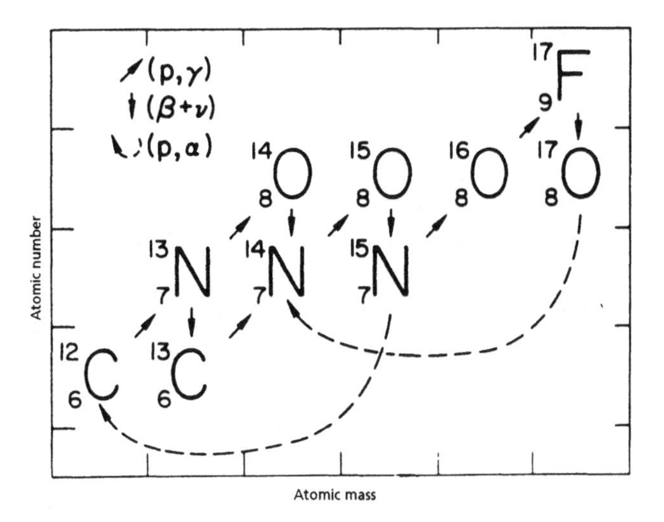

Figure 5.15 Nuclear reaction network in the CNO cycle. From Starrfield (1989).

The inertia of the envelope results in the temperature overshooting $T_F(\sim 8 \times 10^7$ K) by factors of 2–4, following which envelope expansion cools and prevents further nuclear burning. For greatly enhanced CNO abundances all of the possible proton captures occur before envelope expansion and this terminates the runaway. The $\gtrsim 10^{46}$ erg deposited in the envelope by the β^+ unstable nuclei begin to be radiated while the envelope is still only expanding slowly. The $\sim 10^5 L_\odot$ radiated through the small surface area generates $T_{\rm eff} \sim (0.5–1) \times 10^6$ K. At this stage, therefore, novae are predicted to be very luminous EUV and soft X-ray sources, and their luminosities exceed the Eddington limit (equation (5.4b)), which converts mere expansion into a radiatively-driven wind.

A comparison of the energy released in the CNO reactions with the binding energy of the envelope (Truran 1982) shows that only if CNO are considerably overabundant is there enough energy available for *rapid* ejection; otherwise the expansion is relatively slow. In consequence, models predict that fast novae will always be associated with enhancements of the decay products of the β^+ unstable nuclei, namely ^{13}C, ^{14}N, ^{15}N and ^{17}O, and that their abundances will be far from the equilibrium CNO cycle abundances characteristic of solar composition. Sneden & Lambert (1975) show from CN band spectra that ^{13}C and ^{15}N were almost certainly present in considerable abundance during the eruption of DQ Her.

One origin of the CNO enhancements and the high Ne abundances observed in Ne novae is dredge-up from the core of a massive white dwarf. A number of mixing mechanisms have been proposed (reviewed by Fujimoto & Iben (1992), Iben (1992) and Livio (1993b)), including shear mixing and/or diffusion and convective mixing between envelope and core during the time between eruptions, and convective overshoot during the initial phases of the thermonuclear runaway. CO enhancements will thereby occur from CO cores already stripped of their He outer layers by the first eruptions, and Ne enhancements arise from ONeMg cores which have been stripped of both overlying He and CO layers. If this were the only way of making Ne novae, the implications would be that (a) nova eruptions on ONeMg cores should be *relatively common*, which, as ONeMg white dwarfs themselves are the 1.2–1.4$M_\odot$ products of evolution of 8–12$M_\odot$ stars (e.g., Truran & Livio 1989) and are therefore *comparatively rare*, must be the result of the much greater frequency of eruptions on high-mass white dwarfs (equations (5.12) and (5.13)), and (b) part of the core is ejected in each eruption and therefore the more massive *CV primaries* would *secularly decrease in mass* (see Section 9.7.3 for possible exceptions to this conclusion). Computations of the expected relative frequencies of novae as a function of $M(1)$ predict that about one half should occur in the range $1.25 \leq M_1(1) \leq 1.40$ or about one third in $1.35 \leq M_1(1) \leq 1.40$ (Politano *et al.* 1989).

Another way in which modest overabundances of Ne can occur is through production of ^{22}Ne during the He-burning phase in the red giant precursor of the CV primary. Ne enrichments of factors ~3–10 can then occur in lower mass CO white dwarfs (Livio & Truran 1994).

There is, however, a further way of making ONeMg-rich white dwarfs, including ones of low mass. Runaway mass transfer from the secondary (at $\sim 10^{-6}M_\odot$ y^{-1}) for systems that emerge from common envelope evolution with $M(2) > M(1)$ (described in Section 9.3.1) results in mild He flashes which build an outer layer rich in Ne and Mg (Shara & Prialnik 1994). The existence of relatively low mass primaries among the Ne nova is shown by (a) the presence of slow Ne novae (Section 5.3.6) and (b) large ejected shell masses for some Ne novae (e.g., QU Vul: Table 5.7).

The bias towards high $M(1)$ for *observed* novae means that they are not a fair sample of CVs. The majority of CVs have $M_1(1) \sim 0.6$ (Section 2.8.2) and should have relatively infrequent eruptions as slow novae. However, extreme enhancements of CNO can counteract the effect of lower mass: e.g., DQ Her has a well-determined mass $M_1(1) = 0.60 \pm 0.07$ (Horne, Welsh & Wade 1993), $Z = 0.38$ (Martin 1989b) and $t_2 = 67$ d. The existence of the slow novae V450 Cyg ($t_2 = 110$ d) and V2214 Oph ($t_2 = 145$ d) among the Ne novae (Section 5.3.6) also implies very large CNO enhancements on low mass primaries.

The amount of mixing between core and envelope that occurs between eruptions is affected by $\dot{M}(1)$. Chemical diffusion moves heavy elements from the core into the accreted envelope and H inward into the core. The depth of penetration of H into the core and the resulting mass ΔM_H for which $X > 0.005$, is proportional to T_R, which, through equations (5.12) and (5.13), gives $\Delta M_H/M_{env}(\mathrm{crit}) \propto \langle \dot{M}(1) \rangle^{-1}$. This ratio determines Z (i.e., the CNONe abundances) of the ejected envelope. More detailed computations of the Z-profile at the onset of thermonuclear runaway are given by Kovetz & Prialnik (1985).

Iben, Fujimoto & MacDonald (1992b) consider that during quiescence further mixing occurs because of the development of convective zones in the outer part of the CO core. Convective overshoot and diffusion lead to greater mixing of H into the core than for diffusion alone, producing even higher values of Z in the nova eruption. Instead of Z decreasing with increasing $\dot{M}(1)$, as is the case for pure diffusion (Kovetz & Prialnik 1990), Z increases, being a factor of 22 times solar at $\dot{M}(1) = 10^{-10}M_\odot$ y^{-1} and a factor of 30 for $\dot{M}(1) = 10^{-8}M_\odot$ y^{-1} for runaways on a 1.25 $M_\odot$ CO white dwarf.

Fujimoto & Iben (1992; see also Iben 1992); point out that nova abundances fall into two categories: group A (IV Cep, V1500 Cyg, V1668 Cyg, CP Lac, CP Pup and PW Vul) which have roughly solar $Y/X = 0.5 \pm 0.03$, and group B in which $Y/X = 1.0 \pm 0.15$ (only V842 Cen falls between these groupings). There is no correlation between group and t_2. Group A is interpreted as the result of large M_{env} accreted at $\dot{M}(1) \lesssim 2 \times 10^{-10} M_\odot$ y^{-1}, and group B as relatively small M_{env} accreted at $\dot{M}(1) > 1 \times 10^{-9} M_\odot$ y^{-1}. In group A, P_b(crit) is estimated to be large enough for dynamical mass ejection to have taken place; in group B, P_b(crit) $\lesssim 10^{19}$ dyn cm^{-2} and the H flash would be too weak to eject mass dynamically.

Although there is no correlation between t_2 and the Y group, there are correlations between t_2 and M_{env} and between t_2 and the degree of mixing between core and envelope. Fujimoto & Iben deduce that fast novae occur in two ways: among group A stars with massive envelopes and among group B stars with envelopes of very small mass. In the former the fast decline is a result of large P_b(crit) and consequent strong shell flash; in the latter it is a result of the relative ease of ejecting a small envelope mass.

5.8.3 *The Nova Progress*

Maximum bolometric luminosity is reached at the peak of the thermonuclear runaway, typically on a time scale of hours. Visual maximum occurs when the photosphere of the expanding envelope reaches its largest radius, which, from the observed T_{eff} and luminosities (Section 5.2.5), is $\sim 10^{12}$–10^{13} cm. The time difference between the two maxima depends simply on the rate of expansion and is the explanation of the observed correlation between $t_{r,2}$ and t_2 (equation (5.3)).

The envelope undergoes hydrodynamic expansion which, in the most rapid and energetic novae, can be accelerated by propagation of a shock wave (Sparks 1969). Spherically symmetric models produce rise times to visual maximum that are much shorter than observed. A possible reconciliation arises through considering the time scale of propagation of a *local thermonuclear runaway* over the surface of a white dwarf (Shara 1982; Shankar, Arnett & Fryxell 1992).

During the nuclear burning phase, the structure of a nova resembles that of the central stars of planetary nebulae evolving towards white dwarfs – but with the direction of evolution reversed. Degenerate cores with low mass hydrogen envelopes in quasi-hydrostatic equilibrium obey a simple core mass–luminosity relationship (Paczynski 1971):

$$L = 5.92 \times 10^4 [M_1(\text{core}) - 0.522] \quad L_\odot \quad M_1(\text{core}) \geq 0.60. \tag{5.14}$$

Luminosities for lower mass envelopes are given by Schönberner (1983).

Hydrodynamic models of nova eruptions follow this same relationship (Truran 1982), which gives luminosities comparable to L_{Ed} (equation (5.4b)). The invariance of M_1(core) is the underlying cause of the constant bolometric luminosity phase of nova eruptions, which was predicted (Starrfield, Sparks & Truran 1976) before it was observed (Section 5.2.6). Prior to the plateau stage many novae show super-Eddington luminosities (Section 5.2.5).

Numerical models show that the maximum luminosity L(max) is a function of $M(1)$, $\dot{M}(1)$, (X, Y, Z) and the intrinsic luminosity of the white dwarf core. The most important of these is $M(1)$. Livio (1992c) finds that the models may be approximated by

$$\frac{L(\max)}{L_{\rm Ed}} \approx \left[\frac{M_1(1)}{0.6}\right]^3 \tag{5.15a}$$

or

$$\frac{L(\max)}{\mathrm{L}_\odot} = 3.0 \times 10^5 \frac{M_1^4(1)}{1+X} \,. \tag{5.15b}$$

This implies that $M(1)$ is the dominant parameter that generates the observed range of $M_{\rm v}(\max)$ (Section 5.2.3).

To the radiated energy must be added the kinetic contribution $E_{\rm kin} \approx \frac{1}{2} M_{\rm env} v_{\rm exp}^2$, or kinetic luminosity,

$$\begin{aligned} L_{\rm kin} &\approx \frac{1}{2} \dot{M}_{\rm env} v_{\rm exp}^2 \\ &\approx 2 \times 10^4 \left(\frac{v_{\rm exp}}{10^3 \ \mathrm{km\ s^{-1}}}\right)^4 [t(\mathrm{d}) - t_0] \ \mathrm{L}_\odot \end{aligned} \tag{5.16}$$

from equation (5.6). For the fastest novae (e.g., V1500 Cyg, $v_{\rm exp} = 1600$ km s^{-1}), $L_{\rm kin} \gg L_{\rm Ed}$. As explained in Section 5.8.2, the source of energy for the super-Eddington phase is nuclear decay of (overabundant) CNO non-equilibrium cycle products.

If the plateau luminosity is well determined, equation (5.14) provides an estimate of $M(1)$. This is the relationship underlying MacDonald's (1983) simplified model of expanding envelopes. The luminosities given in Section 5.2.5 imply $M_1(1) = 0.94$, 1.0, 0.81 and 1.2 for V1370 Aql, V1500 Cyg, V1668 Cyg and GQ Mus respectively. MacDonald found $1.02 \leq M_1(1) \leq 1.18$ for eight CN.

During the early decline phase of the visual light curve a nova radiates at $\sim L_{\rm Ed}$ and is expelling much of its envelope through a potential $GM(1)/R(1)$. The time scale t_2 should therefore be (Livio 1993b).

$$\begin{aligned} t_2 &\tilde{\propto} \frac{GM(1)M_{\rm env}}{R(1)L_{\rm Ed}} \\ &\tilde{\propto} \frac{M_{\rm env}}{R(1)} \,. \end{aligned}$$

Models show (Fujimoto 1982a,b; MacDonald 1983) that $P_{\rm b}(\mathrm{crit}) \lesssim 1 \times 10^{20}$ dyn cm^{-2} for $Z = 0.02$, and 2×10^{19} dyn cm^{-2} for $Z = 0.5$. Taking $\mathrm{P_b(crit)} \propto Z^{-n}$ (where n ~ 2), equations (2.82) and (5.11) give

$$t_2 = AM_1^{-1}(1)\left[\left(\frac{M(1)}{M_{\rm ch}}\right)^{-2/3} - \left(\frac{M(1)}{M_{\rm ch}}\right)^{2/3}\right]^{3/2} Z^{-n}, \tag{5.17}$$

where A is a constant. Adopting $(M_1(1), Z) = (0.60, 0.38)$ for DQ Her gives $A = 4.2$ d (for $n = 2$).

Figure 5.16 shows the observed correlation between Z and t_2 (from Table 5.9), compared with equation (5.16) with $M_1(1) = 0.91$ (see Table 2.5). The arrows show the

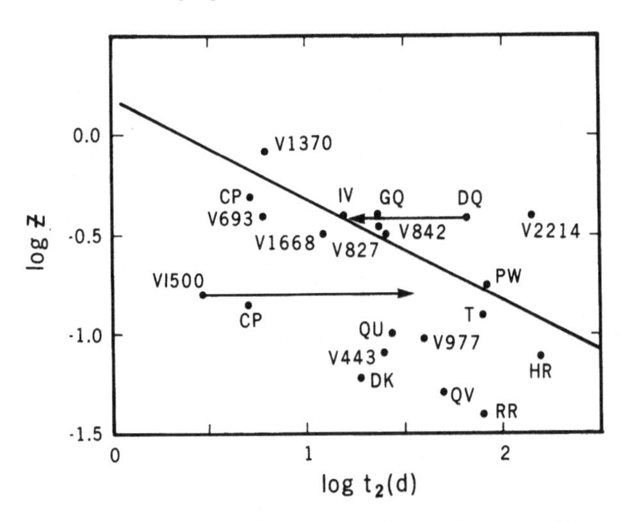

Figure 5.16 Correlation between decay time t_2 and heavy element abundance Z in CN. The diagonal line is equation (5.16) with $A = 4.2$ d and $M_1(1) = 0.91$. The arrows show the effect of adopting $M_1(1) = 0.60$ for DQ Her and $M_1(1) = 1.28$ for V1500 Cyg.

expected departures from the relationship for DQ Her ($M_1(1) = 0.60$: used for normalization) and V1500 Cyg ($M_1(1) \approx 1.28$: Horne & Schneider 1989; Livio 1993b). This is an indication that part of the scatter in Figure 5.15 is caused by a spread in $M(1)$.

Hydrostatic models of envelopes on shell source white dwarfs show a radius – M_{env} relationship (Paczynski 1971a). The initial expansion stage of a nova eruption ejects 10–50% of the accreted envelope, according to speed class, leaving a residual envelope that settles down, in simplified models, to hydrostatic equilibrium with radius of 10^{10}–10^{12} cm (according to M_{env}) with a shell source $T \sim 5 \times 10^7$ K and $L \sim L_{Ed}$. The continuation of mass loss through a wind at this stage steadily decreases M_{env}, which shrinks the radius and leads to an increase of T_{eff} and the observed shift in energy maximum towards shorter wavelengths (Section 5.2.5).

Without a stellar wind the extended envelope would continue until the nuclear fuel was exhausted, taking typically hundreds of years. The integrated luminosities of novae $\sim L(\text{plateau}) \times 1$ y $\sim 10^{46}$ ergs (plus a comparable amount of kinetic energy), requiring the conversion of only $\sim 10^{-6} M_\odot$ of H to He and showing directly that most of the envelope is lost without nuclear processing.

5.8.4 *Continuous Ejection*

The diffuse enhanced and Orion absorption-line systems (Section 5.3) are observational evidence for continued ejection after the initial ejection of gas in the principal system. The development of a nebular spectrum within a year for most novae shows that the central region is cleared of gas (allowing exposure to ionizing radiation from the central source) quite quickly. The requirement to remove almost all of the H-rich envelope in order to turn a nova off on time scales $\ll 100$ y is the theoretical evidence for continued ejection.

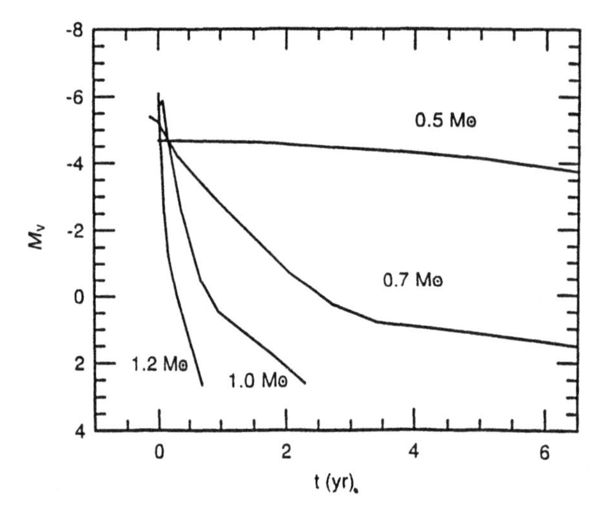

Figure 5.17 Theoretical light curves for slow novae. The curves are labelled with $M(1)$. Solar abundances are adopted. From Kato & Hachisu (1994).

Some success has been obtained from semi-analytical modelling of optically thick steady winds driven by radiation pressure of constant bolometric luminosity (Bath & Shaviv 1976; Bath 1978; Bath & Harkness 1989), but until recently expansion velocities greater than $\sim$100 km s^{-1} were not achievable. However, the mass-loss rates are sufficient to shorten the decay phase of novae to $\sim$20 y even for $M_{\rm env} \sim 10^{-4} M_\odot$ (Kato & Hachisu 1988, 1989) and produce light curves similar to those observed.

Kato & Iben (1992) found that by artificially enhancing wind opacities, velocities up to 1000 km s^{-1} could be obtained, together with nova durations as short as 1 y. New opacity computations (e.g., Rogers & Iglesias 1992) show the required enhancement and lead to good agreement between theory and observation (Kato & Hachisu 1994; Kato 1994: Figure 5.17). Increased opacity drives the wind more effectively, increasing the rate of mass loss, and (equivalently) high opacities reduce $L_{\rm Ed}$ (equation (5.4a), with absorption coefficient $>\sigma_e$), making the nova luminosity even more super-Eddington than previously realized. Optically thick winds occur on erupting white dwarfs even with masses as low as 0.55 $M_\odot$. Fitting model UV and optical light curves to the observations promise great potential for determining $M_1(1)$ and distance (Kato 1994).

5.8.5 *Common Envelope Phase*

The quasi-hydrostatic envelopes remaining after the intial ejection have radii $\gg a$. As pointed out by Paczynski (1976) and MacDonald (1980), provided the envelope density is large enough, the energy released by frictional drag of the secondary as it orbits within the *common envelope* (CE) is sufficient to eject all of the gas lying outside the orbit. The lifetime of the CE is the shorter of the frictional time scale and the wind ejection time scale (of the previous section). Quantitative estimates of the frictional drag are given in Section 9.2.1.3 in the context of CE evolution. Novae in eruption provide perhaps the only opportunity for directly observing the CE phase.

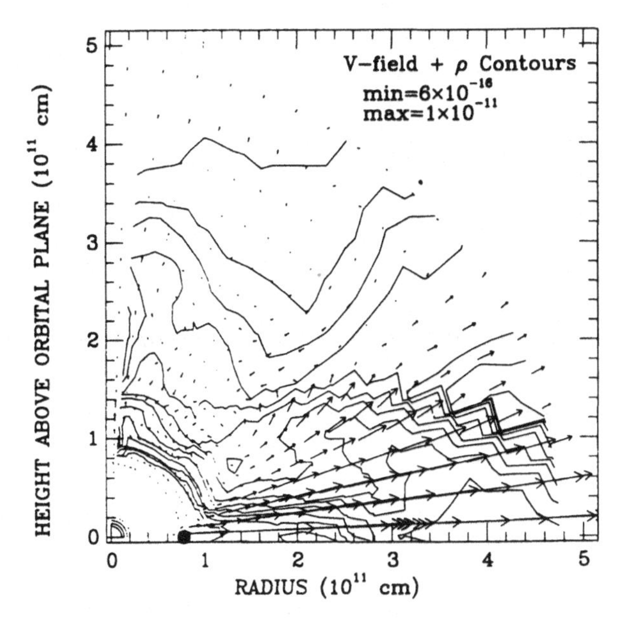

Figure 5.18 Two-dimensional computation of the velocity field and density contours inside a common envelope. Minimum and maximum densities (g cm^{-3}) are given; the largest velocity vector is 1976 km s^{-1}. From Livio *et al.* (1990).

Computations of the effect of frictional drag in nova eruptions have been made by MacDonald (1980), MacDonald, Fujimoto & Truran (1985), Livio *et al.* (1990), Kato & Hachisu (1991) and Livio (1992b). Two-dimensional hydrodynamic calculations show that the orbital angular momentum transferred to the envelope causes an accelerated mass outflow concentrated within 15° of the orbital plane (Figure 5.18). Velocities up to 2000 km s^{-1} can be reached. Weak thermal runaways on low mass primaries, which would otherwise merely expand the radius at less than escape velocity, can be converted into energetic nova ejection provided the maximum radius of the primary exceeds the orbital separation. The greater density in the equatorial plane prevents the later radiation-driven wind from escaping so readily in the plane, resulting in prolate shells when the bulk of mass loss occurs via the wind. This is in agreement with observation (Section 5.6.2). Further discussion is given in Section 9.3.5.

The more rapid expansion of the inner parts of nova envelopes predicted by the new enhanced opacities reduces the density, the interaction time, and hence the deposition of angular momentum, which lessens the effect of CE phases in novae (Kato & Hachisu 1994). The prolate shapes, however, show that some axially symmetric mechanism controls or modifies the wind structure.

5.8.6 Evolution through a Complete Nova Cycle

A white dwarf possessing a hydrogen envelope has two possible quasi-equilibrium structures (Fujimoto 1982a,b): a high luminosity state with a large radius in which a nuclear shell source provides the luminosity (given by equation (5.14)), and a low luminosity state with small radius in which compressional heating replaces the energy loss (for a cold core). Each state has a M_{env} criterion: the high state requires at least a minimum envelope mass, below which nuclear reactions cease, and the low state only exists below a maximum envelope mass, above which nuclear reactions commence. An *accreting* white dwarf is driven between the two states – from low to high by the steady increase in M_{env}, and from high to low by the consumption of nuclear fuel and by mass loss. This constitutes the nova process and its recurrence after time T_R. The instability is suppressed if the white dwarf is too hot or if $\dot{M}(1)$ is very high (both resulting in steady nuclear burning in the high state).

The dependence of M_{env}(crit) on $M(1)$ and $\dot{M}(1)$ for solar abundance envelopes is shown in Figure 5.19. This demonstrates that all accreting white dwarfs, and therefore *all CVs, must undergo nova eruptions.* All CVs not seen in historic times to have a nova eruption (i.e., the DN and NLs) must have had in the past, and will have in the future, very many nova eruptions. The implications of this are presented in more detail in Chapter 9.

All early model computations had to assume initial distributions of composition and physical structure. Only by following the white dwarf structure through several eruptions and subsequent recoveries can the correct initial conditions be discovered. Few such computations have been made (Prialnik 1986, 1990; Livio, Prialnik & Regev 1989; Shara, Prialnik & Kovetz 1993). An example is given in Figure 5.20.

The heavy dashed line in Figure 5.19 shows the limiting $\dot{M}_s(1)$ above which stable nuclear burning occurs; this may be approximated by $\dot{M}_s \approx 2.3 \times 10^{-7}[M_1(1) - 0.19]^{3/2} M_\odot\ y^{-1}$. An example of where $\dot{M}(1) > \dot{M}_s$ has recently been given by GQ Mus, which has turned into an ultrasoft X-ray source 9 y

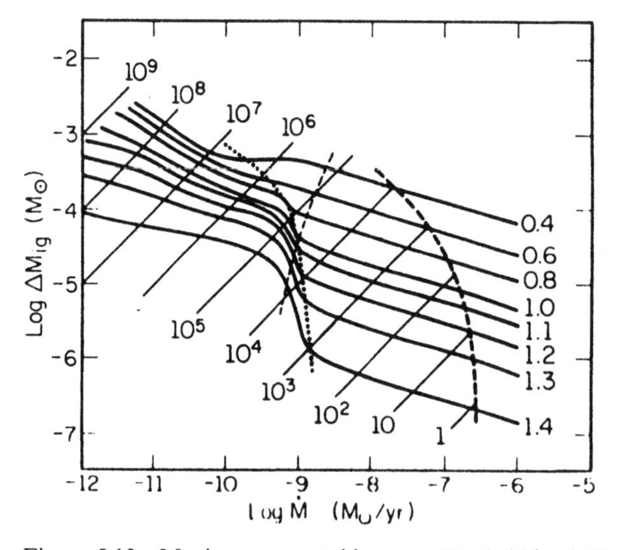

Figure 5.19 Maximum accretable mass M_{env}(crit)($\equiv \Delta M_{ig}$) as a function of $\dot{M}(1)$ and $M(1)$. Thick solid lines give the relationship and are labelled with $M_1(1)$. The heavy dashed line is the upper limit on $\dot{M}(1)$ above which stable H-burning occurs. The light dashed line shows where p–p and CNO reactions contribute equally to energy production. Thin solid lines are labelled with $T_R(y)$ from equation (5.13). From Fujimoto (1982b).

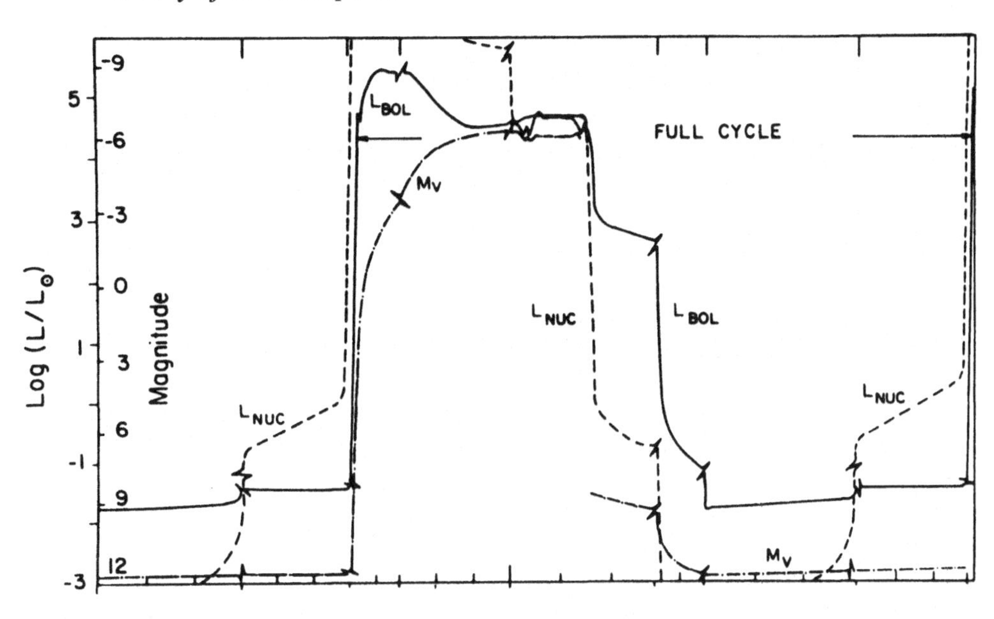

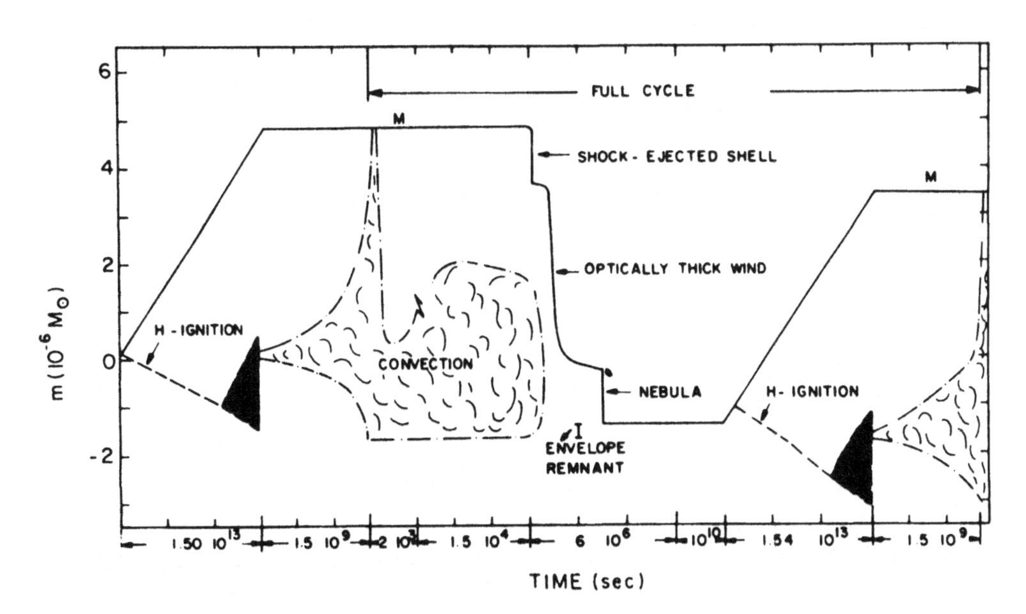

Figure 5.20 Evolution of a nova through a complete cycle. The upper panel shows bolometric luminosity (solid line), nuclear luminosity (dashed line) and visual magnitude (dot-dashed line). The lower panel shows the concomitant changes in the structure of the envelope, above the original white dwarf core ($m > 0$) and below it ($m < 0$). The solid line (labelled M) shows the total mass $M(1) + M_{acc}$ as a function of time. The model has a $M_1(1) = 1.25$ CO core and $\dot{M}(1) = 1 \times 10^{-11}$ $M_\odot y^{-1}$, resulting in $t_3 = 25$ d, ejected mass $= 6.5 \times 10^{-6}$ $M_\odot$ and $v_{exp} = 3800$ km s^{-1}. From Prialnik (1986).

aftereruption(Ögelman*et al.* 1993). With $T_{eff} = 3.5 \times 10^5$ K and $L_{bol} = 1.0 \times 10^{38}$ erg s^{-1} a radius of 3.1×10^9 cm is deduced, and requires hydrogen burning at $\dot{M}(1) \sim 1 \times 10^{-7} M_\odot$ y^{-1}. It is suggested that the reason for the exceptionally high $\dot{M}(1)$ (no other recent nova, among 26 observed, has become a superluminous, ultrasoft soure) is the small P_{orb} (85.5 min) and hence a, which results in exceptional irradiation of the companion.

5.8.7 Novae and the Interstellar Medium

With an average ejected mass $\sim 1 \times 10^{-4}$ $M_\odot$ and a nova rate of 20 y^{-1}, novae return gas to the interstellar medium (ISM) at a rate $\sim$ 0.002 $M_\odot$ y^{-1}. When compared with 0.5 $M_\odot$ y^{-1} for red giant and AGB stars (Knapp, Rachu & Wilcots 1990) and $\sim$ 0.1 $M_\odot$ y^{-1} from supernovae (McKee 1990) it is evident that novae will only be competitive sources of nucleosynthesis for elements with overabundances $\gtrsim 100 \times$ solar.

Williams (1985) estimates that all of the ^{15}N in the ISM may be derived from novae, and Peimbert & Sarmiento (1984) find that ~25% of the ^{13}C may come from novae. A fraction of the ^{7}Li in the ISM could also be produced from nova nucleosynthesis: it is generated from the ^{3}He initially present in the envelope (Truran 1990).

Unstable isotopes are also ejected into the ISM. Ejecta from CNO-enhanced eruptions produce significant amounts of ^{22}Na (half-life = 2.6 y) and ^{26}Al (7.3×10^5 y). Calculations show that nearby ($d < 1$ kpc) ONeMg novae should produce detectable 1.275 MeV ^{22}Na decay γ-rays[4] (Weiss & Truran 1990) and that the amount of ^{26}Al produced is proportional to Z, and the most enriched novae are capable of producing the observed rate of $3 \times 10^{-6} M_\odot$ of ^{26}Al y^{-1} (Clayton & Leising 1987; Nofar, Shaviv & Starrfield 1991; Shara 1994a).

^{22}Na found together with ^{15}N enrichment in interstellar graphite grains is attributed to a nova origin (Amari *et al.* 1990). Excess ^{26}Mg and ^{22}Ne in grains provides the means of assessing ^{26}Al and ^{22}Na production respectively. ^{26}Al is produced in Wolf–Rayet star evolution (Prantzos *et al.* 1986) as well as in novae. Isotopic analyses of interstellar grains indicate admixture from asymptotic giant branch, Wolf–Rayet and nova sources (Amari *et al.* 1993).

5.8.8 QPOs During Eruption

The underlying mechanism that produces quasi-periodic large amplitude luminosity fluctuations in the light curves of some fast novae (Section 5.2.1, Figure 5.2) is still uncertain. Payne-Gaposchkin (1941) studied the V603 Aql oscillations in detail, finding correlations between Orion absorption-line velocities, Balmer emission-line profiles, emission-line intensities and luminosity. An increasing phase lag between emission-line and continuum intensities resulted in cancellation of amplitude in total luminosity, even though both lines and continuum continued as quasi-periodic oscillators.

Friedjung (1966) has stressed that the QPOs must occur in the continuous wind,

[4] Convection of ^{13}N and ^{18}F to the surface in the intial stages of a nova eruption can release positrons and generate a ~0.5 MeV positron-anihilation flux detectable up to several kpc (Leising & Clayton 1987).

rather than any dynamically ejected gas. Sparks, Truran & Starrfield (1976) attribute the phenomenon to oscillations of the quasi-hydrostatic extended envelope of the primary, whose radius greatly exceeds the separation of component stars, which gives pulsation periods of the order of days. In principle a detailed study of the q–p oscillations can provide a means of following the evolution of the cooling remnant. That in practice this may be difficult to achieve is shown by the recent realization that velocity variations in GK Per occurred with twice the period of the luminosity variations (Bianchini, Friedjung & Brinkmann 1990).

No very recent nova has shown prominent long lasting QPOs. This should be a matter for concentrated observations when the next opportunity arises.

5.9 RN

5.9.1 The T Pyx Subclass

T Pyx is unique among the RN in having a short orbital period (2.38 h) and relatively slow decay after eruption. It is the most regular RN, with mean $T_R = 19$ y (and another eruption overdue). In all of its eruptions it has shown a rise to $m_v = 7.8$ in 7 d followed by an irregular rise to $m_v \sim 6.5$ over a time of 20–30 d and then a slow decay (Payne-Gaposchkin 1957; Eggen, Mathewson & Serkowski 1967). An oscillatory behaviour 50–70 d after maximum (Landolt 1970) is similar to the transition stage in CN.

At maximum, T Pyx has an emission-line spectrum, with no A-type absorption (Catchpole 1969), indicative of a small ejecta mass. From an analysis of the nova shell Williams (1982) finds approximately solar abundances. Expansion velocities of ~850 km s^{-1} and ~2000 km s^{-1} were found by Catchpole during eruption. A double nebular shell is observed, the outer of which has a radius of $\sim 5''$ and is expanding at ~350 km s^{-1} and may be associated with the 1944 eruption. The inner shell has a radius $\sim 2''$ and is probably the result of the 1966 eruption (Shara *et al.* 1989). The mass of the ejected shell is $\lesssim 1 \times 10^{-4}$ M$_\odot$.

The distance to T Pyx, and hence M_v(max), is very uncertain. Use of the M_v(max)–t_2 relationship for CN gives $d = 2.1$ kpc. A 1944 origin for the outer shell gives $d \sim 1$ kpc (Shara *et al.* 1989).

5.9.2 The U Sco Subclass

V394 CrA, LMC-RN and U Sco form a distinctive group of RN with He-dominated quiescent discs (Section 4.2.5.2), $P_{orb} \sim 1$ d and, following from the latter, visible secondary spectra in quiescence. From their relatively large P_{orb}, all of these systems must contain evolved secondaries.

The three systems are among the fastest novae known. From their brevity it is quite possible that eruptions of V394 CrA and LMC-RN have been missed: the shortest interval of 8 y between eruptions of U Sco may not be as atypical as at first appears. Only U Sco has a brightness measured on the rise to maximum, but it is clear from immediate pre-eruption magnitudes that all three objects rise very rapidly ($\lesssim 1$ d) to maximum. The declining light curves vary from eruption to eruption: V394 CrA had $t_3 = 10$ d in 1949 and 5.5 d in 1987 (Duerbeck 1988) and U Sco faded more rapidly after its initial 2 mag decline in 1979 than in previous outbursts (Duerbeck & Seitter 1980).

With the accurately known distance and reddening to the LMC, the absolute magnitude of LMC-RN is reliably determined at $M_V(\max) = -7.5$. This is 2 mag fainter than predicted by the $M_V(\max)$–t_2 relationship for normal CN (Figure 5.3: which suggests that LMC 51 in that diagram is also a RN). Adopting this $M_V(\max)$ for all three stars gives the (bracketed) distances in Table 5.3, which put V394 CrA and U Sco 0.7 and 5 kpc above the Galactic plane respectively. These are much greater heights than for typical CN (Section 9.5.3), and place the U Sco subclass as halo objects.

Spectra during outburst show very similar evolution in all these objects. Initially in U Sco and LMC-RN very broad Balmer, HeI, NIII, CIII (FWZI: 10000 km s^{-1}) fade to leave a HeII-dominated spectrum and a narrower (FWZI: 1800 km s^{-1}) component. IUE spectra show strong P Cyg profiles 5 d after maximum. In V394 CrA the broad component is narrower (8000 km s^{-1}), as befits the slightly slower development of its light curve (Barlow *et al.* 1981; Williams *et al.* 1981; Sekiguchi *et al.* 1988,1989,1990b; Shore *et al.* 1991).

Interpretation of the spectra gives $Y/X \sim 8$ and $Z/H \sim$ solar (but enrichment of N relative to C and O) in U Sco and $Y/X \sim 4$ and an N abundance $\sim 30 \times$ solar in LMC-RN. Estimated ejected masses are 10^{-7} $M_\odot$ and $10^{-7.3\pm0.5}$ $M_\odot$ respectively.

Lines from forbidden transitions do not occur during the spectral evolution of the U Sco class. This gives $N_e \gtrsim 10^8$ cm^{-3} and requires the ejecta to consist of a large number ($\sim 10^6$) of dense condensations (Williams *et al.* 1981). Combined with the low shell mass and high v_{exp}, this implies a shell thickness only $\sim 10^7$ cm (Shore *et al.* 1991).

The bolometric luminosity at maximum of LMC-RN was $> 8 \times 10^4$ $L_\odot$, which is super-Eddington (equation (5.4b)) even for a pure He envelope, and even more so when the opacity in the HeII Lyman continuum is included (Shore *et al.* 1991).

5.9.3 The T CrB Subclass

The RN T CrB, RS Oph, V3890 Sgr and V745 Sco (and the RN candidate[5] V723 Sco: Table 5.3) all have M giant secondaries, which require $P_{orb} \gtrsim 100$ d. Their eruptions have very rapid rises ($\lesssim \frac{1}{2}$d) and their light curves repeat quite closely (Figure 5.21). From their brevity it is quite possible that all of these stars have had unobserved eruptions this century.

Distances to the T CrB stars were all considered uncertain until the discovery of V745 Sco, which is near the Galactic Centre (Sekiguchi *et al.* 1990a). This provides $M_V(\max) \approx -8.3$, which is within the spread about $M_V(\max) \sim -8.8$ obtained from Figure 5.3. The estimated distances and reddenings to T CrB and RS Oph in Table 5.3 (see also Weight *et al.* (1994)) give results in excellent agreement with $M_V(\max) = -8.9$ obtained from Figure 5.3. The only discrepant star is V3890 Sgr, which becomes compatible if it too is at the distance of the Galactic Centre and the larger reddening obtained by González-Riestra (1992) is used. Unlike the U Sco subclass, the T CrB

[5] Kopylov *et al.* (1988) suggest that FBS 2351 + 228 also may belong to the T CrB class and Weight *et al.* (1994) suggest V3645 Sgr and EU Sct as candidates.

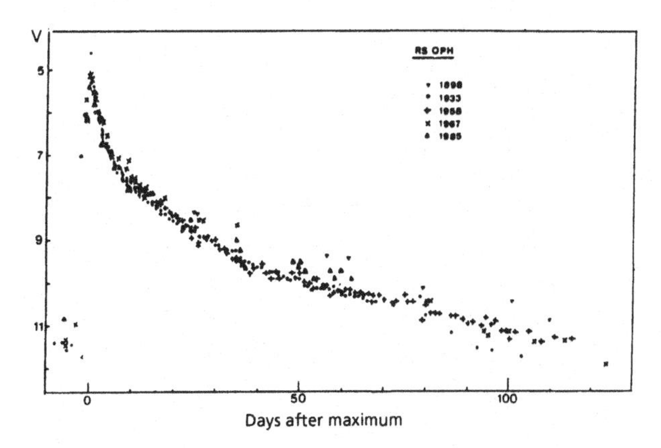

Figure 5.21 Five superposed light curves of RS Oph. From Rosino (1987).

subclass appears to fit the $M_V(\max)$–t_2 CN relationship (cf. González-Riestra (1992) and Harrison, Johnson & Spyromilio (1993)). The luminosities at maximum are $>L_{Ed}$, as in the CN.

RS Oph is the only T CrB star to have been observed (in its 1985 eruption) sufficiently to determine its bolometric evolution. As with CN, the rapid visual decline is deceptive – after an initial peak there was a plateau luminosity $\sim 5.0 \times 10^4$ L$_\odot$ lasting for ~ 50 d (Snijders 1987), which implies $M_1(1) \approx 1.37$ (equation (5.14)). As discussed in Section 4.2.5.2, the mass of the primary in T CrB is also almost certainly close to the Chandrasekhar limit. Harrison, Johnson & Spyromilio (1993) deduce $M_1(1) \sim 1.35$ from IR plateau luminosities of V3890 Sgr and V745 Sco.

The eruption light curves of T CrB are remarkable for showing secondary maxima, $\sim$6 mag below the initial maximum, 106 d after the principal eruption. This has been interpreted as accretion onto the secondary of a ring of gas formed in the principal eruption – the latter being thought to be due to a transfer of $\sim 5 \times 10^{-4}$ M$_\odot$ from the secondary onto a main sequence primary (Webbink 1976; Webbink *et al.* 1987). However, Selvelli, Cassatella & Gilmozzi (1992) point out that the secondary maximum in 1946 coincided with the appearance of a shell spectrum, implying the formation of an optically thick shell which converted the strong UV flux from the primary into an enhanced visible flux. This interpretation, together with the properties of T CrB at quiescence (Section 4.2.5.2), favours a thermonuclear runaway rather than an accretion event model for the eruptions of T CrB.

The spectral development in the T CrB subclass is the reverse of what is seen in CN: initial high velocity ($v_{exp} \sim 5000$ km s^{-1}) emission-line systems are rapidly replaced by lower and lower velocity systems, without any indication of deceleration (Friedjung 1987c; Webbink 1990b). The steady narrowing of individual components, however, speaks for deceleration. Strong coronal lines appear in late decline, as does a strong X-ray flux with $kT \sim 1$ keV; these are interpreted as evidence of a shock wave propagating into the circumstellar wind from the giant secondary (Bode & Kahn 1985; Itoh & Hachisu 1990). Radio mapping of RS Oph shows that ejection occurs in a bipolar pattern (Taylor *et al.* 1989); the multiple components of lines in V3890 also suggest a highly non-spherical ejection (González-Riestra 1992). Optical images of T CrB show a bipolar structure (Williams 1977).

Pottasch (1967) found large He, N and O enhancements in spectra of the 1958 eruption of RS Oph. RS Oph was one of the novae noted to have strong Ne lines by McLaughlin (1943). Ne was also quite prominent in the eruption spectra of T CrB (Payne-Gaposchkin 1957), but no abundances are available. Pottasch (1967) found an ejected mass (corrected to $d = 1.6\,\mathrm{kpc}$) of $5 \times 10^{-7}\ \mathrm{M}_\odot$ for RS Oph from nebular spectra of the 1958 eruption; Bohigas *et al.* (1989) found $M_{\mathrm{env}} = 3.8 \times 10^{-5}\ \mathrm{M}_\odot$ from the 1985 eruption, of which $1.2 \times 10^{-6}\ \mathrm{M}_\odot$ was ejected and the remainder relaxed back onto the primary. O'Brien, Bode & Kahn (1992) find $M_{\mathrm{ej}} = 1.1 \times 10^{-6}\ \mathrm{M}_\odot$ from a model of the X-ray emission.

5.9.4 *Amplitudes and Recurrences*

The amplitude–recurrence relationship for DN (Section 3.3.3.2) was originally thought to extend to RN (Kukarkin & Parengo 1934). Several of the CVs thought to be RN are now known to be DN of long outburst interval, so it is not surprising that some correlation was found. With the discovery that the secondaries of almost all RN are subgiants or giants, the amplitude A'_{RN} of eruption is strongly influenced by the luminosity of the secondary – and whereas in DN it is appropriate to use the magnitude of the disc in quiescence it is not clear which component magnitude should be adopted for RN at minimum. Furthermore, the mean recurrence time $\overline{T}_{\mathrm{R}}$ for RN is strongly dependent on the completeness of coverage – many of the short lived eruptions may have been unobserved. It is no surprise, therefore, that there is no obvious A'_{RN}–$\overline{T}_{\mathrm{R}}$ relationship for RN.

To compare the amplitudes of RN and CN requires subtraction of the contribution from the secondaries of the former. An added complication is that the discs of RN (other than T Pyx) are very much larger and have higher $\dot{M}(\mathrm{d})$ than in CN, and consequently in many cases they contribute relatively more to the visible spectral range. Without correction, all RN except T Pyx lie 2–5 mag below the A'_{CN}–$\log t_2$ relationship for CN (Figure 5.4), which is a useful signature of potential RN. Care, however, must be taken to identify the remnant (misidentification will in general be with too bright an apparent remnant) and to eliminate DN from the candidate list. Some systems having surprisingly low amplitudes are AT Sgr, GR Sgr, HS Sgr, V3888 Sgr, FS Sct and V444 Sct. Six other novae of low amplitude (Warner 1987a) have all been found to have misidentified remnants (Downes & Shara 1993; Duerbeck & Grebel 1993).

5.9.5 *Theory of RN*

From equation (5.13) the short recurrence times for RN require either or both small envelope masses and high $\dot{M}(1)$. Combining equations (5.12) and (5.13) with Figure 5.19 shows that $M_1(1) \geq 1.3$ and $\dot{M}(1) \gtrsim 1 \times 10^{-8}\ \mathrm{M}_\odot\ \mathrm{y}^{-1}$, for $T_{\mathrm{R}} < 100$ y (Livio & Truran 1992). RN therefore represent that fraction of CVs in which accretion is occurring at a high rate onto a primary close to the Chandrasekhar mass.

The U Sco and T CrB subclasses appear to meet these expectations observationally: the short decay times, low M_{ej} and estimates of $M(1)$ are in agreement; the large P_{orb} probably ensures $\dot{M}(2) > 1 \times 10^{-8}$ M$_\odot$ y^{-1} (Section 9.1.2.2). The low M_{ej} results in frictional drag during the brief CE phase being unimportant (Kato & Hachisu 1991), leaving the radiation-driven wind as the dominant determinant of the decline part of the eruption light curve (Kato 1991c), which gives predicted light curves for U Sco and RS Oph which are in good agreement with observation (Kato 1990b, 1991c) and which lead to mass estimates $M_1(1) = 1.38$ and 1.36 and $M_{ej} = 2 \times 10^{-7}$ M$_\odot$ and 1.2×10^{-6} M$_\odot$ respectively. Similarly, Kato (1990c) finds $1.30 \leq M_1(1) \leq 1.37$ for T Pyx, with the prediction that comprehensive observations of the next eruption will lead to very precise deductions of $M(1)$. However, the short P_{orb} and longer t_2 of T Pyx could make the CE phase more influential than in the other RN.

Although initial attempts to produce thermonuclear runaway models with short recurrence times met with difficulties (Starrfield, Sparks & Truran 1985), inclusion of high $\dot{M}(1)$ and observed Y/X give good agreement with observed $M_v(\max)$, M_{ej} and v_{exp} (Starrfield, Sparks & Shaviv 1988; Truran *et al.* 1988). At most 15% of the accreted mass was ejected in the eruption.

5.10 Symbiotic Novae

Symbiotic stars are slowly eruptive variables showing simultaneously absorption-line features of a late-type giant and emission lines of HI, HeI and other ions with ionization potentials >20 eV, which transform to an A- or F-type continuum with additional HI, HeI etc. absorption lines, or to pure nebular emission spectra, during eruptions (Kenyon 1986). The majority of symbiotic stars have shown several eruptions, but a distinct subset have had only one, protracted eruption in a century or more of observation. These are denoted *symbiotic novae (sN)* (Allen 1980) and include objects such as RR Tel which have also been described as very slow novae (e.g., Payne-Gaposchkin (1957), who refers to symbiotic stars in general as symbiotic novae).

The ten currently known sN are listed in Table 5.10 (Kenyon 1986; Viotti 1988, 1990; Munari 1992; Munari *et al.* 1992). Webbink *et al.* (1987) proposed V4074 Sgr (≡AS295B) as a possible RN of the T CrB class, but it is classified as a symbiotic star by Allen (1984) and is probably a sN. Recent reviews of symbiotic stars in general are given by Allen (1984), Kenyon (1986), Mikolajewska *et al.* (1988) and Luthardt (1992b).

There are two types of symbiotic star: S-type, which contains normal M giants, and D-type, which contains heavily dust-reddened Mira variables. The estimated $\dot{M}_{wind}$ in the former is 10^{-8}–10^{-7} M$_\odot$ y^{-1}, and in the latter 10^{-6}–10^{-5} M$_\odot$ y^{-1} (Kenyon 1988a). Representatives of both types are found among the sN.

All symbiotic stars are probably binaries with long orbital periods; most of those measured have $200 \lesssim P_{orb}(\mathrm{d}) \lesssim 1000$, but V1016 Cyg, HM Sge and RR Tel probably need $P_{orb} \gtrsim$ decades for $R(2) \leq R_L(2)$. The hot component typically has $T_{eff}(1) \sim 1 \times 10^5$ K, which may be produced by accretion onto a main sequence star at a rate $\sim 1 \times 10^{-5}$ M$_\odot$ y^{-1}, requiring a Roche lobe-filling M giant (Kenyon 1988a); or by wind accretion onto a white dwarf at a rate $\dot{M}(1) \sim$ few $\times 10^{-6}$ M$_\odot$ y^{-1}, requiring

Table 5.10. *The sN.*

Star	Start of Eruption	Maximum of eruption	m_v^* (min)	m_v (max)	$t_1(y)$	$Sp(2)$	Photom period (d)	Type	P_{orb}(d)
V1016 Cyg	1964	1967	15	10.5	>130	M6–7	450	D	
V1329 Cyg	1964.5**	1964.8	14	11.5	12–20	M5:	955:	S	950
V2110 Oph		1940:		11:	10:	>M3		D	
AG Peg	1855	1871	9	6	40	M3	816.5	S	812±6
HM Sge	1975.3	1975.7	18	11	>65	>M4	540	D	
RT Ser	1909:	1923:	>16	9.5	7:	M5.5		S	
RR Tel	1944.8	1945.3	14	6	9	>M5	387	D	
PU Vel	1977	1982	15	8.8	15–30	M5–6		S	
FG Ser	1988.5	1988.7	12	9.7	5:	M5		S	650
AS 338	1981.5	1982.2	14	10.6	3	M5		S	434.1

* In many cases these are averages over long period variability.
** There was a pre-outburst rise of 1.3 mag during 1956–64.

$M_{wind}(2) \sim$ few $\times 10^{-5}$ $M_\odot$ y^{-1} (Munari & Whitelock 1989), or by steady H-burning on a white dwarf, requiring $\dot{M}(1) \sim$ few $\times 10^{-8}$ $M_\odot$ y^{-1} (Iben 1982; Munari & Whitelock 1989). Probably all of these contribute to the mixture of systems known collectively as symbiotic stars.

The evidence for lobe-filling secondaries varies from star to star. Some symbiotic stars certainly fill their lobes (Kenyon 1986), but the eclipsing sN FG Ser has no IR ellipticity modulation, which implies that it underfills its Roche lobe by at least a factor of 2 (Munari *et al.* 1992) and in AS338 eclipses show no evidence for an accretion disc, indicating wind accretion rather than Roche lobe overflow (Munari 1992).

Stellar evolution in wide binaries requires $M(2) > M(1)$ in the sN if the primary is a white dwarf, in which case Roche overflow is dynamically unstable (Section 9.2). It is probable, therefore, that all sN are *detached* binaries, with wind from the giant secondary supplying the $\dot{M}(1)$. Whitelock (1987) has pointed out that the long periods of Miras found in symbiotic stars imply $0.9 \lesssim M_1(2) \lesssim 1.1$ and a lifetime $\sim 10^4$ y.

The *visual* light curve for AG Peg is shown in Figure 5.22. The presence of an M giant in a sN greatly reduces the apparent amplitude of eruption. As in CN, there is evidence that the *bolometric* luminosity is approximately constant (e.g., Kenyon *et al.* 1993). Mass ejection with $v_{exp} \sim 100$ km s^{-1} is evident in many sN, both from optical and radio observations, often with multiple shells (Kenyon 1986). In FG Ser and AS338, however, there is no evidence for mass loss, and $L(\max) < L_{Ed}$ (Munari 1992; Munari *et al.* 1992). In these objects, therefore, the eruption is simply the expansion and slow contraction of the photosphere of the white dwarf primary. Those systems that do eject mass have $M_{env} \sim 10^{-4}$ $M_\odot$ (e.g., AG Peg: Kenyon *et al.* 1993). The ejected material runs into the circumstellar gas generated by the wind from the M giant, creating a $\sim 10^7$ K shock in V1016 Cyg, HM Sge and RR Tel with $L_x \sim L_\odot$,

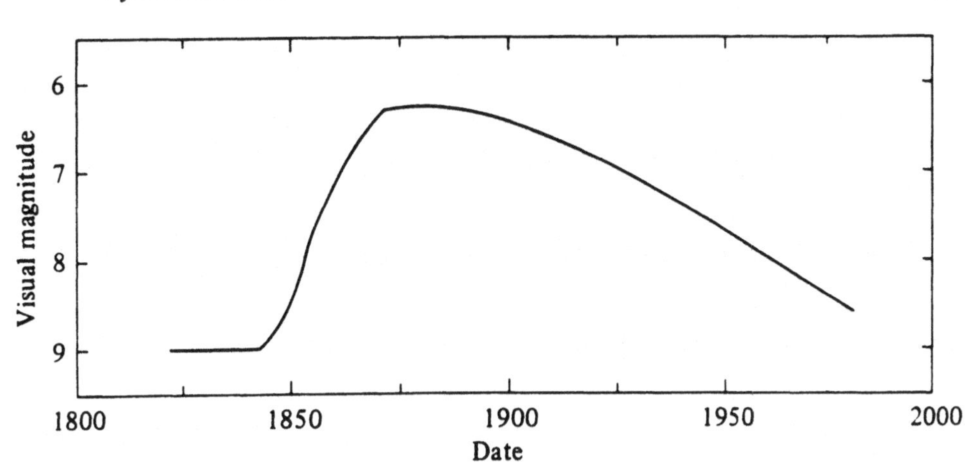

Figure 5.22 Light curve of the sN AG Peg. From Kenyon (1988a).

which requires $v_{exp} \gtrsim 700$ km s^{-1} and $\dot{M} \sim 10^{-8}$ M$_{\odot}$ y^{-1} in the shock region. At least in these systems there must be common envelope acceleration or rapid wind ejection of gas.

Most symbiotic stars show normal CNO abundances (Nussbaumer *et al.* 1988), but PU Vul (and possibly HM Sge) shows typical nova overabundances (Vogel & Nussbaumer 1992).

With their giant secondaries the sN are clearly related to the T CrB subclass of RN. However, their eruptions have very slow rises (0.2–16 y: Table 5.10; Figure 5.22) and are of great duration (t_1 in Table 5.10 is the time to fall 1 mag $\approx$ e-folding time). The slowest (t_2 > 200 d) of the CN have rapid rises to maximum (e.g., DO Aql: Payne-Gaposchkin 1957) although few have observed pre-maximum light curves, this very fact shows that their rises are rapid. Furthermore, no CN are known to have M giant secondaries.

The T CrB subclass and the sN are therefore distinct types, with no transition objects between them. Having already identified RN as eruptions on very massive white dwarfs it is reasonable to conclude that sN constitute the remainder of eruptions in M giant-containing systems, and that their slowness is the result of the absence of any substantial CE phase in such widely separated binaries (Kenyon & Truran 1983). Indeed, many symbiotic star eruptions may be of this nature (Kenyon 1986), with eruptions arising from weak, recurrent H-shell flashes on low mass, or high $\dot{M}(1)$, or hot white dwarfs (Iben 1982; Kenyon 1988b; Sion & Ready 1992; Munari & Renzini 1992). Typical decline time scales observed for these are $200 < \tau_d < 1000$ d.

Reliable direct estimates of $M(1)$ and $M(2)$ are rare, but AG Peg is fairly well determined at $M_1(1) = 0.65 \pm 0.10$, $M_1(2) = 2.5 \pm 0.4$, by Kenyon *et al.* (1993) who also find L(plateau) $\approx 2 \times 10^3$ L$_{\odot}$. Similar luminosities are found for FG Ser, PU Vul and AS338, all of which indicate $M_1(1) \approx 0.55$ (Section 5.8.3).

Computations of thermonuclear runaways on low mass ($\sim$0.5 M$_{\odot}$) white dwarfs have been made by Shara, Prialnik & Kovetz (1993) and Sion (1993a). At

$\dot{M}(1) = 1 \times 10^{-9}$ $M_\odot$ y^{-1} onto a cool white dwarf, slow eruptions with $v_{exp} \sim 100$ km s^{-1} are found, which remain at $L(\max) \sim 1 \times 10^4$ $L_\odot$, fading over few $\times$ 10^3 y and recurring after 4×10^5 y. Most of the envelope (4×10^{-4} $M_\odot$) is ejected. $\dot{M}(1) = 1 \times 10^{-8}$ $M_\odot$ y^{-1} onto a 1×10^5 K white dwarf gives initial multiple eruptions to $L \sim 10^4$ $L_\odot$, but the long-term cyclic behaviour is not known.

Whitelock (1987) points out that the different kinds of eruptions that occur in sN may be attributed partly to the wide range of $\dot{M}_{wind}$ that occurs in Miras and non-variable red giants (in particular, $\dot{M}_{wind} \tilde{\propto} P^{2.17} \Delta L^{1.88}$, where P is the pulsation period and ΔL is the amplitude in the L band), and partly to differences in orbital separation. Both of these strongly affect $\dot{M}(1)$.

6

Polars

A man's friends are his magnetisms.

Emerson. *Conduct of Life: Fate.*

From systems that are weakly or covertly magnetic we turn to ones in which the magnetic field of the primary is strong enough to control the accretion flow, preventing the formation of an accretion disc and generating the signatures of magnetic accretion: large linear and circular optical polarization and strong X-ray emission.

6.1 Historical Development

The discovery of the polars provides a lesson that even relatively familiar objects may reveal exotic phenomena if interrogated in the correct way. The star AM Her had been discovered as a variable in 1924 and listed as a NL on the basis of slow variations in brightness over a range of 3 mag and an emission-line spectrum. In 1976 Berg & Duthie (1977) suggested that AM Her could be the optical counterpart of the Uhuru X-ray source 3U 1809+50 and Hearn, Richardson & Clark (1976) using the SAS-3 satellite found a variable soft X-ray source near the same position. The similarity of this source to the low mass X-ray binaries Sco X-1 and Cyg X-2 stimulated Cowley & Crampton (1977) to obtain spectra, which revealed a 3.09 h orbital period.

The main surprise came, however, when Tapia discovered in August 1976 that AM Her is linearly and circularly polarized at optical wavelengths (Tapia 1977a). Its linear polarization varies from zero up to 7% and its circular polarization from −9% to +3%, both changing smoothly over the period of 3.09 h (Figure 1.12). The high degree of circular polarization, previously only seen in magnetic white dwarfs (Angel 1978), suggested the presence of a strong magnetic field. From the theory of non-relativistic cyclotron radiation, which predicts a fundamental frequency at

$$\nu_c = \frac{\omega_c}{2\pi} = \frac{eB}{2\pi mc} = 2.8 \times 10^{14} B_8 \text{ Hz} \tag{6.1}$$

(Ingham, Brecher & Wasserman 1976), where B is the magnetic field strength in gauss, Tapia concluded that the field strength in AM Her must be $\sim 2 \times 10^8$ G. However, later investigations showed that the polarized emission arises from cyclotron *harmonics*, which reduces the implied field by a factor $\sim$5. Such a field can only be located in the white dwarf primary; the fact that the periods of the polarization variation and the orbit are identical implies that the white dwarf rotates at the same rate as the orbital motion – i.e., the primary rotates *synchronously*, or is *phase-locked* in orbit.

A search for circular polarization in other known NL variables soon revealed two more objects: AN UMa (Krzeminski & Serkowski 1977) and VV Pup (Tapia 1977b); these were later found also to be X-ray sources. As a class, these objects are known after the type star as AM Her stars, or, recognizing the high degree of polarization and following the recommendation of the *Polish* astronomers Krzeminski & Serkowski, as *polars*. The current definition expands on the original concept by requiring that the primary is *secularly* phase-locked: it may, however, be jolted temporarily out of synchronism by a nova eruption.

Optical identifications of sources discovered in later X-ray surveys have increased the number of *confirmed* polars to 42. These are listed[1] in Table 6.1. Additional probable candidates are given in Table 6.2. The derived field strengths (see Sections 6.3.3, 6.5.3 and 6.5.4 for details) lie in the range $11 \leq B \leq 75$ MG. Cropper (1990) estimates that at least 10% of CVs are polars. In contrast, isolated magnetic white dwarfs constitute only about 2% of the white dwarf population and their fields have been measured in the range $1 < B < 1000$ MG, with approximately an equal fraction in each decade of field strength (Angel, Borra & Landstreet 1981; McCook & Sion 1984; Schmidt 1987; Schmidt & Liebert 1987).

6.2 Magnetically-Controlled Accretion for Synchronous Rotation

The rich variety of observed optical, X-ray and radio phenomena in polars may appear perplexing without first acquiring a basis for their interpretation. Accordingly, the physics of accretion onto a magnetic degenerate star is treated in this section, reserving the observations and deductions made from them to later sections. Much of the physics in this and the following chapter applies to accretion onto neutron stars as well as white dwarfs; discoveries in one area have often found application in the other.

6.2.1 The Magnetosphere

Gas which is at least partially ionized and falling towards a magnetized star will at some point have its motion resisted by the magnetic field. The *magnetosphere* is defined as that volume (not in general spherical) within which the field strongly affects the flow of mass, energy and angular momentum (e.g., Lamb 1989). For *spherically symmetric infall*, the radius $r_{\mu,\mathrm{sph}}$ of the magnetosphere is determined from the balance of magnetic pressure $B^2(r)/8\pi$ and ram pressure of the infalling gas (Davidson & Ostriker 1973):

$$\frac{B^2(r)}{8\pi} = \rho(r) v_{\mathrm{in}}^2(r). \tag{6.2}$$

For a dipole field, $B(r) = \mu/r^3$, where $\mu = BR^3$ is the magnetic moment of the star.

For steady state accretion

$$\dot{M} = 4\pi\rho(r) v_{\mathrm{in}}(r) r^2 \tag{6.3}$$

[1] There is a strong tendency in the CV literature to quote polar orbital periods in minutes, rather than hours; this is followed here.

and the infall velocity will be comparable to the free-fall velocity $v_{ff} = (2GM/r)^{1/2}$. The equilibrium radius (also known as the Alfvén radius) is therefore given by

$$r_{\mu,\rm sph} \approx 2^{-3/7}\mu^{4/7}(GM)^{-1/7}\dot{M}^{-2/7} \tag{6.4}$$

$$= 9.9 \times 10^{10}\mu_{34}^{4/7}M_1^{-1/7}\dot{M}_{16}^{-2/7} \text{ cm.} \tag{6.5}$$

Accretion from a stream results in a balance being obtained where

$$\frac{B^2}{8\pi} = \rho v^2 = (\dot{M}/\pi\sigma^2 v)v^2 \tag{6.6}$$

where σ is the radius of the stream and v is the velocity of the stream $\sim v_{ff}$ (Mukai 1988). Then

$$r_\mu = 1.45 \times 10^{10}\mu_{34}^{4/11}\sigma_9^{4/11}M_1^{-1/11}\dot{M}_{16}^{-2/11} \text{ cm.} \tag{6.7}$$

σ can be found from Lubow & Shu (1975). For a given $\dot{M}$, a stream is able to penetrate much closer to the magnetic star than a spherically accreting flow.

The magnetic moment of the primary is

$$\mu(1) = 1.0 \times 10^{34} B_7(1)R_9^3(1) \text{ G cm}^3 \tag{6.8a}$$

$$= 3.9 \times 10^{33} B_7(1)M_1^{-1}(1) \text{ G cm}^3 \qquad 0.4 < M_1(1) < 0.7 \tag{6.8b}$$

which shows that the polars listed in Table 6.1 all have $\mu_{34}(1) \sim 1$ and have $r_\mu \sim$ few $\times 10^{10}$ cm; the latter is comparable with the distance from the primary to the L_1 point (i.e., $R_{L_1}(1)$), so field lines from the primary can readily connect with the secondary which in turn enables various mechanisms to ensure synchronous rotation of the primary (Section 6.6).

The current state of interpretation of polars requires a more general formula for r_μ than the simple form of equation (6.7). This is given by Ferrario, Wickramasinghe & Tuohy (1989):

$$\frac{r_\mu}{R(1)} = 13.4\left[\frac{B_p(1+3\sin^2\beta)^{1/2}}{3\times 10^7 \text{ G}}\right]^{4/7}\left(\frac{f_s}{10^{-3}}\right)^{2/7} M_1^{-8/21}(1)\dot{M}_{16}^{-2/7}, \tag{6.9}$$

where B_p is the polar field strength of the dipole field, β is the inclination of the field axis to the rotation axis of the white dwarf (assumed perpendicular to the plane of the orbit) and $4\pi f_s$ is the solid angle subtended by the stream at r_μ as seen from the primary. Values of $\dot{M}$ can be obtained from the accretion luminosity (emitted mostly in X-rays)

$$L_{acc} = 1.34 \times 10^{33} M_1(1)R_9^{-1}(1)\dot{M}_{16} \text{ erg s}^{-1}. \tag{6.10}$$

For typical values ($10^{-3} < f_s < 10^{-2}, 20 \leq B \leq 50$ MG, $10^{33} < L_{acc} < 10^{34}$ erg s^{-1}) equation (6.9) gives $10R(1) \leq r_\mu \leq 30R(1)$. Note that this mimimum value of r_μ is sufficient to prevent the formation of a disc in the polars – a disc can only exist if the stream can compress the field sufficiently to enable it to pass completely around the primary and collide with itself. To achieve this in CVs with $P_{orb} \lesssim 4$ h and

Table 6.1. *Polars.*

Star	Alias	P_{orb} min	m_v	Distance pc	M_V (max)	i°	$Sp(2)$	References
RXJ 1015+09		78						225
RXJ 0132-65		78	19					241
EV UMa	RE/RX 1307+53	79.69	17.1–20.7	>700		75 ± 5		1,226,232
EF Eri	2A0311-227	81.02	13.5–17.7	94	8.6	65 ± 5		2–26,53,54,78,133–137
RXJ 0153-59		89	17					241
DP Leo	E1114+182	89.80	17.5– > 22	450	9.2	79.6		27–30,53,55,236,248
RE/RX 1844-74		89.91	15–17.6					1,221
EU UMa	RE/RX 1149+28	90.14	17			30–70		1,138,242
RE/RX 0453-42		94:	19					1,241
RXJ 1957-57		99	17					241
VV Pup		100.44	14.5–18	145	8.2	74 ± 4	M4V	31–52,78,139,140
V834 Cen	E1405-451	101.52	14.2–16.6	86	9.5	45 ± 5	M6.5V	24,56–71,141,229
EP Dra	1H1907+690	104.63	17.6–18.4	>300	<10.2	80:		72
RXJ 1002-19		107	17					241
CE Gru	Grus VI	108.6	18.0–20.7			50 ± 5		73–75
V2301 Oph	1H 1752+08	112.97	16.4			80 ± 5		244
RXJ 1802.1+1804		113.0	14			65–80		243
MR Ser	PG 1550+191	113.58	14.5–17	139	8.8	40 ± 5	M5-6V	28,30,76–85,96,120, 142,143,224,233,234
BL Hyi	H0139-68	113.65	14.3–18.5	128	8.8	71 ± 10	M3-4V	86–95,144
ST LMi	CW1103+254	113.89	14.7–17.2	128	9.2	55 ± 5	M5-6V	28,63,96–104,145,237
EK UMa	E1048+542	114.5	17.5–20			56 ± 14		105,106
AN UMa	PG 1101+453	114.84	15.5–20	≥270	<8.3	65 ± 5		30,40,78,110–121
WW Hor	EXO 0234-523	115.49	17.6–21	430	9.4	74 ± 5	M6V:	107-109
AR UMa	ES1113+432	115.9	12.8–16.5					231,235
HU Aqr	RE/RX 2107-0518	125.02	15.3–18.2	190	8.9	85 ± 1	M4.5V	1,122,201,230
UZ For	EXO 0333-255	126.53	17–20.5	230	10.2	86 ± 5	M4.5V	123–130
RE/RX 0531-462		139.8	16.5–17.5					1,245
QS Tel	RE/RX 1938-4612	140.0	15.2–17.2	300:	7.8:			1,131,228
V2008-65.5	Drissen V211b	159.7	18	>400		40		132,239,240

RXJ 0525+45	Paloma	160						225,241
RXJ 0501-03		171	15			80:		241
AM Her	3U 1809+50	185.65	12–15.8	75	7.6	52 ± 5	M4.5V	40,146–194
V1500 Cyg	N Cyg 1975	201.0	17–21	1200	6.6	55		202–210
BY Cam	H0538+608	201.9	13–18	190	8.2	50 ± 10		30,195–200,223
RXJ 1940.2-1025		201.94	16–17	230:	7:	60:	M4V	222,246,247
RX 0929.1-2404		203.4	17.0			75:		1,238
RX 1007-20		207.9:	18					1
RX 2316-05		208.9	18					1,225
QQ Vul	E2003+225	222.51	15.3–18	≥320	<7.8	60 ± 14	M2-4V	96,211–218
EXO 032957-2606.9		228	17–19.2	520	8.4	9 ± 3	M4.5V	106,219–221
RX 1313-32		255	16					1,241
RX 0203+29		275.5	17					1,227

References: 1. Beuermann & Thomas 1993; 2. Charles & Mason 1979; 3. Griffiths *et al.* 1979; 4. Williams *et al.* 1979; 5. Bond, Chanmugam & Grauer 1979; 6. Williams & Hiltner 1980; 7. Verbunt *et al.* 1980; 8. Schneider & Young 1980a; 9. Watson, Mayo & King 1980; 10. Crampton, Hutchings & Cowley 1981; 11. White 1981; 12. Coe & Wickramasinghe 1981; 13. Patterson, Williams & Hiltner 1981; 14. Bailey & Ward 1981b; 15. Allen, Ward & Wright 1981; 16. Young *et al.* 1982; 17. Williams & Hiltner 1982; 18. Hutchings *et al.* 1982; 19. Motch *et al.* 1982; 20. Bailey *et al.* 1982; 21. Cropper 1985; 22. Mukai & Charles 1985; 23. Piirola, Reiz & Coyne 1987a; 24. Rosen, Mason & Cordova 1987; 25. Beuermann, Stella & Patterson 1987; 26. Watson *et al.* 1989; 27. Biermann *et al.* 1985; 28. Szkody, Liebert & Panek 1985; 29. Schaaf, Pietsch & Biermann 1987; 30. Cropper *et al.* 1989; 31. Thackeray, Wesselink & Oosterhoff 1950; 32. Herbig 1960; 33. Walker 1965a; 34. Smak 1971b; 35. Warner & Nather 1972a; 36. Liebert *et al.* 1978b; 37. Bailey 1978; 38. Liebert & Stockman 1979; 39. Visvanathan & Wickramasinghe 1979; 40. Szkody & Capps 1980; 41. Schneider & Young 1980b; 42. Wickramasinghe & Visvanathan 1980; 43. Visvanathan & Wickramasinghe 1981; 44. Allen & Cherepashchuk 1982; 45. Cowley, Crampton & Hutchings 1982; 46. Wickramasinghe & Meggitt 1982; 47. Szkody, Bailey & Hough 1983; 48. Wickramasinghe, Reid & Bessell 1984; 49. Patterson *et al.* 1984; 50. Canalle & Opher 1988; 51. Larsson 1989a; 52. Piirola, Coyne & Reiz 1990; 53. Cropper *et al.* 1990a; 54. Beuermann, Thomas & Pietsch 1991; 55. Cropper & Wickramasinghe 1993; 56. Jensen, Nousek & Nugent 1982; 57. Mason *et al.* 1983a; 58. Nousek & Pravdo 1983; 59. Bailey *et al.* 1983; 60. Visvanathan & Tuohy 1983; 61. Maraschi *et al.* 1985; 62. Tuohy, Visvanathan & Wickramasinghe 1985; 63. Wickramasinghe & Meggitt 1985b; 64. Cropper, Menzies & Tapia 1986; 65. Wickramasinghe, Tuohy & Visvanathan 1987; 66. Takalo & Nousek 1988; 67. Wright *et al.* 1988; 68. Puchnarewicz *et al.* 1990; 69. Schwope & Beuermann 1990; 70. Sambruna *et al.* 1991; 71. Schwope *et al.* 1993a; 72. Remillard *et al.* 1991; 73. Tuohy *et al.* 1988; 74. Cropper *et al.* 1990b; 75. Wickramasinghe *et al.* 1991a; 76. Echevarria, Jones & Costero 1982; 77. Mukai & Charles 1986; 79. Schmidt, Stockman & Grandi 1986; 79. Wilson *et al.* 1986; 80. Echevarria *et al.* 1988; 81. Szkody 1988; 82. Angelini, Osborne & Stella 1990; 83. Wickramasinghe *et al.* 1991c; 84. Schwope *et al.* 1991; 85. Schwope, Jordan & Beuermann 1993; 86.Thorstensen, Schommer & Charles 1983; 87. Pickles & Visvanathan 1983; 88. Visvanathan & Tuohy 1983; 89. Singh, Agrawal & Riegler 1984; 90. Wickramasinghe, Visvanathan & Tuohy 1984; 91. Hutchings, Cowley & Crampton 1985; 92. Piirola, Reiz & Coyne 1987c; 93. Cropper 1987; 94. Schwope & Beuermann 1989; 95. Beuermann & Schwope 1989; 96. Mukai & Charles 1986; 97. Miller 1982; 98. Stockman *et al.* 1983; 99 . Schmidt, Stockman & Grandi 1983; 100. Bailey *et al.* 1985; 101. Beuermann & Stella 1985;

References to Table 6.1 (cont.)

102. Cropper 1986a; 103. Vojkhanskaya *et al.* 1987; 104. Peacock *et al.* 1992; 105. Morris *et al.* 1987; 106. Cropper, Mason & Mukai 1990; 107. Beuermann *et al.* 1987; 108. Bailey *et al.* 1988; 109. Beuermann *et al.* 1987; 110. Krzeminski & Serkowski 1977; 111. Downes & Urbanski 1978; 112. Hearn & Marshall 1979; 113. Schneider & Young 1980b; 114 Gilmozzi, Messi & Natali 1981; 115. Szkody *et al.* 1981a; 116. Liebert *et al.* 1982b; 117. Middleditch 1982; 118. Imamura & Steiman-Cameron 1986; 119. Vojkhanskaya 1986a; 120. Szkody, Downes & Mateo 1988; 121. Bonnet-Bidaud *et al.* 1992; 122. Hakala *et al.* 1993; 123. Berriman & Smith 1988; 124. Beuermann, Thomas & Schwope 1988; 125. Osborne *et al.* 1988; 126. Ferrario *et al.* 1989; 127. Allen *et al.* 1989; 128. Schwope, Beuermann & Thomas 1990; 129. Bailey & Cropper 1991; 130. Ramsay *et al.* 1993; 131. Buckley *et al.* 1993; 132. Drissen *et al.* 1992; 133. van Paradijs *et al.* 1981; 134. Seifert *et al.* 1987; 135. Watson, King & Williams 1987; 136. Oestreicher & Seifert 1988; 137. Oestreicher *et al.* 1990; 138. Mittaz *et al.* 1992; 139. Cropper & Warner 1986; 140. Wickramasinghe, Ferrario & Bailey 1989; 141. Ferrario *et al.* 1992; 142. Liebert *et al.* 1982a; 143. Vojkhanskaya 1985a; 144. Beuermann *et al.* 1985; 145. Shore *et al.* 1982; 146. Berg & Duthie 1977; 147. Chanmugam & Wagner 1977; 148. Cowley & Crampton 1977; 149. Crampton & Cowley 1977; 150. Fabian *et al.* 1977; 151. Hearn & Richardson 1977; 152. Priedhorsky 1977; 153. Stockman *et al.* 1977; 154. Tapia 1977a; 155. King, Raine & Jameson 1978; 156. Priedhorsky, Krzeminski & Tapia 1978; 157. Raymond *et al.* 1979; 158. Young & Schneider 1979; 159. Panek 1980; 160. Szkody & Margon 1980; 161. Szkody *et al.* 1980; 162. Bailey & Axon 1981; 163. Crosa *et al.* 1981; 164. Fabbiano *et al.* 1981; 165. Hutchings, Crampton & Cowley 1981; 166. Latham, Liebert & Steiner 1981; 167. Rothschild *et al.* 1981; 168. Schmidt, Stockman & Margon 1981; 169. Tuohy *et al.* 1981; 170. Young, Schneider & Schectman 1981a; 171. Szkody, Raymond & Capps 1982; 172. Bailey *et al.* 1984; 173. Wickramasinghe & Martin 1985; 174. Mazeh, Kieboom & Heise 1986; 175. Stella, Beuermann & Patterson 1986; 176. Beuermann & Osborne 1985; 177 Cordova, Mason & Kahn 1985; 178. Jablonski & Busko 1985; 179. Bond *et al.* 1987a; 180. Heise *et al.* 1987; 181. Hellier *et al.* 1987; 182. Kaitchuck *et al.* 1987; 183. Beuermann & Osborne 1988; 184. Bond & Freeth 1988; 185. Rosen, Mason & Cordova 1988; 186. Bailey, Hough & Wickramasinghe 1988; 187. Siegel *et al.* 1989; 188. Hill & Watson 1990; 189. Reinsch & Beuermann 1990; 190. Rosen *et al.* 1991; 191. Hellier & Sproats 1992; 192. Vojkhanskaya 1988; 193. Götz 1991; 194. Schaich *et al.* 1992; 195. Remillard *et al.* 1986b; 196. Bonnet-Bidaud & Mouchet 1987; 197. Mason, Liebert & Schmidt 1989; 198. Szkody, Downes & Mateo 1990; 199. Ishida *et al.* 1991; 200. Silber *et al.* 1992; 201. Schwope, Thomas & Beuermann 1993; 202. Kaluzny & Semeniuk 1987; 203. Chlebowski & Kaluzny 1988; 204. Kaluzny & Chlebowski 1988; 205. Stockman, Schmidt & Lamb 1988; 206. Horne & Schneider 1989; 207 Schmidt 1990; 208 Schmidt & Stockman 1991; 209. Katz 1991; 210. Pavlenko & Pelt 1991; 211. Nousek *et al.* 1984; 212. Mukai *et al.* 1985, 1986; 213. McCarthy, Bowyer & Clarke 1986; 214. Mukai & Charles 1986; 215. Osborne *et al.* 1986, 1987; 216. Andronov & Fuhrmann 1987; 217. Mukai, Charles & Smale 1988; 218. Vojkhanskaya 1986b; 219. Beuermann *et al.* 1989; 220. Schmidt & Norsworthy 1989; 221. O'Donoghue *et al.* 1993; 222. Rosen, Done & Watson 1993; 223. Kallman *et al.* 1993; 224. Schwope & Beuermann 1993; 225. Kolb & de Kool 1994; 226. Osborne *et al.* 1994; 227. Silber & Remillard 1993; 228. Warren *et al* 1993; 229. Cropper 1989; 230. Glenn *et al.* 1994; 231. Remillard *et al* 1993; 232. Osborne *et al.* 1993; 233. Schwope, Jordan & Beuermann 1993; 234. Schwope *et al.* 1993b; 235. Wenzel 1993b; 236. Robinson & Cordova 1995; 237. Ferrario, Bailey & Wickramasinghe 1993; 238. Sekiguchi, Nakada & Bassett 1994; 239. Wickramasinghe *et al.* 1993; 240. Drissen *et al.* 1994; 241. Beuermann 1994; 242. Howell *et al.* 1995; 243. Greiner, Remillard & Motch 1994; 244. Barwig, Ritter & Bärnbantner 1994; 245. Buckley *et al.* 1995a; 246. Staubert *et al.* 1994; 247. Watson *et al.* 1995a; 248. Stockman *et al.* 1994.

Table 6.2. *Probable Polars.*

Star	P_{orb}(min)	m_V	References
RXJ 0859+05	80:	18	1
GQ Mus	85.5	17.5	Section 6.7
CP Pup	88.3	15.0	Section 6.7
V2051 Oph	89.90	15.0	2
RXJ 0953+14	90:	19	1
FY Per	93.38	11.0–14.5	3
T Pyx	105:	15.3	Section 6.7
EU Cnc	125.4	20.8	4,5
V348 Pup	146.6	15.5	6,7
V Per	154.3	18.5	Section 4.4.4.2
V2214 Oph	169.2	20.5	Section 6.7
1H 0551–819	200.5	13.4	15
CQ Dra**	238.5		8–10
RXJ 0515+01	478.97	15	1,11
V0252-3037		20.5	12
FBS 1031+590		14.5–15	13
1E 0830.9–2238		17.7	14
RXJ 0512–32		17	1
RXJ 0600–27		19	1
RXJ 2022–39		19	1

** Triple system with 4 Dra (M3III).

References: 1 Beuermann 1994; 2 Warner & O'Donoghue 1987; 3 Sazonov & Shugarov 1992; 4 Gilliland *et al.* 1991; 5 Belloni, Verbunt & Schmitt 1993; 6 Tuohy *et al.* 1990; 7 Bailey 1990; 8 Reimers, Griffin & Brown 1986; 9 Eggleton, Bailyn & Tout 1989; 10 Hvic & Urban 1991; 11 Shafter *et al.* 1995; 12 Drissen *et al.* 1994; 13 Kopylov *et al.* 1988; 14 Hertz *et al.* 1990; 15 Buckley *et al.* 1994.

$0.1 \lesssim q \lesssim 0.50$ requires the stream to survive to $r_{min} \sim 10R(1)$ from the primary (Section 2.4.2).

It is interesting to note that the magnetic moment of the secondary is

$$\mu(2) = 1.0 \times 10^{33} B_3(2) R_{10}^3(2) \text{ G cm}^3, \tag{6.11}$$

which, from the secondary fields required to drive the magnetic braking mechanism (Section 9.1.2), means that $\mu(2)$ is comparable with the magnetic moments of the primaries in polars.

6.2.2 *The Accretion Stream*

If $r_\mu \geq R_{L_1}(1)$ then ionized gas leaving the secondary could be expected to be attached to the field lines of the primary for its entire interstar trajectory. Early calculations assumed this configuration (e.g., Schneider & Young 1980a,b), but Liebert & Stockman (1985) pointed out that for any system in which $r_\mu < R_{L_1}(1)$, gas leaving the L_1 point would follow the same stream trajectory as in non-magnetic systems until the radius r_μ is reached (Figure 6.1). Mukai (1988) shows that in ST LMi $r_\mu \approx R_{L_1}(1)$ and

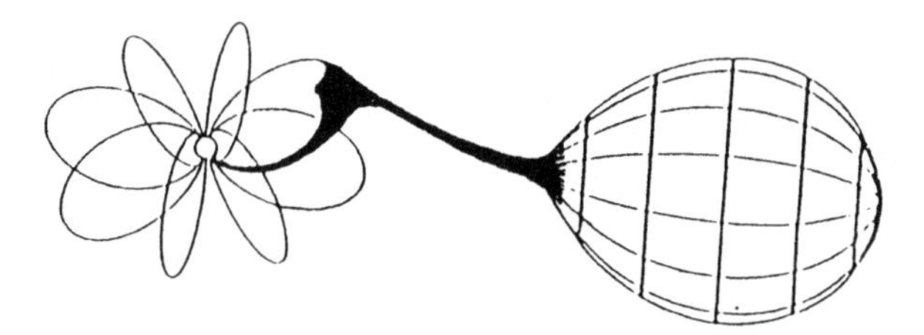

Figure 6.1 Schematic view of a polar. The secondary loses mass via a stream that retains its identity until magnetic forces are able to control the flow. The only field lines shown of the primary are those corresponding to the radius r_μ. Adapted from Cropper (1990).

in QQ Vul $r_\mu < R_{L_1}(1)$, so both configurations actually exist. In general, for polars with $P_{\rm orb} < 2$ h Mukai (see also Hameury, King & Lasota (1986b)) expects $r_\mu > R_{L_1}(1)$, but for $P_{\rm orb} > 2$ h, $r_\mu < R_{L_1}(1)$. Note also that a system for which normally $r_\mu < R_{L_1}(1)$ may switch to the other configuration if $\dot{M}$ drops to a low value (Section 6.3.1).

In polars where $r_\mu < R_{L_1}(1)$ the gas in the stream decreases in density by up to a factor of 30 as it accelerates. Before reaching r_μ the stream will pass through a region, commencing at radius $r_{\rm p}$, where the magnetic pressure exceeds the thermal pressure of the gas. From this point the field determines the density and shape of the stream, but not its trajectory. As the stream probably has a denser core than periphery, the effect of the field will first be felt in its outer parts (Mukai 1988). At $r_{\rm p}$, the field strengths are typically a few hundred gauss.

As the gas accelerates, the magnetic pressure increases faster than the gas can adjust subsonically, so the stream is shattered into small fragments. In this region ($r_\mu \leq r \leq r_{\rm p}$) a number of competing processes result in fine comminution of the stream and threading of the blobs onto magnetic field lines. These processes are poorly understood (Burnard, Lea & Arons 1983; Liebert & Stockman 1985; Hameury, King & Lasota 1986a; Lamb & Melia 1986, 1987, 1988) but include Rayleigh–Taylor instabilities which arise from the opposition of thc magnetic field to the stream – which is equivalent to a heavy fluid being supported against gravity by a light fluid. This produces large blobs which would be only slowly penetrated by the field. The blobs are compressed by the increasing magnetic pressure, but at the same time they are eroded at their surfaces by Kelvin–Helmholtz instabilities – caused by strong shear in a fluid with a density gradient (as with wind blowing over the surface of the sea). These small droplets are rapidly penetrated by the field, are then confined to flow along the field lines, and thus constitute a cross-wind capable of further ablating the larger blobs. As any relative velocities are likely to be supersonic, there will also be shock heating, radiation from which cools the flow.

The gas arriving at r_μ is therefore thought to consist predominantly of a spray of small blobs (typically with radii $<10^8$ cm), with the possibility of some threaded larger blobs that survived intact, and perhaps some unthreaded large blobs which will penetrate further before becoming threaded. There will therefore be a spread in azimuth in the threading region, with the smallest and most rapidly threaded blobs falling first to the surface.

Within the distance r_μ the field redirects the flow and the stream follows the field lines – usually out of the plane of the orbit[2] – down to the white dwarf surface. The region of impact is the zone of footpoints of field lines that pass through the interaction region where the stream material becomes threaded. Computations by Mukai (1988), adopting a number of simplifying assumptions (e.g., no distortion of the field by the stream), show that gas will strike the white dwarf in arcs typically $\sim 10°$ long and displaced from the primary's magnetic pole by $\sim 10°$ (as first pointed out by Lamb (1985)). Because of the way in which blobs are threaded, there may be a correlation of blob size, and hence physical conditions, along the arc.

If $r_\mu > R_{L_1}(1)$, there are no closed field lines starting at L_1 within the Roche lobe of the primary, which inhibits accretion directly from L_1. In observed polars the stream penetrates for a considerable distance into the magnetosphere for even the strongest fields.

The location and form of the accretion region on the white dwarf will depend on the inclination of the magnetic field and on the location of the threading region. Large differences from polar to polar should therefore be expected, and are indeed observed (Section 6.3.2). In particular, depending on the location of the threading region in the magnetic field, matter may be fed towards either one or both polar regions. One aspect of this may be visualized by reference to Figure 6.2, which shows cross-sections perpendicular to the orbital plane for three dipole inclinations ($\beta = 0°$, $45°$ and $90°$) and for the longitude of the magnetic axis with respect to line of centres of the stars $\psi = 0$: the definition of the various angles is shown in Figure 6.3. From energy considerations, gas threaded onto a field line will only be able to reach the stellar surface if in moving along the line it does not pass outside the Roche lobe of the primary (therefore the field lines in Figure 6.2 do not extend beyond the lobe).

For $\beta = 0°$, gas threaded very near to L_1 cannot reach the primary, but a small penetration into the field can reach and feed field lines that would result in equal feeding of two polar regions. In general these regions will not be diametrically opposed on the surface of the primary. For $\beta = 45°$, only feeding of the pole nearest to the secondary is possible, unless the stream penetrates quite deep into the field. For $\beta = 90°$, the stream (moving in its orbit into or out of the plane of the page) does not have to penetrate far before reaching field lines that can feed both poles.

In general, Ferrario, Wickramasinghe & Tuohy (1989) find that this mode of two-pole accretion can occur only if the threading region (r_μ–$r_{\rm p}$) lies at a distance less than $r_{\rm c}$ given by

$$r_{\rm c} \approx \frac{0.85 R_{L_1}(1)}{1 + \tan^2\beta \, \cos^2\psi}. \qquad (6.12)$$

There is an alternative way of feeding both poles: if the threading region lies on the line of intersection of the orbital plane and the magnetic equatorial plane then gas will be channelled equally easily down to footpoints near each pole, provided that the field line lies wholly within the primary's Roche lobe.

[2] Once a particle or blob has become attached to a field line it is constrained to revolve around the primary with the angular velocity of the binary system. This provides much less acceleration than that required for centrifugal balance and the net force on the particle has a component along the field lines.

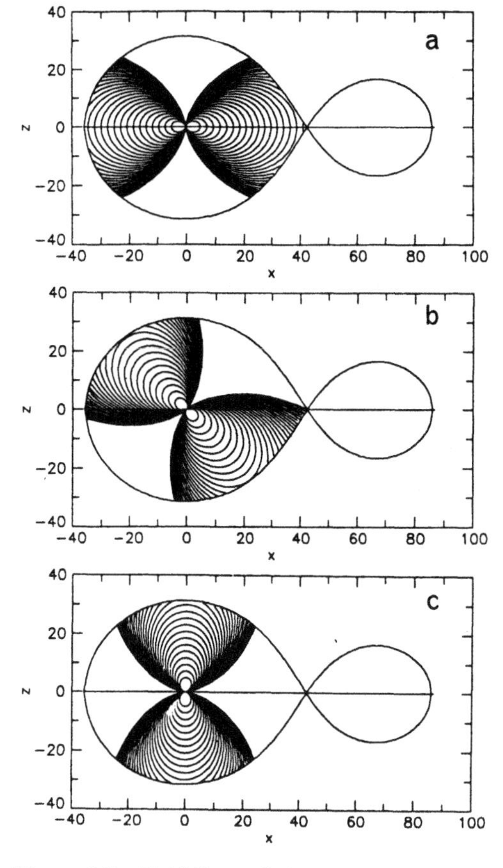

Figure 6.2 Field lines of the primary which are contained within the Roche lobe for different values of dipole inclination: (a) $\beta = 0°$, (b) $\beta = 45°$, (c) $\beta = 90°$. The projection plane is perpendicular to the orbital plane, passing through the centres of the stars. The axes are in units of radii of the primary. From Ferrario, Wickramasinghe & Tuohy (1989).

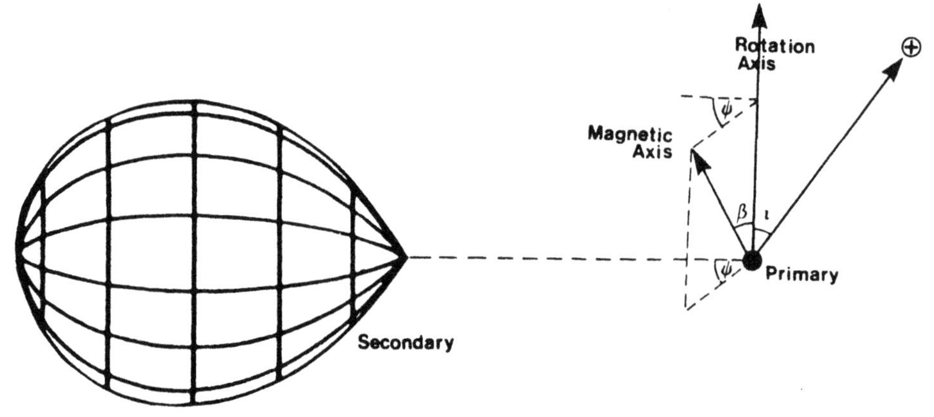

Figure 6.3 Definitions of angles commonly used in polar geometry. The inclination i is the angle between the rotation axis and the line of sight, β is the angle between the rotation and magnetic axes, y is the angle between the line of centres of the stars and the projection of the magnetic axis on the orbital plane. From Cropper (1990).

The situation is further complicated because many polars have *offset dipole fields*, making one pole typically twice as strong as the other and allowing the weaker pole to be fed more easily than the stronger.

What is actually seen from Earth depends sensitively on i and β. For example, a system with small i and small β will have one pole continuously visible and one permanently hidden behind the primary. The system will have very different observed properties according to which of the polar regions is accreting.

6.2.3 The Standard Accretion Column

If it had just started accreting, gas falling supersonically along the field lines would impact the atmosphere of the primary, producing a shock front. As the gas below the front in general would not cool as fast as the heating caused by the shock, it would expand, moving the shock above the stellar surface. The equilibrium height of this standoff shock is determined by the requirement that the post-shock flow must have sufficient time to cool and decelerate to match conditions in the stellar photosphere.

Early models of the resultant accretion column assumed circular cross-section, location at the magnetic pole (i.e., *radial* accretion) and uniform post-shock conditions (Hoshi 1973; Fabian, Pringle & Rees 1976; Masters *et al.* 1977; Lamb & Masters 1979). Although displaced by more realistic models (described below), this simplified picture includes most of the relevant physics.

The reduction in velocity from supersonic to subsonic flow through the thin shock is roughly a factor of 4; from continuity the density increases by a comparable amount. The height of the shock is determined by the efficiency of the cooling mechanisms acting below the shock, which are

(a) bremsstrahlung (free–free) emission (principally in the X-ray region) by free electrons,
(b) cyclotron emission by semi-relativistic electrons spiralling around the magnetic field lines,
(c) Compton cooling through scattering of the relatively lower energy photons by the shocked electrons.

The relative importance of these (and for many calculations only the first two have been included) depends on conditions in the post-shock region, especially on B and on the accretion rate per unit area (also known as the *specific accretion rate*). The latter is related to L/f, where L is the total luminosity (proportional to the accretion rate) and f is the fraction of the stellar surface covered by the accretion column.

Lamb & Masters (1979) delineate three regions in the log B–log L/f plane (Figure 6.4). In the first, above a critical line bremsstrahlung dominates the radiative cooling. Below that cyclotron emission dominates, which cools the electrons but not the ions, so two-temperature plasma calculations are required. In the third region, cyclotron cooling is so effective that even the ions, exchanging energy with the electrons, are driven away from a Maxwellian distribution of velocities and the shock structure collapses.

From equation (2.56) the temperature of the shock is

$$T_{sh} = 3.7 \times 10^8 M_1(1) R_9^{-1}(1) \text{ K} \tag{6.13}$$

and the density in the post-shock gas flow is $4 \times [\dot{M}(1)/4\pi R^2(1) f v_{\rm ff}]$ which gives an electron density

$$N_e = 3.1 \times 10^{15} \dot{M}_{16}(1) M_1^{1/2}(1) R_9^{-3/2}(1) f_{-3}^{-1} \text{ cm}^{-3}. \tag{6.14}$$

To remain below the Eddington luminosity (equation (5.4b)) per unit area requires $\dot{M}(1) \lesssim 1.25 \times 10^{18} R^2(1) f_{-3}$ g s^{-1} and $N_e \lesssim 5 \times 10^{17}$ cm^{-3}.

A column that is cooled by free–free emission has a shock height h_s given by

$$h_s = \text{accretion luminosity per unit area}/j_{\rm ff}$$

$$= 9.6 \times 10^7 M_1(1) R_9^{-1}(1) \left(\frac{N_e}{10^{16} \text{ cm}^{-3}}\right)^{-1} \text{cm}, \tag{6.15}$$

where the free–free emissivity $j_{\rm ff} = 2.0 \times 10^{-27} N_e^2 T_e^{1/2}$ erg cm^{-3} s^{-1}.

The flux distribution from the accretion column comprises three components:

(a) Hard X-ray bremsstrahlung (typically $kT_{\rm br} \sim 30$ keV $\approx 3 \times 10^8$ K) emitted from the column, which is optically thin to hard X-rays.

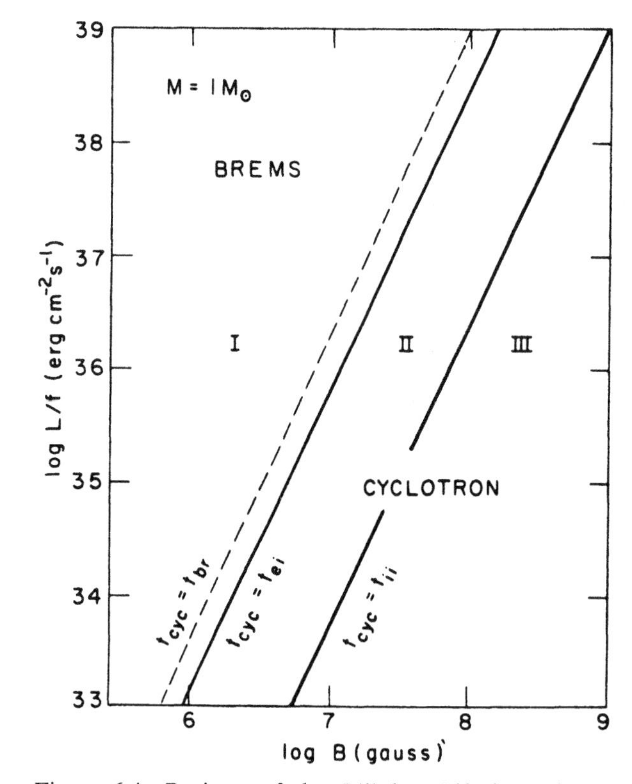

Figure 6.4 Regimes of the $B(1)$–log L/f plane, for $M_1(1) = 1.0$. Bremsstrahlung cooling dominates above the dashed line, cyclotron cooling below. In region I, above the line marked $t_{\rm cyc} = t_{\rm ei}$, a one-temperature plasma obtains. In region II a two-fluid treatment is required. In region III the cyclotron cooling is so efficient that the fluid becomes non-hydrodynamic. From Lamb & Masters (1979).

(b) Cyclotron emission from the column, which is optically thick for low harmonics but optically thin at higher ones.

(c) As the column is situated quite close to the stellar surface, nearly half of the emission is intercepted by the primary. This radiation is mostly reflected for $E \gtrsim 30$ keV but the lower energies are absorbed, thermalized and reemitted as an approximately blackbody spectrum in the UV or soft X-ray region with $kT_{BB} \sim 40$ eV ($\approx 4 \times 10^5$ K) (Milgrom & Salpeter 1975).

The schematic appearance of this standard accretion column is shown in Figure 6.5.

Computed spectral energy distributions for $B = 20$ MG and $\log L/f = 35$ and 37 are shown in Figure 6.6. As seen from Figure 6.4, these examples lie respectively in the cyclotron and bremsstrahlung cooled regimes. Note in particular the behaviour of the cyclotron emission peak: as the cyclotron opacity falls rapidly with increasing energy the result is an optically thick, unpolarized Rayleigh–Jeans flux distribution with $F(\nu) \propto \nu^2$ at low energies, rising to a highly polarized peak near the cyclotron harmonic at which the gas becomes optically thin and then falling at higher energies proportional to the opacity.

Later calculations have introduced a more physically consistent picture: computation of the hydrodynamic flow through the post-shock region to deduce the detailed temperature and density structure of the column, allowance for radiation emitted through the sides of the column, inclusion of electron conduction and of the interaction of the various sources of radiation with the gas itself (Imamura & Durisen 1983; Imamura 1984; Imamura *et al.* 1987). The effect of radiation pressure, Compton scattering and electron thermal conduction on the pre-shock flow is also taken into account (Kylafis & Lamb 1979, 1982; Imamura *et al.* 1987). However, the computations so far performed assume a spherically symmetric shock wave.

The resulting temperature structure is shown schematically in Figure 6.7. In region A the inflowing gas is heated in a *precursor* by hot electrons moving upstream from the post-shock gas (whereas the ions in the post-shock region have thermal velocities comparable with the pre-shock flow, the electrons have velocities $\sim (m_{\rm ion}/m_{\rm electr})^{1/2}$ larger). The ion shock occurs at $R_* + d_{\rm ion}$, below which Compton cooling is largely

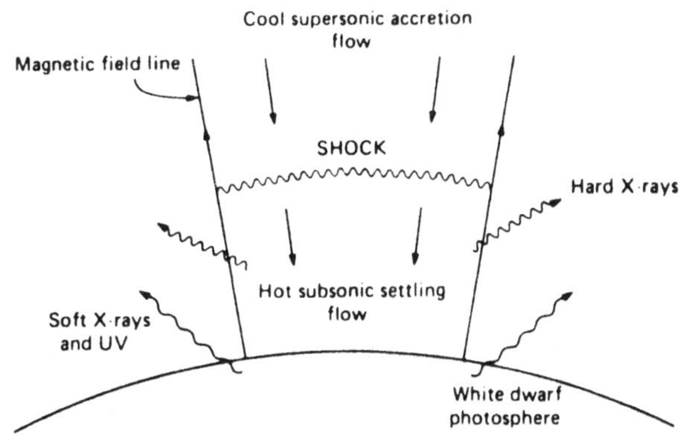

Figure 6.5 Schematic picture of a standard accretion column. From Watson (1986).

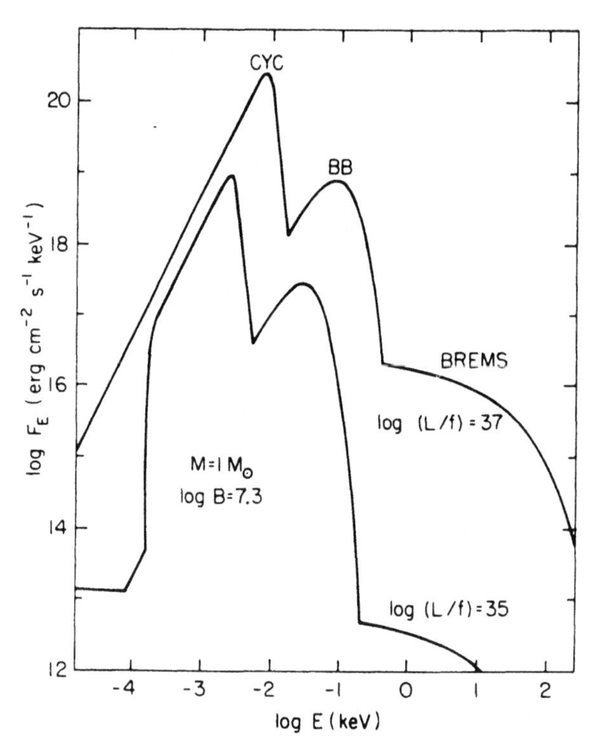

Figure 6.6 Flux distributions produced by two different accretion rates onto a $M_1(1) = 1.0$ primary with $B(1) = 20$ MG. From Lamb & Masters (1979).

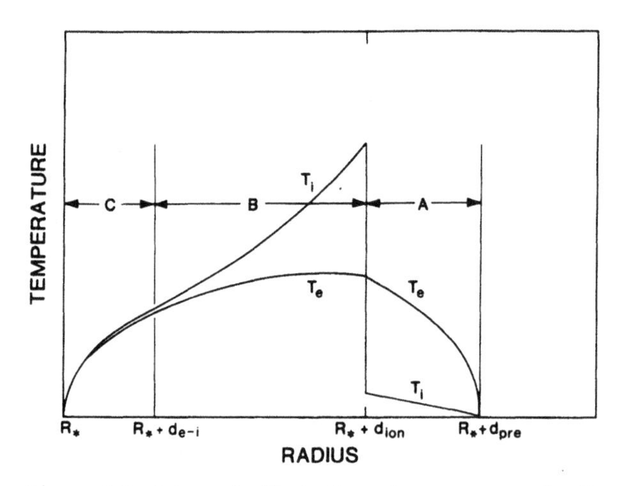

Figure 6.7 Schematic illustration of temperature distribution above the surface of the white dwarf primary (at R_*). The ion shock occurs at $R_* + d_{ion}$; the hot electron precursor starts at $R_* + d_{pre}$. The electron and ion temperatures are designated T_e and T_i respectively. Region A is the precursor. Most of the bremsstrahlung radiation is produced in region C. From Imamura *et al.* (1987).

responsible for keeping $T_{\rm elec} < T_{\rm ion}$. However, near the surface of the star the density of the flow increases rapidly and as the electron–ion energy exchange is $\propto \rho^2$ the temperatures equalize. Also the emission of bremsstrahlung radiation is $\propto \rho^2$, so this is the dominant region of cooling.

Electron scattering in the pre-shock flow results in a greater optical depth $\tau_\parallel$ parallel to the field lines than that $\tau_\perp$ for photons escaping transversely from the post-shock region. Imamura & Durisen (1983) (see also King & Shaviv (1984b)) find

$$\frac{\tau_\perp}{\tau_\parallel} = 0.95\left(\frac{f}{0.1}\right)^{1/2} \tag{6.16}$$

and

$$\tau_\parallel = \frac{0.027}{f}\, L_{35} R_9^{1/2}(1) M_1^{-3/2}(1) \tag{6.17}$$

$$\approx 0.02\left(\frac{L_{35}}{f}\right)\, M_1^{-5/3}(1) \tag{6.18}$$

from equation (2.83a), where L_{35} is the total accretion luminosity. Hence, if $L/f \gtrsim 5 \times 10^{36}$ erg s^{-1} hard X-ray emission will be significantly non-isotropic, emerging as a *fan-beam* perpendicular to the field lines. As electron scattering is independent of energy, the shape of the beam (and hence the pulse shape as the primary rotates) will not depend on energy. However, the Imamura & Durisen models do not include photoabsorption in the pre-shock flow, which may affect the shape of the beam at energies $\lesssim$2 keV.

6.2.4 *Cyclotron Emission*

The observations of continuum polarization and of cyclotron harmonics in the spectra of the polars (Section 6.5.4) carry a wealth of information about the magnetic fields and physical conditions in the accretion regions. Considerable effort has therefore been expended in modelling the transfer of radiation through a standard accretion column (Chanmugam & Dulk 1981; Meggitt & Wickramasinghe 1984; Barrett & Chanmugam 1984; Wickramasinghe & Meggitt 1985a; Cannalle & Opher 1988; Wickramasinghe 1988b: this last reference includes a comprehensive review, as does Schwope 1990).

An individual electron radiates over a range of harmonics, the distribution of power among the harmonics being determined by the energy of the electron. The power shifts to higher harmonics at higher energies, e.g., at $kT_{\rm elec} = 50$ keV only a few per cent is radiated at the fundamental. As a result of the Maxwellian tail of the electron velocity distribution and relativistic mass increase the profiles of the cyclotron emission humps are asymmetric.

The polarized components of the radiation at frequency ω emerging from an homogeneous column are determined by $T_{\rm elec}, \omega/\omega_{\rm c}$ (see equation (6.1)), the angle θ with respect to the magnetic field and a parameter Λ, which is approximately the

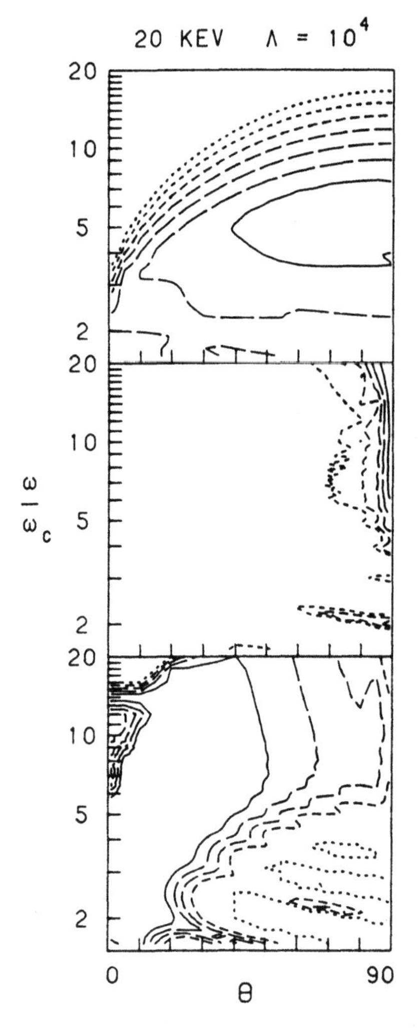

Figure 6.8 Contours of intensity (upper panel), linear polarization (centre) and circular polarization (lower) for $kT_{\rm elec} = 20$ keV, $N_{\rm e} = 10^{16}$ cm^{-3} and $\Lambda = 10^4$. Intensity contours differ by 1 mag, the solid curve being the brightest. Linear polarization contours are at 80 (solid curve), 60, 40, 20 and 10%. Circular polarization contours are at 80 (solid), 60, 40, 20 and 0% (dotted). From Wickramasinghe & Meggitt (1985a).

optical depth at the cyclotron fundamental:

$$\Lambda = 6.06 \times 10^8 \left(\frac{h_{\rm s}}{10^8 \text{ cm}}\right) \left(\frac{N_{\rm e}}{10^{16} \text{ cm}^{-3}}\right) B_7^{-1}(1) \tag{6.19a}$$

$$= 6.0 \times 10^8 M_1(1) R_9^{-1}(1) B_7^{-1}(1) \text{ cm for free–free cooling} \tag{6.19b}$$

from equation (6.15).

An example of results obtained including both cyclotron and free–free opacity (and in conditions where electron scattering is not important) is given in Figure 6.8. The ordinate, $\omega/\omega_{\rm c}$, is labelled in cyclotron harmonic number. For the conditions considered, the flux peaks strongly at roughly 20° from the perpendicular to the magnetic field direction (the *cyclotron beaming* effect) and in the 4–6th harmonics. (Cyclotron beaming becomes stronger with increase in harmonic number.) The flux diminishes slowly towards lower harmonics but rapidly towards high harmonics. Linear polarization is produced only by the higher harmonics and is concentrated strongly in the direction transverse to the field. Circular polarization is a maximum for radiation of low harmonic number travelling along the field lines, changing to $\sim 20°$ to the field for the highest harmonics.

At lower temperatures than that shown in Figure 6.8 ($kT_{\rm elec} \lesssim 10\,{\rm keV}$) the flux distribution shows a strong harmonic structure with linear polarization (whose plane is always parallel to the projected field) seen at larger angles from the perpendicular (Wickramasinghe & Meggitt 1985a). An observed cyclotron spectrum can in principle give the magnetic field directly from the frequency spacing of the harmonics, and the temperature from the width of the features. In practice the spread of B and Λ through the emitting region introduces small ambiguities.

Models from which $T_{\rm elec}$, B and Λ can be deduced from observations of cyclotron harmonics, are given by Wickramasinghe & Meggitt (1982), Barrett & Chanmugam (1985), Wickramasinghe (1988b), Chanmugam *et al.* (1989) and Chanmugam & Langer (1991). Because of the different path lengths through a cylindrical column, strong cyclotron harmonic features are not produced at high accretion rates.

6.2.5 *Inhomogeneous Accretion*

Although the predicted flux distributions (both optical and X-ray) and the polarization properties of the standard accretion column are in broad agreement with the observations of polars, a number of discrepancies have forced a generalization of the model. We have already seen in Section 6.2.2 that an accretion *arc* rather than a cylindrical column is formed – this carries with it the expectation that accretion will not be radial. Furthermore, as already pointed out by Liebert & Stockman (1985) (at a 1983 Conference) and elaborated by Schmidt, Stockman & Grandi (1983), Stockman & Lubenow (1987) and Stockman (1988), the accretion stream from the secondary will not have a uniform transverse section but is more likely to be at least as structured as a high density core with a less dense sheathing. Applied to the simple accretion column, this implies a shock height varying with a minimum at the axis to a maximum on the circumference. Physical conditions then vary considerably in section across the post-shock flow, with higher energy radiation originating from a bremsstrahlung-dominated narrow dense core and lower energy emissions arising from the cyclotron-dominated halo (and overwhelming the core in the optical region because of its much greater volume).

The first inhomogeneous accretion flow models were parameterized with idealized accretion profiles such as Gaussian, exponential or power-law, and had a dramatic consequence on the predicted energy spectrum of the accretion column. Figure 6.9 shows the results for a model with $L = 10^{33}$ erg s^{-1}, $< L/f > = 10^{38}$ erg s^{-1}, $M_1(1) = 0.8$ and $B = 20\,{\rm MG}$ (Stockman & Lubenow 1987). Instead of the cyclotron flux peaking in the UV as in the homogeneous case (Figure 6.6), it now peaks in the red or IR and there is a $F(\nu) \mathrel{\tilde{\propto}} \nu^{-1}$ dependence to the short wavelength side of the peak

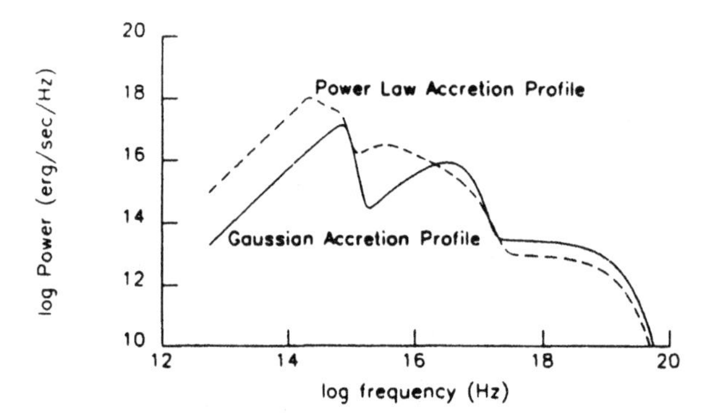

Figure 6.9 Flux distributions for inhomogeneous accretion, using power-law and Gaussian accretion rate profiles across the stream. From Stockman & Lubenow (1987).

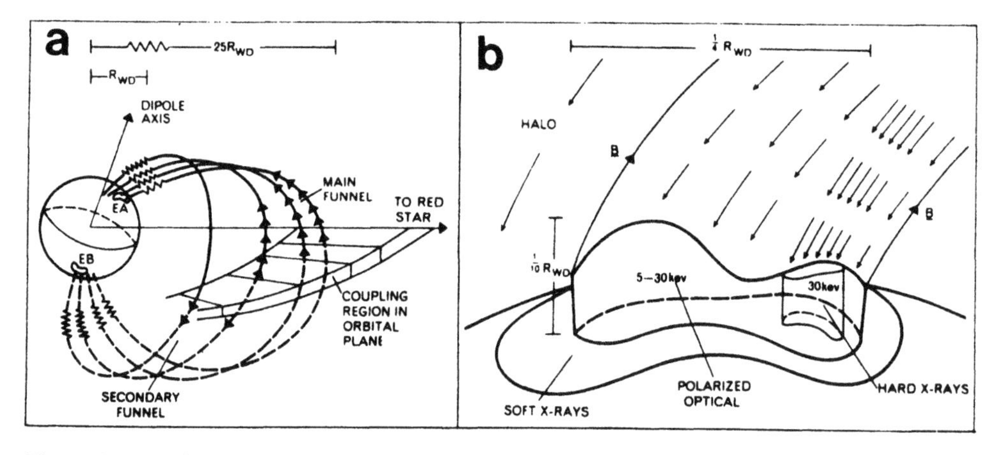

Figure 6.10 Schematic views of (a) the coupling region and the two accretion funnels leading to accretion shocks at EA and EB near the surface of the primary, and (b) the structure of the accretion zone showing a high density compact region emitting hard X-rays and an extended lower density region emitting polarized optical and IR radition. From Wickramasinghe & Meggitt (1985a).

instead of the much steeper $F(\nu) \propto \nu^{-8}$ for optically thin cyclotron emission from a homogeneous region (Dulk 1985). The result has been optimized to give the best agreement with observation, but it shows the capabilities of the model. Further calculations are given by Wu & Chanmugam (1988, 1989) and Wickramasinghe & Ferrario (1988).

The true picture is even more complex; a suggested schematic view is shown in Figure 6.10 (Wickramasinghe 1990). The threading region may funnel gas towards both magnetic poles (Figure 6.10(a)). The extent of the accretion zone results in a spread of inclination of field lines over the emitting area. Thus not only the range of accretion rates but also the variation in field strengths and projected angles to the line of sight must be taken into account.

Modelling of these inhomogeneous accretion zones has gradually become more realistic. Wickramasinghe & Ferrario (1988) assumed a circular cross-section accretion region and allowed for the variable height of the shock caused by different density and temperature profiles. Chanmugam & Wu (1990) and Wu & Chanmugam (1990) computed spectra and polarization properties for columns with an accretion rate varying across the profile, but only applicable for $10^{14} \lesssim N_e \lesssim 10^{17}$ cm^{-3} and thicknesses $> 10^7$ cm. Brainerd (1989) included cyclotron emission from above the accretion shock. Further models for homogeneous arc-shaped accretion zones, taking into account field spread, were given by Ferrario & Wickramasinghe (1990).

The first three-dimensional computations, including self-consistent computation of post-shock conditions through allowance for cyclotron and brehmsstrahlung cooling, were carried out by Wu & Wickramasinghe (1990) and presented the angular dependence of intensity and polarization from axisymmetric shocks. Except for low accretion rates, these models do not produce strong cyclotron harmonic features. Consequently Wu & Wickramasinghe (1992) computed the polarized emission from ridge-shaped regions 2×10^7 cm long and 2×10^6 cm thick in uniform fields with various accretion profiles. Such narrow ridges are less optically thick for cyclotron emission than are axisymmetric cylindrical regions and consequently produce prominent cyclotron harmonics over a wide range of viewing directions. This is in better agreement with observations (Section 6.5.4).

Eventually it should be possible, from the variations of energy distribution and polarization variations (supplemented by X-ray observations) around the orbit of a polar, to model the average structure of the accretion region during that orbit – and to follow the variations of structure on time scales of days or longer and tie these to changes in $\dot{M}$.

There is another aspect of inhomogeneous accretion that can have important observable consequences. The clumpiness of the accretion flow, resulting from conditions in the stream-threading region (Section 6.2.2), will lead to variations in the accretion region on time scales of seconds or minutes. The larger blobs of gas may have sufficient density that the shock front they produce will be within the atmosphere of the white dwarf; if this is deeper than an optical depth of unity then the kinetic energy of the blob will be thermalized and emitted as soft X-rays instead of through the usual accretion column structure (Kuijpers & Pringle 1982).

A spherical blob of gas falling from the threading zone does not retain its shape – it is compressed by the converging field lines and elongated by tidal forces (Kuijpers & Pringle 1982). This elongated structure can be thought of as a mini-accretion column, existing for the length of time it takes for the column to pass into the surface of the star. Calculations (Frank, King & Lasota 1988) show that even an *average* density blob in a polar accretion stream, when squeezed and stretched on its passage through the magnetosphere, will have sufficient density and duration to bury the shockfront in the atmosphere of the white dwarf. However, the further condition must be met that the blob has sufficient momentum to carry the gas in the atmosphere in front of it – this translates into a requirement on blob mass and hence length. Then only the larger blobs produce buried shocks.

Figure 6.10(b) demonstrates the separation of hard and soft X-ray emitting regions consequent upon a range of blob sizes across the accretion flow. (The 'halo' flow is discussed in Section 6.5.3.)

6.3 Photometric and Polarimetric Observations

6.3.1 Long Term Light Curves

As seen from Table 6.1, many polars show ranges in brightness greatly exceeding that expected solely from orbital variations. For those polars where a long enough archival plate record is available, it is found that the system usually hovers around a maximum brightness, but that at intervals typically of ~ years an excursion to a lower brightness occurs. Figure 6.11 illustrates the light curve of AM Her from 1933 to 1976[3] (the photographic record actually extends back to 1890) in which a range of magnitudes from 12.0 to 15.5 occurs. Although there is no preferred minimum magnitude during the fainter phases in this light curve (Feigelson, Dexter & Liller 1978), that for 1982–1993 (Götz 1993; Honeycutt *et al.* 1994[4]) shows that AM Her does not descend below, and can stay for hundreds of days at $m_B \sim 15.6$.

Long term light curves for other polars are given for BY Cam (Remillard *et al.* 1986a), EF Eri (Griffiths *et al.* 1979), BL Hyi (Vojkhanskaya 1989), ST LMi (Götz 1987), MR Ser (Romano 1983), AN UMa (Meinunger 1976) and QQ Vul (Fuhrmann 1984). In many cases isolated observations of low states of other polars have been made.

From the description of the accretion process given earlier it is obvious that significant long term reductions in optical luminosity imply a lowering of the average rate of mass transfer. Similar variations occur in NLs (Section 4.3.2) and they also occur in the intermediate polars (Section 7.3.1).

V834 Cen (Middleditch *et al.* 1991), AN UMa (Meinunger 1976; Garnavich & Szkody 1988) and QQ Vul (Osborne *et al.* 1987) have all shown 'super–high' states at 1–2 mag brighter than in their usual high states.

Garnavich & Szkody (1988) found a strong correlation between the range (from high to low state) and P_{orb}, with smallest range $\sim 1\frac{1}{2}$ mag occurring at $P_{orb} \sim 130$ min and largest $\sim 4\frac{1}{2}$ mag at $P_{orb} \sim 80$ min. However, the more extensive observations now available (Table 6.1) show that large range systems occur also near $P_{orb} \sim 130$ min.

6.3.2 Orbital Variations

The polars show brightness modulation at P_{orb}, with amplitude increasing towards longer wavelength (in some there is little or no variation at U). Consequently the colours are also modulated at P_{orb}. In general, with a few exceptions, they are UV-rich, as can be seen from the mean (U–B, B–V) for VV Pup, EF Eri, V834 Cen and AM Her respectively: (−1.1, 0.1), (−0.8, 0.65), (−0.87, 0.2), (0.05, 0.58)

The first attempts to understand the variations of brightness and polarization observed around the orbital cycles of polars assumed a standard accretion column located at the magnetic pole. Although able to account for gross effects, such models failed to reproduce the asymmetries seen in the observations and the wide wavelength range over which circular and linear polarization are detected. The recent models, which include the effects of inhomogeneous accretion, have resulted in very large improvements in agreement between observation and theory.

[3] An instructive comparison between the photographic and amateur observed visual light curves from 1982 to 1987 is given by Götz (1989).

[4] Note that the Honeycutt *et al.* m_v scale is ~2 mag too bright.

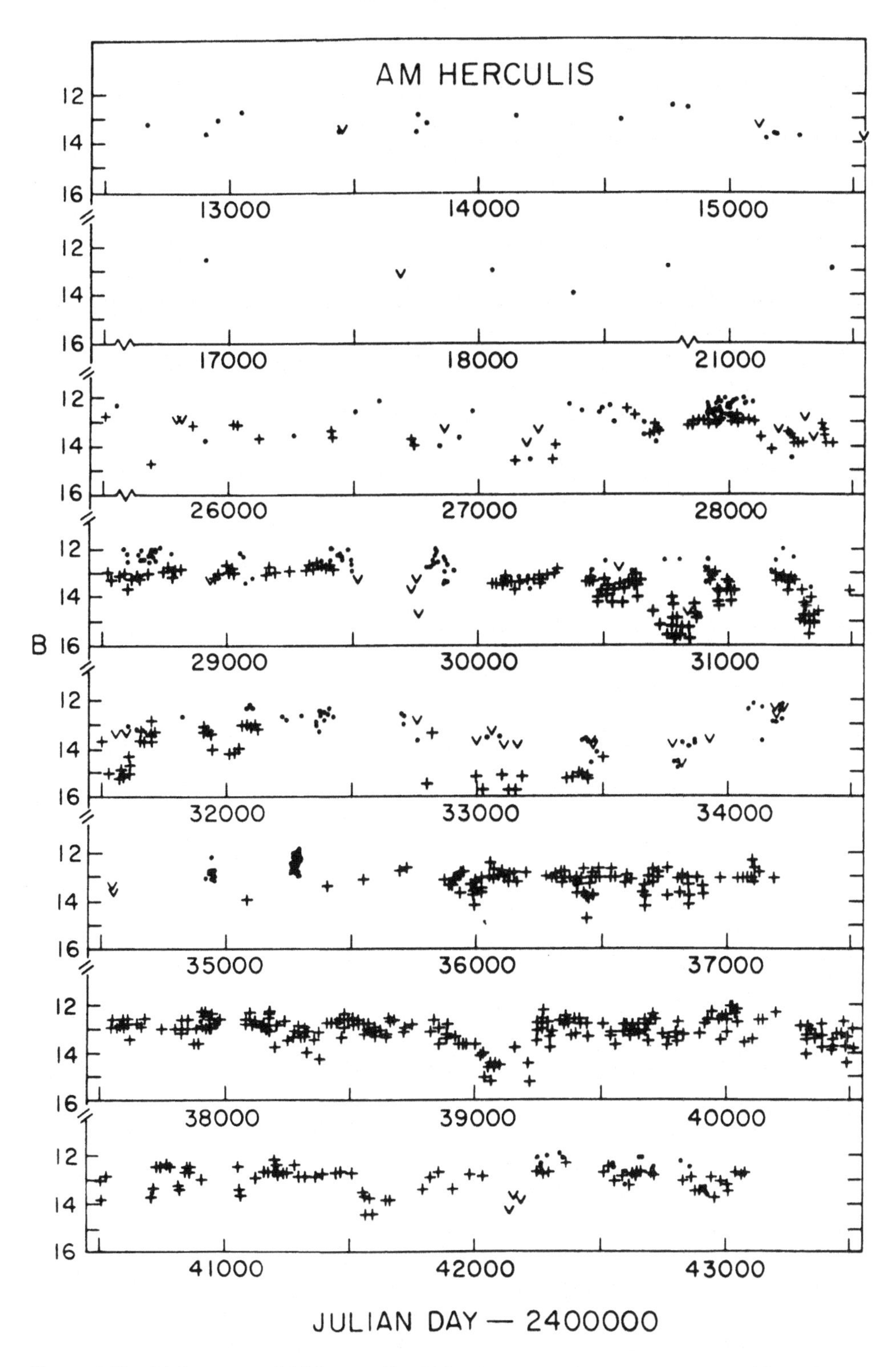

Figure 6.11 Light curve of AM Herculis 1933–76. From Feigelson, Dexter & Liller (1978).

6.3.2.1 Geometrical Considerations

Before illustrating the diversity of photometric and polarimetric observations, the following variety of definitions and relationships should be noted.

As most polars are non-eclipsing, and spectroscopic information on the position in orbit of the secondary is often absent or inaccurate, the usual use of superior conjunction of the primary as the definition of orbital phase zero is frequently not possible. Instead, if there is a prominent linear polarization spike in the observations, this is used to define $\varphi = 0$; otherwise some other photometric signature may be used.

Even among the models of polars there are two definitions of phase zero, neither of which is compatible with the observational definition. One definition (a) uses the moment when the most prominent magnetic pole is nearest to the observer (e.g., Brainerd & Lamb 1985); the other (b) adopts the instant when the magnetic pole is directly beyond the rotation pole, i.e., at its most distant from the observer (e.g., Cropper 1990). The differences between these and the observational definitions should be clear: a linear polarization maximum will occur when an accretion column is viewed most nearly transversely, a circular polarization maximum will occur when looking most nearly along the field lines (Figure 6.8).

The plane of linear polarization is parallel to the magnetic field (Section 6.2.4). For an axially symmetric field, the angle of projection of the field lines, and hence the position angle of the linear polarization is, with definition (b)

$$\cot \theta = \cot \varphi \cos i + \cot \beta \operatorname{cosec} \varphi \sin i. \tag{6.20}$$

If accretion occurs at a single magnetic pole, then the viewing angle α between the direction of the field lines and the observer is given by

$$\cos \alpha = \cos i \cos \beta - \sin i \sin \beta \cos \varphi. \tag{6.21}$$

When a linear polarization spike is observed, the accretion column is seen transversely, at which time $\alpha = 90°$ and

$$\cos \varphi = \cot i \cot \beta. \tag{6.22}$$

Differentiating equation (6.20) and substituting from equations (6.20) and (6.22) gives (Meggitt & Wickramasinghe 1982)

$$d\theta/d\varphi = \cos i \qquad \text{when } \alpha = 90°. \tag{6.23}$$

From the observed variation of θ with φ, equations (6.20) and (6.23) provide a means of estimating i and β. The values listed in Table 6.1 have been obtained in this way (Cropper (1990)).

If the accretion region disappears behind the limb of the primary for a fraction f_d of an orbital cycle then an additional relationship, $\cot i = \tan \beta \cos \pi f_d$ obtains. This can be generalized to allow for the height of the accretion region and its angular extent (Cropper 1986a).

In many polars accretion of different strengths occurs near two poles (Meggitt & Wickramasinghe 1989). At times both accretion regions are visible simultaneously, producing superimposed polarized emissions that are difficult to deconvolve. However, with sufficient time resolution the profiles of some polarization pulses can be seen to be composed of contributions from both poles.

6.3.2.2 V834 Centauri

The optical broad band observations shown in Figure 6.12 for V834 Cen are averages for 10 orbital cycles obtained over 3 d (Cropper, Menzies & Tapia 1986) and are typical for a one-pole system where the pole is continually visible. In this case $\varphi = 0$ corresponds to the minimum in the light curve and is a result of viewing most nearly

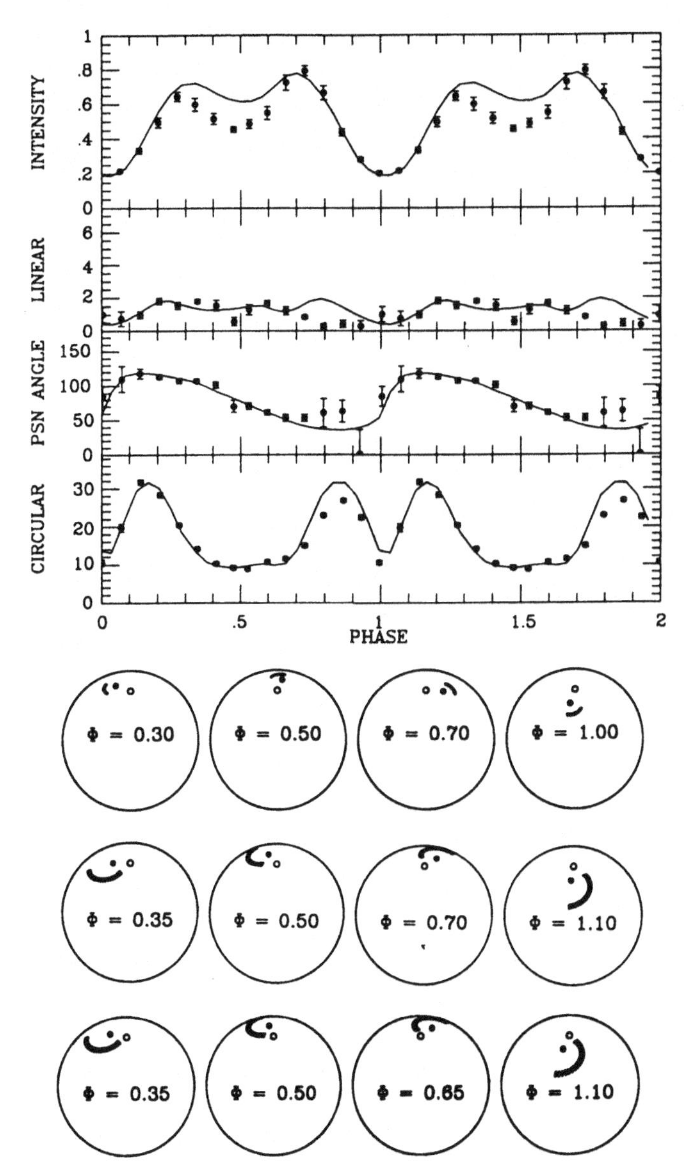

Figure 6.12 Optical broad–band observations of V834 Cen (Cropper, Menzies and Tapia 1986) modelled by accretion arcs. The continuous lines in the upper panels are model fits. The first row of diagrams shows the relative positions of magnetic dipole axis (solid circle), rotation axis (open circle) and accretion arc for four selected orbital phases. The other two rows of diagrams show deduced geometries for other nights of observations. From Ferrario & Wickramasinghe (1990).

down the field lines of the accretion region (intensity is a minimum there: Figure 6.8). The secondary minimum at $\varphi = 0.5$ arises from a projection effect – there the accretion region is closest to the limb of the primary. Occasionally a linear polarization pulse occurs near $\varphi = 0.5$ (Cropper 1989; Cropper, Menzies & Tapia 1986; Tuohy, Visvanathan & Wickramasinghe 1985), showing that the field can be nearly perpendicular to the line of sight there.

The derived model is shown in the first series of diagrams in Figure 6.12, and the fit of the model to the observations is shown by the continuous lines (Ferrario & Wickramasinghe 1990). A short accreting arc, centred 10° from the magnetic pole, extends 80° in magnetic longitude. Observations of V834 Cen made at two other times have also been analysed by Ferrario & Wickramasinghe (1990) and lead to the other series of pictures. As can be seen, in both of these the accretion arc runs over the limb at $\varphi = 0.5$, which produces a linear polarization pulse near that phase.

The light curves of V834 Cen at different wavelengths (Figure 6.13) show the wide variability typical of polars, and in particular the narrower cyclotron beaming at higher frequencies which results in a broader minimum at shorter wavelengths (Bonnett-Bidaud *et al.* 1985; Sambruna *et al.* 1991).

6.3.2.3 *ST Leo Minoris*

The broad band optical observations of ST LMi shown in Figure 6.14 (Cropper 1986a) are an average over 12 orbital cycles. The circular polarization is zero outside of the region of the prominent brightness hump. That there is structure in the accretion zone is seen by the series of brightness and circular polarization steps that occur as the zone appears around the limb of the primary. The linear polarization spike at $\varphi = 0$ shows that as the accretion zone is eclipsed at the primary's limb, we view the field lines transversely – but an asymmetry in the zone results in no linear polarization spike when it reappears.

The position angle of the linear polarization progresses steadily while the accretion zone is visible, but the fact that the linear polarization does not reduce to zero (and the position angle continues to increase, even though noisy because of the much weaker signal) while the zone is eclipsed suggests that there is another accreting region near the opposite, unobserved pole. This is confirmed by polarimetry in the IR: Figure 6.15 shows that strong circular polarization exists in the H band and is roughly 180° out of phase with the R band polarization (Bailey 1990).

The structure of the dominant optical accretion region, as determined by Ferrario & Wickramasinghe (1990), shows a narrow arc 70° in magnetic longitude, inclined slightly about a line of latitude $\sim 10°$ from the magnetic pole. From the fact that the H band polarization is present for a larger fraction of an orbital period than is the visible (Figure 6.15), the second accretion region cannot be diametrically opposite the optically dominant one.

6.3.2.4 *VV Puppis*

Figure 6.16 gives the average brightness and polarization behaviour of VV Pup over 14 orbital cycles (Cropper & Warner 1986). The prominent brightness hump and its attendant circular polarization and linear polarization spike are similar (but from a region of opposite magnetic polarity) to what is seen in ST LMi (Figure 6.14).

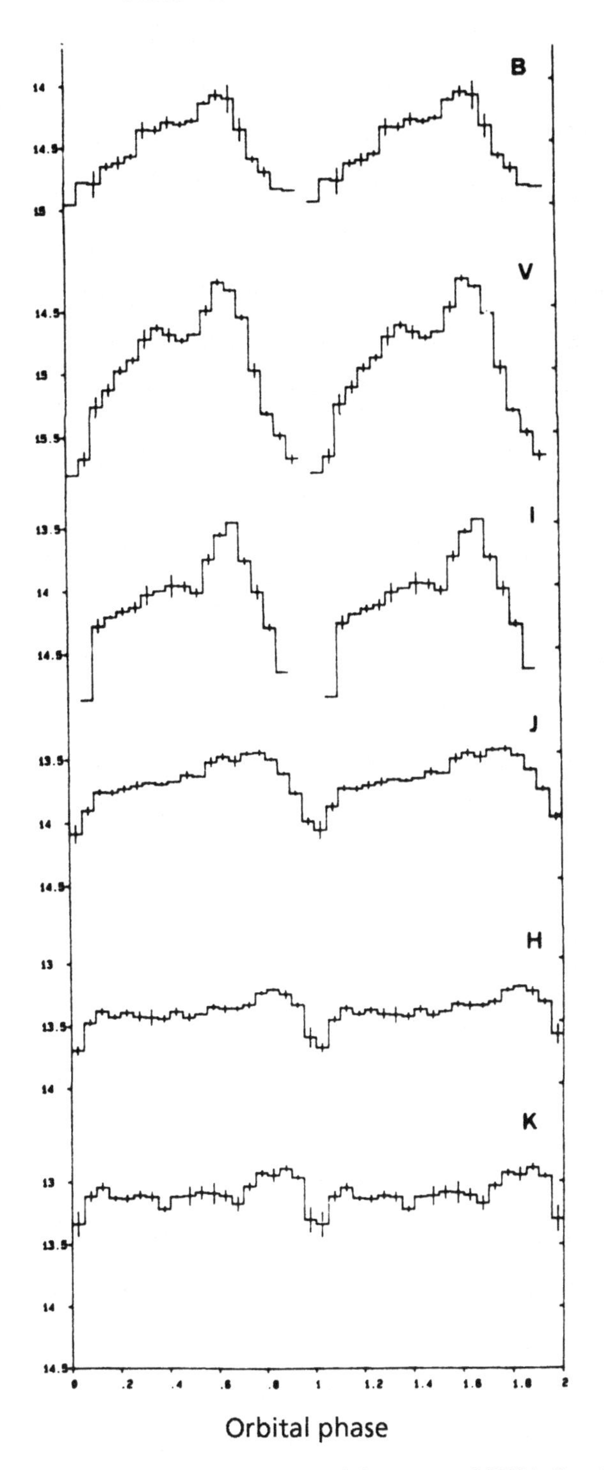

Figure 6.13 Optical and IR light curves of V834 Cen. From Bonnet-Bidaud *et al.* (1985).

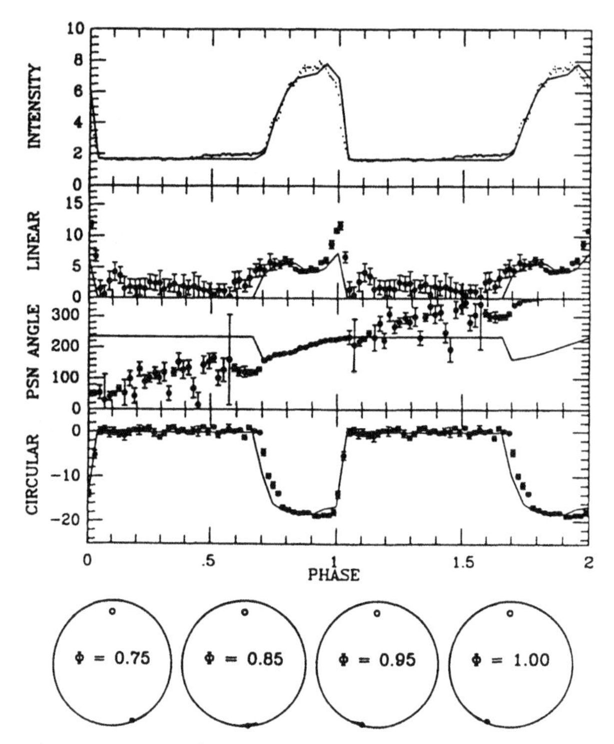

Figure 6.14 Optical broad band observations of ST LMi (Cropper 1986a) modelled with an accretion arc. From Ferrario & Wickramasinghe (1990).

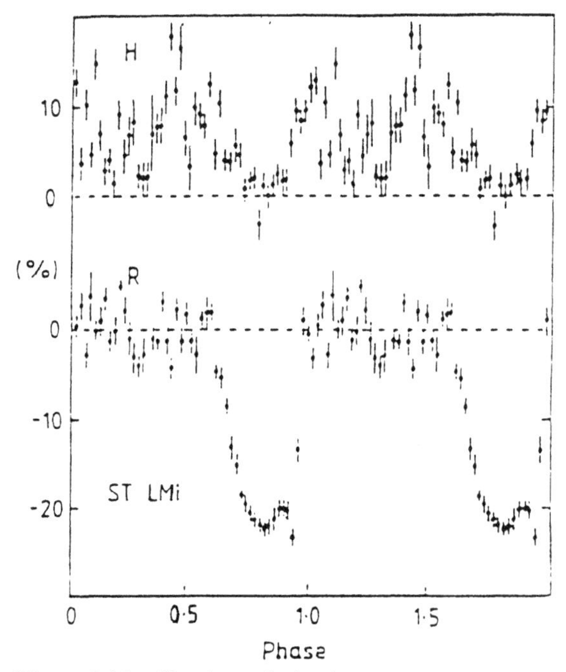

Figure 6.15 Circular polarization curves for ST LMi in the R and H bands, as a function of orbital phase. From Bailey (1988).

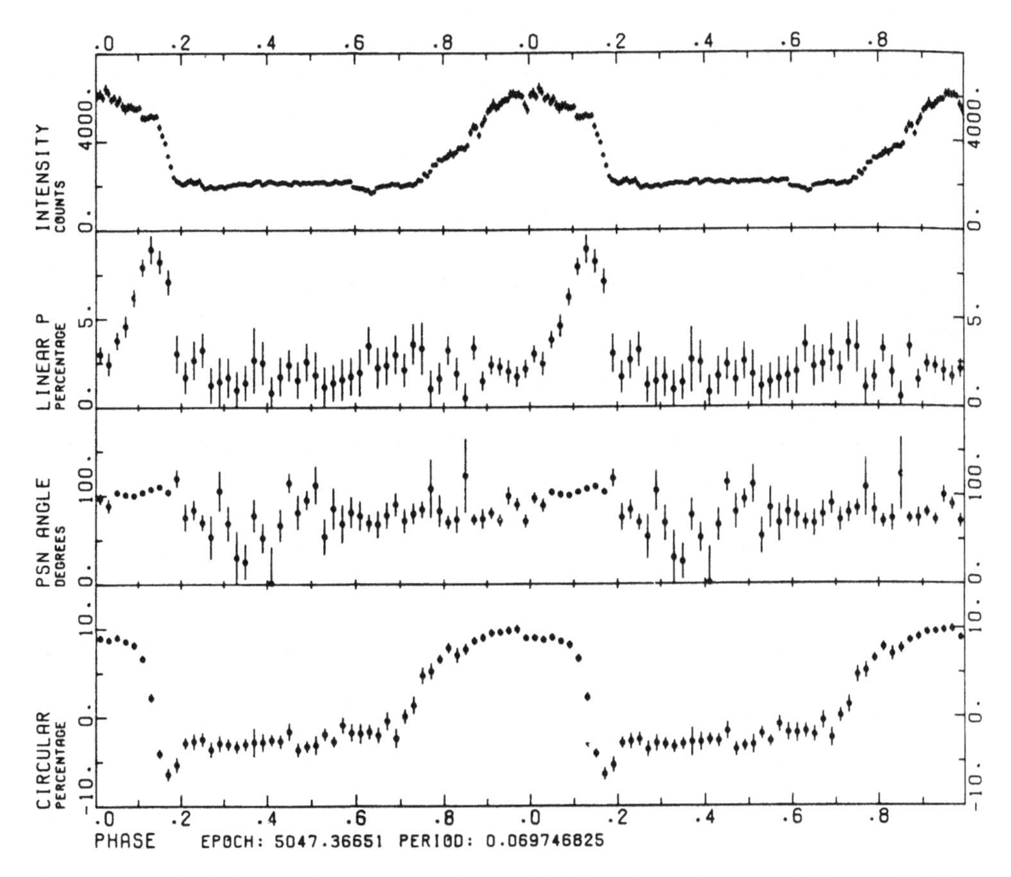

Figure 6.16 Optical broad band observations of VV Pup. From Cropper & Warner (1986).

However, unlike the latter, the circular polarization is non-zero when the dominant pole is hidden behind the primary, which led to an early suggestion that accretion occurs at two poles (Liebert & Stockman 1979).

Initial interpretations (Lamb 1985; Barrett & Chanmugam 1984), however, viewed the reversal of the sign of circular polarization as implying that the accretion column has great extension above the surface of the primary, so that part of it is still visible (and viewed with the field lines directed away from, instead of towards, the observer) even when the bulk of the column is behind the primary. However, modelling of the column in other systems (Wickramasinghe & Meggitt 1985b) showed that the column is not more than $0.1R(1)$ in height. The brief, deep reversal of circular polarization at the end of the brightness hump is still caused by viewing the main accretion region from behind instead of in front: the field lines are non-radial so it is possible to view the accretion zone with the lines inclined away from the observer for a short time before the main accretion zone disappears beyond the limb of the primary.

A new aspect was introduced when Piirola, Reiz & Coyne (1987b) discovered that *three* separate linear polarization pulses are occasionally present in VV Pup. It is then clear that there must be (at least) two active accretion regions, of opposite magnetic polarity (these could produce four polarization spikes, two of which can coincide). These have been modelled by Meggitt & Wickramasinghe (1989), who find that the regions are $\sim 8°$ from the magnetic poles.

Extensive high time resolution polarimetry and multicolour photometry by Piirola, Coyne & Reiz (1990) show that the emission from the two poles has a different wavelength dependence of circular polarization: one is brightest in U and B, but the other is brightest in V. From self-eclipse by the primary it is found that the size of the main emitting region increases by a factor of ~ 2 from the UV to the near IR, indicating inhomogeneous accretion. The emission regions extend 50–100° of longitude.

6.3.2.5 *Two-Pole Accretion in Other Systems*

Three polarization spikes have also been seen in EF Eri, AM Her, in other polars (Piirola *et al.* 1985; Piirola, Reiz & Coyne 1987b; Piirola 1988; Wickramasinghe *et al.* 1991a,b, 1993; Cropper & Wickramasinghe 1993) and also occasionally in V834 Cen (Cropper 1993). Two-pole accretion is therefore a common property of polars. In the case of EF Eri, the two visible accretion zones have the *same* polarity (Piirola, Reiz & Coyne 1987a). Meggitt & Wickramasinghe (1989) show that this can be explained if EF Eri has a *quadrupole* magnetic field.

The angular displacement $\Delta\beta$ of the accretion zones from the magnetic poles can be used to estimate the radius r_μ of the threading region (Meggitt & Wickramasinghe 1989; Wickramasinghe *et al.* 1991b). For a centred dipole

$$r_\mu / R(1) = \mathrm{cosec}^2 \Delta\beta \tag{6.24}$$

which gives $r_\mu \sim (30\text{–}50)R(1)$ for $\Delta\beta \sim 8°$. However, a more detailed analysis of AM Her (Wickramasinghe *et al.* 1991b) reveals an off-set dipole field, and a coupling region for the main accretion zone that starts at $r \sim 18R(1)$ and extends along the stream to $r_{\rm min} \sim 7R(1)$ at the point of deepest penetration into the field. The stream continues beyond this point, but is eventually coupled to field lines that connect to the accretion zone near the second pole.

In EF Eri the two accreting regions have field directions $\sim 90°$ to each other. Similarly, in CE Gru (Wickramasinghe *et al.* 1991a) and DP Leo (Cropper & Wickramasinghe 1993) one accreting region is near the magnetic pole and one near the magnetic equator of an assumed dipole structure. This suggests quadrupolar components for all of these systems.

A list of derived parameters for one-pole and two-pole models is given in Table 6.3 (D. Wickramasinghe: private communication). The former refer to the main accretion region; part of the spread in values is caused by variations in the source. The latter are based on linear polarization observations, or, in the case of MR Ser, the phase dependence of cyclotron lines (Section 6.5.4).

6.3.2.6 *Eclipsing Polars*

The absence of an accretion disc in polars requires $i > i_{\rm min} \approx 74°$ for any eclipse features to appear. About 28% of polars would therefore be expected to eclipse, and indeed 4 out of the 17 pre-ROSAT known polars are eclipsing: DP Leo (Biermann *et al.* 1985; Bailey *et al.* 1993), WW Hor (Bailey *et al.* 1988), UZ For (Ferrario *et al.* 1989)

Table 6.3. *Parameters of Well-Observed Polars.*

Star	One pole		Two pole		References
	β	i	δ_α	i	
V834 Cen	25 ± 5	45 ± 9	15 ± 5	45±5	1–4
EF Eri	33 ± 12	58 ± 18	14 ± 5	65 ± 5	2,5–7
UZ For			32 ± 5	86 ± 5	8,9
CE Gru			15 ± 5	50 ± 5	10
AM Her	50 ± 5	34 ± 5	58 ± 5	52 ± 5	2,6,7,11
WW Hor			40 ± 5	74 ± 5	12
BL Hyi	143 ± 10	71±10			2,7
DP Leo			45 ± 10	80 ± 10	13,14
ST LMi	143 ± 4	65 ± 5	30 ± 5	55 ± 5	1,2,5,7
VV Pup	153 ± 5	78 ± 5	30 ± 5	74 ± 4	2,6,15,16
MR Ser	43 ± 5	38 ± 5	15 ± 5	40 ± 5	17–19
AN UMa	20 ± 5	65 ± 5			18
EK UMa	56 ± 14	56 ± 14			2,20
QQ Vul	19 ± 9	60 ± 14			21
HU Aqr	40 ± 10	80 ± 5			

Note: β is the angle between the *field vector* and the rotation axis. δ_a is the angle between the dipole axis and the rotation axis.

References: 1. Ferrario & Wickramasinghe 1989; 2. Cropper 1987; 3. Cropper, Menzies & Tapia 1986; 4. Tuohy, Visvanathan & Wickramasinghe 1985; 5. Cropper 1985; 6. Meggitt & Wickramasinghe 1989; 7. Piirola 1988; 8. Ferrario *et al.* 1989; 9. Schwope, Beuermann & Thomas 1990; 10. Wickramasinghe *et al.* 1991a; 11. Wickramasinghe *et al.* 1991b; 12. Bailey *et al.* 1988; 13. Cropper & Wickramasinghe 1993; 14. Biermann *et al.* 1985; 15. Piirola, Coyne & Reiz 1990; 16. Cropper & Warner 1986; 17. Wickramasinghe *et al.* 1991c; 18. Brainerd & Lamb 1985; 19. Liebert *et al.* 1982a; 20. Morris *et al.* 1987; 21. Glenn *et al.* 1994.

and EP Dra (Remillard *et al.* 1991), with VV Pup a near-miss. Of the 15 polars discovered by ROSAT 2 are known at present to be eclipsing (HU Aqr: Schwope, Thomas & Beuerman 1993; RXJ0929.1−2404: Sekiguchi, Nakada & Bassett 1994). The light curve of WW Hor (at $m_V \sim 19$) is shown in Figure 6.17 and is one of the most remarkable of any CV; that of EP Dra is very similar. The prominent peaks at $\varphi = 0.2$ and 0.7 are strongly circularly polarized, proving that they are caused by cyclotron beaming; the peaks are broader in the J band because beaming is less effective at low harmonics.

High speed photometry of the eclipses in UZ For (Bailey & Cropper 1991: Figure 6.18) shows the rapid drop (~6 s) as the accretion region disappears, followed by the more leisurely (~40 s) eclipse of the remainder of the white dwarf photosphere. Because of the different angles made by the limb of the secondary at immersion and emersion, the accretion region only appears after most of the white dwarf has emerged. The deduced size of the accretion zone gives $f \sim 5 \times 10^{-3}$. The eclipse duration provides a relationship between i and q; constraints on i from the geometry of the magnetic field

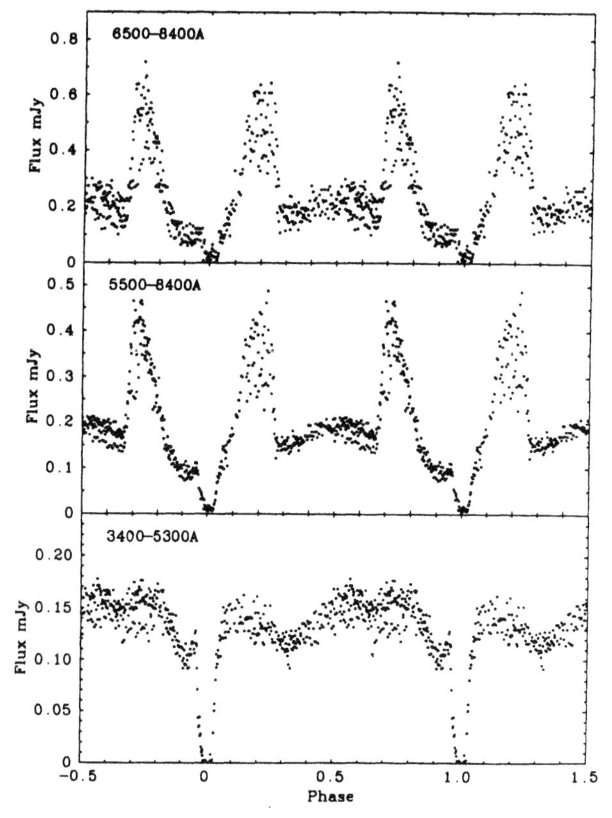

Figure 6.17 Light curves of WW Hor (at $m_V \approx 19$) in different wavelength bands. From Bailey *et al.* (1988).

deduced from polarization measurements then lead to $0.61 \le M_1(1) \le 0.79$ and $M_1(2) \approx 0.17$.

A similar analysis of DP Leo (Bailey *et al.* 1993; Schmidt *et al.* 1993) gives $M_1(1) = 0.71$, $M_1(2) = 0.106$ and shows that IR cyclotron and EUV emission regions are a spot with $f \sim 2 \times 10^{-2}$.

The high resolution observations made during eclipses confirm the standard model of polars that had been deduced from general luminosity and polarization measurements.

6.3.3 Magnetic Field Strengths from Polarization

A remarkable result in the theory of transfer of cyclotron radiation (Meggit & Wickramasinghe 1982; Barrett & Chanmugam 1984) is that the ratio of the linear to the circular polarization is given by

$$R = \frac{\text{Linear}}{\text{Circular}} = -\frac{\sin^2\alpha}{2(\omega/\omega_c)\cos\alpha} \tag{6.25}$$

which is independent of both temperature and optical depth (or Λ). Therefore, if R is observed at frequency ω, and α is deduced from the geometry (equation (6.21)), the

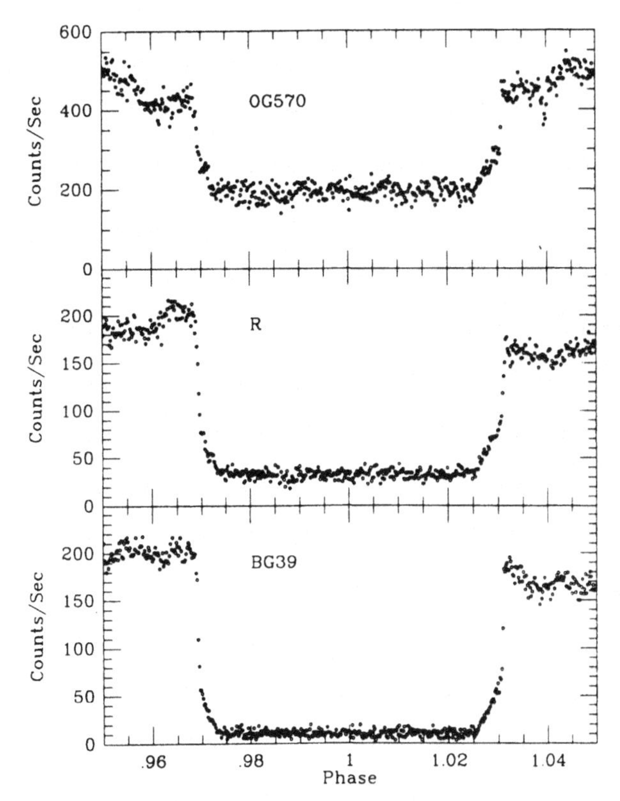

Figure 6.18 Eclipse curves of UZ For in three wavelength regions. From Bailey & Cropper (1991).

cyclotron frequency ω_c and hence the field strength B (equation (6.1)) are immediately measured.

Application of equation (6.25) has been restrained by the fact that the polars are so faint that good signal-to-noise in polarization observations has only been obtained in broad bands, which integrate over one or more cyclotron harmonics. Nevertheless, initial results are promising: Cropper, Menzies & Tapia (1986) found $B \approx 30$ MG for the accretion zone magnetic field in V834 Cen, which agrees satisfactorily with the surface-averaged value of 22 MG found by Beuermann, Thomas & Schwope (1989); and Piirola, Reiz & Coyne (1987b) deduced a mean field of 33 ± 2 MG in BL Hyi from UBVRI polarimetry, in good agreement with 30 ± 3 MG found from Zeeman spectroscopy (Wickramasinghe, Visvanathan & Tuohy 1984). Results for EF Eri (Piirola, Reiz & Coyne 1987a) are less clear cut because of the complicated field geometry (Section 6.3.2.5).

More generally, the distribution of polarized flux as a function of wavelength is obviously dependent on magnetic field (through Λ), but, as the discussions in Sections 6.2.3, 6.2.4 and 6.2.5 make clear, there is also sensitivity to field geometry and the spatial distribution of the accretion flow.

In systems such as ST LMi (Figure 6.14), where the dominant pole is eclipsed for part of the orbit, the spectral flux distribution of the accretion region can be derived by subtracting the distributions near the peak of the hump and that between the humps. In ST LMi (Bailey *et al.* 1985) the distribution from 0.4 to 2.2 μm resembles that from a standard accretion column, with a Rayleigh–Jeans distribution in the IR and a $\sim\nu^{-8}$ dependence from short wavelengths. However, the circular polarization is nearly constant at 20% throughout, which is a much broader wavelength range than that produced by a homogeneous column and implies inhomogeneous accretion.

Fits of homogeneous models to these flux and polarization distributions (Wickramasinghe & Meggitt 1985b) give fields in the range $19 \leq B \leq 25$ MG. In contrast, the inhomogeneous model-fitting of Ferrario & Wickramasinghe (1990) to the orbital variations of polarization gives an effective dipole field strength $B = 40$ MG, but with an offset of $0.1R(1)$ along the dipole axis.

Non-centred dipole magnetic field configurations are well established in isolated white dwarfs (Achilleos & Wickramasinghe 1989), so it is not surprising that they exist also in polars. In the latter the magnetic pole that is associated with the dominant optical/IR accretion region typically has about *half* the strength of the secondary pole (Ferrario *et al.* 1989; Wickramasinghe, Ferrario & Bailey 1989; Schwope, Beuermann & Thomas 1990; Wickramasinghe *et al.* 1991b). Displacements $(0.1\text{–}0.2)R(1)$ along the dipole axis are found (lateral offsets may be detectable with improved observations). The weaker pole accretes at a higher rate, is closest to the orbital plane and points most directly towards the secondary (Wickramasinghe & Wu 1991).

In general, the flux and polarization distributions (Bailey 1988) confirm the results obtained from Zeeman observations (Section 6.5.3) and cyclotron harmonics (Section 6.5.4), listed later in Table 6.7, that $10 \leq B \leq 60$ MG. However, Brainerd (1989) has noted a possibly important selection effect. He finds that, for a wide range of accretion rates and magnetic fields, a stream of gas cooling by cyclotron emission maintains a constant temperature until a field $\sim$20 MG is reached, at which point the gas radiates efficiently and cools quickly. As a result polarized emission and cyclotron harmonics are only associated with local fields in the range $10 \lesssim B \lesssim 50$ MG. The implication is that polars can only be recognized from their continuum polarization if they have fields greater than 10 MG and that the *surface* fields may be considerably larger than those deduced from the cyclotron emission. Brainerd's model needs considerable pre-shock heating, which is not compatible with observed Balmer emission lines from the coupling region nor the cool halo gas.

A general conclusion is that it will be worthwhile examining the white dwarf spectra of *all* cataclysmic variables that drop into states of low $\dot{M}$ in the hope of detecting Zeeman-split lines (Section 6.5.3) from fields *below* or *above* the presently known range of the polars.

6.3.4 Low State Observations

During optical low states of polars the UV and X-ray emissions are also reduced, implying that the fundamental cause is a lowered $\dot{M}$ from the secondary. EUV flares with duration $\sim$1 h have been seen in QS Tel in the low state (Warren *et al.* 1993). The faintness of polars in their low states (Table 6.1) produces technical problems in obtaining high quality observations. With lowered $\dot{M}$ the threading region will move towards, and may be on, the secondary. Consequently the location and structure of the accretion zones on the primary may change significantly from high to low states.

AM Her has been well studied in both high and low states (Priedhorsky & Krzeminski 1978; Priedhorsky, Krzeminski & Tapia 1978; Latham, Liebert & Steiner 1981; Szkody, Raymond & Capps 1982; Bailey, Hough & Wickramasinghe 1988, Wickramasinghe *et al.* 1991b). During the low state, the prominent modulation and circular polarization in the V band, present in the high state, disappear completely. However, strong polarization appears in the H and K IR bands. The effect on the polarized flux is that the peak flux remains unchanged in height but shifts towards longer wavelengths in the low state. Bailey, Hough & Wickramasinghe (1988) find $\Lambda \sim 10^3$ from homogeneous accretion models for the low state, which is about a factor of 10 lower than that found from inhomogeneous models in the high state (Wickramasinghe *et al.* 1991b) and is the result of the drop of $\dot{M}(1)$ by an order of magnitude estimated from total X-ray fluxes (equation (6.10)).

In BL Hyi, on the other hand (Schwope & Beuermann 1987), there is apparently no change of cyclotron temperature from high to low state, so the reduction in cyclotron flux by a factor of 30 implies a reduction in emitting area by that factor. Observations of other polars during low states similarly show that circular polarization at red or IR wavelengths is comparable with or higher than high state levels.

An incidental advantage of low states is that they give an opportunity to study the secondary. A major flare observed in AM Her has characteristics of those seen in flare stars (UV Ceti stars) (Shakhovskoy *et al.* 1993).

6.3.5 IR Ellipsoidal Variations

Despite the substantial contribution from cyclotron emission in the IR, IR photometry of VV Pup (Szkody, Bailey & Hough 1983) and ST LMi (Bailey *et al.* 1985) has successfully detected the double humped 'ellipsoidal' modulation (see Section 2.6.1.4) from the secondary. Observations of AM Her made during an intermediate state (Bailey, Hough & Wickramasinghe 1988) show that the lowered $\dot{M}(1)$ and Λ move the cyclotron emission peak to the IR, so the relative contribution from the secondary does not increase very much. As a result the ellipsoidal contribution is not detectable.

The amplitude of the modulation is a rough indicator of binary orbital inclination, but the principal value of the modulation is to provide the phase in orbit when the secondary is at conjunction. In ST LMi there is excellent agreement between times of conjunction determined from the ellipsoidal variation and from absorption lines of the secondary (Mukai & Charles 1987).

IR colours of the secondaries are typical of M dwarfs (Bailey 1985). The distances given in Table 6.1 have mostly been determined from the K magnitudes of the secondaries (Section 2.9.5; Cropper 1990).

6.3.6 Flickering, Flaring and QPOs

All polars show stochastic optical brightness fluctuations on time scales of seconds to minutes (though three of the high inclination systems – DP Leo, WW Hor and EP Dra – appear less active than others). On occasion some systems have spectacular short lived flares superimposed on the general flickering. The most impressive are seen in VV Pup (Figure 6.19) and were known long before this star was classified as a polar –

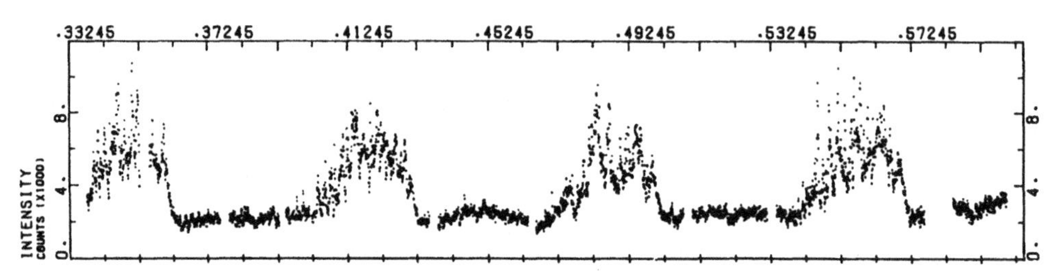

Figure 6.19 Light curve of VV Pup showing rapid flaring. From Cropper & Warner (1986).

indeed, they were first detected by eye at the telescope (Thackeray, Wesselink & Oosterhoff 1950) and only later verified by photoelectric photometry (Walker 1965a; Warner & Nather 1972a). In general, flares are only present when the dominant accretion region is visible. The presence or otherwise of flaring episodes in VV Pup is not correlated with its mean luminosity (Cropper & Warner 1986; Larsson 1989a).

Studies of flickering and other examples of flares have been given by Priedhorsky, Krzeminski & Tapia (1978), Stockman & Sargent (1979), Panek (1980), Szkody *et al.* (1980), Tuohy *et al.* (1981), Cropper (1985) and Watson, King & Williams (1987). Typical flares last for 5–100 s and are probably caused by the survival of large gas blobs in the threading zone, as described in Sections 6.2.2 and 6.2.5. Correlations between optical and X-ray flares and flickering are considered in Section 6.4.5.

In the largest flares in Figure 6.19, VV Pup brightened by a factor of 3 and returned to normal within 25 s. No change in circular polarization was seen during the flares (Cropper & Warner 1986). However, in AM Her a correlation between polarization and intensity has been seen (Stockman & Sargent 1979). As the state of polarization depends on optical thickness, blobs in an accretion column will only produce polarization changes if their optical thickness differs from that of the bulk of the column. The duration of the shortest flares in VV Pup implies a length $\sim 1 \times 10^{10}$ cm, which, before stretching, would have been $\sim 1.5 \times 10^9$ cm. This may be the largest blob size to survive intact through the threading region, or it may be the largest blob size to reach the threading region (the stream diameter at L_1 is $\sim 1.5 \times 10^9$ cm).

Time-dependent calculations by Langer, Chanmugam & Shaviv (1981, 1982) showed that a standard accretion column should be thermally unstable and generate luminosity fluctuations on time scales 1–100 s. A search for these, using higher time resolution than hitherto, was immediately successful: bands of oscillation power (from QPOs) centred on 0.6 Hz were found in V834 Cen and AN UMa (but not the type star AM Her) by Middleditch (1982). QPOs have been found in four polars (Table 6.3), but they are not always present and searches among other polars have been unsuccessful (Ramseyer *et al.* 1993a). They may be present in wavelength ranges not yet investigated.

A train of QPOs observed in V834 Cen is shown in Figure 6.20. As they are not coherent over more than a few tens of seconds, the power spectrum shows a spread of power; this has a well-defined range: Figure 6.21. From the most comprehensive studies of QPOs (references 3,5,8 and 11 of Table 6.4) the following properties emerge:

Table 6.4. *QPOs in Polars.*

Star	P_{orb} (h)	Frequency band (Hz)	Peak (Hz)	Pulse fraction (%)	B^* (MG)	References
EF Eri	1.35	0.3–0.9	0.37	11	11	1–3
VV Pup	1.67	0.4–1.2	0.76	1.5	32	3–5,12
V834 Cen	1.69	0.3–0.9	0.55	1–3	22	1,2,6–8,11
AN UMa	1.91	0.3–1.0	0.65	2–4	36	2–4,5,9,10

* Field of dominant pole.

References: 1. Larsson 1987; 2 Middleditch 1982; 3 Ramseyer *et al.* 1993a; 4. Larsson 1988; 5. Larsson 1989a; 6. Mason *et al.* 1983; 7. Cropper, Menzies & Tapia 1986; 8. Larsson 1992; 9. Imamura & Steiman-Cameron 1986; 10. Larsson 1989b; 11. Middleditch *et al.* 1991; 12. Imamura *et al.* 1993.

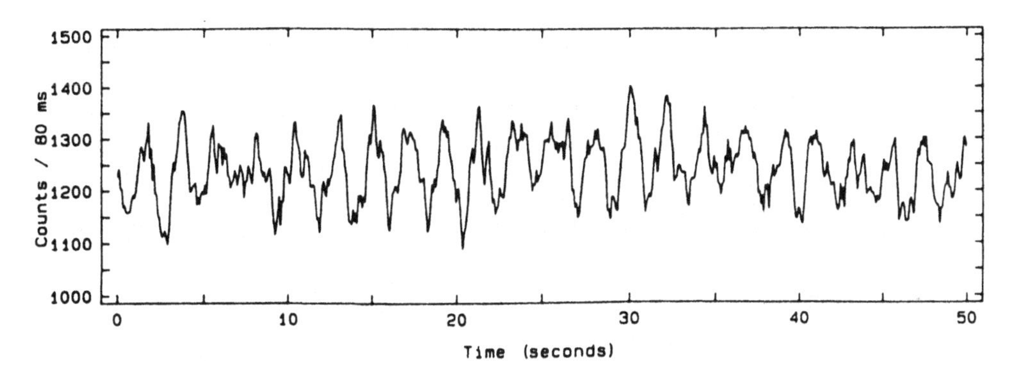

Figure 6.20 Rapid optical QPOs in V834 Cen. From Larsson (1992).

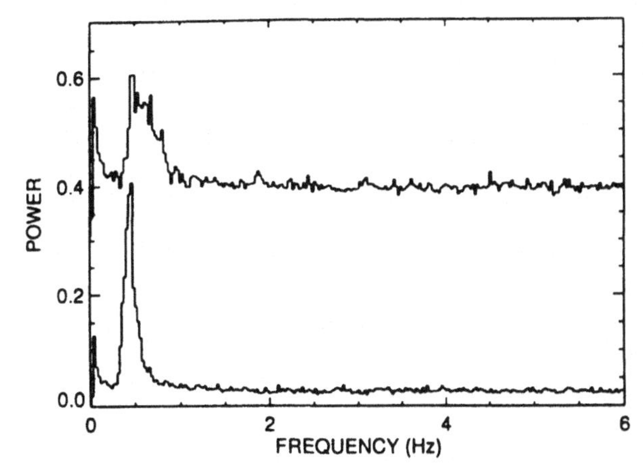

Figure 6.21 Power spectra of QPOs in V834 Cen at different epochs. From Larsson (1992).

(i) The QPOs are clearly associated with the cyclotron emission region: their amplitude follows the emission intensity, and in VV Pup the QPOs disappear when the principal accretion zone passes behind the limb of the primary.

(ii) The colour behaviour of the QPOs is complex: in V834 Cen their highest amplitude has been variously in R, V and B bands, not clearly correlated with changes of overall luminosity; in AN UMa the amplitude is highest in the B band, but there is a variation of colour with orbital phase which give the U band equal oscillation amplitude at peak orbital brightness.

(iii) Apart from one claim in V834 Cen (Larsson 1987b, 1992), not confirmed by Middleditch *et al.* (1991), there is no correlation between QPO frequency and colour. There are no other correlations with system luminosity or colour: in particular the frequency and absolute amplitude in V834 Cen remained constant over a change in luminosity by a factor of 4. There may, however, be a linear relationship between frequency at maximum QPO power and magnetic field strength (Larsson 1992; Table 6.4).

(iv) In a given photometric passband the QPOs have coherence times of 20–40 s, or $Q = |\dot{P}|^{-1} \sim 250$. Comparing simultaneous observations in different passbands, however, shows a correlation time of $\sim$3 s, or 2 cycles. Furthermore, the average frequency of the QPOs changes with wavelength. These observations show that the oscillations seen in one passband are independent of those in another, and that in a broad passband several independent oscillations may be present at once – this is directly evident in V834 Cen where two coherent oscillations can be seen in short light curves (Larsson 1992). Ramseyer *et al.* (1993a) conclude that there are many simultaneous oscillating columns, each generated by an accreting blob of gas.

The thermal instability of accretion columns found by Langer, Chanmugam & Shaviv (1981, 1982) and later elaborated by Chevalier & Imamura (1982), Imamura, Wolff & Durisen (1984), Imamura (1985) and Wolff, Gardener & Wood (1989), results from the following mechanism: an upward movement of the shock increases its velocity relative to the infalling matter, increasing T_{sh} and requiring the column to lengthen even further to allow the gas time to cool by the time it reaches the bottom of the column. Downward perturbations are similarly amplified. It was later found, however, that oscillations excited when bremsstrahlung cooling alone is included are strongly damped if even a small amount of cyclotron cooling is present (Chanmugam, Langer & Shaviv 1985; Imamura, Rashed & Wolff 1991), eliminating self-excitation as a possibility.

The agreement between the estimated cooling time scale of the shocked gas and the periods of QPOs nevertheless encourages further efforts in these directions and it has been found that the columns can readily be excited and sustained by the noise (density fluctuations) in the accretion stream (Wolff, Wood & Imamura 1991; Wu, Chanmugam & Shaviv 1992; Wood, Imamura & Wolff 1992). Even bremsstrahlung-dominated columns can be excited in this way and it is predicted that the QPO frequency should increase with B, as in Table 6.3. The luminosities of the QPOs are a convolution of the response function of the shock and the power spectrum of $\dot{M}(1)$ fluctuations, but the former is linear for conditions typical of polar accretion regions, so it should be possible to extract the $\dot{M}$ spectrum in a straightforward fashion from time series analysis of QPOs. This promises to provide a source of information about conditions in the threading region.

These new models predict that QPOs will be present in X-ray emission with amplitudes similar to those in the optical, and that the two regions will be strongly correlated. QPOs have yet to be observed in the X-ray region, but the upper limits are large (~40%: Cropper 1990).

QPOs at much longer periods are also commonly present in polars. As these are seen also in X-rays they are discussed in Section 6.4.5.

6.4 X-Ray Observations of Polars

The standard accretion column emits from the IR to hard X-rays (Figure 6.6), with three distinct energy regimes. Although inhomogeneous accretion results in a shift of the cyclotron peak to lower energies than in the standard column (Section 6.2.5), the highest density flows within the accretion stream still result in powerful emission in the EUV–soft X-ray region and a substantial hard X-ray flux.

It should be realized, however, that although the hard X-ray bremsstrahlung from polars is a characteristic that distinguishes them from non-magnetic CVs, they are nevertheless only relatively *weak* emitters of hard X-rays. In fact, only AM Her (which appeared as a weak source in the early UHURU survey), EF Eri and BY Cam are sufficiently strong emitters for their spectra to have been well determined above 2 keV.

On the other hand, polars are spectacular emitters of soft X-rays, originating from reprocessed bremsstrahlung, and thermalized subphotospheric deposition (Section 6.2.3). Unfortunately, except for a few nearby objects, interstellar neutral hydrogen is opaque to energies of 13.6–~100 eV where much (perhaps most) of the accretion energy is radiated.

The absence of an accretion disc radiating half of the accretion luminosity reduces the visual flux in polars, compared to non-magnetic CVs, and results in $3 < F_x/F_v < 300$. Consequently, ROSAT has probably detected all polars with $m_v < 20$, apart from those in a low state during the survey (Beuermann & Thomas 1993; cf. Section 2.6.1.5).

6.4.1 The Hard X-Ray Spectrum

The hard X-ray spectrum of AM Her from 2 to 100 keV is shown in Figure 6.22 (Rothschild *et al.* 1981). An optically thin bremsstrahlung spectrum (Tucker 1975) fits the flux distribution well with $kT = 31 \pm 5$ keV. EF Eri has $kT = 38 \pm 3$ keV in its brightest state (Ishida *et al.* 1991).

These temperatures can be used with equation (2.90) to estimate the mass of the primary. There are several reasons, however, to expect the observed temperature to be lower than the shock temperature T_{sh} (Fabian, Pringle & Rees 1976; Kylafis & Lamb 1982; Imamura & Durisen 1983): equipartition of energy between ions and electrons may not be established and the dominant emission may arise from post-shock gas which is already cooled by bremsstrahlung or conduction. The observed temperatures therefore provide the following lower limits on masses: AM Her $M_1(1) \geq 0.71$, EF Eri $M_1(1) \geq 0.55$, BY Cam $M_1(1) \geq 0.80$.

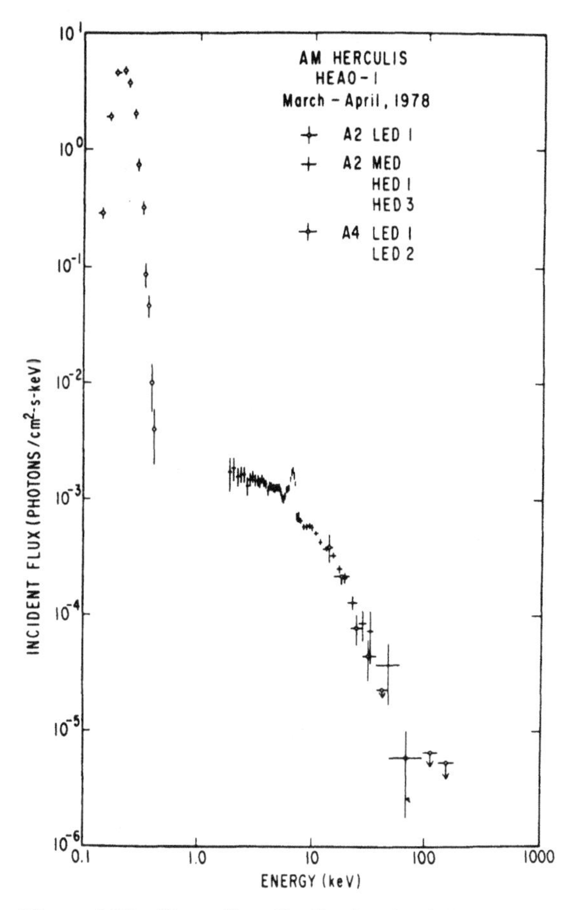

Figure 6.22 X-ray flux distribution in AM Her, obtained by various detectors on HEAO-1. From Rothschild *et al.* (1981).

All three stars show Fe Kα emission at $\sim$6.5 keV (Figure 6.22). This line may be produced (at $\sim$6.9 keV) by thermal emission in the post-shock region or by fluorescence (at $\sim$6.4 keV) from the pre-shock cool region, or from the surface of the white dwarf or (rarely) the surface of the secondary star (Swank, Fabian & Ross 1984; Makishima 1986; Ishida *et al.* 1991). In principle its equivalent width (usually hundreds of keV) can be related to the accretion rate, but the multisite production makes interpretation difficult.

From observed fluxes and upper limits, and the distances listed in Table 6.1, the mean hard X-ray luminosity of polars is 5×10^{30} erg s^{-1} (Chanmugam, Ray & Singh 1991).

6.4.2 *Soft X-Ray and UV Spectrum*

Although often referred to as the blackbody component, the flux in the region 10–300 eV arises from reprocessing of hard X-ray illumination of the atmosphere of the primary (Section 6.2.3) and from radiation of energy deposited subphotospherically (Section 6.2.5); it therefore departs from a blackbody distribution because of the non-greyness of the atmosphere. Williams, King & Brooker (1987) have computed flux distributions for irradiated atmospheres and find that, compared with the usual process of fitting blackbody distributions in the observable regions (10–13 eV and 50–500 eV), the atmospheric fluxes give $T_{\rm eff}$ factors of 2–5 lower than $T_{\rm BB}$ but (because the blackbody fitting process underestimates the emitting area) give total luminosities that are in error by a factor ≤ 2. The higher the contribution from reprocessing, the closer the emerging flux is to a blackbody distribution. Almost all existing estimates of the 10–300 eV flux in polars rely on blackbody fitting.

Interstellar absorption is an acute problem in the 100–300 eV range, where absorption edges of C, N and O in the interstellar gas determine the low energy cut-off (Brown & Gould 1970). Without any use of flux at ~10 eV (i.e., the IUE measurements), a range of correlated $T_{\rm BB}$ and interstellar column densities $N_{\rm H}$ is usually found to fit the soft X-ray distribution adequately. With the constraint of the IUE flux, however, the choice is narrowed and the process becomes more one of interpolating between the 10 eV and 100 eV regions.

Figure 6.23 shows the observed flux distribution of EF Eri in the 70 eV–2 keV region (Beuermann, Thomas & Pietsch 1991). The dashed curves are a blackbody distribution with $kT_{\rm BB} = 19$ eV ($T_{\rm BB} = 2.1 \times 10^5$ K) and a bremsstrahlung distribution with $kT = 20$ keV, both absorbed by interstellar absorption with $N_{\rm H} = 1 \times 10^{19}$ cm^{-2}. Estimates of $T_{\rm BB}$, $N_{\rm H}$ and the luminosity $L_{\rm BB}$ of the blackbody component (for distance estimates listed in Table 6.1) are given in Table 6.5. Because of the large uncertainties in most $T_{\rm BB}$, the values of $L_{\rm BB}$ are probably uncertain by factors ~4.

Many of the estimates of $T_{\rm BB}$ given in Table 6.5 are close to the maximum permitted by stability of a white dwarf atmosphere. A strong upper bound is obtainable from the Eddington limit, which may be written (Mihalas 1978) as an upper limit $T_{\rm E}$ on effective

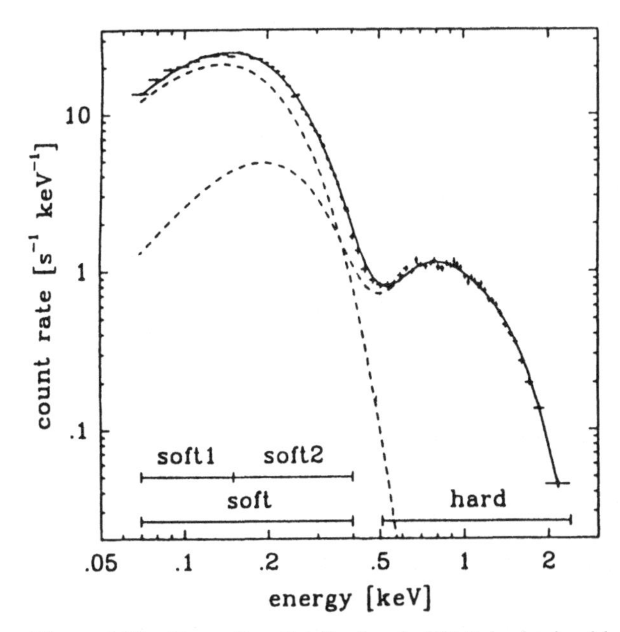

Figure 6.23 X-ray flux distribution in EF Eri, obtained by ROSAT. A deconvolution into two distributions (see text) is shown. From Beuermann, Thomas & Pietsch (1991).

Table 6.5. *Blackbody Emissions in Polars.*

Star	$P_{\rm orb}$ (h)	$kT_{\rm BB}$ (eV)	$N_{\rm H}$ (10^{20} cm^{-2})	$L_{\rm BB}$ (10^{32} erg s^{-1})	References
EF Eri	1.35	19	0.1–0.3	1.8	1,2
VV Pup	1.67	23–43		0.6–4	3
V834 Cen	1.69	15 ± 1	0.5–0.6	3.1	9
BL Hyi	1.89	23–40	1.0	0.4–1.8	4
ST LMi	1.90	18–40	0.06–0.15	1.8	5
UZ For	2.11	19–25	0.7–1.7	$\geq$0.8	10
AM Her	3.09	$\sim$25	0.2–5	3.5	6,7
QQ Vul	3.71	18–29	3.8–9.4	$\geq$1.6	8

References: 1. Beuermann, Stella & Patterson 1987; 2. Beuermann, Thomas & Pietsch 1991; 3. Patterson *et al.* 1984; 4. Beuermann & Schwope 1989; 5. Beuermann, Stella & Krautter 1985; 6. Rothschild *et al.* 1981; 7. Heise *et al.* 1984, 1985; 8. Osborne *et al.* 1986; 9. Sambruna *et al.* 1991; 10. Ramsay *et al.* 1993.

temperature:

$$T_{\rm E} = \left(\frac{gcm_{\rm H}}{\sigma\sigma_{\rm e}}\right)^{1/4}$$

or

$$kT_{\rm E} \approx 68.5 M_1^{5/12}(1)\ {\rm eV} \qquad M_1(1) \lesssim 1 \tag{6.26}$$

from equation (2.83a). Even in very hot white dwarf atmospheres the absorption coefficient is larger than the electron scattering coefficient $\sigma_{\rm e}$, so $T_{\rm E}$ is really about half that given by equation (6.26) (Williams, King & Brooker 1987).

6.4.3 The Ratio of Soft to Hard X-Ray Flux

The standard accretion column requires that the soft X-ray 'blackbody' flux be generated from reprocessed cyclotron and bremsstrahlung radiation intercepted by the stellar photosphere. With allowance for the albedo $a_{\rm x}$ ($\sim$0.3) for hard X-ray scattering from the photosphere, the energy balance should be

$$\frac{L_{\rm BB}}{(L_{\rm brem} + L_{\rm cyc})} = \frac{1 - a_{\rm x}}{1 + a_{\rm x}} \approx 0.56. \tag{6.27}$$

As seen in the previous section, estimates of $L_{\rm BB}$ are uncertain by factors of several, so comparison of equation (6.27) with observation might not seem feasible. However, the first determinations of $L_{\rm BB}$ in AM Her (Raymond *et al.* 1979) and AN UMa (Szkody *et al.* 1981a) gave $L_{\rm BB} \gtrsim 50(L_{\rm cycl} + L_{\rm brem})$. This discordance became known as the 'Soft X-ray Puzzle' (e.g., Lamb 1985).

Later improved analyses (Heise *et al.* 1984; Patterson *et al.* 1984) reduced the discrepancies to within observational (and interpretational) errors, and studies of EF Eri (Beuermann, Stella & Patterson 1987) and VV Pup (Patterson *et al.* 1984) also

Table 6.6. *Energy Budget in ST LMi.*

	Emitting area		Luminosity
	10^{15} cm^2	Fraction of white dwarf surface	(erg s^{-1})
Bremsstrahlung	~0.04	$\sim 5 \times 10^{-6}$	$\sim 5 \times 10^{31}$
Blackbody	0.04–8	5×10^{-6}–9×10^{-4}	12–90 $\times 10^{31}$
Cyclotron	4–32	5×10^{-4}–4×10^{-3}	$\sim 1 \times 10^{31}$

revealed no conflict with theory. However, the later results for QQ Vul (Osborne *et al.* 1986), AM Her (Heise *et al.* 1985), UZ For (Ramsay *et al.* 1993), AN UMa and MR Ser (Ramsay *et al.* 1994) find $L_{BB} \gtrsim 5L_{brem}$, and lead to the conclusion that a soft X-ray excess does exist (Osborne 1988; Beuermann 1988b). A general discussion of ROSAT's results shows that there is a correlation between the amount of soft X-Ray excess and magnetic field strength, in the sense that those with little or no excess have the lowest fields (Ramsay *et al.* 1994).

One proposed source of X-rays – steady nuclear burning in the accretion column (Raymond *et al.* 1979) – has been shown to provide too little luminosity (Papaloizou, Pringle & MacDonald 1982).

The solution to the puzzle lies in the combination of a hard X-ray deficiency and a soft X-ray excess. In the inhomogeneous accretion models the bulk of the hard X-rays comes from the high density region of the stream, which may involve only a small fraction of the total accretion flow. As a result, L_{brem} is much lower than would be judged from any independent knowledge of the total accretion rate. Indeed, King & Watson (1987) point out that the values of $\dot{M}$ inferred from observed L_{brem} are an order of magnitude lower than what is required to drive the orbital evolution of polars (Section 9.3.6.1).

On the other hand, part of the accretion flow may be in the form of blobs that penetrate into the photosphere of the primary (Kuijpers & Pringle 1982; Section 6.2.5) and thus contribute to the soft X-ray flux without generating any hard X-rays. Direct observational evidence for this is given in Section 6.4.5.

The relative rôles of the various emitting regions can be judged from the parameters derived for ST LMi (Beuermann 1987, 1988a,b; similar relative results are obtained for EF Eri, whose distance is less accurately known): Table 6.6.

It is clear from this that the accretion luminosity is dominated by L_{BB}, which (Table 6.5) gives $L_{acc} \sim 3 \times 10^{32}$ erg s^{-1} for polars (see also Chanmugam, Ray & Singh (1991)) and implies (equation 6.10) $\dot{M} \sim 2.5 \times 10^{15}$ g s$^{-1} = 4.0 \times 10^{-11} \mathrm{M}_\odot$ y^{-1}. As seen in Tables 6.1 and 6.5, there is no obvious dependence of M_v or L_{BB} on P_{orb}, so this $\dot{M}$ appears characteristic of polars in general. This is more than an order of magnitude lower than the values of $\dot{M}(2)$ found for most non-magnetic CVs above the period gap, but is similar to that found for DN at quiescence below the gap (Table 2.2 and Section 3.5.4.4). The distances given in Table 6.1 furnish $\overline{M}_v$(high) $= 8.7$, which may be used to obtain distance moduli with an uncertainty ~0.5 mag.

6.4.4 *X-Ray Light Curves*

At first sight, light curves of polars in the X-ray region might be expected to be simpler to interpret than those from the optical: in the standard accretion column the hard X-rays arise only from the shocked gas and the soft X-rays from reprocessing at the stellar surface. In practice, as seen in the previous section, the situation is more complicated – with at least two origins for the soft X-rays (reprocessing of bremsstrahlung and thermalization of buried shocks) and with spatial segregation of high and low density flows.

Despite these complexities, some clear categories exist in the X-ray light curves (Mason 1985). The process of analysis is aided by the nature of the broad band detectors deployed on X-ray satellites: on EXOSAT, for example, the Low Energy Telescope detected X-rays with energies 0.04–2 keV which lie predominantly in the region of the blackbody component, and the Medium Energy Telescope observed 1–10 keV which is an almost uncontaminated measurement of the bremsstrahlung emission. The two telescopes thus separately viewed the heated stellar surface and the shock-heated gas in the accretion flow. In contrast, ROSAT has excellent energy resolution over the range 0.05–2 keV which gives simultaneous coverage of the two regions.

The most clear-cut division of light curves is that caused by the combination of orbital inclination (i) and magnetic axis orientation (β) that results, as seen from the Earth, in either 'one-pole' or 'two-pole' viewing (Mason 1985). In the latter, the accretion zones near the poles are periodically carried behind the primary; if only one pole produces an X-ray emitting region the light curve is particularly simple (Figure 6.24): almost no flux is detected for roughly half of the orbital cycle. The asymmetry of the light curve in its 'on' phase is a repeating phenomenon which is seen also in averaged optical light curves of VV Pup (Patterson *et al.* 1984) and ST LMi (Cropper 1986). This is caused by the variation of density of accretion flow along the arc-like accretion zone.

The soft X-ray light curves of the 'one-pole' systems are more complicated (Figure 6.25). In these objects the X-ray emitting region lies in the rotation hemisphere that is inclined towards us and, visible continuously, may be viewed for part of the orbit through the accretion stream. In the simplest model, with the accretion flow arriving near the magnetic pole after having been lifted out of the orbital plane along the field lines, obscuration by the stream will occur if $\beta < i$ (and $\beta < 90°$). No stream absorption should occur if $\beta > i$, and especially not if $\beta \geq 90°$. The polars listed in Table 6.1 that meet this latter condition indeed do not show any absorption effects. However, in view of the clear indications in the optical that some so-called 'one-pole' systems show evidence of accretion at more than one pole (Section 6.3.2.6), the simple X-ray classification must be used with discrimination.

Some of the polar soft X-ray light curves show two absorption dips (e.g., AN UMa and V834 Cen). This is probably a result of the distortion from a planar trajectory of the accretion stream through the magnetosphere. Different parts of the stream are then separately able to intersect the line of sight to the accretion zone (Rosen 1987).

The general modulation of X-ray flux around orbit in Figure 6.25 is caused by variation of the viewing angle to the accretion zone. The superimposed absorption dips have energy spectra characteristic of absorption by cold matter with a column density

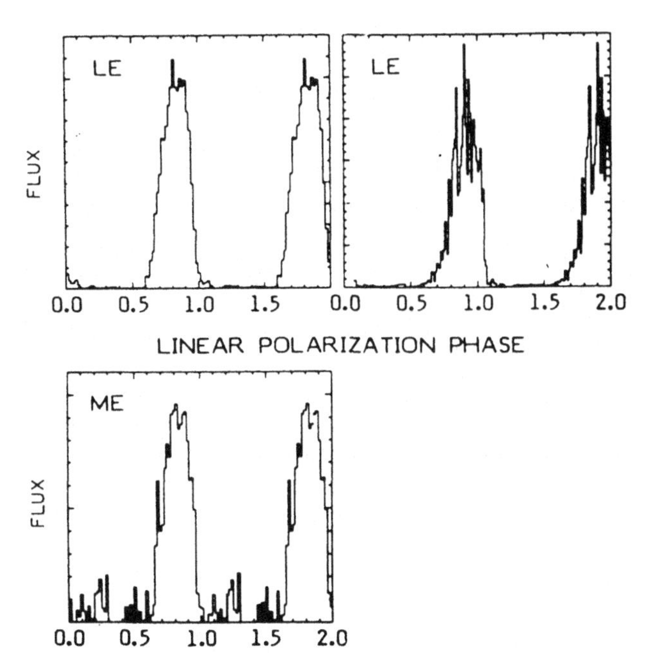

Figure 6.24 X-ray light curves of ST LMi (left hand panels) and VV Pup (right hand panel) as determined by EXOSAT. From Mason (1985).

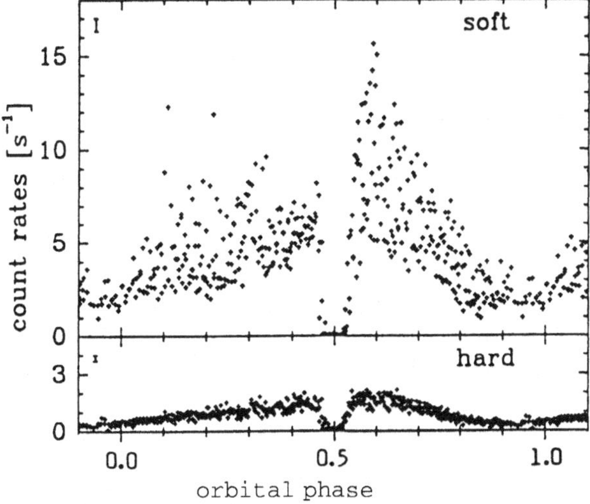

Figure 6.25 X-ray light curves of EF Eri in 'soft' and 'hard' bands (see Figure 6.23) as determined by ROSAT. From Beuermann, Thomas & Pietsch (1991).

$\sim 5 \times 10^{22}$ cm^{-2} (Patterson *et al.* 1984; Beuermann, Stella & Patterson 1987; Beuermann, Thomas & Pietsch 1991; see also King & Williams 1985; Watson *et al.* 1989).

AM Her was early observed to behave like EF Eri (Figure 6.25), with low and medium energy X-ray curves in phase (Swank *et al.* 1977; Tuohy *et al.* 1978; Patterson

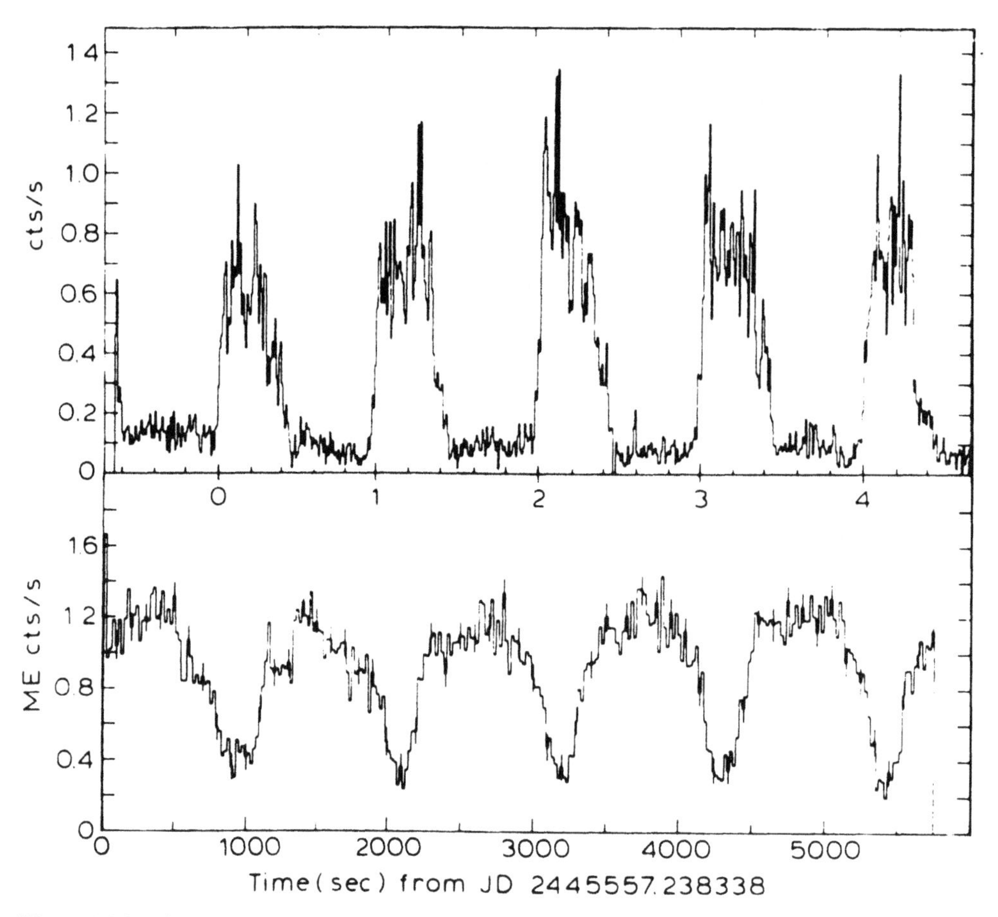

Figure 6.26 Simultaneous hard (lower panel) and soft X-ray light curves of AM Her, obtained by EXOSAT. From Heise *et al.* (1985).

et al. 1984; see also Imamura 1984) and only intermittently displaying absorption dips (Tuohy *et al.* 1978). In 1983, however, Heise *et al.* (1985) observed a complete reversal of the situation, with soft and hard X-rays differing by 180° in phase (Figure 6.26). This 'reversed' X-ray behaviour later became the norm (Mazeh, Kieboom & Heise 1986). Similar changes have been observed in QQ Vul (Osborne *et al.* 1987: Figure 1.21) and V834 Cen (Osborne, Cropper & Cristiani 1986). The very abrupt turn-on of the soft X-rays, and almost square-wave profile, as opposed to the quasi-sinusoidal variations of the hard X-rays, is a common property of the 'reversed' state. There is little difference in optical light curve between the two states (Mazeh, Kieboom & Heise 1986).

It is clear from this behaviour that in these stars the soft X-rays are *not* arising predominantly from reprocessing of bremsstrahlung from the shocked accretion flow. Heise *et al.* (1985) explained the reversal in AM Her by claiming that, despite its current classification as a 'one-pole' system, it must have two accretion poles, one of which can dominate the soft X-rays and the other always dominates the hard X-rays. As seen in Section 6.3.2.6, the optical observers later came to the same conclusion. The region producing hard X-rays is evidently fed by the threading region with small blobs, the other with large blobs. Changes in accretion geometry, caused by variations in average $\dot{M}(2)$ or by changes in alignment of the primary, can evidently have a large effect on the nature of the threading region.

BL Hyi has mostly shown X-ray production at the more distant pole (Beuermann & Schwope 1989), but produced strong soft X-ray flaring from the nearer pole when observed in an unusually bright state (Beuermann & Thomas 1993). RXJ0203+29 shows similar flaring.

In contrast to the complexities of the soft X-ray light curves, the medium and high energy curves are simple. Within the 2–10 keV range there is no dependence of modulation amplitude on energy (Norton & Watson 1989b), just as expected from the Imamura & Durisen (1983) models of hard X-ray beaming described in Section 6.2.3.

6.4.5 X-Ray Variability

Flickering on time scales of seconds to minutes is commonly observed in soft and hard X-rays from polars. Activity varies around the orbit in a manner that shows that it is associated with emission from the accretion zone; there are large changes in activity on longer time scales (Szkody *et al.* 1980; Tuohy *et al.* 1981; Stella, Beuermann & Patterson, 1986; Beuermann, Stella & Patterson 1987; Watson, King & Williams 1987; Beuermann, Thomas & Pietsch 1991).

Studies of correlations between optical, soft X-ray and hard X-ray flickering are included in the references cited; they lead to the conclusion that whereas there is a strong correlation between optical and hard X-rays variations, there is weaker (and often no) correlation between soft X-rays and the other bands. The independence of the soft and hard X-ray light curves in EF Eri is shown in Figure 6.25. This contrasts with an earlier observation (Beuerman, Stella & Patterson 1987) where hard and soft X-ray flickering were well correlated and equation (6.27) was satisfied.

These observations again emphasize that the optical (cyclotron) and hard X-rays are usually generated in one accretion region, and the bulk of the soft X-rays in another (either loosely associated with the first, or at a completely separate accretion pole). An alternative possibility (Watson *et al.* 1989), that the soft X-ray fluctuations are caused by absorption by an inhomogeneous wind driven off the surface of the primary near the base of the accretion region, fails in EF Eri because the observed column density N_H is too small (Beuermann, Thomas & Pietsch 1991).

Quasi-periodic X-ray luminosity variations are also commonly seen in polars, but at much longer periods than the $\sim$1 s QPOs in the optical (Section 6.3.6). Table 6.7 lists examples of longer period QPOs in both X-ray and optical regions.

The simultaneous soft and hard X-ray observations in AM Her and EF Eri show the same QPOs in both. The similar time scales of optical and X-ray QPOs in EF Eri (and in circularly polarized flux: Tapia, cited by Patterson, Williams & Hiltner 1981) indicates they can appear at all energies. Their origin is not yet certain. Imamura & Wolff (1990) find that low accretion rates can produce QPOs of an accretion column with periods in the range 10 s–minutes. The long period QPOs may therefore arise in the lower density accretion flows, with the $\sim$1 Hz optical QPOs coming from the denser flows.

Table 6.7. *Long Period QPOs in Polars.*

Star	P_{orb} (h)	QPO period (min)	Pulsed fraction (%)	Energy*	Reference
EF Eri	1.35	6	30–70	S,H	1
		3.8	10	H	2
		6	20–50	S,H	3
		6	0–30	O	4,11
VV Pup	1.67	3		S	4
BL Hyi	1.89	4.6	50	S	5
ST LMi	1.90	1		S	6
AM Her	3.09	0.3–1	20–50	S	7
		6.7	30–50	S,H	8
		4.2–4.7	20–70	O	10
BY Cam	3.36	31.5,36.2	20	O	12
QQ Vul	3.71	4–11	50	S	9

* S = soft X-rays, H = hard X-rays, O = optical

References: 1. Patterson, Williams & Hiltner 1981; 2. Watson, King & Williams 1987; 3. Beuermann, Stella & Patterson 1987; 4 Cropper 1990; 1984; 5. Singh, Agrawal & Riegler 1984; 6. Beuermann & Stella 1985; 7. Tuohy *et al.* 1981; 8. Stella, Beuermann & Patterson 1986; 9. Osborne *et al.* 1987; 10. Bonnet-Bidaud, Somova & Somov 1991; 11. Williams & Hiltner 1980; 12. Silber *et al.* 1992.

The high coherence oscillations observed in AM Her at an intermediate stage of brightness ($m_v = 14.2$) by Bonnet-Bidaud, Somova & Somov (1991) drifted in period from 250 s to 280 s over a 3.5 h observation. These persistent coherent oscillations suggest that the QPOs usually observed may be a superposition of several quite coherent oscillations, as is the case for the ~1 Hz QPOs (Section 6.3.6).

6.5 Spectroscopic Observations

6.5.1 *The Line Spectrum of the Accretion Flow*

The spectra of polars in their high states consist entirely of emission lines superimposed on a continuum that rises strongly towards the UV and may also rise into the IR. In the far UV (Figure 6.27) the spectra are characterized by strong lines of HeII, CIV, SiIV and NV. The CIV 1500 Å line is always the strongest (Bonnet-Bidaud & Mouchet 1987). There is no evidence of P Cyg or other outflow-associated line profiles in polars (Kallman 1987; Shlosman & Vitello 1993), which is a result of the absence of discs and the generally low $\dot{M}(2)$. In the optical and near IR (Figure 6.28) the Balmer series, the Balmer continuum in emission and strong lines of HeI, HeII and the CIII/NIII blend at 4650 Å are prominent. Typically the strengths of HeII 4686 Å and Hβ are comparable.

In the lowest state the UV emission lines disappear completely (Heise & Verbunt 1987; Szkody, Downes & Mateo 1990; Schmidt *et al.* 1993).

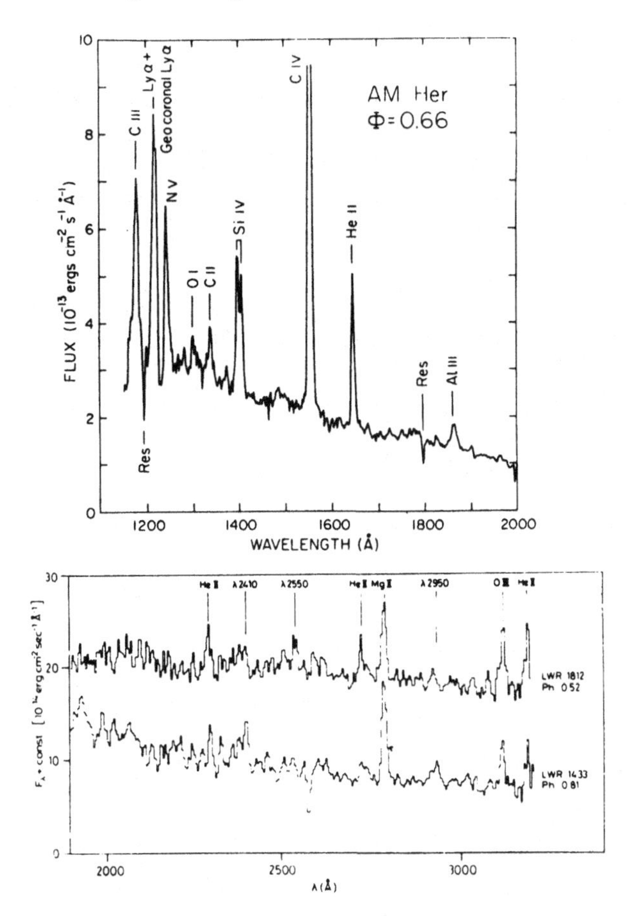

Figure 6.27 UV (IUE) spectra of AM Her. Upper (short wavelength region) from Raymond *et al.* (1979); lower (long wavelength region) from Tanzi *et al.* (1980).

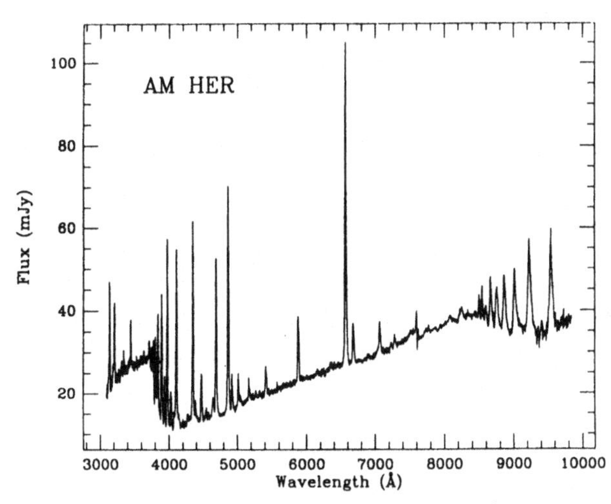

Figure 6.28 Spectrum of AM Her 3000-9800 Å. From Schacter *et al.* (1991). See that paper for an enlarged version, with line identifications.

6.5.1.1 Physical Conditions in the Gas

From the relative intensities of the emission lines some constraints can be placed on the physical conditions in the line emitting regions (Stockman *et al.* 1977; Liebert & Stockman 1985). The ratio Hβ/Hδ is ~1–2, instead of 0.3 for an optically thin case (Hγ or Hδ is usually the strongest Balmer line in polars), and therefore collisions between $n = 2$ and $n = 3$ must be important, from which $N_e \gtrsim 10^{12}$ cm^{-3}. The absence of CIII] λ1909 but presence of CIII λ1176 requires $N_e \gtrsim 10^{13}$ cm^{-3} (Raymond *et al.* 1979). A limit from above can be obtained by noting that Balmer lines are visible to H_{13} in AM Her, which demands $N_e \lesssim 10^{14}$ cm^{-3} for Stark broadening not to blend the lines.

Given the strong UV continuum it is not surprising to find that strong Bowen fluorescence occurs in polars. This is seen not only in the strength of the λ4650 blend but also in the presence of OIII emissions at λλ3123, 3133, 3299, 3312, 3341, 3407, 3429 and 3444 (Raymond *et al.* 1979; Schachter *et al.* 1991).

Bonnet-Bidaud & Mouchet (1987) find that the NV/SiIV and NV/CIV ratios in almost all polars can be accounted for by solar abundances and collisional excitation within a $T \sim 10^5$ K gas photoionized by a ~20 eV blackbody (the radiation processed by the white dwarf surface: BB in Figure 6.6; Liebert & Stockman (1985) show that the gas in the emission-line region is optically thin to the bremsstrahlung radiation emitted by the accretion column). The exception is BY Cam which, assuming the same *physical* conditions as in other polars, requires a C depletion of a factor of ~10 and an O enrichment of ~3 to match the line ratios. Such a CNO redistribution would imply that BY Cam, as with V1500 Cyg (Sections 5.3.9 and 6.7), has suffered nova explosions.

6.5.1.2 Emission-Line Profiles

The first detailed spectroscopic observations of AM Her (Cowley & Crampton 1977) revealed structures in the emission lines that distinguish the polars from CVs with discs. Cowley & Crampton noted a broad 'base' component and a narrow component present throughout an orbital cycle, and the occasional presence of a very narrow component. Modern high resolution spectra (Mukai (1988) estimates that a spectral resolution of at least 2.5 Å and a phase resolution of 0.05 are required) show that line structures and their variation around orbit are quite complex and are variously associated with the stream prior to the line-threading region, the threading region itself, the flow through the magnetosphere, and emission from the secondary star. The locations and extent of these regions can be found from the profiles of the individual line components and their velocity amplitude and γ-velocity.

Zeeman splitting has been observed in the emission lines of only one polar. Although having complex profiles, a persistent splitting $\gtrsim$5Å would probably be detectable, so the magnetic field in the emission-line region is $\lesssim$2.5 MG. This places the emission region at $\gtrsim 4R(1)$ for a dipole surface field of ~30 MG. In EF Eri a possible splitting equivalent to 1.5 ± 0.3 MG is seen in Hα (Seifert *et al.* 1987), implying an emission height $\sim R(1)$ above the primary.

The complex structure in V834 Cen and its decomposition into a broad base (largely lost in the reproduction process), a medium velocity component (MVC), a narrow component and a high velocity component (HVC) are shown in Figure 6.29 (Rosen, Mason & Cordova 1987). These are typical of the best spectral studies, but the relative phasing of the line components, and their phasing with the photometric and polarimetric variations, can change from one object to another.

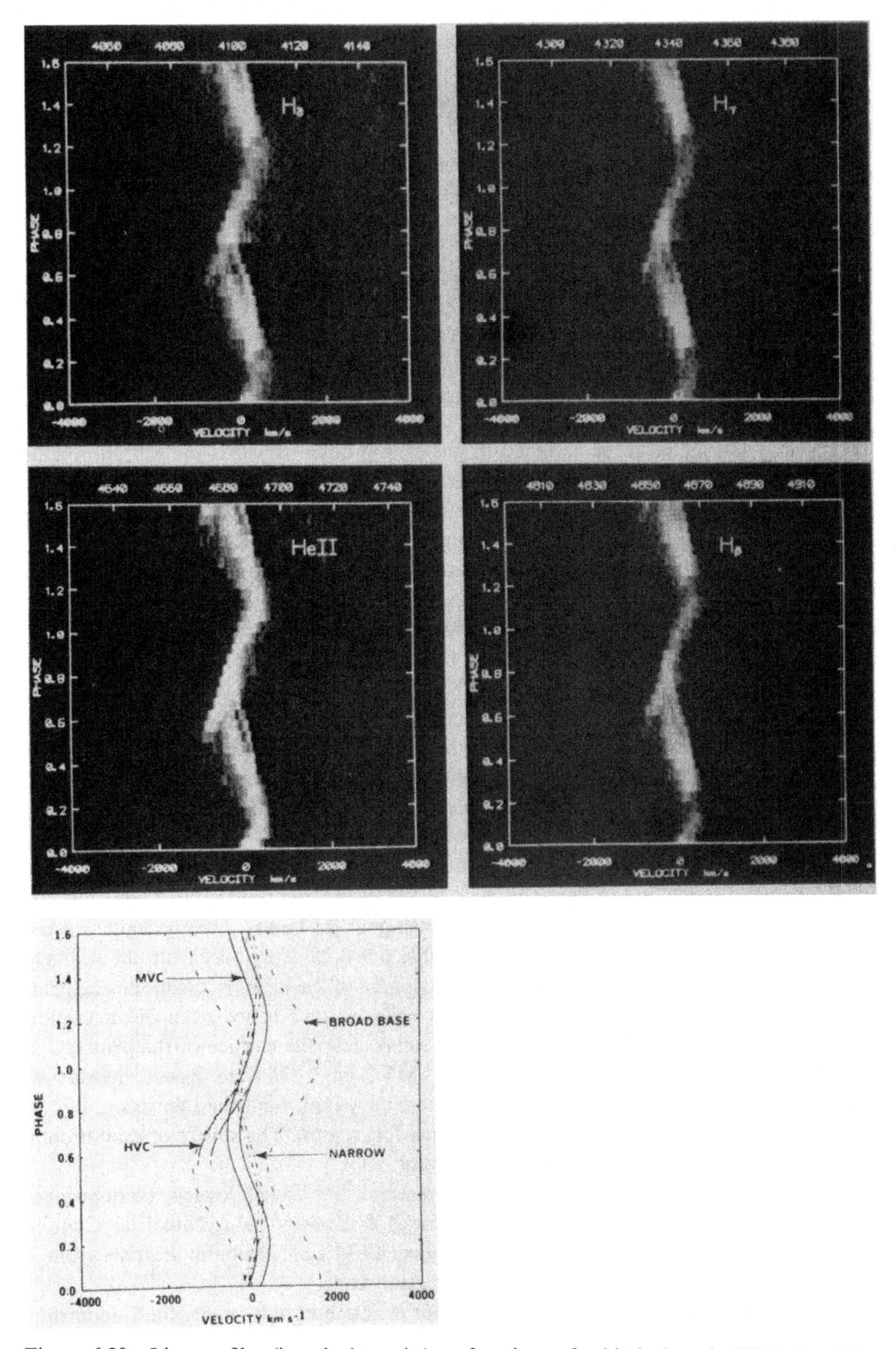

Figure 6.29 Line profiles (in velocity units) as functions of orbital phase in V834 Cen. The lowest diagram is a schematic representation of the multiple components present in the time-resolved spectra. From Rosen, Mason & Cordova (1988).

In general the interpretation of the line profiles has been only semi-quantitative, concentrating on deducing the location of the various line-emitting components from their RV behaviour (e.g., VV Pup: Schneider & Young 1980b; Cowley, Crampton & Hutchings 1982; AN UMa: Schneider & Young 1980b; EF Eri: Mukai & Charles 1985; ST LMi: Bailey *et al.* 1985; Schmidt, Stockman & Grandi 1983; QQ Vul: Mukai *et al.* 1986). However, Schmidt, Stockman & Grandi (1983) and notably Ferrario, Wickramasinghe & Tuohy (1989) have also taken account of the velocity dispersion.

6.5.1.3 Location of the Line Emitting Regions

The broad base component in the polars has a velocity amplitude up to 1000 km s^{-1} and a velocity dispersion almost as large. V834 Cen is representative of most polars, in which maximum redshift of the broad component occurs (Rosen, Mason & Cordova 1988) at phase 0.03 (Figure 6.12), near the centre of the circular polarization maximum – i.e., when we are looking most nearly along the magnetic field lines. The broad component can therefore be assigned to gas falling down towards the accretion zone near the white dwarf surface. The maximum widths (1300 km s^{-1} FWHM) of the broad component occur near phases 0 and 0.5, when we view along the accelerating gas flow and would expect to see the widest range of velocities. Although narrower (750 km s^{-1} FWHM), line widths are large even at phases 0.25 and 0.75, when we view the stream transversely, indicating curvature or convergence of the gas flow.

As the accretion stream leaving the secondary follows a trajectory that takes it ahead of the secondary in the rotating frame of the binary system, the threading region will always lead the secondary in orbit. In V834 Cen the accretion zone on the primary leads the secondary ($\psi = 40°$) so the accretion zone is almost directly below the threading region. This is true for most of the polars, but at least in the case of DP Leo the magnetic pole follows the secondary and a very different flow from threading region to the surface of the primary should be expected. In such systems the phase of maximum redshift of the broad component differs considerably from the phase where the field lines are viewed most nearly end on (Wickramasinghe 1990).

The detailed modelling of the behaviour of the broad components in V834 Cen, ST LMi and UZ For by Ferrario, Wickramasinghe & Tuohy (1989) leads to the conclusion that the threading region lies roughly 0.5–0.75 of the way from the primary to the L_1 point, which is well into the magnetosphere of the primary (Section 6.2.2) and results in accretion over a wide range of longitudes and hence extended accretion emission regions, offset from the magnetic pole(s), near the surface of the primary.

From the amplitude and phasing of the MVC in V834 Cen Rosen, Mason & Cordova deduce that it arises closer to the secondary than the broad emission, i.e., in the stream between the secondary and the threading region. The source of excitation is the EUV and X-ray emission from the accretion zone.

The HVC is not understood: it has been seen in VV Pup (Cowley, Crampton & Hutchings 1982), EF Eri (Crampton, Hutchings & Cowley 1981; Mukai & Charles 1985) and QQ Vul (Mukai *et al.* 1986) as well as V834 Cen. Probably it arises from a heated surface of the stream close to the accretion zone.

Eclipse of the CIV λ1550 emission in UZ For is slow and only briefly total, requiring a location $\sim 2R(1)$ above the accretion shock and a size $\sim R(2)$ (Schmidt *et al.* 1993).

The narrow component seen in Figure 6.29 has a large phase shift ($\sim 110°$) with respect to the broad component; this, its RV amplitude, $\gamma \sim 0$ and its narrow width

($\sim$70 km s^{-1}) place it on the surface of the secondary star (Rosen, Mason & Cordova 1987). From its intensity modulation around orbit, the narrow component can originate in emission distributed over the surface of the secondary illuminated by high energy radiation from the primary (Biermann *et al.* 1985). In AM Her the narrow component and the absorption lines from the secondary show the same phasing (Greenstein *et al.* 1977; Young & Schneider 1979) but with very different amplitudes because the heating that generates the emission also removes the absorption lines, so the emission and absorption lines originate in opposite hemispheres.

Although lines from the secondary may be seen during low states (Section 6.5.2), the faintness of the systems usually precludes accurate RV measurements. The narrow components can therefore be useful sources of information for the location in orbit of the secondaries. For some polars this has been used to assist determination of the longitude (ψ) of the field axis relative to the line of centres (Cropper 1988, 1990). Mukai (1988) points out that in some polars the 'narrow' component may be the same as the MVC of V834 Cen, and therefore not originate on the secondary. Furthermore, Mukai shows that although the narrow components in AM Her and V834 Cen are correctly ascribed to the secondary, in other polars (e.g., QQ Vul, VV Pup, ST LMi) the stream from the L_1 point may instead be the source of narrow emission. In this case, even for the short period systems, the threading region is not at the L_1 point but is situated deep in the magnetosphere of the primary (i.e., $r_\mu < R_{L_1}$: Sections 6.2.1 and 6.2.2).

The long period polar QQ Vul has shown the large structural changes that may be expected from variations in $\dot{M}$. The γ-velocity of the broad line component changed from $\sim$0 to ~ -400 km s^{-1} within a few hours (Mukai *et al.* 1986) and was $\sim +500$ km s^{-1} on another date (Nousek *et al.* 1984).

6.5.2 *Spectrum of the Secondary*

The strength of the cyclotron emission in the red and IR makes it difficult to detect the spectrum of the secondary during the high states of polars – difficult, but not impossible: Young & Schneider (1979) succeeded in detecting NaI 8183, 8194 Å absorption lines in AM Her in high resolution observations. This NaI doublet is the strongest atomic absorption feature in near IR spectra of stars cooler than M2V. Chromospheric emission from the secondary is also occasionally detectable. These lines provide a valuable means of measuring the RV amplitude $K(2)$ in polars. They have been used to this end in ST LMi, MR Ser and QQ Vul (Mukai & Charles 1986, 1987; Schwope *et al.* 1991, 1993b), UZ For (Beuermann, Thomas & Schwope 1988), and V834 Cen (Schwope *et al.* 1993a).

During a low state the optical spectrum of the secondary appears more clearly, although never free of emission lines. Figure 6.30 (Schmidt, Stockman & Margon 1981) shows the spectrum of the secondary in AM Her obtained during the 1980 low state. The TiO band absorption and the NaI lines provide a spectral classification M4.5–M5 (Latham, Liebert & Steiner 1981; Young & Schneider 1981). In the low state the broad component of the emission lines is not present and the Balmer, HeI and FeII line RVs phase with the motion of the secondary. HeII emission does not occur, indicating the absence of strong ionizing radiation, and the Balmer decrement (e.g., Hα/H$\beta \sim 4$) is consistent with pure recombination (cf. Section 6.5.1.1). The IR spectrum of the secondary in AM Her fits a $T_{\rm eff}(2) = 3250$ K model atmosphere (Bailey, Ferrario & Wickramasinghe 1991).

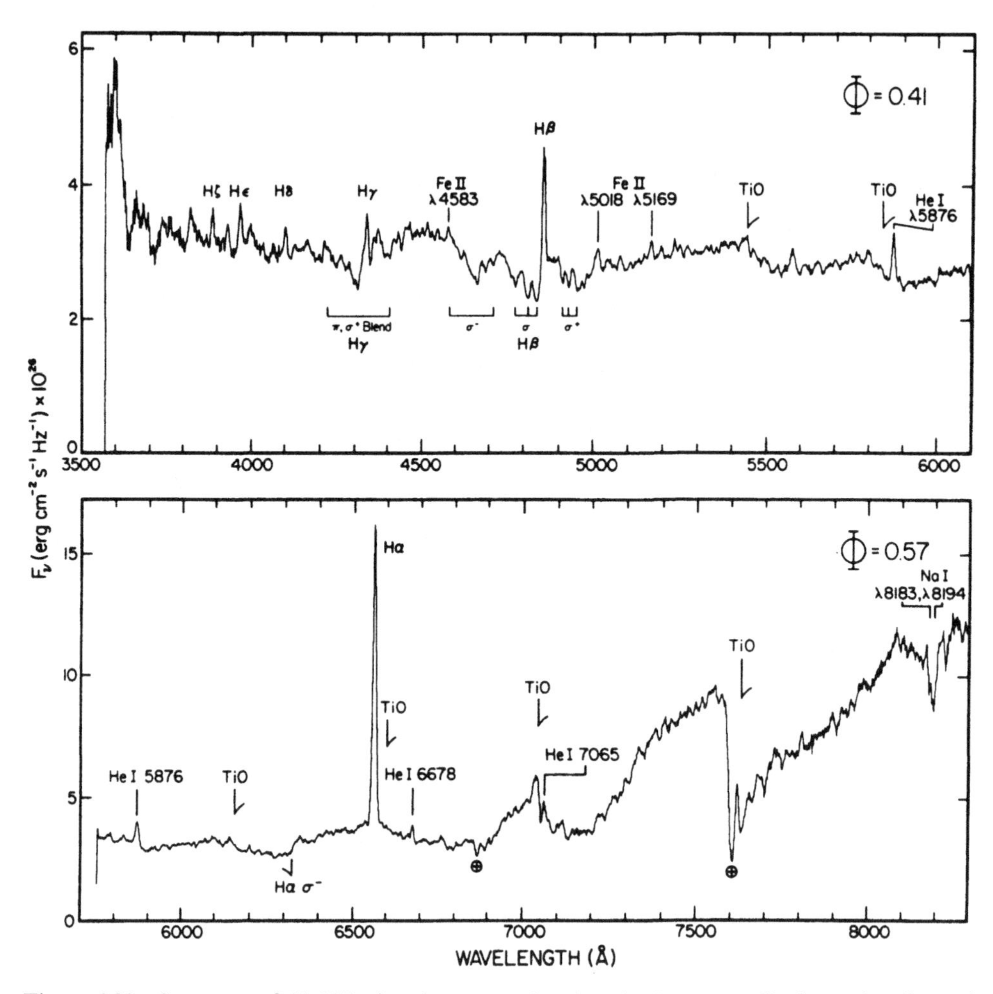

Figure 6.30 Spectrum of AM Her in a low state, showing the Zeeman-split absorption lines of the primary and the TiO absorption spectrum of the secondary. From Schmidt, Stockman & Margon (1981).

Balmer emission from the secondary is seen not only in the low state. Schwope *et al.* (1991,1993b) have observed it in MR Ser in the high state, as have Rosen, Mason & Cordova (1988) and Schwope *et al.* (1993a) in V834 Cen. Modelling the distribution of the Balmer emission over the irradiated secondary (Section 2.9.6), the latter authors deduce q and hence $M_1(1) = 0.66 \pm 0.17$ in V834 Cen.

From five polars in which secondary absorptions are seen Mouchet (1993) deduces $\overline{M}_1(1) = 0.65 \pm 0.24$ if the suspect high mass of $M_1(1) = 1.54$ for UZ For is excluded.

6.5.3 Field Strengths from Zeeman Splitting

The 1980 low state of AM Her revealed not only the spectrum of its secondary, it showed also the absorption spectrum of the photosphere of the primary (Schmidt, Stockman & Margon 1981; Latham, Liebert & Steiner 1981; Patterson & Price 1981a; Hutchings, Crampton & Cowley 1981). Because of the large magnetic field, the Balmer

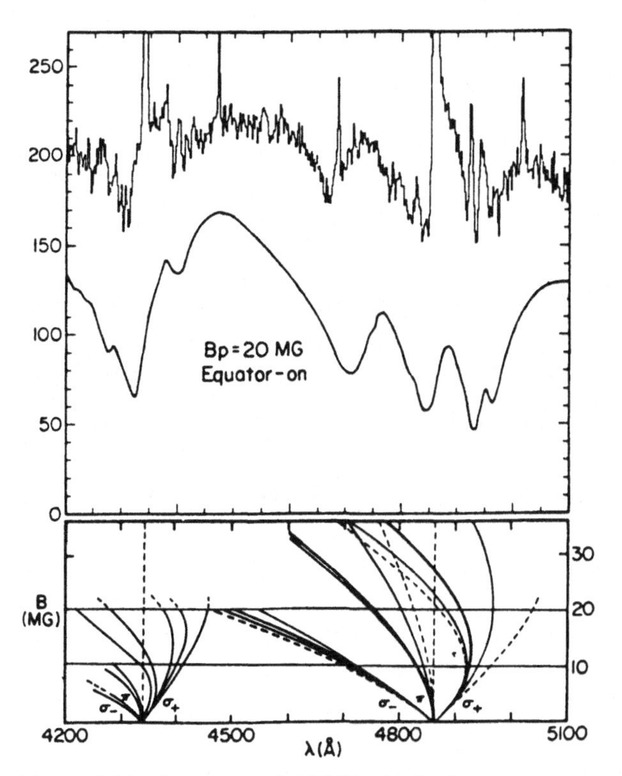

Figure 6.31 Spectrum of AM Her in the low state compared with the synthetic spectrum of a DA white dwarf with $B = 20$ MG. The lower panel shows the shift of Zeeman components of Hβ and Hγ as a function of field strength. From Latham, Liebert & Steiner (1981).

absorption lines in the DA spectrum of the primary are split into Zeeman components (Figure 6.30). This had previously been extensively observed in isolated magnetic white dwarfs (reviewed by Angel (1978)) for which calculations of the field-dependent wavelengths of the Zeeman components were made by Kemic (1974). These are shown for Hγ and Hβ in Figure 6.31, together with computed line profiles at 20 MG (Wickramasinghe & Martin 1979) in a DA white dwarf and the observed spectrum of AM Her (Latham, Liebert & Steiner 1981). Later computations were made specifically for AM Her by Wickramasinghe & Martin (1985) who deduced 22 MG for a centred dipole configuration, but found a significantly better fit to the observed orbital variations of the Zeeman pattern if the field has an effective dipole strength of 14 MG but is offset along the dipole axis by $0.17R(1)$ to give a field of 22 MG at the accreting pole (an updated result is given in Table 6.7). Off-set dipole fields, and sometimes more complex non-axisymmetric fields, are generally inferred from the observations (Wickramasinghe 1990).

Zeeman absorption components have also been observed in ST LMi (Schmidt, Stockman & Grandi 1983), BL Hyi (Wickramasinghe, Visvanathan & Tuohy 1984), V834 Cen (Beuermann, Thomas & Schwope 1989; Ferrario *et al.* 1992; Puchnarewicz *et al.* 1990), EF Eri[5] (Wickramasinghe *et al.* 1990) and MR Ser (Wickramasinghe *et al.*

[5] This result is uncertain (D.T. Wickramasinghe, private communication).

1991c); derived fields are given in Table 6.7. In the case of ST LMi the Zeeman absorption was detected during the faint part of the orbital cycle at high state, when the accretion zone was eclipsed behind the primary, and refers therefore to the non-accreting pole.

A direct comparison between fields derived from photospheric Zeeman-split absorption lines and those derived from cyclotron radiation (Sections 6.3.4 and 6.5.4) is not straightforward. The absorption lines are flux-weighted over the surface of the white dwarf; in the case of an isolated star this is a simple limb-darkening effect, but in the polars even during low states there may be significant extra flux from the weakly active accretion zones.

In a number of cases, however, the Zeeman effect is found to be operating *in association with the accretion zone*. Wickramasinghe, Tuohy & Visvanathan (1987) discovered broad absorption features during the orbital bright phase of V834 Cen which do not change wavelength as the viewing angle varies, showing them not to be cyclotron harmonics. Instead, the features can be attributed to Zeeman-split hydrogen absorptions originating in a field of 20–25 MG. The polarization of the absorptions and the fact that they disappear when the accretion zone passes behind the primary show that it is the accretion zone, and not the photosphere, that produces the spectrum. The absorption arises in the cool *halo* of the inhomogeneous accretion flow (Figure 6.10(b)); the fact that neutral hydrogen absorption is seen implies that the gas has not been shock heated. It is this cool gas that probably loses its kinetic energy subphotospherically and gives rise to the soft X-ray and UV flux.

The temperature of the halo gas is determined by the balance between Compton heating by hard X-rays emitted by the dense parts of the column and cyclotron and bremsstrahlung cooling. Achilleos, Wickramasinghe & Wu (1992) have computed models which show that for a typical hard X-ray luminosity $\sim 10^{33}$ erg s^{-1} and a halo density $N_e \sim 10^{13}$ cm^{-3} placed at a distance $\sim 0.1R(1)$ from the X-ray source, temperatures are $(1\text{–}4) \times 10^4$ K, which are correct to populate the $n = 2$ level of hydrogen and hence produce Hα absorption.

In V834 Cen, MR Ser, EF Eri and DP Leo the fields derived from Zeeman-split halo lines agree with those derived from cyclotron humps (next section) in these stars (Wickramasinghe, Tuohy & Visvanathan 1987; Cropper *et al.* 1990a). Results are presented in Table 6.8.

6.5.4 *Field Strengths from Cyclotron Harmonics*

Visvanathan & Wickramasinghe (1979) discovered cyclotron harmonic structure in the spectrum of VV Pup during the bright phase of its orbital cycle. Further observations were made by Wickramasinghe & Visvanathan (1980) and Stockman, Liebert & Bond (1979). Analysis of the spectra by Wickramasinghe & Meggitt (1982) gave $kT_{\text{elec}} \sim 10$ keV, $B = 3.2 \times 10^7$ G and $\Lambda = 10^5$, which are conditions in which cyclotron cooling dominates.

For a decade VV Pup was the only object in which cyclotron harmonics had been detected, but Cropper *et al.* (1989, 1990a), Bailey, Ferrario & Wickramasinghe (1991) and Wickramasinghe *et al.* (1991c) have shown that good signal-to-noise and a wide wavelength range is generally adequate to detect harmonics in all polars. UZ For and MR Ser show them quite prominently (Beuermann, Thomas & Schwope 1988; Ferrario

Table 6.8. *Magnetic Field Strengths (MG) in Polars.*

		First pole			Second pole			
Star	P_{orb} min	Cyclotron	Halo Zeeman	Photospheric Zeeman	Cyclotron	Photospheric Zeeman	Dipole Photospheric Zeeman	Offset
EF Eri	81.0	12	11				7	
DP Leo	89.8	31	30		59			
VV Pup	100.4	32			56			
V834 Cen	101.5	24	23	22		39	31	−0.1
MR Ser	113.6	24	25				27.3	0.3
BL Hyi	113.6	22					30	
ST LMi	113.9	12					19	
EK UMa	114.5	47						
AN UMa	114.8	36						
HU Aqr	125.0	36		20				
UZ For	126.5	53			75			
V2009-65.5	159.7	20:						
AM Her	185.6	14		14		27	18	−0.17
V1500 Cyg	201.0	25:						
BY Cam	201.9	41						

References: Wickramasinghe (1993); Glenn *et al.* (1994).

et al. 1989; Schwope & Beuermann 1993). The magnetic fields listed in the cyclotron column of Table 6.8 have been determined by this method.

As with the Zeeman absorption lines, the variations around orbit of the cyclotron harmonic positions and their profiles (Figure 6.32) provide a means of measuring the strength and geometry of the field.

The field strengths $1 \lesssim B_7(1) \lesssim 7$ appearing in Table 6.8 imply $0.48 \lesssim \mu_{34}(1) \lesssim 3.4$ from equation (6.8) for $M_1(1) = 0.7$.

6.6 Synchronization of the Primary

The coincidence of the rotation period of the primary, as determined from polarization observations, and the orbital period establishes synchronism of rotation only to the limits of accuracy of the methods employed. For DP Leo, an eclipsing polar, a limit $(P_{\mathrm{orb}} - P_{\mathrm{rot}})/P_{\mathrm{orb}} < 2 \times 10^{-6}$ was obtained by Biermann *et al.* (1985) from a baseline of four years. Further study has shown that the longitude ψ of the magnetic pole in DP Leo and WW Hor can vary by up to 20° over several years (Bailey *et al.* 1993). Yet there is indirect evidence that on long time scales exact synchronism is maintained – if it were not so then ψ would be uniformly distributed, but there is a strong clustering about $\psi \sim 20°$ (Cropper 1988).

This synchronism is maintained despite the accreted angular momentum that tries to increase the angular velocity of the primary. Clearly there are one or more torque mechanisms counteracting the accretion torque, and these must result from the magnetic field of the primary. The possibility of two different interactions must be kept in mind: (a) a braking torque that might act only while the primary is (initially) asynchronous and (b) a torque that can counteract the accretion torque and maintain synchronism even when the primary is already phase-locked.

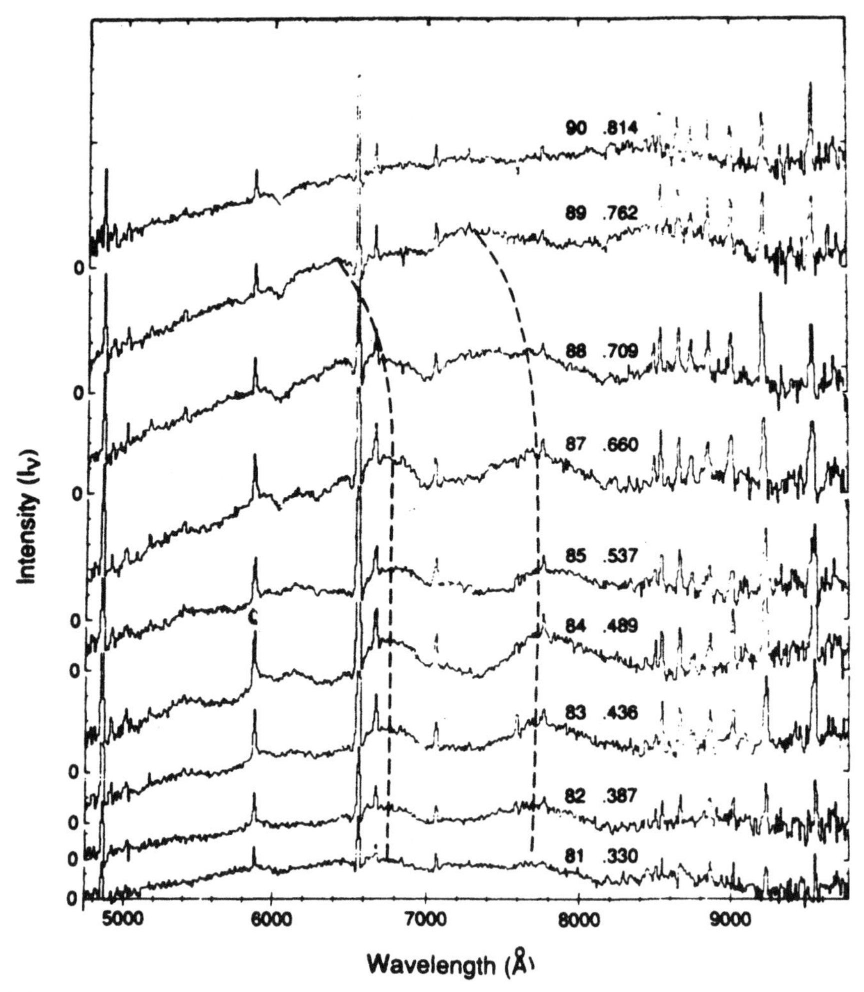

Figure 6.32 Cyclotron harmonics in spectra of MR Ser, showing changes in frequency with orbital phase. From Wickramasinghe *et al.* (1991c).

The first suggestion (Joss, Katz & Rappaport 1979) for the braking torque was that the field of the primary induces a field in the secondary (which is highly conducting and, except in a thin surface layer, excludes the field); this oscillates as the primary rotates and the ohmic dissipation losses in the surface of the secondary must derive their energy from the spin energy of the primary. However, this torque is estimated to be considerably less than the accretion torque and can at most only be effective when the stars are detached (e.g., in the pre-CV stage, or in an orbital period gap). Greater dissipation occurs if the field of the primary is not excluded from the interior of the secondary, but is carried inwards by convection (Campbell 1983); but this is still insufficient to result in synchronization (Lamb 1985; Lamb & Melia 1988).

A stronger braking torque was identified by Lamb *et al.* (1983), who pointed out that in the presence of any plasma of reasonable density ($N_e > 0.01$ cm^{-3}), the space between the primary and secondary does *not* behave like a vacuum. Instead, asynchronous rotation winds up the field lines from the primary that thread the secondary (if the secondary's magnetic moment $\mu(2) = 0$) or connect with the field lines of the secondary (if $\mu(2) \neq 0$), driving large currents along the field lines and generating large magnetic stresses. This MHD torque is sufficient to produce *and (nearly) maintain* synchronization in the presence of an accretion torque (Lamb & Melia 1988), The strength of the MHD torque is $\propto a^{-5} \tilde{\propto} P_{\rm orb}^{-3.3}$ and therefore increases rapidly as $P_{\rm orb}$ decreases.

This, however, cannot be the whole story, for the MHD model requires at least a small degree of asynchronism in order to generate a torque. The additional coupling between primary and secondary needed to maintain synchronism arises in the interaction between the intrinsic fields of the primary and secondary (Campbell 1985, 1986, 1989).

The synchronizing torque is (Hameury, King & Lasota 1987)

$$N_{\rm syn} \approx \frac{\mu(1)\mu(2)}{a^3} \tag{6.28}$$

(this applies also if $\mu(2)$ is induced rather than intrinsic). The moment of inertia of the primary is

$$I(1) = M(1)R^2(1)k^2(1) \tag{6.29}$$

where $k(1)$ is the radius of gyration of the primary, which may be obtained from (Ritter 1985b)

$$k(1) = 0.452 + 0.0853 \log\left[1 - \frac{M(1)}{M_{\rm ch}}\right] \qquad M(1) \leq 0.95\, M_{\rm ch} \tag{6.30a}$$

$$k(1) \to 0.275 \text{ at } M(1) = M_{\rm ch}. \tag{6.30b}$$

Hence the synchronization time scale (see also Zylstra (1989)) is

$$\begin{aligned} \tau_{\rm syn} &= \frac{I(1)\Omega(1)}{N_{\rm syn}} \\ &\approx 3.2 \times 10^4\, \frac{M_1^2(1)R_9^2(1)[k(1)/0.43]^2(1+q)P_{\rm orb}^2(\rm h)}{P_3(1)\mu_{34}(1)\mu_{34}(2)} \text{ y.} \end{aligned} \tag{6.31}$$

Competing against this is the accretion torque $N_{\rm acc} \approx R_{L_1}^2 \Omega_{\rm orb} \dot{M}(2)$, which leads to a spin-up time scale for the primary

$$\tau_{\rm acc} \approx 1.6 \times 10^8\, \frac{M_1^{1/3}(1)R_9^2(1)[k(1)/0.43]^2}{P_3(1)P_{\rm orb}^{1/3}(\rm h)\dot{M}_{15}(2)} \text{ y.} \tag{6.32}$$

Thus $\tau_{\rm syn} \ll \tau_{\rm acc}$ for polars, in accord with their observed synchronism.

The stability of the equilibrium state for dipole, quadrupole and for offset-dipole geometries has been investigated (Lamb & Melia 1988; King, Frank & Whitehurst 1990; Wickramasinghe & Wu 1991; King & Whitehurst 1991; Wu & Wickramasinghe 1993a) with the conclusion that the variable accretion torque will drive oscillations in ψ with amplitudes up to $\sim 50°$ should occur on time scales 30–50 y. These account for the observed variations of ψ.

Another effect has been pointed out by Katz (1989) and Campbell (1990). The Maxwell stresses produced by its internal field distort the primary (with moments of inertia differing by $\sim 10^{-9}$ along the various axes), which allows a gravitational couple to be applied by the secondary. This can be larger than the magnetostatic torque.

6.7 Magnetic Fields and Nova Eruptions

Nova Cygni 1975 (V1500 Cyg) was one of the brightest novae, both intrinsically and apparently. It was also one of the fastest novae, declining 7 mag in 45 d. During the decline a 3.3 h variation was observed in the emission-line velocities, which later appeared in optical photometry. The period decreased from 0.141 d to 0.138 d in the first year and then increased and stabilized at 0.140 d at the end of 1977 (Patterson 1978, 1979a). Photometry of the NR in 1987 produced a light curve reminiscent of a polar (Kaluzny & Semeniuk 1987), which stimulated polarimetric observations and the consequent discovery that V1500 Cyg is a polar (Stockman, Schmidt & Lamb 1988), with $B(1) \sim 25$ MG.

Strictly speaking, V1500 Cyg is *not* a polar – the period of its polarization modulation is 0.137154 d and its photometrically determined orbital period (from the modulation caused by irradiation of the secondary by the primary) is 0.1396129 d. This difference of 1.8% between the rotation period of the primary and the orbital period is ascribed to the effects of the nova explosion in 1975 – until then V1500 was probably synchronized, and at some time in the future it will resynchronize. This system therefore provides a unique opportunity to measure the braking time scale.

As the primary rotates asynchronously (a situation which is treated more fully in the next chapter) the accretion flow is more variable than for a phase-locked polar. Accretion occurs on both magnetic poles, but at any given time accretion onto the pole nearest the secondary is most favoured.

The initial evolution of the photometric period is explained as follows (Stockman, Schmidt & Lamb 1988): the initial expansion of the nova envelope increases the moment of inertia of the primary, decreasing its angular velocity, breaking synchronism and producing the 0.141 d period. Through interaction with the secondary, the envelope is spun up to the binary period; strong coupling between envelope and core ensures that the core also achieves synchronism. When the outer envelope (beyond the secondary) is finally ejected, the primary with its remnant envelope shrinks back to its original white dwarf radius, reducing its moment of inertia and spinning the primary up to its 0.137 d period.

Over the $3\frac{1}{2}$ years from March 1987 to September 1990 the polarization period of V1500 Cyg increased at a rate $\dot{P}_{rot} = 3.6(\pm 0.3) \times 10^{-8}$, corresponding to a time scale for synchronization of $(P_{orb} - P_{rot})/\dot{P}_{rot} = 185 \pm 15$ y (Schmidt & Stockman 1991; Schmidt, Liebert & Stockman 1995). This shows that V1500 Cyg will again become synchronized well before the next nova eruption, and is consistent with the assumption that it was a polar prior to the 1975 eruption.

The observed $\dot{P}_{rot}$ is related to the difference between the braking torque N_{br} and the accretion torque N_{acc} by

$$N_{br} - N_{acc} = I(1)\dot{\Omega}_{rot}(1) = -\frac{2\pi I(1)\dot{P}_{rot}(1)}{P_{rot}^2(1)} . \tag{6.33}$$

The right hand side of equation (6.33) is $\approx 2 \times 10^{35}$ dyn cm. N_{acc} will be $\sim \dot{M}l$, where l is the angular momentum per unit mass of gas leaving the L_1 point. For a probable $\dot{M} \sim 2 \times 10^{16}$ g s^{-1} this gives $N_{acc} \approx 3 \times 10^{34}$ dyn cm, which is an order of magnitude smaller than the observed torque. Therefore $N_{br} \approx 2 \times 10^{35}$ dyn cm. By comparison, the MHD model (Lamb & Melia 1988) gives $N_{br} \sim 10^{32}$–10^{34} dyn cm for $1 \leq \mu_{33}(1) \leq 10$ and a non-magnetic secondary, or $N_{br} \sim 2 \times 10^{35}$ dyn cm for $\mu_{34}(1) = (1)$ and $\mu_{34}(2) = 2$. This implies $B(2) \sim 1$ kG (equation (6.11)).

An alternative suggestion by Katz (1991) is that the large torque is a result of interaction of the primary's magnetic field with an enhanced wind from the secondary (caused by irradiation by the hot primary: Section 4.4.2) which generates a magnetic torque greater than expected (Section 9.1.2.2). However, the discussion in Section 9.3.6.1 indicates that the wind from the secondary is in fact suppressed in polars.

BY Cam is a circularly polarized system showing abrupt transitions in polarization direction with occasional protracted times of constant sign; it has strong flaring activity. These properties suggest non-steady accretion onto a slightly asynchronous rotator (Mason, Liebert & Schmidt 1989). Extended X-ray and optical observations confirm the alternating feeding of different poles, as would occur by stream accretion onto an asynchronous primary (Ishida *et al.* 1991; Silber *et al.* 1992). Optical photometry (Piirola *et al.* 1993, 1994) reveals a period 199.85 min that is 1.0% longer than the spectroscopically determined P_{orb}. BY Cam is therefore similar to V1500 Cyg in its P_{orb} and asynchronicity, but is dramatically more active. This is probably because V1500 Cyg has a lower $B(1)$ (Table 6.7) and a larger $M(1)$ (and therefore separation) than in BY Cam, which allows disc accretion (Section 7.2) in the former but produces stream accretion in the latter. As stated in Section 6.5.1.1, BY Cam has abundances that suggest nova activity, compatible with the possibility that it is a polar knocked out of synchronism by a nova eruption in the past century or two. RXJ 1940.2–1025 has an orbital period almost identical to that of BY Cam and may also be desynchronized (Watson *et al.* 1995).

GQ Mus (Diaz & Steiner 1989), V2214 Oph (Baptista *et al.* 1993), CP Pup (O'Donoghue *et al.* 1989) and possibly T Pyx (Schaefer *et al.* 1992) are novae that have 'orbital' light curves like V1500 Cyg. They are possibly also desynchronized polars, but polarization has yet to be detected in them. Evidence for V Per having a significant magnetic field is given in Section 4.4.4.2. All of these CN have $P_{orb} < 3.4$ h, whereas no CN with $P_{orb} > 3.4$ h are strongly magnetic (an exception may be GK Per with $P_{orb} = 2$ d: Section 7.1). In fact, of the CN with $P_{orb} \lesssim 3.4$ h (Table 4.2) only RW UMi is not known or suspected to be magnetic.

From equation (6.9), the high $\dot{M}(2)$ ($\sim 10^{-8}$ M$_\odot$ y^{-1}) phase subsequent to a nova eruption (Section 4.4.2) will reduce r_μ by a factor of $\sim$5 below that for a polar with the standard $\dot{M}(2) \sim 4 \times 10^{-11}$ M$_\odot$ y^{-1} (Section 6.4.3). In most cases this will result in $r_\mu < r_{min}$, allowing a disc to form, the light from which will dilute that from the accretion zone, making polarization difficult to detect.

V348 Pup (1H 0709–360) is a strong X-ray source whose eclipses establish the orbital period at 2.444 h. There is photometric evidence for a second period $\sim$2% away from P_{orb} (Tuohy *et al.* 1990). The eclipse width shows the presence of a disc or annulus; absence of polarization is probably the result of disc dilution. This star is a strong candidate for a desynchronized polar.

The fraction of polars desynchronized through nova eruptions should be $\sim \tau_{syn}/T_R$. Assuming that the above systems are desynchronized polars, and taking $\tau_{syn} = 185$ y from V1500 Cyg, gives $T_R \sim 37/7 \times 185 \approx 1 \times 10^3$ y.

6.8 Radio Emission

Radio emission at 4.9 GHz from AM Her was first detected by Chanmugam & Dulk (1982) using the Very Large Array. It remained observable as a radio source with a flux ~0.5 mJy during 1982 and 1983 but declined to indetectability (<0.3 mJy) during 1984, after which it reappeared at 0.3–0.7 mJy (Chanmugam 1987). The source is also present at 14.85 GHz (Bastion, Dulk & Chanmugam 1985).

Attempts to detect AN UMa, VV Pup, EF Eri, MR Ser and ST LMi were unsuccessful (Dulk, Bastion & Chanmugam 1983), but those authors observed a 100% circularly polarized flare from AM Her lasting <30 min. ST LMi has since been detected by Pavelin, Spencer & Davis (1994). V834 Cen has been detected at 8.4 GHz with a flux density ~2 mJy (Wright *et al.* 1988).

The high brightness temperature ($\gtrsim 10^{10}$ K) and circular polarization of the AM Her flare imply a coherent emission mechanism such as plasma oscillations (Kuijpers 1985) or cyclotron maser action (Melrose & Dulk 1982). Chanmugam (1987) points out that the former mechanism requires a high electron density and would be absorbed by free–free transitions. Similarly, the fundamental of cyclotron masing is subject to absorption, so the emission probably arises from first harmonic emission. Then $B \sim 10^3$ G from equation (6.1), and the emission must originate in the stream at a distance where the primary's field has dropped to such a value, or on the secondary. As similar highly polarized radio flares have been observed in RS CVn binaries (Brown & Crane 1978), and in a DN (Section 3.3.7), where there is no $\sim 10^7$ G primary, it is most probable that the flare originates in the magnetosphere of the secondary.

The variable quiescent unpolarized fluxes in AM Her and V834 Cen can be explained by gyrosynchrotron emission by electrons with energies of ~500 keV in a magnetic field of ~40 G and harmonic number ~40 (Chanmugam & Dulk 1982; Chanmugam 1987). The origin of these high energy electrons is unknown: they certainly do not escape from the magnetically confined accretion zone near the primary. Chanmugam & Dulk (1982) suggest acceleration mechanisms akin to those operating in Type II bursts in the solar corona, or by electrodynamical processes of the kind operating in the Jupiter–Io magnetic interaction.

6.9 Gamma-Ray Observations

A possible detection of PeV and TeV gamma rays from AM Her has been reported (Bhat 1990; Bhat *et al.* 1991) with time-averaged luminosities of $\sim 2 \times 10^{32}$ erg s^{-1} in both energy ranges. These give an overall flux distribution for AM Her of $F_\nu \propto \nu^{-1}$ from IR to PeV. The gamma-ray light curves appear 100% modulated at period P_{orb}. At MeV energies detections of V834 Cen and VV Pup with COS-B satellite observations have been claimed (Bhat 1990; Bhat, Richardson &Wolfendale 1989). A mechanism for production of TeV photons has been proposed by Kaul, Kaul & Bhat (1993), who envisage the diffusive shock acceleration process (Blandford & Eichler 1987) working in the threading region.

7

Intermediate Polars

...burning for the ancient heavenly connection to the starry dynamo in the machinery of the night.

Allen Ginsberg. *Howe.*

With less powerful magnetic fields on the primaries, synchronism cannot be achieved. This opens the way for a multitude of periodic phenomena and generates the class of magnetic CVs known as intermediate polars (IPs).

7.1 Historical Development

Charles *et al.* (1979) identified the X-ray source 2A0526−328 as a thirteenth magnitude star with a spectrum like AM Her but lacking detectable polarization; this was the first CV to be discovered by means of its X-ray emission. Photometry (Motch 1981) and spectroscopy (Hutchings *et al.* 1981a) revealed an orbital period (from RV modulation) of 5 h 29.2 min and a photometric period of 5 h 11.5 min. The beat period between these, *viz.* 4.024 d, was also present in the light curve.

TV Col (as 2A0525−328 was designated) has turned out to be more complicated than even those initial observations suggested, so although Hutchings *et al.* proposed the correct basic model for the system, it is now applied to TV Col in a different manner (see below). For simplicity we will therefore move attention to the case history of AO Psc, which was being studied almost at the same time as TV Col.

The X-ray source H2252−035 was identified with a thirteenth magnitude previously unrecognized variable (later designated AO Psc) by Griffiths *et al.* (1980). Its spectrum resembled a DN at quiescence and photometry showed an orbital modulation of 3.6 h with a prominent 14.3 min periodic luminosity variation superimposed (Warner 1980; Patterson & Price 1980). There was some surprise when a strong modulation of the X-ray flux was found at a period of 13.4 min (White & Marshall 1980, 1981). In more extended optical photometry this period was also detected at very low amplitude, along with the 14.3 min modulation (Warner, O'Donoghue & Fairall 1981).

The combination of photometric and spectroscopic phenomena led to a model (Patterson & Price 1981b; Warner, O'Donoghue & Fairall 1981; Hassall *et al.* 1981), based on the one proposed by Hutchings *et al.* for TV Col (which itself was derived from the pre-existing model of DQ Her discussed in Section 8.1.4). In this an X-ray emitting region on the white dwarf is carried around by its rotation ($P_{rot}(1) = 13.4$ min) causing periodic illumination of some region *fixed in the rotating frame of the binary.* The latter could, for example, be the secondary star (Patterson & Price 1981b) or an enlarged region of the disc surrounding the bright spot (Hassall *et al.* 1981). The expected period for the X-ray illumination (reprocessed to optical radiation), the synodic period $P_{syn}^{-1} = P_{rot}^{-1}(1) - P_{orb}^{-1}$, was verified to within observational accuracy (Warner, O'Donoghue & Fairall 1981). In contrast to the polars, TV Col and AO Psc showed no detectable polarization at optical wavelengths.

TV Col and AO Psc constituted the first two members of a new class of CV: strong X-ray emitters in which, unlike the polars, the white dwarf rotates *asynchronously*. In anticipation that such a class might exist, in which the primary's magnetic field would not be strong enough to enforce synchronous rotation, Krzeminski had already suggested the term 'intermediate polar' (see Warner & McGraw (1981)). Although their underlying structure is similar to that of DQ Her, the IPs have sufficiently distinctive observational properties to warrant treating them separately from the stars of DQ Her type (Chapter 8). However, the DQ Her stars are to be understood as a *subtype* of the IPs.

To complete the story of TV Col: although at first it appeared that the 5 h 11.5 min photometric period was the rotation period of the white dwarf, and the 4 d period its beat with the orbital period, X-ray observations (Schrijver, Brinkman & van der Woerd 1987) have shown that the true rotation period is 31.9 min. Eclipses show that 5 h 29.2 min is the true orbital period (Hellier, Mason & Mittaz 1991). It thus appears that the 4 d modulation has no immediate connection with the IP nature of TV Col, but is probably a manifestation of the precessing tilted disc or precessing elliptical disc phenomena (Sections 2.6.6 and 4.5). The initial proposal of an asynchronous primary was fortuitously correct.

To be classified as *definitely* an IP a star currently requires to be a hard X-ray emitter and multiperiodic. Table 7.1 lists the objects that meet these requirements. Some examples of candidate asynchronous rotators, mostly without strong X-ray emission, are given in Table 7.2.

Of the stars listed in Table 7.1 all but three were discovered first as X-ray sources. The exceptions were GK Per, the remnant of Nova Persei 1901, which first displayed X-ray pulsations during its infrequent DN outbursts; EX Hya, originally discovered in 1940 from DN-like outbursts, early suggested as an X-ray source candidate (Warner 1973a) but only later recognized to be multiperiodic (Vogt, Krzeminksi & Sterken, 1980) and an IP (Warner & McGraw 1981), and YY Dra, found and misclassified in 1934 as an Algol and wrongly renamed DO Dra when identified as a CV (Patterson *et al.* 1992b).

Reviews of the properties of intermediate polars have been given by Warner (1983a, 1985a), Mason (1985), Berriman (1988), Mason, Rosen and Hellier (1988), Osborne (1988) and Patterson (1990b, 1994).

7.2 Magnetically-Controlled Accretion for Asynchronous Rotation

The existence of an asynchronously rotating primary is an indication that the braking torque is insufficient to produce synchronism. Obvious (non-equilibrium) exceptions are polars in the process of being synchronized for the first time, or having been jolted out of synchronism by a nova explosion (Section 6.7) or a phase of exceptionally high

$\dot{M}$. From the observed $\dot{P}_{rot}$ in V1500 Cyg (Section 6.7), such states are short lived and will be observed only rarely.

There are two distinct modes of mass transfer onto asynchronous rotators:

(i) If the magnetic moment $\mu(1)$ of the primary is too small to prevent formation of an accretion disc, then magnetically-controlled accretion will be restricted to the region from the inner edge of the disc down to the surface of the primary. (Strictly speaking, the term accretion annulus should be used – but that is equally true of the 'complete' discs around non-magnetic primaries, where the disc is truncated by the surface of the white dwarf).

(ii) For sufficiently large $\mu(1)$ no accretion disc can form. The process of mass transfer then is determined by the location of the threading region.

The precise conditions that govern the formation of a disc are not yet known (Hameury, King & Lasota 1986b; Lamb & Melia 1988). The following considerations are relevant:

(i) If $r_\mu < r_{min}$ (Sections 2.4.2 and 6.2.1) a disc will definitely form.

(ii) If $r_\mu > r_{min}$ there are two possible regimes, determined by the radius r_r (Section 2.4.2). If $r_\mu > r_r$ then the transferring matter has insufficient angular momentum to orbit the primary outside the magnetosphere, so a disc cannot form. If $r_{min} < r_\mu < r_r$ a disc may form, depending on the details of how the stream penetrates into the regions of higher field. For example, some slowly threaded large blobs may survive passage past r_{min} and initiate a disc.

(iii) The existence of a disc may also depend on the history of the system. A period of high $\dot{M}$ may reduce r_μ to the point where a disc forms, after which its persistence depends on r_μ(disc) (see Section 7.2.1) rather than r_μ(stream). Lamb & Melia (1986, 1987, 1988) conclude that a disc always forms and survives if r_μ(disc) $< r_r$.

7.2.1 *Accretion from a Disc*

The results obtained from the polars show that the magnetic axes of white dwarfs in CVs are in general inclined to their rotation axes. An asynchronous primary is therefore usually an *oblique rotator*. The *general* pattern of disc accretion onto an oblique rotator has not been solved: the accretion flow is time-dependent, three-dimensional and non-axisymmetric. Consequently, the heuristic model in current use is for an aligned rotator, in which axisymmetric, steady flows can exist (see Anzer & Börner (1980, 1983) and Spruit & Taam (1990) for some initial estimates for oblique rotators). Most of the work in this area has been directed at disc accretion onto neutron stars (reviewed by Lamb (1989) and Lamb & Ghosh (1991)) but applies *mutatis mutandam* to white dwarfs.

The foundations of the understanding of the interaction between the magnetosphere and the disc were laid in a series of papers by Ghosh & Lamb (1978, 1979a,b; see also Wang 1987). A cross-section through their model is shown in Figure 7.1. We will consider the various regions of interaction, starting from that most remote from the magnetic primary. Not shown in Figure 7.1 are the field lines which connect to the

Table 7.1. *IPs.*

Star	Alias	P_{orb} (h)	m_v	$P_{rot}(1)$ (s)	$\dot{P}_{rot}(1)$ $\times 10^{-11}$	P_{syn} (s)	i°	References
EX Hya	4U1228−29	1.633	13.5	4021.61 OX	−3.8± 0.2		78	1–20,24,98 106–108,110–118
RXJ1712.6–2414		2.65 or 3.41	14.2	927 O		1003/1027 O		150
BG CMi	3A0729+103	3.235	14.8	913.49 X	5.7 ± 0.6	913.49 OX	30:	21–27,97,119–124,147
V1223 Sgr	4U1849−31	3.366	13.0–>16.8	745.43 X	2.3 ± 0.3	794.38 OX	<40	24,28–33,95,109, 120,125–128
AO Psc	H2252−035	3.591	13.2–15.3	805.4 XO	−6.6 ± 1.0	858.69 OX:	60:	34–44,95,120, 129–132
YY Dra	3A1148+719	3.910	16.0	529.2 X	<21	550 O	42	45–50,95,133,144
FO Aqr	H2215−086	4.849	13.0–13.8	1254.45 OX	3.4 ± 0.4	1373 OX	70 ± 5	24,51–67,120, 134,145
PQ Gem	RE0751+14	5.18	14.5	833.37 XO		872 O	low?	90–95,148,149
TV Col	2A0526−328	5.487	13.8	1910 OX			70:	68–81,135,136
TX Col	1H0542−407	5.718	15.7	1910.5 X		2106 OX	25:	24,82–84,98
XY Ari	1H0253+193	6.06	$K = 13.5$	206.30 X			>70	85–89
V1062 Tau	1H0459+248	9.952	15.6	3720 OX		3720 OX		96,146
GK Per	A0327+43	47.923	13	351.34 X	−2.5:	351.3 O	65	99–105,137–143

Notes: m_v is the apparent magnitude averaged over the orbit. Where a range of m_v is given, this denotes high and low states. Outburst magnitudes are not given here (see Section 7.7).

O and X denote optical and X-ray determinations respectively.

Norton *et al.* (1992b) give $P_{rot}(1)$ = 847 s for BG CMi - but see Section 7.5.1.3.

References: 1. Kraft 1962; 2. Mumford 1964b,1967b, 1969; 3. Warner 1972a, 1973b; 4. Vogt, Krzeminski & Sterken 1980; 5. Warner & McGraw 1981; 6. Gilliland 1982a; 7. Sterken *et al.* 1983; 8. Jablonski & Busko 1985; 9. Hellier *et al.* 1987; 10. Kaitchuck *et al.* 1987; 11. Cordova, Mason & Kahn 1985; 12. Kruszewski *et al.* 1981; 13. Hill & Watson 1984,1990; 14. Beuermann & Osborne 1985; 15. Beuermann & Osborne 1988; 16.Rosen, Mason & Cordova 1988; 17. Bond & Freeth 1988; 18. Siegel *et al.* 1989; 19. Rosen *et al.* 1991; 20. Singh & Swank 1993; 21. McHardy *et al.* 1984, 1987; 22. Vaidya *et al.* 1988; 23. Patterson 1990a; 24. Hellier & Mason 1990; 25. Singh *et al.* 1991; 26. Augusteijn, van Paradijs & Schwarz 1991; 27. Patterson & Thomas

1993; 28. Steiner *et al.* 1981; 29. Warner & Cropper 1984; 30. Watts *et al.* 1985; 31. Osborne *et al.* 1985; 32. Pakull & Beuermann 1987; 33. van Amerongen, Augusteijn & van Paradijs 1987; 34. Griffiths *et al.* 1980; 35. Warner, O'Donoghue & Fairall 1981; 36. Patterson & Price 1981b; 37. White & Marshall 1981; 38. Motch & Pakull 1981; 39. Wickramasinghe, Stobie & Bessell 1982; 40. Cordova *et al.* 1983; 41. van der Woerd, de Kool & van Paradijs 1984; 42. van Amerongen *et al.* 1985; 43. Kaluzny & Semeniuk 1988; 44. Hellier, Cropper & Mason 1991; 45. Patterson *et al.* 1982; 46. Williams 1983; 47. Szkody 1987a; 48. Friend *et al.* 1988; 49. Mateo, Szkody & Garnavich 1991; 50. Patterson *et al.* 1992b; 51. Shafter & Targan 1982; 52. Patterson & Steiner 1983; 53. Cook, Watson & McHardy 1984; 54. Sherrington, Jameson & Bailey 1984; 55. Berriman *et al.* 1986; 56. Pakull & Beuermann 1987; 57 Shafter & Macry 1987; 58. Chiappetti *et al.* 1988; 59. Semeniuk & Kaluzny 1988; 60. Osborne & Mukai 1989; 61. Hellier, Mason & Cropper 1989; 62. Steiman-Cameron, Imamura & Steiman-Cameron 1989; 63. Norton & Watson 1989b; 64. Norton *et al.* 1990, 1992a; 65. Martell & Kaitchuck 1991; 66. Kruszewski & Semeniuk 1993; 67. de Martino *et al.* 1993; 68. Charles *et al.* 1979; 69. Motch 1981; 70. Hutchings *et al.* 1981a; 71. Mouchet *et al.* 1981; 72. Watts *et al.* 1982; 73. Szkody & Mateo 1984; 74. Bonnet-Bidaud, Motch & Mouchet 1985; 75. Schrivjer *et al.* 1985; 76. Schrivjer, Brinkman & van der Woerd 1987; 77. Barrett, O'Donoghue & Warner 1988; 78. Schwarz *et al.* 1988; 79. Hellier, Mason & Mittaz 1991; 80. Hellier 1993a; 81. Hellier & Buckley 1993; 82. Tuohy *et al.* 1986; 83. Buckley & Tuohy 1989; 84. Mouchet *et al.* 1991; 85. Patterson & Halpern 1990; 86. Kamata, Tawara & Koyama 1991; 87. Koyama *et al.* 1991; 88. Zuckermann *et al.* 1992; 89. Kamata & Koyama 1993; 90. Mason *et al.* 1992; 91. Andronov 1993; 92. Wenzel 1993a; 93. Rosen, Mittaz & Hakala 1993; 94. Piirola, Hakala & Coyne 1993; 95. Beuermann & Thomas 1993; 96. Silber 1992; 97. Norton *et al.* 1992b; 98. Ferrario, Wickramasinghe & King 1993; 99. Watson, King & Osborne 1985; 100. Crampton, Cowley & Fisher 1987; 101. Norton, Watson & King 1988; 102. Eracleous, Patterson & Halpern 1991; 103. Patterson 1991; 104. Ishida *et al.* 1992, 105. Yi *et al.* 1992; 106. Reinsch & Beuermann 1990; 107. Hellier *et al.* 1989; 108. Buckley & Schwarzenberg-Czerny 1993; 109. van Amerongen & van Paradijs 1989; 110. Schwarz *et al.* 1979; 111. Hellier & Sproats 1992; 112. Bath, Pringle & Whelan 1980; 113. Breysacher & Vogt 1980; 114. Sherrington *et al.* 1980; 115. Cowley, Hutchings & Crampton 1981; 116, Frank *et al.* 1981b; 117. Shafter 1984a; 118. Heise *et al.* 1987; 119. Falomo *et al.* 1985; 120. Penning 1985; 121. Penning, Schmidt & Liebert 1986; 122. West, Berriman & Schmidt 1987; 123. Chanmugam *et al.* 1990; 124. Patterson & Thomas 1993; 125. Bonnet-Bidaud, Mouchet & Motch 1982; 126. King & Williams 1983; 127. Jablonski & Steiner 1987; 128. Hassall *et al.* 1981; 129. Clarke, Mason & Bowyer 1983; 130. Kubiak 1984b; 131. Williams, King & Watson 1984; 132. Hutchings & Cote 1986; 133. Mukai *et al.* 1990; 134. Mukai & Hellier 1992; 135. Warner 1980; 136. Coe & Wickramasinghe 1981; 137. Gallagher & Oinas 1974; 138. Bianchini, Hamzaoglu & Sabbadin 1981; 139. Bianchini & Sabbadin 1983b; 140. Mazeh *et al.* 1985; 141. Bianchini *et al.* 1986; 142. Seaquist *et al.* 1989; 143. Wu *et al.* 1989; 144. Patterson & Szkody 1993; 145. Hellier 1993c; 146. Remillard *et al.* 1994a; 147. Garlick *et al* 1994a; 148. Hellier & Ramseyer 1994; 149. Hilditch & Bell 1994; 150. Buckley *et al.* 1995b.

Table 7.2. *Possible Intermediate Polars.*

Star	Alias	Type	Quiescent m_v	P_{orb} (h)	Other periods (h)	Notes	References
SW UMa		DN(SU)	17.0	1.36	0.265 O	1	1–3
T Leo		DN (SU)	15.5	1.41		2	4,5
AL Com		DN(SU)?	19	1.5:	0.67 O		6–8,27,30
HT Cas		DN(SU)	16.5	1.77		2	9,10
VZ Pyx	1H0857−242	DN(SU)	16.8	1.78	0.86 X		11,12,37,39
V348 Pup	1H0709−360	NL	15.6	2.44			40
V795 Her		NL	13.0	2.60	2.796 O	3	13,14,38
TT Ari		NL	10	3.30	3.19 O	4	15
V603 Aql		NR	11.4	3.32	1.02 OX	5	16–18
1H0551–819		NL	13.5	3.34		6	19
S193		NL	13.0	3.6:	0.32: O		30,37
V347 Pup	LB1800	NL	13.4	5.57		7	23
TW Pic	1H0534−581	NL	14.1–15.9	6.5:	1.996 O	8	20,24–26
KO Vel	E1013−477	NL	16.7–19.5	10.12	1.7: O		20–22,31–33
V485 Cen		DN	17.9		0.986 O		28
AH Eri		DN	18.4		0.70 OX		7,37
CP Eri		DN	19.7		0.48 O		35
X Ser		NR	18.3		0.60 O		29
CT Ser		NR	16.6		0.22 O		29
1H0201–029		DN?	14.8		0.485		34
CBS 31		NL					36

Notes: 1. 15.9 min optical period at quiescence, suspected in EXOSAT observations.
2. Spectroscopic observations suggest the presence of a magnetosphere (see Section 7.7).
3. No X-rays (Rosen *et al.* 1989) and no low states (Wenzel, Banny & Andronov 1988), possibly a disc precessor (Section 4.5). Orbital period may be 4.86 h (Section 4.2.1) and the spectroscopic period may be the rotational period of the primary (Haswell *et al.* 1993). Patterson & Skillman (1994) have found an optical quasi-period near 19 min. 4. Multiple periodicities (Section 4.5) and hard X-ray source (Jensen *et al.* 1983) give strong similarity to TV Col, but no spin period has been determined in X-rays. Low state observations (Shafter *et al.* 1985) give $B(1) < 4$ MG.
5. Low amplitude 61.4 min modulation detected in optical, X-ray and UV, but not definitely stable. Strong X-ray source in 1979 and 1981 (Drechsel *et al.* 1983) but very weak in 1984 (Haefner, Pietsch & Metz 1988). Variable circular polarization claimed (Haefner & Metz 1985) but not confirmed (Cropper 1986b). 6. Strong X-ray flux, large L_x/L_{opt} and strong HeII 4686 suggested magnetic system, but no spin period is observed. 7. X-ray and optical characteristics of a magnetic CV, but no circular polarization. 8. 2.0 h periodicity shows in RV measurements, optical photometry, but only possibly in X-ray flux.

References: 1. Shafter, Szkody & Thorstensen 1986; 2. Robinson *et al.* 1987; 3. Szkody, Osborne & Hassall 1988; 4. Shafter & Szkody 1984; 5. Slovak, Nelson & Bless 1988; 6. Howell & Szkody 1988; 7. Szkody *et al.* 1989; 8. Mukai *et al.* 1990; 9. Young, Schneider & Schectman 1981b; 10. Zhang, Robinson & Nather 1986; 11. de Martino *et al.* 1992; 12. D. Buckley, private communication; 13. Kaluzny 1989; 14. Shafter *et al.* 1990; 15. Semeniuk *et al.* 1987; 16. Udalski & Schwarzenberg-Czerny 1989; 17. Eracleous, Patterson & Halpern 1991; 18. Schwarzenberg-Czerny, Udalski & Monier 1992; 19. Buckley *et al.* 1994; 20. Mukai & Corbet 1987, 1991; 21. Mouchet *et al.* 1987; 22. Kubiak & Krzeminski 1989; 23. Buckley *et al.* 1990b; 24. Tuohy *et al.* 1986; 25. Buckley & Tuohy 1990; 26. Patterson & Moulden 1993; 27. Howell & Szkody 1991;

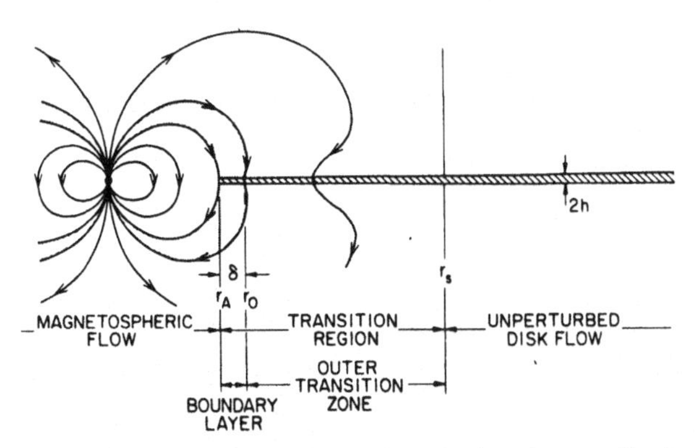

Figure 7.1 Schematic picture of accretion from a disc into a magnetosphere. From Ghosh & Lamb (1978).

secondary. However, as $\mu(1)$ is too small to prevent formation of a disc, any magnetic coupling between the primary and secondary is weak.

An homogeneous plasma is diamagnetic and excludes an external field. Some models of disc–field interactions have used this property to simplify the models (e.g., Pringle & Rees 1972; White & Stella 1987). However, penetration of the disc by the field will occur through convective or turbulent motions where the kinetic energy density of these exceeds the magnetic energy density at the surface of the disc. Also, the interface between the field and the flow in the disc is prone to Kelvin–Helmholtz instabilities which can grow and carry the field into the disc (Ghosh & Lamb 1978). A third process is reconnection of the primary's field to that of the disc. And finally, Livio & Pringle (1992) point out that the process (magnetic or hydrodynamic) that generates the effective disc kinematic viscosity ν_k will probably produce a magnetic diffusivity of similar magnitude. As a result, the field is able to thread the disc out to a radius r_s (Figure 7.1). Ghosh & Lamb (1979a,b) estimate that (for $r < r_s$) at the midplane of the disc the field strength will be reduced by a factor of 5 relative to that of the undisturbed dipole field.

By definition, outside the magnetosphere the field does not determine the gas flow. There can therefore be an extensive region of the accretion disc, with radius $r > r_0$ (Figure 7.1), in which the flow departs insignificantly from Keplerian. Within the region $r_0 < r < r_s$ the gas flow drags the field lines with it, producing a torque on the primary. The sign of this torque depends on the angular velocity $\Omega(1)$ of the primary: if the Keplerian angular velocity is less than that of the primary then the field lines are swept backwards and the primary is slowed down. The radius in the disc at which the Keplerian and primary angular velocities match is called the *corotation radius* (also

References to Table 7.2 (cont.)
28. Augusteijn, van Kerkwijk & van Paradijs 1993; 29. Howell *et al.* 1990; 30. Garnavich, Szkody & Goldader 1988; 31. Sambruna *et al.* 1992; 32. Mason *et al.* 1983a; 33. Cropper 1986b; 34. Silber 1992; 35. Howell *et al.* 1991; 36. Starrfield *et al.* 1991; 37 Szkody *et al.* 1994; 38 Haswell *et al.* 1993.; 39 Remillard *et al.* 1994a; 40 Tuohy *et al.* 1990.

known as the *centrifugal radius*) and is found from

$$r_{co}\Omega^2(1) = \frac{GM(1)}{r_{co}^2}$$

or

$$r_{co} = \left[\frac{GM(1)}{\Omega^2(1)}\right]^{1/3}. \tag{7.1}$$

Therefore, threaded regions of the disc with $r > r_{co}$ act to slow the primary's rotation and regions with $r < r_{co}$ speed it up. Note, however, that gas that is attached to the field lines at $r > r_{co}$ is centrifugally accelerated. *Steady accretion along field lines can occur only for gas attached at radii $r < r_{co}$.*

Note also that if the primary rotates so slowly that $r_{co} \geq r_s$, then all the magnetic torque (and the material torque, produced by the accreted matter striking the surface of the primary) acts to spin the primary up, whereas for a faster rotator, where $r_{co} < r_s$, $\dot{P}_{rot}(1)$ is determined by the net torque and may even be zero. (Ghosh & Lamb show that in general the material torque is small in comparison with the magnetic torques.) This can be made more quantitative as follows:

Set $r_s = f_d r_d$. Then from equations (2.1a), (2.5b), (7.1) and $r_d = 0.75\ R_L(1)$ it follows that $r_{co} > r_s$ and hence $\dot{P}_{rot}(1) < 0$ if

$$P_{rot}(1) \geq 0.20 f_d^{3/2} P_{orb} \quad (f_d \leq 1). \tag{7.2}$$

The only $\dot{P}$ measured for a very slow rotator in Table 7.1 (EX Hya) is compatible with condition (7.2). Furthermore, the spin-*down* of BG CMi requires $f_d > 0.51$ for there to be *any* threaded disc outside of r_{co}, and presumably $f_d \approx 1$ for there to be sufficient slowly rotating threaded outer disc to overwhelm the spin-up torque from the inner disc. Consequently, one observational verification for the Ghosh & Lamb model of the interaction between magnetosphere and disc will be provided if it is found that all $\dot{P}_{rot}(1) < 0$ for $P_{rot}(1) \gtrsim 0.20 P_{orb}$

Within the threaded region $r_0 < r < r_s$ (the outer transition zone in Figure 7.1) the viscous-induced inward radial drift of gas drags the field lines inward. This bending of the field lines decreases at smaller r because $B^2 \propto r^{-6}$, whereas the ram pressure of the inward drift is constant.

Within the boundary layer $r_A < r < r_0$ magnetic forces distort the gas flow so that the Keplerian flow at r_0 is converted to a corotating flow at r_A. The position of r_0 is determined by the need to conserve angular momentum (Lamb 1989) and is dependent on the geometry of the (distorted) field. Estimates (Lamb 1989) give $r_0 \approx (0.4$–$0.95) r_{\mu,sph}$ (see equation (6.4)). The relationship usually adopted (Ghosh & Lamb 1979b) is

$$r_0 = r_{\mu,disc} \approx 0.52 r_{\mu,sph}. \tag{7.3}$$

If r_0 is significantly different from r_{co} a considerable dissipation of angular momentum and energy must take place in the boundary layer. The energy dissipation will occur in shocks which will heat the boundary layer, producing a more luminous annulus and heating of the accreting gas just as it commences its fall along the field lines to the primary.

The width δ of the region $r_A < r < r_0$, where the magnetic field strongly interacts with the Keplerian flow, is expected to be quite narrow because of the rapid inward increase of $B^2(r)$ compared with the comparitively slow variation of Keplerian flow (Lamb 1989). Values of $0.01r_0 < \delta < 0.3r_0$ have been suggested (Lamb 1989), but a possible larger inward extension of the boundary layer by blobs that are threaded only slowly (as in Section 6.2.2) is possible (Scharlemann 1978; Arons & Lea 1980; Aly 1980; Aly & Kuijpers 1990; Spruit & Taam 1990, 1993a,b). Unfortunately, the inward extent of the boundary layer is of great importance to the interpretation of IPs: (a) the torque generated in the boundary layer dominates the spin-up torque acting on the primary and (b) the accretion zone on the primary is determined by the foot points of field lines passing through the boundary layer.

For the field to produce any disruption of the disc at all requires $r_0 > R(1)$, or, using equations (2.83a), (6.5) and (7.3),

$$\mu_{31}(1) > 0.58 M_1^{-1/3}(1) \dot{M}_{16}^{1/2} \tag{7.4a}$$

or

$$B(1) \gtrsim 1.5 \times 10^4 M_1^{2/3}(1) \dot{M}_{16}^{1/2}\ G. \tag{7.4b}$$

This is for a dipole field. Close to the surface of the primary, higher multipole components may become important, which will be more resilient to the accretion pressure. Furthermore, if $r_0 \lesssim 3R(1)$ the disappearance of predominantly dipole geometry may remove much of the simple anisotropy of the emitted radiation and hence reduce the amplitude of any rotationally modulated signal (Lamb 1988). From equations (2.83a) and (7.1) this translates to

$$P_{\rm rot}(1) \lesssim 56 M_1^{-1}(1)\ {\rm s} \tag{7.5}$$

for dipole accretion in equilibrium.

The accretion zones at the surface of the primary are determined by the same considerations as for polars (Sections 6.2.3, 6.2.4 and 6.2.5). As very extended zones are expected, large variations of B and N_e (and, as a result, h_s which is sensitive to the cooling rate) and hence Λ will occur (equation (6.19a)). As the polarization is a strong function of Λ, only detailed modelling can hope to account for the observed polarization properties of IPs (Section 7.4). At the lower fields expected in IPs the peak of the cyclotron flux moves into the IR. In the optical and near IR an additional physical process may then contribute to the flux – *viz.* free–free emission in the magnetic field (Wickramasinghe 1988b). This magneto-bremsstrahlung is weakly circularly polarized ($\sim \omega_c/\omega$).

Preliminary models have been computed by Wickramasinghe, Wu & Ferrario (1991), who allow for variations in specific accretion rate and shock height along the accretion arc. Field lines from the transition zone map onto arcs on the surface with characteristic radius $r_{\rm arc} \sim R(1)[R(1)/r_A]^{1/2}$ and widths $\sim 10^{-2} r_{\rm arc}$ (Lamb 1985; Rosen, Mason & Cordova 1988; Mason, Rosen & Hellier 1988; Lamb 1988; Buckley & Tuohy 1989), which gives $f \sim 10^{-2}$, i.e. ~100 – 1000 times greater accretion areas than in polars (Section 6.4.3). However, as $\dot{M}$ is larger in IPs than polars, the specific accretion rate is not greatly different in the two types. This has interesting consequences for the emission of hard X-rays: the optical thickness can be sufficient to cause significant fan-beaming (Section 6.2.3).

The polarized flux in IPs will be diluted by unpolarized radiation from the primary and from the accretion disc (and from the secondary star for observations in the IR). After inclusion of these, Wickramasinghe, Wu & Ferrario (1990) find that circular polarization should be observable for $B(1) \gtrsim 10$ MG at optical wavelengths and for $B(1) \gtrsim 5$ MG in the IR. The wavelength of the peak of the polarized flux is sensitive to the viewing angle to the dipole axis, so large amplitudes of flux variation are predicted. At a given viewing angle the wavelength of the peak flux gives an estimate of B almost independent of $\dot{M}$.

These calculations show that absence of circular polarization in the optical and IR in IPs (Section 7.4) implies $B(1) \lesssim 2$ MG ($\mu(1) \lesssim 1.0 \times 10^{33} \mathrm{M}_1(1)$ G cm^3 from equation (6.8)). This is a radically different conclusion from earlier models in which it was supposed that the accretion zone covers a large polar cap ($f \sim 0.1$) and the spread of field strengths and orientations, even with $B(1) \geq 20$ MG, depolarizes the optical emission (Chanmugam & Frank 1987). However, Wickramasinghe & Ferrario (1988) show that for $B(1) = 20$ MG the circular polarization at optically thicker frequencies is actually *increased* in this model.

7.2.2 *Accretion without a Disc*

If the radius $r_{\mu,\mathrm{str}}$ at which the gas stream becomes attached to the field lines lies outside the circularization radius (i.e., $r_{\mu,\mathrm{str}} > r_\mathrm{r}$) then no disc can form. In a first approximation this discless IP may be thought of as similar in its properties to a polar (Section 6.2) but with gas accreted onto near-polar zones, alternating between them with period P_syn.

The details of stream penetration into a rotating magnetic field are even less certain than those for the synchronous rotator. For gas that is threaded sufficiently rapidly an approximate value of $r_{\mu,\mathrm{str}}$ is given by (Hameury, King & Lasota 1986b)

$$r_{\mu,\mathrm{str}} = 0.37 r_{\mu,\mathrm{sph}}. \tag{7.6}$$

Accepting the criterion that a disc cannot form if $r_{\mu,\mathrm{str}} > r_\mathrm{r}$, then the condition for *discless* accretion is (Wickramasinghe, Wu & Ferrario 1991)

$$\frac{\mu_{34}(1)}{\dot{M}_{17}^{1/2}} > 0.27\left(\frac{\eta}{0.3}\right)\left(\frac{P_\mathrm{orb}}{4\ \mathrm{h}}\right)^{7/6} M_1^{5/6}(1), \tag{7.7}$$

where $\eta = r_\mathrm{r}/R_{\mathrm{L}_1}(1)$ lies in the range $0.22 < \eta < 0.38$ and can be found from Lubow and Shu (1975) or equation (2.19) and Table 2.1.

As typical mass transfer rates in IPs (Section 7.5.3; Table 7.4) are $\dot{M}_{17} \sim 1$, equation 7.7 shows that discs will form for $P_\mathrm{orb} > 3$ h if $\mu_{34}(1) \lesssim 0.3$. The smallest μ measured for a *polar* above the period gap is $\mu_{34}(1) \approx 0.6$ (for AM Her), so there is only a narrow range of μ in which asynchronous, discless accretion can occur if $\dot{M}_{17} \sim 1$. However, any excursions to states of lowered $\dot{M}$ can convert normally disced IPs to discless ones; and below the period gap, where $\dot{M}$ is low, disclessness may be common (but short lived: Section 9.3.7.2).

A further criterion must be met if a discless system is to be able to accrete steadily: if

$r_{\mu,\mathrm{str}} > r_{\mathrm{co}}$ then threaded material will be moved out radially along the rotating field lines. Conservation of angular momentum results in the primary being spun down, but only on a time scale $\sim 10^5$ y. On a much shorter time scale, the torus of gas, which is partially supported radially by magnetic pressure, will be subject to Rayleigh–Taylor instabilities and other diffusive processes which will enable material to reach radii $< r_{\mathrm{co}}$. This is probably a chaotic rather than a steady accretion process.

Although most currently known IPs probably have discs (see below and Hellier (1991)), a state of lowered $\dot{M}$ (Section 7.3.1) may allow r_μ first to expand beyond r_{r}, thereby destroying the disc, and then beyond r_{co}, preventing steady accretion. From equations (6.4), (7.1) and (7.6), the condition $r_{\mu,\mathrm{str}} > r_{\mathrm{co}}$ transforms to

$$P_{\mathrm{rot}}(1) < 3.29\mu_{33}^{6/7}(1)\dot{M}_{17}^{-3/7}M_1^{-5/7}(1) \quad \mathrm{min.} \tag{7.8}$$

Equation (7.8) shows that steady accretion onto the primary should cease as $\dot{M}_{17}$ drops below ~ 0.03 for a system with $\mu_{33}(1) = 1$, $\mathrm{M}_1(1) = 1.0$ and $P_{\mathrm{rot}}(1) = 15$ min. (Of course, if a system is accreting near equilibrium then even a very small reduction in $\dot{M}$ will result in non-steady accretion. GK Per (Section 7.6.6) may be an example of this.)

The physical appearance of a discless, asynchronous rotator is a matter of some controversy. Hameury, King & Lasota (1986b) consider stream accretion onto asynchronous rotators with $\mu_{34}(1) \sim 1$ and conclude that the threading region will generate hard X-rays from strong shocks. They also envisage that the threaded gas will form a shell (a 'non-accretion disc') around the primary that could simulate the appearance of an accretion disc, and therefore that some or all of the existing IPs could be discless accretors. The notion that IPs have $\mu_{34}(1) \sim 1$ arose from consideration of the evolution of magnetic CVs (discussed more fully in Section 9.3.7.1). Noting that most of the polars were below the period gap and most of the IPs were above it, King, Frank & Ritter (1985) proposed that the polars are the descendants of IPs and therefore both classes should have the same general magnetic field strengths. In that case, all IPs with $P_{\mathrm{orb}} \lesssim 6$ h should be discless (equation (7.7) with $M_1(1) = 0.6$). However, the discovery that BG CMi, which has one of the strongest fields in an IP, has $\mu_{34}(1) \sim 0.1$ (Sections 7.4 and 7.5.3) removes this expectation.

The lower magnetic moments now associated with IPs lead to a different accretion picture (Wickramasinghe, Wu & Ferrario 1991; see also King & Lasota 1990). Because of the rapid motion of the field lines, fragmentation of the stream may lead to a greater clumpiness of the accreted gas than is the case for polars. The fractional area over which accretion occurs, extending in a zone of $\sim 360°$ of magnetic longitude, should be $f \sim 0.01$. This combination of clumpiness and low specific accretion rate between the blobs will result in little hard X-ray production but strong EUV emission from thermalized subphotospheric energy release on the primary (Section 6.2.5). The discless IPs would therefore be of high luminosity in the 0.01–0.1 keV energy range, which is largely obscured from us by interstellar absorption. In the specific case of EX Hya, however, Rosen *et al.* (1991) show that the observed flux above 1 keV already implies an $\dot{M}$ typical of CVs below the orbital period gap, so any exceptional EUV flux would require an abnormal mass transfer rate.

The descriptions given above for accretion with or without a disc are canonical first approximations, perhaps realized by few, if any, IPs. King (1993; see also King &

Wynn 1993) points out that large diamagnetic stream blobs may not be threaded until they have orbited the primary several times. The surface drag on these blobs, caused by the field exciting Alfvén waves, is large enough to make their orbits circularize near the radius r_r, where they fragment into smaller blobs which will then be accreted. This will happen only if an accretion disc has not already been established.

Even when a disc is present, part of the stream can continue its trajectory over the surfaces of the disc (Lubow 1989), colliding with the magnetosphere at r_0. In this case accretion occurs through a hybrid of disc and discless models[1] (Hellier 1993b,c). For this to be possible requires $r_{min} < r_0$. From equations (2.14) and (7.1), for *equilibrium* disc accretion,

$$\frac{r_{min}}{r_0} \approx \frac{r_{min}}{r_{co}} \approx 0.050 \frac{(1+q)^{1/3}}{q^{1/2}} \left[\frac{P_{orb}}{P_{rot}(1)}\right]^{2/3}. \tag{7.9}$$

The condition $r_{min}/r_{co} < 1$ is satisfied by all IPs in Table 7.1 except for 1H0523 and GK Per.

In Sections 6.3.2, 6.3.4, 6.4.4 and 6.5.3 it is seen that the primaries in polars have offset magnetic fields resulting in different polar strengths with different accretion rates and emergent fluxes. In an IP the effects of asymmetric fields are greatly reduced (Rosen *et al.* 1991): over a synodic cycle both poles are equally exposed to the accretion flow (both in disced and discless systems). As any asymmetry in surface field strength is greatly reduced at r_0 where material is threaded by the field lines, there is unlikely to be any strong favouring of mass transfer to one pole rather than the other.

7.2.3 *Accretion Torques and Primary Spin Rate*

In a disc, the specific angular momentum at the outer edge of the boundary layer is $l_0 = [GM(1)r_0]^{1/2}$. As this is transferred from the disc to the primary it exerts a material torque

$$N_0 = \dot{M}[GM(1)r_0]^{1/2}. \tag{7.10}$$

The magnetic torque acting on the primary depends on the details of the boundary layer and the threading of the disc. Following the Ghosh & Lamb model (Ghosh & Lamb 1979a; Lamb 1988,1989), the *fastness parameter* ω_s plays a leading role. ω_s is a dimensionless stellar angular velocity:

$$\omega_s = \frac{\Omega_{rot}(1)}{\Omega_K(r_0)} \tag{7.11}$$

$$= \left(\frac{r_0}{r_{co}}\right)^{3/2} \tag{7.12}$$

where $\Omega_K(r_0)$ is the Keplerian angular velocity at the inner radius r_0 of the disc.

When $\omega_s < 1$ and $r_{co} \sim r_d$ the lines threading the transition zone (Figure 7.1) are all swept forward and produce an additional spin-up torque of $\sim 0.4\ N_0$. If r_{co} lies near to r_0 then the transition zone contributes a negative torque. At some critical value of ω_s,

[1] Norton (1993) refers to this connection as a 'non-accretion disc', but this term should be reserved for the discless model described by Hameury, King & Lasota (1986b): Section 7.2.2.

denoted ω_c, the positive torque (material and magnetic) balances the spin-down torque and *there is no net torque on the primary even though it is accreting.* If $\omega_c < \omega_s < 1$ the negative torque dominates and the primary spins down. The situation $\omega_s > 1$ corresponds to $r_0 > r_{co}$, so steady accretion cannot occur (although there is still a negative torque acting on the primary).

The torque acting on the primary, as calculated from the Ghosh and Lamb model, can be expressed as

$$N = n(\omega_s)N_0, \tag{7.13}$$

where $n(\omega_s)$ is a dimensionless torque function.

From their model, Ghosh & Lamb found $\omega_c \sim 0.35$, but correction for inconsistencies in their model (Wang 1987; Zylstra 1988; Lamb 1989) gives $\omega_c \approx 0.975$. A plot of $n(\omega_s)$ from Wang's analysis is given in Figure 7.2.

From equations (6.4), (7.3), (7.10), (7.13) and the relationship $N = I\dot{\Omega}_{rot}(1)$, the rate of change of rotation is

$$-\dot{P}_{rot}(1) = 1.6 \times 10^{-10} n(\omega_s)\mu_{33}^{2/7}(1)M_1^{3/7}(1)I_{50}^{-1}(1)P_3^2(1)\dot{M}_{17}^{6/7}. \tag{7.14}$$

The same equations give

$$\omega_s = 0.329\mu_{33}^{6/7}(1)M_1^{-5/7}(1)P_3^{-1}(1)\dot{M}_{17}^{-3/7}. \tag{7.15}$$

Equation (7.15) can be rearranged as

$$\mu_{33}(1) = 3.66M_1^{5/6}(1)P_3^{7/6}(1)\dot{M}_{17}^{1/2}\omega_s^{7/6} \quad \text{(disced)} \tag{7.16}$$

which may be used to estimate $\mu(1)$ if it can be established that $P_{rot}(1)$ is close to its equilibrium value ($\omega_s = \omega_c = 0.975$) for a (perhaps secular mean) measured $\dot{M}$.

Equations (7.1) and (7.12) combine to give

$$r_0 = 1.50 \times 10^{10}\omega_s^{2/3}M_1^{1/3}(1)P_3^{2/3}(1) \text{ cm} \tag{7.17}$$

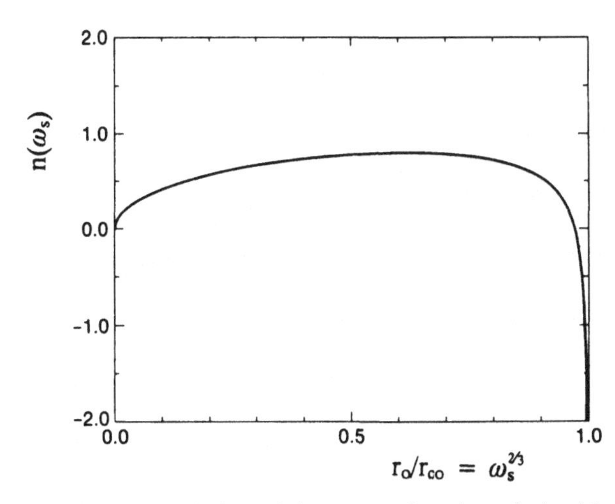

Figure 7.2 Variation of the torque function $n(\omega_s)$ with fastness parameter ω_s. Adapted from Wang (1987).

which, as $\omega_s \leq 1$, gives an upper limit on r_0 (for accretion within r_{co}), or an estimate of r_0 if $P_{rot}(1)$ is near its equilibrium value ($\omega_s = \omega_c \approx 1.0$).

Equation (7.12) has been compared with observations of X-ray pulsars (for which almost all of the luminosity emerges in hard X-rays and therefore $\dot{M}$ can be relatively well estimated from $L_x = GM(1)\dot{M}/R(1)$) with considerable success (Ghosh & Lamb 1979a; Parmar *et al.* 1989; Angelini (cited by Lamb 1989)). In X-ray pulsars, QPOs (Section 8.5) can provide an estimate of $\Omega_K(r_0)$ and the pulse frequency $\Omega_{rot}(1)$ is directly observed. Hence ω_s (equation (7.11)) is known. Variations of $\dot{M}$ and the concomitant changes in $P_{rot}(1)$ then can reveal the form of $n(\omega_s)$ through application of equation (7.14).

In discless accretion, angular momentum conservation leads to

$$N = \dot{M}[GM(1)r_r]^{1/2}. \tag{7.18}$$

However, if $\mu(1)$ is large enough to prevent a disc being formed, then (particularly for small P_{orb} where the separation of the stars is relatively small) there may be a significant braking torque from interaction between the moving field lines and the secondary (and, to a lesser extent, the stream).

From equations (2.1b), (2.19) and (7.18) we have

$$-\dot{P}_{rot}(1) = 2.5 \times 10^{-9} \frac{(1+q)^{1/6}}{q^{0.21}} P_3^2(1) I_{50}^{-1}(1) M_1^{2/3}(1) \dot{M}_{17} \left(\frac{P_{orb}}{4\ \text{h}}\right)^{1/3}. \tag{7.19}$$

This is the maximum rate of spin-up (i.e., with braking torques neglected) expected from discless accretion. For disced accretion the characteristic time scale to change the rotation period is $P_{rot}(1)/\dot{P}_{rot}(1) \sim 1.5 \times 10^5 P_3(1)$ y.

If accretion in a discless system occurs near the equilibrium value, so that $r_{\mu,str} = r_{co}$, then equations (6.5), (7.1), (7.6) give

$$\mu_{33}(1) = 6.6 M_1^{5/6}(1) P_3^{7/6}(1) \dot{M}_{17}^{1/2} \quad \text{(discless)} \tag{7.20}$$

which, compared with equation (7.16), shows that derived magnetic moments differ by a factor of two according to whether the IP is assumed to have a disc or not.

7.2.4 *Instabilities in Truncated Discs*

The magnetosphere of the primary in effect truncates the accretion disc, giving it an inner radius r_0 instead of $R(1)$. In a first approximation, at $r > r_0$ the disc may be represented by the steady state model of Section 2.5.1. However, the absence of the hot inner region of the disc can significantly lower the luminosity of the disc. The radial temperature distribution is given by equation (2.35) with the substitution $R(1) \rightarrow r_0$. The luminosity is then

$$L(\text{d}) = 4\pi\sigma \int_{r_0}^{r_d} T_e^4(r)\ r\ \text{d}r$$

$$= \frac{GM(1)\dot{M}(\text{d})}{2r_0} \left\{1 - \frac{r_0}{r_d}\left[3 - 2\left(\frac{r_0}{r_d}\right)^{1/2}\right]\right\}. \tag{7.21}$$

The remaining energy, liberated in the boundary layer and at the surface of the primary, is

$$L(i) = \frac{GM(1)\dot{M}(\mathrm{d})}{2r_0}\left[2\frac{r_0}{R(1)} - 1\right]. \tag{7.22}$$

If $R(1) \ll r_0 \ll r_\mathrm{d}$ then $L(i) \gg L(\mathrm{d})$, so the disc is a minor contributor to the total luminosity of the system.

Time-dependent computations for the MTI and DI models (Section 3.5) have been made by Angelini & Verbunt (1989; see also Kim, Wheeler & Minshige 1992), who find shorter outburst durations, particularly in the V band, compared to non-truncated discs. This is a result of the shorter time taken for the transition waves to travel from r_d to r_0 and vice versa, not only because $r_0 > R(1)$ but because r_d is smaller in a truncated disc. The latter is a consequence of the reduction of total amount of angular momentum that must be removed from the truncated disc by tidal interaction, which can be achieved at a smaller outer disc radius.

For outside-in outbursts, which are triggered by conditions at the outer edge of the disc, the interval between outbursts is found to be insensitive to r_0. For inside-out outbursts, however, the outburst interval increases greatly with increasing r_0; this is a result of the larger value of Σ_max at r_0 than at $R(1)$ (equation (3.15)), which takes longer to achieve.

From Section 3.5.3.5, a disc is stable if $T_\mathrm{eff}(r)$ is everywhere less than 6000 K. For a steady state truncated disc, the maximum temperature is 0.488 T_*, where T_* is given by equation (2.36b) with the substitution $R(1) \to r_0$. Therefore a truncated disc will stabilize on the lower branch of the S-curve (Figure 3.25) if

$$\dot{M}_{16}(\mathrm{d}) < 8.1 M_1^{-1}(1)\left(\frac{r_0}{10^{10}\ \mathrm{cm}}\right)^3, \tag{7.23}$$

or, using equations (6.4) and (7.3),

$$\dot{M}_{16}(\mathrm{d}) < 5.2 M_1^{-10/13}(1)\mu_{33}^{12/13}(1) \tag{7.24}$$

or, using equation (7.16),

$$\dot{M}_{16}(\mathrm{d}) < 27.5 P_3^2(1)\omega_\mathrm{s}^2. \tag{7.25}$$

Disc instabilities (DN outbursts) should not occur if these inequalities are satisfied. Most IPs (Section 7.5.2) appear to satisfy these conditions.

Livio & Pringle (1992) have considered the effect of field lines from the primary threading the disc. The winding up of the lines generates a torque on the disc which produces an effective α. For a dipole field on the primary this increased viscosity falls off rapidly with radius, so only the innermost parts of the disc are affected. As a result, relatively low fields (10^{4-5} G) rapidly drain the inner disc, leaving a truncated disc. Computations of DN outbursts for these discs show the UV delay observed in some systems (Section 3.5.4.2).

7.3 Photometric Variations

7.3.1 Long Term Light Curves

The brightest IPs fortunately are well above the detection limits of archived plates and thus offer themselves for long term brightness studies. V1223 Sgr has a light curve similar to that of AM Her (Figure 6.11) but with even more time spent below high state brightness. AO Psc is almost always at high state, but showed one drop of ~2 mag in 1946. FO Aqr has shown no distinct low states since 1923 (Belserene 1981; Hudec 1984; Garnavich & Szkody 1988); it has a slow modulation of brightness of amplitude ~0.2 mag which is probably similar in origin to that seen in many other CVs (Section 4.3.1). The possible IP TW Pic has been observed more than a magnitude below its usual brightness (Mouchet *et al.* 1991) and KO Vel has been observed $2\frac{1}{2}$ mag fainter than normal (Cropper 1986b; Mukai & Corbet 1987).

The IPs include within their class some objects from other distinct types. GK Per was Nova Per 1901 and also shows DN of 2–3 mag range every ~3 y (Sections 3.1 and 4.3.3). YY Dra is a DN with very infrequent outbursts (Patterson *et al.* 1992a). EX Hya is classified as a DN with ~3 mag outbursts every ~1 y (but see Section 7.7).

7.3.2 Orbital Modulations

Only three of the IPs discovered have detectable optical eclipses (though the suspected IP, V348 Pup, is an eclipsing system which could increase the count to four: (Tuohy) *et al.* 1990 and XY Ari has an X-ray eclipse). In EX Hya (Warner & McGraw 1981; Hellier *et al.* 1987), FO Aqr (Hellier, Mason & Cropper 1989) and TV Col (Hellier, Mason & Mittaz 1991; Hellier 1993a) the eclipse is shallow and of variable depth in the optical. In FO Aqr and TV Col there are no X-ray eclipses (Chiappetti *et al.* 1988; Hellier, Mason & Mittaz 1991) and in EX Hya there is a partial eclipse of soft X-rays (Cordova, Mason & Kahn 1985), so in all systems it appears that the optical eclipse is at most a partial obscuration of a disc, not affecting the primary: indeed, Martell & Kaitchuck (1991) have questioned whether there is any eclipse at all in FO Aqr, but Kruszewski & Semeniuk (1993) provide further observational evidence. In EX Hya during outburst the eclipse becomes broad and shallow, as of an expanded disc (Reinsch & Beuermann 1990).

Figure 7.3 shows a light curve of BG CMi of about two and a half orbits' duration. The orbital modulation of brightness is conspicuous. This modulation, which is seen in most of the IPs, could originate as an aspect variation of a bright spot or an aspect variation of a region that is heated by X-rays from the central source. If this region is the secondary, then minimum brightness of the orbital modulation would correspond to inferior conjunction; if it is an extended region near the bright spot then maximum brightness would correspond to superior conjunction of the spot. If the modulated brightness arises from self-luminosity of the spot, then, as in DN, maximum brightness would correspond to inferior conjunction of the spot.

These competing processes can be distinguished by noting the relative phases of rotational and orbital modulations (Section 7.3.4). In the case of FO Aqr, eclipse and orbital minimum coincide (Hellier, Mason & Cropper 1989), so the orbital modulation arises predominantly from heating of the side of the secondary facing the primary, but there is evidence for some contribution from reprocessing at the bright spot as well (de

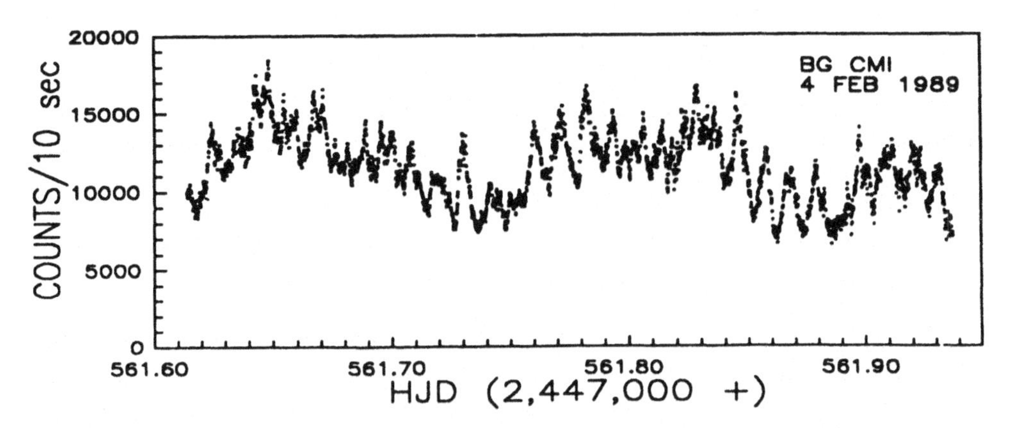

Figure 7.3 Light curve of BG CMi showing orbital modulation at $P_{orb} = 3.2$ h and beat modulation at $P_{syn} = 15.2$ min. From Patterson & Thomas (1993).

Martino *et al.* 1993). In V1223 Sgr and AO Psc it appears to be due to the self-luminous bright spot (Warner 1986a; Hellier, Cropper & Mason 1991).

In the UV a large orbital modulation is seen in V1223 Sgr and possible small amplitude ones in TV Col and FO Aqr (Mouchet & Bonnet-Bidaud 1984), but BG CMi and AO Psc are unmodulated (see also Cordova *et al.* (1983)). These properties are not yet understood – note, for example, that aspect variation of the 12 600 K blackbody component present in AO Psc (Section 7.6.6), which produces much of the flux in the visible, cannot be the source of optical modulation or its effect would easily be detected in the UV. Long baseline flux distributions as a function of orbital phase are required to investigate the solution to this problem.

It is noteworthy that there are no known deeply eclipsing IPs. For high inclination a disc and accretion curtain may prevent X-rays from being seen at sufficient flux to register as an IP. Dhillon, Marsh & Jones (1991) point out that in eclipsing NLs such as V1315 Aql (an SW Sex star) the line profile behaviour *could* be understood with an accretion curtain model. A search for X-ray pulsations in these stars is needed.

7.3.3 Rotational Modulations

7.3.3.1 $P_{rot}(1)$ and P_{syn}

The large amplitude ~15 min modulation seen in Figure 7.3 is typical of most IPs: nearly sinusoidal variations with large cycle-to-cycle variations in amplitude. Power spectra of the light curve for most IPs show no harmonics of the fundamental periods, so their average pulse shape is quite closely sinusoidal (e.g., van der Woerd, de Kool & van Paradijs 1984; van Amerongen, Augusteijn & van Paradijs 1987). In FO Aqr small departures from sinusoidality are seen at shorter wavelengths for $P_{rot}(1)$, and a considerable departure for P_{syn} (de Martino *et al.* 1993).

Among the IPs known so far, there is a fairly distinct division between those having strong optical modulation at $P_{rot}(1)$ and those at P_{syn}. In many IPs the alternative period is weakly and variably present. Only in AO Psc have both $P_{rot}(1)$ and P_{syn} been detected with nearly equal amplitudes in the visible region (van der Woerd, de Kool & van Paradijs 1984; Hellier, Cropper & Mason 1991) but its $P_{rot}(1)$ modulation is more normally weak (Warner, O'Donoghue & Fairall 1981) or even absent (Wickramasinghe, Stobie & Bessell 1982). PQ Gem is unique, however, in possessing a strong, double-peaked $P_{rot}(1)$ modulation in R and I bands, but a sinusoidal P_{syn} in UBV (Piirola, Hakala & Coyne 1993; Rosen, Mittaz & Hakala 1993).

One way of explaining this near dichotomy of behaviour is by postulating that the high energy radiation from the primary is non-isotropic (Warner 1985a). If the radiation is *beamed*, then, according to the location of the accretion zone on the surface of the primary, the beam may or may not sweep across the bright spot bulge and/or secondary, producing the P_{syn} signal. The $P_{rot}(1)$ signal can arise from the beam sweeping across the disc (which, because of its concavity, has a front-to-back asymmetry for large inclinations), or around the inner edge of the accretion disc, or simply from direct viewing of the beam from the accretion zone itself (Hellier *et al.* 1987; Hellier & Mason 1990: see Section 7.5.1.2).

Although this model was originally proposed as a purely phenomenological model, computation of the angular distribution of intensity from the accretion arc near the surface of the primary shows that X-ray emission occurs as a fan-beam, perpendicular to the magnetic field lines (Buckley & Tuohy 1989: Figure 7.4; Hellier & Mason 1990). What that beam illuminates as the primary rotates is determined by the location of the accretion arc – i.e., primarily it depends on the colatitude β of the magnetic pole.

X-ray fan-beams are produced by the accretion columns in polars (Imamura & Durisen 1983; Section 6.2.3); their X-ray light curves show the cross-sections of the beams as viewed from the Earth. In IPs, we have the additional opportunity of

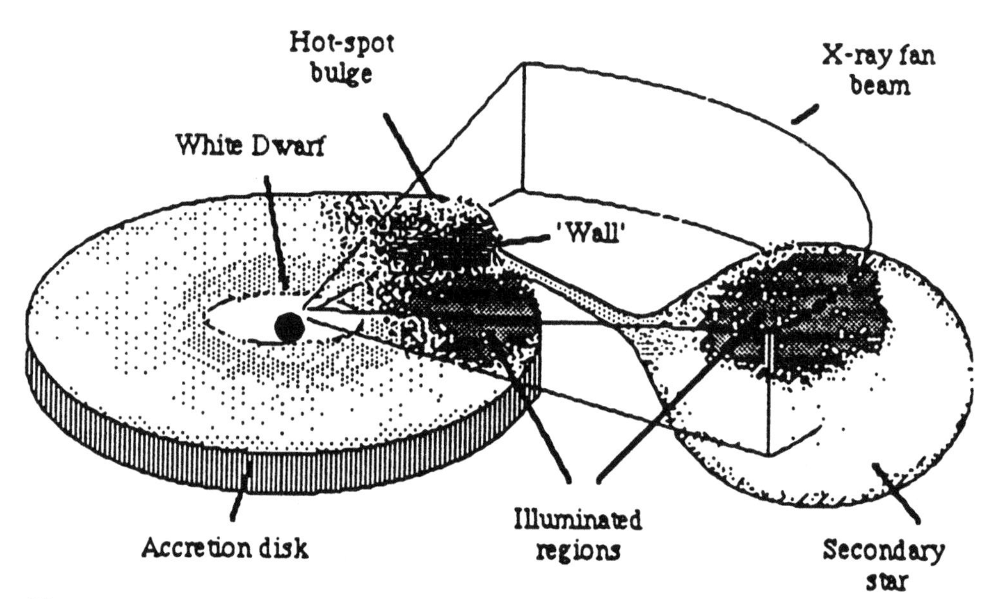

Figure 7.4 Fan-beam illumination of the bright spot region and secondary. From Buckley & Tuohy (1989).

determining the intensity cross-section of the beam as seen from the various reprocessing sites. The changes in amplitude of the optical modulations suggest that the beam shape and direction are variable, as might be expected from movements of the accretion zone resulting from variations of $\dot{M}$ and r_0. There may also be variations in cross-section of the reprocessing sites. An alternative model for the amplitudes of $P_{rot}(1)$ and P_{syn} is considered in Section 7.5.4.

There are several possible ways of determining the locations of the emitting and reprocessing sites. That dependent on phasing of the modulations is discussed in Section 7.3.4; we first look at alternative approaches.

7.3.3.2 Flux Distributions of the Pulsed Sources

In AO Psc, when the $P_{rot}(1)$ and P_{syn} pulses are of comparable amplitude it is possible to deconvolve them to obtain separate average pulse shapes (van der Woerd, de Kool & van Paradijs 1984). Performed over a range of photometric bands, this demonstrates that whereas the $P_{rot}(1)$ (805 s) modulation increases steeply in amplitude towards short wavelengths, the P_{syn} (859 s) modulation is relatively flat – Figure 7.5. Transformed to fluxes and corrected for reddening, the flux distribution of the 805 s component is $F(\nu) \propto \nu^{2.53\pm0.15}$, which is steeper than that expected from the Rayleigh–Jeans tail of a hot blackbody. However, there are complicating factors in the 805 s modulation – it is a composite of two pulses phased 180° apart. Usually the 805 s is

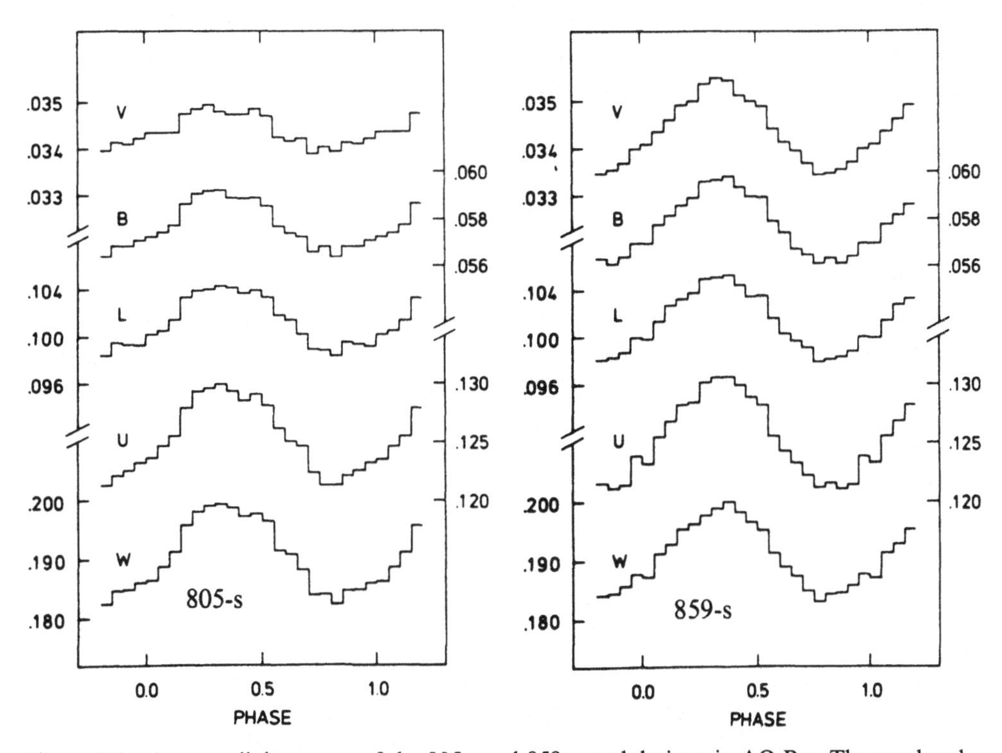

Figure 7.5 Average light curves of the 805 s and 859 s modulations in AO Psc. The passbands are the VBLUW filters of the Walraven system. From van der Woerd, de Kool & van Paradijs (1984).

synchronized with the X-ray pulse, but on one occasion it was observed to be 180° phase-shifted (Pietsch *et al.* 1987). The changing amplitude of the observed 805 s modulation arises from the variable contribution of this second component, which could be either a second accretion zone on the primary or originate from reprocessing of the X-ray beam on the inner edge of the disc. If the latter is the case, it will have a flat spectrum similar to the 859 s modulation and, being antiphased with the main 805 s component, results in the flux distribution becoming steeper than the Rayleigh–Jeans law. The flux distribution of the 859 s modulation, extended into the IR (Hassall *et al.* 1981; Berriman 1988), has a colour temperature of 16 500 K.

This analysis shows that, as anticipated, the principal 805 s component arises from a hot ($\geq 10^5$ K) opaque surface that is probably an extended region of the surface of the white dwarf heated by radiation from the accretion zone. The 859 s and secondary 805 s components arise from opaque regions heated by the radiation beam.

The flux distribution of the dominant (P_{syn}) pulse in V1223 Sgr has a colour temperature of 12 500 K, in agreement with its reprocessed origin (Watts *et al.* 1985). The amplitudes of all of the observed modulations (Section 7.3.4) in FO Aqr have been determined in all photometric bands from U to K (de Martino *et al.* 1993). The $P_{rot}(1)$ pulse is compatible with a 11 300 K blackbody and P_{syn} fits a 7300 K blackbody.

7.3.3.3 Reprocessing Site: Bright Spot or Secondary?

Patterson & Price (1981b) suggested that the atmosphere of the secondary star is the principal reprocessing site for the P_{syn} pulses, but Hassall *et al.* (1981) rejected this on the grounds that if the entire face of the secondary were periodically modulated to colour temperatures ~12 000 K then a much larger amplitude of reprocessed radiation would be expected. However, Wickramasinghe, Stobie & Bessell (1982) showed that when shadowing of the secondary by the disc and the reduction in apparent amplitude because of the luminosity of the disc itself are taken into account, there is no reason to exclude the secondary as a reprocessing site. The alternative site proposed by Hassall *et al.* is the bulge in the accretion disc in the vicinity of the bright spot (see also Elitzur, Clarke & Kallman (1988)).

In reality, in those cases where the beam can strike the disc bulge and the secondary, there is reprocessing from both sites and it is a question of which dominates. Buckley & Tuohy (1989) and Mouchet *et al.* (1991) have modelled TX Col in considerable detail, including the effects of X-ray beam heating. The calculated flux distributions from the various components (Figure 7.6) show that whereas reprocessing from the secondary dominates for $\lambda > 4200$ Å the disc bulge ('hot-spot' in Figure 7.6) dominates at shorter wavelengths. The summed model contributions in Figure 7.6 agree well with the observed continuum distribution of TX Col.

7.3.4 Optical Orbital Sidebands

The observed light curves of IPs have been routinely Fourier analysed in order to search for coherent periodicities. In this way the detection and measurement of $P_{rot}(1)$ and P_{syn} are achieved. In carrying out this process, occasional additional periodicities were found which were difficult to interpret (Patterson & Steiner 1983; Warner & Cropper 1984) and even dismissed as spurious (Penning 1985; Watts *et al.* 1985).

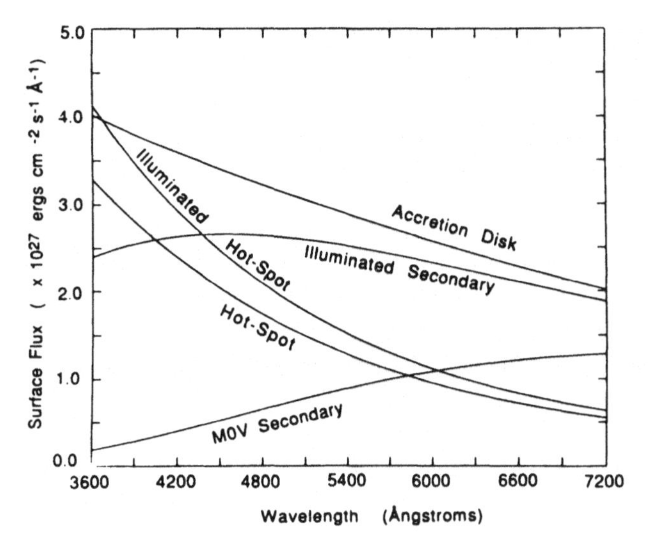

Figure 7.6 Flux distributions of the various components of the reprocessing model of TX Col. From Buckley & Tuohy (1989).

Consideration of the variable geometry of the reprocessing sites shows, however, that multiple orbital sidebands of $P_{rot}(1)$ are naturally produced (Warner 1986a).

The projected area of the principal processing site, whether it be secondary star or disc bulge, varies around the orbit. As an illustration, suppose that the reprocessed light curve is sinusoidal with frequency $2\pi/P_{syn} = \omega_{rot} - \Omega_{orb}$ where $\omega_{rot} = 2\pi/P_{rot}(1)$ and $\Omega_{orb} = 2\pi/P_{orb}$. If the orbital modulation is also sinusoidal then the time variable part of the observed intensity $I(t)$ is

$$I(t) = I_0(1 + A \sin \Omega_{orb}t) \sin (\omega_{rot} - \Omega_{orb})t \tag{7.26}$$

where $t = 0$ when the orbital and reprocessed modulations are in phase. This can be rewritten

$$I(t) = I_0 \sin(\omega_{rot} - \Omega_{orb})t + \frac{I_0 A}{2}\cos(\omega_{rot} - 2\Omega_{orb})t + \frac{I_0 A}{2}\cos \omega_{rot}\, t \tag{7.27}$$

which shows that a power spectrum of $I(t)$ will have component frequencies ω_{rot}, $\omega_{rot} - \Omega_{orb}$ and $\omega_{rot} - 2\Omega_{orb}$. This is, of course, simple amplitude modulation which produces sidebands of equal amplitude. (Orbital sidebands were first discussed by Katz (1975) in connection with a proposed model for DQ Her.)

The reprocessing projected area will probably also vary with a $2\Omega_{orb}$ component: for example, although the Roche lobe-shaped secondary has the same cross-sectional area when seen at each point of quadrature, there is a difference in visible reprocessing area when the secondary is viewed at superior conjunction from when it is viewed at inferior conjunction. (The magnitude of this effect is a strong function of orbital inclination.) This will introduce components at frequencies $(\omega_{rot} + \Omega_{orb})$ and $(\omega_{rot} - 3\Omega_{orb})$ into the power spectrum.

An added source of $\omega_{rot} - \Omega_{orb}$ and $\omega_{rot} + \Omega_{orb}$ Fourier components is any orbital modulation of the amplitude of the ω_{rot} signal(s). Such a modulation is observed in AO

Psc (van der Woerd, de Kool & van Paradijs 1984); its origin is uncertain but it is probably an obscuration effect associated with the disc bulge (it is present also in the X-ray emission – Section 7.6.2). Amplitude modulation at frequencies Ω_{orb} and $2\Omega_{orb}$ will produce Fourier components from $\omega_{rot} - 2\Omega_{orb}$ to $\omega_{rot} + 2\Omega_{orb}$.

The various contributions to each of the sidebands can be summed, making allowance for phase differences (Warner 1986a). The observed relative amplitudes and phases of ω_{rot} and its orbital sidebands and the orbital modulation therefore contain information on the geometry of the reprocessing sites, which can be used to distinguish between the disc bulge and the secondary as the principal reprocessing site (only, however, when the position of the secondary in orbit is known).

In FO Aqr, Osborne & Mukai (1989) have deduced that the disc bulge is the reprocessing site, but Hellier, Mason & Cropper (1989) show that the orbital modulation arises from X-ray heating of the secondary, so that some reprocessing on the secondary does occur. Semeniuk & Kaluzny (1988) find that the coherence point of the ω_{rot} and $\omega_{rot} - \Omega_{orb}$ modulations varies between orbital phases 0.5 and 0.75 over several nights, so there is evidently competition between the reprocessing sites. This is confirmed by de Martino *et al.* (1993).

The set of $\omega_{rot} + \Omega_{orb}$, ω_{rot}, $\omega_{rot} - \Omega_{orb}$ and $\omega_{rot} - 2\Omega_{orb}$ frequencies has been seen in AO Psc (Hellier, Cropper & Mason 1991), TX Col (Buckley & Tuohy 1989) and FO Aqr (de Martino *et al.* 1993, who also detect $\omega_{rot} + 2\Omega_{orb}$. In TX Col the average pulse shapes of the sideband modulations show distinctly non-sinusoidal profiles (Figure 7.7). Modelling of such profiles, produced by convolution of the beam profile with the projected reprocessing areas (as intercepted at the reprocessing site, and then reprojected towards Earth) should lead to better understanding of the geometry of IPs.

In EX Hya, reprocessing from any site fixed in the rotating frame of the binary is too weak to detect, but there is a strong $2\Omega_{orb}$ modulation of ω_{rot}, so the $\omega_{rot} \pm 2\Omega_{orb}$ components are observed but not the $\omega_{rot} \pm \Omega_{orb}$ components (Siegel *et al.* 1989).

7.3.5 *Period Changes*

7.3.5.1 *Variation of Orbital Period*

Among the IPs, only FO Aqr, EX Hya and TV Col have the eclipses necessary for accurate orbital phase measurements and potential detection of period variations. Only EX Hya has been observed over a sufficiently large number of orbital cycles to carry this into effect.

Vogt, Krzeminski & Sterken (1980) were the first to detect a significant $\dot{P}_{orb}$ in EX Hya. Later eclipse timings (Jablonski & Busko 1985; Bond & Freeth 1988) show that P_{orb} varies irregularly; the observations to date are well represented by a sinusoid with period 19 y and amplitude O–C $= 1.9 \times 10^{-4}$ d. As there does not appear to be a significant bright spot in EX Hya (Sterken *et al.* 1983; Siegel *et al.* 1989), and the eclipse is of a region within $\sim 2R(1)$ of the primary (Siegel *et al.* 1989), the orbital O–C variations must signify real changes in P_{orb}. This is discussed further in Section 9.4.2.

7.3.5.2 *Variations of Rotation Period*

Changes of primary rotation rate have been determined in six IPs: FO Aqr (Shafter & Macry 1987; Steiman-Cameron, Imamura & Steiman-Cameron 1989; Osborne &

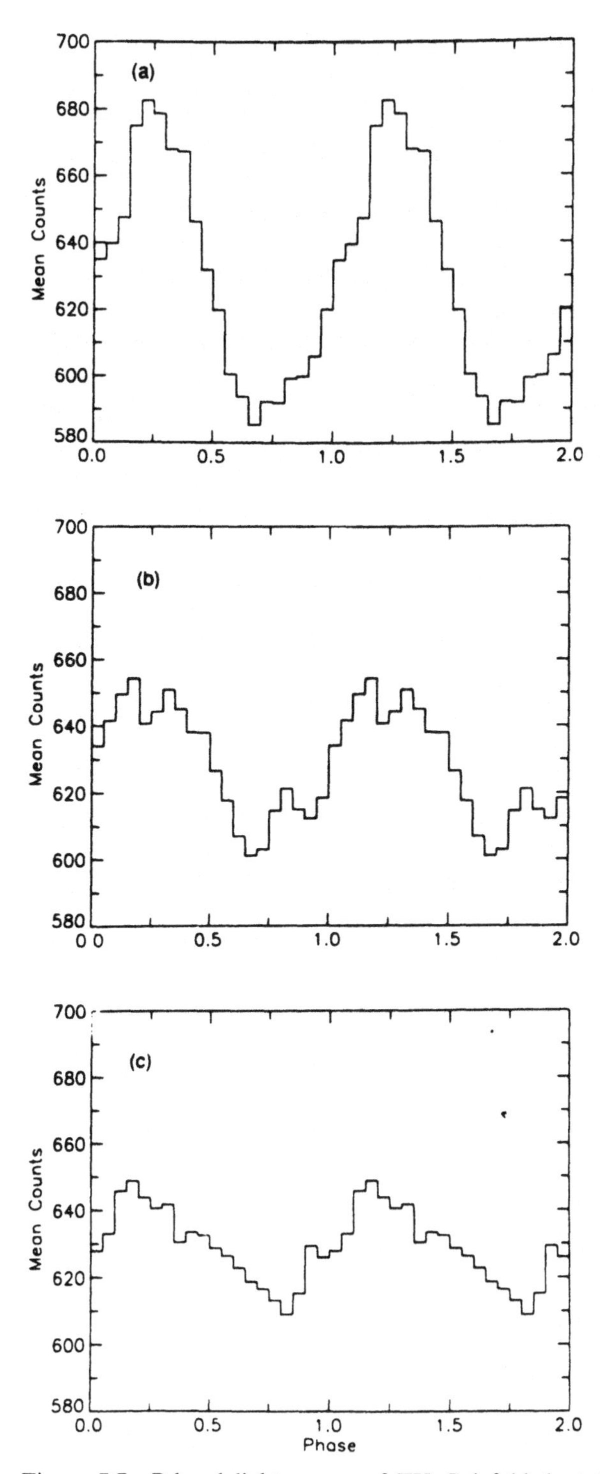

Figure 7.7 B-band light curves of TX Col folded at (a) the $\omega_{rot} - \Omega_{orb}$ frequency, (b) the $\omega_{rot} - 2\Omega_{orb}$ frequency after removing the $\omega_{rot} - \Omega_{orb}$ component, and (c) the $\omega_{rot} + \Omega_{orb}$ frequency after removing the other two frequencies. From Buckley & Tuohy (1989).

Mukai 1989; Kruszewski & Semeniuk 1993), AO Psc (van Amerongen *et al.* 1985; Kaluzny & Semeniuk 1988), V1223 Sgr (van Amerongen, Augusteijn & van Paradijs 1987), EX Hya (Vogt, Krzeminski & Sterken 1980; Bond & Freeth 1988), BG CMi (Patterson 1990a; Singh *et al.* 1991; Augusteijn, van Paradijs & Schwarz 1991; Patterson & Thomas 1993) and GK Per (Patterson 1991).

The quality of the determination in BG CMi is shown by the small scatter about the parabolic variation of O–C in Figure 7.8. For IPs where there are several Fourier components of comparable amplitude, it is important to determine O–C from Fourier transforms or whole-orbit averages (Warner 1986a; Osborne & Mukai 1989).

FO Aqr shows evidence of variable $\dot{P}_{rot}(1)$. Shafter & Macry (1987) derived $\dot{P}_{rot}(1) = 8.0\ (\pm\ 1.1) \times 10^{-11}$ from observations over the period 1981–5. Steiman-Cameron, Imamura & Steiman-Cameron (1989) added observations from 1985–7 and found $\dot{P}_{rot}(1) \sim 2 \times 10^{-11}$. Osborne & Mukai (1989) have extended the baseline to 1988 and in a discussion of all available observations conclude that the O–C diagram departs significantly from a parabolic relationship; the value of $\dot{P}_{rot}(1)$ quoted in Table 7.1 is their best quadratic fit to the observations, a better fit is given by $\dot{P}_{rot}(1) = 3.7 \times 10^{-11}$ and $\ddot{P}_{rot}(1) = -2.5(\pm 0.8) \times 10^{-11}$ y^{-1}. Zhang & Robinson (1994) have observations extending to 1992. Similar results have been found for BG CMi (Garlick *et al.* 1994).

The fact that both positive and negative values of $\dot{P}_{rot}(1)$ are observed, and that $\dot{P}_{rot}(1)$ for FO Aqr appears to be changing quite rapidly (and may even have changed sign in the past few years), indicates that the $P_{rot}(1)$ are close to their equilibrium values, where relatively small changes in $\dot{M}$ can carry them from one side of equilibrium to the other (Warner 1990). Consequently, it is not possible to apply equation (7.14) with any conviction as $n(\omega_s)$ depends so sensitively on ω_s (Figure 7.2). Some general conclusions may, however, be drawn.

First, note that because $n(\omega_s)$ changes sign at $\omega_s = \omega_c$, $P_{rot}(1)$ is always driven towards a value appropriate for equilibrium at the current $\dot{M}$. Therefore the sign of $\dot{P}_{rot}(1)$ is determined by the difference between the average (over $\sim 10^5$ y: Section 7.2.3) and the current $\dot{M}$.

In Table 7.1 the absolute values of the $\dot{P}$ that are < 0 are similar to those that are > 0. From the very rapid decrease in $n(\omega_s)$ with ω_s for $\omega_s > \omega_c$, compared to the slower

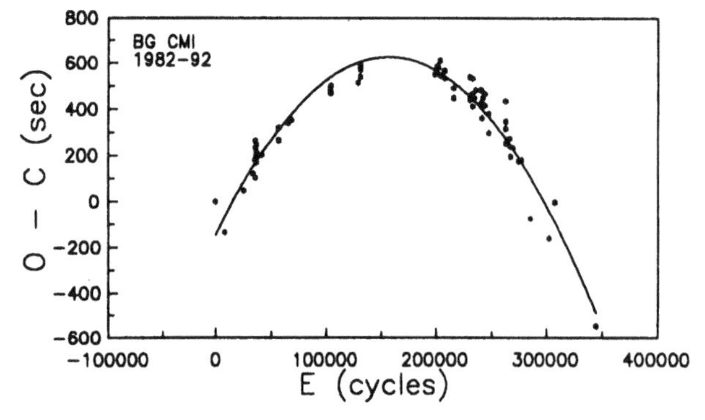

Figure 7.8 O–C diagram of variations in pulse timing in BG CMi. The abscissa is the number of 913 s cycles passed since the initial epoch. From Patterson & Thomas (1993).

increase for $\omega_s < \omega_c$ (Figure 7.2), we deduce that the individual values of ω_s must be *quite* close to ω_c. However, it is unlikely that they are *extremely* close because the $\Delta m_v \sim \pm 0.15$ slow variations in FO Aqr (Section 7.3.1) probably arise from variations in mass transfer $\Delta \dot{M}/\dot{M} \sim \pm 0.20$ which, through equation (7.15), imply $\Delta\omega_s/\omega_s \sim 0.09$. This suggests (Figure 7.2) that $n(\omega_s)$ lies in the range $-1 < n(\omega_s) < 0.3$.

Inserting this derived range into equation (7.14) (with standard values $\mu_{33} = M_1 = \dot{M}_{17} = P_3 = I_{50} = 1$) gives $-2 \times 10^{-10} < \dot{P}_{\rm rot}(1) < 5 \times 10^{-11}$, which matches the observed range quite well. Applications of equation (7.16), which does not depend on $n(\omega_s)$, can then be made with some confidence. These are given in Section 7.5.3.

It must be noted, however, that there is one star for which equation (7.14) can be tested with some acuteness. The primary of EX Hya rotates so slowly that $r_{\rm co} > R_{L_1}$ and therefore accretion from the disc (or without a disc) is far from equilibrium. In this case $n(\omega_s) = 0.6 \pm 0.1$ from Figure 7.2. The mass transfer rate appears well established at $\dot{M}_{17} = 0.2$ (Table 7.4) or 0.1 (Cordova, Mason & Kahn 1985) and the smallest that μ can reasonably be assumed is $\mu_{33}(1) = 0.03$ (see Section 7.5.3.). Then equation (7.14) or (7.19) gives $\dot{P}_{\rm rot}(1) \approx -1.0 \times 10^{-10}$, which is a factor of $\sim$3 larger than observed (Table 7.1). Thus for the one system which is demonstrably far from equilibrium (and for which, since $r_{\rm co} > r_{\rm d}$, the spin-up torque has no uncertain negative contribution) there is a substantial disagreement between theory and observation.

7.4 Polarimetric Observations

Broad band polarimetry of intermediate polars has detected significant circular polarization only in BG CMi, PQ Gem and RXJ1712.6–2414 (Cropper 1986b; Berriman 1988; Stockman *et al.* 1992; Buckley *et al.* 1995b). Typical errors for the more intensive studies are ±0.02%. The low, non-variable linear polarization seen in FO Aqr and AO Psc (Cropper 1988) is probably interstellar in origin.

Until the detection of -0.24 ± 0.03% circular polarization in the I band of BG CMi (Penning, Schmidt & Liebert 1986) there was no observational evidence that the asynchronous magnetic rotator model of IPs was correct. Subsequent observations by West, Berriman & Schmidt (1987) gave unmodulated circular polarization percentages of -0.053 ± 0.051 in a passband 3200–8600 Å, -0.253 ± 0.062 in the I band, -1.74 ± 0.26 in the J band and -4.24 ± 1.78 in an H band. This rapid increase from 0.8 to 1.6 μm is what is expected for cyclotron emission in magnetic fields somewhat lower than those in the polars, and thus provided a more solid foundation for the model.

The polarization behaviour of BG CMi can be modelled by accretion onto a large magnetic polar cap and a field $\sim$4 MG (Chanmugam *et al.* 1990) or, more realistically, with an accretion arc covering a fractional area $f \sim 0.01$ of the primary's surface and a field $2 < B < 5$ MG (Wickramasinghe, Wu & Ferrario 1991). The polarization in the red arises from magneto-bremsstrahlung and that in the IR from cyclotron emission. As the wavelength of the peak of the circular polarization has not yet been determined, a more precise value of the magnetic field cannot be estimated. Dhillon & Marsh (1993) have detected a cyclotron hump in 2.0–2.5 μm spectra of BG CMi. Extended baselines for this and other IPs promise hope of direct measurements of $B(1)$.

PQ Gem has prominent circular and linear polarization in the red and IR, modulated at the primary's rotation period of 13.9 min. The amplitudes of the circular polarization are 1.1% at R and 2.1% at I. Linear polarization spikes reach 1% in R and 3% in I (Piirola, Hakala & Coyne 1993; Rosen, Mittaz & Hakala 1993). Based on the calculations for accretion arcs by Wickramasinghe, Wu & Ferrario (1991), these results suggest $B(1) \sim 8$ MG, which is close to the lowest field measured in a polar (Table 6.8). The occurrence of both negative and positive circular polarization in the rotation cycle, associated with double humps in the red and IR light curve, indicates that cyclotron emission occurs near two opposite magnetic poles.

RXJ 1712.6–2414 has the largest polarization yet detected in an IP, with amplitudes ~1% in V and R, and up to 5% in I (Buckley *et al.* 1995b). The circular polarization has a negative sign over the entire spin cycle, implying that only one of the accreting regions is visible, in turn suggesting that this is a low inclination system. From the wavelength dependence of the circular polarization, Buckley *et al.* estimate a field ~8 MG.

Unless these systems are special (e.g., lower $\dot{M}$) in some way that makes their cyclotron emission easily visible, it would appear that the sequence RXJ1712.6−2414, PQ Gem, BG CMi, other IPs, is one of diminishing $B(1)$, and that most IPs have $B(1) \lesssim 2$ MG.

Since the intensive study of IPs began, none of the brighter ones has descended to a very low state of $\dot{M}$; it has therefore not yet been possible to obtain a spectrum of the photosphere of any IP primary to search for Zeeman components.

7.5 X-Ray Observations

The high mass transfer rate in most IPs gives them $\sim 10^2$ times the hard X-ray luminosity of polars; any X-ray flux-limited survey therefore probes a volume $\sim 10^3$ larger for IPs than polars. However, the greater optical flux in IPs results in their L_X/L_V being 10–100 times smaller than for polars (Beuermann & Thomas 1993), so many of the IPs remaining to be identified in the ROSAT X-Ray survey will be quite faint. The ROSAT EUV survey detected only EX Hya and PQ Gem.

Reviews of pointed observations of IPs by EXOSAT are given by Norton & Watson (1989b) and Hellier, Garlick & Mason (1993), and by GINGA by Ishida (1991).

7.5.1 *X-Ray Light Curves*

7.5.1.1 Orbital Modulation

In all of the IPs for which sufficient observations are available an orbital modulation in the form of dips near orbital phase 0.8 is seen (Figure 7.9: Hellier, Garlick & Mason 1993). The depth of modulation increases at lower energies, indicating photoelectric absorption by obscuring gas. Similar dips occur in the X-ray light curves of low mass X-ray binaries, Figure 7.9, where they are caused by material above the orbital plane subtending angles 10–15° at the primary. In IPs the obscuring gas could be raised to such heights above the plane either by impact of the stream on the disc, and/or by impact on the magnetosphere of that part of the stream that runs over the face of the disc. McHardy *et al.* (1987) proposed that in BG CMi the latter region is responsible for a considerable fraction of the X-ray emission, and that the orbital dips arise from self-obscuration of this region. However, as pointed out by Hellier, Garlick & Watson

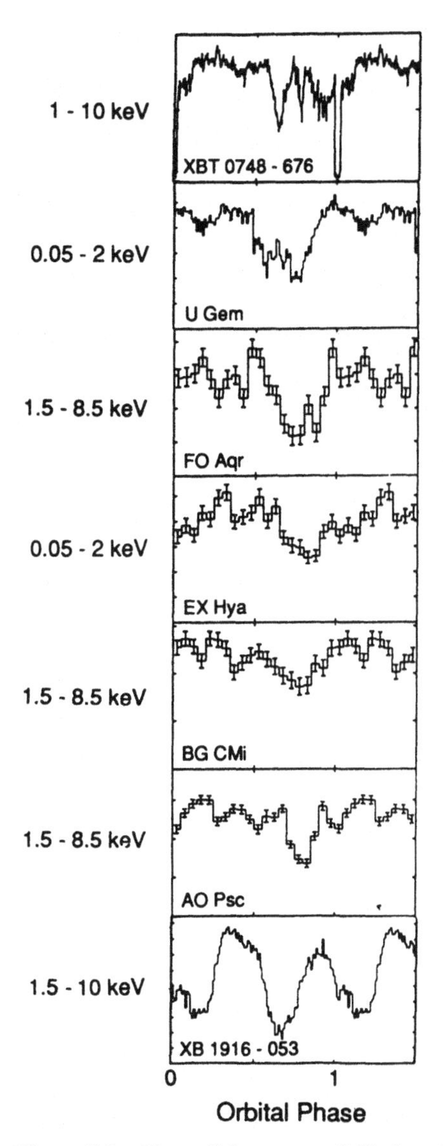

Figure 7.9 X-ray light curves, folded at the orbital period, for four IPs, the DN U Gem, and two low-mass X-ray binaries. These are from the EXOSAT medium energy detector, except for EX Hya and U Gem where the modulation is seen only at low energy. From Hellier, Garlick & Mason (1993).

(1993), 90% of the gravitational energy has still to be released by the time that r_μ is reached, so it is unlikely to be a major contributor of X-ray emission.

7.5.1.2 Rotational Modulation

The copious X-ray emission of IPs in the range 2–20 keV enables good light curves (Figure 1.23) and spectra to be derived for the rotational modulation. Early indications were that the light curves are approximately sinusoidal (Watson 1986; Osborne 1988;

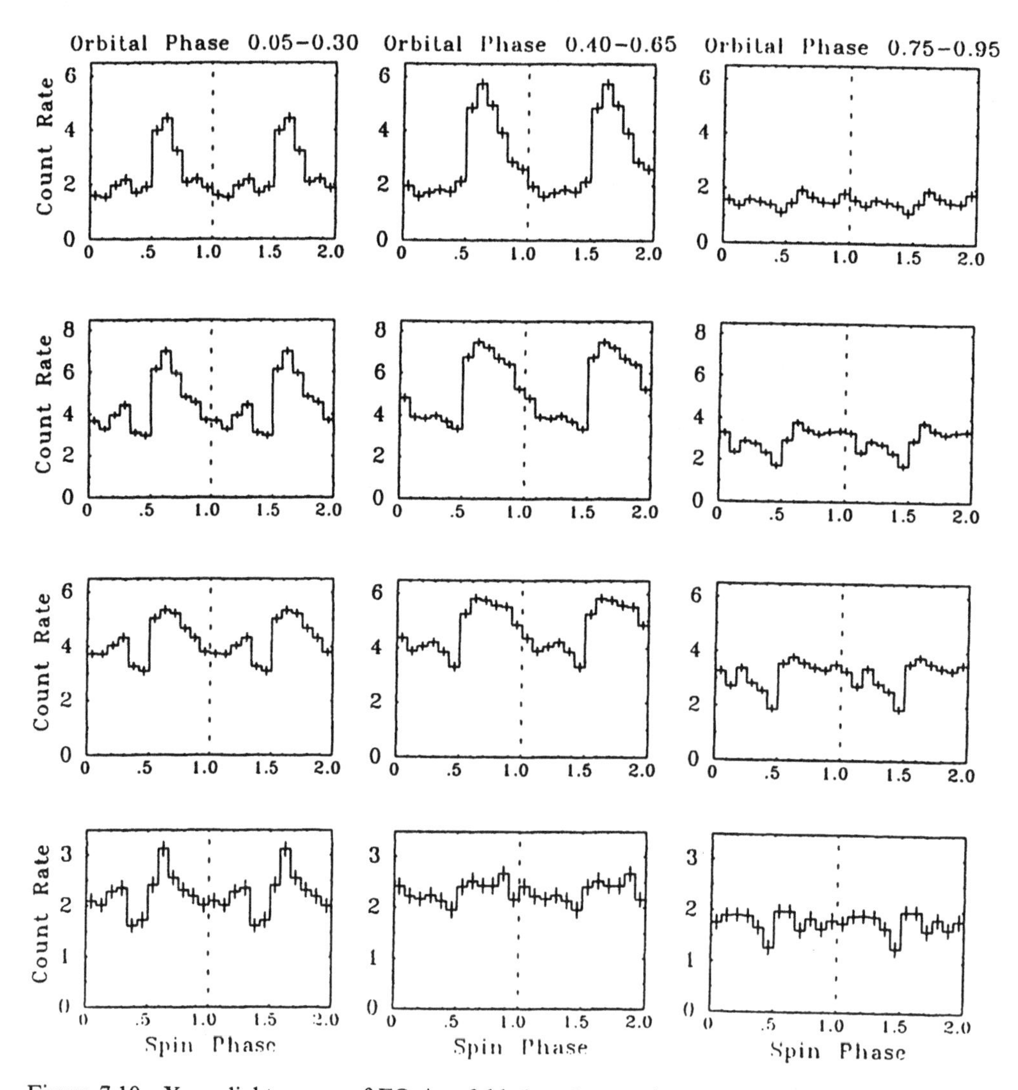

Figure 7.10 X-ray light curves of FO Aqr folded at the rotational period for different orbital phases. From top to bottom the energy ranges are 2–4 keV, 4–6 keV, 6–10 keV and 10–20 keV respectively. From Norton *et al.* (1992a).

Norton & Watson 1989b), but observations with higher resolution are better described as sinusoidal with superimposed dips, the strengths of which vary with orbital phase (Norton *et al.* 1990, 1992a,b; Norton 1993). Figures 1.22 and 7.10 illustrate this for FO Aqr and show the strong dependence of modulation amplitude with energy. Average amplitudes are given in Table 7.3 (Norton & Watson 1989b; Shafter, Szkody & Thorstensen 1986; Mason *et al.* 1992). Individual light curves often contain short lived flares (e.g., EX Hya: Rosen *et al.* 1991). XY Ari is peculiar in showing two maxima per rotation period, which may be a geometrical result of its large inclination (Kamata & Koyama (1993).

The general smoothness and nearly sinusoidal variation of the X-ray fluxes led to the

Table 7.3. *Average X-Ray and Optical Modulation Amplitudes (%) and Column Densities.*

	Energy range (keV)				Optical		
Star	0.05–2.0	2–4	4–6	6–10	$P_{rot}(1)$	P_{syn}	$N_H(10^{23}cm^{-2})$
GK Per		22	19	18			2.5
TX Col	34	12	9	8		7	0.02:
TV Col	18	16	10	9			0.9
PQ Gem		35	36	23			
FO Aqr		42	28	17	21	8	2.0
AO Psc	40	39	29	18	1.5	3	0.9
V1223 Sgr	21	15	10	4		7	1.0
BG CMi	14	22	5	2	22		1.0
EX Hya	29	13	7	6	20		0.007

suggestion (King & Shaviv 1984a) that this must be a geometrical effect, and could most simply be produced by postulating accretion over a large ($f \gtrsim 0.25$) polar cap (so that oblique rotation in general never leads to complete occultation of the cap). However, accretion zones that are long arcs also produce the necessary slowly varying light curves and with $f \sim 0.01$ produce shock heights that are $< R(1)$ (whereas with $f \sim 0.25$ the shock height is $> R(1)$ and the X-ray emitting region could not be modulated with the large amplitudes seen in Table 7.3 (Kylafis & Lamb 1982; Lamb 1988)). In addition, unless there are very large variations of accretion properties (i.e., L/f and blob size) along an arc, purely geometric effects cannot give an energy-dependent pulse profile – and variations of properties and geometries are in any case unlikely to lead to the systematic dependence seen in Table 7.3.

Estimates of f, for a centred dipole configuration and a range of r_0 and δ (Section 7.2.1), give $0.001 \lesssim f \lesssim 0.02$ for disc accretion (Rosen 1992). With typical specific accretion luminosities $L/f \sim 3 \times 10^{36}(0.01/f)$ erg s^{-1} (Section 7.5.2) the accretion zones of IPs are just into the regime where hard X-ray beaming occurs (Section 6.2.3). However, this is independent of energy and so cannot account for the energy-dependent modulations. Rosen *et al.* (1991), Hellier, Cropper & Mason (1991) and Hellier (1991) discuss other objections to the large polar cap model.

A solution to the problem was proposed by Rosen, Mason & Cordova (1988) (see also Mason, Rosen & Hellier (1988) and Lamb (1985)) and further developed by Norton & Watson (1989b), Rosen *et al.* (1991), Rosen (1992) and Norton (1993). The accretion flow at $r < r_0$, falling towards the stand-off shock, produces an absorbing 'curtain'; the intensity modulation of the X-ray flux from the post-shock region is caused by the varying path lengths through the curtain as well as by passage of part of the accretion arc behind the primary (Figure 7.11). The modulation thus has two components – an energy-dependent photoelectric absorption and energy-independent electron scattering optical depth effect plus optical geometric occultation. It is these varying optical thicknesses that produce the dips in the otherwise sinusoidal rotational light curves (Section 7.5.1.2).

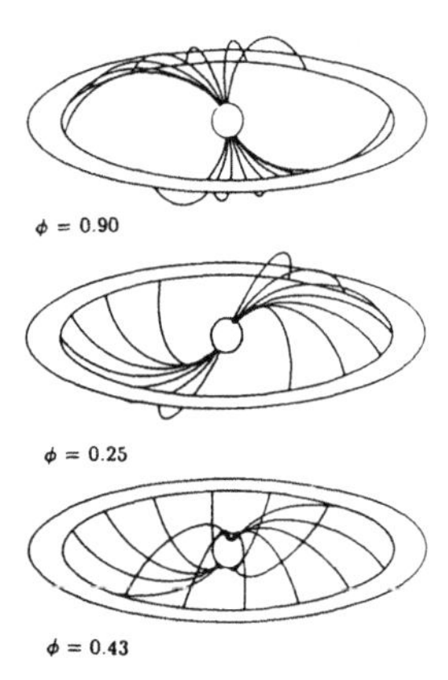

Figure 7.11 Schematic view of accretion curtains at three rotational phases. From Hellier, Cropper & Mason (1991).

Simulated light curves for this model, for discless accretion and for combinations of both, have been generated by Norton (1993). He points out that accretion via a disc will result in each accretion region on the primary always receiving exactly half of the accreted gas, whereas in the discless model accretion onto each region is modulated by 100%. Furthermore, in the latter case the accretion region 'migrates' around the magnetic pole to follow the incoming stream of gas. Norton finds that complex pulse profiles, similar to those observed, result whenever at least part of the accretion flow is channelled to the magnetosphere via a stream.

In the accretion curtain model the $P_{\rm rot}(1)$ optical modulation arises from varying aspects of the optically thick accretion curtain, rather than from reprocessing of a beam by the accretion disc – Section 7.3.3.1. This can explain why the X-ray and optical pulses are in phase and have similar profiles (Hellier & Mason 1990). The X-ray emission is produced closer than the optical to the primary, where the variations in obscuration with spin phase are most effective, which may explain the persistence of X-ray modulation even at low inclination (Hellier & Mason 1990; Hellier, Cropper & Mason 1991).

EX Hya provides a particularly sensitive means of probing the structure of an IP. X-ray observations by GINGA (Rosen *et al.* 1991) have given greater range and resolution of the energy spectra in and out of eclipse. These results can be understood if there are two accretion sites on the primary, one above and one below the orbital plane. The eclipse geometry is such that only the latter accretion site is obscured by the limb of the secondary, explaining the partial eclipse. The depth of eclipse thereby depends on the visibility (dependent on phase in rotation of the primary) of that site. Furthermore, rotational modulation of the X-ray flux is a result of occultation during rotation of the primary of the lower site but not the upper one.

In FO Aqr the rotational modulation is separable into a sinusoidal component, caused by optical depth effects in an accretion curtain, and an energy-independent constant component with a superimposed 'notch' of phase width ~0.22. The latter is thought to be caused by self-occultation of a compact emission region on the surface of the primary, indicative of a magnetic field geometry more complex than a simple dipole (Norton *et al.* 1992a).

Norton & Watson (1989b) were able to fit the amplitudes and energy dependences of the pulsations listed in Table 7.3 with the hydrogen column densities also listed there. (Their low column density for TV Col is confirmed by Buckley & Tuohy (1989).) The column densities are mostly ~100 times greater than those deduced for polars (Table 6.5). This is partly due to the generally lower accretion rates in polars – note that EX Hya, which has an $\dot{M}$ *similar* to that of polars, as expected from its position among the polars below the orbital period gap, has an N_H of the same order as the polar (Table 6.5), and about half is interstellar (Singh & Swank 1993). An additional factor will be that the accretion column effectively obscures a greater fraction of the accretion zone than the accretion funnel in polars.

7.5.1.3 X-Ray Orbital Sidebands

Hellier (1991) pointed out that the relative amplitudes of ω_{rot}, Ω_{orb} and the orbital sidebands of X-ray light curves should be used to discriminate between disced and discless accretion. In the simplest model of disced accretion ω_{rot} should dominate over $\omega_{rot} - \Omega_{orb}$ and a strong Ω_{rot} modulation should appear only at large inclinations. This was observed to be the case for all IPs except TX Col, in which the P_{syn} and P_{orb} modulations have a larger amplitude than that of $P_{rot}(1)$; TX Col therefore was suggested as a possible case of discless accretion (see also Mason, Rosen & Hellier (1988)). This is in accord with accretion via a stream (fixed in orbital phase) striking the rotating magnetosphere, which would be expected to generate the strongest modulations at P_{syn} and P_{orb}.

Wynn & King (1992) found, however, that prominent $P_{rot}(1)$ and low P_{syn} modulations can be generated by discless accretion if the two accretion zones differ greatly in their properties. These models predict a strong $2\omega_{rot} - \Omega_{orb}$ component to be present. Hellier (1992,1993c) was unable to find such a modulation in any IP and concluded that accretion in IPs is dominantly through discs, but that TX Col and FO Aqr (which occasionally shows a strong P_{syn} modulation: Norton *et al.* 1992a) have some stream overflow.

Norton *et al.* (1992b) found a weak ~847 s X-ray modulation in BG CMi, as well as the strong 913 s period seen in both X-ray and optical. They suggested that 847 s = $P_{rot}(1)$ and 913 s = P_{syn}, which would be a clear sign that BG CMi is a predominantly discless system. However, Hellier, Garlick & Mason (1993) point out that such a pattern can arise from orbital modulation of $P_{rot}(1)$ = 913 s plus cancellation of the $\omega_{rot} - \Omega_{orb}$ sideband by an intrinsic P_{syn} modulation of the kind observed in FO Aqr (Norton *et al.* 1992a, but the situation is not yet resolved (Garlick *et al.* 1994)).

The pulse-fraction of the ω_{rot} modulation decreases during the orbital dips. This cannot be due simply to obscuration; it is probably due to (unresolved) interference between ω_{rot} and its orbital sidebands, which implies some discless accretion as in the hybrid model (Hellier, Garlick & Mason 1993).

7.5.2 X-Ray Fluxes and Spectra

Typical X-ray fluxes F_x(2–10 keV) in the 2–10 keV region are listed in Table 7.4 (Norton & Watson 1989a). To convert these to *detected* luminosity in that energy range

Table 7.4. *X-ray Fluxes, Accretion Rates and Magnetic Moments of IPs.*

Star	d (pc)	Reference	F_X(2–10 keV) (10^{11} erg cm^{-2} s^{-1})	L_X (10^{33} erg s^{-1})	L_{acc} (10^{34} erg s^{-1})	$\dot{M}_{17}(1)$	μ_{33}
EX Hya	105	1	8.4	0.057	0.29	0.11	0.06
BG CMi	700	2,3	2.3	0.70	3.5	1.4	3.9
V1223 Sgr	600	4,5	5.0	1.1	5.5	2.3	4.0
AO Psc	420	9	4.0	0.44	2.2	0.90	2.7
YY Dra	155	6	4.5	0.067	3.3	1.4	2.1
FO Aqr	325	7	3.4	0.22	1.1	0.45	3.2
TV Col	$\geq$500	8	4.1	$\geq$0.6	$\geq$3	>1.2	>8.3
TX Col	550	8	1.9	0.36	1.8	0.74	6.5
XY Ari	200	10		0.2	1.0	0.4	0.35
GK Per	340	1 quiesc	3	0.2	1.0	0.4	0.71
		outburst	15	1.1	5.5	2.3	1.6

References to distance estimates: 1. Warner 1987a; 2. Berriman 1988; 3. McHardy *et al.* 1987; 4. Watts *et al.* 1985; 5. Bonnet-Bidaud, Mouchet & Motch 1982; 6. Mateo, Szkody & Garnavich 1991; 7. de Martino *et al.* 1993; 8. Buckley & Tuohy 1989; 9. Hellier, Cropper & Mason 1991; 10. Zuckerman *et al.* 1992.

requires knowledge of distances. For almost all of the IPs distances are very uncertain (e.g., Berriman 1987a); those given in Table 7.4, derived from the references listed, are adopted for reason of necessity.

The conversion to X-ray luminosity L_X(2–10 keV) is made using $L_X = 2\pi d^2 F_X$, which allows for the fact that only $\sim 1/2$ of the emitted flux avoids the stellar surface. A bolometric correction must be made to this bremsstrahlung emission component. The limited energy range (most IPs are not detected above 10 keV), the significant absorption, and the problem that a single temperature source often does not give a good fit to the spectra (Norton & Watson 1989a), make the bolometric correction uncertain. King & Watson (1987) find that a factor of 3.3 increase to L_X(2–10 keV) is required to give the total bremmstrahlung emission at $kT = 30$ keV. An additional increase is required to remove the effect of absorption at the lower end of the energy range.

To derive the *total* accretion luminosity would require an exact knowledge of the spectrum in the 0.01–1 keV range. As one half of the bremsstrahlung radiation is absorbed and reemitted by the photosphere at $T_{BB} \sim 40$ eV a factor of $\sim$2 increase over the hard X-ray emission is immediately obvious. A further factor depends on whether the IPs have a 'soft X-ray excess' (see Section 6.4.3), which is even less detectable in IPs than in polars because of the larger local absorption by the accretion stream. Indeed, most IPs are not detected in soft X-rays at all (Watson 1986; Osborne 1988). The radiation absorbed by the accretion curtain is of course not lost to the system – it preheats the stream prior to arriving at the stand-off shock or in the stellar photosphere and appears eventually as enhanced luminosity of one or more of the emission components.

There is every reason to expect that accretion in IPs will be at least as inhomogeneous as in polars. The large cycle-to-cycle variations in intensity of the rotational modulation (Figure 7.3) must arise from such inhomogeneity. Therefore EUV and soft X-ray emissions exceeding the bremsstrahlung emission by factors of 5 or 10, but hidden from us both by local and interstellar absorption, are probable. Much of the EUV radiation will escape without being absorbed by the accretion curtain (it must escape, to maintain energy balance) and will be to some extent fan-beamed in the process (but not necessarily with the same profile as the hard X-rays). It may be this radiation, rather than the hard X-rays, which energizes the reprocessing sites.

Without allowing for beaming effects, Hassall *et al.* (1981) estimated that the total accretion luminosity L_{acc} of AO Psc is $\sim$100 times the detected L_x(2–10 keV). In the model of TX Col constructed by Buckley & Tuohy (1989) L_x(2–10 keV) $\sim 10L_x$(2–10 keV). Allowing for the bolometric correction discussed earlier, these models indicate $L_{acc} \sim 50 \times$ the detected L_x(2–10 keV) given in Table 7.4, with an uncertainty of at least a factor of 2. Norton & Watson (1989a), from analogous considerations, deduce a factor of 33.

The values of L_{acc} given in Table 7.4 are simply 50 L_x(2–10 keV).The deduced rates of mass transfer assume $M_1(1) = 1.0$.

The effects of the usually large absorption column densities N_H (Table 7.3) on the spectra of IPs are seen in Figure 7.10 (Norton & Watson 1989b). The varying line of sight through the accretion curtain results in a large variation of spectrum with orbital phase. N_H may vary by up to a factor of 2 around orbit. Models using only bremsstrahlung or power-law continua passing through an homogeneous absorber do not fit the observations – the absorber must either be only partially covering the source or there is a second continuum source (Norton *et al.* 1992b). The bremsstrahlung temperature is usually $\gtrsim$30 keV and is poorly confined at the upper limit (Norton *et al.* 1992b); EX Hya, however, has a two-component model, with a $\sim$9 keV thermal source and a $\sim$0.74 keV optically thin plasma. The hotter component arises in the accretion zone; the softer component may be any of pre-shock, or post-shock regions or transition zone (Singh & Swank 1993).

The presence of a strong Fe Kα emission line at 6.4–6.5 keV is evidence of the importance of significant electron scattering in the accretion column (Norton, Watson & King 1991; Section 6.4.1). In EX Hya, however, the line is at 6.7 ± 0.05 keV, which indicates a thermal origin (Rosen *et al.* 1991); there is also an emission line of Si Kα at 1.72 keV (Singh & Swank 1993).

A source of 0.5 MeV positron annihilation radiation was identified with V1223 Sgr (Briggs *et al.* 1991) but has since been shown to arise in a nearby low mass X-Ray binary (Briggs *et al* 1994)

7.5.3 *Magnetic Moments from Rates of Mass Transfer*

On the assumption, to a large extent justified in Section 7.3.5.2, that most IPs accrete close to their equilibrium values, equations (7.16) or (7.20) can be used, together with the values of $\dot{M}$ listed in Table 7.4, to give estimates of the magnetic moments $\mu(1)$ (Norton & Watson 1989a). We adopt $\omega_s = 1.0$ and $M_1(1) = 1.0$. The magnetic moments for discless accretion will be twice those deduced for disced systems, so only the latter are listed in Table 7.4. It may be noted that, since the deduced $\mu \propto \dot{M}^{1/2}$, the uncertainties in $\dot{M}$ are not too harmful. It should be noted, however, that the $\dot{M}_{17}(1)$

estimates given in Table 7.4 are the *current values*. As discussed further in Section 8.5, the observed $P_{rot}(1)$ is determined by some long term average $< \dot{M}(1) >$ which may depart by small factors from the current value.

The field of $2 \lesssim B \lesssim 5$ MG derived from polarization measurements of BG CMi (Section 7.4) transforms to $1.3 \lesssim \mu_{33}(1) \lesssim 3.3$, which is just compatible with the independently derived value of $\mu_{33}(1) \approx 3.9$ in Table 7.4. It may be noted that $\mu \propto r_{\mu}^{7/4}$ (equation (6.4)); therefore the value of μ derived using equation (7.16) depends very sensitively on the constant of proportionality in equation (7.3). A definitive measurement of $\mu(1)$ for BG CMi from polarization observations will serve to calibrate the technique.

In Table 7.4 the value of $\mu(1)$ derived for EX Hya does not come from application of equation (7.6) as it is clearly far from the equilibrium accretion condition (Sections 7.3.5.2 and 7.6.2). Instead, we make use of equations (6.5) and (7.3) and estimates of r_0 deduced from line profiles (see Section 7.6.1). In EX Hya Ferrario, Wickramasinghe & King (1993) find $r_0 \approx 5R(1)$, which, together with the $\dot{M}$ given in Table 7.4, provides the evaluation of $\mu(1)$. In EX Hya, Cordova, Mason & Kahn (1985) independently find $\dot{M}_{17} \approx 0.1$, in reasonable agreement with Table 7.3.

The two values of $\mu(1)$ given for GK Per in Table 7.4 should bracket the actual $\mu(1)$ as this star appears to go from $\omega_s > 1$ to $\omega_s < 1$ during outburst (Section 7.6.6).

7.6 Spectroscopic Observations

Examples of the UV and optical spectra of IPs are given in Figures 7.12 and 7.13. There is a general resemblance to the polars (Figures 6.21 and 6.22) but HeII and SiIV are less prominent in the UV relative to CIV, and HeII is generally somewhat weaker relative to Hβ in the optical. Other examples of spectra can be found among the references accompanying Table 7.1 and in the compilation by la Dous (1990).

In AO Psc the observed ratios of NV, SiIV, CIV and MgII intensities are in agreement with collisional excitation at high temperature ($T \gtrsim 10^5$ K) and the ratio HeII $\lambda1640/\lambda4686$ is consistent with a recombination spectrum (Cordova *et al.* 1983). However, abnormally large ratios of NV/CIV are observed in TX Col which have not been modelled successfully (Bonnet-Bidaud & Mouchet 1987; Mouchet *et al.* 1991).

7.6.1 Line Profiles and Modulations: Theory

Before reviewing the observations, it is helpful to consider what might be expected from the accretion processes described in Section 7.2.

A Keplerian disc exists from the inner boundary at radius r_0 to an outer boundary that should be similar to that seen in non-magnetic CVs (but see Section 7.2.4). The line profile of such a truncated disc will differ from that of a complete disc, especially by the absence of any rotational broadening with velocities greater than those at r_0. From equations (2.16), (2.23) and (7.12), at r_0 the disc velocity is

$$v_d(r_0) = 943\left[\frac{M_1(1)}{\omega_s P_3(1)}\right]^{1/3} \text{km s}^{-1}. \qquad (7.28)$$

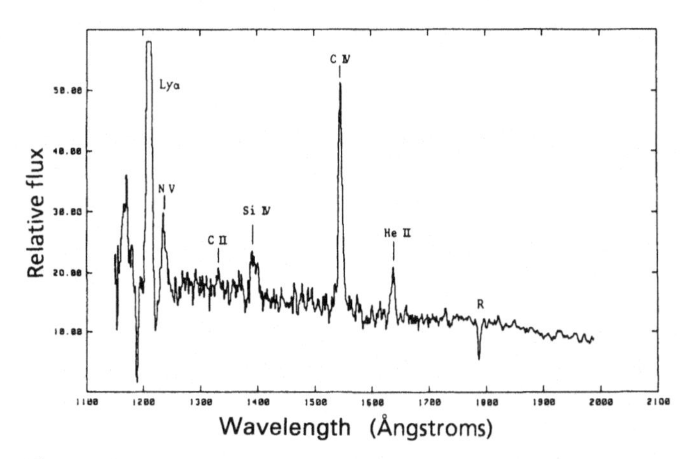

Figure 7.12 UV spectrum of V1223 Sgr. From Bonnet-Bidaud, Mouchet & Motch (1982).

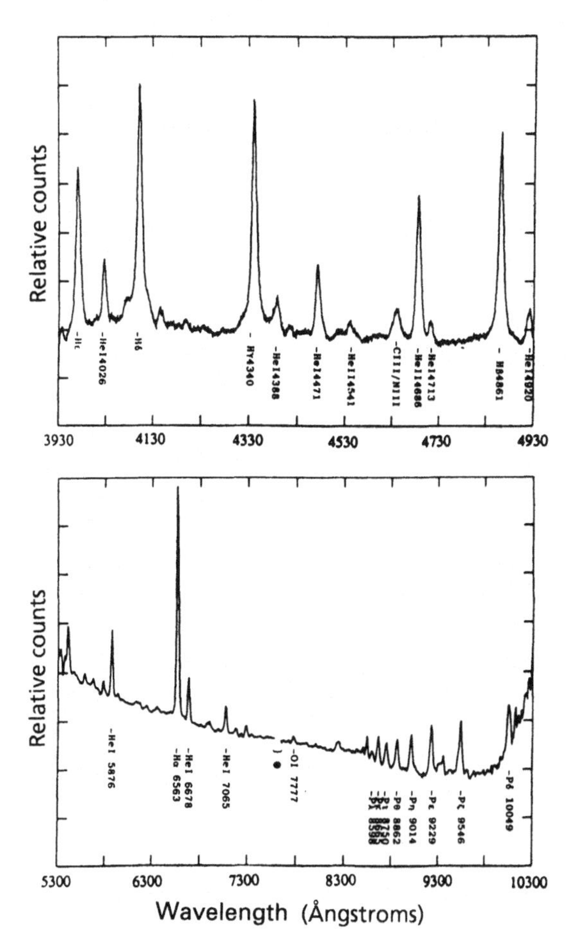

Figure 7.13 Spectra of TX Col. From Buckley & Tuohy (1989).

Therefore, for IPs accreting near their equilibrium condition ($\omega_s \sim \omega_c = 0.97$) and $500 \lesssim P_{rot}(1) \lesssim 2000$ s, the disc emission-line profiles should have FWZI $\sim 2000 \sin i$ km s^{-1}.

However, if we are able to see to the surface of the primary (without obscuration by the disc or accretion curtain) there will be an additional source of line emission: *viz.*, the same source that produces the broad emission-line component in polars (Section 6.5.1.3), i.e., gas falling into the accretion zone. In polars this has a FWZI ~ 2000 km s^{-1} and a velocity amplitude up to 1000 km s^{-1} with period P_{orb}. In the IPs we might expect a similar width and amplitude (which is determined by velocity in the stream, not by the rotational velocity of the surface of the primary), but modulated with period $P_{rot}(1)$. (The modulation of this contributor to the line profile will not be easily detectable if there are two accretion zones active, spaced $\sim 180°$ apart in rotational longitude.)

Therefore, spectra taken with long integrations ($\gtrsim P_{rot}(1)$) could be expected to show emission lines with FWZI up to 4000 km s^{-1} rather than that appropriate for a truncated disc (note that only the disc velocity projects simply with a sin i factor; projection of the infall velocity depends on the geometry of the flow). However, spectra with sufficient time resolution should show that the wings of the emission lines beyond the velocity $v_d(r_o) \sin i$ are modulated with period $P_{rot}(1)$.

Detailed models of optically thin emission from the accretion curtains have been computed by Ferrario, Wickramasinghe & King (1993; see also Ferrario & Wickramasinghe 1993). By comparison with observation these provide estimates of r_0 and, with lower accuracy, i. TX Col is found to have $r_0 \sim 8 \times 10^9$ cm which, from equations (6.5) and (7.3), gives $\mu_{33}(1) = 1.0$, in good agreement with Table 7.4.

7.6.2 *Line Profiles and Modulations: Observation*

High quality, time-resolved spectroscopic studies have been made of EX Hya (Kaitchuck *et al.* 1987; Hellier *et al.* 1987), TX Col (Buckley & Tuohy 1989), FO Aqr (Hellier, Mason & Cropper 1990; Martell & Kaitchuck 1991), AO Psc (Hellier, Cropper & Mason 1991) and GK Per (Reinsch 1994; Garlick *et al.* 1994b).

In EX Hya, Hellier *et al.* were able to separate the broad emission-line profiles into two components: a double-peaked profile with separation of peaks $\sim$1300 km s^{-1} and FWZI of 2500 km s^{-1}, and a component with a maximum FWZI of $\sim$7000 km s^{-1} which is modulated in strength and V/R (and probably velocity) at a period of 67 min = $P_{rot}(1)$. There is also an S-wave component with amplitude 600 km s^{-1} and period = P_{orb} which is attributed to a bright spot at the edge of the disc.

The 67 min period also appears in the equivalent widths of the emission lines; the pulsed fractions are 0.06 in Hα, 0.18 in Hβ and 0.25 in Hγ. This systematic variation is a result of a steady increase in width of the broad component along the Balmer series (FWZI of Hα is 4000 km s^{-1}). The inner disc regions evidently radiate Hα only weakly (Hellier *et al.* 1987, 1989).

The unpulsed, double-peaked broad profile is very strong evidence for an accretion disc in EX Hya. The maximum observed $v \sin i$ of 1250 km s^{-1}, together with $M_1(1) = 0.9$ and $i = 78°$, give, from equations (7.12) and (7.28), $r_0/r_{co} = \omega_s^{2/3} = 0.10$, so EX Hya is far from accretion equilibrium. From equation (7.17), $r_0/R(1) \approx 6$. The pulsed very broad profile in EX Hya is the equivalent of the broad component in polars. See Section 7.7 for spectroscopic observations made during outburst.

In FO Aqr, Hellier, Mason & Cropper (1990) find that phasing their spectra on the $P_{rot}(1)$ cycle shows the presence of a strong V/R modulation caused by an S-wave feature with an amplitude $\sim$800 km s^{-1} and FWZI $\sim$800 km s^{-1}. The S-wave is only strongly present in its blueward excursions and only becomes an obvious structure when the underlying broad profile is subtracted (Figure 7.14). Penning (1985) measured a $P_{rot}(1)$ S-wave for HeII in FO Aqr with an amplitude $\sim$80 km s^{-1}; a similar amplitude can be seen in the raw profile in Figure 7.14. This misleadingly small amplitude is a result of the strong V/R variation, which practically eliminates the redward half of the S-wave, and blends with the disc emission line.

Similar low amplitude S-waves have been observed in V1223 Sgr, BG CMi and AO Psc (Penning 1985) and TX Col (Buckley & Tuohy 1989) and probably also arise from blending of a much larger amplitude S-wave with the disc emission.

In FO Aqr, as in EX Hya, there is therefore evidence for an emission-line region in the vicinity of the accretion zone. The phasing of the $P_{rot}(1)$ S-wave with the continuum variations shows that maximum blueshift corresponds to photometric and X-ray maximum. This implies that at maximum light the accretion 'pole' points away from the observer where the material falling in the accretion curtain towards the primary has its maximum component towards the observer. At the same time, the continuum-emitting accretion zone is seen transversely, in the direction of minimum optical thickness and hence maximum brightness.

Hellier, Mason & Cropper do not extract a profile for the disc emission in FO Aqr, but from inspection of their published spectra the FWZI $\lesssim$ 1500 km s^{-1}, which is compatible with equation (7.28) (for $i \sim 70°$) and therefore with equilibrium accretion. Line widths for other IPs are listed in Table 7.4. Many of the inclinations (Table 7.1) are very uncertain and must be used with caution – the reasoning (a small $K(1)$) that leads to low inclinations in AO Psc, V1223 Sgr and BG CMi also applied to FO Aqr until it was discovered to be an eclipsing system (Hellier, Mason & Cropper 1989). Taken at face value, the line widths in many IPs imply that there are substantial rotational S-wave contributions to their line profiles.

In AO Psc, although the dominant continuum variation is at P_{syn}, the principal variation in RV has a period $P_{rot}(1)$. The amplitude is 260 km s^{-1}, implying that the emission lines are formed at $\leq 7R(1)$. Unlike EX Hya, there is pronounced asymmetry in the line profile, indicating that emission comes predominantly from only one pole.

In GK Per Reinsch (1994) has observed a spectral S-wave at the 351 s spin period, similar to that seen in FO Aqr. The Balmer and HeII line fluxes are pulsed at fractional amplitudes $\sim$2% and $\sim$4% respectively.

7.6.3 *Orbital Sidebands*

As in the continuum and X-ray emissions, the emission-line profiles in IPs can be expected to show a range of orbital sidebands which can aid identification of the emission-line sites. For FO Aqr and AO Psc intensive studies have been made (Hellier & Mason 1990; Hellier, Mason & Cropper 1990; Hellier, Cropper & Mason 1991). In FO Aqr the $\omega_{rot} - \Omega_{orb}$ (P_{syn}) component is the strongest observed sideband and reaches maximum intensity at inferior conjunction of the secondary. This sideband

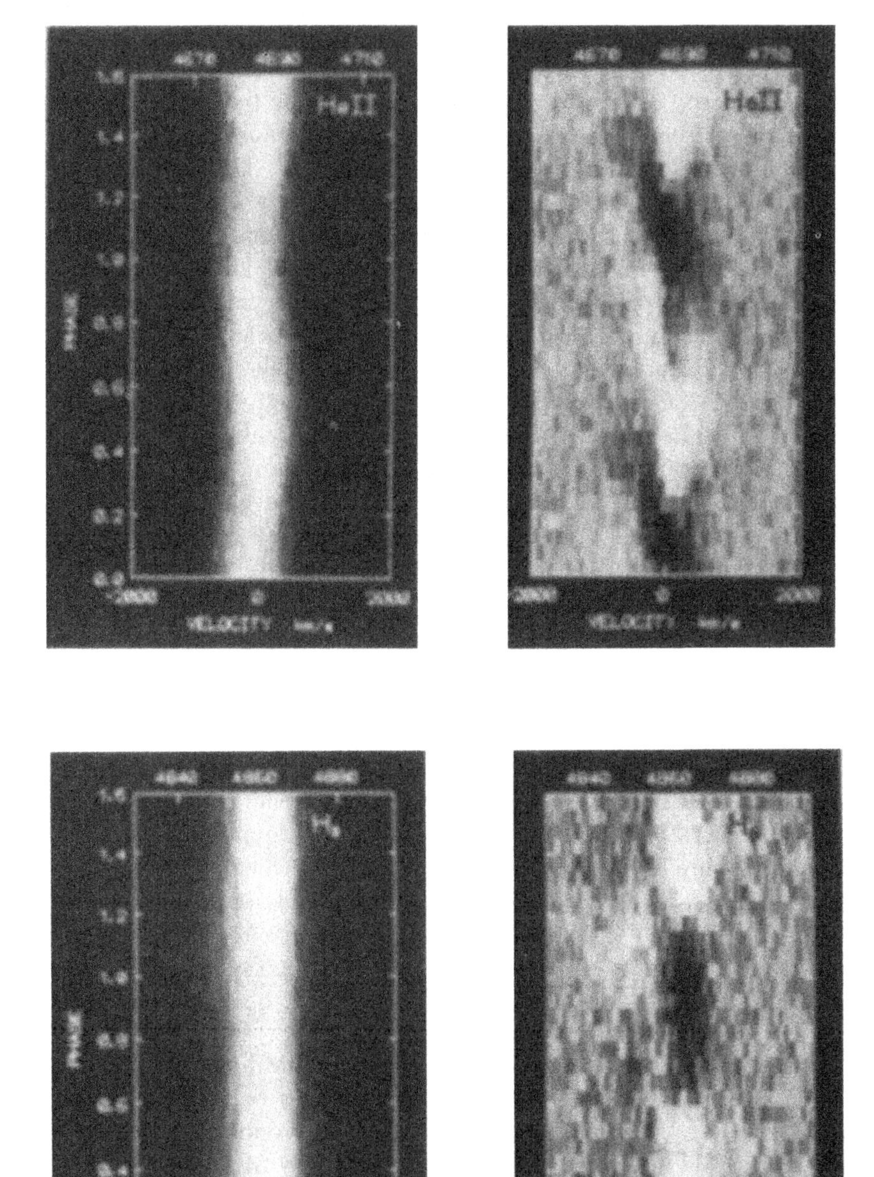

Figure 7.14 Optical spectra of FO Aqr folded on the rotation cycle. Rotational phase is plotted on the ordinate. The left hand illustration is the raw image; the right hand is an image with the mean line profile subtracted. From Hellier, Mason & Cropper (1990).

Table 7.5. *Emission-Line Widths in IPs.*

Star	FWZI ($km\ s^{-1}$)	References
GK Per	2900	1
TX Col	2800	2
TV Col	3000	3
FO Aqr	1800	4
AO Psc	1100	5
V1223 Sgr	1100	6
BG CMi	2500	7
EX Hya	7000	8

References: 1 Gallagher & Oinas 1974; 2 Buckley & Tuohy 1989; 3 Hutchings *et al.* 1981a; 4 Hellier, Mason & Cropper 1990; 5 Clarke, Mason & Bowyer 1983; 6 Watts *et al.* 1985; 7 McHardy *et al.* 1987; 8 Hellier *et al.* 1987.

modulation is confined to the low velocity core of the emission line, showing that its origin is in the outer regions of the system. In both AO Psc and FO Aqr the HeII $\lambda4686$ line (but not the Balmer emission) shows also $\omega_{rot} - 2\Omega_{orb}$ and $\omega_{rot} + \Omega_{orb}$ modulations.

7.6.4 Eclipse Disturbance

In EX Hya (Cowley, Hutchings & Crampton 1981; Hellier *et al.* 1987) there is a strong rotational disturbance in the V/R ratio of the emission lines which lasts for $\sim$0.3 of the orbit. From the geometry of the orbit the implied line-emitting region has a radius $\sim$0.8RL(1), which is in good accord with disc radii in non-magnetic systems (Section 2.6.2). In FO Aqr a V/R disturbance in HeII $\lambda4686$ led to the realization that its inclination must be quite high and that an eclipse could be discerned in many published light curves which had gone unnoticed because of the prominent $P_{rot}(1)$ modulation (Hellier, Mason & Cropper 1989). However, Martell & Kaitchuck (1991) claim that the classical rotational disturbance is not present, and the V/R effect is due to the presence of an S-wave component that probably originates in a wind.

7.6.5 Orbital S-Wave and Conditions in the Bright Spot

Orbitally modulated S-waves are seen in almost all of the IPs. The velocity amplitude of the wave in the studies with high spectral resolution can provide estimates of the radius vector of the line-emitting region. However, as the S-wave is phased differently than the orbital motion, care must be used to obtain the correct amplitude. In FO Aqr Hellier, Mason & Cropper (1990) find that the true amplitude is $\sim$400 km s^{-1} and maximum intensity occurs 20° ahead of the line of centres. This is typical of accretion disc bright spots in non-magnetic CVs. If the S-wave originated in the impact of the stream onto the magnetosphere in a discless system, then an entirely different phasing would be expected (Hellier, Mason & Cropper 1990). Similar results are obtained for EX Hya (Hellier *et al.* 1989).

Elitzur, Clarke & Kallman (1988) exploit the special circumstances offered by an IP to compute a detailed model for the Balmer emission in AO Psc. They assign the observed component modulated at P_{orb} to the bright spot region. The absence of any $P_{rot}(1)$ or P_{syn} modulations in the Balmer lines (Clarke, Mason & Bowyer 1983) shows that it is UV from the disc, rather than X-rays from the primary, that ionizes the discward face of the disc bulge. The deduced density in the line emitting bulge region is $\sim 10^{13}$ cm^{-3}.

7.6.6 Flux Distributions

Derivations of the overall flux distributions in IPs have been made by several workers (Sherrington *et al.* 1980; Hassall *et al.* 1981; Cordova *et al.* 1983; Mouchet 1983; Mouchet & Bonnet-Bidaud 1984; Szkody & Mateo 1984; Szkody 1985a,b; Cordova & Howarth 1987; Verbunt 1987; Cordova & Cropper 1988). Standard accretion disc models with inner radii $r_0 \sim$ few white dwarf radii fit the flux distributions quite well (e.g., Mouchet 1983) but derived mass transfer rates cannot be trusted because of their great sensitivity to $M(1)$ (Verbunt 1987). Examples of flux distributions for discs with three different inner radii are shown in Figure 7.15.

Verbunt (1987) has compared the UV fluxes of IPs with NLs and finds that the IPs generally have lower short wavelength fluxes. The spectral index $F_u = \log F_{1460}/F_{2280}$ lies in the range $0.03 \leq F_u \leq 0.57$ for six IPs, whereas the range for NLs is $0.20 < F_u < 0.87$. Furthermore, whereas a single power index $\alpha \sim 2.0$ fits the flux distribution quite well over the range 1200–3000 Å of the NLs, the IPs require a two–index fit, with the shorter wavelength region requiring the larger spectral index.

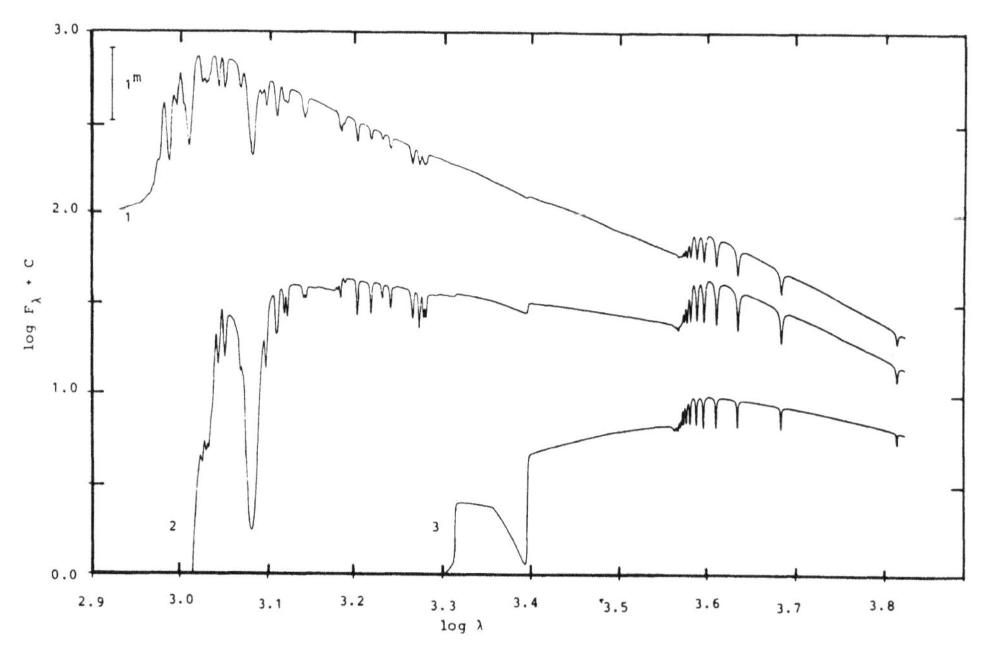

Figure 7.15 Theoretical flux distributions for truncated discs. Model 1 has log r_d = 10.6, log r_0 = 9.2; model 2 has log r_d = 10.5, log r_0 = 9.9; Model 3 has log r_d = 10.6, log r_d = 10.3, in units of cm. From la Dous (1989).

Table 7.6. *UV Fluxes from Model Accretion Discs.*

$\dot{M}$(d) ($10^{-10}M_{\odot}y^{-1}$)	Standard disc			Truncated disc		
	log F_{1460}	F_u	T_{max}(K)	log F_{1460}	F_u	T_{max}(K)
1	0.46	0.66	26 700	−0.52	0.25	11 700
10	1.4	0.77	47 500	0.92	0.58	20 900
100	2.3	0.84	84 500	2.0	0.77	37 100

This relative deficiency of short wavelength flux is readily explained by the absence of the inner regions of the disc. Verbunt compares values of F_u computed for standard accretion discs ($M_1(1) = 1$, $r_d = 3 \times 10^{10}$ cm, $i = 60°$) with ones computed for truncated discs with $r_0 = 3R(1)$. The results are shown in Table 7.6. There is a large reduction in F_{1460}, F_u and the maximum temperature T_{max} in the disc for $\dot{M}_{17}(d) \lesssim 3$ (i.e., $\dot{M}(d) < 3 \times 10^{-9} M_{\odot}$ y^{-1}). The standard disc results agree well with observations for most of the NLs, and the truncated disc values are lower by amounts similar to those seen in the IPs. Even smaller values of F_{1460} and F_u would be expected for the equilibrium accretion truncations at r_0 given by equation (7.17).

Detailed fitting of model flux distributions to individual stars reveals a variety of results that are not yet understood in terms of structural differences between systems. Decompositions of the observed flux distributions are not unique, so care must be taken when comparing one star with another. The descriptions of the AO Psc flux distribution provide a good example. The optical continuum is roughly fitted with spectral index $\alpha = 2.4$, but the IUE spectral distribution requires $\alpha = 1.2$. A good fit to the optical and most of the IUE fluxes can be obtained by a blackbody with $T = 12\,600 \pm 1500$ K (Hassall *et al.* 1981). However, an excellent fit is obtained with an $\alpha = 2.74$ distribution, reddening of $E(B–V) = 0.10$ and the addition of a blackbody contribution with $T = 13\,400$ K Figure 7.16 (Cordova *et al.* 1983). $\alpha = 2.3$ and a 9000 K blackbody fit the TV Col flux distribution (Mouchet *et al.* 1981).

Fits to the flux distributions of BG CMi and V1223 Sgr also require $\alpha \sim 2.0$ in the IUE region, but FO Aqr has almost level flux in the UV (i.e., $\alpha = 0$) which diminishes below $\sim$ 1400 Å. A standard accretion disc with $\dot{M}_{17}(d) \sim 0.06$ fits the observed flux distribution (Szkody 1985a), but it is probable that a much higher $\dot{M}$ through a truncated disc will fit equally well. A simple Planckian disc without any truncation fits the EX Hya flux distribution from 1200 Å to 2 μm very well (Sherrington *et al.* 1980), although $r_0 \approx 5R(1)$ (Sections 7.5.3 and 7.6.2).

Before GK Per was known to be an IP, Bianchini & Sabbadin (1983a,b) pointed out that its X-ray to IR continuum distribution could be fitted only with a truncated disc, from which they deduced that the primary in GK Per must possess a magnetic field. They found $r_0 = 1.2 \times 10^{10}$ cm from their fit to the flux distribution, which, from equation (7.17), gives $\omega_s = 1.8$ for $M1(1) = 1.2$. This is for quiescence, when GK Per has irregular, low-amplitude X-ray pulsations (Norton, Watson & King 1988; Ishida *et al.* 1992), and suggests that steady accretion does not occur from the inner disc to the

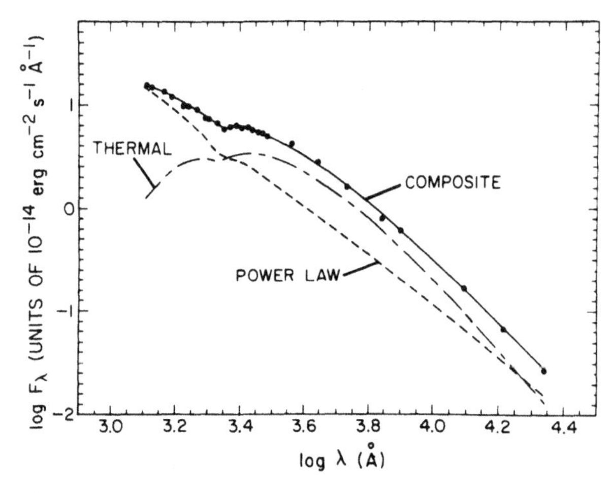

Figure 7.16 Observed flux distribution of AO Psc fitted with a combination of thermal and power-law distributions (see text). From Cordova *et al.* (1983).

primary. During an outburst of GK Per, $\dot{M}$(d) increases by a factor ~10 (Watson, King & Osborne 1985), which, through equation (7.15), reduces ω_s below unity where stable accretion can occur and accounts for the observed strongly modulated X-ray emission. If the $\dot{M}$(d) can be acertained at which unsteady accretion changes to steady accretion, then the magnetic moment $\mu(1)$ can be found from equation (7.8).

Mauche *et al.* (1990) have computed flux distributions for GK Per in outburst, integrating over empirical spectra for individual annuli. The temperature distribution in the truncated disc is found from a treatment which allows for the reduction in viscous heating in the region $r_0 < r < r_{co}$, caused by the angular momentum carried in the gas flowing along field lines, and the increase in heating for $r > r_0$ caused by dissipation of electrical currents flowing within the disc.

Whereas the maximum temperature T_{max} in an appropriate complete standard disc is 125 000 K, a magnetic model with $\dot{M}_{17}(d) = 20$ and $\mu_{33}(1) = 2$ (both considerably greater than is indicated by Table 7.4) has $T_{max} = 20\,000$ K, which fits the observed flux distribution quite well.

7.7 Outbursts

GK Per has a long orbital period (2 d) and therefore has adequate room for a large accretion disc, even though the inner regions are kept clear by the magnetosphere of the primary. The DN outbursts of GK Per (Bianchini, Sabbadin & Hamzaoglu 1982; Sabbadin & Bianchini 1983) show a very slow rise (~35 d) to maximum, which can be modelled by an instability in a large non-truncated disc (Cannizzo & Kenyon 1986) but Kim, Wheeler & Mineshige (1992) fit both outburst light curve and recurrence time with a disc having $r_0 = (2.5\text{–}3.5) \times 10^{10}$ cm and $\dot{M}(1) = 6\text{–}9 \times 10^{17}$ g s^{-1} during outburst. The spectroscopic behaviour during outburst (Bianchini, Sabbadin & Hamzaoglu 1982) differs from ordinary DN: all lines remain strongly in emission; HeII 4686 Å becomes much stronger than Hβ, the Balmer lines are narrower (625 km s^{-1}) than at quiescence. These changes have not been modelled, but are evidently related to the magnetically controlled accretion process, which produces enhanced hard X-ray emission during outburst (Table 7.4; Yi *et al.* 1992). Similar spectroscopic behaviour occurs in BV Cen and V1017 Sgr (Sekiguchi 1992b).

TV Col (Szkody & Mateo 1984; Schwarz *et al.* 1988; Hellier & Buckley 1993) and V1223 Sgr (van Amerongen & van Paradijs 1989) have both shown outbursts with a range of ~2 mag and lasting only 6–12 h. In the case of TV Col there were two outbursts observed within 10 d, which suggests that they may be generally quite frequent (and would make IPs a useful target for intensive amateur study). In both stars the spectra during outburst show greatly enhanced emission-line fluxes.

One of the TV Col outbursts was observed simultaneously with IUE and optically and showed a ten-fold increase in strength of the UV emission lines and the development of P Cygni profiles. The continuum power-law component steepened to $\alpha \approx 3.0$ with a concomitant increase of ~2000 K in the blackbody component. This shows that the outburst was an inner disc event (with associated additional heating of the reprocessing site), which is in accord with the brevity of the outburst and the *increase* in emission-line strength, which are completely unlike DN outbursts that involve the whole disc. Hellier (1993a) points out that enhanced accretion during outburst of an IP will arrive near the poles, rather than at the equator as in a DN, and lead to more efficient irradiation of the disc.

Outbursts of EX Hya provide a rare opportunity to investigate the effects of greatly increased $\dot{M}$ in an IP. EX Hya is normally at $m_V \sim 13.0$, with quite frequent short lived (~days) minor outbursts to ~12.2 and major outbursts to $m_V \sim 9.3$–9.6 with a mean interval ~600 d. Their brevity, however, almost certainly results in some having occurred unrecorded. The rise to maximum takes ~12 h, and recovery to quiescence takes at most 2 d. As such, the outbursts in EX Hya resemble neither normal nor superoutbursts of DN (Hellier *et al.* 1990; Reinsch & Beuermann 1990; Buckley & Schwarzenberg-Czerny 1993).

The absence of DN-like outbursts in the IPs other than the four listed above could be a result of either $\dot{M} > \dot{M}_{crit2}$ (Section 3.5.3.4) or that their magnetic nature suppresses the effect. The (rather uncertain) values of $\dot{M}$ listed in Table 7.4 are all $\sim \dot{M}_{crit2}$, so in reality many IPs may well have $\dot{M} > \dot{M}_{crit2}$. Taken at face value, Table 7.4 indicates that EX Hya and GK Per have $\dot{M} < \dot{M}_{crit2}$, so their outbursts are not unexpected. TV Col and V1223 Sgr, on the other hand, may have $\dot{M} > \dot{M}_{crit2}$, in which case disc instabilities would not be expected.

At maximum light in EX Hya there is little or no $P_{rot}(1)$ modulation of brightness, but at $m_V \sim 11\frac{1}{2}$ the modulation is quite evident (Bond *et al.* 1987a; Hellier *et al.* 1989; Reinsch & Beuermann 1990). The effect is one of adding light from an unmodulated outbursting accretion disc to a $P_{rot}(1)$ modulation of increased brightness (because of the higher $\dot{M}$ in the accretion zone). In contrast, the spectroscopic changes during outburst (cf. Section 7.6.2) are unexpected. The emission lines increase greatly in equivalent width (to 125 Å for Hα near maximum light) but their $P_{rot}(1)$ modulation is diminished or absent. The increase in strength of the lines is confined to regions near the core; this and the absence of modulation demonstrate that the central regions of the disc are emitting relatively less line radiation during outburst.

A broad component to the line profiles, producing an S-wave at P_{orb} (not $P_{rot}(1)$ as at quiescence), appears during outburst. This is in addition to an S-wave which originates in the bright spot on the rim of the disc, as at quiescence. The broad line S-wave is interpreted by Hellier *et al.* (1989) to be the result of the stream flow over the face of the disc (see Sections 2.4.3 and 7.5.1), striking the rotating magnetosphere to produce a second bright spot (Figure 7.17). As this effect does not occur at quiescence, it appears that the *major* outbursts in EX Hya have associated higher mass transfer rates from the secondary and are therefore allied to superoutbursts (Section 3.6.6.4). As all DN below the period gap are SU UMa stars, this is a satisfactory conclusion, but the low amplitude and short duration of outbursts in EX Hya, which are *major* perturbations of the usual superoutburst behaviour, may provide a clue to the nature of the superoutburst itself.

The polars are not definitely known to show short lived outbursts of the kind evinced by TV Col and V1223 Sgr, but the 'superhigh' states (Section 6.3.1) may be a manifestation of enhanced $\dot{M}(2)$, which appears also to be required in at least some of the IP outbursts.

The one definite SU UMa star that is a suspected IP, SW UMa, shows no $P_{rot}(1)$ modulation in either optical or X-ray emission during outburst (Robinson *et al.* 1987; Szkody, Osborne & Hassall 1988). Furthermore, unlike EX Hya in outburst, there is no HeII λ1640 emission in SW UMa. The explanation is probably straightforward: during superoutburst $\dot{M}(d)$ in SW UMa increases over its quiescent value by a factor ~300 (Szkody, Osborne & Hassall 1988) and therefore the inner edge of the accretion disc, which is at ~2R(1) at quiescence and whose radius is $\propto \dot{M}^{-2/7}$ (equations (6.5) and (7.3)), is pushed down to the surface of the primary, eliminating magnetically-controlled accretion. In contrast, the $\dot{M}$ increase of a factor ~20 during an EX Hya outburst is evidently insufficient to reduce its $r_0(\sim 3 \times 10^9$ cm : Section 7.5.3) to $R(1)$.

With the insight gained from the EX Hya outburst observations it is possible to suggest a reinterpretation of T Leo and HT Cas observations. Young, Schneider & Schectman (1981b) found that the wings of emission lines of HT Cas in quiescence have maximum RV at orbital phase 0.84, instead of the 0.75 expected from orbital motion.

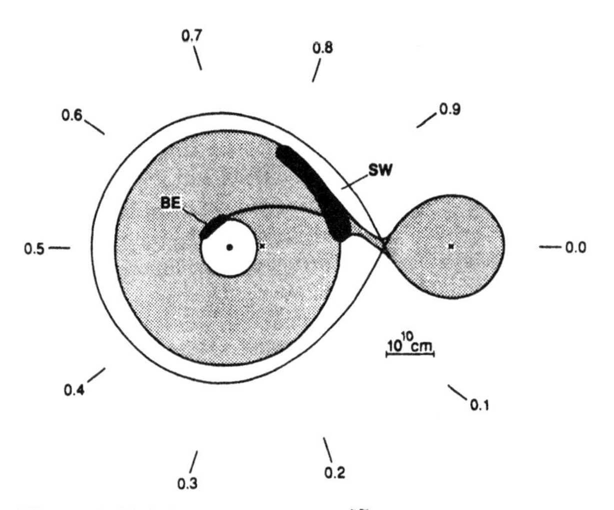

Figure 7.17 Schematic view of EX Hya during outburst showing stream overflow and the location of the S-wave (SW) and broad line (labelled BE) regions. From Hellier *et al.* (1989)

They correctly state that no distortion from disc bright spot emission should be expected in the wings; but if most of the emission is coming from the second bright spot (as in EX Hya during outburst) then the anomalous phasing is understood (see Figure 7.17). Similarly, Shafter & Szkody (1984) find an anomalous velocity phase shift in T Leo which they liken to HT Cas. In both HT Cas and T Leo, therefore, the spectroscopic observations at *quiescence* are suggestive of stream overflow and impact onto a magnetosphere. As there is clear evidence for very extensive discs in these two stars (e.g., both are SU UMa type dwarf novae) but the overflowing stream is able to strike the magnetosphere, they probably have magnetic fields comparable to that of EX Hya.

7.8 Radio Emission

BG CMi has been observed to flare at 5 GHz (Pavelin, Spencer & Davis 1994), reaching a flux of 2 mJy.

8

DQ Her Stars and Other Rapid Rotators

These flashes on the surface are not he. He has a solid base of temperament.
Alfred Lord Tennyson. *The Princess.*

The DQ Her stars constitute a subset of the IPs. Observationally they are defined as rapidly rotating IPs lacking hard X-ray emission. The latter property probably results from the proximity of the magnetospheric radius to the surface of the primary. For such systems the accretion shock may be almost at r_0 (Wu & Wickramasinghe 1991) and the infall from r_0 to the surface is unlikely to be radial. For equilibrium accretion equation (7.5) shows that any IP with $P_{rot}(s) \lesssim 120$ s is likely to be a DQ Her star. The three IPs without hard X-rays do indeed satisfy this condition.

Rapid quasi-periodic luminosity variations seen in DN during outburst, and in some NLs, are shown in this chapter to be probably an extension of the DQ Her class to lower field strengths.

8.1 DQ Her

As described in Section 1.3, DQ Her, the remnant of Nova Herculis 1934, was for many years unique in possessing an observable highly coherent brightness modulation with a remarkably short period.

8.1.1 *Photometry of the 71 s Oscillations*

Since their discovery in 1954, the 71 s oscillations in brightness of DQ Her have undergone some seventeen million cycles. This is sufficient, not only for accurate determination of P and $\dot{P}$, but also to investigate possible variations in $\dot{P}$. Phase measurements of the 71 s cycles have been contributed principally by Walker (1961), Nather & Warner (1969), Warner *et al.* (1972), Herbst, Hesser & Ostriker (1974), Nelson (1975, 1976), Lohsen (1977), Patterson, Robinson & Nather (1978) and Balachandran, Robinson & Kepler (1983). The last named authors, using all of these published observations, give the following emphemeris (after correction (Robinson 1990)) for barycentric times of maxima:

$$T_{max} = \text{HJED}\,2434954.78589(\pm 31) + 8.2252168(\pm 2) \times 10^{-4}\text{E} - 3.23(\pm 4) \times 10^{-16}\text{E}^2 + 3.9(\pm 2) \times 10^{-24}\text{E}^3. \quad (8.1)$$

There is some evidence (Warner 1988a) that the residuals from this ephemeris follow a ~14 y cycle.

The average pulse profile is nearly sinusoidal (Kiplinger & Nather 1975). Investigation of the possibility that the true period of DQ Her is 142 s, with two nearly equal maxima per cycle, has had a varied history. Nelson (1975) found some evidence, of low statistical signficance, which was contradicted by the power spectrum of a very extensive set of observations by Kiplinger & Nather (1975). The question was reopened by Schoembs & Rebham (1989) who found different average amplitudes for odd and even cycles in the U band, although BVRI do not show the effect.

An unexpected diagnostic tool for probing the nature of the 71s oscillations and the structure of DQ Her was found by Warner *et al.* (1972) who, setting out to use phase measurements of the 71s cycles to measure the light travel time across the orbit of the primary, found the effect masked by very large phase shifts through eclipse. This was studied in more detail by Patterson, Robinson & Nather (1978), with more extensive observations made at a time when the 71 s oscillations were particularly prominent (Figure 1.10). Further observations, in UBVRI photometry, have been made by Schoembs & Rebham (1989).

The phase shifts through eclipse (Figure 8.1) are difficult to measure because they are largest where the oscillation amplitude is least. Nevertheless, it is clear that the phase

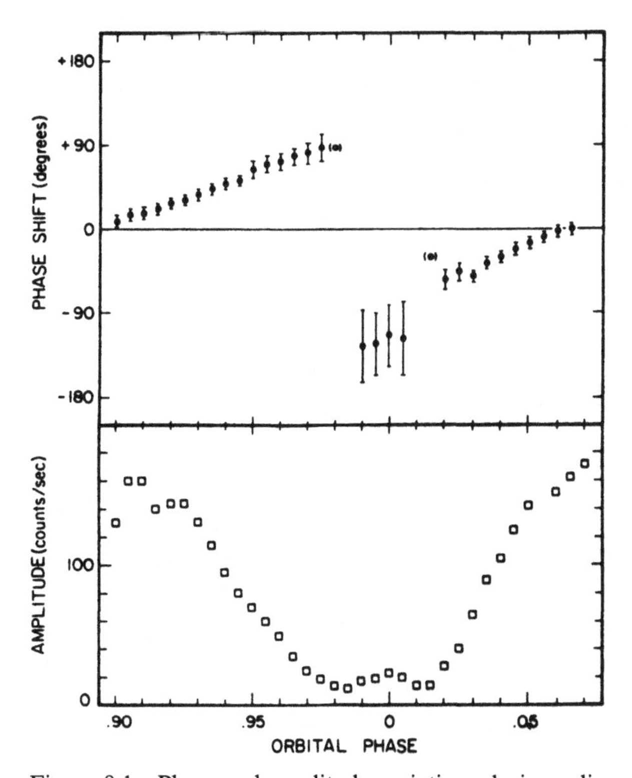

Figure 8.1 Phase and amplitude variations during eclipse of the 71 s oscillations in DQ Her. This is a weighted average of seven eclipses. Bracketed points are of low weight. From Patterson, Robinson & Nather (1978).

shift is a smooth ~90° change from orbital phase 0.9 (which coincides with the start of eclipse) to phase 0.0, followed by a rapid swing of ~180° and subsequent recovery to zero phase shift at the end of eclipse. There are also smaller systematic phase shifts and substantial systematic amplitude variations around orbit outside of eclipse (Patterson, Robinson & Nather 1978; Schoembs & Rebham 1989).

The colours of the 71s pulsed component have been measured by Chanan, Nelson & Margon (1978) who find that for $\lambda \gtrsim 4000$ Å the flux distribution can be fitted by a 7700 ± 600 K blackbody. There may, however, be a non-thermal distribution at $\lambda < 4000$ Å which has not yet been observed (cf. the flux distribution in DNOs: Section 8.5.2).

8.1.2 Polarimetric Observations of DQ Her

Broad (3700–5800 Å) circular polarimetry and linear polarimetry of DQ Her were obtained by Swedlund, Kemp & Wolstencroft (1974) and Kemp, Swedlund & Wolstencroft (1974). They found 0.1% amplitude circular and 0.3% linear polarization (largest in the UV), modulated at twice the 71 s photometric period. These 142 s variations show no phase shift through eclipse, which led the authors to suggest that the polarized light comes directly from the white dwarf, which must then *remain visible* at mid-eclipse (i.e., a *partial* eclipse of the disc is required). The low values of the polarizations and the results drawn from them, which conflict with current models of the system (Section 8.1.4), make it difficult to accept them at face value. There is an urgent need to repeat these measurements with the largest telescopes and efficient polarimeters, to average over many more than the few hundred 71s cycles employed by Swedlund *et al.* and to explore any wavelength dependence.

8.1.3 Spectroscopic Observations of DQ Her

The emission-line spectrum of DQ Her is that of a normal NR (Section 4.2). Greenstein & Kraft (1959) demonstrated the presence of a classic rotational disturbance through eclipse of the HeII 4686 Å line, which provides direct evidence for the existence of an accretion disc in DQ Her and shows that the 4686 Å line originates in the disc.

Chanan, Nelson & Margon (1978) obtained time-resolved spectrophotometry of DQ Her which they summed modulo the 71 s period to show that the HeII 4686 Å line and NIII–CIII 4640 Å blend are pulsed at 71 s with ~2.4% pulsed fraction – considerably greater than the ~0.9% pulsed fraction of the continuum. The Balmer lines are not pulsed.

The HeII line, resolved at ~5 Å, shows a strong variation of pulse phase across its profile, with pulsations at the centre of the line in phase with those of the continuum (Figure 8.2). In the short wavelength wing of the line the pulsations lag behind the continuum pulsations. This is exactly the behaviour that would be expected if a beam of ionizing radiation sweeps around the disc in the same direction as the disc rotation, with the maximum of continuum brightness coinciding with the beam pointed away from the observer – i.e., with the 71 s continuum oscillations arising from the front-to-back asymmetry of the concave disc. There is no systematic variation of pulse phase across the 4640 Å profile because, being a blend, there is no simple mapping of the Doppler velocity to wavelength.

The absence of pulsations in the Balmer lines is a result of the different ionization

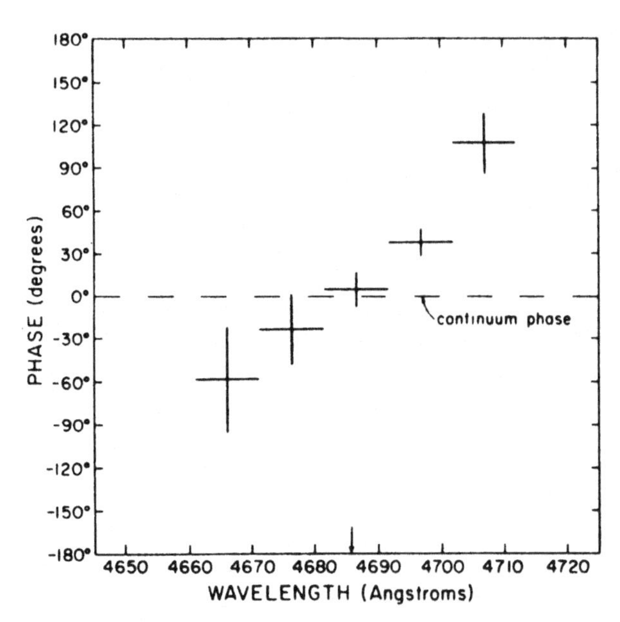

Figure 8.2 71 s pulsation phase across the profile of HeII λ4686 in DQ Her. The arrow indicates the central wavelength of the line. From Chanan, Nelson & Margon (1978).

conditions for hydrogen and heavier ions. The recombination time τ_{rec} is $\sim (\alpha N_e)^{-1}$, where α is the *total* recombination coefficient. For HII at 20 000 K, $\alpha = 2.5 \times 10^{-13}$ $cm^3\,s^{-1}$ and for HeIII, $\alpha = 1.4 \times 10^{-12}$ $cm^3\,s^{-1}$ (Osterbrock 1989). With typically $N_e > 10^{13}$ cm^{-3} (Section 2.7.2) we have $\tau_{rec} < 2$ s for hydrogen and $\tau_{rec} < 0.4$ s for helium. The emission-line strengths from both species will therefore respond to variations in ionization conditions with a time lag that is very small compared with the 71 s modulation caused by the rotating beam in DQ Her. However, if H is almost *totally* ionized, but He, C and N are only *partially* doubly ionized, then only the ionization conditions for the latter elements are significantly modulated. (See Section 7.6.5 for an alternative interpretation used for an IP.)

8.1.4 Models of the 71 s Oscillations in DQ Her

The early explanations for the 71s oscillations in the DQ Her light curve invoked radial oscillations (Walker 1961; Kraft 1963) or non-radial oscillations (Baglin & Schatzman 1969; Warner *et al.* 1972) of the primary. In analogy with models proposed for pulsing X-ray sources (e.g., Cen X-3, Her X-1 (Pringle & Rees 1972)), Bath, Evans & Pringle (1974) suggested an accreting oblique dipole rotator for DQ Her. This was also proposed by Lamb (1974) and Katz (1975). (Both the non-radial oscillator and oblique rotator produce anisotropic emission from a heated region rotating around the primary. Strictly speaking, the non-radial oscillator model has not been eliminated from consideration for the DQ Her stars, but it certainly is not correct for the IPs and should, by analogy, be at least deemed unnecessary for the DQ Her stars.)

As the phase shift through eclipse shows that the whole disc is in some way involved in the 71 s modulation, is relevant that Kato (1978) has demonstrated that an accretion disc cannot oscillate coherently (nor with the high stability of period possessed by DQ Her).

The phase shift through eclipse has been modelled as the effect of reprocessing of a high energy beam: Patterson, Robinson & Nather (1978), Chester (1979), Petterson (1980) and O'Donoghue (1985). The $\pm 90°$ phase shift is very easily understood as the progressive eclipse of a disc illuminated by a rotating beam – the effect of the eclipse is gradually to obscure part of the sinusoidal pulse profile, causing a shift in phase (obtained from least squares fitting of a sinusoid at such low signal-to-noise ratios) which flips from $+90°$ to $-90°$ as the final parts of the following lune of the disc disappear and the first parts of the leading lune reappear. The predicted variation of the 71 s oscillation phase is very sensitive to inclination; the fit to the phases shown in Figure 8.3 (the amplitude variations are fitted simultaneously) gives $i \approx 89°$ (Petterson 1980). As the steady state disc used in the model has a rim which subtends a half-angle of 1.4° as seen from the primary, the entire nearside inner surface of the disc *and the primary* are obscured by the disc. This accounts for the absence of any detectable X-ray emission from DQ Her ($< 1 \times 10^{-13}$ erg cm^{-2} s^{-1} in 0.16–4.5 keV, or $L_X \lesssim 1 \times 10^{30}$ erg s^{-1} (Cordova, Mason & Nelson 1981)).

O'Donoghue (1985) has shown that the variations of phase and amplitude of the 71 s optical pulse around the orbital cycle can be modelled by reflection from, and obscuration by, an outer disc rim of irregular height. With its extreme inclination, the pulse shapes in DQ Her are particularly sensitive to the azimuthal profile of the rim.

8.1.5 Radio Emission

A flare at 5 GHz, reaching 4 mJy and falling to 1 mJy within 20 min has been observed by Pavelin, Spencer & Davis (1994).

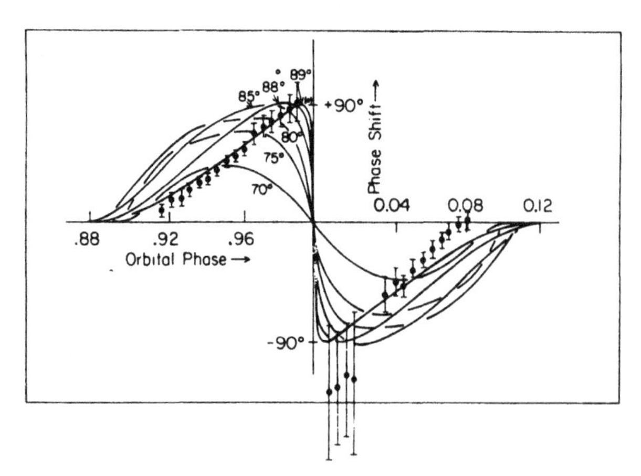

Figure 8.3 Observed DQ Her 71 s phase shift (Figure 8.1) compared with model predictions for different inclinations. From Petterson (1980).

8.2 AE Aqr

The NL AE Aqr, one of the brightest CVs, played an important rôle in early studies of the nature of interacting binaries (Section 1.4.2). It continues as a uniquely bright source of coherent rapid luminosity variations. Its light curve shows extended intervals (1–2 h) of flaring activity, during which it brightens by factors of ~3–5 (Zinner 1938; Lenouvel & Daguillon 1954; Walker 1965b; Chincarini & Walker 1974; Patterson 1979b). With its long orbital period (9.9 h) the K4 secondary contributes substantially to optical continuum and emission-line radiation.

8.2.1 Photometry of AE Aqr

In 1978 Patterson (1979b) discovered that AE Aqr has a persistent brightness modulation with a period of 33.08 s and amplitude of 0.2–0.3%, increasing to ~1% during flares. Comparable amplitudes are seen at the first harmonic (16.54 s). Although there is a small amount of 'phase wandering' on a time scale of hours, which is correlated with orbital phase (Patterson 1979b), the 33 s oscillations have been shown to be stable in the long term (Patterson, Beuermann & Africano 1988). The latter authors found $\dot{P} < 1 \times 10^{-14}$, but De Jager *et al.* (1994) show that there has been a cycle count ambiguity and that in fact $\dot{P} = 5.64 \times 10^{-14}$.

The 16 s oscillations show greater stability in phase than the 33 s fundamental. The formula for barycentric ephemeris time of maximum of the 16.5 s oscillations is (De Jager *et al.* 1994)

$$T_{\max} = \text{BJED } 2445172.000042 + 1.914163192 \times 10^{-4}\text{E} + 2.70(\pm 0.01) \times 10^{-18}\text{E}^2. \quad (8.2)$$

To this must be added an orbital period modulation with amplitude 2.04 ± 0.13 s (De Jager *et al.* 1994).

By analogy with DQ Her, Patterson proposed an oblique dipole rotator model for AE Aqr, with reprocessed light able to reach us from both poles. With an optically thick disc, reprocessed light from the concave surface of the disc reaches us from the illumination of only one beam; but beams from both poles can illuminate the inner edge of the disc (and as $i \sim 55°$ we may receive some radiation directly from the accretion zones on the surface of the white dwarf). This could account for the different stabilities of the fundamental and the harmonic.

Welsh, Horne & Oke (1993) find that the two maxima per 33 s cycle are not quite equally spaced. The flux distribution of the modulated component can be fitted equally well with an $F_\nu \propto \nu^{0.9}$ law or a blackbody with $T = (1.2\text{–}5.7) \times 10^4$ K and area comparable to that of the primary.

During episodes of optical flaring the observed amplitude of the periodic signals increases (runs 2202 and 2227 of Figure 8.4) but, as a fraction of the contribution of luminosity from the primary and its accretion disc *alone*, the fractional amplitude remains constant. This is similar in behaviour to EX Hya during outburst (except at its peak: Section 7.7), and shows that both disc and accretion zone luminosities increase proportionally during the increase of $\dot{M}$ that causes the flares. From multicolour photometry van Paradijs, Kraakman & van Amerongen (1989) find that the time-average flux in flares is three times that from the disc in quiescence. The latter is seen as

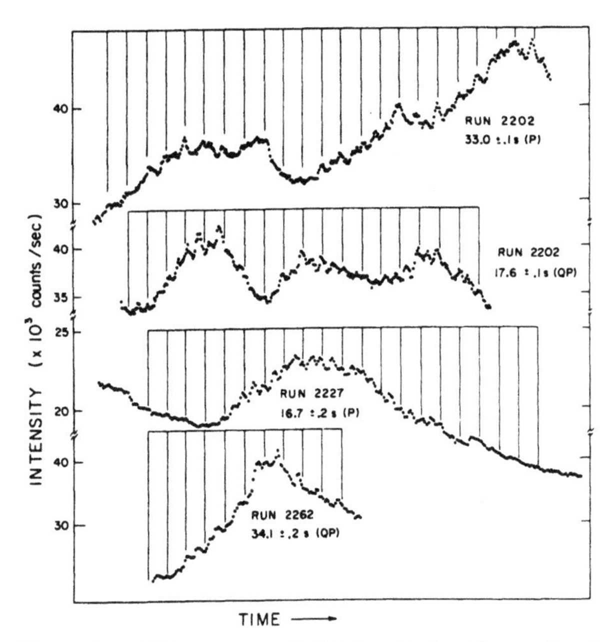

Figure 8.4 Light curves of AE Aqr during flares, showing the enhanced periodic and quasi-periodic oscillations. Vertical lines mark the pulse maxima. From Patterson (1979b).

a blue continuum (contributing only 7% in the V band) with integrated flux $\sim 8 \times 10^{31}$ erg s^{-1}. The total time-averaged $\dot{M}(2)$ is therefore $\sim 1.5 \times 10^{15}$ g s^{-1}, which is very low for such a long period CV. (Patterson (1984) and Warner (1987a) deduced $\dot{M}(\mathrm{d}) \sim 1 \times 10^{16}$ g s^{-1}, but this assumes a complete steady state disc, whereas in AE Aqr the disc is certainly truncated.)

Walker (1981) has observed Balmer line flares, occurring at a frequency ~0.13 h^{-1}, of great width (~7000 km s^{-1} FWZI). These are quite separate from the frequent narrow and medium breadth emission-line flares that are associated with the continuum flaring (Chincarini & Walker 1974). From their great width, these broad flares must occur close to the primary and may be the result of instabilities in the magnetic threading zone at the inner radius of the disc. Welsh, Horne & Oke (1993) find the spectrum of flares indicates optically thin gas with $T \sim 1.0(\pm 0.2) \times 10^4$ K.

HST observations show strong flaring activity and a pulsed fraction of up to 40% for the 33 s oscillations. Here the oscillation amplitude is not proportional to flare luminosity, which suggests that the flares are not simply unstable accretion onto the primary. The oscillation light curve can be deconvolved to produce an intensity map of the surface of the primary. This shows two bright spots, 180° apart in longitude and 20° above and below the equator. These are the accretion zones of the dipole accretor (Horne 1993b; Eracleous *et al.* 1994).

There has been considerable confusion over the relative phasing of the emission-line and absorption-line orbits and the optical and X-ray pulse-timing orbits (Feldt & Chincarini 1980; Robinson, Shafter & Balachandran (1991); Eracleous, Patterson & Halpern 1991; De Jager 1991a; Marsh 1992). This has been eliminated by more recent work (Welsh, Horne & Oke 1993; Welsh, Horne & Gomer 1993; De Jager *et al.* 1994;

Reinsch & Beuermann 1994) in which it is shown that (a) because of contamination from gas with independent motion, the emission-line orbit is not a reliable tracer of the primary and (b) the pulse-timing orbits are correctly anti-phased with the absorption line orbit. The 33 s optical and X-ray oscillations therefore originate in the vicinity of the primary, and not at some reprocessing site as had earlier been deduced.

8.2.2 *Polarization of AE Aqr*

Szkody, Michalsky & Stokes (1982) found linear polarization in AE Aqr, varying from 0.1 to 0.9 ($\pm$0.1)% in the B and V bands over a few days. Cropper (1986b) found 0.15–0.20% in a similar position angle and in addition detected 0.05($\pm$0.01)% time-averaged circular polarization. The slowly varying linear polarization may originate in scattering from the disc, but the circular polarization is indicative of a magnetic field in AE Aqr probably $> 10^6$ G. From time-resolved polarimetry, Cropper was able to set a limit $\leq$40% for the amplitude of the circular polarization in the 33 s pulsed component.

8.2.3 *X-Ray Observations of AE Aqr*

AE Aqr was detected in the 0.1–4.0 keV X-ray band by the Einstein Observatory (Patterson *et al.* 1980). The 33 s modulation is seen at the same period as the optical pulse, without any first harmonic, with amplitude $\sim$25% independent of energy (Patterson, Beuermann & Africano 1988; Eracleous, Patterson & Halpern 1991; De Jager 1991a). De Jager (1991a) found evidence for the presence of a 33.11 s period, as well as 33.08 s. Marsh (1992) attributes this to an orbital sideband, which implies an X-ray scattering site in addition to the direct beam.

From EXOSAT observations Osborne (1990) finds $kT < 1.8$ keV for optically thin emission, and a bolometric luminosity $\sim 3 \times 10^{31}$ erg s^{-1}. The low temperature illustrates the difference between this DQ Her star and the intermediate polars, all of which have $T \gtrsim 10$ keV (Osborne 1988): accretion onto the primary in AE Aqr is evidently not quasi-radial.

8.2.4 *Radio Observations of AE Aqr*

Bookbinder & Lamb (1987), Bastian, Dulk and Chanmugam (1988) and Pavelin, Spencer & Davis (1994) observed radio emission from AE Aqr in the frequency range 1.5–22.5 GHz, with strong flaring behaviour on time scales $\sim$20 min–1 h (similar to the optical flares) and flux enhancements by factors as large as 12 at 15 GHz; the degree of variability increases with frequency. The flux increases up to at least 240 GHz (Abado-Simon *et al.* 1993). From the high brightness temperature and lack of circular polarization the emission probably arises as incoherent synchrotron radiation. Dimensions of the emitting region 1×10^9–3×10^{10} cm and variable magnetic field $B \sim 25$–250 G are consistent with the observations. The origin of the activity is uncertain: as the secondary in AE Aqr is observed to flare in the Balmer lines (Chincarini & Walker 1981) it is a possible source of the required relativistic electrons; but the special character of AE Aqr as a magnetic accreting system is more probably involved.

The suggestion (Bastian, Dulk & Chanmugam 1988) that the non-thermal radio emission is located in expanding plasma clouds has been confirmed in VLBI observations, where expansion velocities $\sim 0.01c$ are seen during a flare (Neill 1992).

8.2.5 Gamma-Ray Observations of AE Aqr

Chanmugam, Brecher (1985) and Ruderman (1987), arguing by analogy with Cyg X-3 and Her X-1 (e.g., Dowthwaite *et al.* 1984), suggested that polars and IPs could be a source of TeV gamma-rays. The radio emission observed in AE Aqr led Bastian, Dulk & Chanmugam (1988) to describe it as 'low-power Cyg X-3'. Although intrinsically 10^5 times less luminous than Cyg X-3, AE Aqr is $\sim$100 times closer. The lower rotational frequency (factor of $\sim$30) than the neutron star in Cyg X-3 facilities observational procedures.

The production of gamma-rays in compact magnetic binaries through magnetic wind-up and reconnection (Treves 1978), production of potential gaps in the magnetosphere of the rapidly rotating primary (Eichler & Vestrand 1984) or accretion onto a unipolar inductor (Chanmugam & Brecher 1985) has been proposed, and includes predictions of the possibility of gamma-rays from CVs – based on the similar magnetic moment of neutron stars and magnetized white dwarfs. Only in AE Aqr have TeV gamma-rays so far been detected with some certainty.

Meintjes (1990), Brink *et al.* (1990, 1991) De Jager *et al.* (1990, 1991), Raubenheimer *et al.* (1991), Bowden *et al.* (1991,1992) and Meintjes *et al.* (1992,1994) have observed the 33.08 s modulation in TeV emission from AE Aqr, with a luminosity $\sim 1.5 \times 10^{32}$ erg s^{-1}. The TeV emission shows flaring enhancements of factors up to 60. Simultaneous observations with independent gamma-ray telescopes at one site, and with gamma-ray and optical telescopes, show similar enhancements of the 33 s signal during flares but indicate that TeV emission occurs immediately before or after optical flares. The 33.08 s TeV pulse is in phase with the optical and X-ray pulses during quiescent emission; at the time of flares the TeV pulses appear at longer periods ($\Delta P/P \lesssim 0.01$) and are quasi-periodic, just as with the optical pulses (Section 8.7).

8.2.6 The Energetics of AE Aqr

The observed rate of spin-down of the primary implies an energy supply $I(1)\Omega(1)\dot{\Omega}(1) \sim 6 \times 10^{33}$ erg s^{-1}, which is $\sim$ 75 times the quiescent accretion luminosity. With such a large spin-down AE Aqr is far from spin equilibrium: the strong braking torque arises because in quiescence the accretion disc lies entirely outside the corotation radius (eliminating any spin-up torque) and is very large in this long P_{orb} system. Storage of mass in the disc increases the pressure on the magnetosphere, ultimately allowing accretion onto the primary from the vicinity of r_{co}. The luminosity of the strongest TeV flares is $\sim 10^{34}$ erg s^{-1}, which shows that at times the entire spin-down power can be channeled into particle acceleration.

Suggested proton acceleration mechanisms in AE Aqr include the effect of differential rotation between star and disc (Cheng & Ruderman 1991) and the Fermi mechanism in unstable magnetospheric flow with development of collisionless shocks (De Jager 1991b, Meintjes *et al.* 1992). Subsequent impact of protons with energies $\gg 1$ TeV onto regions of surface density $\gtrsim 50$ g cm^{-2} generates neutral pions which emit gamma-rays in their fast decay.

8.3 V533 Her

V553 Her, the remnant of Nova Herculis 1963, was found by Patterson (1979c) in 1978 to possess a brightness modulation with period 63.63309 s and mean amplitude 1%.

Although the amplitude was variable (0.4–2% in less than an hour) the oscillation was quite stable ($\dot{P} < 2 \times 10^{-11}$; Lamb & Patterson (1983) later gave $\dot{P} < 3 \times 10^{-13}$). In 1980, V533 Her still oscillated and was coherent at its fainter ($m_V \sim 15.7$) phases but some phase variation was present at a bright ($m_V \sim 15.0$) excursion (Middleditch & Nelson 1980). In 1982, with $m_V \sim 14.7$, no oscillations to a limiting amplitude of 0.05% were detectable (Robinson & Nather 1983); this brighter state has persisted and no oscillations have been seen since.

The disappearance of the V553 Her oscillations led Robinson & Nather to propose that their cause was non-radial pulsation of the primary, rather than an oblique rotator. Since then, however, even AE Aqr has been found with its 33 s oscillations absent (Steiman-Cameron & Imamura 1988). It remains an open question which model is correct, but noting the large short time scale amplitude variations, which cannot be intrinsic to a single-mode non-radial pulsator, the presence of variable obscuration (in the case of the pulsator) or variable $\dot{M}(1)$ (in the case of the rotator) is indicated and a long term change in either could result in disappearance of the modulated signal. In particular, the increased phase noise at brighter magnitude (i.e., at higher $\dot{M}$) could mean that the magnetosphere of the primary is being compressed so close to the surface of the primary that complexities of the magnetic field destroy most of the anisotropy of the emission (Section 7.2.1). Indeed, equation (7.17) gives $r_0/R(1) = 3$ for equilibrium accretion, so modest increases in $\dot{M}$ above the mean may well destroy the source of pulsed radiation.

The inclination of V533 Her is ~60°, so the non-detection of X-rays, which gives $L_x \lesssim 1.8 \times 10^{31}$ erg s^{-1} (Cordova, Mason & Nelson 1981), implies at most the same X-ray luminosity as AE Aqr, which probably has a very much lower $\dot{M}$.

Other NRs have been examined for rapid oscillations (Robinson & Nather 1983) without success. The intermittent behaviour of V533 Her suggests that all NRs should be regularly monitored.

8.4 WZ Sge

The SU UMa star WZ Sge may be a magnetic accretor but there is no model which accounts for all of the observations; it will therefore be described only briefly. Brightness modulations at 28.952 s and 27.868 s were discovered by Robinson, Nather & Patterson (1978); in general either one or the other oscillation is present. Patterson (1980) found that the 27.87 s mode was dominant prior to the 1979 superoutburst of WZ Sge and that other oscillations were present that might be orbital sidebands or QPOs (Section 8.7). The 27.87 s modulation maintains coherence on a long time scale, but there are typically ~30° departures from predicted phase from night to night. Over a one-year baseline there was a change of period of $\sim 2 \times 10^{-4}$ s, or $\dot{P} \sim 10^{-11}$. Subsequent to the 1979 superoutburst the 28.95 s oscillation was for a while dominant and had $\dot{P} = 2 \times 10^{-10}$ (Middleditch & Nelson 1979; Cordova & Mason 1983), but no clear periodicity has been present in the last decade.

The colours at $\lambda \gtrsim 4000$ Å of the 27.87 s pulsed component can be fitted with an 8900 ± 1600 K blackbody (Middleditch, Nelson & Chanan 1978), but there could be an undetected strong UV flux.

As there are two prominent periodicities, and they do not form a simple relationship with the orbital period, one or both may be non-radial oscillations of the primary.

Table 8.1. *Properties of DQ Her Stars.*

Star	P_{orb} (h)	m_v (quiesc)	$P_{rot}(1)$ (s)	$\dot{P}_{rot}(1)$	$< \dot{M}_{17}(1) >$	$\mu_{32}(1)$
DQ Her	4.65	14.2	71.07	-8.1×10^{-13}	2.5	3.0
V533 Her	5.04	14.3	63.63	$<3 \times 10^{-13}$	4	3.0
AE Aqr	9.88	10.0–12.5	33.08	5.64×10^{-14}	40	4.1

Alternatively, WZ Sge may be both a DQ Her star and a non-radial oscillator. An upper limit of 15% has been set on the pulsed fraction of the hard X-ray emission from WZ Sge

8.5 Magnetic Fields in the DQ Her Stars

The general properties of the DQ Her stars are given in Table 8.1. Equation (7.14), with appropriate estimation of current values of $\dot{M}(2)$, leads to $n(\omega) \approx -22\mu_{32}^{-2/7}(1)$ for AE Aqr and $n(\omega) \approx 0.25\mu_{32}^{-2/7}$for DQ Her, showing that whereas DQ Her and V533 Her accrete nearly in equilibrium, AE Aqr is far from equilibrium. This is understood in terms of the observed $\dot{M}(2)$ – whereas DQ Her and V533 Her are NRs and probably have $\dot{M}(2)$ factors of a few above their long–term value (when the effects of irradiation from the primary have died away), the $\dot{M}(2)$ in AE Aqr is very low for such a large P_{orb} system. In all three cases the observed $P_{rot}(1)$ reflects the long term average $\dot{M}(2)$, not the current value. Thus DQ Her is currently spinning up because it has an above average $\dot{M}(2)$ from the after effects of the 1934 nova eruption, whereas AE Aqr is spinning down from the effect of an anomalous (medium term) low $\dot{M}(2)$.

The equilibrium accretion equation (7.16) should be moderately well approximated for all three systems if $<\dot{M}(2)>$ can be estimated. In Section 9.1.2.2 theoretical values of $\dot{M}(2)$ are derived, which provide the rates given in Table 8.1; for DQ Her and V533 Her these are a factor of a few lower than ones derived from M_v(d) (Patterson 1984; Warner 1987a). Equations (7.16) and (9.18b) combined give $P_{rot}(1) \propto \mu^{6/7}(1)P_{orb}^{-1.4}$, which, provided $\mu(1)$ is restricted in range, accounts for the inverse relationship between $P_{rot}(1)$ and P_{orb} seen in Table 8.1.

8.6 DNOs

8.6.1 *Introduction*

The discovery of low amplitude periodic brightness modulations in DN (DNOs) during outburst (Section 1.4; Warner & Robinson 1972) introduced a phenomenon with a time scale that was not only shorter than that of DQ Her but was also less stable. As already seen, the DQ Her stars have a period stability parameter $Q = |\dot{P}|^{-1} > 10^{12}$ and the IPs have $Q > 10^{10}$. In contrast, for DN $10^4 \lesssim Q \lesssim 10^6$ in optical observations (e.g., Nather 1978; Patterson 1981) and Q is even lower in X-ray observations (Section 8.6.4). From the large stability gap between DQ Hers and DNOs it might be thought that the underlying mechanisms are quite different. However, the observational phenomena associated with the DNOs are very similar in other respects, which justifies associating the DNOs with the magnetic CVs.

DNOs are seen among all subclasses of DN and are present also in some NLs. A list of the stars in which DNOs have been detected, and their range of periods, is given in Table 8.2. In some DN, despite good observational coverage, DNOs have never been observed (e.g., YZ Cnc: Patterson 1981); in others they are seen in one but not in other outbursts (e.g., KT Per: Nevo & Sadeh 1978; Robinson & Nather 1979). Similarly, the bright NL RW Sex has been well observed without any DNO having been found. DNOs have never been seen in DN at quiescence (Patterson 1981).

In general, the DNO cannot be seen directly in the optical light curve (amplitude $<$ 0.5%) but must be detected through power spectrum or periodogram analyses (Figure 8.5: see Warner & Robinson (1972)b; Warner & Brickhhill (1978) and Warner (1988b) for explanations of such techniques). However, very rarely they can be seen directly (e.g., VW Hyi: Warner & Harwood 1973, Robinson & Warner 1984; TY PsA: Warner, O'Donoghue & Wargau 1989 – Figure 8.6). In other stars, if Q is sufficiently high during the time of observation, amplitudes as low as 0.02% may be measured (e.g., SS Cyg: Patterson, Robinson & Kiplinger 1978). In all cases the pulse profile of DNOs, as judged by the absence of harmonics to the fundamental, is sinusoidal within limits of measurement (Figure 8.5: note the logarithmic scale of the ordinate; see also Stiening, Hildebrand & Spillar (1979)).

Broadband photometry has shown that the DNO components in AH Her (Hildebrand, Spillar & Stiening 1981b) and SY Cnc (Middleditch & Cordova 1982) are much bluer than the disc continuum, in particular the pulsed component has strong emission for $\lambda \lesssim 3600$ Å. A very hot thermal component augmented with a small amount of recombination spectrum is the only flux distribution that matches the observations (Middleditch & Cordova 1982).

A more or less distinct class having very low Q, known as QPOs, is discussed in Section 8.7. Reviews of DNOs and QPOs can be found in Warner (1976a, 1979, 1986b, 1988b), Robinson (1976a), Patterson (1981) and Cordova & Mason (1982a, 1983).

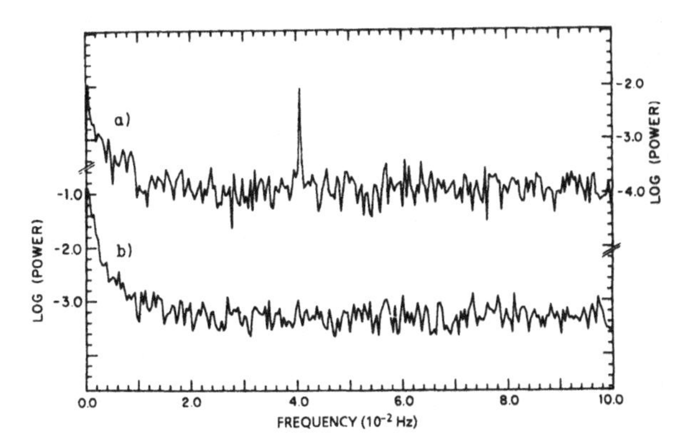

Figure 8.5 Power spectra of light curves of SY Cnc: (a) during outburst, (b) in quiescence. The spike shows the presence of DNOs at a 24.6 s period during outburst. From Robinson (1973b).

Table 8.2. *Optical Oscillations in Cataclysmic Variables.*

Star	Type	P_{orb} (h)	DNO periods (s)	Quasi-periods (s)	References
SS Cyg	DN	6.60	7.3–10.9	32–36,730	1–3,47,65
RU Peg	DN	8.99	11.6–11.8	~51	4,5
TT Ari	NL	3.30		~12,~32,~40 50–1600	6–8,62–64
EM Cyg	DN(Z)	6.98	14.6–21.2		9–11
Z Cam	DN(Z)	6.96	16.0–18.8		12,31
OY Car	DN(SU)	1.50	18–28.0		42,61
V436 Cen	DN(SU)	1.50	19.5–20.1		13
HL Aqr	NL	3.25	19.6		58
VW Hyi	DN(SU)	1.78	20–36	23,88,253,413	14,15,23,51,52
HT Cas	DN(SU)	1.77	20.2–20.4	~100	10
RR Pic	N	3.48		20–40	16,17
KT Per	DN(Z)	3.92	22.4–29.3	82–147	5,18,19
SY Cnc	DN(Z)	9.12	23.3–33.0		9,10,19,20
V1159 Ori	DN(SU)	1.50	24		70
AH Her	DN(Z)	5.93	24.0–38.8	~100	10,21,22,48
CN Ori	DN	3.91	24.3–32.6		23,24,31
IX Vel	NL	4.65	24.6–29.1		25
Z Cha	DN(SU)	1.79	27.7		23,27
TY PsA	DN(SU)	2.02	25.5–30	245	49
PG0859+415	NL	3.67	25.5		60
WZ Sge	DN(SU)	1.36	27.87,28.97		28,29
UX UMa	NL	4.72	28.5–30.0		30,31,50
V3885 Sgr	NL	4.94	29–32		32,33
AE Aqr	DQ	9.88	33.08	~36	34,35
RX And	DN(Z)	5.08		36	19,36
V2051 Oph	DN(?)	1.50		40	26
V533 Her	N(DQ)	5.3	63.63		37,46
DQ Her	N(DQ)	4.65	71.07		38–40
U Gem	DN	4.25		73–146	5
YZ Cnc	DN(SU)	2.1		75–95	10
LX Ser	NL	3.80		~140	59
X Leo	DN	3.95		~160	20
SW UMa	DN(SU)	1.36		280–370	68,69
GK Per	N,DN	1.99d		360–390	10,41,45,67
KR Aur	NL	3.91		400–800	66
RW Sex	NL	5.93		620,1280	43
V442 Cen	DN	11.0:		925	44
WX Hyi	DN(SU)	1.80		1140,1560	53
V426 Oph	DN(Z)	6.85		1680	54
GO Com	DN(?)	1.6:		1980	55
SU UMa	DN(SU)	1.83		2280	56
MV Lyr	NL	3.20		2820	57

References: 1 Hildebrand, Spillar & Stiening 1981a; 2 Horne & Gomer 1980; 3 Patterson, Robinson & Kiplinger 1978; 4 Patterson, Robinson & Nather 1977; 5 Robinson & Nather 1979;

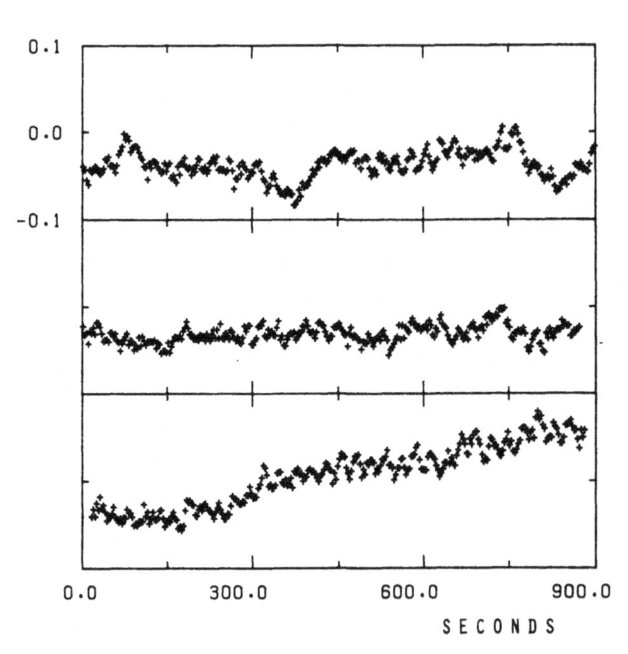

Figure 8.6 DNOs visible in the light curve of TY PsA during outburst. The three panels are contiguous, starting at the top. The vertical scale is fractional intensity. From Warner, O'Donoghue & Wargau (1989).

References to Table 8.2 (cont.)

6 Jensen *et al.* 1983; 7 Mardirossian *et al.* 1980; 8 Sztanjo 1978; 9 Nevo & Sadeh 1978; 10 Patterson 1981; 11 Stiening, Dragovan & Hildebrand 1982; 12 Robinson 1973a; 13 Warner 1975a; 14 Warner & Brickhill 1974; 15 Robinson & Warner 1984; 16 Warner 1981; 17 Schoembs & Stolz 1981; 18 Nevo & Sadeh 1976; 19 Robinson 1973b; 20 Middleditch & Cordova 1982; 21 Hildebrand, Spillar & Stiening 1981b; 22 Hildebrand *et al.* 1980; 23 Warner & Brickhill 1978; 24 Schoembs 1982; 25 Warner, O'Donoghue & Allen 1985; 26 Warner & O'Donoghue 1987; 27 Warner 1974a; 28 Patterson 1980; 29 Robinson, Nather & Patterson 1978; 30 Nather & Robinson 1974; 31 Warner & Robinson 1972a; 32 Warner 1973b; 33 Hesser, Lasker & Osmer 1974; 34 Patterson 1979b; 35 Patterson *et al.* 1980; 36 Szkody 1976a; 37 Patterson 1979c; 38 Walker 1958; 39 Warner *et al.* 1972; 40 Patterson, Robinson & Nather 1978; 41 Watson, King & Osborne 1985; 42 Schoembs 1986; 43 Hesser, Lasker & Osmer 1972; 44 Marino & Walker 1984; 45 Mazeh *et al.* 1985; 46 Robinson & Nather 1983; 47 Arévalo, Solheim & Lazaro 1989; 48 Stiening, Hildebrand & Spillar 1979; 49 Warner, O'Donoghue & Wargau 1989; 50 Warner & Nather 1972b; 51 Haefner, Schoembs & Vogt 1979; 52 Schoembs & Vogt 1980; 53 Kuulkers *et al.* 1991b; 54 Szkody, Kii & Osaki 1990; 55 Kato & Hirata 1990; 56 Udalski 1988b; 57 Borisov 1992; 58 Haefner & Schoembs 1987; 59 Eason *et al.* 1984; 60 Ringwald 1992; 61 Horne 1993b; 62 Hollander & van Paradijs 1992; 63 Semeniuk *et al.* 1987; 64 Udalski 1988a; 65 Bartolini *et al.* 1985; 66 Singh *et al.* 1993; 67 Patterson 1991; 68 Robinson *et al.* 1987; 69 Kato, Hirata & Mineshige 1992; 70 Patterson *et al.* 1994.

8.6.2 DNO Behaviour During Outburst

From the very first observations (Warner & Robinson 1972a) it was suspected that there is a relationship between DNO period and luminosity in outburst. This is illustrated for AH Her in Figure 8.7, which shows the typically ~10–30% variation in DNO period, P_D, during outburst (see Table 8.2 for observed ranges).

Most DNOs appear about half way up the rising branch of the outburst, are present through maximum and disappear about half way down the declining branch. There are exceptions: in AH Her the DNOs disappear near maximum but reappear later (Hildebrand *et al.* 1980; Figure 8.7); in VW Hyi the DNOs appear only towards the end of an outburst or superoutburst (Warner & Brickhill 1978; Robinson & Warner 1984).

The periods P_D decrease on the rising branch, reach a minimum about one day after maximum visual brightness, and increase again on the declining branch (Warner & Robinson 1972a; Patterson 1981). The time of P_D minimum probably coincides with the time of maximum of the EUV and X-ray flux (Bailey 1980; Hassall *et al.* 1983; Section 3.3.5.2), but a direct comparison has not been made. As P_D differs at equal

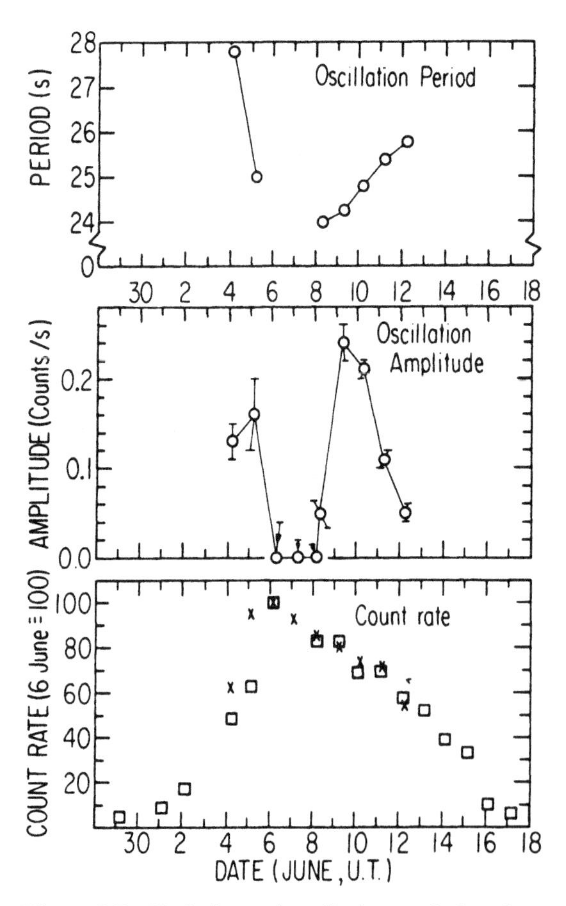

Figure 8.7 Variations of oscillation period and amplitude and of total brightness during an outburst of AH Her. From Hildebrand *et al.* 1980.

visual brightnesses on the rising and falling branch of outburst, a plot of P_D versus m_v produces a loop around which the star travels during outburst (Patterson 1981); however, it is probable that P_D has a monotonic relationship with $\dot{M}(1)$ (but this has not been tested).

On the declining branch, and in the variations of mean brightness of the NLs, the period-luminosity relationship in the visible (where 'white light' observations are made using S-11 response photomultiplier) has a slope dlog P_D/dlog $L_{opt} = -0.25 \pm 0.1$ (Warner & Robinson 1972a; Nevo & Sadeh 1976; Patterson 1981; Warner, O'Donoghue & Allen 1985; Warner, O'Donoghue & Wargau 1989).

As can be seen from Table 8.2, the periods of DNOs cover the range $7.3 < P_D < 40$ s, but each star has a characteristic value. Many of the ranges listed in Table 8.2 for individual stars are undoubtedly too small: although the minimum value of P_D is probably well observed (because it occurs when an outburst is near its brightest), the maximum values often occur after observing runs have terminated. The minimum values of P_D range from 7.5 s up almost to the rotation period of AE Aqr. There is a strong clustering in the range 19–29 s.

There is no dependence of amplitude of DNOs on the inclination of the system: they occur at their highest amplitudes in low (e.g., V3885 Sgr, IX Vel) and intermediate (e.g., VW Hyi, TY PsA) as well as high inclination systems (e.g., UX UMa).

8.6.3 Phase Shifts and Stability of DNOs

Although quite rapid period changes are observed in some of the DNOs, higher stability fortunately is often present which gives an opportunity to investigate eclipse-related phase shifts as was done for DQ Her. In the NL UX UMa, in which DNOs are seen in about half of the observations (Warner & Nather 1972b; Nather & Robinson 1974) and appear to vary in period smoothly on a time scale of weeks, a monotonic 360° phase shift is observed during eclipse (Figure 8.8). This occurs in the opposite

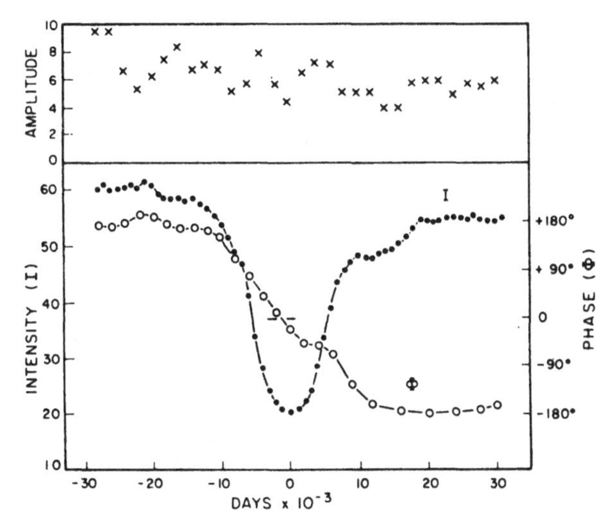

Figure 8.8. Variation of oscillation amplitude (crosses) and phase (open circles) through eclipse (dots) of UX UMa. From Nather & Robinson (1974).

sense to that in DQ Her (Figure 8.1), but also begins and ends at the same phases as eclipse of the disc. Similarly, during its outbursts HT Cas shows an eclipse-related phase shift of the same nature as UX UMa (Patterson 1981). Z Cha also has phase shifts in eclipse during superoutbursts, but these are more complicated and have yet to be modelled (Warner & Brickhill 1978; Warner & O'Donoghue, unpublished). OY Car, late in a superoutburst, showed DNOs which were eclipsed but without any phase shift (Schoembs 1986).

At first sight the direction of phase shift in UX UMa and HT Cas could be thought to imply retrograde rotation of the primary; however, all that is required is that the observer receives more forward directed light than light reprocessed from the back of the accretion disc (as is the case in DQ Her). As this is not possible with the projection geometry of the disc alone, it requires a boost from light emitted directly from the stellar surface. If this is a substantial fraction of the total pulsed radiation then a rapid drop in amplitude and change of phase will be seen as the primary is eclipsed and emerges.

The phase shift and amplitudes through eclipse of UX UMa (Figure 8.8) are adequately fitted with the above model with only a small contribution from the primary, and imply an inclination of 75° (Petterson 1980). The observations of HT Cas (Patterson 1981) have not been modelled in detail, but there are rapid variations indicative of substantial contribution from the primary.

Petterson's models are axisymmetric, but he points out that the disc bulge and other structural asymmetries could account for the lack of symmetry in the observations of phase and amplitude variations. In the case of DQ Her this has been confirmed by O'Donoghue (1985). In OY Car, Schoembs (1986) finds that the time of mideclipse of the DNOs is delayed by 0.022 orbital phase with respect to the middle of optical eclipse, indicating that a large contribution to the reprocessing occurs in the trailing lune, presumably from the disc bulge.

Phase stability is sufficient in UX UMa to examine orbital dependence. Although this has been seen (Nather & Robinson 1974) it has not been modelled (and many more observations are required). In OY Car, Schoembs (1986) finds that the amplitude of the DNOs follows the superhump profile, suggesting that an increase in reprocessing area accompanies the superhump modulation.

In general, however, the DNOs have unstable periods. Clearly, if the DNOs are both unstable and of low amplitude they are liable to be undetectable in a Fourier analysis (see the discussion on QPOs in Section 8.8). Therefore a bias is introduced into the observability of DNOs: *at low amplitude only the most stable oscillations can be recorded.* It is quite possible, therefore, that DN without observed DNOs, or intervals when DNOs are unobserved (including the disappearance of DNOs towards the end of an outburst) are connected with states of low stability rather than complete absence of oscillation. A study of the coherence of DNOs in SS Cyg and AH Her during outbursts confirms this: coherence is maximal at maximum light and, in SS Cyg, fell by a factor of 8 until oscillations were no longer observable at the end of outburst (Hildebrand, Spillar & Stiening 1981a,b; Cordova *et al.* 1984).

The few occasions when DNOs of unusually large amplitude occur can be used to examine phase and period stability. In these circumstances a phase and amplitude may be obtained from fitting only a few oscillation cycles (usually to a sinusoid with pre-

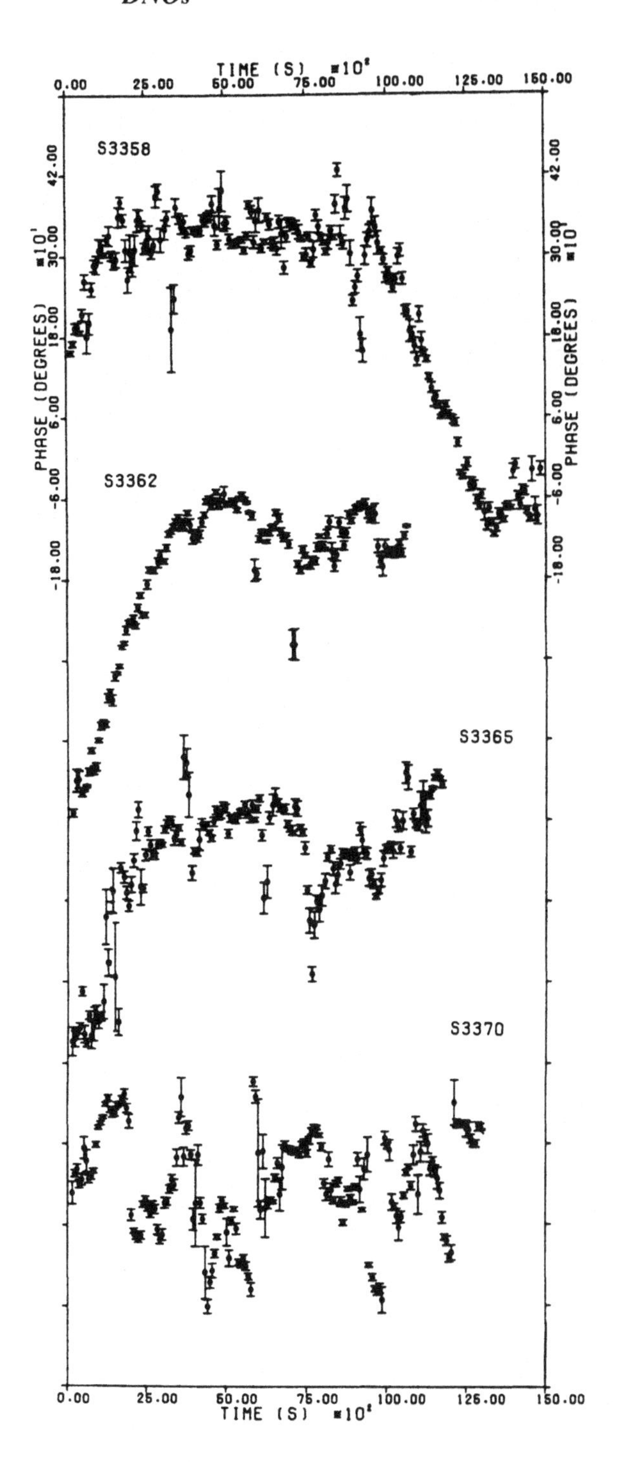

Figure 8.9 Phase diagram of the DN oscillations seen in TY PsA during outburst (see Figure 8.6) on consecutive nights. The phases are determined relative to constant mean periods (25.2, 26.6, 28.0 and 30.2 s respectively). Each point represents a fit to 150 s of light curve (i.e., ~ 5 or 6 oscillations), with a 50% overlap. From Warner, O'Donoghue & Wargau (1989).

determined period) instead of the dozens or hundreds required at low amplitude. This has been accomplished in TY PsA (see Figure 8.6 for one of the light curves) with the result shown in Figure 8.9. Each point is a phase measurement averaged over 5 cycles, and there is 50% overlap for adjacent measurements. Near maximum light (of a superoutburst) there are intervals of ~1–2 h when the oscillation period remains almost constant (i.e., constant slope in Figure 8.9) with only a small amount of phase wandering. As the star declines in brightness (lower two curves of Figure 8.9) the phase becomes so irregular that a well-defined period hardly exists (although even in the lowest curve there is clearly a preference for the phase to remain, even over $3\frac{1}{2}$ h, near to that of the average period of 30.2 s. No oscillations were detectable on the following night, compatible with an even greater amount of phase noise.

8.6.4 X-Ray Observations of DNOs

Rapid flux modulations are also observed in the X-ray emission from some CVs and, being clearly associated, at least for SS Cyg, with the optical DNOs, show that the origin of the phenomenon is in the inner regions of the accretion disc, close to the surface of the primary. The observations are listed in Table 8.3. In the case of SS Cyg, VW Hyi, U Gem, YZ Cnc and AB Dra the modulations were seen only during outbursts.

8.6.4.1 SS Cyg

Pulsed soft X-ray (0.1–0.5 keV) emission has been observed during several outbursts of SS Cyg (Cordova *et al.* 1980a, 1984; Watson, King & Heise 1985; Jones & Watson 1992). The hard X-rays are not pulsed (Swank 1979). The soft X-ray pulsed fraction varies from 0 to 100%, so it is evident that soft X-ray production and the modulation mechanism are intimately related. During an outburst there is an increase in soft X-ray flux by a factor of ~10 at maximum light (Section 3.6).

Table 8.3. *X-ray Modulations in CVs.*

Star	Type	P_{orb}(h)	Periods (s)	References
SS Cyg	DN	6.60	8–12	1–3,9
WZ Sge	DN(SU)	1.36	9.557	4
TT Ari	NL	3.30	9,12,32	5
VW Hyi	DN(SU)	1.78	14.06, 14.2–14.4	6
HT Cas	DN(SU)	1.77	21.85	4,7
U Gem	DN	4.25	20–30,121,135,585	4,7,8
SU UMa	DN	1.83	1.00434, 33.93	4
YZ Cnc	DN(SU)	2.21	222	7
RW Sex	NL	5.93	254	7
AB Dra	DN	3.65	290	7

References: 1 Cordova *et al.* 1980a; 2 Cordova *et al.* 1984; 3 Watson, King & Heise 1985; 4 Eracleous, Patterson & Halpern 1991; 5 Jensen *et al.* 1983; 6 van der Woerd *et al.* 1986; 7 Cordova & Mason 1984a; 8 Mason *et al.* 1988; 9 Jones & Watson 1992.

Although simultaneity of optical and X-ray observations has not been achieved, it is obvious from the similarity of the optical and X-ray periodicities (Tables 8.2 and 8.3) that different aspects of a single modulation process are being observed. As with optical observations of DNOs, the average pulse profile in the X-ray region is strikingly sinusoidal (Figure 8.10a). The period is well behaved ($\dot{P} = -1 \times 10^{-5}$ for the 6 h observation (Cordova *et al.* 1980a) analysed in Figures 8.10, 8.11 and 8.12) but there are substantial stochastic variations in phase relative to the mean period. Figure 8.11 gives an example of the phase wanderings and changes of amplitude, displayed as pulse profiles averaged over ~14 pulses. These same observations, with sine waves fitted to four consecutive pulses to determine phases, produce the variations seen in Figure 8.12. This has an almost identical appearance to that seen in the optical DNOs of TY PsA (Figure 8.9) and underlines the similar statistical behaviour of the phases during large amplitude optical and X-ray pulsations.

Analysis of the amplitude variations again shows similar behaviour in the optical and X-ray. Furthermore, the amplitude and phase variations are *independent* (Cordova *et al.* 1980a).

8.6.4.2 U Gem

No optical DNOs have been observed in U Gem; low coherence soft X-ray oscillations have been found during one outburst but not in two others (Cordova *et al.* 1984; Cordova & Mason 1983).

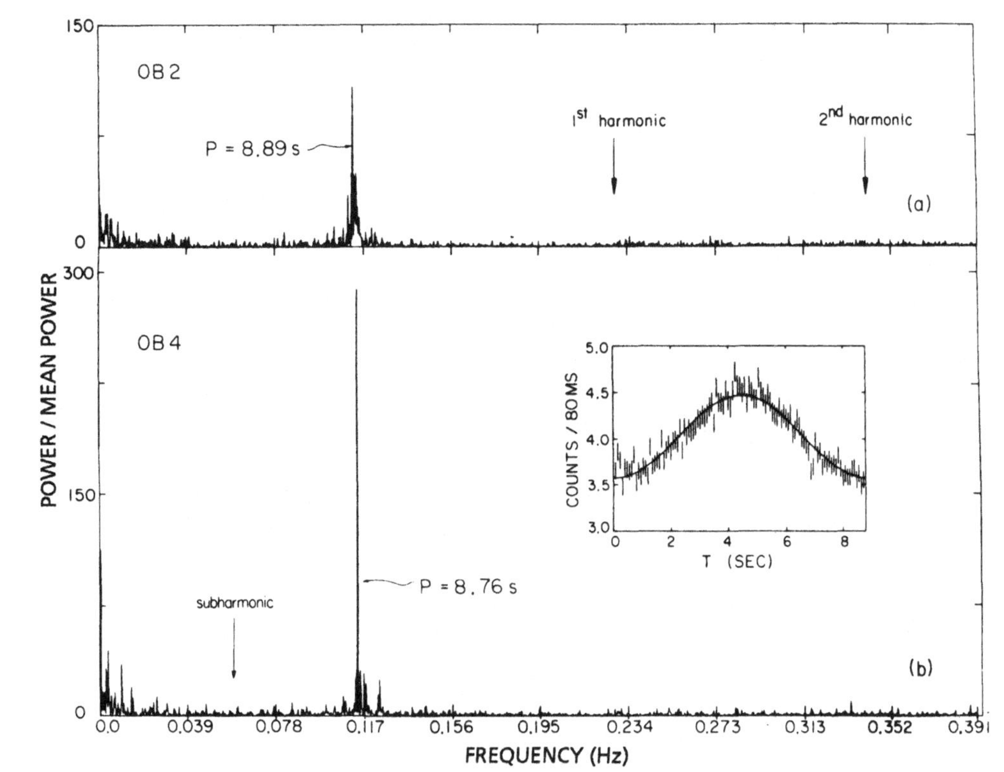

Figure 8.10 Power spectra of soft X-ray emission of SS Cyg during outburst. Inset is the mean pulse profile with the best fitting sinusoid. From Cordova *et al.* (1980a).

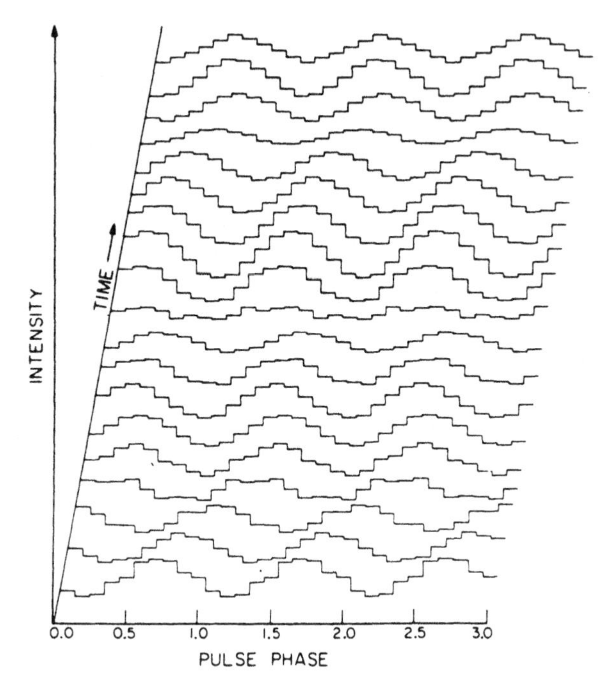

Figure 8.11 Mean X-ray light curves of SS Cyg in outburst. Each trace is 128 s of light curve folded modulo 8.76 s. There is 50% overlap between traces. From Cordova *et al.* (1980a).

8.6.4.3 VW Hyi

Soft X-ray (0.04–0.12 keV) modulations in VW Hyi are occasionally present during the 'plateau' section of superoutbursts of VW Hyi (van der Woerd *et al.* 1987). Near the maximum of the 27 October 1984 superoutburst an oscillation with asymmetric pulse profile and of low coherence was detected, showing rapid changes of amplitude and period varying over the range 14.2–14.4 s. At the end of the plateau of the November 1983 superoutburst a sinusoidal oscillation with period 14.06 $\pm$ 0.02 s and $Q > 2.5 \times 10^3$ was observed. (A claim (Heise, Paerels & van der Woerd 1984) that this period was present for several consecutive days was later withdrawn (van der Woerd *et al.* 1987).)

Table 8.4 lists optical and X-ray observations of the DNOs in VW Hyi. The X-ray periodicities occur earlier in outburst than any optical DNOs, but are in agreement with the general period–luminosity law (Section 8.6.2) on the declining branch: from Table 8.4, $d\log P_D/d\log L_{opt} \approx -0.22$.

8.6.5 Models of DNOs

8.6.5.1 Statistical Models

A number of different approaches have been made to the mathematical description of the rapid oscillations in CVs (both the DNOs under discussion here and the QPOs of Section 8.8). The purpose of these investigations is to provide a quantitative evaluation

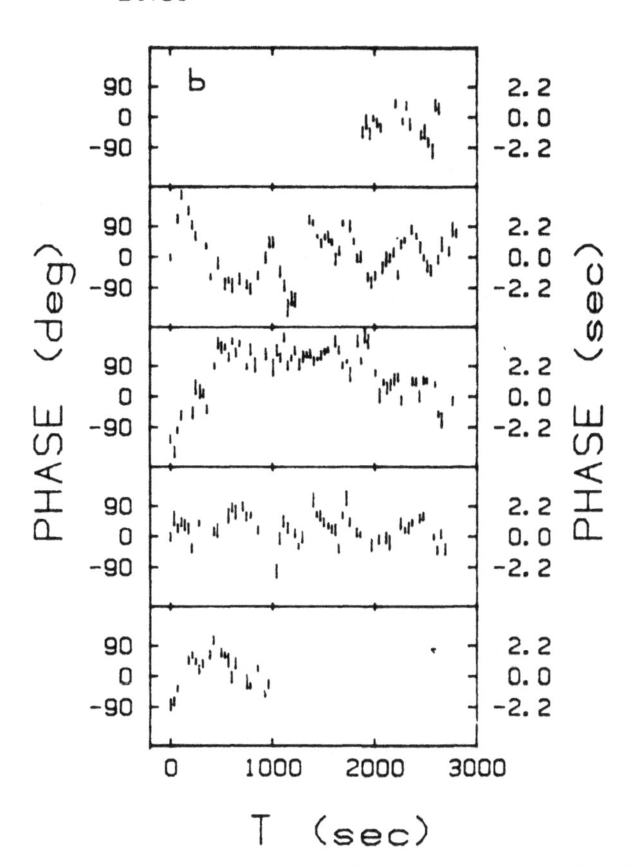

Figure 8.12 Phase diagram of the X-ray pulses in SS Cyg. The phases are determined relative to constant mean periods of 8.893, 8.876, 8.785, 8.757 and 8.745 s respectively. From Cordova *et al.* (1980a).

Table 8.4. *DNOs in VW Hyi.*

m_v	$P_D(s)$	X-ray or optical	Reference
8.8	14.2–14.42*	X	1
9.4	14.06	X	1
13.0:	23.6	O	2
12.2–12.4	28–32	O	3
12.8	32–34*	O	3
13.1	28-32*	O	3
13.3	28	O	4
13.6	36	O	4

* Very low Q

References: 1. van der Woerd *et al.* 1987; 2. Robinson & Warner 1984; 3. Warner & Brickhill 1978; 4. Haefner, Schoembs & Vogt 1979.

of the coherence of the oscillations and then to see if that correlates with properties of the system (e.g., the time since outburst, as described in Section 8.6.3).

The statistical models and analysing processes that have been applied include:

(i) Noise-excited damped harmonic oscillator (Robinson & Nather 1979; Hildebrand, Spillar and Stiening 1981a). This autoregressive process produces a Lorentzian profile in the power spectrum; the FWHM measures the decay time of the oscillation. Within the profile, the standard deviation of the power is equal to the power, which accounts for the large noise in the power spectra of low coherence DNOs and QPOs.

(ii) Random walk in phase of a sinusoidal oscillator (Horne & Gomer 1980; Cordova *et al.*, 1980a, 1984).

(iii) Autocorrelation of the observed oscillations, which, for an autoregressive process, leads to a cosine function with an exponentially decaying amplitude (Robinson & Nather 1979).

The quantitative relationships between the various measures of noise or coherence length derived from these analyses are given by Cordova *et al.* (1984). Although the observations can be fitted with any of these mathematical models, some restrictions on their nature do emerge. Pure amplitude modulation (Hildebrand, Spillar & Stiening 1981a) is excluded by the analysis (Cordova *et al.* 1984); the large amplitude of the X-ray modulation limits the number of independent oscillations that might be excited simultaneously; harmonics are more affected (spread out in the power spectrum) than the fundamental, which could account for the apparent sinusoidal character of the DNOs. None of the statistical analyses, however, throw much light on the underlying physical mechanism of the oscillations.

8.6.5.2 Physical Models

The variety of models that have been proposed as possible explanations for the DNOs all use the fact that near the surface of a white dwarf there is a gravitationally determined time scale $\tau_G \sim (G\bar{\rho})^{-1/2}$, where $\bar{\rho}$ is the mean density of the primary; i.e., $\tau_G \sim$ few seconds. All revolutionary or pulsational phenomena will have a fundamental period near τ_G. These include inhomogeneities in the accretion disc, travelling waves excited in the surface layers of the primary, rotation near break-up of the primary and oscillations near the inner edge of the disc. It is feasible, by general arguments, to eliminate only a few possibilities.

First, the low Q of the oscillations excludes a simple DQ Her model from consideration: the rotational energy in the primary is $\frac{1}{2}I\omega^2 \sim 10^{49}$ erg; to enact period changes of up to 25% would therefore require $\sim 3 \times 10^{48}$ erg, whereas only $\sim 10^{39-40}$ erg is available in a DN outburst (Section 3.7). However, if the outer $\sim 10^{-10}$M$_\odot$ of the white dwarf is only loosely coupled to the interior (which implies a relatively weak field) then a variable rotation magnetic accretor becomes possible (Paczynski 1978).

Second, luminous blobs in the inner disc, resulting from some unknown instability, have periods of revolution of the correct size. To obtain a luminosity modulation through eclipses by the primary (Bath 1973) does not appear tenable because the DNOs are observed even in low inclination systems.

Third, non-radial pulsations of the primary, which have periods in the observed range (e.g., Brickhill 1975), also cannot change their periods on the time scales observed (Bath *et al.* 1974). However, *r*-modes or trapped *g*-modes, both of which have periods $\sim P_{\rm rot}(1)/m$, where m is the azimuthal harmonic index, and which are concentrated in the surface layers of a star, have the correct range of periods (Papaloizou & Pringle 1978b). The difficulties with this model are: that m bright spots should be produced; that more than one m-mode would be expected to be excited, whereas the observations show no evidence for multiple periods; that the coupling between the oscillation in the star and the X-ray producing region, which must be capable of modulating the X-rays by up to 100% (which itself suggests that only one or two high energy regions can be involved), is difficult to envisage (a similar problem arises with the low amplitude surface oscillations driven by viscous instabilities studied by Papaloizou & Stanley (1986)); and, finally, that the statistical properties of the DNOs seem too variable and 'random' to be produced by just a few oscillation modes.

Fourth, oscillations of the accretion disc (e.g., van Horn, Wesemael & Winget 1980) are not confined to one particular annulus, so the oscillation power is spread over a wide range of periods which produces a signal much less coherent than the DNOs of even moderate coherence.

The observation that the DNOs have periods smaller than that of AE Aqr suggests that the DNO phenomenon is in some way an extension of the oblique magnetic rotator to lower magnetic fields: the large amplitude of modulation of the soft X-rays (as in AE Aqr) shows that much or most of the accreting material is passing through the modulating regions just as in magnetically controlled accretion; the phase shifts in eclipse show that a beaming mechanism similar to that in DQ Her is operating; the absence of circular polarization in the DN and NLs is unsurprising when considered in relation to the situation in the IPs which presumably have higher magnetic fields; the presence of Balmer line flares with FWZI $\sim$7000 km s^{-1} in the DN SS Cyg (Walker 1981), similar to those observed in AE Aqr, indicates analogous instabilities in the inner disc. Furthermore, DQ Her, AE Aqr and V533 Her all have equilibrium $r_0/R(1) \approx 3$; if the distribution of magnetic fields of the primaries continues to even lower values than in the DQ Her stars, it is reasonable to expect some expression of magnetically controlled accretion for the systems with $1 < r_0/R(1) < 3$. And it is in just this range that more complex behaviour is expected.

However, recognising the *similarities* of behaviour is much easier than finding a model for DNOs that accounts for the *differences* from DQ Her stars. There appear to be two possibilities for explaining the range of periods seen in the DNOs of a particular star: either there is slippage of the exterior layers of the primary (Paczynski 1978; see also van der Woerd *et al.* 1987 and Jones & Watson 1992), or the observed periods are in some way related to the period at the inner edge of the accretion disc.

King (1985a) has proposed a model in which the differentially rotating surface layers, caused by accretion of angular momentum during outburst, amplify existing small fields and generate a sufficiently strong field to channel electrons from a hot corona (Section 2.6.1.5) through magnetic tubes to produce accretion zones on the primary. In this model the field survives only while differential rotation exists; it seems inappropriate for the NL systems in which DNOs are persistently present even in the absence of the significant variations in $\dot{M}$. And DNOs might be expected in all DN outbursts. The model does, however, have the ability to generate a noisy periodic signal because of the short lifetime of individual flux tubes.

Warner (1983a) suggested a model which combines the ideas of inhomogeneities in the inner disc with magnetic accretion. The magnetosphere rotates with period $P_{\rm rot}(1)$, and the inner edge of the disc will have a different period $P_{\rm K}(r_0)$ during the non-equilibrium accretion which must occur in the high $\dot{M}$ outburst phase, so any inhomogeneity at the inner edge of the disc threaded by the field most efficiently at some preferred magnetic longitude will cause mass transfer to the accretion zones to be modulated at the period $P_{\rm D}^{-1} = P_{\rm K}^{-1}(r_0) - P_{\rm rot}^{-1}(1)$. From equations (2.23), (6.5) and (7.3),

$$P_{\rm K}(r_0) = 6.3\mu_{31}^{6/7}(1)M_1^{-5/7}(1)\dot{M}_{17}^{-3/7}(\rm d)\ s. \tag{8.3}$$

Therefore $P_{\rm D}$ will reach a minimum at maximum $\dot{M}$ and will behave qualitatively in accordance with the observed period–luminosity relationship of DNOs. The noisy phase properties also follow from the replacement of one inhomogeneity by another at a random position.

Although this 'beat-period' model has had considerable success in explaining one kind of QPO in X-ray binaries (e.g., Lamb *et al.* 1985) it fails to account for the phase shifts during eclipse of DNOs, which require a simple rotating beam, not a beam rotating with period $P_{\rm rot}(1)$ modulated in intensity at $P_{\rm D}$ (Warner 1987b).

In view of what is now known about polars, IPs and DQ Her stars, and in the spirit of *Pluralitas non est ponenda sine necessitate* (used, but not invented, by William of Ockham, 1300?–1349), an elaboration of Paczynski's (1978) model seems most in harmony with the global properties of magnetic CVs, *viz.*,

> DNOs occur in those systems where the magnetic field of the primary is low enough to allow differential rotation of its outer layers – hence the difference in Q of DQ Her and UX UMa, for example.
>
> In high $\dot{M}$(d) (NL) systems, $P_{\rm D}$ is determined by the dynamic balance between accretion torque and viscous dissipation in the outer layers. Thus small variations in $\dot{M}$ can lead to the changes in DNO period seen in UX UMa, IX Vel etc.
>
> DN with weak fields ($\lesssim 5 \times 10^4$ G: equation (7.4b)) are not capable of producing DNOs during outburst because there is no magnetically controlled accretion.
>
> For DN with fields $\gtrsim 5 \times 10^4$ G, magnetically controlled accretion, with attendant DNOs, occurs as long as $r_0 > R(1)$. If the magnetosphere is crushed to the surface during the highest values of $\dot{M}$(d), then the DNOs will disappear as in Figure 8.7. At the same time the hard X-ray flux is quenched (Section 3.3.6).
>
> The periods of the DNOs follow the accretion of angular momentum during the high $\dot{M}$ phase of outburst, with subsequent relaxation to $P_{\rm rot}(1)$ as the outer layers of the primary interact with the interior.

In this picture, the minimum observed $P_{\rm D}$ provides a lower limit on the mass of the primary. Noting that the outer layers are only bound if they rotate at less than Keplerian velocity gives[1]

$$M_1(1) \geq 296\frac{R_9^3(1)}{P_{\rm D,min}^2(\rm s)}. \tag{8.4}$$

[1] This is for a non-rotating primary. See Section 9.3.6 for the minimum period of a primary rotating at break-up velocity.

Thus for SS Cyg, $M_1(1) > 1.0$ and in AH Her, as the DNOs disappear and reappear at $P_{D,min} = 24.0$ s, $M_1(1) = 0.49$. The former is in agreement with $M_1(1) = 1.20 \pm 0.10$ found by Kiplinger (1979b); the latter requires $i = 62°$ instead of $i = 45°$ as adopted in the analysis by Horne, Wade Szkody (1986).

The anisotropy of radiation in DNOs arises from the same mechanism as in IPs and DQ Her stars. The complexity of the magnetic field (e.g., multipole geometry) near the surface of the primary results in alternate feeding of more than one accretion zone, thereby introducing phase instability to an underlying slowly varying periodic mechanism.

Differential rotation may enhance the field during outburst, thus leading to simpler field geometry and DNOs that are more coherent. Some evidence exists for viscous heating by shearing during outburst (Section 3.2.1).

8.7 QPOs

Brightness oscillations of a kind different from the DNOs were noted by Patterson, Robinson & Nather (1977) during an outburst of RU Peg (Figure 8.13). They designated these *quasi-periodic oscillations* (QPOs). These had an amplitude ~0.005 mag, a period near 50 s and persisted for at least 5 d during outburst. The very low Q of the oscillations (~1–10) results in them appearing as a very low amplitude, noisy, broad band in the power spectrum, which would possibly be overlooked were it not for the prominence of the oscillations in the light curve (Figure 8.13). The fact that the QPOs in RU Peg coexisted with a DNO at 11 s shows that they are distinct phenomena.

Robinson & Nather (1979) and Patterson (1981) later found QPOs in other DN during outburst. Others have been found (Table 8.2) in NL systems and in DN in which no DNOs have yet been detected. The NR RR Pic also possesses QPOs (Warner 1981; Schoembs & Stolz 1981). In general, the average period of QPOs is ~2–3 times that of DNOs in the same stars. On occasion the power in QPOs is very high (Warner 1987b; Robinson & Warner 1984), but the power is spread over such a wide range of

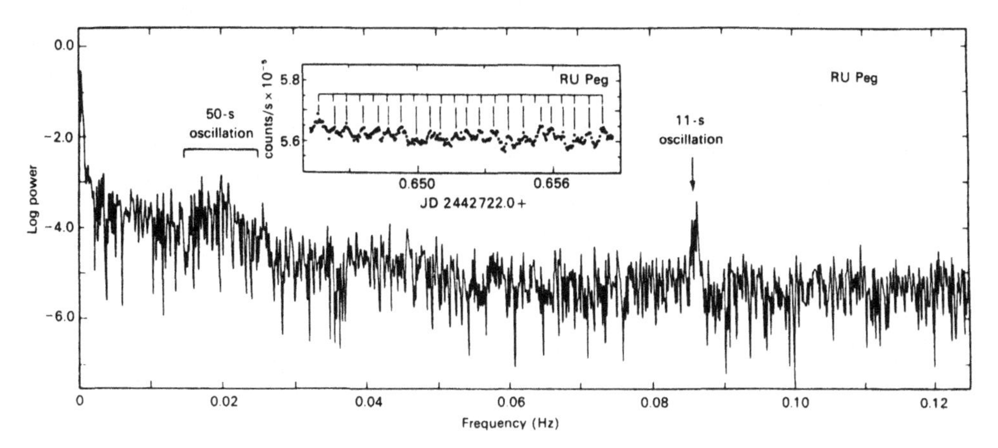

Figure 8.13 Power spectra and (inset) part of the light curve of RU Peg during outburst showing the simultaneous presence of DNOs with a period 11 s and QPOs with periods near 50 s. Adapted from Robinson & Nather (1979).

frequencies and is so noisy (Section 8.6.5.1) it is difficult to detect; QPOs probably occur more frequently than is realized. It is not yet known whether there is any systematic change of mean period of QPOs during outburst of a DN.

The longer periods, and the observation that QPOs in U Gem persist through eclipse and are therefore not associated with the bright spot region (Robinson & Nather 1979), suggest that they are produced further out in the disc than the DNOs. Vertical or radial oscillations of accretion discs have been proposed as the probable origin of QPOs (Kato 1978; Van Horn, Wesemael & Winget 1980; Cox 1981; Cox & Everson 1982; Carroll *et al.* 1985). Because of the weak coupling in the radial direction, individual annuli oscillate independently. The period of oscillation is proportional to, and not very different from, the Keplerian period at the radius of the annulus. Superposition of the contributions from the visually brightest annuli produces a spread of power with periods in the range 10–150 s, very similar to what is observed in the QPOs.

More general treatment, performing perturbation analyses on thin discs, produces several spectra of oscillations, analogous to p-mode and g-mode non-radial oscillations in stars (Abramowicz *et al.* 1984; Blumenthal, Yang & Lin 1984; Carroll *et al.* 1985). Carroll *et al.* found that there are three groups of oscillations: (i) $2 \lesssim P \lesssim 140$ s, which overlap the DNOs and some of the QPOs; (ii) $49 \lesssim P \lesssim 640$ s, which overlap the QPOs, and (iii) $P \gtrsim 480$ s which, if spread over a wide range of periods, would be difficult to detect observationally unless of large amplitude.

The radial oscillations of steady state α-disc models have been extensively computed by Okuda *et al.* (1992). The driving process is analogous to the ε-mechanism of stellar pulsation: as the viscous energy generation is proportional to gas pressure, oscillations are amplified. Local oscillation amplitudes of $\dot{M}$ can be several times the mean transfer rate $\dot{M}(\mathrm{d})$; however, these generate only $\sim$1% variations of T_{eff} and luminosity. Oscillation amplitudes are low in the inner regions of all discs and in the outer regions of large ($r_\mathrm{d}/r_0 \sim 100$) discs. However, in systems with small discs ($P_{\mathrm{orb}} < 2$ h) the oscillation is large even at r_d. Decay times are $\gtrsim 40 P_\mathrm{K}(r)$. In general, the region contributing to the luminosity modulation is the radius range $(4\text{–}12)r_0$, giving QPO time scales

$$\begin{aligned} P_\mathrm{Q} &\approx (8\text{–}40) P_\mathrm{K}(r_0) \\ &\approx (86\text{–}430) M_1^{-1}(1) \text{ s} \end{aligned} \qquad (8.5)$$

from equations (2.16) and (2.83a), if $r_0 \equiv R(1)$. Therefore, if magnetic accretion is always overwhelmed during the high $\dot{M}(\mathrm{d})$ phase of DN outbursts, there would be a significant dependence of P_Q on $M(1)$. This is not evident in the observations.

Observationally it would be useful to investigate the power spectra of QPOs as a function of wavelength. If the QPOs are the result of the superposition of many oscillations of disc annuli then the peak power should move to shorter periods at shorter optical wavelengths.

The longest quasi-periods listed in Table 8.2 are comparable with the rotation period $P_\mathrm{K}(r_\mathrm{d})$ at the outer edge of the disc. The latter may be estimated from equations (2.1a), (2.5b) and (2.16) as

$$\frac{P_\mathrm{K}(r_\mathrm{d})}{P_{\mathrm{orb}}} \approx 0.31 \left[\frac{r_\mathrm{d}}{R_\mathrm{L}(1)}\right]^{3/2} \approx 0.20 \qquad (8.6)$$

for $r_d = 0.75R_L(1)$ (Section 2.6.2). These QPOs may therefore be associated with long-lived inhomogeneities ('blobs') created by impact of the accretion stream (Section 2.5.3), or with tidal resonance with the secondary.

A completely different origin has been proposed for QPOs that occur (Figure 8.4) during the flaring stages of AE Aqr (Patterson 1979b). These appear at periods ~1 s longer than the 16.5 and 33 s coherent periodicities (Section 8.2.1) and may be caused by 'blobs' in the disc (which may themselves be the effect of multimode vertical oscillations) on which the rotating beam in AE Aqr is reprocessed (Patterson 1979b). Alternatively, the 'beat-period' model (Section 8.6.5.2) may apply to the QPOs in AE Aqr (Abado-Simon *et al.* 1992).

The beat-period model has also been invoked (Hollander & van Paradijs 1992) in the case of TT Ari, which has shown a remarkable steady decline in QPO period from ~27 min to ~15 min over 25 y (Semeniuk *et al.* 1987). During that time the average brightness has not changed by more than 0.08 mag; the resulting limit on the range of $\dot{M}(2)$ eliminates most models (especially an IP model with variable r_μ). A spin-up of the surface layers of the primary, following some unobserved extended low state (last century) is another possibility.

The occurrence in GK Per of ~380 s QPOs near its $P_{rot}(1) = 351.34$ s (Patterson 1981,1991) shows that the underlying regular rotation of the primary may appear only as quasi-periodic modulation of the optical light curve. This opens the possibility that some of the QPO systems in Table 8.2 may be intermediate polar candidates (Table 7.2).

8.8 Distribution of Magnetic Moments in CVs

Combining the information in this and the prevous two chapters provides a view of the over all distribution of magnetic moments among the CVs.

In Table 6.8 magnetic fields are given for 15 out of the total of 34 known polars. Adopting $M_1(1) = 0.7$ these cover a range $33.7 < \log \mu(1) < 34.6$; the 19 polars with unknown field strengths probably fall within the same range.

Because of their greater intrinsic luminosity, the observed IPs average about twice the distance of polars, but are probably much less completely known. Of 12 known IPs, 10 have estimates of $\mu(1)$ in Table 7.4, to which maybe added $\mu_{33}(1) \approx 8$ for PQ Gem (Section 7.4). To these should be added T Leo and HT Cas (Section 7.7) with $\mu(1)$ similar to that of EX Hya. There are probably many more similar objects, but they will be counted among the CVs with DNOs.

The magnetic moments for the DQ Her stars (Table 8.1) are larger than that of EX Hya, but are a factor of 10 lower than most IPs of similar P_{orb}.

No direct measurements of $\mu(1)$ are available for the CVs that show DNOs, but from applying the approximate criteria $r_0 \lesssim 3R(1)$ and equation (7.4b), and noting that $\dot{M}(1) \sim (1\text{–}10) \times 10^{17}$ g s^{-1} when DNOs are observed, gives $30.7 \lesssim \log \mu(1) \lesssim 31.7$. There are many stars with no observed DNOs, for which $\log \mu(1) \lesssim 31.0$ if the interpretation offered here is correct.

The above results are summarized in Table 8.5. It is impossible to make convincing allowances for the differences of volumes of space surveyed and of completeness for the different types of star, but the general appearance is that the statistics are compatible with a uniform distribution per decade in μ as found for isolated magnetic white dwarfs

Table 8.5. *Distribution of Magnetic Moments.*

Type	Number known	Range of log $\mu(1)$
Polar	34	33.7–34.6
IP	1	33.9
	10	32.5–33.9
	3	31.8–32.1
DQ Her	3	32.5–32.6
DN	19	30.7–31.7

(Section 6.1). However, among the latter, magnetic moments up to log $\mu = 35.5$ are known (Schmidt & Liebert 1987). The question of the missing high-μ CVs is addressed in Section 9.3.7.1.

9

Evolution of CVs

...he questioned the dwarves about their doings, and where they were going to, and where they were coming from.

J.R.R. Tolkien. *The Hobbit*

And is old Double dead?

W. Shakespeare. *King Henry IV, Part 2*

Thus far the structures of CVs have been described without much necessity for knowing whence the CVs came, how they live, or whither they go. The occurrence of mass transfer has been taken as given, without enquiry as to the driving mechanism. The long term consequences of mass transfer and loss, with their attendant exchange and loss of angular momentum, have now to be described. The common ancestors and the exotic descendants of CVs have also to be identified.

Although this chapter is mostly about the theory and computation of structural and orbital evolution, two further observational classes of stars are introduced: the pre-cataclysmic systems and the double degenerate binaries.

9.1 Transfer and Loss of Mass and Angular Momentum

9.1.1 Mass and Angular Momentum Transfer

The general case, allowing $\dot{M}(1) \neq -\dot{M}(2)$ and the total angular momentum to vary, is as follows. The total orbital angular momentum is

$$J = \frac{G^{1/2} M(1) M(2)}{[M(1) + M(2)]^{1/2}} \, a^{1/2} \tag{9.1a}$$

$$= \frac{M(1) M(2)}{M(1) + M(2)} \, a^2 \Omega_{\mathrm{orb}}. \tag{9.1b}$$

The angular momentum $j(2)$ of the secondary is

$$j(2) = M(2) k^2(2) R^2(2) \Omega_{\mathrm{orb}} \tag{9.2}$$

which gives, from equations (2.1a), (2.5b) and (9.1a),

$$j(2)/J \approx 0.21 k^2(2)(1 + q)^{1/3} q^{2/3}, \tag{9.3}$$

where $k(2)$ is the radius of gyration of the secondary, which is ~0.45 for low mass main sequence stars: therefore $j(2)/J \lesssim 0.03$ and $j(2)$ can usually be neglected.

Then, from equations (2.1a) and (9.1)

$$\frac{\dot{P}_{orb}}{P_{orb}} = 3\frac{\dot{J}}{J} - \frac{2+3q}{(1+q)}\frac{\dot{M}(1)}{M(1)} - \frac{3+2q}{1+q}\frac{\dot{M}(2)}{M(2)} \tag{9.4a}$$

$$\frac{\dot{a}}{a} = 2\frac{\dot{J}}{J} - \frac{1+2q}{1+q}\frac{\dot{M}(1)}{M(1)} - \frac{2+q}{1+q}\frac{\dot{M}(2)}{M(2)} \tag{9.4b}$$

$$= 2/3\frac{\dot{P}_{orb}}{P_{orb}} + \frac{1}{3(1+q)}\frac{\dot{M}(1)}{M(1)} + \frac{q}{3(1+q)}\frac{\dot{M}(2)}{M(2)} \tag{9.4c}$$

(and here $\dot{M}(2) < 0$).

For the fully conservative case, $M(1) + M(2) = \text{const}$ and $\dot{J} = 0$, we then have

$$\frac{\dot{a}}{a} = 2(q-1)\frac{\dot{M}(2)}{M(2)} \tag{9.5a}$$

$$\frac{\dot{P}_{orb}}{P_{orb}} = 3(q-1)\frac{\dot{M}(2)}{M(2)}\,. \tag{9.5b}$$

These equations show that, as $M(2)$ is steadily reduced by mass transferred to the primary, a and P_{orb} *decrease only if* $q > 1$. (The reason is that, if $q < 1$, mass is moved nearer to the centre of mass of the system, so the separation must increase in order to conserve angular momentum.) However, because the Roche lobe around the secondary continues to shrink as q decreases further, mass transfer does not terminate at $q = 1$. For example, from equations (2.5b) and (9.5a),

$$\frac{\dot{R}_L(2)}{R_L(2)} = \left(2q - \frac{5}{3}\right)\frac{\dot{M}(2)}{M(2)} \tag{9.6}$$

and therefore the minimum size of the Roche lobe occurs for $q = 5/6$. More general formulae, allowing for angular momentum or mass loss, can be found in Kruszewski (1966), Hadjidemetriou (1967), Plavec (1968) and Warner (1978).

As seen from Section 2.8.2, most CVs have $q < 5/6$, *yet steady mass transfer is observed.* This implies that either or both of the conservancy equations are incorrect. The observed rates of *mass loss* from winds (Section 2.7.3) are insufficient to provide the solution; therefore, in order to maintain a low mass main sequence secondary star in steady contact with its Roche lobe, *there must be a mechanism of orbital angular momentum loss from CVs.*

If the secondary is evolved, however, sustainable mass transfer may be viable as the secondary expands on a nuclear time scale. The radius (and luminosity) of a subgiant or giant is a function only of the core mass M_c (Refsdal & Weigert 1970; Paczynski 1971a; Kippenhahn 1981: equation (5.14)) and therefore $\dot{R}(2)$ is a function of $\dot{M}_c(2)$ which in turn gives $\dot{M}(2)$ as a function of $\dot{M}_c(2)$ through equation (9.6). Detailed computations (Webbink, Rappaport & Savonije 1983; Iben & Tutukov 1984) and analytic approximations (King 1988) *for the fully conservative case* give

$$\dot{M}(2) \approx 5.3 \times 10^{-10} P_i(\mathrm{d}) \qquad \mathrm{M}_\odot\ \mathrm{y}^{-1}, \tag{9.7}$$

where P_i is P_{orb} at the time when the secondary first fills its Roche lobe.

In the above models, P_{orb} and a increase with time. There may, however, be competition from orbital angular momentum loss by magnetic braking (Section 9.1.2.1). To be evolved at all, within the age of the Galaxy, requires $M_1(2) \gtrsim 1.0$, and therefore $q \gtrsim 0.7$ at the start of mass transfer. As $R_1(2) = 1.75$ at the base of the giant branch (Webbink, Rappaport & Savonije 1983), equations (2.1a) and (2.5a) give $P_i \geq 24.2\,\mathrm{h}$.

9.1.2 Angular Momentum Loss

9.1.2.1 Magnetic Braking in Single Stars

For the Sun it was early recognized that the ionized solar wind would be forced to corotate on magnetic field lines out to the Alfvén radius $R_A \sim 100\ \mathrm{R}_\odot$, thereby exerting a braking torque (through Maxwell stresses) sufficient to account quantitatively for the slowness of the solar rotation (Weber & Davis 1967). The possibility that this mechanism accounts for slow rotation in all late-type dwarfs (Wilson 1966a) is supported by (a) the existence of coronae in stars later than type F as demonstrated by their X-ray emission, the strength of which increases with rotational velocity and decreases with age (Cruddace & Dupree 1984; Rosner, Golub & Vaiana 1985; Schrijver & Zwaan 1991), (b) the decrease in rotational velocity with age (as measured by membership of clusters or groups) (Skumanich 1972; Soderblom 1991), (c) the reduction in (field-related) H and K emission line strengths with age (Wilson 1966b; Baliunas & Vaughan 1985) and (d) the similar correlations with chromospheric and transition region emission-line strengths (Basri, Laurent & Walter 1985; Zwaan 1986). These are reviewed in detail by Baliunas & Vaughan (1985), Rosner, Golub & Vaiana (1985) and Hartmann & Noyes (1987).

The observed (Kraft 1967) dramatic decrease in rotational velocities of stars on the main sequence in passing from F0 to F5 coincides with the development of a deep convective envelope for $M \lesssim 1.3\ \mathrm{M}_\odot$. This suggests a causal connection. Two models are commonly considered: the *distributed dynamo* acting throughout the convective zone, in which differential rotation in the star stretches the field lines and creates a strong toroidal field (the ω effect) and the initial poloidal field is regenerated from this by the cyclonic motions in the convective cells (the α *effect*), and the *shell dynamo* in which the ω effect is situated in the transition region between the convective envelope and the radiative core (Rosner 1980; Schussler 1983). The latter has usually been favoured because Parker (1975) showed that, unless the field lines are anchored in the radiative core, flux tubes would be buoyed to the surface too quickly for the α effect to operate. However, Tout & Pringle (1992a) consider that in a rapidly rotating star the α effect can operate in a small but finite way. The result is that the dynamo mechanism continually generates a magnetic field, the equilibrium value of which is determined by the steady loss of flux from the star to the base of the corona, where it incidentally provides the energy that drives the stellar wind.

The angular momentum loss rate caused by the wind is

$$\dot{J}_{\rm wind} = \frac{2}{3}\dot{M}_{\rm wind}R^2\Omega_{\rm rot}\left[\frac{R_{\rm A}}{R}\right]^n, \tag{9.8}$$

where $\Omega_{\rm rot}$ is the angular velocity of the star, R is the stellar radius and n parameterizes the field geometry ($n = 1$ for a radial field, $n = 2$ for a dipole field in which the field strength $B(r) \propto r^{-3}$). The value of $R_{\rm A}/R$ depends on the field geometry (Kawaler 1988).

Estimation of $\dot{M}_{\rm wind}$ is very model-dependent. Verbunt (1984) argued that the wind flows along radial field lines; Tout & Pringle (1992a) use a dipolar field and Mestel & Spruit (1987) use an inner dipole configuration with outer radial field lines, both studies allowing for the fact (seen in X-ray pictures of the solar corona) that not all of the surface of the star is connected to open field lines (Mestel 1968,1984). As an example of the complications of *ab initio* calculations, there is observational evidence that the fraction of the surface thus connected (the 'filling factor') is related exponentially to the Rossby number $R_{\rm o}(= P_{\rm rot}/\tau_{\rm con}$, where $\tau_{\rm con}$ is the convective turnover time at the base of the convection zone) and increases non-linearly with the fractional coverage of the surface with field lines (Saar 1991; Stepién 1991).

The connection between the rate of magnetic energy release $L_{\rm mag}$ by the dynamo process and $\dot{M}_{\rm wind}$ is made on the assumption that the wind velocity equals the escape velocity from the star; then $L_{\rm mag} \sim GM\dot{M}_{\rm wind}/R$. For example, Tout & Pringle (1992a) find $L_{\rm mag} \sim 0.1\ Mv_{\rm A}^3/R$, where $v_{\rm A}$, the Alfvén speed in the star, is $\sim B/(4\pi\bar{\rho})^{1/2}$.

There is no accepted fundamental theory that connects the field strength of a star with its rotation rate. It is commonly supposed that $B \propto \Omega_{\rm rot}^p$, where p is usually taken as unity, although directly measured surface fields in cool stars show no correlation with $\Omega_{\rm rot}$ (Saar 1991). This is possibly because non-linear effects saturate the dynamo mechanism in rapidly rotating stars (Durney & Robinson 1982); saturation effects are certainly visible in magnetic activity indices (Vilhu & Walter 1987). Furthermore, as the filling factor increases with angular velocity (Saar 1987), the mean surface field will increase even if the field itself does not. Then

$$B = {\rm B}_\odot({\rm P}_\odot/P_{\rm rot})^p \tag{9.9a}$$

$$\approx 650P_{\rm rot}^{-1}({\rm h})\ {\rm G} \quad {\rm for}\ p = 1. \tag{9.9b}$$

For the Sun, with ${\rm P}_\odot \sim 25\,{\rm d}$, observations give $\dot{M}_{\rm wind}(\odot) \sim 2 \times 10^{-14}\ {\rm M}_\odot\ {\rm y}^{-1}$. Evaluation of the Mestel & Spruit theory gives $\dot{M}_{\rm wind} = 2.6 \times 10^{-11}\ {\rm M}_\odot\ {\rm y}^{-1}$ for $P_{\rm rot} = 4\,{\rm h}$ (Hameury *et al.* 1988).

A question of some importance in the evolution of CVs is what happens to the magnetic fields and winds for stars that are fully convective? This occurs on the main sequence for $M \lesssim 0.25\ {\rm M}_\odot$ (Kippenhahn & Weigert 1990), corresponding to spectral type later than M4V. Active chromospheres in this mass range are evident from the existence of dMe and flare stars of type M5 and later (Giampapa & Liebert 1986). Fleming *et al.* (1992, 1993) find that although X-ray *fluxes* diminish at later types, coronal heating efficiencies are similar ($L_{\rm x}/L_{\rm bol} \sim 0.0016$) over the whole range $8 < M_{\rm v} < 20$. There is therefore no observational evidence for any strong differences in *field strengths* above and below 0.25 $M_\odot$. Nevertheless, the *wind strength*, and, through increased complexity of the field configuration, the number of open field lines, *could* be less in lower masses (Taam & Spruit 1989).

9.1.2.2 Magnetic Braking in Binaries

Tidal forces in close binaries are able to maintain equality of rotational and orbital periods (Section 9.1.3). Therefore, a lobe-filling secondary that is losing rotational angular momentum via a magnetic wind is kept in synchronous rotation by spin–orbit coupling, and angular momentum instead drains from the orbit (usually, and paradoxically, decreasing $P_{rot}(2)$ in the process: cf. the increase in angular velocity of a satellite losing energy to atmospheric friction).

It is not yet clear if the rapid rotation forced upon the secondary in a short period binary causes it to differ from a single star with the same P_{rot}. Basri (1987: see also Rutten 1987; Young, Ajir & Thurman 1989) found that the indicators of magnetic activity in late-type dwarfs and giants have the same dependence on P_{rot} and R_o, whether or not they are single or in synchronized binaries. These studies use stars with $P_{rot} \gtrsim 1$ d. There are indications that magnetic activity could be reduced in rapidly rotating stars (Rucinski 1983; Vilhu & Moss 1986), perhaps being the result of saturation of activity. However, Schrijvcr & Zwaan (1992) find that the presence of a companion star enhances magnetic activity in rapidly rotating stars, perhaps because differential rotation is affected or because the axis for rotational effects passes through the centre of the gravity of the system rather than the centre of the star, causing a different dynamo structure.

The orbital evolution of a CV is largely determined by the value of $\dot{J}$ at each stage of its life (equations (9.4)). Empirical estimates of $\dot{J}$ may be obtained from the evolution of Ω_{rot} in late-type stars, using $\dot{J} = \dot{J}_{rot}$ for a synchronously rotating secondary.

For a rotating star

$$\dot{J}_{rot} = Mk^2R^2\dot{\Omega}_{rot}, \tag{9.10}$$

where k is the radius of gyration of the part of the star coupled to the magnetic wind. This may be only the outer convective zone, but is usually assumed to be the entire star, in which case $k^2 \approx 0.10$ for low mass dwarfs.

In a study of slowly rotating G dwarfs, Skumanich (1972; see also Smith 1979 and Iben, Fujimoto & MacDonald 1992a) deduced that the equatorial rotation velocities vary with age, t, according to

$$v_{eq} = R\Omega_{rot} = 7.3 \times 10^{13} t^{-1/2} \text{ cm s}^{-1}. \tag{9.11}$$

This may be differentiated to give

$$\dot{\Omega}_{rot} = -4.6 \times 10^{-7} R_1^2(2)\Omega_{rot}^3 \text{ rad s}^{-2}. \tag{9.12}$$

Following Verbunt & Zwaan (1981), assuming that this relationship holds for rapidly rotating late-type dwarfs in general, equations (2.1a), (9.10) and (9.12) give

$$\dot{J} = \dot{J}_{rot} = -3.4 \times 10^{36} \left(\frac{k^2(1)}{0.1}\right) M_1^{5/2}(2) R_1^{-1/2}(2) \text{ dyn cm.} \tag{9.13a}$$

If the secondary does not depart greatly from the empirical mass–radius–period relationship found for CVs (Section 2.8.3), equations (2.99) and (2.100) can be used to obtain

$$\dot{J}_{rot} = -1.2 \times 10^{34} \left(\frac{k^2(2)}{0.1}\right) P_{orb}^{31/12}(\mathrm{h}) \text{ dyn cm} \tag{9.13b}$$

(it should be noted that the exponent of P_{orb} in this equation is very sensitive to the exponent of the mass–radius relationship – see McDermott & Taam (1989)).

From a study of $\dot{P}_{orb}$ in CVs (but see Section 9.4.2 for the problems associated with interpretation of $\dot{P}_{orb}$), Patterson (1984) derived $\dot{J} = -1 \times 10^{37}\ M_1^4(2)$, which may be written, with the empirical mass–radius relationship,

$$\dot{J} = -1.8 \times 10^{32} P_{orb}^5(\mathrm{h}) \text{ dyn cm} \qquad (P_{orb} \lesssim 9 \text{ h}). \tag{9.14}$$

These, or similar, formulae have been used extensively in computations of orbital evolution of CVs (Section 9.3). There are indications, however, that the correct formulations may differ considerably from these. For example, Maceroni (1993; see also van't Veer & Maceroni 1988) finds for rapidly rotating G stars (v_{eq} up to 200 km s^{-1}) that

$$\dot{\Omega}_{rot} = 7 \times 10^{-20} \text{ rad s}^{-2} \tag{9.15}$$

independent of Ω_{rot} and applicable in both single and binary stars. This is lower by factors of 10–100 than the extrapolated Skumanich relationship.

Because of uncertainties in the braking law, $\dot{J}$ is often parameterized as a function of powers of R, M or Ω_{rot} (e.g., Rappaport, Verbunt & Joss 1983; Kawaler 1988; McDermott & Taam 1989) and the consequences for CV evolution computed for different values of the powers. The interaction between theory and observation in detached binaries is reviewed by Charbonneau (1992).

The rate of orbital angular momentum loss required to drive the observed rates of mass transfer in CVs is easily estimated. From equations (2.5b), (2.99), (2.100), (9.1a) and (9.4b), for main sequence secondaries

$$\dot{J} \approx 3.5 \times 10^{17} \frac{(19 - 15q)}{15q^{2/3}(1+q)^{1/3}} P_{orb}^{7/6}(\mathrm{h})\dot{M}(2) \text{ dyn cm} \quad (P_{orb} \lesssim 9 \text{ h}). \tag{9.16}$$

Taking $M_1(1) = 1$, $P_{orb} = 4\,\mathrm{h}$ and $\dot{M}(2) = -3 \times 10^{17}$ g s^{-1} (see, e.g., Section 4.3.3) gives $\dot{J} \sim -8 \times 10^{35}$ dyn cm. The Tout & Pringle (1993) magnetic braking theory predicts $\dot{J}_{MB} \sim -5 \times 10^{35}$ dyn cm and the Verbunt & Zwaan empirical formula (equation (9.13)) gives $\dot{J} \sim -4.3 \times 10^{35}$ dyn cm. The Mestel & Spruit theory integrates to give (McDermott & Taam 1989)

$$\dot{J}_{MB} = -9.06 \times 10^{35}\ M_1^{7/6}(2)\ R_1^{1/6}(2) \text{ dyn cm}, \tag{9.17a}$$

which, for the empirical mass–radius–period relationship, becomes

$$\dot{J}_{MB} = -2.52 \times 10^{34}\ P_{orb}^{1.64}(\mathrm{h}) \text{ dyn cm} \tag{9.17b}$$

and $\dot{J}_{MB} \sim -2.4 \times 10^{35}$ dyn cm for the above example.

Equation (9.16) can be reversed to give $\dot{M}(2)$ for a chosen $\dot{J}$, but the relationship to

P_{orb} is sensitive to the mass–radius relationship. From an evolutionary calculation using the Mestel & Spruit law McDermott & Taam (1989) find

$$\dot{M}(2) = 2.00 \times 10^{-11} P_{orb}^{3.7}(\mathrm{h})\ \mathrm{M}_\odot\ \mathrm{y}^{-1} \quad (P_{orb} > 2.7\ \mathrm{h}) \tag{9.18a}$$

Rapport, Verbunt & Joss (1983) obtain

$$\dot{M}(2) = 2.0 \times 10^{-11} P_{orb}^{3.2\pm0.2}(\mathrm{h})\ \mathrm{M}_\odot\ \mathrm{y}^{-1} \quad (P_{orb} > 2.7\ \mathrm{h}) \tag{9.18b}$$

from the Verbunt & Zwaan braking law.

9.1.2.3 Braking by Gravitational Radiation

As pointed out by Kraft, Matthews & Greenstein (1962; see also Paczynski 1967, Faulkner 1971), the small separation of component stars in the shortest period CVs makes gravitational quadrupole radiation a possible mechanism for loss of orbital angular momentum. General Relativity predicts that a binary will have (Landau & Lifschitz 1958)

$$\frac{\dot{J}_{GR}}{J} = -\frac{32G^3}{5c^5}\,\frac{M(1)M(2)[M(1)+M(2)]}{a^4}\,. \tag{9.19a}$$

With equations (2.1a), (2.5b) and (9.1a) this gives, for a lobe-filling secondary,

$$\dot{J}_{GR} = -8.50 \times 10^{35}\,\frac{q^{5/3}}{(1+q)^{2/3}}\,\frac{M_1^3(1)}{P_{orb}^2(\mathrm{h})}\,M_1^{1/2}(2)R_1^{-1/2}(2)\ \mathrm{dyn\ cm}, \tag{9.19b}$$

and *for a main sequence secondary*, through equations (2.100), (2.101) and (9.16),

$$\dot{M}(2) = -2.4 \times 10^{15}\,\frac{M_1^{2/3}(1)P_{orb}^{-1/6}(\mathrm{h})}{\left(1-\frac{15}{19}q\right)(1+q)^{1/3}}\ \mathrm{g\ s^{-1}} \quad (P_{orb} \lesssim 9\ \mathrm{h}). \tag{9.20}$$

The suprising weakness of dependence on P_{orb} results from the effect of decrease in P_{orb} (or a) being almost cancelled by the decrease of $M(2)$.

Braking by emission of gravitational radiation therefore falls far short of what is required for driving the observed rate of mass transfer in NLs and NRs (Section 4.3.3), but may be the dominant mechanism for at least some short period DN at quiescence (Table 2.2) and in polars (Section 6.4.3).

9.1.3 Tidal Synchronization and Circularization

In a non-synchronously rotating binary component the outer parts of its envelope travel through the tidal bulge raised by its companion. If the flow were inviscid the bulge would always point towards the companion, but the presence of viscosity results in the bulge lagging behind the substellar point if $P_{rot} > P_{orb}$ and leading it if $P_{rot} < P_{orb}$. The off-axis bulge results in a tidal torque, acting to synchronize rotation and to reduce any eccentricity of orbit (e.g., Zahn 1966; Hut 1981).

Ignoring for the moment the effects of mass transfer and loss of angular momentum discussed above, $J + j(2)$ must remain constant through the tidal evolution. The characteristic time for circularization $\tau_{cir} = J/\dot{J}$ and for synchronization

$\tau_{syn} = -\Omega_{rot}(2)/\dot{\Omega}_{rot}(2) = j(2)/\dot{j}(2)$, from a generalization of equation (9.2). Hence

$$\tau_{cir} = \frac{J}{j(2)}\tau_{syn} \tag{9.21}$$

(Lecar, Wheeler & McKee 1976; Tassoul 1988).

These characteristic times are inversely proportional to the effective viscosity in the stellar envelope. Three competing mechanisms have been proposed for evaluating the viscosity in convective envelopes of late-type stars. Zahn (1977; see also Campbell & Papaloizou 1983; Savonije & Papaloizou 1985) estimated the eddy-viscosity from a phenomenological argument and found

$$\tau_{syn} \sim 2.5 \times 10^3 (1+q)^2 P_{orb}^4(\mathrm{d})\ \mathrm{y}. \tag{9.22}$$

Tassoul (1988) and Tassoul & Tassoul (1992) propose a hydrodynamical mechanism, involving transient meridional currents induced by the asymmetry of the rotational motion, which gives

$$\tau_{syn} \sim 34\left(\frac{\mathrm{L}_\odot}{L}\right)^{1/4} M_1^{5/4}(2) R_1^{-3}(2) P_{orb}^{11/4}(\mathrm{d})\ \mathrm{y}. \tag{9.23a}$$

For a lobe-filling system this becomes, from equations (2.3a,b),

$$\tau_{syn} \sim 0.43\left(\frac{\mathrm{L}_\odot M(2)}{L(2)\mathrm{M}_\odot}\right)^{1/2} P_{orb}^{3/4}(\mathrm{h})\ \mathrm{y}. \tag{9.23b}$$

Zahn (1989) and Zahn & Bouchet (1989) note that the time scale of tidal deformation is similar to, or shorter than, the convective turn-over time, which makes the effective viscosity smaller than that adopted earlier and used in equation (9.22).

Attempts to decide between these mechanisms by appeal to observations of the incidence of synchronism in detached binaries have not been conclusive (Mazeh *et al.* 1990; Goldman & Mazeh 1991). As the time scales given by equations (9.22) and (9.23b), applied to CVs, are so short this may appear only of academic interest. However, the time τ_{syn} applies only to systems which are *already close to synchronisation* – for large departures ($|\,\Omega_{rot} - \Omega_{orb}\,|/\Omega_{orb} \gtrsim 0.1$) the viscous dissipation model breaks down and τ_{syn} can be factors 10^{5-6} greater (Campbell & Papaloizou 1983), whereas the hydrodynamical model continues to be effective (Tassoul & Tassoul 1992).

Adding the effect of magnetic braking demonstrates the importance of knowing τ_{syn}. If its time scale, $\tau_{MB} = J/\dot{J}_{MB}$, is significantly shorter than τ_{syn} then the secondary will be spun-down as for a solitary star, resulting in strong asynchronism. In that case a much smaller $|\,\dot{J}_{orb}\,|$ results, with lower values of $|\,\dot{M}(2)\,|$. This might be thought an explanation (van Paradijs 1986) of the spread in $\dot{M}(2)$ deduced from the M_v of discs (e.g., Figure 3.9). However, Czerny & King (1986) point out that a *bimodal* distribution of $\dot{M}(2)$ would result, with $\dot{M}(2)$ high when synchronous rotation obtains and $\dot{M}(2)$ low if the secondary spins down. There is no observational evidence for such a distribution.

The evidence from observed values of $v_{eq}(2)\sin i$ (Section 2.8.1) points to synchronous rotation for CVs containing dwarf secondaries. However, there may be asynchronous rotation of the giant secondary in RS Oph (Section 4.2.5.2) and the possibility also exists in the giant components of symbiotic stars (Kenyon 1986).

Another reason for requiring an accurate knowledge of τ_{syn} is that, as the tidal couple is $\propto (\Omega_{orb} - \Omega_{rot})$ (Zahn 1977), and therefore goes to zero at exact synchronism, magnetic braking of the secondary results in non-synchronous rotation at an equilibrium value

$$\frac{\Delta\Omega}{\Omega_{orb}} = \frac{\Omega_{orb} - \Omega_{rot}}{\Omega_{orb}} \sim \frac{\tau_{syn}}{\tau_{MB}} . \tag{9.24}$$

In the rotating frame of the binary, the secondary is then rotating with an equatorial velocity deficit of $\Delta\Omega R(2) = (\tau_{syn}/\tau_{MB})v_{eq}$. Provided $\tau_{syn} < 0.01\tau_{MB}$, which appears certain in CVs, the velocity deficit is $< c_s$ and no account of asynchronism need be taken on the flow from the L_1 point.

If the magnetic field of the secondary produces a significant triaxility of its shape, tidal interaction may induce *exact* synchronism (Applegate 1989).

The energy dissipation implicit in tidal braking heats the envelope of the secondary. This tidal luminosity is (Verbunt & Hut 1983)

$$L_{tid} = (\Omega_{orb} - \Omega_{rot}) \mid \dot{J} \mid \tag{9.25a}$$

$$\sim 2 \times 10^{31} \left(\frac{\tau_{syn}}{\tau_{MB}}\right) P_{orb}^{19/12}(\mathrm{h}) \text{ erg s}^{-1} \tag{9.25b}$$

from equations (9.13) and (9.24). Therefore, provided $\tau_{syn} \ll \tau_{MB}$, then $L_{tid} \ll L(2)$. However, Rieutord & Bonazzola (1987), who compute in detail the flow inside the secondary induced by tidal interaction, point out that if viscous dissipation is very much smaller (as in the Zahn (1989) formulation) then $L_{tid} \sim L(2)$ and the outer structure of the secondary would be affected accordingly.

No CVs have been found with a detectable orbital eccentricity. The observational limit is $e \lesssim 0.01$ in the most favourable case, but is confused by the effect of non-uniform irradiation of the secondary which induces an apparent eccentricity in the RV curve (Section 2.7.6).

A small eccentricity $e \sim 10^{-4}$ may be introduced at the time of a nova eruption (Edwards & Pringle 1987b) and could be maintained by tidal resonance with the disc (Papaloizou & Pringle 1979). This would lead to modulated transfer of the kind discussed in Section 3.6.6.1, for which there is no convincing observational evidence. Furthermore, Livio, Govarie & Ritter (1991) point out that the common envelope (Section 9.3.4) formed during the nova eruption would lead to a rapid decay of eccentricity. Observation of a finite e and its decrease after a nova eruption would give a direct measurement of τ_{cir}.

9.2 Pre-Cataclysmic Evolution

The theory of evolution of a single star demonstrates that a white dwarf forms as the core of a giant star which has a radius of typically 50–500 $R_\odot$ (see, e.g., Kippenhahn & Weigert (1990)). The existence, in CVs, of white dwarfs in binaries with separations $< 1R_\odot$ implies interesting events in their past evolution. It is clear that any extrapolation backwards in time will require that the entire binary be engulfed in the giant envelope surrounding the nascent white dwarf. The white dwarf must have formed in a wide binary ($a \gtrsim 50\mathrm{R}_\odot$) by the usual single star processes, with subsequent loss of orbital angular momentum (and energy) to draw the stars together. With $J \approx 10^{53}$ g cm^2 s^{-1} for the wide binary, it is evident that the braking mechanisms discussed in the previous section are inadequate to provide the reduction of J by a factor $\sim\sqrt{50}$ (equation (9.1a)) within the age of the Galaxy – a much more efficient braking mechanism is required. This is thought to be the frictional braking that occurs in a CE binary, a phenomenon already encountered in Section 5.8.5.

9.2.1 CE Binaries

9.2.1.1 Evolution into Contact

The more massive of the components in a binary evolves most rapidly off the main sequence, becoming a giant. The radius of a red giant star is determined almost entirely by the mass M_c of its degenerate core (see, e.g., Kippenhahn (1981)). From stellar models, the relationship for a carbon-oxygen core is (Ritter 1976)

$$\log R_1 = 1.48 + 1.30\frac{M_c}{\mathrm{M}_\odot} \; . \tag{9.26}$$

To grow white dwarfs in the mass range $0.5 < M_1 < 1.4$ therefore requires binaries with initial separations $130 < a/\mathrm{R}_\odot < 2000$, which have $0.3 < P_{\mathrm{orb}}(\mathrm{y}) < 165$ (for $M(1) + M(2) = 3\mathrm{M}_\odot$). Shorter period systems can lead to the primary filling its Roche lobe before helium ignition in the core, giving rise to helium white dwarfs in the mass range 0.18–0.46 $\mathrm{M}_\odot$. The variety of possible internal structures at the start of mass transfer, as functions of initial mass and separations, is described by Webbink (1979b). The initial masses of the primaries lie in the range $0.95 < M_1(1) \lesssim 10$: lower masses have not had time to evolve, higher masses lead to the formation of neutron stars.

9.2.1.2 Dynamical Mass Transfer

Because the more massive component evolves first, it is inevitable that $q > 1$ when mass transfer begins (the definition of q here temporarily reverses its previous rôle, but is consistent with the principle that it is the ratio of the mass of the lobe-filling component to that of its companion). This results, even for the conservative case, in shrinkage of the Roche lobe as soon as mass transfer begins (equation (9.6)). But, as seen from equation (9.26), the equilibrium radius of a cool giant does not depend on its total mass, and therefore no stabilizing reduction in radius takes place. The result is a runaway of mass transfer, limited only by the inability of the envelope to expand faster than the sonic velocity. Then $\dot{M} \sim M/\tau_{\mathrm{dyn}}$, where $\tau_{\mathrm{dyn}} \sim [R^3/GM]^{1/2}$ is the dynamical time scale (Section 9.3.1). Therefore mass transfer rates from the giant $\sim 0.1\mathrm{M}_\odot$ y^{-1} may be reached (cf. equation (2.12) with $\Delta R/R \sim 1$).

9.2.1.3 The CE

The accreting secondary is unable to adjust its structure at the rate at which mass is arriving (its Kelvin–Helmholtz time scale $\tau_{\mathrm{KH}} = GM^2/RL \gg \tau_{\mathrm{dyn}}$). If it has a radiative envelope it will expand towards a giant structure (Kippenhahn & Meyer-Hofmeister

1977). A convective envelope does not react so drastically (Webbink 1985) but in any case the Eddington limit establishes a rate $\dot{M} \lesssim 2 \times 10^{-3} R_1$ M$_\odot$ y^{-1} (see equation (9.35)) above which mass could not reach the surface. The result is that the transferring gas initially fills the outer Roche lobe (on which L_2 lies in Figure 2.2) and then grows to form an extended CE around the stellar components.

The CE binary as a nursery for CVs was first suggested by Ostriker (1976) and Paczynski (1976; see also Meyer & Meyer-Hofmeister 1979). The drag force experienced by the dwarf secondary orbiting within the giant envelope is

$$D \sim R^2(2) v_{\mathrm{orb}}^2 \rho \sim R_{\mathrm{L}}^2(2) v_{\mathrm{orb}}^2 \rho \sim a^2 v_{\mathrm{orb}}^2 \rho \sim GM(1) a \rho. \tag{9.27}$$

The luminosity deposited within the envelope is then

$$L_{\mathrm{D}} \sim D v_{\mathrm{orb}} \sim D \frac{a}{P_{\mathrm{orb}}} \sim \frac{GM(1) a^2}{P_{\mathrm{orb}}} \rho \tag{9.28}$$

and this is taken from the binding energy of the orbit; therefore

$$L_{\mathrm{D}} \sim \frac{\mathrm{d}}{\mathrm{d}t} \left[\frac{GM^2(1)}{a} \right] \sim \frac{GM^2(1)}{\tau_{\mathrm{D}} a}, \tag{9.29}$$

where τ_{D} is the time scale of the spiralling-in process. From equations (9.28) and (9.29)

$$\tau_{\mathrm{D}} \sim \frac{\overline{\rho}}{\rho} P_{\mathrm{orb}} \tag{9.30}$$

where $\overline{\rho} \sim M(1)/a^3$ is the mean density of the system within the binary core. In general $\overline{\rho} \gg \rho$, so $\tau_{\mathrm{D}} \gg P_{\mathrm{orb}}$; typically $\tau_{\mathrm{D}} \sim 10^3$ y.

Eventually the heat deposited in the envelope exceeds its binding energy and the envelope is ejected as a planetary nebula, exposing the central detached binary which now consists of the main sequence secondary orbiting a hot subdwarf primary.

Detailed models of this process have been computed (reviewed by Taam (1988, 1989) and Livio (1989, 1993b)). The most recent results show that (a) most of the ejection takes place close to the orbital plane (Livio & Soker 1988; Taam & Bodenheimer 1989: Figure 9.1), (b) although the secondary may accrete a substantial amount of mass, this expands and is transferred back to the CE, leaving the secondary only ~0.01 M$_\odot$ more massive than it started (Hjellming & Taam 1990, 1991) and (c) the CE can indeed be removed before the stellar components coalesce – e.g., a 2 M$_\odot$ red giant and a 1 M$_\odot$ dwarf emerge as a 29.2 h binary (Taam & Bodenheimer 1991). The latter is a non-trivial point: if the time scale τ_{E} for mass loss from the envelope were $< \tau_{\mathrm{D}}$ then coalescence would occur. However, because the envelope is spun-up by the orbiting secondary, which reduces the relative velocity and hence the viscous dissipation, τ_{D} becomes $\ll \tau_{\mathrm{E}}$. Also, there is a steep decline in density just outside the degenerate core; as a result, as soon as the orbiting secondary approaches this region and much of the envelope mass has already been driven off by centrifugal effects, the nuclear burning shells die out and the matter interior to the orbit contracts onto the white dwarf, increasing τ_{D} dramatically and effectually terminating further reduction in the orbital dimensions. The occurrence or not of this steep density gradient in different stages of evolution of giants of various masses may be the selection process that decides between coalesced stars and pre-cataclysmic candidates.

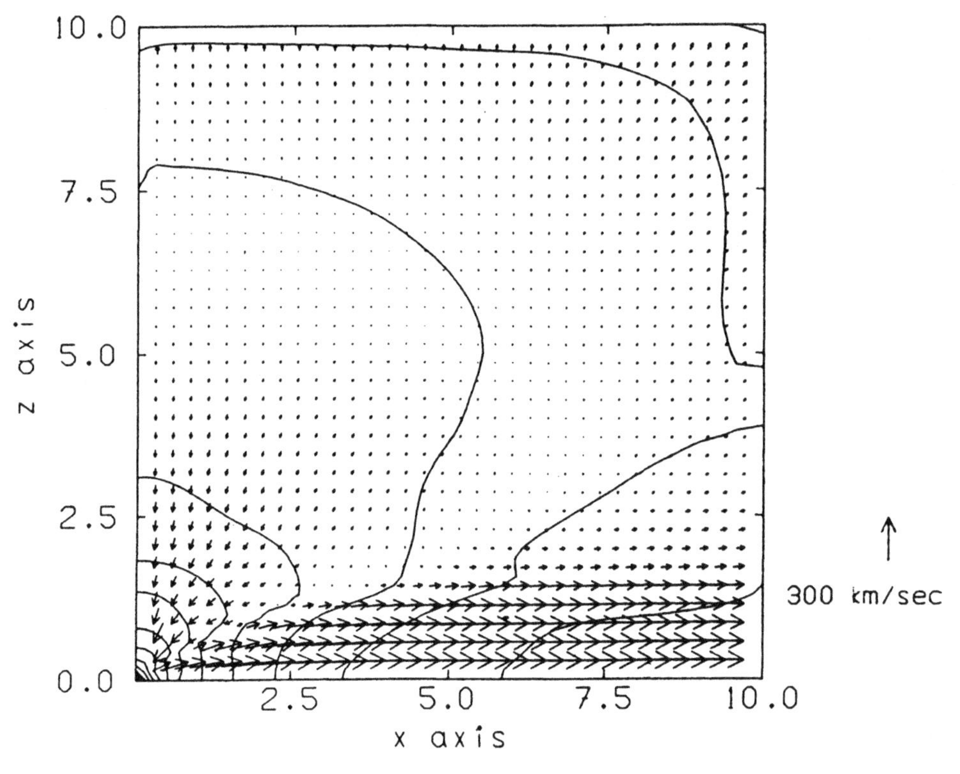

Figure 9.1 Velocity field and density distribution in the CE of a 5 $M_\odot$ red giant and a 1 $M_\odot$ companion. The axes are in units of 10^{12}cm, the z-axis is perpendicular to the orbital plane. Density contours are log ρ spaced logarithmically over $-6.6 \leq \log \rho \leq -3.6$. The gas within 12° of the equatorial plane is moving at velocities approaching escape velocity. From Taam & Bodenheimer (1989).

Another possible origin of CV precursors is through engulfment of a planet or very low mass star by expansion of a red giant. It is possible to find circumstances in which the planet will accrete sufficient gas to produce a low mass dwarf in a short period orbit, perhaps resembling the DN WZ Sge (Livio 1982; Soker, Horpaz & Livio 1984).

9.2.2 *Observations of Pre-Cataclysmic Binaries*

No object has yet been certainly identified as being in the CE phase of evolution[1] (which has a duration $\sim 10^3$–10^4 y). There are, however, several planetary nebulae (PN) known with short period binary nuclei. As the lifetime of a PN is $\sim 10^4$ y it is evident that these nuclei have recently emerged from a CE.

9.2.2.1 PN

From the observed incidence of binarity among late F and G stars, Livio (1993c) estimates that at least 17% will eventually undergo CE evolution. Observationally, ~13% of PN are known to have binary nuclei; the small difference is in agreement with the expected ~25% fraction of coalesced nuclei.

[1] But see the case for the carbon star V Hya (Kahane *et al.* 1993)

About 80% of PN show axial symmetry (Zuckerman & Aller 1986), partly due to CE evolution and partly to the influence of more distant orbiting companions on the PN ejected from undisturbed evolution. However, and ironically, only about half of the PN with known binary nuclei actually show axial symmetry (Bond & Livio 1990), so CE evolution may not always result in non-spherical ejection (some will appear symmetrical because they are observed along the axis of symmetry, the shapes of others may be modified by interaction with the interstellar medium).

9.2.2.2 PN Nuclei

A list of short period binaries containing hot subdwarf or white dwarf primaries with known or suspected late dwarf companions, including those in PN nuclei, is given[2] in Table 9.1. There are ten PN nuclei; three more (LW Hya, IN Com and the nucleus of LoTr1) have observed periods that are probably caused by (non-synchronous) rotation of the secondaries, so their orbital periods are not yet known (Livio 1993c). Discussion of the parameters and evolutionary status of most of the systems in Table 9.1 has been given by Ritter (1986a) and de Kool & Ritter (1993).

9.2.2.3 Other Detached Binaries

The remainder of the binaries in Table 9.1 are pre-cataclysmic candidates with no visible PN. Some of these are quite bright, but have been discovered only in the last decade; undoubtedly there are many more to be found, especially the older systems where the primary has low intrinsic luminosity. In the Palomar–Green Survey about two dozen white dwarfs and hot subdwarfs were found to have composite spectra revealing the presence of cool companions. Although an initial investigation of 19 systems (Ferguson, Green & Liebert 1984) did not generate any new orbital periods, later work has been more successful (Saffer *et al.* 1993). The Edinburgh–Cape Survey has begun to produce new candidates in the southern sky (O'Donoghue, private communication; Table 9.1). A significant number of faint white dwarfs are found to have composite spectra (Berg *et al.* 1992), but ~90% of such systems are expected to be wide and non-interacting (de Kool & Ritter 1993).

No systems with strong magnetic fields, which would be the precursors of the magnetic CVs, are yet known.

9.2.3 Evolution into Contact

As the pre-cataclysmic binaries are detached systems, to first order $\dot{M}(1) = \dot{M}(2) = 0$ and equations (9.4a) and (9.4b) give

$$\frac{\dot{J}}{J} = 1/2\frac{\dot{a}}{a} = 1/3\frac{\dot{P}_{\text{orb}}}{P_{\text{orb}}} \;. \tag{9.31}$$

For a given $\dot{J}$ these can be integrated to yield P_{orb} and a as functions of time, and in particular to give the time t_{sd} taken to reach the semi-detached state, i.e., for the secondary to make contact with its Roche lobe. The importance of magnetic braking in the evolution of these systems was first pointed out by Eggleton (1976).

[2] As these stars are not themselves CVs, only the most recent reference to substantial discussion of each object is given. More complete references are given by Ritter (1990, 1992)

Table 9.1. *Pre-Cataclysmic Binaries.*

Star	Alias	Type	m_V	P_{orb}(h)	Spectral type	$i°$	$M_1(1)$	$M_1(2)$	$R_1(2)$	P_f(h)	Reference
V651 Mon	AGK3−0°965	PN	11.3	15.991d	A5V	60:	0.40 ± 0.05	1.8±0.3	2.2±0.2		1
FF Aqr	BD −3°3537		9.3	9.2078d	G8III+sdOB	81	0.5	2.0	5.7		2
Feige 24			12.4	4.2318d	M3.5V + DA1	>44	0.54±0.20	0.41		4.4	3
HD 33959C			8.0	71.8h	F4V+DA						26
SP1		PN	13.9	69.8							4
PG 1538+269	Ton 245		13.9	60.02	sdB			>0.68		>6.5	5
BE UMa	PG 1155+492		14.1	54.988	K5V + DO	>84	0.6±0.1	0.34±0.13	0.47±0.15		6
HtTr 4		PN		41.0							4
HD 49798			8.3	37.1441	F4-K0+sdO		1.75±1.0	1.75:	1.45±0.25		7
RE 0720-318			14.8	24.5	M3 + DAO						27
Abell 65		PN	15.9	24.00							4
RE 2013+400			14.6	17.04	M4+DAO						26
VW Pyx		PN	16.5	16.09							8
EG UMa	Case 1		13.3	16.023	M4V + DA	<72	0.38±0.07	0.26±0.04	0.5	3.0	9
PG 1026+002			13.8	14.334	M4V + DA2	>49	0.68±0.23	0.22±0.05	0.25±0.03	2.6	10
V664 Cas	HFG 1	PN	14	13.96							11
UX CVn	HZ 22		13.1	13.769	B2 + B3	90	0.39±0.05	0.42	<1.23	4.4	12
HZ 9			14.0	13.544	M5eV + DA2		0.51±0.10	0.28±0.04		3.2	9

V471 Tau	BD +16°516		9.2	12.508	K2V + DA2	79.5	0.71±0.02	0.73±0.03	0.80± 0.03	6.9	6
V477 Lyr		PN	14.6	11.321			0.51±0.07	0.12±0.02		1.6	13
UU Sge		PN	14.2	11.162	KV + sdO		0.63±0.06	0.29±0.03	0.53±0.02	3.3	14
KV Vel	LSS 2018	PN	12.3	8.571	sdO	62±10	0.55±0.15	0.25± 0.06	0.35±0.05	2.9	15
GK Vir	PG 1413+015		17.0	8.264	M3-5V+DAO	90	0.51±0.04	0.10	0.15	1.4	16
PG 0900+401			12.9	8.116	K3						17
EC 11575-1845			12.8	7.86	sdO +						18
RR Cae	LFT 349		14.4	7.289	MeV + DAwk		0.28	0.15:		1.9:	19
PG 0308+096			15.3	6.823	M4-5V+DA2	42±9	0.40±0.11	0.18±0.05	0.21±0.02	2.3	10
AA Dor	LB 3459		11.2	6.277	sdO		0.25±0.05	0.043±0.005	0.08±0.01	0.72	25
Feige 36	PG 1101+249		12.7	4.94	sdB		0.5:				20
BPM 71214				4.33							21
EC 13471-1258			14.6	3.617	dM4-5V+DA						18
NN Ser	PG 1550+131		16.6	3.122	M5-6V	>81	0.54±0.07	0.12±0.03		1.6	22
HW Vir	BD −7°3477		10.5	2.801	sdB	80.6±0.9	0.54±0.09	0.16 ± 0.03	0.19 ± 0.01	2.0	23
MT Ser		PN	15.4	2.717	sdO		0.6:	0.2±0.1		2.5:	24

References: 1. Mendez *et al.* 1985; 2. Dworetsky *et al.* 1977; 3. Vennes *et al.* 1991; 4. Bond & Livio 1990; 5. Ritter 1992; 6. Ferguson *et al.* 1987, Liebert *et al.* 1994; 7. Kudritzki & Simon 1978; 8. Bond & Grauer 1987; 9. Stauffer 1987; 10. Saffer *et al.* 1993; 11. Acker & Stenholm 1990; 12. Schoenberner 1978; 13. Pollaco & Bell 1993b; 14. Pollacco & Bell, 1993a; 15. Landolt & Drilling 1986; 16. Fulbright *et al.* 1993; 17. Lipunova & Shugarov 1991; 18. Chen 1994; 19. Bragaglia *et al.* 1990; 20. Reid, Saffer & Liebert 1993; 21. Livio & Shara 1987; 22. Wood & Marsh 1991; 23. Wood, Zhang & Robinson 1993; 24. Grauer & Bond 1983; 25. Kudritzki *et al.* 1982; 26. Barstow *et al.* 1995a; 27. Barstow *et al.* 1995b.

For systems with $M_1(2) \gtrsim 1$ there will of course be a race between shrinkage of the Roche lobe and expansion of the secondary on a nuclear time scale. The following treats only the situation where t_{sd} is shorter than the latter (see Section 9.3.4 for the alternative).

Taking $\dot{J} = \dot{J}_{GR}$, which may be appropriate for $M_1(2) \lesssim 0.25$ (and for strongly magnetic systems: Section 9.3.7), equations (9.18) and (9.31) yield, for a lifetime in the detached state,

$$t_{sd} = 4.73 \times 10^{10} \frac{(1+q)^{1/3}}{q} M_1^{-8/3}(1)[P_i^{8/3}(d) - P_f^{8/3}(d)] \text{ y}, \tag{9.32}$$

where P_i is the initial orbital period and P_f is the final orbital period, i.e., at contact. The latter is found from (2.100) as

$$P_f(d) = 0.37 M_1^{4/5}(2) = 0.37 q^{4/5} M_1^{4/5}(1) \quad (P_f(d) \lesssim 0.38). \tag{9.33}$$

Equation (9.32) may also be used to estimate the time remaining in a detached state (for GR braking), for example, HW Vir (Table 9.1) will become a CV after $\sim 1.7 \times 10^9$ y.

Because of uncertainty in the braking law, the situation with magnetic braking is less clear cut.[3] Using equations (9.10), (9.12), (9.31) gives

$$t_{sd} \approx \frac{8.26 \times 10^6}{(1+q)^{1/3} q^{2/3} M_1^{2/15}(2)[k^2(2)/0.1]} \left[\left(\frac{P_i}{P_f} \right)^{10/3} - 1 \right] \text{ y}. \tag{9.34}$$

For example, V471 Tau (Table 9.1) would be expected to become a CV with $P_{orb} = 6.9$ h after a time $t_{sd} \sim 3 \times 10^7$ y. There is indeed evidence from eclipse timings that P_{orb} is decreasing on a time scale $\sim 10^7$ y, on which are superimposed quasi-periodic variations of the kind described in Section 9.4 (Skillman & Patterson 1988).

Many of the systems in Table 9.1 have $t_{sd} >$ Hubble time, and are therefore not representative of the population from which the currently observed CVs have evolved. Only a few have $t_{sd} < 10^9$ y and only V471 Tau has $t_{sd} < 10^8$ y. V471 Tau is a member of the Hyades cluster, which has an evolutionary age $\sim 7 \times 10^8$ y; this membership implies an original mass of ≥ 2 M$_\odot$ for the primary in V471 Tau, which is the present mass at the main sequence turn-off in the HR diagram. Clearly the probability of observing a system in the detached state is $\tilde{\propto}\ t_{sd}$ (see Ritter (1986a) for more detailed discussion), so the observed systems are strongly biased towards large t_{sd}. In particular, if the CE phase commonly delivers binaries that are only barely detached, few if any of these would be observed. However, of the PN nuclei in Table 9.1 none other than MT Ser (and perhaps V664 Cas – see below) appear close to becoming semi-detached.

At contact the primary can only accept mass up to a rate which generates L_{Ed}. From equation (5.4a) this requires

$$\dot{M}(1) < \dot{M}_{Ed} = 1.8 \times 10^{-5} R_9(1) \quad \text{M}_\odot \text{ y}^{-1}. \tag{9.35}$$

[3] Long term studies should be made of the eclipsing systems GK Vir, NN Ser and HW Vir to look for secular (or quasi-periodic) changes in P_{orb} which would elucidate the braking mechanism in detached binaries. In HW Vir Kilkenny, Marang & Menzies (1994) find $P/\dot{P} \sim 1.1 \times 10^6$ y, which is an order of magnitude shorter time scale than predicted by magnetic braking. Perhaps cyclical effects, similar to those in CVs discussed in Section 9.4.2, are dominating over secular changes.

If $\dot{M}(1) > \dot{M}_{\rm Ed}$ then a second CE phase will ensue. If $\dot{M}_s(1) < \dot{M}(1) < \dot{M}_{\rm Ed}(1)$ (Section 5.8.1), stable nuclear burning will occur (for examples of this see Sections 5.8.1 and 9.6).

The weighted mean $\overline{M}_1(1)$ in Table 9.1 is 0.57, similar to that of isolated white dwarfs (Bergeron, Saffer & Liebert 1994). Selection effects strongly favour discovery of CVs with high $M(1)$ (Section 9.5.6.3), so the pre-CVs in Table 9.1 are not representative of *observed* CVs for this reason as well. The values of $M(2)$ in Table 9.1 will result in the majority of systems having convective secondaries on first contact. As can be seen from the tabulated values of $P_{\rm f}$(h), the observed population of pre-CVs is not at all representative of that from which CVs with $P_{\rm orb} \gtrsim 5$ h is drawn.

Surprisingly, there is evidence for at least two CVs being associated with planetary nebulae. BZ Cam is a NL (Table 4.1) with a surrounding nebula that was thought to be an ejected nova shell (Ellis, Grayson & Bond 1984) but which has too low an expansion velocity and may instead be a PN (Krautter, Klaas & Radons 1987). GK Per has an IR dust shell of $\sim$0.7 pc radius which may be an ancient PN (Bode *et al.* 1987; Scott, Rawlings & Evans 1994; cf. Hessman 1989b). The PN nucleus V664 Cas (Table 9.1) shows some signs of already being a CV, with a spectrum similar to the IP TW Pic (Acker & Stenholm 1990). Bond (1989) has pointed out that Barnard's (1908) observations of >3 mag variations of the nucleus of the PN NGC 7662 are possible DN outbursts. These systems suggest that CE evolution is able in some cases to leave a mass-transferring binary at the start of its CV evolution.

Although detached, and formally pre-cataclysmic, it is possible that pre-contact systems can have nova eruptions. V471 Tau has circumstellar gas expanding at 1200 km s^{-1} (Sion *et al.* 1989) which could be from a nova eruption in AD396 (Pskovski 1979). If V471 Tau had such a nova event it would have been as the result of wind-accreted mass, which is estimated to occur at $\dot{M}(1) \sim 4 \times 10^{-13}$ M$_\odot$ y^{-1} (Sion *et al.* 1989).

9.3 Secular Evolution

From the moment of contact with its Roche lobe, the secondary has a mass-losing outer boundary. This can have a profound effect on its internal structure, driving it out of thermal equilibrium. In some circumstances mass transfer may commence on a dynamical time scale, leading to a second CE phase (see Webbink & Iben (1987) for the response of a white dwarf to dynamical mass transfer rates).

Long term mass transfer, driven by magnetic or gravitational braking, changes the orbital elements through equations (9.4). The population of CVs at each $P_{\rm orb}$ is generated from the initial $P_{\rm orb}$ distribution acted upon by secular changes. The period gap in non-magnetic CVs, and its absence for magnetic CVs, and the presence of a minimum $P_{\rm orb}$ for hydrogen-rich CVs, should be explicable from these inputs.

Recent reviews of CV evolution are given by Ritter (1986b) and King (1988).

9.3.1 Response of the Secondary Star

The secondary star has two time scales on which to respond to mass loss. Adjustments to hydrostatic equilibrium (except near L_1) occur adiabatically on a dynamical time scale, which is the characteristic sound-crossing time of the region affected (or, equivalently, the pulsational time scale). In the interior of a star the dynamical time scale is

$$\tau_{dyn} = [G\bar{\rho}(r)]^{-1/2} \tag{9.36a}$$

where $\bar{\rho}(r)$ is the mean density interior to r. For the gas lying outside the Roche lobe,

$$\tau_{dyn} = \left[G\bar{\rho}(r)\frac{\Delta R}{R}\right]^{-1/2} \tag{9.36b}$$

(Webbink 1985).

Thermal adjustment, on the other hand, is determined by the rate at which heat can diffuse through the star, and occurs on the thermal, or Kelvin–Helmholtz, time scale

$$\tau_{KH} = \left(\frac{3}{10-2n}\right)\frac{GM^2(2)}{R(2)L(2)} \text{ for a polytrope of index } n. \tag{9.37}$$

For $M_1(2) = 0.5$, $R_1(2) = 0.5$, $n = 3$ and $L(2) = 0.1\mathrm{L}_\odot$, $\tau_{dyn} = 0.45\,\mathrm{h}$ and $\tau_{KH} = 1.1 \times 10^8\,\mathrm{y}$. For rates of mass transfer $\dot{M}(2) \sim 1 \times 10^{-8}\,\mathrm{M}_\odot\,\mathrm{y}^{-1}$, the time scale of mass loss $\tau_{ML} = M(2)/\dot{M}(2) < 10^8\,\mathrm{y}$ and such a secondary star will not be able to maintain thermal equilibrium. This reduces $L(2)$ and pushes the star even further from equilibrium.

If, when first making contact with its Roche lobe, the secondary star is unable to adjust its radius rapidly enough to remain within the lobe, mass transfer will proceed on the dynamical time scale, saturating at $\dot{M}(2) \sim M(2)/\tau_{dyn}$; otherwise transfer is stable at a rate $\dot{M}(2) \sim M(2)/\tau_{ML}$ (where $\tau_{ML} = \tau_{MB}$, τ_{GR}, or τ_{nuc} for expansion on a nuclear time scale, as appropriate).

The criterion for avoiding mass transfer on a dynamical time scale is

$$\xi = \frac{\partial \ln R(2)}{\partial \ln M(2)} > \frac{d \ln R_L(2)}{d \ln M(2)} \equiv \xi_L \tag{9.38}$$

where $\xi = \xi_{ad}$ represents the *adiabatic* response of the secondary, i.e., its response on a dynamical time scale, or, if $\tau_{ML} \gg \tau_{KH}$ then $\xi = \xi_{KH}$, which can be obtained from the exponent in the mass–radius relationship for equilibrium models (e.g., $\xi_{KH} = 0.87$ for main sequence stars, from equation (2.99)), whereas if $\tau_{ML} \leq \tau_{KH}$ the adiabatic response is required. Paczynski (1965a) found $\xi = -1/3$ for the dynamic response of stars with convective envelopes.

For secondaries with $M_1(2) \gtrsim 1$, equation (9.37) shows that $\tau_{KH} < \tau_{ML}$. For conservative mass transfer, equations (9.6) and (9.38) require

$$q < \frac{1}{2}\,\xi_{KH} + \frac{5}{6} \tag{9.39}$$

and therefore $q < 1.26$ for stability of mass transference from main sequence stars. As the maximum mass of the primary is M_{ch}, this gives an upper limit $M_1(2) \lesssim 1.8$; systems not obeying $q < 1.26$ would transfer mass on a thermal time scale, generating stable nuclear burning or passage through a second phase of CE evolution, according to $\dot{M}_s(1) < \dot{M}(2) < \dot{M}_{Ed}(1)$.

In general, few primaries have initial masses close to M_{ch} (Politano & Webbink

1989), so a more realistic observational cut off is $M_1(2) \lesssim 1.5$, which gives $P_{orb} \lesssim 12.5$ h. This is the reason that main sequence secondaries of $M_1(2) \gtrsim 1.5$ and of spectral types earlier than ~G0 are not found in CVs (but see Section 9.6).

Secondaries with $M_1(2) \lesssim 0.8$ have $\tau_{KH} > \tau_{ML}$, for which $\xi_{ad} = -1/3$ is appropriate (Faulkner 1976; Tutukov, Fedorova & Yungelson 1982). From equation (9.39) this demands $q < 2/3$ for stability. However, if $\xi_{KH} > \xi_{ad} > \xi_L$, which can occur for $M_1(2) \sim 0.4$, $\dot{M}(2)$ rises briefly to a moderate rate until the envelope is driven out of thermal equilibrium (raising ξ to ~1) whereupon stability is achieved (Rappaport, Joss & Webbink 1982). Hjellming & Webbink (1987) have considered the stability of mass transfer in detail, finding that ξ_{ad} for convective stars depends on the mass of the radiative core (cf. Paczynski 1965a), with $q < 0.634$ for stability of transfer from a fully convective secondary ($M_1(2) \approx 0.25$) but steadily increasing for higher masses (see also Cameron & Iben (1986)). A more detailed evaluation of the critical q for stable mass transfer is given in Figure 2 of de Kool (1992).

Following contact, the evolution of the secondary star may be computed (assuming hydrostatic but not thermal or nuclear equilibrium) as mass is stripped from it (usually assumed to be in a spherically symmetrical fashion). The rotational and tidal distortion has only a very small effect on the luminosity and radius (Chan & Chau 1979; Nelson, Chau & Rosenblum 1985). The value of $\dot{M}(2)$ is found from, e.g., equation (9.16), with an adopted $\dot{J}$ from gravitational and magnetic braking, together with an assumption about mass loss, e.g., conservancy, or $\dot{M}(1) = 0$ because of nova eruptions. In the latter case it may be assumed that the ejecta carry away further orbital angular momentum at a time-averaged rate (Warner 1978; see also Section 9.3.4)

$$\frac{\dot{J}}{J} = \frac{q^2}{1+q} \frac{\dot{M}(2)}{M(2)} . \tag{9.40}$$

Extensive computations of CV evolution have been carried out by Faulkner (1971), Chau & Lauterborn (1977), Taam, Flannery & Faulkner (1980), Whyte & Eggleton (1980), Paczynski & Sienkiewicz (1981, 1983), Rappaport, Joss & Webbink (1982), D'Antona & Mazzitelli (1982a,b), Taam (1983), Spruit & Ritter (1983), Joss & Rappaport (1983), Rappaport, Verbunt & Joss (1983), Rappaport & Joss (1984), Iben & Tutukov (1984), Verbunt (1984), Sienkiewicz (1984), Ritter (1985a), Nelson, Rappaport & Joss (1986), Hameury *et al.* (1987), Pylyser & Savonije (1988a,b), Hameury *et al.* (1988), McDermott, Taam & Ringwald (1988), D'Antona, Mazzitelli & Ritter (1989), McDermott & Taam (1989), Hameury (1991) and Kolb & Ritter (1990, 1992).

An example (Verbunt 1984) of the effect of departure from thermal equilibrium on the radius and temperature (and hence luminosity) of the secondary, caused by steady mass transfer, is shown in Figure 9.2. The initial model had $M_1(1) = 1.2$, $M_1(2) = 1.0$, $P_{orb} = 7.2$ h. Two different laws of magnetic braking were used, giving 'strong' and 'intermediate disrupted' (see Section 9.3.3) braking. The effective temperature as a function of mass is hardly affected by braking, but the (non-equilibrium) radius increases greatly with stronger braking relative to that of an equilibrium main sequence star of the same mass. Therefore, at a given P_{orb}, the mass, radius and luminosity are *reduced* relative to those of an equilibrium lobe-filling dwarf. At $P_{orb} \sim 3$ h, Verbunt

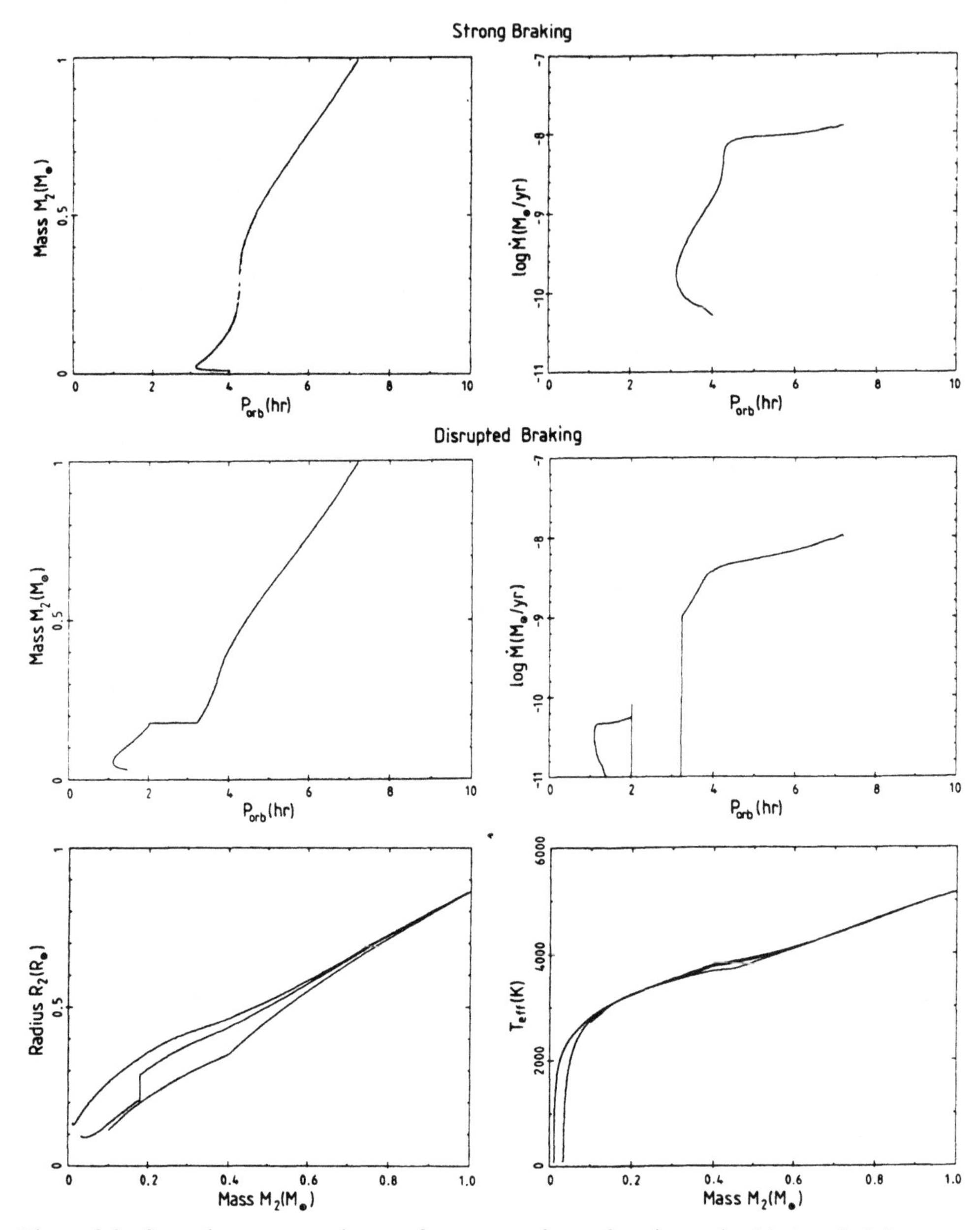

Figure 9.2 Secondary mass and rate of mass transfer as functions of orbital period for two models of angular momentum loss. In the temperature and radius graphs the lower lines are for equilibrium main sequence stars, the middle are for intermediate disrupted braking and the upper are for strong braking. Adapted from Verbunt (1984).

found ($M_1(2)$, $R_1(2)$, $L(2)/L_\odot$) = (0.40, 0.35, 0.02), (0.18, 0.29, 0.0072), (0.02, 0.14, 2.1×10^{-4}) respectively for equilibrium structure, intermediate and strong braking. The fact that secondaries have $M_V(2)$ that do not depart significantly from those of equilibrium stars (Section 2.9.5, Figure 2.46) then excludes any kind of strong braking. From Figure 9.2 this in turn implies that the *long term average* $\dot{M}(2)$ for $P_{orb} \lesssim 5\,h$ cannot be much greater than $\sim 3 \times 10^{-9}\,M_\odot\,y^{-1}$. At larger P_{orb} $(M_1(2) \gtrsim 0.6)$ the departures from thermal equilibrium are insufficient to set any limits on $\dot{M}(2)$. The Verbunt & Zwaan (1981) braking law and other prescriptions given in Section 9.1.2.2 satisfy these conditions. The higher values of $\dot{M}(2)$ *observed* in some systems with $P_{orb} < 5\,h$ are discussed in Section 9.5.1.

The nuclear luminosity in the secondary is generated by the p–p chain, in which the $^2D + p \rightarrow {}^3He + \gamma$ rate is much greater than that of $^3He + {}^3He \rightarrow {}^4He + p$ (Clayton 1968). As a result, the time taken to reach equilibrium abundances of ^3He and ^4He is $> \tau_{ML}$ over a considerable part of the star's energy generating region (D'Antona & Mazzitelli 1982a; Joss & Rappaport 1983; Iben & Tutukov 1984; Nelson, Chau & Rosenblum 1985; McDermott, Taam & Ringwald 1988). Typically ^3He is in equilibrium in the central regions, rising to a maximum abundance $Y(^3He)$ of $\sim$ several $\times$ equilibrium in an intermediate region, and falling in the convective envelope to the initial abundance. As the energy generation rate for $^3He + {}^3He$ is proportional to $Y^2(^3He)$ the luminosity is sensitive to the $Y(^3He)$ profile.

A second effect is that convective mixing takes place more rapidly than ^3He can find equilibrium. The result of departure from thermal equilibrium is both to drive convection in the envelope to greater depths and to create a convective core (d'Antona & Mazzitelli 1982a). ^3He is mixed through the core, increasing $Y(^3He)$ at the centre. When the outer and inner convection zones meet, ^3He is mixed throughout the star, causing a decrease in $Y(^3He)$ and a fall in nuclear luminosity (McDermott, Taam & Ringwald 1988).

The response of the secondary to irradiation by the companion has been considered by several authors and is used in one model of superoutbursts (Section 3.6.6.4). Although most of the soft X-ray and EUV illumination may be scattered away (Section 2.9.6), hard X-rays and photons with $E < 13.6$ eV can penetrate to subphotospheric regions (Hameury, King & Lasota 1986a; Sarna 1990). Provided the irradiation energy is $\gtrsim 10^{11}$ erg cm^{-2} s^{-1} the outermost convective zone is modified and the star becomes radiative throughout (Podsiadlowski 1991), giving a large equilibrium radius. The increase in radius has been estimated to be $\sim$2% for a secondary heated by a $\sim 5 \times 10^4$ K white dwarf subsequent to a nova eruption (Kovetz, Prialnik & Shara 1988) or $\sim$10% from high boundary layer luminosities (Podsiadlowski 1991). The actual BL luminosities (Section 2.5.4) are much lower, which will reduce the latter estimate. Short time scale effects of irradiation are discussed in Section 9.4.3.

Suppression of convection on the irradiated side of the secondary could largely destroy its magnetic dynamo, leading to reduced braking and a strongly asymmetric wind from the secondary. This should be looked for in the detached systems (Section 2.9.6).

9.3.2 *The Minimum Orbital Period*

The shortest observed orbital period for a CV with normal hydrogen-rich composition is $\sim$75 min (Section 2.2). As pointed out independently by Paczynski & Sienkiewicz (1981) and Rappaport, Joss & Webbink (1982) (see also D'Antona & Mazzitelli 1982b), this is readily understood from the response of very low mass secondaries to continuing mass transfer.

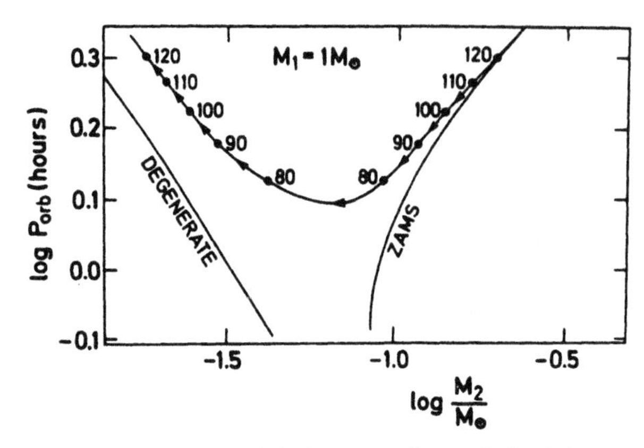

Figure 9.3 Secular evolution near the period minimum. The equilibrium main sequence and white dwarf sequence are shown. Orbital periods in minutes are marked on the evolutionary sequence. From Ritter (1986b).

There are three factors involved. First, for $M_1(2) \lesssim 0.08$ the secondary becomes fully degenerate and ceases to burn hydrogen. It becomes in effect a very low mass hydrogen-rich white dwarf and, if in thermal equilibrium, would obey the mass–radius relationship (Chandrasekhar 1939)

$$R(2) \approx 8.96 \times 10^8 (1+X)^{5/3} M_1^{-1/3}(2) \text{ cm} \tag{9.41}$$

where X is the fractional hydrogen content. Therefore, decrease of mass leads to *expansion* of the secondary. Secondly, from equations (2.1b),(2.5b) and (9.41)

$$P_{\text{orb}}(\text{h}) \approx 1.41 \times 10^{-2}(1+X)^{5/2} M_1^{-1}(2), \tag{9.42}$$

so mass transfer results in an *increasing* orbital period. This implies a period minimum somewhere during the transformation from a shrinking non-degenerate secondary to an expanding degenerate secondary (Figure 9.3). (This may seem a surprising result – that the orbital separation is able to increase to accommodate an expanding secondary. It can be understood by thinking of the transfer of mass as a two-stage event: since $q < 1$, moving mass from the secondary to the primary increases the separation, cutting off mass transfer. Orbital angular momentum loss reduces the separation, restoring mass transference at a larger separation than before if the secondary expanded on losing mass, or at smaller separation if it shrank. In practice the process is of course a continuous one.) From the previous section it is seen that, since $q \ll 0.67$, mass transfer from a degenerate secondary is stable.

For a low mass degenerate He secondary in thermal equilibrium, equations (2.3b) and (2.83d) give the more accurate relationship

$$M_1(2) = 1.20 \times 10^{-2} P_{\text{orb}}^{-1}(\text{h})[1 - 8.65 \times 10^{-2} P_{\text{orb}}^{2/3}(\text{h})]^{15/2} \quad P_{\text{orb}} \lesssim 2.5 \text{ h}. \tag{9.43}$$

Finally, for small $M(2)$, even while still on the main sequence, thermal equilibrium cannot be maintained: the lowest possible *average* $\dot{M}(2)$ is given by gravitational braking which, from equations (2.1a), (2.100) and (9.18) has

$$\tau_{\rm GR} \sim 1.2 \times 10^9 \frac{(1+q)^{1/3}}{M_1^{2/3}(1)} P_{\rm orb}^{17/12}(\rm h)\ y, \tag{9.44}$$

whereas, using $L = L_\odot M_1^{3.5}(2)$ for low mass dwarfs, equations (2.100), (2.101) and (9.37) give

$$\tau_{\rm KH} \sim 1.95 \times 10^{10} P_{\rm orb}^{-71/24}(\rm h)\ y. \tag{9.45}$$

Therefore $\tau_{\rm GR} < \tau_{\rm KH}$ for $P_{\rm orb} \lesssim 1.9\,\rm h$. On being driven out of equilibrium, $L(2)$ is reduced, giving an even larger $\tau_{\rm KH}$.

The value, $P_{\rm orb,min}$, at which the orbital period reaches a minimum therefore depends sensitively on the actual $\dot{M}(2)$ (which, through the operation of magnetic braking, may be greater than that from GR alone) and on the internal structure of the secondary. Unfortunately, opacities are not accurately known for stellar surface temperatures $\lesssim$2500 K. However, the calculated values of $P_{\rm orb,min}$ (Paczynski & Sienkiewicz 1981; Rappaport, Joss & Webbink 1982; Rappaport, Verbunt & Joss 1983) are ~80 min for moderate $\dot{M}(2)$, in agreement with observations. The effect of *strong* magnetic braking is seen in Figure 9.2: the secondary is far from thermal equilibrium and degeneracy only sets in when its mass is very low, giving a smaller $M(2)$ at $P_{\rm orb,min}$ and a *larger* value of $P_{\rm orb,min}$. In general, $P_{\rm orb,min}$ occurs at the point where $\tau_{\rm KH}/\tau_{\rm ML} \sim 1.5$ (Nelson, Rappaport & Joss 1986).

$P_{\rm orb,min}$ is sensitive to chemical composition of the secondary (Sienkiewicz 1984; Nelson, Rappaport & Joss 1986; Pylyser & Savonije 1988a,b). $P_{\rm orb,min}$ decreases monotonically with X. This is because equilibrium radii of low mass stars decrease with decreasing X (equation 9.41). At $X \sim 10^{-3}$, $P_{\rm orb,min} \sim 35$ min.

Assuming purely gravitational braking, $\dot{M}(2)$ can be estimated for late evolution when the secondary is degenerate and $P_{\rm orb}$ is increasing (Nelson, Rappaport & Joss 1986; King, 1988). From equations (9.4b), (9.19a) and (9.38), for conservative mass transfer

$$\dot{M}(2) = \frac{1.27 \times 10^{-8} q^2 M_1^{8/3}(1)}{(1+q)^{1/3}\left(\frac{5}{6} + \frac{\xi_{\rm ad}}{2} - q\right)} P_{\rm orb}^{-8/3}(\rm h)\ \ M_\odot\ y^{-1}. \tag{9.46a}$$

To proceed further, both $\xi_{\rm ad}$ and a mass–radius relationship for the secondary (out of thermal equilibrium) are required. In thermal equilibrium $\xi_{\rm ad} = -1/3$ and equations (9.42) and (9.46a) give

$$\dot{M}(2) = 2.5 \times 10^{-12} \frac{(1+X)^5 M_1^{2/3}(1)}{\left(\frac{2}{3} - q\right)(1+q)^{1/3}} P_{\rm orb}^{-14/3}(\rm h)\ \ M_\odot\ y^{-1}. \tag{9.46b}$$

For example, with $X = 0.6$ and $P_{\rm orb} = 1.5\,\rm h$, $\dot{M}(2) \sim 3 \times 10^{14}$ g s^{-1}, which is much smaller than what is observed in typical CVs (Table 2.2) and implies that they are on the decreasing branch of $P_{\rm orb}$ evolution. On the other hand, the large interval between outbursts of DN like WZ Sge (Section 3.5.4.5) suggests much lower $\dot{M}(2)$ and the possibility that such stars have increasing P_{orb}.

From equations (9.42) and (9.46b) it is seen that, for a given $M(2)$, $\dot{M}(2) \propto (1+X)^{-20/3}$ and therefore $\dot{M}(2)$ and the rate of evolution on the rising branch of P_{orb} increase greatly for hydrogen-deficient systems (Nelson, Rappaport & Joss 1986).

The orbital evolution of pure He systems is discussed in Section 9.7.2.4.

9.3.3 The Period Gap

The deficiency of non-magnetic CVs in the period range $2.3 \lesssim P_{orb}(\mathrm{h}) \lesssim 2.8$ is quite marked (Section 2.2). There are, however, detached systems already in that range (HW Vir, MT Ser: Table 9.1), or which will pass through the region on the way to making contact, or which will reach contact within the gap. From equation (9.33), contact will be made *within* the gap if $0.18 \lesssim M_1(2) \lesssim 0.24$ and *below* it if $M_1(2) \lesssim 0.18$. Because of the stability criterion $q < 2/3$, He white dwarf primaries (with $M \lesssim 0.46\,\mathrm{M}_\odot$: Section 9.2.1.1) can only give rise to CVs with $P_{orb} \lesssim 3.4\,\mathrm{h}$, whereas CO (or ONeMg) primaries can populate the whole P_{orb} range.

The exact location of the 'mass gap' depends on the efficiency of ejection of the CE (de Kool 1992), but it is tempting to suspect that the period gap is in some way a consequence of the mass gap – with primaries below the period gap having $M_1(1) \lesssim 0.46$ and primaries above having $M_1(1) \gtrsim 0.56$ (Webbink 1979a). The well-determined masses below the gap do not support this hypothesis (Table 2.6). Furthermore, to sustain any such natal dichotomy would require $\dot{M}(2) \lesssim 5 \times 10^{-11}\,\mathrm{M}_\odot\,\mathrm{y}^{-1}$ in order to prevent orbital evolution *into and through* the gap, which is also at variance with observations (Verbunt 1984).

The sudden mixing of ^{3}He throughout the secondary when it becomes fully convective lowers the nuclear luminosity (Section 9.3.1) and causes the star to contract (d'Antona & Mazzitelli 1982a). This can be sufficient to detach the secondary from its Roche lobe for times up to few $\times\ 10^7\,\mathrm{y}$ (McDermott, Taam & Ringwald 1988). For conservative mass transfer there is no difficulty in reestablishing stability when contact is regained (see below), but if angular momentum is not conserved then dynamical mass transfer may occur. Taam & McDermott (1987) suppose that the accretion disc can have a low viscosity ($t_\nu \sim 10^5\,\mathrm{y}$), in which case the angular momentum of accreting gas is stored in the disc and the criterion for stable mass transfer is $q < 0.26$, which is not satisfied for a $\sim 0.4\,\mathrm{M}_\odot$ secondary. In the Taam & McDermott model, the binary evolves to longer P_{orb} and eventually detaches. However, as pointed out by Melia & Lamb (1987), (a) the dynamical mass transfer phase will in reality lead to CE evolution with coalescence of the stars or reemergence as a CV below the period gap and (b) systems coming into contact for the first time (from the pre-cataclysmic phase) would undergo the same instability, resulting in no CVs with $P_{orb} \gtrsim 3\,\mathrm{h}$. Furthermore, from Sections 3.5.4.1 and 3.5.4.5 it is clear that even low mass accretion discs have $t_\nu \lesssim 10^2\,\mathrm{y}$, and discs with $M(\mathrm{d}) \gtrsim 10^{-9}\,\mathrm{M}_\odot$ have $t_\nu \sim$ days.

As seen in Figure 9.2, strong braking can produce a $P_{orb,min}$ in the vicinity of 3 h. A possible explanation of the period gap could therefore be that systems born with P_{orb} above the gap 'bounce' back at 3 h, whereas those born below have weak braking and bounce off $P_{orb,min} \sim 75$ min (Eggleton 1983b). However, for $P_{orb,min} \sim 3$ h the secondaries must be degenerate structures with $M_1(2) < 0.1$ and very depressed luminosities (Verbunt 1984). The simple fact that the secondaries are visible at normal main sequence luminosities in systems with P_{orb} just above $P_{orb,min}$ (e.g., MV Lyr, AM Her and see Figures 2.45 and 2.46) makes this scheme untenable.

Until recently, the generally accepted model has been one that arose from the suggestion by Robinson *et al.* (1981) in connection with the temporary cessation of $\dot{M}(2)$ in MV Lyr. They noticed that the period gap coincides with the range of secondary masses over which the internal structure is changing from a deep convective envelope to total convection, and suggested that contraction of the secondary, due to the internal restructuring, would terminate mass transfer until braking caused contact again at the lower edge of the period gap. The current models of 'disrupted braking' are based on these initial suggestions.

Even quite low rates of magnetic braking result in $\tau_{KH} > \tau_{ML}$ at $P_{orb} \sim 3$ h. At this stage, therefore, the secondary has a radius (for its mass) in excess of the equilibrium value. Any reduction of braking that gives $\tau_{KH} < \tau_{ML}$ will allow the secondary to shrink to the equilibrium radius on a time scale $\tau_{KH}L(2)/\Delta L(2)$ (where $\Delta L(2)$ is the luminosity deficit from the equilibrium value), detaching itself from the Roche lobe and terminating mass transfer. The lowered efficiency of braking is hypothesized to be caused by the rearrangement of field structure when the convective envelope becomes very deep, resulting in a lowered stellar wind (Section 9.1.2.1). *As there is no complete theory of this process, the disrupted braking model is only an hypothesis*, the consequences of which may be tested by comparing predicted and observed distributions of CVs as functions of P_{orb}, and by the existence and location of the period gap.

The consequences of the assumption of termination of magnetic braking when the secondary becomes fully convective are shown in Figure 9.2. For the magnetic braking model shown, this occurs at $M_1(2) = 0.18$, but more recent computations give

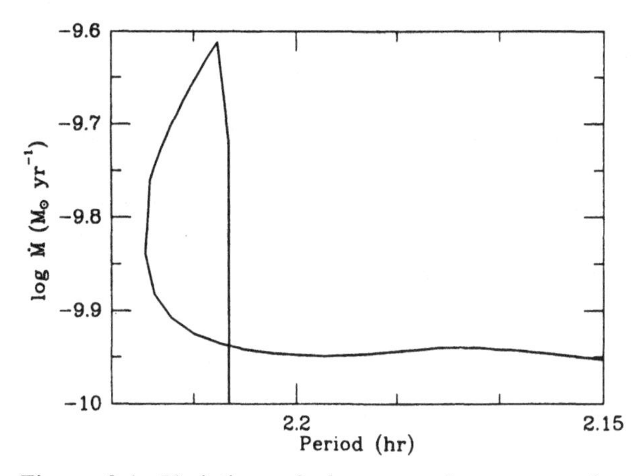

Figure 9.4 Variation of the rate of mass transfer and orbital period following the reestablishment of contact with the Roche lobe at the lower end of the period gap. From McDermott & Taam (1989).

$M_1(2) \approx 0.26$ (McDermott & Taam 1989). The secondary shrinks to its equilibrium radius and the system remains as a low luminosity detached binary until sufficient orbital angular momentum has been lost (by gravitational plus any residual magnetic braking) to shrink the orbit and reestablish contact. As there is negligible loss of mass in the period gap, the P_{orb}s at the upper and lower end of the gap are related by $P_{orb}(\text{low})/P_{orb}(\text{up}) = [R_{low}(2)/R_{up}(2)]^{2/3}$(equations (2.3)). The value of $P_{orb}(\text{up})$ depends sensitively on the magnetic braking law (Hameury *et al.* 1987; McDermott & Taam 1989) and on the physics of the stellar models (Hameury 1991). The value of $P_{orb}(\text{low})$ is not sensitive to the braking law but requires accurate knowledge of stellar opacities. The observed gap is reproduced by model calculations in which $\dot{M}(2) \sim (1\text{–}2) \times 10^{-9}\ \text{M}_\odot\ \text{y}^{-1}$ at the upper edge of the gap. The braking laws of equations (9.13) and (9.17) generate just such transfer rates. Lower rates result in the incursion of systems into the gap; higher rates move $P_{orb}(\text{up})$ to longer periods. A possible alternative explanation of the period gap is given in Section 9.4.3.

The time t_g taken to cross the gap is

$$t_g = \frac{P_{orb}(\text{up}) - P_{orb}(\text{low})}{\dot{P}_{orb}}$$

$$= \frac{2}{3}\frac{P_{orb}(\text{up}) - P_{orb}(\text{low})}{P_{orb}(\text{up}) + P_{orb}(\text{low})} \cdot \frac{J}{\dot{J}} \quad \text{from equation (9.31)} \qquad (9.47\text{a})$$

$$\approx 0.08\frac{J}{\dot{J}}, \qquad (9.47\text{b})$$

where $\dot{J}$ includes both gravitational and any magnetic braking. For typical evolutionary sequences the time scale of stellar contraction at turn-off of magnetic braking is $\sim 3\tau_{KH} \sim 1 \times 10^9$ y, increasing rapidly to $\sim 10^{11}$ y as the luminosity deficit is removed. The shrinking Roche lobe therefore catches up with the secondary's surface when equilibrium is almost fully established (McDermott & Taam 1989). For gravitational braking, equation (9.32) is therefore suitable for calculating t_g, giving typically $t_g \sim (2\text{–}5) \times 10^8$ y. Any significant residual magnetic braking will shorten this.

On resuming mass transfer at the lower end of the period gap, the secondary is fully convective and satisfies the condition $\xi_{KH} > \xi_{ad} > \xi_L$ of Section 9.3.1. Consequently, the secondary expands, increasing $\dot{M}(2)$ and P_{orb} until, after $\sim 2 \times 10^7$ y, departure from thermal equilibrium increases ξ_{ad} and stable mass transfer is established. The behaviour of $\dot{M}(2)$ in the vicinity of $P_{orb}(\text{low})$ for purely gravitational braking is shown in Figure 9.4, where P_{orb} increases by 0.65 min before again decreasing (McDermott & Taam 1989). Inclusion of residual magnetic braking can cause a ~3 min increase in P_{orb} (Hameury *et al.* 1988).

9.3.4 Evolved Secondaries

The existence of CVs with P_{orb} in the range 0.5–6 d (Tables 3.1, 4.1, 4.2) shows that stable mass transfer can occur from subgiant secondaries, which must have had initial masses $\gtrsim 1.0\ \text{M}_\odot$. Evolution of systems for which Roche lobe contact is made when the secondary is between the main sequence and the initial ascent of the giant branch has been computed by Pylyser & Savonije (1988a,b).

For mass conservancy and magnetic braking, equation (9.4b) can be written as

$$\frac{1}{2}\frac{\dot{a}}{a} = (1 - q)\tau_{\mathrm{ML}}^{-1} - \tau_{\mathrm{MB}}^{-1}. \tag{9.48}$$

Therefore, for $q > 1$ the orbital separation must decrease, but for $q < 1$ the orbit may expand if $\tau_{\mathrm{MB}} > \tau_{\mathrm{ML}}$. As the effectiveness of magnetic braking decreases with increasing P_{orb}, there can exist an initial orbital period $P_{\mathrm{orb}}(\mathrm{crit})$ above which the separation increases, but mass transfer is sustained by expansion of the secondary on a time scale τ_{nuc} (and $\tau_{\mathrm{ML}} \approx \tau_{\mathrm{nuc}}$).

$P_{\mathrm{orb}}(\mathrm{crit})$ is a function of $M(1)$, $M(2)$ and τ_{MB}. For braking according to equation (9.13a) and mass conservation, Pylyser & Savonije find $P_{\mathrm{orb}}(\mathrm{crit}) \sim 12\,\mathrm{h}$; i.e., *all* evolved secondaries will drive diverging systems. This is because as soon as any He core $\gtrsim 0.01\ \mathrm{M}_\odot$ has developed, τ_{nuc} falls drastically from the main sequence value, causing relatively rapid expansion of the secondary.

Pylyser & Savonije (1988a) show that GK Per (Table 3.1), with $P_{\mathrm{orb}} = 2\,\mathrm{d}$ and a secondary of K0 IV spectral type and probable mass $M_1(2) \approx 0.25$ (Watson, King & Osborne 1985), can be generated from initial masses $M_1(1) = 0.70$, $M_1(2) = 1.00$ and initial $P_{\mathrm{orb}} > 0.73\,\mathrm{d}$.

The ultimate fate of an expanding system is a detached state when all of the envelope of the secondary has been lost. Any inclusion of additional angular momentum loss or mass loss from the system (as in the next section) will increase $P_{\mathrm{orb}}(\mathrm{crit})$. This has still to be explored in detail.

Systems that begin mass transfer when the secondary has already become a giant ($M_{\mathrm{c}} \gtrsim 0.15\ \mathrm{M}_\odot$) will usually be dynamically unstable, leading to the CE evolution of Section 9.2.1.2. The exceptions are (a) if the giant is very evolved, so that the envelope mass is smaller than the core mass (Hjellming & Webbink 1987) or (b) the mass ratio satisfies the stability criterion $q \lesssim 0.65$ (Section 9.3.1). Iben & Tutukov (1984) have explored the latter criterion and find that there is only a narrow range of initial masses for which stable mass transfer occurs; this gives equation (9.7) for $\dot{M}(2)$.

De Kool, van den Heuvel & Rappaport (1986) point out that the stellar wind (irrespective of whether or not it is magnetic) from a giant carries away orbital and rotational angular momentum from its surface. With

$$\dot{M}_{\mathrm{wind}}(2) = -5.5 \times 10^{-13} R_1(2) \left(\frac{L}{\mathrm{L}_\odot}\right) M_1^{-1}(2) \quad \mathrm{M}_\odot\ \mathrm{y}^{-1} \tag{9.49}$$

(Kudritzki & Reimers 1979), wind losses can be larger than the transfer rate through the Roche lobe and significantly affect the orbital evolution.

9.3.5 *The Effects of Nova Eruptions*

A nova eruption creates a temporary CE during which friction transmits orbital momentum from the secondary to the envelope, which is then expelled from the system (Section 5.8.5). This is in addition to the angular momentum lost from the primary (equation (9.40)). A high mass, slowly expanding envelope clearly maximizes the effect.

If, as is currently believed, CVs eject most or all of the mass accreted onto the primary in nova eruptions, this can constitute an important additional loss of angular momentum, affecting their secular evolution.

The time-averaged $\dot{J}$ produced by this mechanism has been estimated by MacDonald (1986), Shara *et al.* (1986a) and Livio, Govarie & Ritter (1991) to be

$$\frac{\dot{J}}{J} = \frac{1}{4}(1+q)\frac{R^2(2)}{a^2}\frac{(1+U^2)^{1/2}}{U}\frac{\dot{M}_{env}}{M(2)}, \tag{9.50}$$

where $U = v_{exp}/v_{orb}(2)$ is the ratio of the expansion velocity of the envelope at the position of the secondary to the orbital velocity of the secondary in the primary's frame of reference and $\dot{M}_{env}$ is the rate of mass flow past the secondary. Because the envelope accelerates as it expands, U typically increases from ~ 0.05 at $P_{orb} = 2\,\text{h}$ to ~ 0.5 at $P_{orb} = 10\,\text{h}$. The interplay of mass loss and angular momentum loss results in an increase of orbital separation for $P_{orb} \gtrsim 8\,\text{h}$, but a decrease for shorter periods. This is consistent with the observed increase of P_{orb} for BT Mon after an eruption (Schaefer & Patterson 1983).

9.3.6 *Spin Evolution of the Primary*

The effect of a stellar wind during post-main-sequence evolution removes >99% of the angular momentum, leaving a white dwarf rotating with $P_{rot} \gtrsim 10^3\,\text{s}$ (Villata 1992). Isolated non-magnetic white dwarfs have $P_{rot} \sim 1\,\text{d}$, as determined from asteroseismology (Kepler 1990). White dwarfs in CVs will also probably start as slow rotators, but will decrease $P_{rot}(1)$ through the accretion torque applied by mass transfer.

For an average $\dot{M}(1) \sim \text{few} \times 10^{-9}\,\text{M}_\odot\,\text{y}^{-1}$, the time taken to spin up the primary to break-up angular velocity is relatively short: $\tau_{acc} \ll \tau_{MB}$ (equation (6.32)). However, the evolution of $P_{orb}(1)$ depends on a number of competing and uncertain factors. Although the viscosity in the envelope is similar to that in non-degenerate stars, the degenerate core has a time scale for angular momentum transfer $> 10^{10}\,\text{y}$ (Durisen 1973). The presence of small initial magnetic fields will, however, produce greater rigidity after shear amplification by differential rotation. (There is recent evidence that isolated white dwarfs do rotate differentially (Winget *et al.* 1994).) The internal field configuration in white dwarfs is not known: in particular, whether the field is generated and confined to the envelope or whether a 'fossil' field permeates the entire star. Furthermore, if the interval between nova eruptions is much less than the coupling time scale between envelope and core, then little of the accreted angular momentum will be transmitted to the core.

That it is possible to accrete angular momentum into the cores of *magnetic* primaries is shown by the examples of DQ Her and AE Aqr, whose stable $\Omega_{rot}(1)$ imply full-body rotation (Chapter 8). The equatorial radius of an $n = 1.5$ polytrope at the point of break-up is 1.255 times its average radius (Tassoul 1978). From equations (2.23) and (2.83a) this gives for the minimum period of rotation of a white dwarf

$$P_{rot,min}(1) = 15.1 M_1(1)\ \text{s} \qquad (0.4 \le M_1(1) \le 0.7). \tag{9.51}$$

Therefore DQ Her and AE Aqr have $P_{rot}(1)/P_{rot,min}(1) = 7.8$ and 2.9 respectively.

The *maximum* periods of DNOs (Section 8.6.5.2) may be more indicative of the general rotation periods of the outer envelopes of CV primaries, in which case $P_{\mathrm{rot}}(1) \sim 30\,\mathrm{s}$ and $\Omega_{\mathrm{rot}}(1)/\Omega_{\mathrm{K}}[R(1)] \sim 1/3$. From equation (2.54a) this would produce boundary layer luminosities of ~40% those of non-rotating primaries. King, Regev & Wynn (1991) show that if more mass is lost in a nova eruption than is accreted between eruptions, this will prevent the primary reaching $P_{\mathrm{rot,min}}(1)$.

Interesting limits may be placed on $P_{\mathrm{rot}}(1)$ from observations of the spectra of VY Scl stars in the low state. Shafter *et al.* (1985) note that the steep-sided Hγ absorption line profile in TT Ari requires $v \sin i \lesssim 1000$ km s^{-1}, or $P_{\mathrm{rot}}(1) \gtrsim 40\,\mathrm{s}$. Sion *et al.* (1994) find $v \sin i \approx 200$ km s^{-1} in U Gem at quiescence, which translates to $P_{\mathrm{rot}}(1) \sim 110\,\mathrm{s}$ if $M_1(1) = 1.2$.

9.3.7 *Evolution of Magnetic Systems*

9.3.7.1 Orbital Evolution

The observed low value of $\dot{M}(2)$ for all polars (Section 6.4.3) is sufficient reason for the absence of a period gap in their distribution (Section 2.2). Prior to the recent realization that the strongly magnetic systems have little or no gap, the standard theory of magnetic braking was applied (King, Frank & Ritter 1985), particularly for explaining an apparent clustering of polars near $P_{\mathrm{orb}} = 114\,\mathrm{min}$ (Hameury *et al.* 1988, 1989), which was assigned to the hesitation in orbital evolution on emergence from the lower end of the period gap (Section 9.3.3; Figure 9.4). It would require a very small range of $M(1)$ among polars to produce such a narrow spike in their P_{orb} distribution. The later discovery of UZ For, with $P_{\mathrm{orb}} = 126.5\,\mathrm{min}$, in the context of this model of evolution required this star to have $M_1(1) > 1.0$ (Hameury *et al.* 1989), which is contradicted by observation (Section 6.3.2.6).

With ROSAT having doubled the number of known polars, only one of which falls at the 114 min spike, and the general filling of previous gaps in the P_{orb} distribution, the distribution now looks more uniform (Figure 2.11); the cluster near 114 min is probably no more significant than that near 200 min. This has led to a reassessment of the secular evolution of magnetic CVs (Wickramasinghe & Wu 1993), based on the realization that, ironically, the magnetic braking model probably does not apply to strongly magnetic systems (Li, Wu & Wickramasinghe 1993).

It had previously been speculated (Liebert & Stockman 1985; King 1985b) that coupling of the secondary's wind onto the field of the primary could *enhance* magnetic braking and hence the rate of evolution of polars. However, noting that this would lead to a *wider* orbital period gap than for non-magnetic CVs (Hameury *et al.* 1987), Hameury, King & Lasota (1988) suggested that such enhanced magnetic braking would only apply if $B(1) \gtrsim 40$ MG, thus explaining the absence of fields in polars as high as those seen in some isolated white dwarfs (Section 6.1).

Detailed computation of the combined field structure of polars shows, however, that for sufficiently large B the field lines of the secondary are either closed or connect on to field lines of the primary, thus preventing any outflow of wind from the system as a whole (Li, Wu & Wickramasinghe 1993a). At $P_{\mathrm{orb}} = 5\,\mathrm{h}$, as $B(1)$ is increased from 10 MG to 70 MG the magnetic braking efficiency drops from nearly 100% to zero. For the smaller P_{orb} typical of most polars, less extreme fields will shut off magnetic braking, leaving only gravitational radiation as a mechanism to drive mass transfer. The values of $\dot{M}(2)$ derived for polars (Section 6.4.3) are indeed just at the level predicted by gravitational radiation (equation (9.20)). The secondaries in polars will not depart greatly from thermal equilibrium.

Applied to the pre-cataclysmic evolution of high-B systems, Wickramasinghe & Wu (1993) point out that diminution of magnetic braking will occur as soon as the primary's rotation becomes synchronized with the orbital rotation. For large B, therefore, only gravitational radiation acts and equations (9.32) and (9.33) show that $t_{sd} > 1.0 \times 10^{10}$ y if $P_i > 9$ h (for $M_1(1) = 0.7$ and $q \leq 0.8$). Therefore many, perhaps the majority, of high-field binaries emerging from CE evolution during the lifetime of the Galaxy remain in a detached state. For this reason, there are no polars with $B \gtrsim 70$ MG.

Among the definite and probable intermediate polars (Tables 7.1 and 7.2) there are none in the period gap. Mass transfer rates (Section 7.5.2) are higher than in polars, requiring magnetic braking as a driving mechanism. These are indicators that the orbital evolution of IPs is similar to that for non-magnetic CVs. Wu & Wickramasinghe (1993b) conclude that polars and IPs are essentially unrelated systems, following different orbital evolutions.

9.3.7.2 Spin Evolution

For sufficiently strong primary fields $P_{rot}(1) = P_{orb}$, with brief episodes of unlocking due to nova eruptions (Sections 6.6 and 6.7). At lower $\mu(1)$, for constant $\dot{M}(1)$ the spin of the primary will evolve towards the equilibrium value determined by equation (7.16):

$$P_{rot,eq}(1) = 1510 \mu_{33}^{6/7}(1) M_1^{-5/7}(1) \dot{M}_{17}^{-3.7}(1) \omega_s^{-1} \quad \text{s.} \tag{9.52}$$

Long term departures of $\dot{M}(2)$ from the average will result in $P_{rot}(1)$ deviating from $P_{rot,cq}$, as appears to be the case in EX Hya (Section 7.5.3) and AE Aqr (Section 8.2.3).

The log $P_{rot}(1)$–log P_{orb} diagram[4] in not uniformly populated with magnetic CVs (Figure 9.5(a)). Apart from the obvious relationship for the synchronized polars, the diagram shows a clustering of IPs in the region of $P_{rot}(1) \sim 0.07 P_{orb}$ (Barrett, O'Donoghue & Warner 1988). From equations (2.14) and (7.1) this means that $P_{rot}(1)$ is in the vicinity of the value expected for a primary that rotates with the same angular velocity as gas at the radius r_r. This has been interpreted in two ways:

(a) accretion onto the primary occurs from a disc whose inner edge $r_o \sim r_r$ (Warner & Wickramasinghe 1991; Wickramasinghe, Wu & Ferrario 1991); or
(b) discless accretion in which an equilibrium is reached when centrifugal effects expel angular momentum at the rate it arrives (King & Lasota 1991).

In view of the evidence for discs in IPs (Section 7.2.2) only the first of these options is discussed further. Warner & Wickramasinghe (1991) proposed that large long term variations in $\dot{M}(2)$ (discussed more fully in Section 9.4.3) will cause $P_{rot}(1)$ to prefer values near that where discs first form (i.e., with the specific angular momentum at r_r).

[4] Patterson & Thomas (1993) include the detached binary V471 Tau (Table 9.1) in this diagram. But only Roche lobe-filling systems, rather than wind-accreting, systems are discussed here.

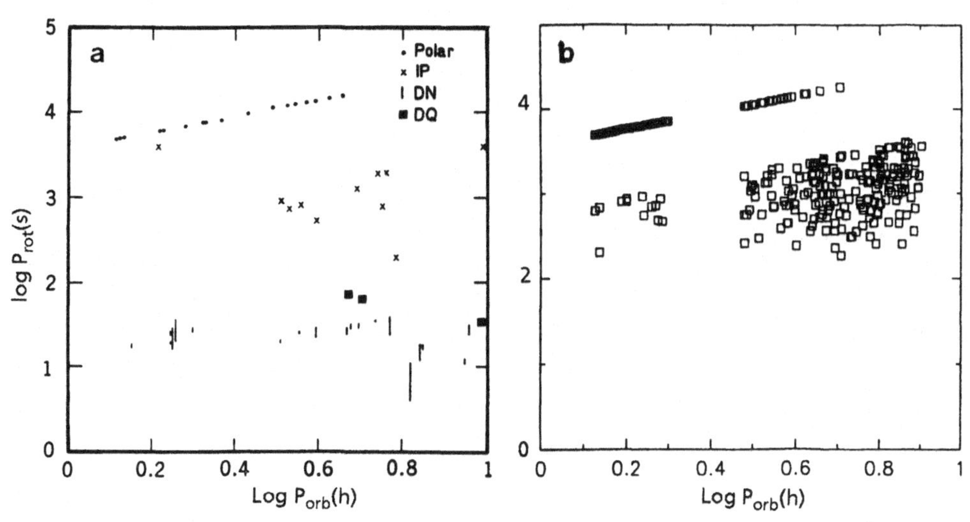

Figure 9.5 The log P_{rot}–log P_{orb} diagram as observed (a) and computed (b) by Wu & Wickramasinghe (1991).

A more generalized approach has been made by Wu & Wickramasinghe (1991,1993c), who deduce the distribution of $P_{rot,eq}(1)$ from Gaussian distributions (or from appropriate evolutionary models – as in Section 9.5.6.2) of $M(1)$, log $\dot{M}(1)$ and log $\mu(1)$. Their simulated log $P_{rot}(1)$–log P_{orb} distribution is shown[5] in Figure 9.5(b) (the DQ Her subclass is not included). This general agreement with the observed distribution is obtained for $\overline{\mu}_{33}(1) = 5$ with a dispersion in log $\mu(1)$ of 0.3. The gap between the polars and the IPs arises because no equilibrium exists there, and these (very rare) systems evolve relatively rapidly towards the polars or disced IPs.

9.4 Departures from Evolutionary Equilibrium

The mechanisms driving CV evolution, described in Section 9.1 and applied in Section 9.3, result in mass transfer rates that are monotonic (except for $P_{orb} \lesssim 1\frac{1}{2}$h) functions of orbital period. Yet the observed spread in disc absolute magnitudes at a given P_{orb} above the gap (Figures 3.9, 4.16 and 4.20; see also Patterson (1984)) implies a range of $\dot{M}(2) \sim 10^{16}$–10^{18} g s^{-1}. This section describes the explanations that have been offered for the dispersion in $\dot{M}(2)$.

9.4.1 Evolved Secondaries

Pylyser & Savonije (1988b) show that $\dot{M}(2)$ is very sensitive to the state of evolution of the secondary at the time it first makes contact with its Roche lobe. This is because, from equation (9.13a), the magnetic braking torque is proportional both to k^2, which depends on the internal mass distribution of the secondary, and to $\sim M^2(2)$, which is much smaller (at given $R(2)$) for evolved systems. The result is that at $P_{orb} \sim 4$ h the

[5] A number of conditions were imposed, including a period gap for polars, which would no longer be thought appropriate.

moment of inertia of the secondary, and in consequence $\dot{M}(2)$, shows a range of a factor ~100 for core hydrogen fractions X_c in the range 0.1–0.7.

There are several reasons to reject this proposed explanation of the $\dot{M}(2)$ dispersion. First, it would imply a spread in $M_1(2)$ from 0.2 to 0.6 at $P_{orb} \sim 5$ h (Figure 3 of Pylyser & Savonije (1988b)), which is not shown by the observations (Section 2.8.3). In particular, taking U Gem as an example of low $\dot{M}(2)$ (Figure 3.9) and UX UMa and DQ Her as examples of high $\dot{M}(2)$ (Figures 4.16 and 4.18), the derived masses are $M_1(2) = 0.53 \pm 0.06$, 0.40 ± 0.05 and ~0.50 respectively (Zhang & Robinson 1987; Horne, Welsh & Wade 1993; Shafter 1984b). Secondly, secondaries with low X_c have permanently low $\dot{M}(2)$, but from Section 9.3.3 it is seen that this would populate the period gap. In the context of disrupted magnetic braking *the time-averaged $\dot{M}(2)$ must be* $\gtrsim 1 \times 10^{-9}$ $M_\odot$ y^{-1}. Thirdly, from Section 9.3.2, substantial numbers of CVs would be expected with $40 < P_{orb} < 75$ min, which are not observed.

An important general point, not addressed by any of the secular evolution driving mechanisms, is that in many CVs $\dot{M}(2)$ is observed to drop by orders of magnitude and recover again on time scales of days to years – this is found in non-magnetic CVs (the VY Scl stars: Section 4.3.2, and post-novae: Section 4.4.2) as well as magnetic CVs (Sections 6.3.1 and 7.3.1). There are also quasi-periodic variations in $\dot{M}(2)$ on time scales of years (e.g., Section 4.3.1). These demonstrate that there is at least one additional parameter that determines the instantaneous $\dot{M}(2)$. Before pursuing further the observed (long term) dispersion in $\dot{M}(2)$ it is necessary to consider possible causes of temporary variations.

9.4.2 Magnetism in the Secondaries

The magnetic moment of a secondary star, required by the magnetic braking mechanism, is sufficient to affect flow in the vicinity of the L_1 point (e.g., Barrett, O'Donoghue & Warner 1988). Magnetic activity (e.g., prominences, dead zones) on a range of time scales may therefore be expected to appear as concomitant variations in $\dot{M}(2)$.

Studies of chromospheric activity in single late-type stars show that magnetic cycles of the same kind as the solar sunspot cycle with quasi-periods ~3–20 y are common (Baliunas & Vaughn 1985; Baliunas 1988). As with the Sun, about one third of the time is spent in 'magnetic minima' in which there is no apparent variability (Baliunas & Jastrow 1990). Similar behaviour is seen in close binaries of the RS CVn and BY Dra types (Maceroni *et al.* 1990). If the secondaries in close binaries undergo similar magnetic cycling then variations in internal magnetic pressure $B^2/8\pi$ (which in the Sun is probably largest at the base of the convective envelope) could produce changes in radius which, through the great sensitivity to $\Delta R(2)$ shown by equation (2.12), would modulate $\dot{M}(2)$ and hence many of the observable properties.

Quasi-periodic variations on time scales of 4–30 y have indeed been found in CVs, with median period ~6 y as in single low mass main sequence stars (Bianchini 1987, 1988, 1990a,b; Warner 1988a, Maceroni *et al.* 1990). These appear in (a) the mean brightness $\overline{m}_v$, (b) the mean interval T_n between outbursts of DN and (c) the orbital periods. A list of these is given in Tables 9.2 and 9.3 and examples are illustrated in Figures 4.19, 9.6 and 9.7. A quantitative connection between (a) and (b) has been established and requires $\Delta R(2)/R(2) \sim 10^{-4}$ over the cycle (Warner 1988a). Variations

Table 9.2. *Orbital Period Modulations.*

Star	Type	P_{orb} (h)	Quasi-period (y)	Amplitude O–C (min)	References
V4140 Sgr	DN(SU)?	1.47	5.4:	0.14	20
V2051 Oph	Polar?	1.50	6.9	0.22	21
EX Hya	IP	1.64	17.5	0.40	1,2
Z Cha	DN(SU)	1.78	>15	>0.3	3,4
IP Peg	DN	3.80	4.7	1.5	5,6,19
U Gem	DN	4.17	8:	1.0:	7,8
DQ Her	N	4.65	6	1.2	9,10
UX UMa	NL	4.72	7–30	2.5	11–15
T Aur	N	4.91	23:	3:	16
RW Tri	NL	5.57	8 or 14	1.2	17,18

References: 1. Bond & Freeth 1988; 2. Hellier & Sproats 1992; 3. Cook & Warner 1981; 4. Wood *et al.* 1986; 5. Wood *et al.* 1989a; 6. Wood 1990a; 7. Eason *et al.* 1983; 8. Warner 1988a; 9. Patterson, Robinson & Nather 1978; 10. Dmitrienko 1992; 11. Mandel 1965; 12. Kukarkin 1977; 13. Quigley & Africano 1978; 14. Panek & Howell 1980; 15. Rubenstein, Patterson & Africano 1991; 16. Beuermann & Pakull 1984; 17. Africano *et al.* 1978; 18. Robinson, Shetrone & Africano 1991; 19. Wolf *et al.* 1993; 20. Baptista, Jablonski & Steiner 1992; 21. Echevarria & Alvarez 1993.

Table 9.3. *Quasi-Periodic Long-Term Brightness Variations*

Star	Type	P_{orb}(h)	Quasi-period (y)	Amplitude* Δm_v	$\Delta T_n/T_n$	References
UU Aql	DN(SU)	1.88	12.9		0.18	1–3
TT Ari	NL(IP?)	3.30	12.6	min		2–4
MV Lyr	NL	3.30	11–12	min		3–5
V603 Aql	N	3.32	11–12	0.2		6
RR Pic	N	3.48	14	0.1		7,8
U Gem	DN	4.17	6.9		0.26	3,4
SS Aur	DN	4.33	2.3		0.33	3,4
DQ Her	N	4.65	14:	0.3:		8
RX And	DN(ZC)	5.08	0.8		0.26	3
SS Cyg	DN	6.63	7.3	0.075	0.32	3,4,9,10
Z Cam	DN(ZC)	6.95	8.8:		0.26	3
RU Peg	DN	8.99	6		0.26	3,11
V841 Oph	N	14.5	3.3	0.15		2–4
GK Per	N,DN	48	7 or 18	0.2:		2–4,6
Q Cyg	N		6.6	0.2		2–4
V446 Her	N		12.2	0.4		3
IV Cep	N		0.75	0.2		12

* min indicates that the star is usually in a high state but has low states that according to the references cited appear to occur quasi-periodically.

References: 1. Shakun 1987; 2. Bianchini 1990a; 3. Bianchini 1990b; 4. Bianchini 1988; 5. Andronov & Shugarov 1982; 6. Richman 1991; 7. Warner 1986d; 8. Warner 1988a; 9. Kiplinger 1988; 10. Cannizzo & Mattei 1992; 11. Saw 1983; 12. Della Valle & Calvani 1990.

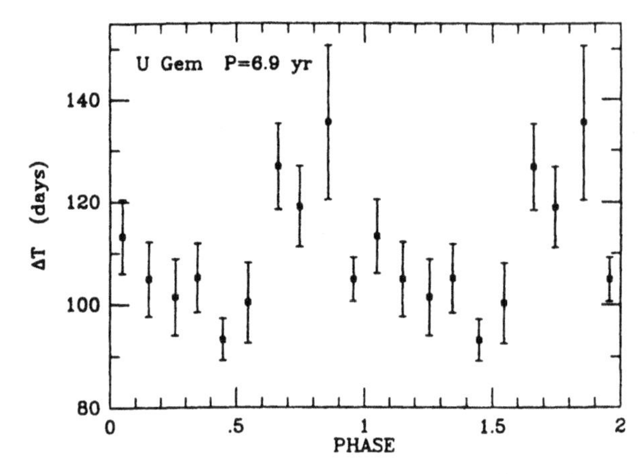

Figure 9.6 Variations in interval between outbursts in U Gem, folded onto a 6.9 y period. From Bianchini (1987).

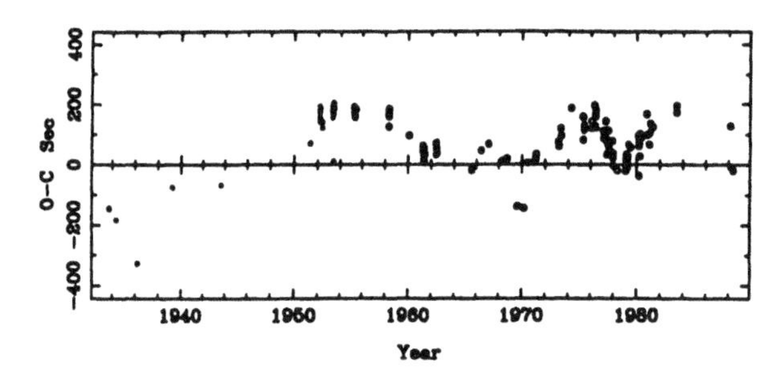

Figure 9.7 Orbital O–C diagram of UX UMa. From Rubenstein, Patterson & Africano (1991).

in radius of this magnitude have been claimed for the Sun during its magnetic cycle (Thomas 1979; Gilliland 1981, 1982b) and helioseismological studies show correlations with sunspot cycle equivalent to fractional changes in resonator diameter $\sim 10^{-4}$ (Elsworth *et al.* 1990; Libbrecht & Woodard 1990).

In those CVs for which very long light curves are available (e.g., U Gem, SS Cyg) intervals with no quasi-periodic variations are found, as in 'magnetic minima' of single stars. It appears, therefore, that the light curves of CVs, constructed from many decades of patient observation by amateur astronomers, may play an unanticipated rôle in elucidating solar-type magnetic cycles.

There have been a number of proposals to account for the P_{orb} modulations, Smak (1971b, 1972) suggested that angular momentum is stored in the disc and then transferred back to the orbit by tidal interaction. This requires disc masses $\sim 10^{-6}$–10^{-4} $M_{\odot}$ which are now known to be too large by factors $\gtrsim 10^4$ (Section 3.5.4.1). In systems where eclipse is of a luminous disc (e.g., RW Tri, UX UMa) some O–C variations could arise from changes in intensity distribution across the disc (though why these should be quasi-periodic on a long time scale would remain to be explained);

however, some of the P_{orb} variations (Z Cha, IP Peg) are deduced from eclipses of the primary which are free from such uncertainty.

In general at most one or two cycles of the P_{orb} variation have been observed. Z Cha has an apparently parabolic variation in the O–C diagram which formally gives $P_{orb}/\dot{P}_{orb} \sim 10^{5-6}$ y. This is too short by factors of ~100 to be the result of even the highest observed $\dot{M}(2)$, and in any case $\dot{P}_{orb} > 0$, which disagrees with ideas of secular evolution. The observed variation is almost certainly a section of long term cyclical behaviour. The absence of some CVs from Table 9.2 is also noteworthy (e.g., OY Car, for which no P_{orb} variation is detectable: Wood *et al.* 1989b).

There was a possibility that the P_{orb} variations in UX UMa are caused by a third body in orbit (Africano & Wilson 1976; Bois, Lanning & Mochnacki 1988) until the apparent sinusoidal variation of O–C (Africano & Wilson 1976) was found not to repeat (Rubenstein, Patterson & Africano 1991: Figure 9.7). This is also seen in DQ Her (Dmetrienko 1992). There are in any case too many incidences of O–C modulations to attribute all to third bodies. It remains to be seen whether the sinusoidal O–C variation in IP Peg, which is interpreted as a ~0.1 $M_\odot$ third body in a 4.7 y orbit (Wolf *et al.* 1993), continues to repeat faithfully.

Applegate & Patterson (1987) and Warner (1988a) (see also Matese & Whitmire (1983) and Applegate (1989)) have suggested that the P_{orb} variations are generated by the changes in quadrupole moment of the secondary that accompany the $B^2/8\pi$ pressure variation during a magnetic cycle. Although the original model has been criticized energetically by Marsh & Pringle (1990), Applegate (1992) has resurrected it by showing that the distortions arise from internal redistribution of angular momentum during a magnetic cycle, and not from changes in hydrostatic equilibrium. The mechanism requires the secondary to be variable in luminosity by about 10% over the magnetic (and, therefore, P_{orb}) cycle; this may be too small to detect in CVs but has been confirmed in the eclipsing binary CG Cyg (Hall 1991). More generally, the mechanism can explain quantitatively the P_{orb} variations seen in Algol variables and other eclipsing binaries containing convective secondaries (Hall 1989; Applegate 1992).

As the time scale ~10 y for P_{orb} variation is shorter than the synchronization time in polars (section 6.6), P_{spin} and P_{orb} may often be slightly different. This appears to be the case in DP Leo, where $P_{orb} - P_{spin} \simeq 5 \times 10^{-3}$ s (Robinson & Cordova 1994b).

9.4.3 Cyclical Evolution and Hibernation

In Figure 9.8 the disc absolute visual magnitudes for non-magnetic CVs from Figures 3.9, 4.16 and 4.20 are combined to illustrate the distribution of $M_V(d)$ as a function of P_{orb}. An approximate calibration in terms of $\dot{M}(2)$ (= $\dot{M}$(mean)) is indicated, based on the best determinations for individual systems given in Sections 2.6.5, 3.5.4.4 and 4.3.3. The $\dot{M}(2)$ expected from gravitational radiation (equation (9.20)) is also shown. There are strong indications that some of the shortest period DN have $M_V > 10$ (Szkody 1992), which would extend the diagram downwards, well below the gravitational radiation level.

The large spread in $\dot{M}(2)$, already evident in Patterson's (1984) work, emphasizes the need to supplement the magnetic and gravitational braking mechanisms with another process, capable of generating a range of $\dot{M}(2)$ at a given P_{orb}. It is misleading to put a

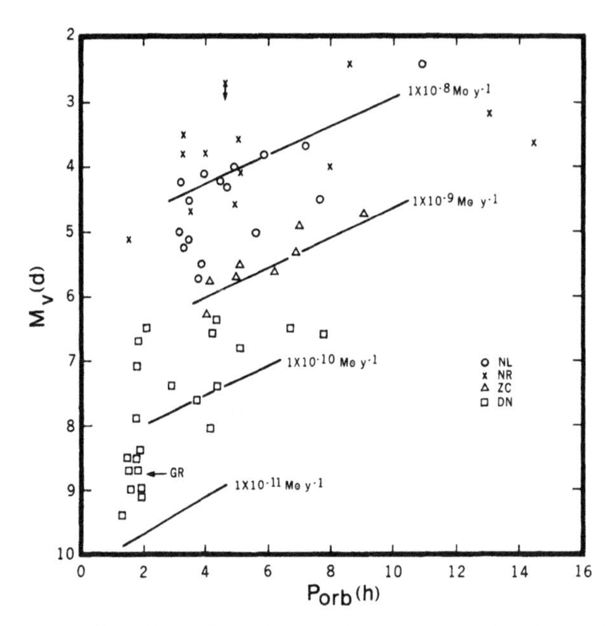

Figure 9.8 Disc absolute magnitudes versus orbital periods for CVs: a superposition of Figures 3.9, 4.16 and 4.20. An approximate calibration of $\dot{M}$(d) is given. The GR arrow indicates the steady state brightness of a short period system driven by gravitational radiation braking.

mean $\dot{M}(P_{orb})$ relationship through the systems plotted in Figure 9.8 (cf. Patterson 1984): rather, the diagram is better described as $\dot{M}(2) \leq 2 \times 10^{-8}$ M$_\odot$ y^{-1} for P_{orb}(h) > 3 and $\dot{M}(2) \leq 1 \times 10^{-9}$ M$_\odot$ y^{-1} for P_{orb}(h) < 2. It is noticeable that some systems below the gap have $\dot{M}(2)$ much larger than the gravitational radiation value. This may imply substantial magnetic braking even in fully convective secondaries.

Although variations of $\dot{M}(2)$ on time scales of ~ years can be attributed to magnetic secondaries (Section 9.4.2), it appears unlikely that the >100 y stability of $\dot{M}(2)$ in such low $\dot{M}$ systems as U Gem can be so explained. Instead, a mechanism that allows CVs to move up and down in Figure 9.8 on time scales >100 y (*cyclical evolution*) should be sought. This would imply that all types of non-magnetic CVs (and perhaps some or all of the mildly magnetic CVs) are interchangeable – a system's characteristics are determined by its current $\dot{M}(2)$, but DN can become CN and *vice versa*, evidence for which has already been led in Sections 4.4.1 and 4.4.2.

The general fading of post-novae (Section 4.4.2) led to a proposal that, following nova eruption, a CV may 'hibernate' for hundreds or thousands of years (Shara *et al.* 1986a), during which it would have very low or zero $\dot{M}(2)$ and would in general be inactive and difficult to recognize at $m_v \gtrsim 20$ (but the system LTT 329 at $m_v = 14.5$ is a possible candidate (Bragaglia, Greggio & Renzini 1988)).

The mechanism originally proposed to induce the state of low $\dot{M}(2)$ was an expected increase of orbital separation a during nova eruption (Livio & Shara 1987), but inclusion of the effect of the CE phase (MacDonald 1986; Section 9.3.4) showed that (a) a *decrease* in a is expected for $P_{orb} \lesssim 8$ h and (b) the changes in a are at most one or two scale heights $H(2)$ of the secondary's atmosphere, so $\dot{M}(2)$ is not dramatically affected (Livio, Govarie & Ritter 1991). The current proposal (Kovetz, Prialnik & Shara 1988; Section 4.4.2) is that irradiation of the secondary during nova eruption leads to sufficient enhanced mass loss that, after the primary has cooled down (~100 y) the secondary relaxes and underfills its Roche lobe. Braking then reduces a, gradually increasing $\dot{M}(2)$, carrying the system up through the DN region to the NL region. If the amount of underfill is $nH(2)$, then, from equations (2.5b), (2.8) and (9.5a), the time taken to return to the full $\dot{M}(2)$ is

$$t_f \approx 3.3 \times 10^{-5} n\tau_{\mathrm{MB}}(1-q)^{-1}\left(\frac{T_s(2)}{4000\ \mathrm{K}}\right) M_1^{-2/3}(2) P_{\mathrm{orb}}^{2/3}(\mathrm{h}) \tag{9.53}$$

$$\approx 2.6 \times 10^{-4} n\tau_{\mathrm{MB}} \text{ for } P_{\mathrm{orb}} = 4\ \mathrm{h}$$

$$\gtrsim 10^4\ \mathrm{y}.$$

An interesting example is GQ Mus, a nova with $P_{\mathrm{orb}} = 85$ min, which has been transferring mass at $\sim 1 \times 10^{-7}\ \mathrm{M}_\odot\ \mathrm{y}^{-1}$ for the past 10 y in essentially a fully conservative manner (Section 5.8.6). As a result, P_{orb} and a will have *increased* fractionally by $\sim 2 \times 10^{-5}$ (equation (9.5)), which will take $\sim 2 \times 10^4$ y of gravitational radiation to reduce (however, the long term evolution is not delayed by this). If $\dot{M}(2)$ continues at its present rate it will generate a measurable $\dot{P}_{\mathrm{orb}}$ (0.013 s y^{-1}, or an orbital phase delay of $\sim 40t^2(\mathrm{y})$ s). Monitoring of the X-ray flux from GQ Mus will provide a valuable comparison for models of irradiation-induced mass transfer.

An important feature of Figure 9.8 is the dominance of high $\dot{M}(2)$ systems in the range $3 \lesssim P_{\mathrm{orb}}(\mathrm{h}) \lesssim 4$, with the result that DN (with the exception only of AB Dra, a Z Cam star) are absent at these periods. Shafter, Wheeler & Cannizzo (1986) and Shafter (1992b) have pointed out that the DN orbital period distribution is sensitive to $\dot{M}(P_{\mathrm{orb}})$, and deduce that the latter cannot be as strongly dependent on P_{orb} as magnetic braking models (equations (9.13b), (9.14), (9.16) and (9.17b) predict. In fact, Figure 9.8 shows that for $P_{\mathrm{orb}} \gtrsim 3$ h there is an approximate upper limit of $\dot{M}(2) \sim 2 \times 10^{-8}\ \mathrm{M}_\odot\ \mathrm{y}^{-1}$ *independent* of P_{orb}. Note, however, that the highest values are for recent post-novae, so $\sim 1 \times 10^{-8}\ \mathrm{M}_\odot\ \mathrm{y}^{-1}$ is the maximum for long term equilibrium. For magnetic braking, this rate is only achievable for $P_{\mathrm{orb}} \gtrsim 6$ h (Shafter 1992b). There are therefore two questions to ask: what limits the mass transfer rate to $\dot{M}(2) \lesssim 1 \times 10^{-8}\ \mathrm{M}_\odot\ \mathrm{y}^{-1}$, and what mechanism supplements magnetic braking at $3 < P_{\mathrm{orb}}(\mathrm{h}) \lesssim 6$ to enable this high rate to be attained?

The answer to both questions lies in the effects of irradiation of the secondary (Wu, Wickramasinghe & Warner 1994). From equations (2.1b) and (2.110) the energy per unit area falling on the secondary is $\propto a^{-2} \propto (1+q)^{-2/3} P_{\mathrm{orb}}^{-4/3}$, which increases by a factor 3 as P_{orb} falls from 6 h to 3 h. The CVs with $3 < P_{\mathrm{orb}}(\mathrm{h}) < 4$ are mostly high $\dot{M}(2)$ systems with $T_{\mathrm{eff}}(1) \sim 5 \times 10^4$ K (Table 2.8) and boundary layer temperatures (5–10) $\times 10^4$ K (Section 2.7.3), thereby maximizing the effect of irradiation and its attendant increase of $\dot{M}(2)$ (Section 9.3.1). The high $T_{\mathrm{eff}}(1)$ and BL luminosity are generated by the high $\dot{M}(2)$. A runaway situation is avoided by (a) the increase in shielding by the disc as $\dot{M}(\mathrm{d})$ rises (equation (2.52)), and (b) the local Eddington limit for gas arriving at the surface of the primary. The rate $\dot{M}(2) \sim 2 \times 10^{-8}\ \mathrm{M}_\odot\ \mathrm{y}^{-1}$ is the saturation value found by these processes. Just that rate is attained in young post-novae; the above scenario explains why NLs in general and some pre-novae also have large $\dot{M}(2)$ and why that is much greater than that produced by magnetic braking alone (and is not otherwise explicable by the hibernation hypothesis).

When the irradiation-induced enhancement of $\dot{M}(2)$ is a substantial fraction of that driven by magnetic braking the mass flow becomes bistable, preferring either the high state or the low state, with no long term stable state in between. Models show this to occur for $3 \lesssim P_{orb}(h) \lesssim 4$, which accounts for the behaviour of the VY Scl stars. In this range, the low state $\dot{M}(2)$ is such that no (or only very infrequent) DN outbursts would occur, and the high state $\dot{M}(2)$ is above $\dot{M}_{crit2}(d)$; so very few DN are expected in that orbital period range. At $P_{orb} = 3$ h the system, once having stayed in the low state for a moderate length of time, allowing the primary to cool, is unable to reach the meta-stable high state again, permitting the secondary to relax towards thermal equilibrium and creating the orbital period gap. No speculations on the behaviour of magnetic fields are required. Magnetic braking can continue through the period gap and below, though at a monotonically decreasing rate.

The few NLs in and below the orbital period gap (Table 4.1; Section 4.5) must possess very hot primaries – remnants of prehistoric nova eruptions – that hold $\dot{M}(2)$ stably in the high state.

9.4.4 Temperature Evolution of the Primaries

Isolated white dwarfs cool to $T_{eff} = 5 \times 10^4$ K in about 3×10^6 y (D'Antona & Mazzitelli 1990), so it is clear that the three NLs in Table 2.8 with these temperatures are no longer simply radiating their original core energies. To reach $T_{eff} = 1 \times 10^4$ K takes $\sim 6 \times 10^8$ y, which sets a *lower limit* to the age of the coolest observed CV primaries. This is compatible with expected durations (t_{sd}) as detached systems, plus the evolutionary times as CVs (with $< \dot{M} > \sim$ few $\times 10^{-9}$ M$_\odot$ y^{-1}).

The surface temperature of a CV primary is determined by its history, depending on the time since it last had a nova eruption (and if any mass was stripped from the core in the process), heating by $\dot{M}(1)$ averaged over a suitable time, and the efficiency with which hot accreted matter is distributed from the equator over the surface.

The effect of an eruption is seen in V1500 Cyg, which currently has $T_{eff}(1) = 1.1 \times 10^5$ K (Sion 1991; Schmidt, Liebert & Stockman 1995). Prialnik (1986) finds that the temperature of a 1.25 M$_\odot$ primary after eruption is

$$T_{eff}(1) = 4.8 \times 10^5 \left[\frac{t(\mathrm{y})}{0.1\ \mathrm{y}}\right]^{-0.28} \mathrm{K}, \tag{9.54}$$

which agrees well with that observed for V1500 Cyg.[6] The NLs with $T_{eff}(1) = 5 \times 10^4$ K could therefore have experienced eruptions $\sim$300 y ago, and cooler primaries must have erupted at much earlier epochs. If this were the only source of heating, then those systems with $T_{eff}(1) \sim 1 \times 10^4$ K would imply $T_R \geq 1 \times 10^5$ y (see

[6] V1500 Cyg may be used to test models of atmospheres of irradiated secondaries (section 2.9.6). There is no shielding by a disc; $T_{eff}(1)$ will vary sufficiently over the next 25 y to produce a detectable decrease in the temperature of the irradiated surface. Currently the irradiated face has $T \simeq 8000$ K (Schmidt, Liebert & Stockman 1995).

also Sion (1985a)). However, it is also possible that $T_{\rm eff}(1)$ is determined (except for very recent novae) by accretion heating. In this case,

$$T^4_{\rm eff}(1) = f\frac{GM(1)\dot{M}(1)}{8\pi\sigma R^3(1)}\ , \tag{9.55}$$

where f is the fraction of accretion energy deposited in the outer envelope.

The $T_{\rm eff}(1)$ given in Table 2.8 are 'quiescent' values, determined after cooling from outburst (Section 3.3.5.3) or during temporary low states. (Sion (1993a) has modelled the asymptotic approach of $T_{\rm eff}(1)$ to its quiescent value between DN outbursts.) The appropriate $\dot{M}(1)$ is therefore the quiescent value of $\dot{M}$(d) in DN but the high state $\dot{M}$(d) for NLs that spend most of their time in that state.

The observed relationship is shown in Figure 9.9, where the M_V(d)–$\dot{M}(1)$ relationship has been taken from Figure 9.8. An average $T_{\rm eff}(1)$ and $< \dot{M}(1) >= 4 \times 10^{-11}$ M$_\odot$ y^{-1} for polars (Section 6.4.3) are adopted. The solid line is drawn from equations (2.82) and (9.55) with $M_1(1) = 0.6$ and $f = 0.40$. Because a high mass primary is heated more for a given $\dot{M}(1)$, but has less area to radiate, the expected $T_{\rm eff}(1)$ is sensitive to $M(1)$. U Gem (Marsh *et al.* 1990) and SS Cyg (Yoshida, Inoue & Osaki 1992; Robinson, Zhang & Stover 1986, with $i = 38°$) have $M_1(1) \approx 1.2$. The effect of reducing these masses to 0.6 M$_\odot$ (through equation (9.55)) is shown in Figure 9.9.

Figure 9.9 constitutes evidence that $T_{\rm eff}(1)$ is governed by $\dot{M}(1)$, (and/or vice versa, according to Section 9.4.3) and that, within errors, approximately half of the available accretion energy is distributed over the surface layers of the primary.

9.5 Space Densities and Population Characteristics

Estimates of the local space density of CVs are bedevilled by incompleteness of surveys and by selection effects (Patterson 1984; Shara *et al.* 1986a). However, all studies indicate that the local population of CVs is dominated by low $\dot{M}(2)$ systems, giving

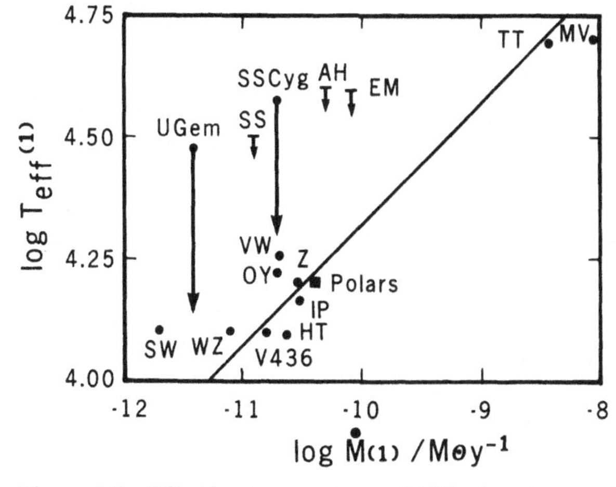

Figure 9.9 Effective temperatures of CV primaries as a function of accretion rate. The arrows show the effect of allowing for the larger masses of U Gem and SS Cyg.

Table 9.4. *Population Characteristics of CVs.*

Type	No.	$\bar{b}^0$	z_0(pc)	$\rho(0)(\mathrm{pc}^{-3})$
U Gem	305	18.9	103	4.0×10^{-7}
SU UMa	47	27.7	176	2.5×10^{-7}
Novae (Fast)	105	5.7	100:	
Novae (Slow)	35	7.6	> 200:	
Novae (Unclass.)	91	8.4		
NL	76	27.6		1.6×10^{-6}
Polars	34	40.5		3.0×10^{-7}
RN (T CrB type)	5	14	660	4×10^{-11}

special importance to DN and, unfortunately, systems with such low $\dot{M}(2)$ that they are infrequent outbursters, the majority of which probably remain unknown.

9.5.1 DN and other low $\dot{M}(2)$ Systems

In dealing with systems below the orbital period gap (i.e., mostly polars and SU UMa stars), where gravitational braking is probably the principal driver of evolution, the time scales required to evolve down to $P_{\mathrm{orb}} \sim 100$ min become $\gg 10^9$ y (Section 9.3.2) and these CVs may be drawn from an old Galactic disc population. Evidence for this is seen in the mean Galactic latitude $\bar{b}$ of CVs in the Downes & Shara (1993) catalogue: whereas U Gem and Z Cam subtypes show no difference in $\bar{b}$ (and have consequently been consolidated in Table 9.4), the SU UMa stars (including the WZ Sge category of Downes & Shara) have a significantly larger $\bar{b}$, indicating either an intrinsically fainter population observed nearby or a greater scale height z_0 perpendicular to the Galactic Plane.

To investigate the latter point, Duerbeck's (1984) analysis has been reworked, using later classifications (Chapter 3 and Downes & Shara (1993)) and improved distance measurements (Warner 1987a and supplementary observations). Only DN with distance in the Plane ≤ 250 pc have been included; Duerbeck shows there is significant incompleteness at greater distances. The resulting vertical density distribution $\rho(z)$ is shown in Figure 9.10, and is fitted by the equation

$$\rho(z) = \rho(0)\mathrm{e}^{-z^2/2z_0^2} \tag{9.56}$$

The scale height of the SU UMa systems (or, equivalently, non-magnetic CVs below the period gap) is nearly twice that of the U Gem stars. (Although the local density $\rho(0)$ is different, the surface density $\Sigma_z = 2\int_0^\infty \rho(z)\mathrm{d}z$ is approximately 2.5×10^{-4} pc^{-2} for both groups.) This result is in general agreement with the investigations of Howell & Szkody (1990), Szkody & Howell (1992b) and Howell (1993), who have found that high-z CVs are predominantly systems with large outburst amplitudes below the P_{orb} gap (note, however, that this is probably a selection effect: Drissen *et al.* (1994)). As would be expected of an old disc population, these stars are distributed at all heights above the Plane (i.e., they are not confined to high z, some of them reach high z because

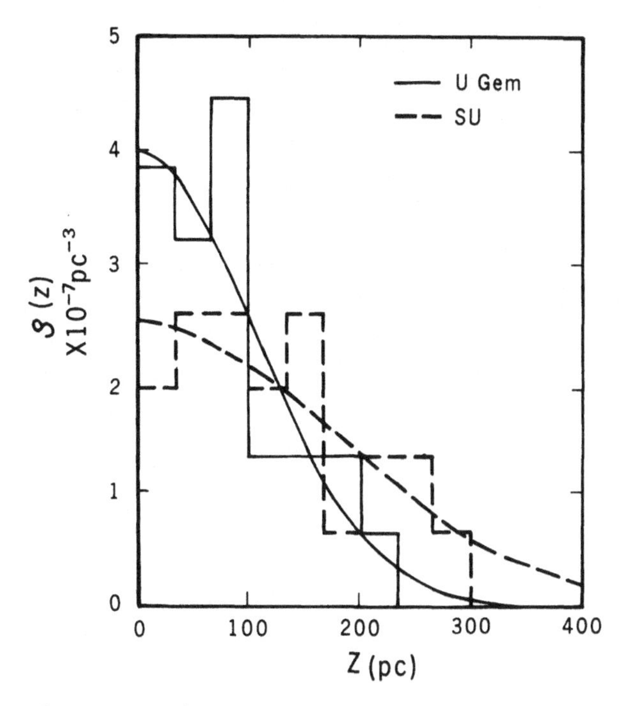

Figure 9.10 Histogram and theoretical fit of the distribution of DN in height above the Galactic Plane.

the population has a large velocity *dispersion* in the z direction, cf. Howell (1993)). WZ Sge is an example of a low z member of the group.

The distribution of DN in Galactic coordinates is shown in Figure 9.11(a). The right–left asymmetry shows that many more DN remain to be discovered in the southern celestial hemisphere.

The polars are another group with low $\dot{M}(2)$ and almost all nearby ones should have been found by ROSAT. Their $\bar{b}$ is characteristic of an intrinsically faint population extending to a considerable height above the plane. From the distances given in Table 6.1, supplemented with others deduced from $\overline{M}_V(\text{high}) = 8.7$, there is a strong grouping at $z \leq 150$ pc, but with several outliers at $z \sim 400$ pc. The numbers are insufficient to determine z_0. The vertical cylinder of radius 250 pc around the Sun contains 20 polars, giving $\rho(0) \approx 3.0 \times 10^{-7}$ pc^{-3} for an assumed $z_0 = 300$ pc, in agreement with Patterson (1984). The older population nature of polars is seen in ~25% having RVs ≥ 100 km s^{-1} (Mouchet 1993).

The total space density for *active* low $\dot{M}$ systems is $\sim 9.5 \times 10^{-7}$ pc^{-3}. However, the large T_n or T_s for very low $\dot{M}(2)$ CVs will introduce an unknown and possibly overwhelming bias against detection of these by their cataclysmic signature. Instead, unbiased surveys are looked to; these may be objective prism, UV-rich object, or all-sky X-ray searches. Patterson (1984) deduced $\rho(0) \sim (3\text{–}6) \times 10^{-6}$ pc^{-3} from the Palomar–Green, Michigan–Tololo optical surveys and the Einstein serendipitous X-ray sources. However, evidence is mounting that even these figures are underestimates.

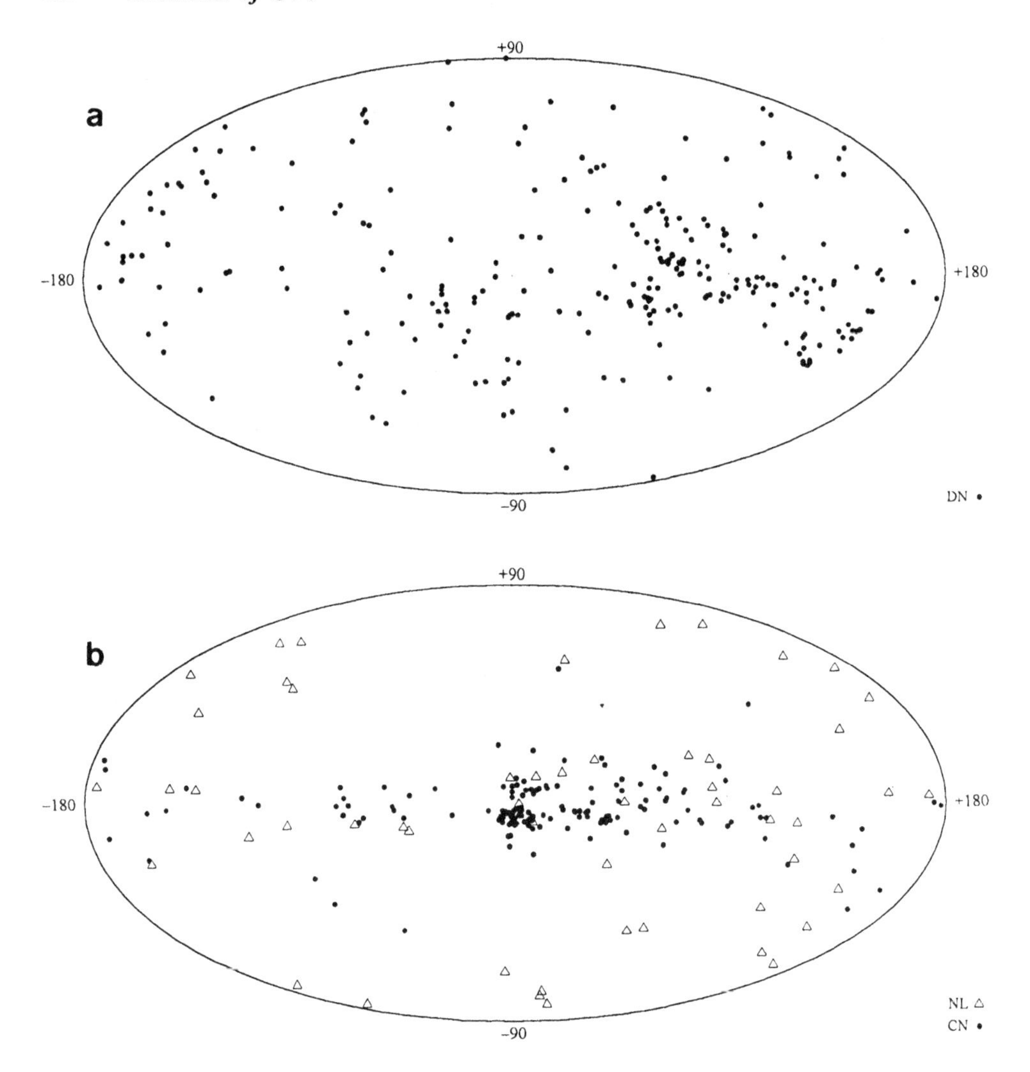

Figure 9.11 The distribution in Galactic Coordinates of (a) DN and (b) CN and NLs. Only CVs with certain classifications from the Downes & Shara (1993) catalogue are plotted.

The EXOSAT medium energy survey detected many DN, with $1 \times 10^{30} \lesssim L_X \lesssim 1 \times 10^{32}$ erg s^{-1}. Combining the total observed diffuse X-ray flux with an assumed $\overline{L}_X = 4 \times 10^{31}$ erg s^{-1} gives $\rho(0) \lesssim 1 \times 10^{-5}$ pc^{-3} (Mukai & Shiokawa 1993). However, this $\overline{L}_X$ corresponds to $\dot{M}(1) \sim 10^{16}$–10^{17} g s^{-1} (Patterson & Raymond 1985a); if very low $\dot{M}$ CVs dominate then $\overline{L}_X \sim 5 \times 10^{30}$ erg s^{-1} and $\rho(0) \lesssim 2 \times 10^{-4}$ pc^{-3}. Hertz *et al.* (1990) identified three Einstein sources in 144 square degrees with faint ($m_V \sim 17.5$–19.0) CVs which have $\overline{L}_X \sim 4 \times 10^{30}$ erg s^{-1} ($\dot{M}(1) \sim 1.5 \times 10^{14}$ g s^{-1}) and give $\rho(0) \gtrsim (2\text{–}3) \times 10^{-5}$ pc^{-3}. These objects are comparable with WZ Sge, which has $L_X \sim 4 \times 10^{30}$ erg s^{-1} (Patterson & Raymond 1985a); however, their spectra show no signs of absorption lines from the primary, which excludes them from having very short P_{orb}. On the other hand, Figure 3.5 shows that emission lines will become progressively more difficult to detect against the secondary's spectrum for $P_{orb} \gtrsim 3\,h$ in CVs with $\dot{M}(d) \sim 10^{14}$ g s^{-1}, and such low $\dot{M}$ objects will not be found in UV surveys. A deep optical search for CVs (Shara *et al.* 1990,1993) in a 2 square degree field in Scorpius has produced 13 CVs at $B \sim 19$. Their blue colour limits their distance to $\lesssim 1$ kpc and implies $\rho(0) \sim 1 \times 10^{-4}$ pc^{-3}. One apparently has $P_{orb} \sim 3.6\,h$.

The Hertz *et al.* (1990) estimate of $\rho(0)$ is currently the most reliable. It implies a mean separation between low $\dot{M}$ CVs of ~35 pc, or ~13 CVs within 50 pc of the Sun. These will have $F_x \gtrsim 1 \times 10^{-11}$ erg cm^{-2} s^{-1} (i.e., already have been detected in the ROSAT X-ray survey) and be uniformly distributed over the sky. By analogy with the systems already found by Hertz *et al.* these CVs should have $m_v \lesssim 16$ and will appear in any moderately deep optical survey. In general most will have T_n (or T_s) $\gtrsim 100$ y and will therefore not be known as DN. Of the SU UMa stars studied so far, only WZ Sge with $m_v = 14.9$ and $d = 48 \pm 10$ pc (Smak 1993a) satisfies these criteria. Identification of the other low $\dot{M}$ nearby CVs is a challenge to be met, and is crucial to establishing the correctness of the high estimate of $\rho(0)$.

9.5.2 Nova-Like Systems

Applying the statistical technique of Allen (1954), which is based on the tenth brightest star in an all-sky sample, using the NLs in Downes & Shara (1993) and $\overline{M}_v = 4.1$ (Figure 4.16, with allowance for the fact that the brightest listed magnitudes are extrema, rather than average high state magnitudes), gives $\rho(0) \approx 1.6 \times 10^{-6}$ pc^{-3}, which is three times that previously deduced (Warner 1974b). This is largely a result of discoveries of more bright NLs in the past 20 y, which has changed $m_v(10)$ from 12.5 to 11.9. The Galactic distribution of NLs is shown in Figure 9.11(b). Fourteen NLs with known systemic velocities all have RVs under 40 km s^{-1} (Shara 1994b). The contribution from IPs and DQ Her stars (not already included in other subclasses) is $\rho(0) \approx 3 \times 10^{-8}$ pc^{-3}.

9.5.3 CN

The annual rate of production of observed novae has been estimated by Patterson (1984) and Duerbeck (1984; see also Downes 1986) as $\rho(0, y) = (2.6\text{–}3.8) \times 10^{-10}$ pc^{-3} y^{-1}. To deduce $\rho(0)$ requires the mean interval between eruptions:

$$\rho(0) = \rho(0, y)\overline{T}_R, \tag{9.57}$$

which is, in principle, given by equations (5.12) and (5.13) but in practice is uncertain to order of magnitude because of the dependence of $< \dot{M}(1) >$ on the duration and depth of hibernation.

If the large space density of low $\dot{M}(2)$ CVs described in Section 9.5.1 is a population of hibernating CVs, then equation (9.57) gives $\overline{T}_R \approx 1 \times 10^5$ y from the Hertz *et al.* (1990) estimate of $\rho(0)$. Equations (2.82), (5.12) and (5.13) then give (for $\overline{M}_1(1) = 0.65$) $< \dot{M}(1) > \approx 1.5 \times 10^{-9}$ M$_\odot$ y^{-1}, which is compatible with the braking requirements for generating the period gap (Section 9.3.3). If this $< \dot{M}(1) >$ is representative (at least, above the period gap), then T_R is $\sim 2.1 \times 10^4$ y for $M_1(1) = 1.0$ and $\sim 2.5 \times 10^3$ y for $M_1(1) = 1.3$. The observed novae are therefore strongly biased towards CVs with high mass primaries. Below the period gap, where

$\dot{M}(1)$ is an order of magnitude lower, recurrence times ten times longer are expected, which explains the relative rarity of novae at short P_{orb}.

Duerbeck (1990) has suggested that there are two nova populations in the Galaxy, a disc population containing mostly very fast novae and a bulge population containing mostly slower novae. This is seen in $\bar{b}$ (Table 9.4) and in the frequency distribution of t_2, which is bimodal with peaks at $t_2 \sim 22$ d and ~ 8 d (Della Valle *et al.* 1992). In M31 novae are mostly produced in the bulge (Ciardullo *et al.* 1987; Capaccioli *et al.* 1989), only 15–20% appearing in the disc. The fraction of very fast novae differs from system to system, being high (30%) in bulgeless galaxies (M33, LMC) compared with M31 (10%). (Note, however, that Sharov (1993) finds no differences in surface distribution of fast and slow novae in M31.)

The bimodality of t_2 probably indicates a different average white dwarf mass for the two populations. The lower masses for the higher z_0 population are in harmony with the predominance of CVs below the orbital gap found at high z (Howell & Szkody 1990), though from the above discussion it is unlikely that these are significant contributors to observed novae.

The distribution of novae in Galactic coordinates is shown in Figure 9.11(b). As with the DN, fewer CN are known in the southern than in the northern celestial hemisphere. The effect of interstellar absorption within a few degrees of the plane is evident.

In a sample of 12 relatively nearby novae, none have RVs in excess of 100 km s^{-1} (Shara 1994b).

Attempts to estimate the nova rate in our Galaxy by scaling from rates in others (Della Valle 1988; van den Bergh 1991; Della Valle & Duerbeck 1993; Della Valle & Livio 1994) suffer from the unknown effects of Galactic optical depths (Section 5.1).

The five RN of T CrB type in Table 5.3 have $\bar{z} = 525$ pc, or $z_0 = 660$ pc, which is similar to that for symbiotic stars (Duerbeck 1984); their space density is only 5% that of the symbiotics. The U Sco-type RN are even rarer and are members of the halo population.

9.5.4 *Population Models*

In principle, the observed space densities of the CV subtypes and the frequency distributions of the P_{orb} (Figure 2.1) should be understandable in terms of evolutionary processes applied to an initial population of wide binaries. In practice, some of the physical processes (CE evolution, magnetic braking) are not well quantified, and there are strong observational selection effects in operation.

9.5.4.1 Pre-Cataclysmic Systems

The population of close binaries emerging from CE evolution has been computed by Iben & Tutukov (1986) and Politano & Webbink (1990). This can be integrated over the age of the Galaxy, assuming some (usually constant) rate of star formation. De Kool & Ritter (1993) have used a Monte Carlo simulation (de Kool 1992), drawing binary components at random from an appropriate initial mass function restricted to $0.95 \leq M_1(1) \leq 8$ and $M_1(2) > 0.08$, and follow these through the CE phase to generate a distribution of P_{orb}, $M(1)$ and $M(2)$ for detached (pre-CV) systems. The space density of the initial wide binaries is normalized to be compatible with the observed birth rate of single white dwarfs. About 10% of the initial binaries both pass through and survive (as binaries) the CE phase, leading to a birth rate of $\sim 10^{-13}$ pc^{-3} y^{-1} for post-CE systems.

To compare with the observed pre-CVs (Table 9.1) requires allowance for observational selection. This can be quite pronounced: e.g., the nuclei of PN have been especially studied, but are recognizable as such for only the first $\sim 10^4$ y of their lives; white dwarf + main sequence binaries are more difficult to discover but have very long lifetimes.

The results obtained by de Kool & Ritter show great sensitivity to certain assumptions – e.g., the efficiency of the CE phase, which determines whether low mass He cores merge or survive. Nevertheless the space density of post-CE systems is dominated by the hot primaries with He cores because their lifetimes are ~100 times that of those with CO cores, which have higher luminosities.

After applying corrections for observational selection there is tolerable agreement with the statistics of the systems in Table 9.1. In order to distinguish between choices of theoretical parameters, a larger sample of pre-CVs is necessary and determination of $M(2)$ for those already known is a priority.

9.5.4.2 CV Populations

Starting with the population of pre-CVs, an evolved population of CVs can be generated. The form of this depends on the nature of the braking mechanisms (including the effect of nova eruptions) and the fate of systems which have thermally unstable mass transfer on first contact. As with the pre-CV population models, the CV model can be independently investigated with varying assumptions about the efficiency of the CE phase, possible correlations of initial $M(1)$ and $M(2)$, etc. De Kool (1992) has performed Monte Carlo computations to generate CV populations, and finds that 0.5–2.0% of original binaries with mass > 0.95 $M_{\odot}$ evolve into CVs, which predicts a CV formation rate of (0.5–2.0) $\times 10^{-14}$ pc^{-3} y^{-1}. Over the age t_G of the Galaxy this gives ~(0.5–2) $\times 10^{-4}$ pc^{-3}, of which ~80% have evolved beyond the orbital period minimum (Ritter & Burkert 1986) and are no longer recognizable as CVs. The predicted space density is therefore $\rho(0) \sim$ (1–4) $\times 10^{-5}$, in agreement with the conclusions of Section 9.5.1 and 9.5.3.

Kolb (1993) has explored the influences of different magnetic braking laws on CV populations generated from the Politano and de Kool pre-CV populations. This intrinsic population, i.e., not corrected for selection effects, has ~99% of CVs below the period gap and ~70% beyond the period minimum. Such a result is expected because $\tau_{GR}/\tau_{MB} \sim 100$ and $\tau_{GR}/t_G \sim 0.1$.

Kolb finds 2–4% of the *total* CV population should be crossing the orbital period gap – this is on the assumption of gravitational radiation braking in the gap; the probable continuation of magnetic braking (Section 9.4.3) will shorten the lifetime and reduce the fraction by $\sim \dot{J}_{GR}/\dot{J}_{MB} \sim 0.18(1+q)^{-2/3} M_1^{4/3}(1) P_{orb}^{-2.42}$(h) for the Verbunt & Zwaan braking law (equations (9.13a), (9.19) and (2.100)), i.e., by a factor f ~0.011. The expected space density of gap-crossing systems is then $\sim 3 \times 10^{-6}$ pc^{-3} for gravitational radiation braking and $\sim 3 \times 10^{-8}$ for magnetic braking.

9.5.4.3 Selection Effects

Comparison of the intrinsic population of CVs with observation requires knowledge of the distortions caused by observational selection. CVs discovered through their

Table 9.5. *Average Masses in a CV Population.*

Observed sample	$\overline{M}_1(1)$
$m_v \leq 10$	0.812
$m_v \leq 12.5$	0.797
$m_v \leq 15.0$	0.764
DN only	0.811
NL only	0.818
$P_{orb} < 2$ h	0.756
$P_{orb} > 3$ h	0.876
DN with $P_{orb} > 3$ h	0.905
CN	1.12–1.24

accretion luminosities (NLs, polars, DN at maximum or quiescence) have bolometric luminosities $L_{bol} \sim L_{acc} \tilde{\propto} M(1)R^{-1}(1)\dot{M}(2)$. A bolometric magnitude-limited survey will then expect to reveal a number $\propto L_{bol}^{3/2} \tilde{\propto} M^2(1)\dot{M}^{3/2}(2)$. Novae, on the other hand, have $L_{bol} \propto M^4(1)$ at maximum (equation 5.15(b)) and recurrence times $T_R \tilde{\propto} R^{2.8}(1)M^{0.7}(1)$ (equations (5.12) and (5.13)), giving an extremely strong selection in favour of large $M(1)$.

Detailed calculations of the selection effects (Truran & Livio 1986; Ritter & Burkert 1986; Ritter & Özkan 1986; Ritter 1986c; Ritter *et al.* 1991; Dünhuber & Ritter 1993) are able to account for the high mean mass found in novae (Table 2.5) compared with other CVs, and of the high $\overline{M}(1)$ for CVs compared with isolated white dwarfs for which $\overline{M}_1 = 0.56$. An example of theoretical predictions for a visual magnitude (m_v)-limited survey of the Kolb (1993) CV population is given in Table 9.5 (Ritter *et al.* 1991; Dünhuber & Ritter 1993).

9.6 CVs in Clusters

Multiplicity in field CVs is not common. The DN RU Peg has a common proper motion companion (Eggen 1968), the NL EC19314−5915 has a G-type companion (Buckley *et al.* 1992) and the polar CQ Dra is a triple with the M3III 4 Dra (Reimers, Griffin & Brown 1988).

Membership in galactic clusters is known for the DN SS Aur, which is in the Hyades moving group (Eggen 1968), the DN BX Pup in NGC 2482 (Moffat & Vogt 1975) and the probable polar EU Cnc in M67 (Gilliland *et al.* 1991). Belloni, Verbunt & Schmitt (1993) identify EU Cnc as a soft X-ray source and suspect that there may be other CVs in M67.

Two novae have long been recognized as members of globular clusters. The first, T Sco, reached $m_v = 7.0$ in 1860 and is only a few arc seconds from the centre of M80 (Pogson 1860). The second, Nova Oph 1938, erupted ~0.8 core radii from the centre of M14 (Hogg & Wehlau 1964) and had been identified as a blue object at $B = 20.2$ ($M_B \approx 2.7$) by Shara *et al.* (1986b; see also Shara, Moffat & Potter 1990a), but HST images show that the candidate is in fact a grouping of several objects (Margon *et al.* 1991). More recently, a DN has been found in M5 (Oosterhoff 1941;

Margon, Downes & Gunn 1981) which is confirmed as a member, probably with $P_{\rm orb} > 3$ h, varying between $4.3 \leq M_{\rm B} \leq 7.6$ (Shara, Moffat & Potter 1987,1990a; Naylor *et al.* 1989). Another candidate, near M30, is not thought to be a cluster member (Machin *et al.* 1991).

Several globular clusters were observed by the Einstein satellite to have X-ray sources with $L_{\rm x} \sim 10^{33}$ erg s^{-1} in their cores. These are thought to be CVs (Hertz & Grindlay 1983), but initial optical searches in ω Cen failed to confirm this (Margon & Bolte 1987; Shara *et al.* 1988); however, one has been identified in 47 Tuc (Paresce, De Marchi & Ferraro 1992). On the basis of (variable) $L_{\rm x} \sim 7 \times 10^{33}$ erg s^{-1}, $kT \sim 2$ keV, $L_{\rm x}/L_{\rm opt} \sim 10$, a possible 6 h orbital period and intermittent 120.200 s periodicity (Auriére, Koch-Miramond & Ortolani 1989), this is thought to be an IP, like GK Per (Section 7.6.6). Richer & Fahlman (1988) have found seven UV-rich objects in M71 with $6 \lesssim M_{\rm v} \lesssim 10$ which they think may be CVs.

The extension to lower luminosities made possible by ROSAT has revealed many sources down to $L_{\rm x} \sim 3 \times 10^{31}$ erg s^{-1} in globular clusters (Grindlay 1993a,b), including some identified as UV-rich at $M'_{\rm v} \sim 6$, or Hα emitting in NGC 6397 (Cool *et al.* 1993). As predicted by Hertz & Grindlay (1983), the luminosity function of these sources resembles that of CVs, and removes the possibility that they could be low mass X-ray binaries (i.e., accreting neutron stars) in low states (which have $L_{\rm x} \sim 10^{32-33}$ erg s^{-1}). The total number of CVs in NGC 6397 is estimated to be $\sim$20; more generally globular clusters are estimated to have $\sim$5 CV members with $L_{\rm x} > 5 \times 10^{31}$ erg s^{-1} (Grindlay 1993b).

Double stars in globular clusters are thought to be produced by three-body collisions and tidal capture (e.g., Hut & Verbunt 1983). The latter process can produce only very close binaries, of which neutron star binaries are more probable than white dwarf binaries because the neutron stars are more massive. In addition, the neutron stars will have gravitated to the centre of the cluster, where the space density of stars is highest and binary formation is most likely. The resulting numbers and ratios of low mass X-ray binaries and CVs are difficult to calculate; predictions range from a few CVs per cluster (Verbunt & Meylan 1988) to more than 100 in ω Cen and 47 Tuc (Stefano & Rappaport 1994).

The importance of hard binaries in the dynamical evolution of a cluster may be appreciated from the fact that a dozen or so CVs have as much binding energy as an entire globular cluster.

A new window on high $\dot{M}(2)$ CVs has been opened by the discovery of 'ultrasoft' low luminosity sources in the Large Magellanic Cloud (van den Heuvel *et al.* 1992), in the Small Magellanic Cloud (Kahabka & Pietsch 1993), in the globular cluster M3 (Hertz, Grindlay & Bailyn 1993), in M31 (see Greiner, Hasinger & Thomas 1994) and in the Galaxy (Motch, Hasinger & Pietsch 1994). These sources (Table 9.6) have flux distributions that peak at $kT = 30$–50 eV and have $L_{\rm x} \sim 10^{36-38}$ erg s^{-1}. All are variable in X-ray flux over a large range. CAL 83 and CAL 87 are eclipsing systems.

The M3 source during an outburst had $L_{\rm bol} = 1.2 \times 10^{36}$ erg s^{-1} with $kT \approx 45$ eV, which gives a radius $\sim 7.7 \times 10^8$ cm, typical of a white dwarf (Hertz, Grindlay & Bailyn 1993). If entirely powered by accretion, this would require $\dot{M}(1) \sim 6.6 \times 10^{-7}$ M$_\odot$ y^{-1}, which is above the rate required for stable nuclear burning (Figure 5.19). Van den Heuvel *et al.* (1992) model the LMC sources as CVs in

Table 9.6. *Ultrasoft X-ray Sources.*

Star	System	P_{orb}(h)	m_v	M_v(max)	References
CAL 83	LMC	25.0464	16.2–17.1	−2.4	1
CAL 87	LMC	10.624	19.0–20.1	0:	2,3
RXJ0527−6954	LMC		16–17:	−2.5:	4
1E1339.8+2837	M3		>18.5	>3	5
RXJ0439−6809	LMC		>19	>0	6
RXJ0513−6915	LMC		16.7	−1.9	7
RXJ0925.7−4754	Galaxy	3.5:	17.1		8
IE0035.4−7230	SMC		19.9–20.2		9

References: 1. Smale *et al.* 1988; 2. Cowley *et al.* 1990; 3. Pakull *et al.* 1988; 4. Trümper *et al.* 1991; 5. Hertz, Grindlay & Bailyn 1993; 6. Schaeidt, Hasinger & Trümper 1993; 7. Greiner, Hasinger & Thomas 1994.; 8. Motch, Hasinger & Pietsch 1994; 9 Orio *et al.* 1994.

which $0.7 \leq M_1(1) \leq 1.2$ and $1.0 \times 10^{-7} \leq \dot{M}(1) \leq 4.0 \times 10^{-7}\ \mathrm{M_\odot\ y^{-1}}$. The secondaries in CAL 83 and CAL 87 have $M_1(2) \sim 1.5$, which gives $q \gtrsim 1.5$ and shows why mass transfer is so high (equation (9.39)). The computations by Pylyser & Savonije (1988a,b) generate $\dot{M}(2)$ in the above range for $1.5 < M_1(2) < 2.0$.

It appears, therefore, that the sources in Table 9.6 are those in which contact with the Roche lobe has been made by main sequence secondaries in systems with $q > 1.26$, resulting in mass transfer on a thermal time scale (Section 9.3.1). These are the systems thought capable of building Ne-rich layers which later generate relatively low mass Ne novae (Shara & Prialnik 1994: Section 5.8.2).

9.7 Ultimate Evolution

A number of end-points of CV evolution are possible. Most CVs decline into lonely isolation, but some may go out with a bang.

9.7.1 *Hydrogen-Rich Systems*

Beyond the P_{orb} minimum, equations (9.42) and (9.46b) for degenerate secondaries apply. Ignoring the effects of nova eruptions and departure from thermal equilibrium these may be integrated to give the secondary mass as a function of time:

$$M_1(2,t) \approx M_1(2,t_0)[1 - 1.43 \times 10^{-3} M_1^{11/3}(2,t_0)(1+X)^{-20/3} M_1^{2/3}(1)(t-t_0)]. \tag{9.58}$$

Taking $M_1(2,t_0)$ as the secondary's mass when $P_{orb} = P_{orb,min}$ then gives a lifetime on the increasing P_{orb} branch of evolution of

$$t_{inc} \approx 4.3 \times 10^9 P_{orb,min}^{11/3}(\mathrm{h})(1+X)^{-5/2} M_1^{-2/3}(1)\ \mathrm{y}. \tag{9.59}$$

The smallest practical $M_1(2)$ is $\sim 2.5 \times 10^{-3}$ for $X = 0.7$ and 1.1×10^{-3} for $X = 0$ (Zapolsky & Salpeter 1969), below which the secondary becomes planet-like and shrinks in size as it decreases in mass. This does not reduce t_{inc} significantly. The orbital periods for such planets are ~0.7 d.

As seen from equation (9.46b), CVs with degenerate secondaries and $P_{orb} \gtrsim 2$ h have $< \dot{M}(2) > \lesssim 1 \times 10^{-12}$ M$_\odot$ y^{-1} (or $L_x \lesssim 3 \times 10^{29}$ erg s^{-1}) and will be difficult to detect (unless in a state of higher $\dot{M}(2)$ stimulated by (an extremely rare) recent nova eruption). The *observable* lifetime is therefore only $\sim 0.1 t_{inc}$, which implies that the majority of CVs formed during the life t_G of the Galaxy will now appear as single white dwarfs. This is in agreement with the conclusions of Section 9.5.6.2. The space density of white dwarfs generated through this channel is $\sim(0.4\text{–}1.5) \times 10^{-4}$ pc^{-3}, which is ~1% of the space density of isolated white dwarfs.

9.7.2 *Helium-Rich Systems*

Amid the variety of evolutionary possibilities available to the initial distribution of detached binary systems some result in helium secondaries. One such route is followed by initial primaries with $3.5 < M_1 < 7$ which fill their Roche lobes before helium ignition has occurred. Following the CE phase a helium core is exposed which is sufficiently massive to ignite helium as a main sequence star and transfer mass via a wind to its companion (Webbink 1979b). This may be the state of the 12.34 h binary helium–rich Wolf–Rayet star V Sge, which has sometimes been included among the CVs.

Another route is followed by systems in which the initial >3 M$_\odot$ primary has a radiative envelope at the time of filling its Roche lobe, leading first to thermal time scale mass transfer until the mass ratio is reversed and then transfer on a nuclear time scale until the hydrogen envelope is removed. This leaves a $0.21 < M_1 < 0.46$ helium white dwarf orbiting a $1.5 < M_1 < 6$ main sequence star (Webbink 1979b). When the latter evolves, a CE phase ensues which leaves two white dwarfs in a binary of short period (Tutukov & Fedorova 1989). Gravitational radiation braking brings these together until the lower mass white dwarf (which in general is the helium core of the original primary) fills its Roche lobe, forming a double degenerate helium-transferring close binary. Although no nova eruptions would be expected of these systems, they should exhibit most or all of the accretion disc phenomena made familiar by the hydrogen-rich CVs.

9.7.2.1 *The AM Canis Venaticorum Stars: Discovery and Nature*

The six known members (Table 9.7) of the AM CVn class have been found by a variety of techniques. The type star, formerly known as HZ 29 from the catalogue of faint blue stars by Humason & Zwicky (1947), was found by Greenstein & Matthews (1957) to have broad, shallow absorption lines of HeI and was classified as a DB white dwarf. Smak (1967) discovered that the light curve has a double hump with a period of ~18 min. This was confirmed by Ostriker & Hesser (1968) who derived a period of 1051.118 ± 0.015 s. The star was thought to be a pulsating hot subdwarf or a rotating magnetic white dwarf. High speed photometry showed the presence of flickering, which pointed to a mass-transferring binary (Warner & Robinson 1972b).

The second object in the class was recorded first as a high proper motion star, G61–29, found to have a helium emission-line spectrum and thought to be a white dwarf (Burbidge & Strittmatter 1971). Again the presence of flickering suggested a

Table 9.7. *AM CVn Stars.*

Star	Alias	Principal period and harmonics (s)	Other periods (s)	m_V	HeI λ4471 FWHM(Å)	References
AM CVn	HZ 29	1050(3)	1011,1028	14.1–14.2	40	1–16,38,41–44
EC 15330–1403		1117(3)		13.6	32	37
CR Boo	PG 1346+082	1493(3)	1471	13.0–18.0	45	17–19,40
V803 Cen	AE 1	1611(5)	175	13.2–17.4	21	20–22,35,36,39
CP Eri		1724(4)		16.5–19.7		23–25
GP Com	G61-29	2970		15.7–16.0	31	26–34

References: 1. Smak 1967; 2. Ostriker & Hesser 1968; 3. Faulkner, Flannery & Warner 1972; 4. Krzeminski 1972; 5. Warner & Robinson 1972b; 6. Smak 1975a; 7. Patterson *et al.* 1979; 8. Elsworth, Grimshaw & James 1982; 9. Vojkhanskaya 1982; 10. Solheim *et al.* 1984; 11. Patterson *et al.* 1992a; 12. Kruszewski & Semeniuk 1992; 13. Robinson & Faulkner 1975; 14. Wampler 1967; 15. Kovacs 1980; 16. Solheim 1989, 1993; 17. Wood *et al.* 1987; 18. Provencal *et al.* 1989; 19. Westin 1980; 20. O'Donoghue, Menzies & Hill 1987a,b; 21. O'Donoghue & Kilkenny 1989; 22. O'Donoghue *et al.* 1990; 23. Szkody *et al.* 1989; 24. Howell *et al.* 1991; 25. Abbott *et al.* 1992; 26. Burbidge & Strittmatter 1971; 27. Warner 1972b; 28. Richer *et al.* 1973; 29. Smak 1975b; 30. Greenstein, Arp & Schectman 1977; 31. Lambert & Slovak 1981; 32. Nather, Robinson & Stover, 1981; 33. Stover 1983; 34. Marsh, Horne & Rosen 1991; 35. Elvius 1975; 36. Ulla & Solheim 1990,1991; 37. O'Donoghue *et al.* 1994; 38. Lázaro, Solheim & Arévalo 1989; 39. Kepler, Steiner & Jablonski 1989; 40. Nather 1989; 41. Solheim & Kjeldseth-Moe 1987; 42. Patterson, Halpern & Shambrook 1993; 43. Solheim & Sion 1994; 44. Solheim *et al.* 1993.

binary nature (Warner 1972b). Spectroscopy subsequently revealed a Doppler shift modulation with a period of 46.5 min (Nather, Robinson & Stover 1981).

The third object, PG1346+082, was found in the PG survey for blue objects (Green, Schmidt & Liebert 1986) and shown to be a rapid variable (Nather 1989) later designated as CR Boo. At the same time, an object originally discovered on survey plates as a blue object varying between $m_V \sim 13$ and $m_V \sim 17$ (Elvius 1975) and found to be a helium absorption-line star, confirmed by Westin (1980), was found to be a short period variable by O'Donoghue, Menzies & Hill (1987b) and subsequently designated V803 Cen.

More recently, CP Eri, noted by Luyten & Haro (1959) as a high Galactic latitude blue star of large brightness variation, was found to be a short period photometric variable by Howell *et al.* (1991) and to be a helium line system, in absorption when bright and in emission when faint, by Abbott *et al.* (1992). Finally, the star EC 15330–1403 was discovered in the Edinburgh–Cape survey for blue stars, and shown spectroscopically and photometrically also to belong to the AM CVn class (O'Donoghue *et al.* 1994).

The common properties that bind the stars into a distinct subclass of CVs are (a) they are all helium-rich objects (no hydrogen lines have been seen in any of them), (b) they are all ultrashort period variables – in five objects this is demonstrated by photometry but not by spectroscopy, in GP Com the reverse is the case (but note that there appear to be four cycles of the 46.5 min period in the light curve, Fig. 3 of Warner (1972b)), (c) the large breadths of the absorption lines suggest rapid rotation (of disc or star) and/or pressure broadening by compact stars, implying luminosities substantially less than He main sequence stars (in the case of GP Com, the proper motion $\mu = 0''.35$ y^{-1}, requires $M_V \geq 9.2$ if it is to be bound to the Galaxy (Burbidge & Strittmatter 1971); AM CVn has $\pi = 0''.012 \pm 0''.007$ and $\mu = 0''.023$ y^{-1}, which give $M_V = 9.5(+1.0, -2.0)$ and $M_V \geq 5.8$ respectively (Solheim *et al.* 1984)), (d) in some of the stars there is evidence for flickering – in others the evidence is less convincing because of the presence of many oscillation modes, (e) the fact that at least some of the group have high and low states is a signature of mass transfer.

These group properties lead strongly to the conclusion that the AM CVn stars have structures like H-rich CVs, with accretion luminosity domination of their high states, but not necessarily in the low states. If the observed periods are indicative of orbital periods, these systems are very compact (equation (2.1b)), must have degenerate secondaries (equation (2.3b)), and will have accretion discs with radii $\lesssim 1.5 \times 10^{10}$ cm.

A crucial question, which was for long unanswered, is whether the periodicities reveal binary motion, disc oscillation, stellar oscillation or rotation of the primary. In GP Com, where a spectroscopic periodicity *is* present, and the system appears to be in a permanent low state, a magnetic primary could plausibly be responsible. In others, oscillations of the accretion disc or the DB primary *could* be possible. Only for AM CVn itself has there recently been a strong deduction from observations that its orbital period is indeed ultrashort, and by analogy so are those of the other members of the class (Patterson, Halpern & Shambrook 1993). The location of the AM CVn stars among low $\dot{M}(1)$, non-magnetic CVs in their X-ray properties (Beuermann & Thomas 1993) excludes strongly magnetic primaries ($B \lesssim 1$ MG for these very compact systems), but not IPs.

9.7.2.2 Spectroscopy of the AM CVn Stars

By analogy with the H-rich CVs, AM CVn and EC 15330 having absorption spectra appear to be in a stable state of high $\dot{M}(2)$, GP Com with strong line emission is in a state of low $\dot{M}(2)$ and the others move between high and low states, showing predominantly absorption spectra in the high and emission line spectra (or plain continuum) in the lowest states.

The optical spectrum of AM CVn (Figure 9.12) shows shallow, asymmetric HeI absorption lines (Greenstein & Matthews 1957). The line profiles are of variable depth and profile, which suggests varying amounts of in-filling by emission contributions (Robinson & Faulkner 1975; Vojkanskaya 1982). This is supported by the occasional presence of HeII emission (Patterson *et al.* 1992a: Figure 9.12). However, part or most of the variation in profile may be intrinsic to the absorption process and not to addition of emission (O'Donoghue & Kilkenny 1989). CR Boo and V803 Cen in their (short lived) high states can have very similar spectra to AM CVn, but with deeper absorption lines (Figure 9.12) as though less filled with emission. However, line profiles and depths are also variable in these stars, leading to reversal of asymmetries and to lines as shallow as those of AM CVn (Wood *et al.* 1987).

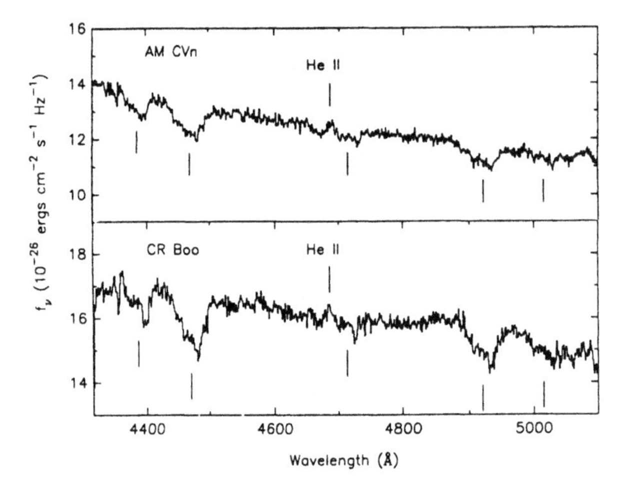

Figure 9.12 Spectra of AM CVn and CR Boo. From Patterson *et al.* (1992a).

The HeI absorption-line profiles are quite different from those of DB white dwarfs: the lines in the latter are almost symmetrical, have broader wings, are much deeper with relatively sharp cores, and have larger equivalent widths (comparisons are given by Patterson *et al.* (1992a)). The physical widths are similar to those in $\log g = 6$ hot subdwarf model atmospheres, rather than $\log g = 8$ DB atmospheres (O'Donoghue & Kilkenny 1989).

All high state spectra show anomalous line strengths relative to DB spectra (Robinson & Faulkner 1975; Wood *et al.* 1987; O'Donoghue & Kilkenny 1989): the 2^1P–n^1S and 2^3P–n^3S HeI series and lowest members of the remaining series are weak or absent. This is more probably a general non-LTE efect in the line-forming region, rather than superposition of an optically thin recombination spectrum (O'Donoghue & Kilkenny 1989).

Line widths show a considerable range over the different members of the AM CVn class (Table 9.7). Full widths at continuum level are ~ 100 Å ($v \sin i \sim 3500$ km s^{-1}), which are similar to H-rich CVs. In V803 Cen, which has the narrowest lines, O'Donoghue & Kilkenny (1989) find that the line widths scatter widely and therefore the Doppler effect is not the dominant broadening mechanism. The lines have widths typical of the Stark broadening in a $\log g = 6$ atmosphere. Equations (2.38b) and (2.51a) give an effective gravity $GM(1)z/r^3 \sim 2 \times 10^6$ cm s^{-2}; therefore the absorption lines must be formed in the inner regions ($r \lesssim 3 \times 10^9$ cm $\lesssim 6R(1)$) of the disc. The broader line AM CVn stars are deduced to be of higher inclination than V803 Cen.

As an AM CVn star falls from high state to low state the spectrum changes from absorption to emission-line, passing through a stage of nearly continuous spectrum (Kepler, Steiner & Jablonski 1989) to one of generally weak emission. The same He lines appear in high and low states, with some changes in relative strength. The weakness of the emission lines suggests that the continuum has considerable optical thickness.

The optical flux gradients and broad band colours in the high state are well fitted by blackbody distributions with $T = 20\,000$ K (AM CVn: Wampler 1967) or

$T = 26\,000$ K (V803 Cen: Kepler, Steiner & Jablonski 1989). The absence of HeII absorption requires $T \lesssim 30\,000$ K. For discs of small radii blackbody distributions would be expected to fit well because the range of $T_{\mathrm{eff}}(r)$ is comparatively small. Patterson *et al.* (1992a) show that the flux distributions of blackbody accretion discs with $r_{\mathrm{d}} \sim 10^{10}$ cm mimic single temperatures and fit the AM CVn flux distribution well for $\dot{M}(\mathrm{d}) \sim 3 \times 10^{-9}$ M$_\odot$ y^{-1} ($M_1(1) = 0.9$ and $r_{\mathrm{d}} = 1.5 \times 10^{10}$ cm).

In falling to the low state, the optical flux distribution in V803 Cen reddens and becomes almost flat, resembling a blackbody with $T \sim 7000$ K (Kepler, Steiner & Jablonski 1989); but in CR Boo the flux distribution remains very blue (Wood *et al.* 1987).

GP Com has had no known outbursts; its emission-line spectrum is similar to others at a low state, except that the lines are very much stronger. The optical flux distribution is flat and indicates $T \sim 10\,000$ K. The optical and near IR spectra are shown in Figure 9.13. Lines of HeI and HeII dominate, but there are also lines from N,O and Mg (Nather, Robinson & Stover 1981; Marsh, Horne & Rosen 1991). In the UV, only NV 1240 Å and HeII 1640 Å emission lines are present – the CIV 1550 Å line is absent (Lambert & Slovak 1981). Modelling the emission spectrum as for a uniform gas in LTE gives $T \sim 11\,000$ K, abundances H/He < 10^{-5}, N/O $\sim$ 50, N/C > 100; the absence of CaII, SiII and FeII lines requires these elements to have abundances $< 10^{-3}$solar (Marsh, Horne & Rosen 1991).

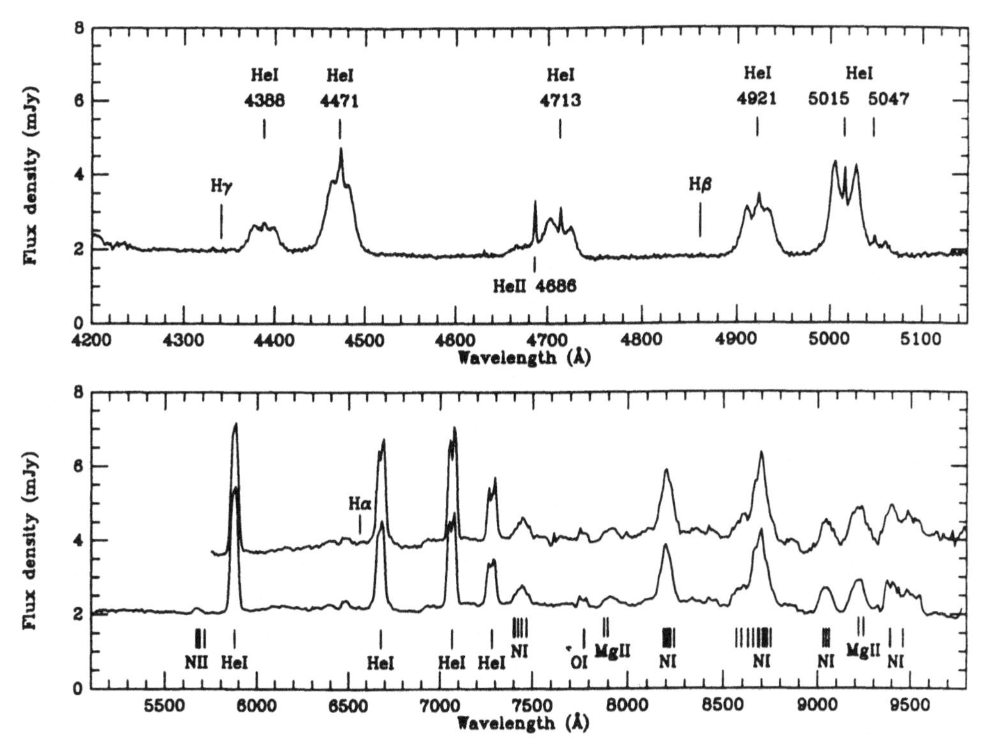

Figure 9.13 Spectra of GP Com. From Marsh, Horne & Rosen (1991).

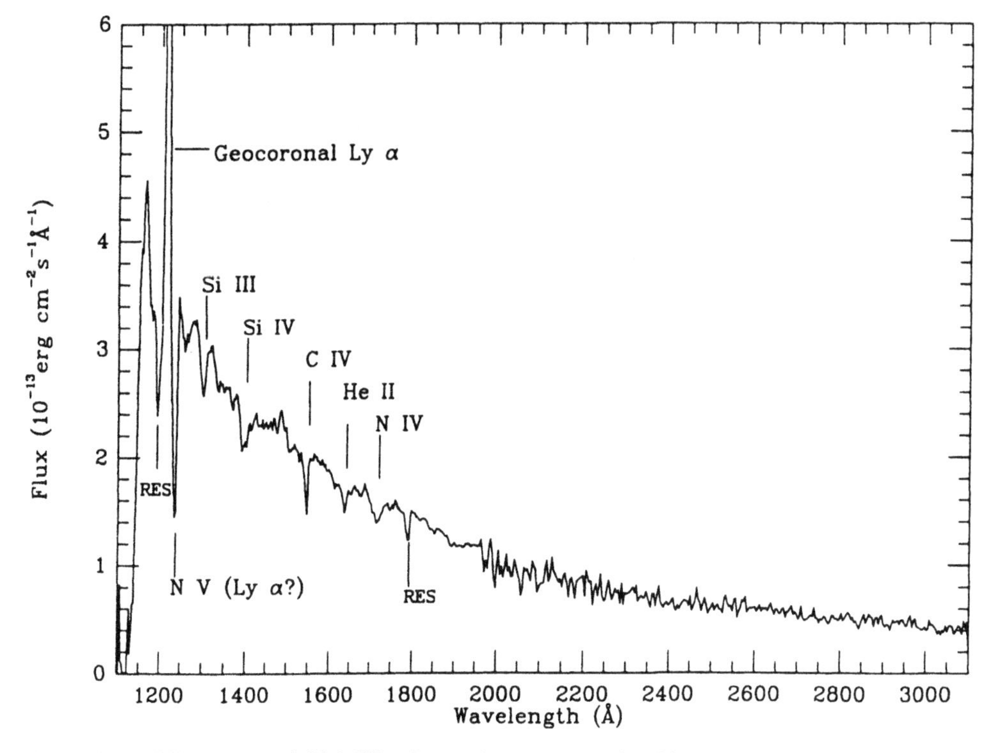

Figure 9.14 UV spectra of AM CVn. From Patterson *et al.* (1992a).

UV spectra of the other AM CVn stars show slopes in the high and low states which are in general bluer than those of H-rich CVs (Solheim 1992). Unlike in the optical region, a blackbody distribution does not fit the UV well – in V803 Cen an $F_\lambda \propto \lambda^{-2.44}$ fits the short wavelength region in the high state and $F_\lambda \propto \lambda^{-1.7}$ in the low state (Ulla & Solheim 1991). AM CVn has shown both $\lambda^{-2.0}$ and $\lambda^{-2.3}$ distributions (Solheim & Kjeldseth-Moe 1987). No models for the discs are available for comparison.

The spectrum of AM CVn (Figure 9.14) is characteristic of the high state spectra of other objects (V803 Cen: Ulla & Solheim 1991; CR Boo: Wood *et al.* 1987). Absorption lines of HeII, CIV, NIV,V, SiIII,IV are present at strengths similar to those seen in NLs (Figure 4.8). The lines have blueshifted cores, varying with time and from line to line over the range 200–1800 km s^{-1}. The UV absorption line widths are < 10 Å, i.e., much smaller than in the optical (Table 9.6), and are generally deeper and more variable in equivalent width than the optical lines. Although Greenstein (1979) claimed strong P Cyg emission, this is disputed (Solheim & Kjeldseth-Moe 1987). However, the conclusion is that the UV lines are formed in a hot wind, not in a disc.

The variable lines profiles in optical spectra lead to apparent changes in RV (e.g., Westin 1980), which have long defied attempts to reveal periodicities (Robinson & Faulkner 1975; Wood *et al.* 1987; O'Donoghue & Kilkenny 1989). However, Patterson *et al.* (1992a) have discovered that the absorption-line asymmetries in AM CVn follow a well-defined 13.38 h period (Figure 9.15), which will undoubtedly open the way to

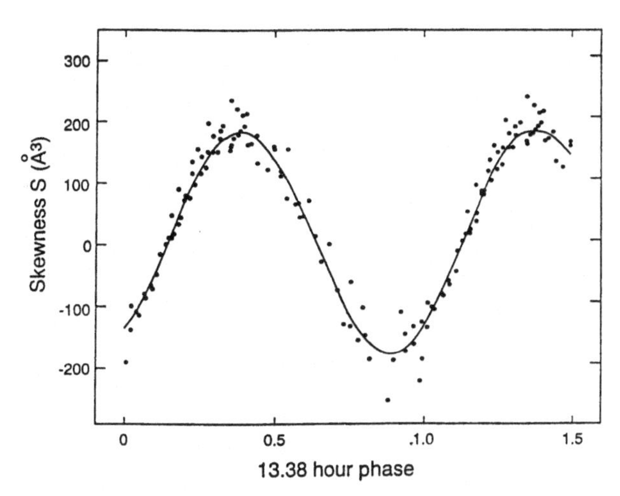

Figure 9.15 Periodicity in skewness of absorption-line profiles in AM CVn. From Patterson, Halpern & Shambrook (1993).

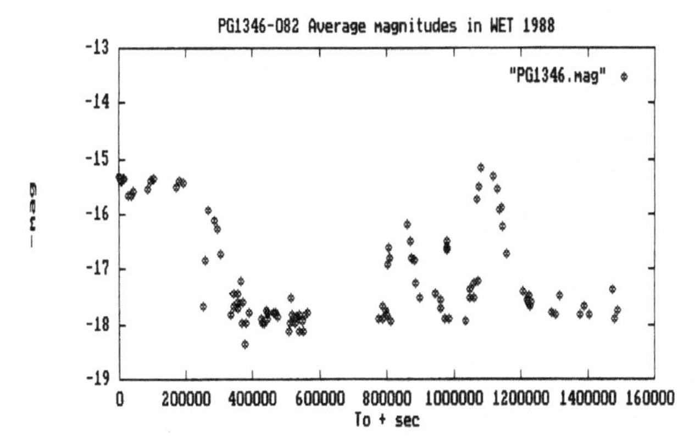

Figure 9.16 Light curves of CR Boo over an 18 d period. From Provencal (1993).

similar results in other AM CVn systems and is seen below to be of pivotal importance in the understanding of the structure of these systems.

In GP Com there is an S-wave (Section 1.4.2) component in all emission lines which has $K(1) = 14.6$ km s^{-1} and a period of 2790 s (Nather, Robinson & Stover 1981). This period is confirmed by the X-ray light curve (Beuerman & Thomas 1993).

9.7.2.3 Photometry of AM CVn Stars

AM CVn has never been seen to dip from its $m_v \sim 14.1$ brightness and GP Com has no known outburst. The observational coverage of the latter, however, is too sparse to exclude occasional brief outbursts. Wood *et al.* (1987) used archival meteor search films to study the light curve of CR Boo over the interval 1952–7 and found $m_v < 14.0$ about 75% of the time, with strong evidence for alternation between bright and faint magnitudes with mean interval 4–5 d. The light curve from an intensive study of CR Boo over 18 d is shown in Figure 9.16. During this time CR Boo was never brighter

Table 9.8. *Mean Colours of AM CVn Stars.*

	High state		Low state	
	B–V	U–B	B–V	U–B
AM CVn	−0.20	−1.06		
EC 15330	−0.12	−0.95		
CR Boo	−0.20	−1.00	0.05	−0.95
V803 Cen	−0.12	−0.90	0.15	−0.95
GP Com			−0.10	−0.97

Table 9.9. *Periodicities in AM CVn.*

Period (s)	Description	Amplitude (mmag)
1051.23	Fundamental	0–20
525.618	1st harmonic	11.6
350.407	2nd harmonic	3.2
262.799	3rd harmonic	1.1
210.244	4th harmonic	1.4
175.200	5th harmonic	1.0
1011.3		1.2–15
1028.8	Orbital period	0.12

than $m_V = 15$. Two outbursts from a distinct base level at $m_V = 18.0$ with rise times $\sim$0.5 d and duration $\sim$1–2 d are present, as is a very brief $1\frac{1}{2}$ mag brightening. V803 Cen is more erratic: Elvius (1975) in 1971, O'Donoghuc, Menzies & Hill (1987b) and O'Donoghue & Kilkenny (1989) found it mostly near maximum brightness; O'Donoghue *et al.* (1990) found it at minimum light for 9 d, with bright states at each end of the interval. Photometry by Westin (1980) showed smooth rise and falls on time scales $\sim$5 h and range $\sim$1 mag about a mean magnitude $m_V \approx 14.0$. No long term light curve is available.

The light curves of the unstable AM CVn stars are very similar to those of the VY Scl stars (cf. Figures 9.16 and 4.15), suggesting the operation of irradiation – enhanced mass transfer (Section 9.4.3).

The mean colours of the absorption-line systems are very similar to each other: Table 9.8; there is some reddening of B–V in the low state but U–B hardly changes.

High speed photometry of the AM CVn stars has revealed an abundance of periodicities and a complexity of behaviour. Unfortunately, none are known to be eclipsing systems. (If $q \sim 0.05$ (next section), then $i_{min} \sim 80°$ and a fraction $\sim \cos i_{min} \sim 0.17$ would be expected to eclipse.) A short section of the light curve of AM CVn itself is shown in Figure 9.17. Fourier analysis of light curves of AM CVn obtained in different years shows variable amplitudes of a small group of periodicities, which are listed in Table 9.9. The 'fundamental' period is not always present; its

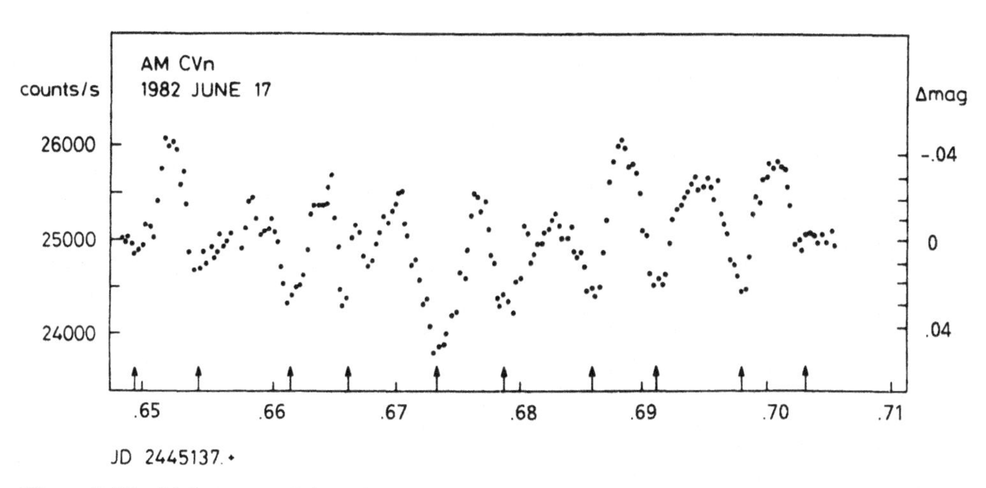

Figure 9.17 Light curve of AM CVn at 25 s time resolution. Arrows point to light minima, the irregular spacings emphasize the complexity of the light curve. From Solheim *et al.* (1984).

harmonics probably vary in amplitude also, but independently of the amplitude of the fundamental. The amplitudes in Table 9.9 are those derived in 1990 (Solheim *et al.* 1993). At that time the harmonics had sidebands 20.7 μHz separated from them, i.e., at a modulation period of 13.4 h.

Long term observation of AM CVn shows variations in the fundamental period which has been interpreted as $\dot{P} = -3.2 \times 10^{-12}$ (Solheim *et al.* 1984), but Patterson *et al.* (1992a) point out that such large phase shifts occur over times of weeks that the fundamental (and its harmonics) is better considered as unstable and no $\dot{P}$ is derivable.

Similar behaviour is seen in the other AM CVn stars (with the exception of GP Com); the periods and the numbers of observed harmonics are listed in Table 9.6. CR Boo, V803 Cen and CP Eri provide the additional dimension of high and low states, with dramatic changes in short time scale photometric behaviour between them. The transformation in CR Boo is shown in Figure 9.18. The observed frequencies change by 6–7 μHz from low to high state (Provencal 1993).

9.7.2.4 *The Structure of AM CVn Stars*

The variability of both line profiles and γ-velocities in the AM CVn stars closely resembles what is observed in superoutbursts of DN (Section 3.6.5.1). The 13.38 h periodicity of profile skewness in AM CVn (Patterson, Halpern & Shambrook 1993) can then be interpreted as the beat period P_b of a superhump phenomenon, which implies that the light from AM CVn stars in their high states is dominated by an optically thick He disc. The fundamental period of 1051 s in AM CVn is a superhump period P_s which, through equation (3.34), provides $P_{orb} = 1028.77 \pm 0.18$ s; there is a weak signal at this period in the light curve (Solheim *et al.* 1993: Table 9.9), The harmonics show that the pulse shape is highly non-sinusoidal, which is typical of superhump profiles in SU UMa stars. The 1011 s period in AM CVn may be retrograde precession caused by a slight tilt to the disc (in addition to its prograde precession as an elliptical disc (Patterson, Halpern & Shambrook 1993). The sidebands to the

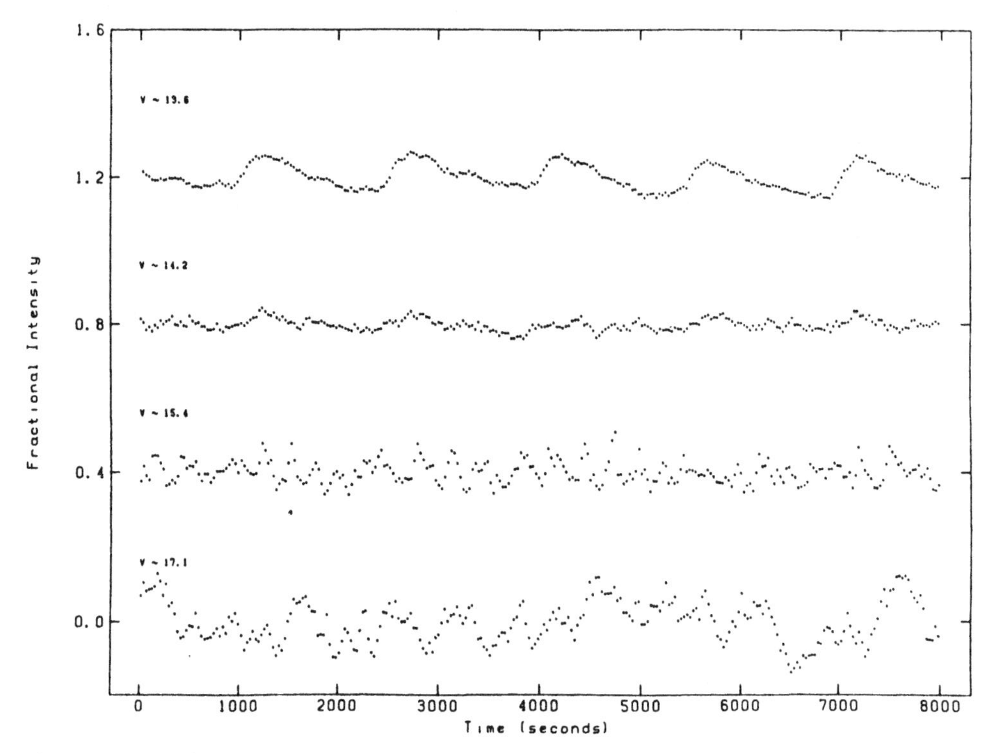

Figure 9.18 Light curves of CR Boo at high, low and two intermediate states. From Wood *et al.* (1987)

harmonics seen in the light curve are probably amplitude modulation caused by varying aspects of the disc.

For AM CVn, application of equation (3.42) gives $q = 0.084$. If the secondary were in thermal equilibrium then equation (9.43) would give $M_1(2) = 0.033$ and hence $M_1(1) = 0.39$. This gives $v_{max} = 2186 \sin i$ km s^{-1} (equation (2.81)) and $K(1) = 55 \sin i$ km s^{-1} (equation (2.79)). The first of these figures would be in conflict with the observed $v_{max} = 2340$ km s^{-1} (Patterson, Halpern & Shambrook 1993) if i is low. The prediction for $K(1)$ is compatible with the observed limit $K(1) < 50$ km s^{-1} (Robinson & Faulkner 1975), but this was set for $P_{orb} = 1051$ s and a new determination for $P_{orb} = 1028$ s is required. Quite different values are obtained when allowance is made for departure from thermal equilibrium. Savonije, de Kool & van den Heuvel (1986) have computed the evolution for a He secondary and find that after passing $P_{orb,min}$ the mass–radius relationship is

$$R_9(2) = 2.03\ M_1^{-0.19}(2) \tag{9.60a}$$

or

$$M_1(2) = 1.86 \times 10^{-2} P_{orb}^{-1.274}(\mathrm{h}) \tag{9.60b}$$

(i.e., $\xi_{ad} = -0.19$), which gives $M_1(2) = 0.092$ and $M_1(1) = 1.09$. These values are compatible with the observed $K(1)$, v_{max}, and $v_d(I_{max})$, and (taking $r_d = r_{32}$) give $i = 31 \pm 6°$.

For GP Com, the observed $K(1) = 14.6$ km s^{-1} and $v_{max} = 2000$ km s^{-1} (Nather, Robinson & Stover 1981), together with $M_1(2) = 0.024$ from equation (9.60b), give $q = 0.039$, $M_1(1) = 0.63$ and $i \approx 30°$. Assuming thermal equilibrium (equation (9.43)) does not give realistic results. The limit $K(1) < 16$ km s^{-1} for V803 Cen (O'Donoghue & Kilkenny 1989), with equation (9.60b) gives $q \leq 0.043$ and $i \leq 26°$ for $M_1(1) \leq 1.2$. The low inclination is in agreement with the narrower lines in V803 Cen (Table 9.7).

The 1471.35 s coherent period in CR Boo (Provencal *et al.* 1989) and the variable period ~1493 s invite identification as P_{orb} and P_s, in which case $q = 0.057$, $M_1(1) = 1.00$ and $M_1(2) = 0.058$. The observed $v_{max} \sim 2650$ km s^{-1} (Wood *et al.* 1987) gives $i = 32°$ and the predicted $K(1)$ is 24 km s^{-1}.

Less is known about the other two AM CVn stars, but it may be assumed that their dominant photometric modulation in the high state is also P_s; P_s and P_{orb} are related through equation (3.42); as $q \mathrel{\tilde{\propto}} M^{-1}(1)P_{orb}^{-1}$ (equation (9.42)) the ratio P_{orb}/P_s should increase towards unity with increasing P_{orb}.

It should be stressed that the parameters derived above are entirely dependent on the correctness of the adopted precessing disc model.

From equations (9.46a) and (9.60a) the mass transfer rate is

$$\dot{M}(2) = \frac{4.4 \times 10^{-12} M_1^{2/3}(1)}{(1+q)^{1/3}(0.74-q)} P_{orb}^{-5.21}(\text{h}) \quad \text{M}_\odot \ \text{y}^{-1} \tag{9.61}$$

giving $\dot{M}(2) = 3.9 \times 10^{-9}$ M$_\odot$ y^{-1} for AM CVn, diminishing down the entries in Table 9.7 to $\dot{M}(2) = 1.0 \times 10^{-11}$ M$_\odot$ y^{-1} for GP Com.

GP Com has a well-defined modulation in the 0.1–2 keV X-ray region (Beuermann & Thomas 1993), with $L_x \sim 5 \times 10^{30}$ erg s^{-1} if $d = 100$ pc. The X-ray light curve is very similar to that for low $\dot{M}$ systems like SU UMa. By analogy with other low $\dot{M}$ CVs, the primary should have $T_{eff} \lesssim 1 \times 10^4$ K (Table 2.8), which gives $m_v(1) \gtrsim 17.2$ at $d = 100$ pc, which is compatible with the absence of any white dwarf absorption spectrum in GP Com.

9.7.2.5 *Helium Accretion Discs*

Helium accretion discs differ from hydrogen-rich ones in the effects of lowered opacity, higher ionization temperature and smaller c_s. From integrations of the vertical structure for discs with composition $Y = 0.98$ and $Z = 0.02$, Smak (1983) found a typical S-curve (Section 3.5.3.4) with critical temperatures $T(\text{crit1}) = 9000$ K and $T(\text{crit2}) = 12\,500$ K, which are higher than for hydrogen discs; the corresponding surface densities (for $\alpha = 0.1$, $M_1(1) = 1$, $r = 1 \times 10^{10}$ cm) are $\Sigma_{min} = 330$ g cm^{-2}, and $\Sigma_{max} = 680$ g cm^{-2} A considerably larger Σ_{max} would obtain if $\alpha_C \lesssim 0.02$ on the lower branch of the S-curve (e.g., equation (3.15)), but $T(\text{crit1})$ is probably independent of α_C (see Figure 3.24). Cannizzo (1984) has computed some disc instability light curves for pure helium discs of dimensions characteristic of the AM CVn stars.

In their high states the AM CVn stars have $T_{eff} > T(\text{crit2})$ and GP Com has $T_{eff} < T(\text{crit2})$. Smak (1983) suspected that GP Com could be stabilized on the lower branch of the S-curve (Section 3.5.3.5), but this is for $\alpha_C = 0.1$; the values of $\alpha_C < 0.01$ deduced for the low $\dot{M}(\text{d})$ hydrogen-rich CVs (Section 3.5.4.5) would change this conclusion to an expectation that GP Com should show DN outbursts with recurrence intervals of tens of years.

The long term light curves suggest that AM CVn and EC 15330 are equivalent to NLs permanently in the high state, CR Boo, V803 Cen and CP Eri are the analogues of VY Scl stars, and GP Com is probably a SU UMa star. This ordering is one of increasing P_{orb} and, therefore, through equation (9.61), one of diminishing $\dot{M}(2)$, in agreement with these classifications.

The irradiative flux $F_{irr}(2)$ on the surface of the secondary is $\propto L(1)/a^2$, which from equations (2.1b), (2.83a) and (9.55) is

$$F_{irr}(2) \propto \frac{M^{2/3}(1)\dot{M}(2)}{(1+q)^{2/3}} P_{orb}^{-4/3}. \tag{9.62}$$

The VY Scl stars have $P_{orb}(h) \sim 3.5$ in the middle of their range, $q \sim 0.3$ and an observed $\dot{M}(2) \sim 1.0 \times 10^{-8}$ M$_\odot$ y^{-1} (Figures 4.16 and 9.8). The AM CVn stars have $q \sim 0.05$ and $\dot{M}(2)$ given approximately by equation (9.61). From equation (9.62) the latter stars will therefore experience the same $F_{irr}(2)$ as the former when $P_{orb} \sim 1550$ s, which is compatible with the observed range of $\dot{M}(2)$ instabilities (Table 9.7). Whereas the VY Scl stars evolve through their instability range towards shorter P_{orb}, the AM CVn stars evolve with increasing P_{orb}. However, having entered the period gap they will evolve back to shorter P_{orb}, to be recycled at decreased $M(2)$. There is no instability region on the decreasing P_{orb} branch for helium systems because $\dot{M}(2)$ is always too high (see Tutukov & Fedorova (1989)). Similarly there is no instability region on the increasing P_{orb} branch of hydrogen-rich systems because $P_{orb,min}$ is ~80 min and $\dot{M}(2)$ is too low thereafter.

The outbursts from $m_v \sim 18.0$ in CR Boo (Figure 9.16) look distinctly like DN outbursts. Taking $r_d = r_{32} \approx 9.3 \times 10^9$ cm, $\alpha_H = 0.2$ and $c_s = 10.1$ km s^{-1} (for helium in a 5×10^4 K transition front) equation (3.20) gives $t_r \approx 0.5$ d, which is in agreement with observation (Section 9.7.2.3). Equation (3.25a) predicts a decay time $t_d \sim 10t_r \sim 5$ d, which is also similar to what is observed (Figure 9.16).

9.7.2.6 *Space Density of AM CVn Stars*

Four AM CVn stars are of high Galactic latitude, characteristic of a population of nearby stars which have been discovered mostly from high latitude surveys. Extension to lower latitudes may be expected to double or triple the number known with $m_v \lesssim 14.5$. Cannizzo (1984) found $M_v = 8.5$ for a face-on helium disc of radius 4×10^9 cm and $\dot{M}(d) \sim 3 \times 10^{-9}$ M$_\odot$ y^{-1}. This suggests $M_v \sim 9.5$ for the AM CVn stars in their high states, in which case probably five of the six stars in Table 9.7 are within 100 pc. With allowance for incompleteness (and the expectation that, since these objects are believed to be evolved from progenitors with mass ~2.5 M$_\odot$ (Iben & Tutukov 1991), they will have Galactic scale heights $\lesssim$100 pc), this gives a local space density $\rho(0) \sim 3 \times 10^{-6}$ pc^{-3}.

Iben & Tutukov (1991) estimate that the birth rate of progenitors is ~0.01 y^{-1} in the entire Galaxy, i.e., a local rate $\sim 1 \times 10^{-13}$ pc^{-3} y^{-1}. From equation (9.59), with $P_{orb,min} \approx 13$ min (Iben & Tutukov 1991) the total lifetime is $t_{inc} \sim 3 \times 10^7$ y and $\rho(0) \sim 3 \times 10^{-6}$ pc^{-3}, in agreement with observation. (The higher $\dot{M}(2)$ in the AM

CVn stars enables them to be seen for a much greater fraction of t_{inc} than hydrogen-rich CVs, as is shown by the very low $M(2)$ deduced for GP Com.) The space density of descendants is $\sim\rho(0)t_G/t_{inc} \sim 1 \times 10^{-3}$ pc^{-3}, which is an order of magnitude greater than descendants of hydrogen-rich CVs (Section 9.7.1) and is comparable to the observed space density of DB white dwarfs (Fleming, Liebert & Green 1986).

9.7.3 *Supernovae from CVs*

Type Ia supernovae are observed in all types of galaxy and appear to avoid spiral arms. The association with an older stellar population, for which current masses are $\lesssim 1$ M$_\odot$, excludes single stars as progenitors of these supernovae. A number of binary star models have been proposed, of which pushing a CV primary over the Chandrasekhar limit is of interest here (Wheeler 1991).

To reach M_{ch} a primary must eject less mass in its nova eruptions than it receives during the quiescent accretion phase. The presence of enhanced CNONe abundances in nova ejecta shows that part of the core is being removed, which eliminates most CVs from the possibility of attaining supernova status. However, the abundances in RN (Section 5.9) do not require dredge up from the core and in theory (Section 5.9.5) the primaries appear already to be close to M_{ch}. The RN, therefore, offer the only currently recognized possibility of generating supernovae from CVs (MacDonald 1984; Livio & Truran 1992; Livio 1993b).

For a CV space density $\rho(0) \sim 1 \times 10^{-4}$ pc^{-3}, Livio & Truran (1992) estimate a supernova production rate $\sim 1 \times 10^{-4}$ y^{-1} from the above evolution, which is only a few per cent of the observed Galactic rate. It does, however, agree with the observed birth rate of low mass X-ray binaries (Kulkarni & Narayan 1988).

References

Abado-Simon, M., Lecacheux, A., Bastian, T.S., Bookbinder, J.A. & Dulk, G.A. 1993, *Astrophys. J.*, **406**, 692.
Abbott, T.M.C., Shafter, A.W., Wood, J.H., Tomaney, A.B. & Haswell, C.A. 1990, *Publ. astr. Soc. Pacific*, **102**, 558.
Abbott, T.M.C., Robinson, E.L., Hill, G.J. & Haswell, C.A. 1992, *Astrophys. J.*, **399**, 680.
Abramowicz, M.A., Livio, M., Piran, T. & Wiita, P.J. 1984, *Astrophys. J.*, **279**, 367.
Achilleos, N. &. Wickramasinghe, D.T. 1989, *Astrophys. J.*, **346**, 444.
Achilleos, N., Wickramasinghe, D.T. & Wu, K. 1992, *Mon. Not. R. astr. Soc.*, **256**, 80.
Acker, A. & Stenholm, B. 1990, *Astr. Astrophys.*, **233**, L21.
Adam, J., Störzer, H., Shaviv, G. & Wehrse, R. 1988, *Astr. Astrophys.*, **193**, L1.
Adams, W.S. 1922, *Trans. Int. Astr. Union*, **1**, 95.
Adams, W.S. & Joy, A.H. 1921, *Pop. Astr.*, **30**, 102.
Adams, F.C. & Shu, F.H. 1986, *Astrophys. J.*, **308**, 836.
Africano, J. & Wilson, J. 1976, *Publ. astr. Soc. Pacific*, **88**,8.
Africano, J.L., Nather, R.E., Patterson, J., Robinson, E.L. & Warner, B. 1978, *Publ. astr. Soc. Pacific*, **90**, 568.
Allen, C.W. 1954, *Mon. Not. R. astr. Soc.*, **114**, 387.
Allen, C.W. 1976, *Astrophysical Quantities*, Athlone Press, London, 3rd edit.
Allen, D.A. 1980, *Mon. Not. R. astr. Soc.*, **190**, 75.
Allen, D.A. 1984, *Proc. astr. Soc. Australia*, **5**, 369.
Allen, D.A. & Cherepashchuk, A.M. 1982, *Mon. Not. R. astr. Soc.*, **201**, 521.
Allen, D.A., Ward, M.J. & Wright, A.E. 1981, *Mon. Not. R. astr. Soc.*, **195**, 155.
Allen, R.G., Berriman, G., Smith, P.S. & Schmidt, G.D. 1989, *Astrophys. J.*, **347**, 426.
Aller, L.H. 1987, in *Spectroscopy of Astrophysical Plasmas*, eds. A. Dalgarno & D. Layzer, Cambridge University Press, Cambridge, p. 89.
Aly, J.J. 1980, *Astr. Astrophys.*, **86**, 192.
Aly, J.J. & Kuijpers, J. 1990, *Astr. Astrophys.*, **227**, 473.
Amari, S., Anders, E., Virag, A. & Zinner, E. 1990, *Nature*, **345**, 238.
Amari, S., Hoppe, P., Zinner, E. & Lewis, R.S. 1993, *Nature*, **365**, 806.
Anderson, N. 1988, *Astrophys. J.*, **325**, 266.
Andrea, G. 1992, Ph.D. Thesis, Univ. Nuremborg.
Andreä, J. 1992, *Rev. Mod Astr.*, **5**, 58.
Andrillat, Y. 1985, *Int. Astr. Union Circ.*, No. 4026.
Andronov, I.L. 1986a, *Astr. Tsirk.*, No. 1417, p. 5.
Andronov, I.L. 1986b, *Astr. Tsirk.*, No. 1432, p. 7.
Andronov, I.L. 1991, *Inf. Bull. Var. Stars*, No. 3645.
Andronov, I.L. 1993, *Inf. Bull. Var. Stars*, No. 3828.
Andronov, I.L. & Fuhrmann, B. 1987, *Inf. Bull. Var. Stars*, No. 2976.
Andronov, I.L. & Shugarov, S. Ya. 1982, *Astr. Tsirk.*, No. 1218, 3.
Andronov, I.L., Fuhrmann, B. & Wenzel, W. 1988, *Astr. Nach.*, **309**, 39.
Andronov, I.L., Kimeridze, G.N., Richter, G.A. & Smykov, V.P. 1989, *Inf. Bull. Var. Stars*, No. 3388.
Angel, J.R.P. 1978, *Ann. Rev. Astr. Astrophys.*, **16**, 487.
Angel, J.R.P., Borra, E.F. & Landstreet, J.D. 1981, *Astrophys. J. Suppl.*, **45**, 457.

Angel, J.R.P., Borra, E.F. & Landstreet, J.D. 1981, *Astrophys. J. Suppl.*, **45**, 457.
Angelini, L. & Verbunt, F. 1989, *Mon. Not. R. astr. Soc.*, **238**, 697.
Angelini, L., Osborne, J.P. & Stella, L. 1990, *Mon. Not. R. astr. Soc.*, **245**, 652.
Antipova, L.I. 1987, *Astrophys. Sp. Sci.*, **131**, 453.
Anupama, G.C. & Prabhu, T.P. 1993, *Mon. Not. R. astr. Soc.*, **263**, 335.
Anzer, U. & Börner, G. 1980, *Astr. Astrophys.*, **83**, 133.
Anzer, U. & Börner, G. 1983, *Astr. Astrophys.*, **203**, 183.
Applegate, J.H. 1989, *Astrophys. J.*, **337**, 865.
Applegate, J.H. 1992, *Astrophys. J.*, **385**, 621.
Applegate, J.H. & Patterson, J. 1987, *Astrophys. J.*, **322**, L99.
Arévalo, M.J., Solheim, J.E. & Lazaro, C. 1989, *Lect. Notes Phys.*, **328**, 462.
Arons, J. & Lea, S.M. 1980, *Astrophys. J.*, **235**, 1016.
Augusteijn, T. 1993, in *Cataclysmic Variables and Related Physics*, eds. O. Regev & G. Shaviv, Inst. Phys. Publ., Bristol, p. 272.
Augusteijn, T. 1994, personal communication.
Augusteijn, T., van Kerkwijk, M.H. & van Paradijs, J. 1993, *Astr. Astrophys.*, **267**, L55.
Augusteijn, T., van Paradijs, J. & Schwarz, H.E. 1991, *Astr. Astrophys.*, **247**, 64.
Auriére, M., Koch-Miramond, L. & Ortolani, S. 1989, *Astr. Astrophys.*, **214**, 113.
Baglin, A. & Schatzman, E. 1969, in *Low Luminosity Stars*, ed. S. Kumar, Gordon & Breach, New York, p. 385.
Baidak, A.V. & Shugarov, S. Yu. 1986, *Sov. Astr.*, **30**, 76.
Baidak, A.V., Lipunov, N.A., Shugarov, S.Yu., Moshkalev, V.G. & Volkov, I.M. 1985, *Inf. Bull. Var. Stars*, No.2676.
Bailey, J. 1975a, *J. Brit. astr. Assoc.*, **86**, 30.
Bailey, J. 1975b, *J. Brit. astr. Assoc.*, **85**, 217.
Bailey, J. 1978, *Mon. Not. R. astr. Soc.*, **185**, 73P.
Bailey, J.A. 1979a, *Mon. Not. R. astr. Soc.*, **187**, 645.
Bailey, J. 1979b, *Mon. Not. R. astr. Soc.*, **189**, 41P.
Bailey, J. 1979c, *Mon. Not. R. astr. Soc.* **188**, 681.
Bailey, J. 1980, *Mon. Not. R. astr. Soc.*, **190**, 119.
Bailey, J.A. 1981, *Mon. Not. R. astr. Soc.*, **197**, 31.
Bailey, J. 1985, *Recent Results on Cataclysmic Variables*, ESA SP-236, Paris, p. 139.
Bailey, J.A. 1988, in *Polarized Radiation of Circumstellar Origin*, eds. G.V. Coyne *et al.*, Vatican Obs., Vatican, p. 105.
Bailey, J. 1990, *Mon. Not. R. astr. Soc.*, **243**, 57.
Bailey, J. & Axon, D.J. 1981, *Mon. Not. R. astr. Soc.*, **194**, 187.
Bailey, J. & Cropper, M. 1991, *Mon. Not. R. astr. Soc.*, **253**, 27.
Bailey, J. & Ward, M. 1981a, *Mon. Not. R. astr. Soc.*, **194**, 17P.
Bailey, J. & Ward, M. 1981b, *Mon. Not. R. astr. Soc.*, **196**, 425.
Bailey, J., Ferrario, L. & Wickramasinghe, D.T. 1991, *Mon. Not. R. astr. Soc.*, **251**, 37P.
Bailey, J., Hough, J.H. & Wickramasinghe, D.T. 1988, *Mon. Not. R. astr. Soc.*, **233**, 395.
Bailey, J., Sherrington, M.R., Giles, A.B. & Jameson, R.F. 1981, *Mon. Not. R. astr. Soc.*, **196**, 121.
Bailey, J., Hough, J.H., Axon, D.J., Gatley, I., Lee, T.J., Szkody, P., Stokes, G. & Berriman, G. 1982, *Mon. Not. R. astr. Soc.*, **199**, 801.
Bailey, J., Axon, D.J., Hough, J.H., Watts, D.J., Giles, A.B. & Greenhill, J.G. 1983, *Mon. Not. R. astr. Soc.*, **205**, 1P.
Bailey, J., Hough, J.H., Gilmozzi, R. & Axon, D.J. 1984, *Mon. Not. R. astr. Soc.*, **207**, 777.
Bailey, J.A., Watts, D.J., Sherrington, M.R., Axon, D.J., Giles, A.B., Hanes, D.A., Heathcote, S.R., Hough, J.H., Hughes, S., Jameson, R.F. & McLean, I. 1985, *Mon. Not. R. astr. Soc.*, **215**, 179.
Bailey, J.A., Wickramasinghe, D.T., Hough, J.H. & Cropper, M.S. 1988, *Mon. Not. R. astr. Soc.*, **234**, 19P.
Bailey, J., Wickramasinghe, D.T., Ferrario, L., Hough, J.H. & Cropper, M. 1993, *Mon. Not. R. astr. Soc.*, **261**, L31.
Balachandran, S., Robinson, E.L. & Kepler, S.O. 1983, *Publ. astr. Soc. Pacific*, **95**, 653.
Balbus, S.A. & Hawley, J.F. 1991, *Astrophys. J.*, **376**, 214.
Balbus, S.A. & Hawley, J.F. 1992a, *Astrophys. J.*, **392**, 662.
Balbus, S.A. & Hawley, J.F. 1992b, *Astrophys. J.*, **400**, 610.

Balbus, S.A. & Hawley, J.F. 1994, *Mon. Not. R. astr. Soc.*, **266**, 769.
Baliunas, S.L. 1988, in *Formation & Evolution of Low Mass Stars*, eds. A.K. Dupree & M.T. Lago, Kluwer, Boston, p. 319.
Baliunas, S.L. & Jastrow, R. 1990, *Nature*, **348**, 520.
Baliunas, S.L. & Vaughn, A.H. 1985, *Ann. Rev. Astr. Astrophys.*, **23**, 379.
Baptista, R. & Steiner, J.E. 1991, *Astr. Astrophys.*, **249**, 284.
Baptista, R. & Steiner, J.E. 1993, *Astr. Astrophys.*, **277**, 331.
Baptista, R., Jablonski, F.J. & Steiner, J.E. 1989, *Mon. Not. R. astr. Soc.*, **241**, 631.
Baptista, R., Jablonski, F.J. & Steiner, J.E. 1992, *Astr. J.*, **104**, 1557.
Baptista, R., Steiner, J.E. & Cieslinski, D. 1994, *Astrophys. J.*, **433**, 332.
Baptista, R., Jablonski, F.J., Cieslinski, D. & Steiner, J.E. 1993, *Astrophys. J.*, **406**, L67.
Baptista, R., Steiner, J.E., Cieslinski, D. & Horne, K. 1994, *Astr. Soc. Pacific Conf. Ser.*, **56**, 264.
Barden, S.C. & Wade, R.A. 1988, *Astr. Soc. Pacific Conf. Ser.*, **3**, 113.
Barlow, M.J., Brodie, J.P., Brunt, C.C., Hanes, D.A., Hill, P.W., Mayo, S.K., Pringle, J.E., Ward, M.J., Watson, M.G., Whelan, J.A.J. & Willis, A.J. 1981, *Mon. Not. R. astr. Soc.*, **195**, 61.
Barnard, E.E. 1908, *Mon. Not. R. astr. Soc.*, **68**, 465.
Barnes, T.G. 1976, *Mon. Not. R. astr. Soc.*, **177**, 53P.
Barnes, T.G. & Evans, D.S. 1976, *Mon. Not. R. astr. Soc.*, **174**, 489.
Barrera, L.H. & Vogt, N. 1989a, *Astr. Astrophys.*, **220**, 99.
Barrera, L.H. & Vogt, N. 1989b, *Rev. Mex. Astr. Astrof.*, **19**, 99.
Barrett, P.E. & Chanmugam, G. 1984, *Astrophys. J.*, **278**, 298.
Barrett, P.E. & Chanmugam, G. 1985, *Astrophys. J.*, **298**, 743.
Barrett, P.E., O'Donoghue, D. & Warner, B. 1988, *Mon. Not. R. astr. Soc.*, **233**, 759.
Barstow, M.A., Fleming, T.A., Diamond, C.J., Finley, D.S., Sansom, A.E., Rosen, S.R., Koester, D., Marsh, M.C., Holberg, J.B. & Kidder, K. 1993, *Mon. Not. R. astr. Soc.*, **264**, 16.
Barstow, M.A., Burleigh, M.R., Fleming, T.A., Holberg, J.B., Koester, D., Marsh, M.C., Rosen, S.R., Rutten, R.G.M., Sakai, S., Tweedy, R.W. & Wegner, G. 1995a, *Mon. Not. R. astr. Soc.*, **272**, 531.
Barstow, M.A., O'Donoghue, D., Kilkenny, D., Burleigh, M.R. & Fleming, T.A. 1995b, *Mon. Not. R. astr. Soc.*, **273**, 711.
Bartolini, C., Guarnieri, A., Lolli, M., Piccioni, A., Giovannelli, F., Guadenzi, S. & Lombardi, R. 1985, in *Multifrequency Behaviour of Galactic Accreting Sources*, ed. F. Giovannelli, Frascati, Inst. Astrofis. Spaziale, p. 50.
Barwig, H. & Schoembs, R. 1983, *Astr. Astrophys.*, **124**, 287.
Barwig, H. & Schoembs, R. 1987, *The Messenger*, No. 47, 19.
Barwig, H., Mantel, K.H. & Ritter, H. 1992, *Astr. Astrophys.*, **266**, L5.
Barwig, H., Ritter, H. & Bärnbarnter, O. 1994, *Astr. Astrophys.*, **268**, 204.
Barwig, H., Hunger, K., Kudritski, R.P. & Vogt, N. 1982, *Astr. Astrophys.*, **114**, L11.
Basri, G. 1987, *Astrophys. J.*, **316**, 377.
Basri, G., Laurent, R. & Walter, F.M. 1985, *Astrophys. J.*, **298**, 761.
Bastian, T.S., Dulk, G.A. & Chanmugam, G. 1985, in *Radio Stars*, eds. R. Hjellming & D.M. Gibson, Reidel, Dordrecht, p. 225.
Bastian, T.S., Dulk, G.A. & Chanmugam, G. 1988, *Astrophys. J.*, **324**, 431.
Bateson, F.M. 1977a, *Pub. Var. Star Sect. Roy. astr. Soc. New Zealand*, **5**, 27.
Bateson, F.M. 1977b, *New Zealand J. Sci.*, **20**, 73.
Bateson, F.M. 1977c, *Pub. Var. Star Sect. R. astr. Soc. New Zealand*, **5**, 1.
Bateson, F.M. 1979a, *Int. Astr. Union Colloq. No. 46*, p. 89.
Bateson, F.M. 1979b, *Pub. Var. Star Sect. R. astr. Soc. New Zealand*, **7**, 5.
Bateson, F.M. 1981, *Pub. Var. Star Sect. R. astr. Soc. New Zealand*, **9**, 2.
Bateson, F.M. 1988, *Vistas*, **31**, 301.
Bateson, F.M. 1990a, *Pub. Var. Star Sect. R. astr. Soc. New Zealand*, **17**, 1.
Bateson, F.M. 1990b, *Pub. Var. Star Sect. R. astr. Soc. New Zealand*, **17**, 83.
Bateson, F.M. & Dodson, A.W. 1985, *Pub. Var. Star Sect. Roy. astr. Soc. New Zealand*, **12**, 1.
Bath, G.T. 1969, *Astrophys. J.*, **158**, 571.
Bath, G.T. 1972, *Astrophys. J.*, **173**, 121.
Bath, G.T. 1973, *Nature Phys. Sci.*, **246**, 84.
Bath, G.T. 1974, *Mon. Not. R. astr. Soc.*, **169**, 456.

Bath, G.T. 1975, *Mon. Not. R. astr. Soc.*, **171**, 311.
Bath, G.T. 1976, *Int. Astr. Union Symp. No.73*, p. 173.
Bath, G.T. 1978, *Mon. Not. R. astr. Soc.*, **182**, 35.
Bath, G.T. & Harkness, R.F. 1989, in *Classical Novae*, eds. M.F. Bode & A. Evans, Wiley, Chichester, p. 61.
Bath, G.T. & Pringle, J.E. 1982, *Mon. Not. R. astr. Soc.*, **199**, 267.
Bath, G.T. & Pringle, J.E. 1985, in *Interacting Binary Stars*, eds. J.E. Pringle & R.A. Wade, Cambridge University Press, Cambridge, p. 177.
Bath, G.T. & Shaviv, G. 1976, *Mon. Not. R. astr. Soc.*, **175**, 305.
Bath, G.T. & van Paradijs, J. 1983, *Nature*, **305**, 33.
Bath, G.T., Edwards, A.C. & Mantle, V.J. 1983, *Int. Astr. Union Colloq. No. 72*, p. 55.
Bath, G.T., Evans, W.D. & Pringle, J.E. 1974, *Mon. Not. R. astr. Soc.*, **166**, 113.
Bath, G.T., Pringle, J. & Whelan, J.A.J. 1980, *Mon. Not. R. astr. Soc.*, **190**, 185.
Bath, G.T., Evans, W.D., Papaloizou, J. & Pringle, J. 1974, *Mon. Not. R. astr. Soc.*, **169**, 447.
Baxendall, J. 1902, *Astr. J.*, **22**, 127.
Beardsley, W.R., King, M.W., Russell, J.L. & Stein, J.W. 1975, *Publ. astr. Soc. Pacific*, **87**, 943.
Beatty, J.K. 1990, *Sky & Tel.*, **80**, 128.
Becker, R.H. 1981, *Astrophys. J.*, **251**, 626.
Becker, R.H. 1989, in *Classical Novae*, eds. M.F. Bode & A. Evans, Wiley & Sons, Chichester, p. 215.
Becker, R.H. & Marshall, F.E. 1981, *Astrophys. J.*, **244**, L93.
Becker, R.H., Wilson, A.S., Pravdo, S.H. & Chanan, G.A. 1982, *Mon. Not. R. astr. Soc.*, **201**, 265.
Beer, A. 1974, *Vistas*, **16**, 179.
Belakov, E.T. & Shulov, O.S. 1974, *Trud. Ast. Obs. Len.*, **30**, 103.
Bell, M. & Walker, M.F., 1980, *Bull. Amer. astr. Soc.* **12**, 63.
Belloni, T., Verbunt, F. & Schmitt, J.H.M.M. 1993, *Astr. Astrophys.*, **269**, 175.
Belloni, T., Verbunt, F., Beuermann, K., Bunk, W., Izzo, C., Kley, W., Pietsch, W., Ritter, H., Thomas, H.-C. & Voges, W. 1991, *Astr. Astrophys.*, **246**, L44.
Belserene, E.P. 1981, *Bull. Amer. astr. Soc.*, **13**, 524.
Belvedere, G. (ed.) 1989, *Accretion Disks and Magnetic Fields in Astrophysics*, Kluwer, Dordrecht.
Benjamin, R.A. & Dinerstein, H.L. 1990, *Astr. J.*, **100**, 1588.
Benz, A.O. & Güdel, M. 1989, *Astr. Astrophys.*, **218**, 137.
Benz, A.O., Fürst, E. & Kiplinger, A.L. 1983, *Nature*, **302**, 45.
Benz, A.O., Fürst, E. & Kiplinger, A.L. 1985, in *Cataclysmic Variables and Low-Mass X-Ray Binaries*, eds. D.Q. Lamb & J. Patterson, Reidel, Holland, p. 331.
Berg, R.A. & Duthie, J.G. 1977, *Astrophys. J.*, **211**, 859.
Berg, C., Wegner, G., Foltz, C.B., Chaffee, F.H. & Hewett, P.C. 1992, *Astrophys. J. Suppl.*, **78**, 409.
Bergeron, P., Saffer, R.A. & Liebert, J. 1994, *Astrophys. J.*, in press.
Berriman, G. 1984a, *Mon. Not. R. astr. Soc.*, **207**, 783.
Berriman, G. 1984b, *Mon. Not. R. astr. Soc.*, **210**, 223.
Berriman, G. 1987a, *Astr. Astrophys. Suppl.*, **68**, 41.
Berriman, G. 1987b, *Mon. Not. R. astr. Soc.*, **228**, 729.
Berriman, G. 1988, in *Polarized Radiation of Circumstellar Origin*, eds. G.V. Coyne *et al.*, Vatican Obs., Vatican, p. 281.
Berriman, G. & Smith, P.S. 1988, *Astrophys. J.*, **329**, L97.
Berriman, G., Kenyon, S. & Bailey, J. 1986, *Mon. Not. R. astr. Soc.*, **222**, 871.
Berriman, G., Kenyon, S. & Boyle, C. 1987, *Astr. J.*, **94**, 1291.
Berriman, G., Szkody, P. & Capps, R.W. 1985, *Mon. Not. R. astr. Soc.*, **217**, 327.
Berriman, G., Beattie, D.H., Gatley, I., Lee, T.J., Mochnacki, S.W. & Szkody, P. 1983, *Mon. Not. R. astr. Soc.*, **204**, 1105.
Berriman, G., Bailey, J., Axon, D.J. & Hough, J.J. 1986, *Mon. Not. R. astr. Soc.*, **223**, 449.
Bertola, F. 1964, *Ann. d'Astrophys.*, **27**, 298.
Bertout, C., Collin-Souffrin, S., Lasota, J.P. & Van, J.T.T. (eds.), 1991, *Structure & Emission Line Properties of Accretion Disks*, Ed. Frontières, Paris.
Beuermann, K. 1987, *Astrophys. Sp. Sci.*, **131**, 625.
Beuermann, K. 1988a, *Adv. Sp. Res.*, **8**, 283.
Beuermann, K. 1988b, in *Polarized Radiation of Circumstellar Origin*, eds. G.V. Coyne *et al.*, Vatican Obs., Vatican, p. 125.

Beuermann, K. 1994, private communication.
Beuermann, K. & Osborne, J.O. 1985, *Sp. Sci. Rev.*, **40**, 117.
Beuermann, K. & Osborne, J. 1988, *Astr. Astrophys.*, **189**, 128.
Beuermann, K. & Pakull, M.W. 1984, *Astr. Astrophys.*, **136**, 250.
Beuermann, K. & Schwope, A.D. 1989, *Astr. Astrophys.*, **223**, 179.
Beuermann, K. & Stella, L. 1985, *Sp. Sci. Rev.*, **40**, 139.
Beuermann, K. & Thomas, H.-C. 1990, *Astr. Astrophys.*, **230**, 326.
Beuermann, K. & Thomas, H.-C. 1993, *Adv. Sp. Res.*, **13**, 115.
Beuermann, K., Stasiewski, U. & Schwope, A.D. 1992, *Astr. Astrophys.*, **256**, 433.
Beuermann, K., Stella, L. & Krautter, J. 1985, in *X-Ray Astronomy '84*, eds. M. Oda & R. Giacconi, Inst. Sp. & Astronaut. Sci., Tokyo, p. 27.
Beuermann, K., Stella, L. & Patterson, J. 1987, *Astrophys. J.*, **316**, 360.
Beuermann, K., Thomas, H.-C. & Pietsch, W. 1991, *Astr. Astrophys.*, **246**, L36.
Beuermann, K., Thomas, H.-C. & Schwope, A. 1988, *Astr. Astrophys.*, **195**, L15.
Beuermann, K., Thomas, H.-C. & Schwope, A. 1989, *Int. Astr. Un. Circ.*, No. 4775.
Beuermann, K., Schwope, A., Weierßsieker, H. & Motch, C. 1985, *Sp. Sci. Rev.*, **40**, 135.
Beuermann, K., Thomas, H.-C., Giommi, P. & Tagliaferri, G. 1987, *Astr. Astrophys.*, **175**, L9.
Beuermann, K., Thomas, H.-C., Giommi, P., Tagliaferri, G. & Schwope, A.D. 1989, *Astr. Astrophys.*, **219**, L7.
Beuermann, K., Thomas, H.-C., Schwope, A., Giommi, P. & Tagliaferri, G. 1990a, *Astr. Astrophys.*, **238**, 187.
Beuermann, K., Schwope, A.D., Thomas, H.-C., & Jordan, S. 1990b, in *Accretion-Powered Compact Binaries*, ed. C.W. Mauche, Cambridge University Press, Cambridge, p. 265.
Beuermann, K., Thorstensen, J.R., Schwope, A.D., Ringwald, F. & Sahin, H. 1992, *Astr. Astrophys.*, **256**, 442.
Beyer, M. 1977, *Verof. Remeis Stern. Bamberg*, **12**, No. 123.
Bhat, C.L. 1990, *Bull. astr. Soc. India,* **18**, 215.
Bhat, C.L., Richardson, K.M. & Wolfendale, A.W. 1989, *Proc. GRO Science Workshop*, ed. W.N. Johnson, Nav. Res. Lab., Washington, 4/57.
Bhat, C.L., Kaul, R.K., Rawat, H.S., Senecha, V.K., Rannot, R.C., Sapru, M.L., Tickoo, A.K. & Razdan, H. 1991, *Astrophys. J.*, **369**, 475.
Bianchini, A. 1980, *Mon. Not. R. astr. Soc.*, **192**, 127.
Bianchini, A. 1987, *Mem. Soc. astr. Ital.*, **58**, 245.
Bianchini, A. 1988, *Inf. Bull. Var. Stars*, No. 3136.
Bianchini, A. 1990a, in *Accretion-Powered Compact Binaries*, ed. C.W. Mauche, Cambridge University Press, Cambridge, p. 149.
Bianchini, A. 1990b, *Astr. J.*, **99**, 1941.
Bianchini, A. & Middleditch, J. 1976, *Inf. Bull. Var. Stars*, No. 1151.
Bianchini, A. & Sabbadin, F. 1983a, *Int. Astr. Union Colloq. No. 72*, p. 127.
Bianchini, A. & Sabbadin, F. 1983b, *Astr. Astrophys.*, **125**, 112.
Bianchini, A., Friedjung, M. & Brinkmann, W. 1990, *Int. Astr. Union Colloq. No. 122*, p. 155.
Bianchini, A., Friedjung, M. & Sabbadin, F. 1985a, *Inf. Bull. Var. Stars*, No. 2650.
Bianchini, A., Friedjung, M. & Sabbadin, F. 1985b, in *Recent Results on Cataclysmic Variables*, ed. W. Burke, ESA SP-236, Paris, p. 77.
Bianchini, A., Friedjung, M. & Sabbadin, F. 1989, *Int. Astr. Union Colloq. No. 122*, p. 61.
Bianchini, A., Hamzaoglu, E. & Sabbadin, F. 1981, *Astr. Astrophys.*, **99**, 392.
Bianchini, A., Sabbadin, F. & Hamzaoglu, E. 1982, *Astr. Astrophys.* **106**, 176.
Bianchini, A., Sabbadin, F., Favero, G.C. & Dalmeri, I. 1986, *Astr. Astrophys.*, **160**, 367.
Bianchini, A. Della Valle, M., Orio, M., Ögelman, H. & Bianchi, L. 1991, *The Messenger*, No. 64, p. 32.
Bianchini, A., Della Valle, M., Duerbeck, H. & Orio, M. 1992, *The Messenger*, No. 69, p. 42.
Biermann, P. Schmidt, G.D., Liebert, J., Stockman, H.S., Tapia, S., Kühr, H., Strittmatter, P.A., West, S. & Lamb, D.Q. 1985, *Astrophys. J.*, **293**, 303.
Bignami, G. F. & Hermsen, W. 1983, *Ann. Rev. Astr. Astrophys.*, **21**, 67.
Billington, I., Marsh, T.R., Horne, K., Cheng, F., Thomas, G., Bruch, A., O'Donoghue, D. & Eracleous, M. 1995, *Mon. Not. R. astr. Soc.*, in press.
Binnendijk, L. 1974, *Vistas*, **16**, 61.

Blandford, R.D. 1976, *Mon. Not. R. astr. Soc.*, **176**, 465.
Blandford, R.D. 1989, in *Theory of Accretion Disks*, eds. F. Meyer, W.J. Duschl, J. Frank & E. Meyer-Hofmeister, Kluwer, Dordrecht, p. 35.
Blandford, R.D. & Eichler, D. 1987, *Phys. Rep.*, **154**, 1.
Blandford, R.D. & Payne, D.G. 1982, *Mon. Not. R. astr. Soc.*, **199**, 883.
Blazit, A., Bonneau, D., Koechlin, L. & Labeyrie, A. 1977, *Astrophys. J.*, **214**, L79.
Blumenthal, G.R., Yang, L.T. & Lin, D.N.C. 1984, *Astrophys. J.*, **287**, 774.
Bode, M.F. 1982, *Vistas*, **26**, 369.
Bode, M.F. 1987, *RS Ophiuchi (1985) and the Recurrent Nova Phenomenon*, ed. M.F. Bode, VNU Sci. Press, Utrecht.
Bode, M.F. & Evans, A. 1983, *Quart. J. Roy. astr. Soc.*, **24**, 83.
Bode, M.F. & Evans, A. 1989, in *Classical Novae*, eds. M.F. Bode & A. Evans, Wiley & Sons, Chichester, p. 163.
Bode, M.F. & Kahn, F.D. 1985, *Mon. Not. R. astr. Soc.*, **217**, 205.
Bode, M.F., Evans, A., Whittet, D.C.B., Aitken, D.K., Roche, P.F. & Whitmore, B. 1984, *Mon. Not. R. astr. Soc.*, **207**, 897.
Bode, M.F., Seaquist, E.R., Frail, D.A., Roberts, J.A., Whittet, D.C.B., Evans, A. & Albinson, J.S. 1987, *Nature*, **329**, 519.
Boeshaar, P.C. 1976, Ph.D. Thesis, Ohio State Univ.
Boggess, A. & Wilson, R. 1987, in *Exploring the Universe with the IUE Satellite*, ed. Y. Kondo, Reidel, Dordrecht, p. 3.
Bohigas, J., Echevarria, J., Diego, F. & Sarmiento, J.A. 1989, *Mon. Not. R. astr. Soc.*, **238**, 1395.
Bois, B., Lanning, H.H. & Mochnacki, S.W. 1988, *Astr. J.*, **96**, 157.
Bolick, U., Beuermann, K., Bruch, A. & Lenzen, R. 1987, *Astrophys. Sp. Sci.*, **130**, 175.
Bond, H.E. 1978, *Publ. astr. Soc. Pacific*, **90**, 216.
Bond, H.E. 1989, *Int. Astr. Union Symp. No. 131*, p. 310.
Bond, H.E. & Grauer, A.D. 1987, *Int. Astr. Union Colloq. No. 95*, p. 221.
Bond, H.E. & Landolt, A.U. 1971, *Publ. astr. Soc. Pacific*, **83**, 485.
Bond, H.E. & Livio, M. 1990, *Astrophys. J.*, **355**, 568.
Bond, H.E., Chanmugam, G. & Grauer, A.D. 1979, *Astrophys. J.*, **234**, L113.
Bond, H.E., Kemper, E. & Mattei, J. 1982, *Astrophys. J.*, **260**, L79.
Bond, H.E. Grauer, A.D., Burnstein, D. & Marzke, R.O. 1987b, *Publ. astr. Soc. Pacific*, **99**, 1097.
Bond, I.A. & Freeth, R.V. 1988, *Mon. Not. R. astr. Soc.*, **232**, 753.
Bond, I.A., Freeth, R.V., Marino, B.F. & Walker, W.S.G. 1987a, *Inf. Bull. Var. Stars*, No. 3037.
Bonnet-Bidaud, J.M. & Mouchet, M. 1987, *Astr. Astrophys.*, **188**, 89.
Bonnet-Bidaud, J.M. & Mouchet, M. 1988, in *A Decade of UV Astronomy with the IUE Satellite*, eds. W. Longdon & E.J. Rolfe, ESA SP-281, Paris, p. 271.
Bonnet-Bidaud, J.M., Motch, C. & Mouchet, M. 1985, *Astr. Astrophys.*, **143**, 313.
Bonnet-Bidaud, J.M., Mouchet, M. & Motch C. 1982, *Astr. Astrophys.*, **112**, 355.
Bonnet-Bidaud, J.M., Somova, T.A. & Somov, N.N. 1992, *Astr. Astrophys.*, **251**, L27.
Bonnet-Bidaud, J.M., Beuermann, K., Charles, P., Maraschi, L., Motch, C., Mouchet, M., Osborne, J., Tanzi, E. & Treves, A. 1985, in *Recent Results on Cataclysmic Variables,* ESA SP-236, Paris, p. 155.
Bonnet-Bidaud, J.M., Mouchet, M., Somova, T.A. & Somov, N.N. 1992, *Int. Astr. Union Circ.*, No. 5673.
Bookbinder, J.A. & Lamb, D.Q. 1987, *Astrophys. J.*, **323**, L131.
Borisov, G.V. 1992, *Astr. Astrophys.*, **261**, 154.
Bowden, C.C.G., Bradbury, S.M., Brazier, K.T.S., Carriminana, A., Chadwick, P.M., Dipper, N.A., Edwards, P.J., Lincoln, E.W., McComb, T.J.L., Orford, K.J., Rayner, S.M. & Turver, K.E. 1991, *Proc. 22nd Int. Cosmic Ray Conf.*, M. Cawley *et al.* (eds.), Dublin Inst. Adv. Stud., Dublin, **1**, 356.
Bowden, C.C.G., Bradbury, S.M., Chadwick, P.M., Dickinson, J.E., Dipper, N.A., Edwards, P.J., Lincoln, E.W., McComb, T.J.L., Orford, K.J., Rayner, S.M. & Turver, K.E. 1992, *Astropart. Phys.*, **1**, 47.
Bowyer, S., Lampton, M., Paresce, F., Margon, B. & Stern, R. 1976, *Bull. Amer. astr. Soc.*, **8**, 447.
Boyarchuk, A.A. & Antipova, L.I. 1990, *Lect. Notes Phys.*, **369**, 97.
Boyarchuk, A.A., Galkina, T.S., Krasnobabtsev, V.I., Rachkovskaya, T.M. & Shakhovskaya, N.I. 1977, *Sov. Astr. J.*, **21**, 257.

Boynton, P.E., Crosa, L.M. & Deeter, J.E. 1980, *Astrophys. J.*, **237**, 169.
Bragaglia, A., Greggio, L. & Renzini, A. 1988, *The Messenger*, No. 52, 35.
Bragaglia, A., Greggio, L., Renzini, A. & D'Odorico, S. 1990, *Astrophys. J.*, **365**, L13.
Brainerd, J.J. 1989, *Astrophys. J.*, **345**, 978.
Brainerd, J.J. & Lamb, D.Q. 1985, in *Cataclysmic Variables and Low-Mass X-ray Binaries*, eds. D.Q. Lamb & J. Patterson, Reidel, Dordrecht, p. 247.
Brett, J.M. & Smith, R.C. 1993, *Mon. Not. R. astr. Soc.*, **264**, 641.
Breysacher, J. & Vogt, N. 1980, *Astr. Astrophys.*, **87**, 349.
Brickhill, A.J. 1975, *Mon. Not. R. astr. Soc.*, **170**, 405.
Briggs, M., Matteson, J., Gruber, D. & Peterson, L. 1991, *Int. Astr. Union Circ.*, No. 5229.
Briggs, M., Gruber, D.E., Matteson, J.L. & Peterson, L.E. 1994, in *The Second Compton Symposium*, Amer. Inst. Phys., New York, p. 255.
Brink, C., Cheng, K.C., de Jager, O.C., Meintjes, P.J., Nel, H.I., North, A.R., Raubenheimer, B.C. & van der Walt, D.J. 1990, *Proc. 21st Int. Cosmic Ray Conf.*, R.J. Protheroe (ed.), Univ. Adelaide, Adelaide, **2**, 283.
Brink, C., Raubenheimer, B.C., van Urk, G., van der Walt, D.J., de Jager, O.C., Meintjes, P.J., Nel, H.I., van Wyk, J.P. & Visser, B. 1991. *Proc. 22nd Int. Cosmic Ray Conf.*, M. Cawley *et al.* (eds.), Dublin Inst. Adv. Studies, Dublin, **2**, 622.
Broadfoot, A.L., Sandel, B.R., Shemansky, D.E., Atreya, S.K., Donahue, T.M., Moos, H.W., Bertaux, J.L., Blamont, J.E., Ajello, J.M., Strobel, D.F., McConnell, J.C., Dalgarno, A., Goody, R., McElroy, M.B. & Yung, Y.L. 1977, *Sp. Sci. Rev.*, **21**, 183.
Brown, R.L. & Crane, P.C. 1978, *Astr. J.*, **83**, 1504.
Brown, R.L. & Gould, R.J. 1970, *Phys. Rev. D.*, **1**, 2252.
Bruch, A. 1982, *Publ. astr. Soc. Pacific*, **94**, 916.
Bruch, A. 1984, *Astr. Astrophys. Suppl.*, **56**, 441.
Bruch, A. 1986, *Astr. Astrophys. Suppl.*, **167**, 91.
Bruch, A. 1987, *Astr. Astrophys.*, **172**, 187.
Bruch, A. 1989, *Astr. Astrophys. Suppl.*, **78**, 145.
Bruch, A. 1990, *Acta Astr.*, **40**, 369.
Bruch, A. 1992, *Astr. Astrophys.*, **266**, 237.
Bruch, A., Duerbeck, H.W. & Seitter, W. 1982, *Astr. Gesell. Mitt.*, **52**, 34.
Bruch, A., Fischer, F.J. & Wilmsen, U. 1987, *Astr. Astrophys. Suppl.*, **70**, 481 and **74**, 351.
Bruhweiler, F.C., Kondo, Y. & McCluskey, G.E. 1981, *Astrophys. J. Suppl.*, **46**, 255.
Brun, A. & Petit, M. 1952, *Bull. Assoc. Franc. d'Obs. d'Etoiles Var.*, Vol. 12.
Bryan, R.K. & Skilling J. 1980, *Mon. Not. R. astr. Soc.*, **191**, 69.
Buckley, D. &. Schwarzenberg-Czerny, A. 1993, in *Cataclysmic Variables and Related Physics*, eds. O. Regev & G. Shaviv, Inst. Phys. Publ., Bristol, p. 178.
Buckley, D.A.H. & Tuohy, I.R. 1989, *Astrophys. J.*, **344**, 376.
Buckley, D.A.H. & Tuohy, I.R. 1990, *Astrophys. J.*, **349**, 296.
Buckley, D.A.H., O'Donoghue, D., Kilkenny, D. & Stobie, R.S. 1990a, *Int. Astr. Union Colloq. No. 129*, p. 389.
Buckley, D.A.H., Sullivan, D.J., Remillard, R.A., Tuohy, I.R. & Clark, M. 1990b, *Astrophys. J.*, **355**, 617.
Buckley, D.A.H., O'Donoghue, D., Kilkenny, D., Stobie, R.S. & Remillard, R.A. 1992, *Mon. Not. R. astr. Soc.*, **258**, 285.
Buckley, D.A.H., O'Donoghue, D., Hassall, B.J.M., Kellett, B.J., Mason, K.O., Sekiguchi, K., Watson, M.G., Wheatley, P.J. & Chen, A. 1993, *Mon. Not. R. astr. Soc.*, **262**, 93.
Buckley, D.A.H., Remillard, R.A., Tuohy, I.R., Warner, B. & Sullivan, D.J. 1994, *Mon. Not. R. astr. Soc.*, **265**, 926.
Buckley, D.A.H., Barrett, P., Wargau, W., Sekiguchi, K., O'Donoghue, D. & Wheatley, P.J. 1995a, in preparation.
Buckley, D.A.H., Sekiguchi, K., Motch, C., O'Donoghue, D., Chen, A-L., Schwarzenberg-Czerny, A., Pietsch, W. & Harrop-Allin, M. 1995b, *Mon. Not. R. astr. Soc.*, in press.
Burbidge, E.M. & Strittmatter, P.A. 1971, *Astrophys. J.*, **170**, L39.
Burm, H. 1985, *Astr. Astrophys.*, **143**, 389.
Burm, H. 1986, *Astr. Astrophys.*, **165**, 120.
Burm, H. & Kuperus, M. 1988, *Astr. Astrophys.*, **192**, 165.

Burnard, D.J., Lea, S.M. & Arons, J. 1983, *Astrophys. J.*, **266**, 175.
Burrell, J.F. & Mould, J.R. 1973, *Publ. astr. Soc. Pacific*, **85**, 627.
Buscombe, W. & de Vaucouleurs, G. 1955, *Observatory*, **75**, 170.
Cabot, W., Canuto, V.M., Hubicky, O. & Pollack, J.B. 1987a, *Icarus*, **69**, 387.
Cabot, W., Canuto, V.M., Hubicky, O. & Pollack, J.B. 1987b, *Icarus*, **69**, 423.
Caillault, J.-P. & Patterson, J. 1990. *Astr. J.*, **100**, 825.
Cameron, A.G.W. & Iben, I. 1986, *Astrophys. J.*, **305**, 228.
Campbell, W.W. 1892, *Publ. astr. Soc. Pacific*, **4**, 233.
Campbell, L. 1934, *Ann. Harvard Coll. Obs.*, **90**, 93.
Campbell, C.G. 1983, *Mon. Not. R. astr. Soc.*, **205**, 1031.
Campbell, C.G. 1985, *Mon. Not. R. astr. Soc.*, **215**, 509.
Campbell, C.G. 1986, *Mon. Not. R. astr. Soc.*, **219**, 589.
Campbell, C.G. 1989, *Mon. Not. R. astr. Soc.*, **236**, 475.
Campbell, C.G. 1990, *Mon. Not. R. astr. Soc.*, **244**, 367.
Campbell, C.G. 1992, *Astrophys. & Geophys. Fluid Dyn.*, **63**, 197.
Campbell, L. & Jacchia, L. 1941, *The Story of Variable Stars*, Blakiston, Philadelphia.
Campbell, G.C. & Papaloizou, J.C.B. 1983, *Mon. Not. R. astr. Soc.*, **204**, 473.
Campbell, R.D. & Shafter, A.W. 1992, in *Proc. 12th N. Amer. Workshop on Cataclysmic Variables*, ed. A.W. Shafter, Mnt Laguna Obs., California, p. 4.
Campbell, R.D. & Shafter, A.W. 1995, *Astrophys. J.*, in press.
Campolonghi, F., Gilmozzi, R., Guidoni, U., Messi, R., Natali, G. & Wells, J. 1980, *Astr. Astrophys.*, **85**, L4.
Canalle, J.B.G. & Opher, R. 1988, *Astr. Astrophys.*, **189**, 325.
Cannizzo, J.K. 1984, *Nature*, **311**, 443.
Cannizzo, J.K. 1992, *Astrophys. J.*, **385**, 94.
Cannizzo, J. 1993a, in *Accretion Disks in Compact Stellar Systems*, ed. J.C. Wheeler, World Sci. Publ. Co., Singapore, p. 6.
Cannizzo, J.K. 1993b, *Astrophys. J.*, **419**, 318..
Cannizzo, J.K. & Cameron, A.G.W. 1988, *Astrophys. J.*, **330**, 327.
Cannizzo, J.K. & Goodings, D.A. 1988, *Astrophys. J.*, **334**, L31.
Cannizzo, J.K. & Kaitchuck, R.H. 1992, *Sci. Amer.*, **266**, No. 1, 42.
Cannizzo, J.K. & Kenyon, S.J. 1986, *Astrophys. J.*, **309**, L43.
Cannizzo, J.K. & Kenyon, S.J. 1987, *Astrophys. J.*, **320**, 319.
Cannizzo, J.K. & Mattei, J.A. 1992, *Astrophys. J.*, **401**, 642.
Cannizzo, J.K. & Pudritz, R.E. 1988, *Astrophys. J.*, **327**, 840.
Cannizzo, J.K. & Reiff, C.M. 1992, *Astrophys. J.*, **385**, 87.
Cannizzo, J.K. & Wheeler, J.C. 1984, *Astrophys. J. Suppl.*, **55**, 367.
Cannizzo, J.K., Ghosh, P. & Wheeler, J.C. 1982, *Astrophys. J.*, **260**, L83.
Cannizzo, J.K., Shafter, A.W. & Wheeler, J.C. 1988, *Astrophys. J.*, **333**, 227.
Cannizzo, J.K., Wheeler, J.C. & Polidan, R.S. 1986, *Astrophys. J.*, **301**, 634.
Cannon, A.J. 1912, *Harv. Coll. Obs. Ann.*, **56**, 65.
Cannon, A.J. 1916, *Harv. Coll. Obs. Ann.*, **76**, 3.
Cannon, A.J. & Pickering, E.C. 1901, *Harv. Coll. Obs. Ann.*, **28**, 131.
Canuto, V.M. & Goldman, I. 1985, *Phys. Rev. Lett.*, **54**, 430.
Canuto, V.M., Goldman, I. & Hubickyj, O. 1984, *Astrophys. J.*, **280**, L55.
Capaccioli, M., Della Valle, M., D'Onofrio, M. & Rosino, L. 1989, *Astr. J.*, **97**, 1622.
Capaccioli, M., Della Valle, M., D'Onofrio, M. & Rosino, L. 1990, *Astrophys. J.*, **360**, 63.
Carone, T.E., Polidan, R.S. & Wade, R.A. 1986, in *New Insights in Astrophysics*, ed. E.J. Rolfe, Noordwijk, ESA SP-263, Paris, p. 493.
Carroll, B.W., Cabot, W., McDermott, P.N., Savedoff, M.P. & Van Horn, H.M. 1985, *Astrophys. J.*, **296**, 529.
Cassatella, A. & Gonzalez-Riestra, R. 1990, *Int. Astr. Union Colloq. No. 122*, p. 115.
Cassatella, A. & Viotti, R. (eds.) 1990, *Physics of Classical Novae, Int. Astr. Colloq. No. 122.*
Cassatella, A., Gilmozzi, R. & Selvelli, P.L. 1985, *Proc. ESA Workshop: Recent Results on Cat. Var.*, ESA SP-236, Paris, p. 213.
Cassatella, A., Benvenuti, P., Clavel, J., Heck, A., Penston, M., Selvelli, P.L. & Macchetto, F. 1979, *Astr. Astrophys.*, **74**, L18.

Cassatella, A., Selvelli, P.L., Gilmozzi, R., Bianchini, A. & Friedjung, M. 1990, in *Accretion-Powered Compact Binaries*, ed. C.W. Mauche, Cambridge University Press, Cambridge, p. 373.
Catchpole, R.M. 1969, *Mon. Not. R. astr. Soc.*, **142**, 119.
Cecchini, G. & Gratton, L. 1942, *Le Stelle Nuove, Pub. Osserv. Astr. Milano*, Milan, Italy.
Chan, K.L. & Chau, W.Y., 1979, *Astrophys. J.*, **233**, 950.
Chanan, G.A., Middleditch, J. & Nelson, J.E. 1976, *Astrophys. J.*, **208**, 512.
Chanan, G.A., Nelson, J.E. & Margon, B. 1978, *Astrophys. J.*, **226**, 963.
Chandrasekhar, S. 1939, *An Introduction to the Study of Stellar Structure*, Univ. Chicago Press, Chicago.
Chanmugam, G. 1987, *Astrophys. Sp. Sci.*, **130**, 53.
Chanmugam, G. & Brecher, K. 1985, *Nature*, **313**, 767.
Chanmugam, G. & Dulk, G.A. 1981, *Astrophys. J.*, **244**, 569.
Chanmugam, G. & Dulk, G.A. 1982, *Astrophys. J.*, **255**, L107.
Chanmugam, G. & Frank, J. 1987, *Astrophys. J.*, **320**, 746.
Chanmugam, G. & Langer, S.H. 1991, *Astrophys. J.*, **368**, 580.
Chanmugam, G. & Wagner, R.L. 1977, *Astrophys. J.*, **213**, L13.
Chanmugam, G. & Wu, K. 1990, in *Accretion Powered Compact Binaries*, ed. C.W. Mauche, Cambridge University Press, Cambridge, p. 347.
Chanmugam, G., Langer, S.H. & Shaviv, G. 1985, *Astrophys. J.*, **299**, L87.
Chanmugam, G., Ray, A. & Singh, K.P. 1991, *Astrophys. J.*, **375**, 600.
Chanmugam, G., Barrett, P.E., Wu, K. & Courtney, M.W. 1989, *Astrophys. J. Suppl.*, **71**, 323.
Chanmugam, G., Frank, J., King, A.R. & Lasota, J.-P. 1990, *Astrophys. J.*, **350**, L13.
Charbonneau, P. 1992, *Astr. Soc. Pacific Conf. Ser.* **26**, 416.
Charles, P.A. & Mason, K.O. 1979, *Astrophys. J.*, **232**, L25.
Charles, P.A., Thorstensen, J., Bowyer, S. & Middleditch, J. 1979, *Astrophys. J.*, **231**, L131.
Chau, W. & Lauterborn, D. 1977, *Astrophys. J.*, **214**, 540.
Chen, A.-L. 1994, Ph.D. Thesis, Univ. Cape Town.
Chen, J.-S., Liu, X.-W. & Wei, M.-Z. 1991, *Astr. Astrophys.*, **242**, 397.
Cheng, F.H. & Lin, D.N.C. 1989, *Astrophys. J.*, **337**, 432.
Cheng, K.S. & Ruderman, M. 1991, *Astrophys. J.*, **373**, 187.
Cheng, F.H., Shields, G.A., Lin, D.N.C. & Pringle, J.E. 1988, *Astrophys. J.*, **328**, 223.
Chester, T.J. 1979, *Astrophys. J.*, **230**, 167.
Chevalier, R.A. & Imamura, J.N. 1982, *Astrophys. J.*, **261**, 543.
Chiapetti, L., Tanzi, E.G. & Treves, A. 1980, *Space Sci. Rev.*, **27**, 3.
Chiapetti, L., Maraschi, L., Tanzi, E.G. & Treves, A. 1982, *Astrophys. J.*, **258**, 236.
Chiappetti, L., Maraschi, L., Belloni, T., Bonnet-Bidaud, J.M., de Martino, D., Mouchet, M., Osborne, J., Tanzi, E.G. & Treves, A. 1988, *Adv. Sp. Res.*, **8**, 309.
Chincarini, G. & Walker, M.F. 1974, in *Electrography and Astronomical Applications*, eds. G. Chincarini, P.J. Griboval & H.J. Smith, Univ. Texas Press, Austin, p. 249.
Chincarini, G. & Walker, M.F. 1981, *Astr. Astrophys.*, **104**, 24.
Chlebowski, T. & Kaluzny, J. 1988, *Acta Astr.*, **38**, 329.
Chlebowski, T., Halpern, J.P. & Steiner, J.E. 1981, *Astrophys. J.*, **247**, L35.
Ciardullo, R., Ford, H., Neill, J.D., Jacoby, G.H. & Shafter, A.W. 1987, *Astrophys. J.*, **318**, 520.
Ciardullo, R., Shafter, A., Ford, H.C., Neill, J.D., Shara, M.M. & Tomaney, A.B. 1990, *Astrophys. J.*, **356**, 472.
Cisneros-Parra, J.U. 1970, *Astr. Astrophys.*, **8**, 141.
Clark, D.H. & Stephenson, F.R. 1976, *Quart. J. Roy. astr. Soc.*, **17**, 290.
Clark, D.H. & Stephenson, F.R. 1977, *The Historical Supernovae*, Pergamon, Oxford.
Clarke, C.J. 1988, *Mon. Not. R. astr. Soc.*, **235**, 881.
Clarke, J.T. & Bowyer, S. 1984, *Astr. Astrophys.*, **140**, 345.
Clarke, J.T., Capel, D. & Bowyer, S. 1984, *Astrophys. J.*, **287**, 845.
Clarke, C.J., Mantle, V.J. & Bath, G.T. 1985, *Mon. Not. R. astr. Soc.*, **215**, 149.
Clarke, J.T., Mason, K.O. & Bowyer, B. 1983, *Astrophys. J.*, **267**, 726.
Clayton, D.D. 1968, *Principles of Stellar Evolution and Nucleosynthesis*, McGraw-Hill, New York, pp. 372-378.
Clayton, D.D. & Leising, M.D. 1987, *Phys. Rept.*, **144**, 1.
Clayton, K.L. & Osborne, J.P. 1994, *Mon. Not. R. astr. Soc.*, **268**, 229.

Coe, M.J. & Wickramasinghe, D.T. 1981, *Nature*, **290**, 119.
Cohen, J.G. 1985, *Astrophys. J.*, 292, 90.
Cohen, J.G. 1988, *Astr. Soc. Pacific Conf. Ser.*, **4**, 114.
Cohen, J.G. & Rosenthal, A.J. 1983, *Astrophys. J.*, **268**, 689.
Cohen, J.M., Lapidus, A. & Cameron, A.G.W. 1969, *Astrophys. Sp. Sci.*, **5**, 113.
Collin-Souffrin, S. 1977, in *Novae & Related Stars*, ed. M. Friedjung, Reidel, Dordrecht, p. 123.
Colpi, M., Nannurelli, M. & Calvani, M. 1991, *Mon. Not. R. astr. Soc.*, **253**, 55.
Cook, M.C. 1981, *Mon. Not. R. astr. Soc.*, **195**, 51P.
Cook, M.C. 1985a, *Mon. Not. R. astr. Soc.*, **215**, 211.
Cook, M.C. 1985b, *Mon. Not. R. astr. Soc.*, **215**, 81P.
Cook, M.C. 1985c, *Mon. Not. R. astr. Soc.*, **216**, 219.
Cook, M.C. & Brunt, C.C. 1983, *Mon. Not. R. astr. Soc.*, **205**, 465.
Cook, M.C. & Warner, B. 1981, *Mon. Not. R. astr. Soc.*, **196**, 55P.
Cook, M.C. & Warner, B. 1984, *Mon. Not. R. astr. Soc.*, **207**, 705.
Cook, M.C., Watson, M.G. & McHardy, I.M. 1987, *Mon. Not. R. astr. Soc.*, **210**, 7P.
Cool, A.M., Grindlay, J.E., Krockenberger, M. & Bailyn, C.D. 1993, *Astr. Soc. Pacific Conf. Ser.*, **50**, 307.
Copeland, E. 1882, *Observatory*, **5**, 100.
Copeland, H., Jensen, J.O. & Jorgensen, H.E. 1970, *Astr. Astrophys.*, **5**, 12.
Cordova, F.A. & Cropper, M.S. 1988, in *Multiwavelength Astrophysics*, ed. F. Cordova, Cambridge University Press, Cambridge, p. 109.
Cordova, F.A. & Howarth, I. 1987, in *Exploring the Universe with the IUE Satellite*, ed. Y. Kondo, Reidel, Dordrecht, p. 395.
Cordova, F.A. & Mason, K.D. 1982a, in *Pulsations in Classical and Cataclysmic Variable Stars*, eds. J.P. Cox & C.J. Hansen, JILA, Boulder, p. 23.
Cordova, F.A. & Mason, K.O. 1982b, *Astrophys. J.*, **260**, 716.
Cordova, F.A. & Mason, K.O. 1983, in *Accretion-Driven Stellar X-ray Sources*, eds. W.H.G. Lewin & E.P.J. van den Heuvel, Cambridge University Press, Cambridge, p. 147.
Cordova, F.A. & Mason, K.O. 1984a, *Mon. Not. R. astr. Soc.*, **206**, 879.
Cordova, F.A. & Mason, K.O. 1984b, in *Future of Ultraviolet Astronomy Based on Six Years of IUE Research*, NASA Conf. Pub. No. 2349, p. 377.
Cordova, F.A. & Mason, K.O. 1985, *Astrophys. J.*, **290**, 671.
Cordova, F.A., Jensen, K. & Nugent, J.J. 1981, *Mon. Not. R. astr. Soc.*, **196**, 1.
Cordova, F.A., Ladd, E.F. & Mason, K.O. 1986, in *Magnetospheric Phenomena in Astrophysics*, eds. R.I. Epstein & W.C. Feldman, Amer. Inst. Phys., New York, p. 250.
Cordova, F.A., Mason, K.O. & Hjellming, R.M. 1983, *Publ. astr. Soc. Pacific*, **95**, 69.
Cordova, F.A., Mason, K.O. & Kahn, S.M. 1985, *Mon. Not. R. astr. Soc.*, **212**, 447.
Cordova, F.A., Mason, K.O. & Nelson, J.E. 1981, *Astrophys. J.*, **245**, 609.
Cordova, F.A., Chester, T.J., Tuohy, I. & Garmire, G.P. 1980a, *Astrophys. J.*, **235**, 163.
Cordova, F.A., Nugent, J.J., Klein, S.R. & Garmire, G.P. 1980b, *Mon. Not. R. astr. Soc.*, **190**, 87.
Cordova, F.A., Fenimore, E.E., Middleditch, J. & Mason, K.O. 1983, *Astrophys. J.*, **265**, 363.
Cordova, F.A., Chester, T.J., Mason, K.O., Kahn, S.M. & Garmire, G.P. 1984, *Astrophys. J.*, **278**, 739.
Cornu, M.A. 1876, *Compt. Rend.*, **83**, 1172.
Coroniti, F.V. 1981, *Astrophys. J.*, **244**, 587.
Couderc, P. 1939, *Ann. d'Astrophys.*, **2**, 271.
Cowley, A.P. & Crampton, D. 1977, *Astrophys. J.*, **212**, L121.
Cowley, A.P. & MacConnell, D.J. 1972, *Astrophys. J.*, **176**, L27.
Cowley, A.P., Crampton, D. & Hesser, J.E. 1977a, *Astrophys. J.*, **214**, 471.
Cowley, A.P., Crampton, D. & Hesser, J.B. 1977b, *Publ. astr. Soc. Pacific*, **89**, 716.
Cowley, A.P., Crampton, D. & Hutchings, J.B. 1980, *Astrophys. J.*, **241**, 269.
Cowley, A.P., Crampton, D. & Hutchings, J.B. 1982, *Astrophys. J.*, **259**, 730.
Cowley, A.P., Hutchings, J.B. & Crampton, D. 1981, *Astrophys. J.*, **246**, 489.
Cowley, A.P., Crampton, D., Hutchings, J.B. & Marlborough, J.M. 1975, *Astrophys. J.*, **195**, 413.
Cowley, A.P., Schmidtke, P.C., Crampton, D. & Hutchings, J.B. 1990, *Astrophys. J.*, **350**, 288.
Cox, J.P. 1981, *Astrophys. J.*, **247**, 1070.
Cox, J.P. & Everson, B.L. 1982, in *Pulsations in Classical and Cataclysmic Variable Stars*, eds. J.R.

Cox & C.J. Hansen, JILA, Boulder, p. 42.
Cox, J.P. & Giuli, R.T. 1968, *Principles of Stellar Structure*, Gordon & Breach, New York.
Crampton, D. & Cowley, A.P. 1977, *Publ. astr. Soc. Pacific*, **89**, 374.
Crampton, D., Cowley, A.P. & Fisher, W.A. 1987, *Astrophys. J.*, **300**, 788.
Crampton, D., Cowley, A.P. & Hutchings, J.B. 1983, *Int. Astr. Union Colloq. No. 72*, p. 25.
Crampton, D., Hutchings, J.B. & Cowley, A.P. 1981, *Astrophys. J.*, **243**, 567.
Crawford, J.A. & Kraft, R.P. 1956, *Astrophys. J.*, **123**, 44.
Cropper, M.S. 1985, *Mon. Not. R. astr. Soc.*, **212**, 709.
Cropper, M.S. 1986a, *Mon. Not. R. astr. Soc.*, **222**, 853.
Cropper, M.S. 1986b, *Mon. Not. R. astr. Soc.*, **222**, 225.
Cropper, M.S. 1987, *Mon. Not. R. astr. Soc.*, **228**, 389.
Cropper, M.S. 1988, *Mon. Not. R. astr. Soc.*, **231**, 597.
Cropper, M.S. 1989, *Mon. Not. R. astr. Soc.*, **236**, 935.
Cropper, M.S. 1990, *Sp. Sci. Rev.*, **54**, 195.
Cropper, M.S. & Horne, K. 1994, *Mon. Not. R. astr. Soc.*, **267**, 481.
Cropper, M.S. & Warner, B. 1986, *Mon. Not. R. astr. Soc.*, **220**, 633.
Cropper, M.S. & Wickramasinghe, D.T. 1993, *Mon. Not. R. astr. Soc.*, **260**, 696.
Cropper, M.S., Mason, K.O. & Mukai, K. 1990, *Mon. Not. R. astr. Soc.*, **243**, 565.
Cropper, M.S., Menzies, J.W. & Tapia, S. 1986, *Mon. Not. R. astr. Soc.*, **218**, 201.
Cropper, M., Mason, K.O., Allington-Smith, J.R., Branduardi-Raymont, G., Charles, P.A., Mittaz, J.P.D., Mukai, K., Murdin, P.G. & Smale, A.P. 1989, *Mon. Not. R. astr. Soc.*, **236**, 29P.
Cropper, M., Mukai, K., Mason, K.O., Smale, A.P., Charles, P.A., Mittaz, J.P.D., Machin, G., Hassall, B.J.M., Callanan, P.J., Naylor, T. & van Paradijs, J. 1990a, *Mon. Not. R. astr. Soc.*, **245**, 760.
Cropper, M.S., Bailey, J.A., Wickramasinghe, D.T. & Ferrario, L. 1990b, *Mon. Not. R. astr. Soc.*, **244**, 34P.
Crosa, L., Szkody, P., Stokes, G., Swank, J. & Wallerstein, G. 1981, *Astrophys. J.*, **247**, 984.
Cruddace, R.G. & Dupree, A.K. 1984, *Astrophys. J.*, **277**, 263.
Czerny, M. & King, A.R. 1986, *Mon. Not. R. astr. Soc.*, **221**, 55P.
D'Antona, F. & Mazzitelli, I. 1982a, *Astrophys. J.*, **260**, 722.
D'Antona, F. & Mazzitelli, I. 1982b, *Astr. Astrophys.*, **113**, 303.
D'Antona, F. & Mazzitelli, I. 1990, *Ann. Rev. Astr. Astrophys.*, **28**, 139.
D'Antona, F., Mazzitelli, I. & Ritter, H. 1989, *Astr. Astrophys.*, **225**, 391.
Davey, S. & Smith, R.C. 1992, *Mon. Not. R. astr. Soc.*, **257**, 476.
Davidsen, A.F., Durrance, S.T., Long, K.S., Kimble, R.A. & Bowers, C.W. 1991, in *Berkeley Colloq. on Extreme UV Astr.* eds. R.F. Malina & S. Bowyer, Pergamon, New York, p. 427.
Davidson, K. & Ostriker, J.P. 1973, *Astrophys. J.*, **179**, 585.
Deeter, J.E., Crosa, L., Gerend, D. & Boynton, P. 1976, *Astrophys. J.*, **206**, 861.
de Freitas Pacheco, J.A. 1977, *Mon. Not. R. astr. Soc.*, **181**, 421.
de Freitas Pacheco, J.A., da Costa, R.D.D. & Codina, S.J. 1989, *Astrophys. J.*, **347**, 483.
De Jager, O.C. 1991a, *Astrophys. J.*, **378**, 286.
De Jager, O.C. 1991b, *Proc. 22nd Int. Cosmic Ray Conf.*, M. Cawley *et al.* (eds.), Dublin Inst. Adv. Stud., Dublin, **2**, 463.
De Jager, O.C., Brink C., Meintjes, P.J., Nel, H.I., North, A.R., Raubenheimer, B.C. & van der Walt, D.J. 1990, *Nucl. Phys. B. Suppl.*, **14A**, 169.
De Jager, O.C., Meintjes, P.J., Raubenheimer, B.C., Brink, C., North, A.R., Visser, B., van Urk, G. & Buckley, D. 1991, *Lect. Not. Phys.*, **391**, 173.
De Jager, O.C., Meintjes, P.J., O'Donoghue, D. & Robinson, E.L. 1994, *Mon. Not. R. astr. Soc.*, **267**, 577.
de Kool, M. 1992, *Astr. Astrophys.*, **261**, 188.
de Kool, M. & Ritter, H. 1993, *Astr. Astrophys.*, **267**, 397.
de Kool, M., van den Heuvel, E.P.J. & Rappaport, S.A. 1986, *Astr. Astrophys.*, **164**, 73.
Della Valle, M. 1988, *Astr. Soc. Pacific Conf. Ser.*, **4**, 73.
Della Valle, M. 1991, *Astr. Astrophys.*, **252**, L9.
Della Valle, M. 1992, *Astr. Soc. Pacific Conf. Ser.*, **29**, 292.
Della Valle, M. & Augusteijn, T. 1990, *The Messenger*, No. 61, 41.
Della Valle, M. & Calvani, M. 1990, *Int. Astr. Union Colloq. No. 122*, p. 48.

Della Valle, M. & Duerbeck, H.W. 1993, *Astr. Astrophys.*, **271**, 175.
Della Valle, M. & Livio, M. 1994, *Astr. Astrophys.*, **286,** 786.
Della Valle, M. & Rosino, L. 1987, *Inf. Bull. Var. Stars*, No. 2995.
Della Valle, M., Bianchini, A., Livio, M. & Orio, M. 1992, *Astr. Astrophys.*, **266**, 232.
de Martino, R., Gonzalez-Riestra, R., Rodriguez, P., Buckley, D., Dickson, J. & Remillard, R.A. 1992, *Int. Astr. Union Circ.*, No. 5481.
de Martino, D., Buckley, D.A.H., Mouchet, M. & Mukai, K. 1994, *Astr. Astrophys.*, **284**, 125.
De Vaucouleurs, G. 1978, *Astrophys. J.*, **223**, 351.
De Young, J.A. & Schmidt, R.E. 1994, *Astrophys. J.*, **431**, L47.
Dgani, R., Livio, M. & Soker, N. 1989, *Astrophys. J.*, **336**, 350.
Dhillon, V.S. & Marsh, T.R. 1993, in *Cataclysmic Variables and Related Physics*, eds. O. Regev & G. Shaviv, Inst. Phys. Publ., Bristol, p. 34.
Dhillon, V.S., Jones, D.H.P. & Marsh, T.R. 1994, *Mon. Not. R. astr. Soc.*, **266**, 859.
Dhillon, V.S., Marsh, T.R. & Jones, D.H.P. 1990, in *Accretion-Powered Compact Binaries*, ed. C.W. Mauche, Cambridge University Press, Cambridge, p. 127.
Dhillon, V.S., Marsh, T.R. & Jones, D.H.P. 1991, *Mon. Not. R. astr. Soc.*, **252**, 342.
Dhillon, V.S., Jones, D.H.P., Marsh, T.R. & Smith, R.C. 1992, *Mon. Not. R. astr. Soc.*, **258**, 225.
Diaz, M.P. & Steiner, J.E. 1989, *Astrophys. J.*, **339**, L41.
Diaz, M.P. & Steiner, J.E. 1990a, *Rev. Mex. Astr. Astrof.*, **21**, 369.
Diaz, M.P. & Steiner, J.E. 1990b, *Astr. Astrophys.*, **238**, 170.
Diaz, M.P. & Steiner, J.E. 1991a, *Astr. J.*, **102**, 1417.
Diaz, M.P. & Steiner, J.E. 1991b, *Publ. astr. Soc. Pacific*, **103**, 964.
Diaz, M.P. & Steiner, J.E. 1994, *Astr. Astrophys.*, **283**, 508.
Dinerstein, H.L. 1986, *Astr. J.*, **92**, 1381.
Disney, M., Davies, J. & Philipps, S. 1989, *Mon. Not. R. astr. Soc.*, **239**, 939.
Dmitrienko, E.S. 1988, *Astrophys. J.*, **29**, 587.
Dmitrienko, E.S. 1992, *Izv. Krym. Astrofiz. Obs.*, **84**, 57.
Dmitrienko, E.S. & Cherapashchuk, A.M. 1980, *Sov. Astr. J.*, **24**, 432.
Dmitrienko, E.S., Matvienko, A.N., Cherepashchuk, A.M. & Yagola, A.G. 1983, *Sov. Astr.*, **28**, 180.
Dobrzycka, D. & Howell, S.B. 1992, *Astrophys. J.*, **388**, 614.
Dobrzycka, D. & Kenyon, S.J. 1994, *Astr. J.*, in press.
Downes, R.A. 1982, *Publ. astr. Soc. Pacific*, **94**, 950.
Downes, R.A. 1986, *Astrophys. J.*, **307**, 170.
Downes, R.A. 1990, *Astr. J.*, **99**, 339.
Downes, R.A. & Margon, B. 1981, *Mon. Not. R. astr. Soc.*, **197**, 35P.
Downes, R.A. & Shara, M.M. 1993, *Publ. astr. Soc. Pacific*, **105**, 127.
Downes, R.A. & Szkody, P. 1989, *Astr. J.*, **97**, 1729.
Downes, R.A. & Urbanski, J.L. 1978, *Publ. astr. Soc. Pacific*, **90**, 485.
Downes, R.A., Mateo, M., Szkody, P., Jenner, D.C. & Margon, B. 1986, *Astrophys. J.*, **301**, 240.
Dowthwaite, J.C., Harrison, A.B., Kirkman, I.W., Macrae, H.J., Orford, K.J., Turver, K.E. & Walmsley, M. 1984, *Nature*, **309**, 691.
Doyle, J.G. & Butler, C.J. 1990, *Astr. Astrophys.*, **235**, 335.
Drechsel, H., Rahe, J., Holm, A. & Krautter, J. 1981, *Astr. Astrophys.*, **99**, 166.
Drechsel, H., Rahe, J., Seward, F.D., Wang, Z.R. & Wargau, W. 1983, *Astr. Astrophys.*, **126**, 357.
Drew, J.E. 1986, *Mon. Not. R. astr. Soc.*, **218**, 41P.
Drew, J.E. 1987, *Mon. Not. R. astr. Soc.*, **224**, 595.
Drew, J.E. 1990, *Int. Astr. Union Colloq. No. 122*, p. 228.
Drew, J.E. & Verbunt, F. 1985, *Mon. Not. R. astr. Soc.*, **213**, 191.
Drew, J.E. & Verbunt, F. 1988, *Mon. Not. R. astr. Soc.*, **234**, 341.
Drew, J.E., Hoare, G. & Woods, J.A. 1991, *Mon. Not. R. astr. Soc.*, **250**, 144.
Drew, J.E., Jones, D.H.P. & Woods, J.A. 1993, *Mon. Not. R. astr. Soc.*, **260**, 803.
Drissen, L., Shara, M., Dopita, M., Wickramasinghe, D.T., Bell, J., Bailey, J. & Hough, J. 1992, *Int. Astr. Union Circ.*, No. 5609.
Drissen, L., Shara, M.M., Dopita, M. & Wickramasinghe, D.T. 1994, *Astr. J.*, **107**, 2172.
Drury, L. O'C. 1980, *Mon. Not. R. astr. Soc.*, **193**, 337.
Duerbeck, H.W. 1981, *Publ. astr. Soc. Pacific*, **93**, 165.
Duerbeck, H.W. 1984, *Astrophys. Sp. Sci.*, **99**, 363.

Duerbeck, H.W. 1986, *Mitt. Astr. Gesell.*, **65**, 207.
Duerbeck, H.W. 1987, *Sp. Sci. Rev.*, **45**, 1.
Duerbeck, H.W. 1988, *Astr. Astrophys.*, **197**, 148.
Duerbeck, H.W. 1990, *Int. Astr. Union Colloq. No. 122*, p. 34.
Duerbeck, H.W. 1992, *Mon. Not. R. astr. Soc.*, **258**, 629.
Duerbeck, H.W. 1993, in *Cataclysmic Variables and Related Physics*, eds. O. Regev & G. Shaviv, Inst. Phys. Publ., Bristol, p. 77.
Duerbeck, H.W. & Grebel, E.K. 1993, *Mon. Not. R. astr. Soc.*, **265**, L9.
Duerbeck, H.W. & Seitter, W.C. 1980, *Inf. Bull. Var. Stars*, No. 1738.
Duerbeck, H.W. & Seitter, W.C. 1987, *Astrophys. Sp. Sci.*, **131**, 467.
Duerbeck, H.W. & Seitter, W.C. 1990, *Int. Astr. Union Colloq. No. 122*, p.425.
Duerbeck, H.W., Seitter, W.C. & Duemmler, R. 1987, *Mon. Not. R. astr. Soc.*, **229**, 653.
Duerbeck, H.W., Klare, G., Krautter, J., Wolf, B., Seitter, W.C. & Wargau, W. 1980, *Proc. Second European IUE Conf.*, ESA SP-157, Paris, p. 91.
Duerbeck, H.W., Duemmler, R., Seitter, W.C., Leibowitz, E.M. & Shara, M.M. 1993, *The Messenger*, No. 71, p. 19.
Dulk, G.A. 1985, *Ann. Rev. Astr. Astrophys.*, **23**, 169.
Dulk, G.A., Bastion, T.S. & Chanmugam, G. 1983, *Astrophys. J.*, **273**, 249.
Dünhuber, H. & Ritter, H. 1993, in *White Dwarfs: Advances in Observation and Theory*, ed. M.A. Barstow, Kluwer, Dordrecht, p. 359.
Durisen, R.H. 1973, *Astrophys. J.*, **183**, 215.
Durney, B.R. & Robinson, R.D. 1982, *Astrophys. J.*, **253**, 290.
Duschl, W.J. 1986, *Astr. Astrophys.*, **163**, 56.
Duschl, W. 1989, *Astr. Astrophys.*, **225**, 105.
Duschl, W.J. & Livio, M. 1989, *Astr. Astrophys.*, **209**, 183.
Duschl, W.J. & Tscharnuter, W.M. 1991, *Astr. Astrophys.*, **241**, 153.
Dwek, E. 1983, *Astrophys. J.*, **274**, 175.
Dworetsky, M.M., Lanning, H.H. Etzel, P.B. & Patenaude, D.J. 1977, *Mon. Not. R. astr. Soc.*, **181**, 13P.
Dyck, G.P. 1989, *J. Amer. Assoc. Var. Star Obs.*, **18**, 105.
Eardley, D.M. & Lightman, A.P. 1975, *Astrophys. J.*, **200**, 187.
Eason, E.L.E., Africano, J.L., Klimke, A., Quigley, R.J., Rogers, W. & Worden, S.P. 1983, *Publ. astr. Soc. Pacific*, **95**, 58.
Eason, E.L.E., Worden, S.P., Klimke, A. & Africano, J.L. 1984, *Publ. astr. Soc. Pacific*, **96**, 372.
Echevarria, J. 1983, *Rev. Mex. Astr. Astrophys.*, **8**, 109.
Echevarria, J. 1984, *Rev. Mex. Astr. Astrophys.*, **9**, 99.
Echevarria, J. 1988, *Rev. Mex. Astr. Astrophys.*, **16**, 37.
Echevarria, J. & Alvarez, M. 1993, *Astr. Astrophys.*, **275**, 187.
Echevarria, J., Jones, D.H.P. & Costero, R. 1982, *Mon. Not. R. astr. Soc.*, **200**, 23P.
Echevarria, J., Jones, D.H.P., Wallis, R.E., Mayo, S.K., Hassall, B.J.M., Pringle, J.E. & Whelan, J.A.J. 1981, *Mon. Not. R. astr. Soc.*, **197**, 565.
Echevarria, J., Pocock, A.S., Penston, M.V. & Blades, J.C. 1983, *Mon. Not. R. astr. Soc.*, **205**, 559.
Echevarria, J., Schwarzenberg-Czerny, A., Jones, D.H.P., Dick, J.S.B., Ward, M., Costero, R. & Gilmozzi, R. 1988, *Rev. Mex. Astr. Astrof.*, **16**, 87.
Echevarria, J., Diego, F., Tapia, M., Costero, R., Ruiz, E., Salaz, L., Guttierrez, L. & Enriquez, R. 1989, *Mon. Not. R. astr. Soc.*, **240**, 975.
Edwards, D.A. 1985, *Mon. Not. R. astr. Soc.*, **212**, 623.
Edwards, D.A. 1987, *Mon. Not. R. astr. Soc.*, **226**, 95.
Edwards, D.A. 1988, *Mon. Not. R. astr. Soc.*, **231**, 25.
Edwards, D.A. & Pringle, J.E. 1987a, *Mon. Not. R. astr. Soc.*, **229**, 383.
Edwards, D.A. & Pringle, J.E. 1987b, *Nature*, **328**, 505.
Efimov, Yu.S., Tovmasyan, G.G. & Shakhovskoi, N.M. 1986, *Astrofiz.*, **24**, 227.
Efremov, Yu.N. & Kholopov, P.N. 1966, *Astr. Tsirk.*, No. 384, 1.
Eggen, O.J. 1968, *Astrophys. J. Suppl.*, **16**, 97.
Eggen, O.J. & Niemela, V.S. 1984, *Astr. J.*, **89**, 389.
Eggen, O.J., Mathewson, D.S. & Serkowski, K. 1967, *Nature*, **213**, 1216.
Eggleton, P.P. 1976, *Int. Astr. Union Symp. No. 73*, p. 209.

Eggleton, P.P. 1983a, *Astrophys. J.*, **268**, 368.
Eggleton, P.P. 1983b, *Int. Astr. Union Colloq. No. 72*, p. 239.
Eggleton, P.P., Bailyn, C.D. & Tout, C.A. 1989, *Astrophys. J.*, **345**, 489.
Eichler, D. &. Vestrand, W.T. 1984, *Nature*, **307**, 613.
Elitzur, M., Clarke, J.T. & Kallman, T.R. 1988, *Astrophys. J.*, **324**, 405.
Ellis, G.L., Grayson, E.T. & Bond, H.E. 1984, *Publ. astr. Soc. Pacific*, **96**, 283.
Elsner, R.F. & Lamb, F.K. 1977, *Astrophys. J.*, **215**, 897.
Elstner, D., Rüdiger, G. & Tschäpe, R. 1989, *Geophys. Astrophys. Fluid Dyn.*, **48**, 235.
Elsworth, Y.P. & James, J.F. 1982, *Mon. Not. R. astr. Soc.*, **198**, 889.
Elsworth, Y.P. & James, J.F. 1986, *Mon. Not. R. astr. Soc.*, **220**, 895.
Elsworth, Y., Grimshaw, L. & James, J.F. 1982, *Mon. Not. R. astr. Soc.*, **201**, 45P.
Elsworth, Y., Howe, R., Isaak, G., McLeod, C.P. & New, R. 1990, *Nature*, **345**, 322.
Elvey, C.T. & Babcock, H.W. 1940, *Pub. Amer. astr. Soc.*, **10**, 51.
Elvey, C.T. & Babcock, H.W. 1943, *Astrophys. J.*, **97**, 412.
Elvius, A. 1975, *Astr. Astrophys.*, **44**, 117.
Ennis, D., Becklin, E.E., Beckwith, S., Elias, J., Gatley, I., Mathews, K., Neugebauer, G. & Willner, S.P. 1977, *Astrophys. J.*, **214**, 478.
Eracleous, M., Halpern, J.P. & Patterson, J. 1991a, *Astrophys. J.*, **382**, 290.
Eracleous, M., Halpern, J.P. & Patterson, J. 1991b, *Astrophys. J.*, **370**, 330.
Eracleous, M., Horne, K., Robinson, E.L., Zhang, E.-H., Marsh, T.R. & Wood, J.H. 1994, *Astrophys. J.*, **433**, 313.
Escalante, V. & Dalgarno, A. 1991, *Astrophys. J.*, **369**, 213.
Evans, A. 1986, in *RS Ophiuchi and the Recurrent Nova Phenomenon*, ed. M.F. Bode, VNU Sci. Press, Utrecht, p. 117.
Evans, A., Bode, M.F., Duerbeck, H.W. & Seitter, W.C. 1992, *Mon. Not. R. astr. Soc.*, **258**, 7P.
Fabbiano, G., Hartmann, L., Raymond, J., Steiner, J., Branduardi-Raymont, G. & Matilsky, T. 1981, *Astrophys. J.*, **243**, 911.
Fabian, A.C., Pringle, J.E. & Rees, M.J. 1976, *Mon. Not. R. astr. Soc.*, **173**, 43.
Fabian, A.C., Pringle, J.E., Rees, M.J. & Whelan, J.A.J. 1977, *Mon. Not. R. astr. Soc.*, **179**, 9P.
Fabian, A.C., Lin, D.N.C., Papaloizou, J., Pringle, J. & Whelan, J.A.J. 1978, *Mon. Not. R. astr. Soc.* **184**, 835.
Fabian, A.C., Pringle, J.E., Stickland, D.J. & Whelan, J.A.J. 1980, *Mon. Not. R. astr. Soc.*, **191**, 457.
Falomo, R., Charles, P.A., Corbet, R., Maraschi, L., Tagliaferri, G., Tanzi, E.G. & Treves, A. 1985, in *Recent Results on Cataclysmic Variables*, ed. W. Burke, ESA SP-236 Paris, p. 239.
Faulkner, J. 1971, *Astrophys. J.*, **170**, L99.
Faulkner, J. 1976, *Int. Astr. Union Symp. No. 73*, p. 193.
Faulkner, J. & Ritter, H. 1982, *Int. Astr. Union Colloq. No. 69*, p. 483.
Faulkner, J., Flannery, B.P. & Warner, B. 1972, *Astrophys. J.*, **175**, L79.
Faulkner, J., Lin, D.N.C. & Papaloizou, J. 1983, *Mon. Not. R. astr. Soc.*, **205**, 359.
Feigelson, E., Dexter, L. & Liller, W. 1978, *Astrophys. J.*, **222**, 263.
Feinswog, L., Szkody, P. & Garnavich, P. 1988, *Astr. J.*, **96**, 1702.
Feldt, A.N. & Chincarini, G. 1980, *Publ. astr. Soc. Pacific*, **92**, 528.
Ferguson, D.H., Green, R.H. & Liebert, J. 1984, *Astrophys. J.*, **287**, 320.
Ferguson, D.H, Liebert, J., Cutri, R., Green R.F., Willner, S.P., Steiner, J.E., & Tokarz, S. 1987, *Astrophys. J.*, **316**, 399.
Ferland, G.J. 1979, *Astrophys. J.*, **231**, 781.
Ferland, G.J. 1980, *Observatory*, **100**, 166.
Ferland, G.J. & Shields, G.A. 1978a, *Astrophys. J.*, **226**, 172.
Ferland, G.J. & Shields, G.A. 1978b, *Astrophys. J.*, **224**, L15.
Ferland, G.J., Lambert, D.L. & Woodman, J. 1977, *Astrophys. J.*, **213**, 132.
Ferland, G.J., Lambert, D.L. & Woodman, J.H. 1986, *Astrophys. J. Suppl.*, **60**, 375.
Ferland, G.J., Lambert, D.L., Netzer, H., Hall, D.N.B. & Ridgway, S.T. 1979, *Astrophys. J.*, **227**, 489.
Ferland, G.J., Lambert, D.L., McCall, M.L., Shields, G.A. & Slovok, M.H. 1982, *Astrophys. J.*, **260**, 794.
Ferland, G., Williams, R.E., Lambert, D.L., Shields, G.A., Slovak, M., Gondhalekar, P.M. & Truran, J.W. 1984, *Astrophys. J.*, **281**, 194.
Ferrario, L. & Wickramasinghe, D.T. 1989, *Astrophys. J.*, **357**, 589.

Ferrario, L. & Wickramasinghe, D.T. 1990, *Astrophys. J.*, **357**, 582.
Ferrario, L. & Wickramasinghe, D.T. 1993, *Mon. Not. R. astr. Soc.*, **265**, 605.
Ferrario, L., Bailey, J. & Wickramasinghe, D.T. 1993, *Mon. Not. R. astr. Soc.*, **262**, 285.
Ferrario, L., Wickramasinghe, D.T. & King, A.R. 1993, *Mon. Not. R. astr. Soc.*, **260**, 149.
Ferrario, L. Wickramasinghe, D.T. & Tuohy, I.R. 1989, *Astrophys. J.*, **341**, 327.
Ferrario, L., Wickramasinghe, D.T., Bailey, J., Tuohy, I.R. & Hough, J.H. 1989, *Astrophys. J.*, **337**, 832.
Ferrario, L., Wickramasinghe, D.T., Bailey, J., Hough, J. & Tuohy, I. 1992, *Mon. Not. R. astr. Soc.* **256**, 252.
Fiedler, S.H. 1994, Dipl. Thesis, Munich University.
Flannery, B. 1975, *Mon. Not. R. astr. Soc.*, **170**, 325.
Fleet, R. 1992, in *Variable Star Research: An International Perspective*, eds. J.R. Percy, J.A. Mattei & C. Sterken, Cambridge University Press, Cambridge, p. 311.
Fleming, W.P. 1912, *Harv. Coll. Obs. Ann.*, **56**, 165.
Fleming, T.A., Gioia, I.M. & Maccacaro, T. 1989, *Astrophys. J.*, **340**, 1011.
Fleming, T.A., Liebert, J. & Green, R.F. 1986, *Astrophys. J.*, **308**, 176.
Fleming, T.A., Giampapa, M.S., Schmitt, J. & Bookbinder, J.A. 1992, *Astr. Soc. Pacific Conf. Ser.*, **26**, 93.
Fleming, T.A., Giampapa, M.S., Schmitt, J.H.M.M. & Bookbinder, J.A. 1993, *Astrophys. J.*, **410**, 387.
Ford, H.C. & Ciardullo, R. 1988, *Astr. Soc. Pacific Conf. Ser.* **4**, 128.
Frank, J. & King, A.R. 1981, *Mon. Not. R. astr. Soc.*, **195** 227.
Frank, J., King, A.R. & Lasota, J.-P. 1988, *Astr. Astrophys.*, **193**, 113.
Frank, J., King, A.R. & Raine, D.J. 1985, *Accretion Power in Astrophysics*, Cambridge University Press, Cambridge.
Frank, J., King, A.R., Sherrington, M.R., Jameson, R.F. & Axon, D.J. 1981a, *Mon. Not. R. astr. Soc.*, **195**, 505.
Frank, J., King, A.R., Sherrington, M.R., Giles, A.B. & Jameson, R.F. 1981b, *Mon. Not. R. astr. Soc.*, **196**, 921.
Fraser, G.W. 1989, *X-Ray Detectors in Astronomy*, Cambridge University Press, Cambridge.
Friedjung, M. 1966, *Mon. Not. R. astr. Soc.*, **131**, 447.
Friedjung, M. (ed.) 1977, *The Novae and Related Stars*, Reidel, Dordrecht.
Friedjung, M. 1981, *Astr. Astrophys.*, **99**, 226.
Friedjung, M. 1985, *Astr. Astrophys.*, **146**, 366.
Friedjung, M. 1987a, *Astr. Astrophys.*, **179**, 164.
Friedjung, M. 1987b, *Astr. Astrophys.*, **180**, 155.
Friedjung, M. 1987c, in *RS Ophiuchi (1985) and the Recurrent Nova Phenomenon*, ed. M.F. Bode, VNU Sci. Press, Utrecht, p. 77.
Friedjung, M. 1989, in *Classical Novae*, eds. M.F. Bode & A. Evans, Wiley, Chichester, p. 187.
Friedjung, M., Andrillat, Y. & Puget, P. 1982, *Astr. Astrophys.*, **114**, 351.
Friedjung, M., Bianchini, A. & Sabbadin, F. 1988, *The Messenger*, No. 52, 49.
Friend, M.T., Martin, J.S., Smith, R.C. & Jones, D.H.P. 1988, *Mon. Not. R. astr. Soc.*, **233**, 451.
Friend, M.T., Martin, J.S., Smith, R.C. & Jones, D.H.P. 1990a, *Mon. Not. R. astr. Soc.*, **246**, 637.
Friend, M.T., Martin, J.S., Smith, R.C. & Jones, D.H.P. 1990b, *Mon. Not. R. astr. Soc.*, **246**, 654.
Fuhrmann, B. 1984, *Mitt. verand. Sterne*, **10**, 97.
Fuhrmann, B. & Wenzel, W. 1990, *Inf. Bull. Var. Stars*, No. 3513.
Fujimoto, M. 1982a, *Astrophys. J.*, **257**, 752.
Fujimoto, M. 1982b, *Astrophys. J.*, **257**, 767.
Fujimoto, M. & Iben, I. 1992, *Astrophys. J.*, **399**, 646.
Fujino, S., Nakai, M., Iida, M., Moriyama, M., Makiguchi, N. & Kato, T. 1989, *Var. Star Bull. Japan*, **10**, 37.
Fulbright, M.S., Liebert, J., Bergeron, P., & Green, R. 1993, *Astrophys. J.*, **406**, 240.
Fürst, E., Benz, A., Hirth, W., Kiplinger, A. & Geffert, M. 1986, *Astr. Astrophys.*, **154**, 377.
Galeev, A.A., Rosner, R. & Vaiana, G.S. 1979, *Astrophys. J.*, **229**, 318.
Gallagher, J.S. & Code, A.D. 1974, *Astrophys. J.*, **189**, 303.
Gallagher, J.S. & Holm, A.V. 1974, *Astrophys. J.*, **189**, L123.
Gallagher, J.S. & Ney, E.P. 1976, *Astrophys. J.*, **204**, L35.

Gallagher, J.S. & Oinas, V. 1974, *Publ. astr. Soc. Pacific*, **86**, 952.
Gallagher, J.S. & Starrfield, S. 1976, *Mon. Not. R. astr. Soc.*, **176**, 53.
Gallagher, J.S. & Starrfield, S. 1978, *Ann. Rev. Astr. Astrophys.*, **16**, 171.
Gallagher, J.S., Hege, E.K., Kopriva, D.A., Williams, R.E. & Butcher, H.R. 1980, *Astrophys. J.*, **237**, 55.
Gaposchkin, S. 1946, *Harv. Coll. Obs. Bull.*, No. 918.
Garcia, M.R. 1986, *Astr. J.*, **91**, 1400.
Garlick, M.A., Rosen, S.R., Mittaz, J.P.D., Mason, K.O. & de Martino, D. 1994a, *Mon. Not. R. astr. Soc.*, **267**, 1095.
Garlick, M.A., Mittaz, J.P.D., Rosen, S.R. & Mason, K.O. 1994b, *Mon. Not. R. astr. Soc.*, **269**, 517.
Garnavich, P. & Szkody, P. 1988, *Publ. astr. Soc. Pacific*, **100**, 1522.
Garnavich, P.M., Szkody, P. & Goldader, J. 1988, *Bull. Amer. astr. Assoc.*, **20**, 1020.
Garnavich, P.M., Szkody, P., Mateo, M., Feinswog, L., Booth, J., Goodrich, B., Miller, H.R., Carini, M.T. & Wilson, J.W. 1990, *Astrophys. J.*, **365**, 696.
Garrison, R.F., Schild, R.E., Hiltner, W.A. & Krzeminski, W. 1984, *Astrophys. J.*, **276**, L13.
Geertsema, G.T. & Achterberg, A. 1992, *Astr. Astrophys.*, **255**, 427.
Gehrz, R.D. 1988, *Ann. Rev. Astr. Astrophys.*, **26**, 377.
Gehrz, R.D. 1990, *Int. Astr. Union Colloq. No. 122*, p. 138.
Gehrz, R.D., Grasdalen, G.L. & Hackwell, J.A. 1986, *Astrophys. J.*, **306**, L49.
Gehrz, R.D., Hackwell, J.A. & Jones, T.W. 1974, *Astrophys. J.*, **191**, 675.
Gehrz, R.D., Harrison, T.E., Ney, E.P., Matthews, K., Neugebauer, G., Elias, J. Grasdalen, G.L. & Hackwell, J.A. 1988, *Astrophys. J.*, **329**, 894.
Gehrz, R.D., Jones, T.J., Woodward, C.E., Greenhouse, M.A., Wagner, R.M., Harrison, T.E., Hayward, T.L. & Benson, J. 1992, *Astrophys. J.*, **400**, 671.
Geisel, S.L., Kleinmann, D.E. & Low, F.J. 1970, *Astrophys. J.*, **161**, L101.
Gerasimovic, B.P. 1936, *Pop. Astr.*, **44**, 78.
Geyer, F., Herold, H. & Ruder, H. 1990, in *Accretion-Powered Compact Binaries*, ed. C.W. Mauche, Cambridge University Press, Cambridge, p. 307.
Ghosh, P. & Lamb, F. 1978, *Astrophys. J.*, **223**, L83.
Ghosh, P. & Lamb, F.K. 1979a, *Astrophys. J.*, **234**, 296.
Ghosh, P. & Lamb, F.K. 1979b, *Astrophys. J.*, **232**, 259.
Giampapa, M.S. & Liebert, J. 1986, *Astrophys. J.*, **305**, 784.
Giannone, P. & Weigert, A. 1967, *Z. Astrophys.*, **67**, 41.
Gibb, D. 1914, *Mon. Not. R. astr. Soc.*, **74**, 678.
Gicgar, A. 1987, *Acta Astr.*, **37**, 29.
Gilliland, R.L. 1981, *Astrophys. J.*, **248**, 1144.
Gilliland, R. 1982a, *Astrophys. J.*, **258**, 576.
Gilliland, R.L. 1982b, *Astrophys. J.*, **253**, 399.
Gilliland, R.L. 1982c, *Astrophys. J.*, **254**, 653.
Gilliland, R.L. 1982d, *Astrophys. J.*, **263**, 302.
Gilliland, R.L. 1983, *Int. Astr. Union Colloq. No. 72*, p. 29.
Gilliland, R.L. 1985, *Astrophys. J.*, **292**, 522.
Gilliland, R.L. & Kemper, E. 1980, *Astrophys. J.*, **236**, 854.
Gilliland, R.L. & Phillips, M.M. 1982, *Astrophys. J.*, **261**, 617.
Gilliland, R.L., Kemper, E. & Suntzeff, N. 1986, *Astrophys. J.*, **301**, 252.
Gilliland, R.L., Brown, T.M., Duncan, D.K., Suntzeff, N.B., Lockwood, G.W., Thompson, D.T., Schild, R.E., Jeffrey, W.A. & Penprase, B.E. 1991. *Astr. J.*, **101**, 541.
Gilmozzi, R., Messi, R. & Natali, G. 1981, *Astrophys. J.*, **245**, L119.
Gingold, R.A. & Monaghan, J.J. 1977, *Mon. Not. R. astr. Soc.*, **181**, 375.
Giovannelli, F. & Martinez-Pais, I.G. 1991, *Sp. Sci. Rev.*, **56**, 313.
Giovannelli, F., Martinez-Pais, I.G., Gaudenzi, S., Lombardi, R., Rossi, C. & Claudi, R.U. 1990, *Astrophys. Sp. Sci.*, **169**, 125.
Glasby, J. 1970, *The Dwarf Novae,* Constable, London.
Glatzel, W. 1992, *Mon. Not. R. astr. Soc.*, **257**, 572.
Glenn, J., Howell, S.B., Schmidt, G.D., Liebert, J., Grauer, A.D. & Wagner, R.M. 1994, *Astrophys. J.*, **424**, 967.
Goetz, W. 1985, *Inf. Bull. Var. Stars*, No.2734.
Goetz, W. 1986, *Inf. Bull. Var. Stars*, No.2918.

Goldman, I. & Mazeh, T. 1991, *Astrophys. J.*, **376**, 260.
Goldreich, P. & Narayan, R. 1985, *Mon. Not. R. astr. Soc.*, **213**, 7P.
González-Riestra, R. 1992, *Astr. Astrophys.*, **265**, 71.
Goodman, J. 1993, *Astrophys. J.*, **406**, 596.
Goranskij, V.P., Lyutyi, V.M. & Shugarov, S.Yu. 1985, *Sov. Astr. Letts.*, **11**, 293.
Gorbatskii, V.G. 1965, *Sov. Astr. J.*, **8**, 680.
Götz, W. 1987, *Inf. Bull. Var. Stars*, No. 3067.
Götz, W. 1989, *Comm. Konkoly Obs.*, **10**, 279.
Götz, W. 1991, *Astr. Nach.*, **312**, 103.
Götz, W. 1993, *Astr. Nach.*, **314**, 69.
Grant, G. 1955, *Astrophys. J.*, **122**, 566.
Granzlo, B.H. 1992, in *Variable Star Research: An International Perspective*, eds. J.R. Percy, J.A. Mattei & C. Sterken, Cambridge University Press, Cambridge, p. 314.
Grasdalen, G.L. & Joyce, R.R. 1976, *Nature*, **259**, 187.
Gratton, L. 1953, *Astrophys. J.*, **118**, 568.
Grauer, A.D. & Bond, H.E. 1983, *Astrophys. J.*, **271**, 259.
Green, R.F., Schmidt, M. & Liebert, J. 1986, *Astrophys. J. Suppl.*, **61**, 305.
Green, R.F., Ferguson, D.H., Liebert, J. & Schmidt, M. 1982, *Publ. astr. Soc. Pacific*, **94**, 560.
Greenhouse, M.A., Grasdalen, G.L., Hayward, T.L., Gerhz, R.D. & Jones, T.J. 1988, *Astr. J.*, **95**, 172.
Greenhouse, M.A., Grasdalen, G.L., Woodward, C.E., Benson, R.D., Rosenthal, E. & Skrutskie, M.F. 1990, *Astrophys. J.*, **352**, 307.
Greenstein, J.L. 1954, *Publ. astr. Soc. Pacific*, **66**, 79.
Greenstein, J.L. 1979, *Int. Astr. Union Colloq. No. 53*, p. 74
Greenstein, J.L. 1960, in *Stellar Atmospheres*, ed. J.L. Greenstein, Univ. Chicago Press, Chicago, p. 676.
Greenstein, J.L. & Kraft, R.P. 1959, *Astrophys. J.*, **130**, 99.
Greenstein, J.L. & Matthews, M.S. 1957, *Astrophys. J.*, **126**, 14.
Greenstein, J.L. & Oke, J.B. 1982 , *Astrophys. J.*, **258**, 209.
Greenstein, J.L., Arp, H.C. & Schectman, S. 1977, *Publ. astr. Soc. Pacific*, **89**, 741.
Greenstein, J.L., Sargent, W.L., Boronson, T.A. & Boksenberg, A. 1977, *Astrophys. J.*, **218**, L127.
Greep, P. 1942, Ph.D. Thesis, Univ. Utrecht.
Greiner, J., Hasinger, G. & Thomas, H.-C. 1994, *Astr. Astrophys.*, **281**, L61.
Greiner, J., Remillard, R.A. & Motch, C. 1994, in press.
Griffiths, R.E., Ward, M.J., Baldes, J.C., Wilson, A.S., Chaisson, L. & Johnston, M.D. 1979, *Astrophys. J.*, **232**, L27.
Griffiths, R.E., Lamb, D.Q., Ward, M.J., Wilson, A.J., Charles, P.A., Thorstensen, J., McHardy, I.M. & Lawrence, A. 1980, *Mon. Not. R. astr. Soc.*, **193**, 25P.
Grindlay, J.E. 1993a, *Astr. Soc. Pacific Conf. Ser.*, **48**, 156.
Grindlay, J.E. 1993b, *Astr. Soc. Pacific Conf. Ser.*, **50**, 285.
Grossman, A.S., Hays, D. & Grasboske, H.C. 1974, *Astrophys. J.*, **181**, 457.
Guinan, E.F. & Sion, E.M. 1982a, *Astrophys. J.*, **258**, 217.
Guinan, E.F. & Sion, E.M. 1982b, *Advances in Ultraviolet Astronomy*, NASA Conf. Pub. No. 2238, p. 465.
Gull, S.F. & Daniell, G.J. 1978, *Nature*, **272**, 686.
Hachenberg, V.O. & Wellmann, P. 1939, *Z. Astrophys.*, **17**, 246.
Hacke, G. 1987a, *Inf. Bull. Var. Stars*, No. 2979.
Hacke, G. 1987b, *Mitt. Verand. Sterne*, **11**, 40.
Hacke, G. & Andronov, I.L. 1988, *Mitt. Verand. Sterne*, **11**, 74.
Hacke, G. & Richert, M. 1990, *Verof. Sternw. Sonn.*, **10**, 336.
Hadjidemetriou, J. 1967, *Adv. Astr. Astrophys.*, **5**, 131.
Haefner, R. & Metz, K. 1982, *Astr. Astrophys.*, **109**, 171.
Haefner, R. & Metz, K. 1985, *Astr. Astrophys.*, **145**, 311.
Haefner, R. & Schoembs, R. 1987, *Mon. Not. R. astr. Soc.*, **224**, 231.
Haefner, R., Barwig, H. & Mantel, K.-H. 1993, *Astrophys. Sp. Sci.*, **204**, 199.
Haefner, R., Pietsch, W. & Metz, K. 1988, *Astr. Astrophys.*, **200**, 75.
Haefner, R., Schoembs, R. & Vogt, N. 1977, *Astr. Astrophys.*, **61**, L37.
Haefner, R., Schoembs, R. & Vogt, N. 1979, *Astr. Astrophys.*, **77**, 7.

Hakala, P.J., Watson, M.G., Vilhu, O., Hassall, B.J.M., Kellett, B.J., Mason, K.O. & Piirola, V. 1993, *Mon. Not. R. astr. Soc.*, **263**, 61.
Hall, D.N.B. 1982, *The Space Telescope Observatory*, ed. D.N.B. Hall, Sp. Tel. Sci. Inst., Baltimore, NASA CP-2244.
Hall, D.S. 1989, *Sp. Sci. Rev.*, **50**, 219.
Hall, D.S. 1991, *Astrophys. J.*, **380**, L85.
Hamada, T. & Salpeter, E.E. 1961, *Astrophys. J.*, **134**, 683.
Hameury, J.M. 1991, *Astr. Astrophys.*, **243**, 419.
Hameury, J.M., King, A.R. & Lasota, J.-P. 1986a, *Astr. Astrophys.*, **162**, 71.
Hameury, J.M., King, A.R. & Lasota, J.-P. 1986b, *Mon. Not. R. astr. Soc.*, **218**, 695.
Hameury, J.M., King, A.R. & Lasota, J.P. 1987, *Astr. Astrophys.*, **171**, 140.
Hameury, J.M., King, A.R. & Lasota, J.P. 1988, *Astr. Astrophys.*, **195**, L12.
Hameury, J.M., King, A.R., Lasota, J.-P. & Ritter, H. 1987, *Astrophys. J.*, **316**, 275.
Hameury, J.M., King, A.R., Lasota, J.-P. & Ritter, H. 1988, *Mon. Not. R. astr. Soc.*, **231**, 535.
Hameury, J.M., King, A.R., Lasota, J.P. & Livio, M. 1989, *Mon. Not. R. astr. Soc.*, **237**, 835.
Hamuy, M. & Maza, J. 1986, *Inf. Bull. Var. Stars*, No. 2867.
Hanawa, T. 1988, *Astr. Astrophys.*, **206**, 1.
Hanes, D.A. 1985, *Mon. Not. R. astr. Soc.*, **213**, 443.
Harlaftis, E., Naylor, T., Sonneborn, G., Hassall, B.J.M. & Charles, P.A. 1990, In *Accretion-Powered Compact Binaries*, ed. C. Mauche, Cambridge University Press, Cambridge, p. 105.
Harlaftis, E.T., Hassall, B.J.M., Naylor, T., Charles, P.A. & Sonnenborn, G. 1992a, *Mon. Not. R. astr. Soc.*, **257**, 607.
Harlaftis, E.T., Naylor, T., Hassall, B.J.M., Charles, P.A., Sonnenberg, G. & Bailey, J. 1992b, *Mon. Not. R. astr. Soc.*, **259**, 593.
Harlaftis, E.T., Marsh, T.R., Dhillon, V.S. & Charles, P.A. 1994a, *Mon. Not. R. astr. Soc.*, **267**, 473.
Harlaftis, E.T., Lazaro, C., Charles, P.A., Dhillon, V.S., Casares, J., Marsh, T.R., Martinez-Pais, I.G. & Mukai, K. 1994b, *Mon. Not. R. astr. Soc.*, in press.
Harlaftis, E.T., Casares, J., Charles, P.A., Dhillon, V.S., Lazaro, C., Marsh, T.R., Martinez-Pais, I.G. & Mukai, K. 1994c, *Mon. Not. R. astr. Soc.*, in press.
Harrison, T.E. 1992, *Mon. Not. R. astr. Soc.*, **259**, 17P.
Harrison, T.E. & Gehrz, R.D. 1988, *Astr. J.*, **96**, 1001.
Harrison, T.E. & Gehrz, R.D. 1991, *Astr. J.*, **101**, 587.
Harrison, T.E. & Gehrz, R.D. 1992, *Astr. J.*, **103**, 243.
Harrison, T.E., Johnson, J.J. & Spyromilio, J. 1993, *Astr. J.*, **105**, 320.
Hartmann, L.W. & Noyes, R.W. 1987, *Ann. Rev. Astr. Astrophys.*, **25**, 271.
Hartwick, F.D.A. & Hutchings, J.B. 1978, *Astrophys. J.*, **226**, 203.
Harwood, J. 1973, M.Sc. Thesis, Univ. Cape Town.
Hassall, B.J.M. 1985a, *Mon. Not. R. astr. Soc.*, **216**, 335.
Hassall, B.J.M. 1985b, in *Cataclysmic Variables & Low-Mass X-Ray Binaries*, eds. D.Q. Lamb & J. Patterson, Reidel, Holland, p. 287.
Hassall, B.J.M., Pringle, J.E. & Verbunt, F. 1985, *Mon. Not. R. astr. Soc.*, **216**, 353.
Hassall, B.J.M., Pringle, J.E., Ward, M.J., Whelan, J.A.J., Mayo, S.K., Echevarria, J., Jones, D.H.P., Wallis, R.E., Allen, D.A. & Hyland, A.R. 1981, *Mon. Not. R. astr. Soc.*, **197**, 275.
Hassall, B.J.M., Pringle, J.E., Schwarzenberg-Czerny, A., Wade, R.A., Whelan, J.A.J. & Hill, P.W. 1983, *Mon. Not. R. astr. Soc.*, **203**, 865.
Hassall, B.J.M., Snijders, M.A.J., Harris, A.W., Cassatello, A., Dennefeld, M., Friedjung, M., Bode, M., Whittet, D., Whitelock, P., Menzies, J., Lloyd Evans, T. & Bath, G.T. 1990, *Int. Astr. Union Colloq. No. 122*, p. 202.
Haswell, C.A., Horne, K., Thomas, G., Patterson, J. & Thorstensen, J.R. 1994, *Astr. Soc. Pacific Conf. Ser.* **56**, 268.
Haug, K. 1987, *Astrophys. Sp. Sci.*, **130**, 91.
Haug, K. 1988, *Mon. Not. R. astr. Soc.*, **235**, 1385.
Haug, K. & Drechsel, H. 1985, *Astr. Astrophys.*, **151**, 157.
Hawkins, N.A., Smith, R.C. & Jones, D.H.P. 1990, in *Accretion-Powered Compact Binaries*, ed. C.W. Mauche, Cambridge University Press, Cambridge, p. 113.
Hawley, J.F. & Balbus, S.A. 1991, *Astrophys. J.*, **376**, 223.
Hawley, J.F. & Balbus, S.A. 1992, *Astrophys. J.*, **400**, 595.

Hayward, T.L., Gehrz, R.D., Miles, J.W. & Houck, J.R. 1992, *Astrophys. J.*, **401**, L101.
Hazen, M. 1985, *Inf. Bull. Var. Stars*, No. 2880.
Hearn, D.R. & Marshall, F.J. 1979, *Astrophys. J.*, **232**, L21.
Hearn, D.R. & Richardson, J.A. 1977, *Astrophys. J.*, **213**, L115.
Hearn, D.R., Richardson, J.A. & Clark, G.W. 1976, *Astrophys. J.*, **210**, L23.
Hearn, D.R., Richardson, J.A. & Li, F.K. 1976, *Bull. Amer. astr. Soc.*, **8**, 447.
Hearnshaw, J.B. 1986, *The Analysis of Starlight*, Cambridge University Press, Cambridge.
Heiles, C. 1975, *Astr. Astrophys. Suppl.*, **20**, 37.
Heise, J. & Verbunt, F. 1987, *Astr. Astrophys.*, **189**, 112.
Heise, J., Paerels, F. & van der Woerd, H. 1984, *Int. Astr. Union Circ.*, No. 3939.
Heise, J., Mewe, R., Brinkman, A.C., Gronenschild, E.H.B.M., den Boggende, A.J.F., Shrijver, J., Parsignault, D.R. & Grindlay, J.E. 1978, *Astr. Astrophys.*, **63**, L1.
Heise, J., Kruszewski, A., Chlebowski, T., Mewe, R., Kahn, S. & Seward, F.D. 1984, *Phys. Scripta*, **T7**, 115.
Heise, J., Brinkman, A.C., Gronenschild, E., Watson, M., King, A.R., Stella, L. & Kiebrom, K. 1985, *Astr. Astrophys.*, **148**, L14.
Heise, J., Mewe, R., Kruszewski, A. & Chlebowski, T. 1987, *Astrophys. J.*, **183**, 73.
Hellier, C. 1991, *Mon. Not. R. astr. Soc.*, **251**, 693.
Hellier, C. 1992, *Mon. Not. R. astr. Soc.*, **258**, 578.
Hellier, C. 1993a, *Mon. Not. R. astr. Soc.*, **264**, 132.
Hellier, C. 1993b, *Publ. astr. Soc. Pacific*, **105**, 966.
Hellier, C. 1993c, *Mon. Not. R. astr. Soc.*, **265**, L35.
Hellier, C. & Buckley, D. 1993, *Mon. Not. R. astr. Soc.*, **265**, 766.
Hellier, C. & Mason, K.O. 1990, in *Accretion-Powered Compact Binaries*, ed. C.W. Mauche, Cambridge University Press, Cambridge, p. 185.
Hellier, C. & Ramseyer, T.F. 1994, *Astr. Soc. Pacific Conf. Ser.*, **56**, 334.
Hellier, C. & Robinson, E.L. 1994, *Astrophys. J.*, **431**, L107.
Hellier, C. & Sproats, L.N. 1992, *Inf. Bull. Var. Stars*, No. 3724.
Hellier, C., Cropper, M. & Mason, K.O. 1991, *Mon. Not. R. astr. Soc.*, **248**, 233.
Hellier, C., Garlick, M.A. & Mason, K.O. 1993, *Mon. Not. R. astr. Soc.*, **260**, 299.
Hellier, C., Mason, K.O. & Cropper, M.S. 1989, *Mon. Not. R. astr. Soc.*, **237**, 39P.
Hellier, C., Mason, K.O. & Cropper, M.S. 1990, *Mon. Not. R. astr. Soc.*, **242**, 250.
Hellier, C., Mason, K.O. & Mittaz, J.P.D. 1991, *Mon. Not. R. astr. Soc.*, **248**, 5P.
Hellier, C., Ringwald, F.A. & Robinson, E.L. 1994, *Astr. Astrophys.*, **289**, 148.
Hellier, C., Mason, K.O., Rosen, S.R. & Cordova, F.A. 1987, *Mon. Not. R. astr. Soc.*, **228**, 463.
Hellier, C., Mason, K.O., Smale, A.P., Corbet, R.H.D., O'Donoghue, D., Barrett, P.E. & Warner, B. 1989, *Mon. Not. R. astr. Soc.*, **238**, 1107.
Hellier, C., O'Donoghue, D., Buckley, D. & Norton, A. 1990, *Mon. Not. R. astr. Soc.*, **242**, 32P.
Hempelmann, A. & Kurths, J. 1990, *Astr. Astrophys.*, **232**, 356.
Hempelmann, A. & Kurths, J. 1993, *Astrophys. J.*, **412**, L41.
Hendry, E.M. 1983, *Inf. Bull. Var. Stars*, No. 2381.
Henize, K.G. 1949, *Astr. J.*, **54**, 89.
Henrichs, H.F. 1983, in *Accretion-driven Stellar X-Ray Sources*, eds. W.H.G. Lewin & E.P.J. van den Heuvel, Cambridge University Press, Cambridge, p. 393.
Henriksen, R.N. & Rayburn, D.R. 1971, *Mon. Not. R. astr. Soc.*, **152**, 323.
Henry, T.J. & McCarthy, D.W. 1990, *Astrophys. J.*, **350**, 334.
Henry, P., Cruddace, R., Lampton, M., Paresce, F. & Bowyer, S. 1975, *Astrophys. J.*, **197**, L117.
Hensler, G. 1982, *Astr. Astrophys.*, **114**, 309.
Herbig, G.H. 1944, *Publ. astr. Soc. Pacific*, **56**, 230.
Herbig, G.H. 1960, *Astrophys. J.*, **132**, 76.
Herbig, G.H. & Smak, J. 1992, *Acta Astr.*, **42**, 17.
Herbst, W., Hesser, J.E. & Ostriker, J.P. 1974, *Astrophys. J.*, **193**, 679.
Herter, T., Lacasse, M.G., Wesemael, F. & Winget, D.E. 1979, *Astrophys. J. Suppl.*, **39**, 513.
Hertz, P. & Grindlay, J.E. 1983, *Astrophys. J.*, **267**, L83.
Hertz, P. & Grindlay, J.E. 1984, *Astrophys. J.*, **278**, 137.
Hertz, P. & Grindlay, J.E. 1988, *Astr. J.*, **96**, 233.
Hertz, P., Grindlay, J.E. & Bailyn, C.D. 1993, *Astrophys. J.*, **410**, L87.

Hertz, P., Bailyn, C.D., Grindlay, J.E., Garcia, M.R., Cohn, H. & Lugger, P.M. 1990, *Astrophys. J.*, **364**, 251.
Hertzog, K.P. 1986, *Observatory*, **106**, 38.
Hesser, J.E., Lasker, B.M. & Osmer, P.S. 1972, *Astrophys. J.*, **176**, L31.
Hesser, J.E., Lasker, B.M. & Osmer, P.S. 1974, *Astrophys. J.*, **189**, 315.
Hessman, F.V. 1986, *Astrophys. J.*, **300**, 794.
Hessman, F.V. 1987, *Astr. Sp. Sci.*, **130**, 351.
Hessman, F.V. 1988, *Astr. Astrophys. Suppl.*, **72**, 515.
Hessman, F.V. 1989a, *Astr. J.*, **98**, 675.
Hessman, F.V. 1989b, *Mon. Not. R. astr. Soc.*, **239**, 759.
Hessman, F.V. 1990a, *Int. Astr. Union Circ.*, No. 4971.
Hessman, F.V. 1990b, in *Accretion-Powered Compact Binaries*, ed. C.W. Mauche, Cambridge University Press, Cambridge, p. 123.
Hessman, F.V. & Hopp, U. 1990, *Astr. Astrophys.*, **228**, 387.
Hessman, F.V., Robinson, E.L., Nather, R.E. & Zhang, E.-H. 1984, *Astrophys. J.*, **286**, 747.
Hessman, F.V., Koester, D., Schoembs, R. & Barwig, H. 1989, *Astr. Astrophys.*, **213**, 167.
Hessman, F.V., Mantel, K.H., Barwig, H. & Schoembs, R. 1992, *Astr. Astrophys.*, **263**, 147.
Hildebrand, R.H., Spillar, E.J. & Stiening, R.F. 1981a, *Astrophys. J.*, **243**, 223.
Hildebrand, R.H., Spillar, E.J. & Stiening, R.F. 1981b, *Astrophys. J.*, **248**, 268.
Hildebrand, R.H., Spillar, E.J., Middleditch, J., Patterson, J. & Stiening, R.F. 1980, *Astrophys. J.*, **238**, L145.
Hilditch, R.W. & Bell, S.A. 1994, *Mon. Not. R. astr. Soc.*, **266**, 703.
Hill, K.M. & Watson, R.D. 1984, *Proc. Astr. Soc. Australia*, **5**, 532.
Hill, K.M. & Watson, R.D. 1990, *Astrophys. Sp. Sci.*, **163**, 59.
Hillier, R. 1984, *Gamma Ray Astronomy*, Oxford Univ. Press, Oxford.
Hiltner, W.A. & Gordon, M.F. 1971, *Astrophys. Letts.*, **8**, 3.
Hind, J.R. 1848, *Mon. Not. R. astr. Soc.*, **8**, 146, 155, 192 and **9**, 18.
Hind, J.R. 1856, *Mon. Not. R. astr. Soc.*, **16**, 56.
Hinderer, F. 1949, *Astr. Nach.*, **277**, 193.
Hirose, M. & Osaki, Y. 1989, in *Theory of Accretion Discs*, eds. F. Meyer, W.J. Duschl, J. Frank & E. Meyer-Hofmeister, Kluwer, Dordrecht, p. 207.
Hirose, M. & Osaki, Y. 1990, *Pub. astr. Soc. Japan*, **42**, 135.
Hirose, M. & Osaki, H. 1993, *Pub. astr. Soc. Japan*, **45**, 1993
Hirose, M., Osaki, Y. & Mineshige, S. 1991, *Pub. astr. Soc. Japan*, **43**, 809.
Hjellming, R.M. 1974, in *Galactic and Extragalactic Radio Astronomy*, eds. G.L. Verschuur & K.I. Kellermann, Springer-Verlag, New York, p. 159.
Hjellming, R.M. 1990, *Lect. Notes Phys.*, **369**, 169.
Hjellming, M.S. & Taam, R.E. 1990, in *Accretion-Powered Compact Binaries*, ed. C.W. Mauche, Cambridge University Press, Cambridge, p. 417.
Hjellming, M.S. & Taam, R.E. 1991, *Astrophys. J.*, **370**, 709.
Hjellming, R.M. & Wade, C.M. 1970, *Astrophys. J.*, **162**, L1.
Hjellming, M.S. & Webbink, R.F. 1987, *Astrophys. J.*, **318**, 794.
Hjellming, R.M., Wade, C.M., Vandenberg, N.R. & Newell, R.T. 1979, *Astr. J.*, 84, 1619.
Hoard, D.W. & Szkody, P. 1994, *Astr. Soc. Pacific Conf. Ser.* in press.
Hoare, M.G. & Drew, J.E. 1991, *Mon. Not. R. astr. Soc.*, **249**, 452.
Hoare, M.G. & Drew, J.E. 1993, *Mon. Not. R. astr. Soc.*, **260**, 647.
Hogg, H.S. & Wehlau, A. 1964, *J. Roy. Astr. Soc. Canada*, **58**, 163.
Hollander, A. & van Paradijs, J. 1992, *Astr. Astrophys.*, **265**, 77.
Hollander, A., Kraakman, H. & van Paradijs, J. 1993, *Astr. Astrophys. Suppl.*, **101**, 87.
Holm, A.V. 1988, in *A Decade of UV Astronomy with the IUE Satellite*, ESA SP-281, Paris, p. 229.
Holm, A.V. & Gallagher, J.S. 1974, *Astrophys. J.*, **192**, 425.
Holm, A.V., Panek, R.J. & Schiffer, F.H. 1982, *Astrophys. J.*, **252**, L35.
Holm, A.V., Lanning, H., Mattei, J. & Nelan, E. 1993, *Bull. Amer. astr. Soc.*, **25**, 909.
Honey, W.B., Charles, P.A., Whitehurst, R., Barrett, P.E. & Smale, A.P. 1988, *Mon. Not. R. astr. Soc.*, **231**, 1.
Honey, W.B., Bath, G.T., Charles, P.A., Whitehurst, R., Jones, D.H.P., Echevarria, J., Areval, O.M.J., Solheim, J.-E., Tovmassian, G. & Takagishi, K. 1989, *Mon. Not. R. astr. Soc.*, **236**, 727.

Honeycutt, R.K. & Schlegel, E.M. 1985, *Publ. astr. Soc. Pacific*, **97**, 1189.
Honeycutt, R.K., Cannizzo, J.K. & Robertson, J.W. 1994, *Astrophys. J.*, **425**, 835.
Honeycutt, R.K., Kaitchuck, R.H. & Schlegel, E.M. 1987, *Astrophys. J. Suppl.*, **65**, 451.
Honeycutt, R.K., Livio, M. & Robertson, J.W. 1993, *Publ. astr. Soc. Pacific*, **105**, 922.
Honeycutt, R.K., Schlegel, E.M. & Kaitchuck, R.H. 1986, *Astrophys. J.*, **302**, 388.
Honeycutt, R.K., Robertson, J.W., Turner, G.W. & Vesper, D.N. 1993, *Publ. astr. Soc. Pacific*, **105**, 919.
Honeycutt, R.K., Robertson, J.W., Turner, G.W. & Vesper, D.N. 1994, *Astr. Soc. Pacific Conf. Ser.* **56**, 277.
Ho Peng Yoke 1962, *Vistas*, **5**, 127.
Ho Peng Yoke 1970, *Oriens Extremus*, **17**, 63.
Hopp, U. & Wolk, C. 1984, *Astrophys. Sp. Sci.*, **98**, 237.
Horne, K. 1980, *Astrophys. J.*, **242**, L167.
Horne, K. 1983, Ph.D. Thesis, Calif. Inst. Tech.
Horne, K. 1984, *Nature*, **312**, 348.
Horne, K. 1985a, *Mon. Not. R. astr. Soc.*, **213**, 129.
Horne, K. 1985b, in *Interacting Binaries*, eds. P.P. Eggleton & J.E. Pringle, Reidel, Dordrecht, p. 327.
Horne, K. 1990, Statement made at Accretion Disc Conference, Paris 1990.
Horne, K. 1991, *Int. Astr. Union Colloq. No. 129*, p. 3.
Horne, K. 1993a, in *Accretion Disks in Compact Stellar Systems*, ed. J.C. Wheeler, World Sci. Pub. Co., Singapore, p. 117.
Horne, K. 1993b, in *Theory of Accretion Disks II*, eds. W.J. Duschl *et al.*, Kluwer, Dordrecht, p. 77.
Horne, K. & Cook, M.C. 1985, *Mon. Not. R. astr. Soc.*, **214**, 307.
Horne, K. & Gomer, R. 1980, *Astrophys. J.*, **237**, 845.
Horne, K. & Marsh, T.R. 1986, *Mon. Not. R. astr. Soc.*, **218**, 761.
Horne, K. & Saar, S.H. 1991, *Astrophys. J.*, **374**, L55.
Horne, K. & Schneider, D.P. 1989, *Astrophys. J.*, **343**, 888.
Horne, K. & Stiening, R.F. 1985, *Mon. Not. R. astr. Soc.*, **216**, 933.
Horne, K., la Dous, C.A. & Shafter, A.W. 1990, in *Accretion-Powered Compact Binaries*, ed. C.W. Mauche, Cambridge University Press, Cambridge, p. 109.
Horne, K., Lanning, H.H. & Gomer, R.H. 1982, *Astrophys. J.*, **252**, 681.
Horne, K., Wade, R.A. & Szkody, P. 1986, *Mon. Not. R. astr. Soc.*, **219**, 791.
Horne, K., Welsh, W.F. & Wade, R.A. 1993, *Astrophys. J.*, **410**, 357.
Horne, K., Wood, J.H. & Stiening, R.F. 1991, *Astrophys. J.*, **378**, 271.
Horne, K., Marsh, T.R., Cheng, F.H., Hubeny, I. & Lanz, T. 1994, *Astrophys. J.*, **426**, 294.
Hoshi, R. 1973, *Prog. Theor. Phys.*, **49**, 776.
Hoshi, R. 1979, *Prog. Theor. Phys.*, **61**, 1307.
Howarth, I.D. 1978a, *J. Brit. astr. Assoc.*, **88**, 458.
Howarth, I.D. 1978b, *J. Brit. astr. Assoc.*, **89**, 47.
Howell, S.B. 1993, in *Cataclysmic Variables and Related Physics*, eds. O. Regev & G. Shaviv, Inst. Phys. Publ., Bristol, p. 67.
Howell, S.B. & Blanton, S.A. 1993, *Astr. J.*, **106**, 311.
Howell, S.B. & Fried, R. 1991, *Int. Astr. Union Circ.*, No. 5372.
Howell, S.B. & Hurst, G.M. 1994, *Inf. Bull. Var. Stars*, No. 4043.
Howell, S.B. & Szkody, P. 1988, *Publ. astr. Soc. Pacific*, **100**, 224.
Howell, S.B. & Szkody, P. 1990, *Astrophys. J.*, **356**, 623.
Howell, S.B. & Szkody, P. 1991, *Inf. Bull. Var. Stars*, No. 3653.
Howell, S.B. & Szkody, P. 1995, *Astrophys. J.*, in press.
Howell, S.B., Szkody, P. & Cannizzo, J.K. 1994, *Astrophys. J.*, in press.
Howell, S.B., Mason, K.O., Reichert, G.A., Warnock, A. & Kreidl, T.J. 1988, *Mon. Not. R. astr. Soc.*, **233**, 79.
Howell, S.B., Szkody, P., Kreidl, T.J., Mason, K.O. & Puchnarewicz, E.M. 1990, *Publ. astr. Soc. Pacific*, **102**, 758.
Howell, S.B., Szkody, P., Kreidl, T.J. & Dobryzycka, D. 1991, *Publ. astr. Soc. Pacific*, **103**, 300.
Howell, S.B., Schmidt, R., De Young, J.A., Freid, R., Schmeer, P. & Gritz, L. 1993, *Publ. astr. Soc. Pacific*, **105**, 579.
Howell, S.B. Sirk, M.M., Malina, R.F., Mittaz, J.P.D. & Mason, K.O. 1995, *Astrophys. J.*, **439**, 991.

Hric, L. & Urban, Z. 1991, *Inf. Bull. Var. Stars*, No. 3683.
Huang, S.S. 1972, *Astrophys. J.*, **171**, 549.
Hubble, E. 1929, *Astrophys. J.*, **69**, 103.
Hubeny, I. 1989, in *Algols*, ed. A.H. Batten, Kluwer, Dordrecht, p. 117.
Hudec, R. 1984, *Bull. astr. Inst. Czech.*, **35**, 203.
Hudec, R. & Wenzel, W. 1976, *Bull. astr. Inst. Czech.*, **27**, 325.
Hudec, R., Huth, H. & Fuhrmann, B. 1984, *Observatory*, **104**, 1.
Huggins, W. 1866, *Mon. Not. R. astr. Soc.*, **26**, 275.
Huggins, W. & Lady Huggins 1892a, *Proc. R. Soc.*, **50**, 465.
Huggins, W. & Lady Huggins 1892b, *Proc. R. Soc.*, **51**, 486.
Humason, M.L. 1938, *Astrophys. J.*, **88**, 228.
Humason, M.L. & Zwicky, F. 1947, *Astrophys. J.*, **105**, 85.
Hunger, K., Heber, U. & Koester, D. 1985, *Astr. Astrophys.*, **149**, L4.
Hunter, C. 1972, *Ann. Rev. Fluid Mech.*, **4**, 219.
Hut, P. 1981, *Astr. Astrophys.*, **99**, 126.
Hut, P. & Verbunt, F. 1983, *Nature*, **301**, 587.
Hutchings, J.B. 1972, *Mon. Not. R. astr. Soc.*, **158**, 177.
Hutchings, J.B. 1979, *Astrophys. J.*, **232**, 176.
Hutchings, J.B. 1980, *Publ. astr. Soc. Pacific*, **92**, 458.
Hutchings, J.B. 1987, *Publ. astr. Soc. Pacific*, **99**, 57.
Hutchings, J.B. & Cote, T.J. 1985, *Publ. astr. Soc. Pacific*, **97**, 847.
Hutchings, J.B. & Cote, T.J. 1986, *Publ. astr. Soc. Pacific*, **98**, 104.
Hutchings, J.B. & Cowley, A.P. 1984, *Publ. astr. Soc. Pacific*, **96**, 559.
Hutchings, J.B. & McCall, 1977, *Astrophys. J.*, **217**, 775.
Hutchings, J.B., Cowley, A.P. & Crampton, D. 1979, *Astrophys. J.*, **232**, 500.
Hutchings, J.B., Cowley, A.P. & Crampton, D. 1985, *Publ. astr. Soc. Pacific*, **97**, 423.
Hutchings, J.B., Crampton, D. & Cowley, A.P. 1981, *Astrophys. J.*, **247**, 195.
Hutchings, J.B., Link, R. & Crampton, D. 1983, *Publ. astr. Soc. Pacific*, **95**, 264.
Hutchings, J.B., Thomas, B. & Link, R. 1986, *Publ. astr. Soc. Pacific*, **98**, 507.
Hutchings, J.B., Crampton, D., Cowley, A.P., Thorstensen, J.R. & Charles, P.A. 1981a, *Astrophys. J.*, **249**, 680.
Hutchings, J.B., Cowley, A.P., Crampton, D. & Williams, G. 1981b, *Astrophys. J.*, **93**, 741.
Hutchings, J.B., Cowley, A.P., Crampton, D., Fisher, W.A. & Liller, M.H. 1982, *Astrophys. J.*, **252**, 690.
Hyland, A.R. & Neugebauer, G. 1970, *Astrophys. J.*, **160**, L177.
Iben, I. 1982, *Astrophys. J.*, **254**, 244.
Iben, I. 1992, *Astr. Soc. Pacific Conf. Ser.* **30**, 307.
Iben, I. & Tutukov, A.V. 1984, *Astrophys. J.*, **284**, 719.
Iben, I. & Tutukov, A.V. 1986, *Astrophys. J.*, **311**, 742.
Iben, I. & Tutukov, A.V. 1991, *Astrophys. J.*, **370**, 615.
Iben, I., Fujimoto, M.Y. & MacDonald, J. 1992a, *Astrophys. J.*, **384**, 580.
Iben, I., Fujimoto, M.Y. & MacDonald, J. 1992b, *Astrophys. J.*, **388**, 521.
Ichikawa, S. & Osaki, Y. 1992, *Pub. astr. Soc. Japan*, **44**, 15.
Ichikawa, S. & Osaki, Y. 1994, in *Theory of Accretion Discs II*, eds. W.J. Duschl *et al.*, Kluwer, Dordrecht, p. 169.
Ichikawa, S., Hirose, M. & Osaki, Y. 1993, *Pub. astr. Soc. Japan*, **45**, 243.
Ichimura, S. 1976, *Astrophys. J.*, **208**, 701.
Iijima, T. 1990, *J. Amer. Assoc. Var. Star Obs.*, **19**, 28.
Ilovaisky, S.A., Chevalier, C., Motch, C., Pakull, M., van Paradijs, J. & Lub, J. 1984, *Astr. Astrophys.*, **140**, 251.
Imamura, J.N. 1984, *Astr. J.*, **285**, 223.
Imamura, J.N. 1985, *Astrophys. J.*, **296**, 128.
Imamura, J.N. & Durisen, R.H. 1983, *Astrophys. J.*, **268**, 291,
Imamura, J.N. & Steiman-Cameron, T.Y. 1986, *Astrophys. J.*, **311**, 786.
Imamura, J.N. & Wolff, M.T. 1990, *Astrophys. J.*, **355**, 216.
Imamura, J.N., Rashed, H. & Wolff, M.T. 1991, *Astrophys. J.*, **378**, 665.
Imamura, J.N., Wolff, M.T. & Durisen, R.H. 1984, *Astrophys. J.*, **276**, 667.

Imamura, J.N., Durisen, R.H., Lamb, D.Q. & Weast, G.J. 1987, *Astrophys. J.*, **313**, 298.
Imamura, J.N., Middleditch, J., Scargle, J.D., Steiman-Cameron, T.Y., Whitlock, L.A., Wolff, M.T. & Wood, K.S. 1993, *Astrophys. J.*, **419**, 793.
Ingham, W.H., Brecher, K. & Wasserman, I. 1976, *Astrophys. J.*, **207**, 518.
Ingram, D., Garnavich, P., Green, P. & Szkody, P. 1992, *Publ. astr. Soc. Pacific*, **104**, 402.
Ishida, M. 1991, Ph.D. Thesis, Univ. Tokyo. *ISAS Research Note*, No. 505.
Ishida, M., Silber, A., Bradt, H.V., Remillard, R.A., Makashima, K. & Ohashi, T. 1991, *Astrophys. J.*, **367**, 270.
Ishida, M., Sakao, T., Makishima, K., Ohashi, T., Watson, M.G., Norton, A.J., Kawada, M. & Koyama, K. 1992, *Mon. Not. R. astr. Soc.*, **254**, 647.
Isles, J.E. 1974, *J. Brit. astr. Assoc.*, **84**, 203.
Isles, J.E. 1976, *J. Brit. astr. Assoc.*, **86**, 327.
Itoh, H. & Hachisu, I. 1990, *Astrophys. J.*, **358**, 551.
Ivison, R.J., Hughes, D.H., Lloyd, H.M., Bang, M.K. & Bode, M.F. 1993, *Mon. Not. R. astr. Soc.*, **263**, L43.
Jablonski, F. & Busko, I.C. 1985, *Mon. Not. R. astr. Soc.*, **214**, 219.
Jablonski, F.J. & Cieslinski, D. 1992, *Astr. Astrophys.*, **259**, 198.
Jablonski, F.J. & Steiner, J.E. 1987, *Astrophys. J.*, **313**, 376.
James, P.A. & Puxley, P.J. 1993, *Nature*, **363**, 240.
Jameson, R.F. & Sherrington, M.R. 1983, *Mon. Not. R. astr. Soc.*, **205**, 265.
Jameson, R.F., King, A.R. & Sherrington, M.R. 1980, *Mon. Not. R. astr. Soc.*, **191**, 559.
Jameson, R.F., King, A.R. & Sherrington, M.R. 1981, *Mon. Not. R. astr. Soc.*, **195**, 235.
Jameson, R.F., King, A.R. & Sherrington, M.R. 1982, *Mon. Not. R. astr. Soc.*, **200**, 455.
Jameson, R.F., Sherrington, M.R., King, A.R. & Frank, J. 1982, *Nature*, **300**, 152.
Jameson, R.F., King, A.R., Bode, M.F. & Evans, A. 1987, *Observatory*, **107**, 72.
Jensen, K.A., Nousek, J.A. & Nugent, J.J. 1982, *Astrophys. J.*, **261**, 625.
Jensen, K.A., Cordova, F.A., Middleditch, J., Mason, K.O., Grauer, A.D., Horne, K. & Gomer, R. 1983, *Astrophys. J.*, **270**, 211.
Johnson, H.L. 1965, *Astrophys. J.*, **141**, 923.
Johnson, H.L., Perkins, B. & Hiltner, W.A. 1954, *Astrophys. J. Suppl.*, **1**, 91.
Johnston, H.M. & Kulkarni, S.R. 1992, *Astrophys. J.*, **396**, 267.
Jones, M.H. & Watson, M.G. 1992, *Mon. Not. R. astr. Soc.*, **257**, 633.
Jones, C., Forman, W. & Liller, W. 1973, *Astrophys. J.*, **182**, L109.
Joss, P.C. & Rappaport, S. 1983, *Astrophys. J.*, **270**, L73.
Joss, P.C., Katz, J.I. & Rappaport, S.A. 1979, *Astrophys. J.*, **230**, 176.
Joy, A.H. 1938, *Publ. astr. Soc. Pacific*, **50**, 300.
Joy, A.H. 1940, *Publ. astr. Soc. Pacific*, **52**, 324.
Joy, A.H. 1943, *Publ. astr. Soc. Pacific*, **55**, 283.
Joy, A.H. 1954a, *Publ. astr. Soc. Pacific*, **66**, 5.
Joy, A.H. 1954b, *Astrophys. J.*, **120**, 377.
Joy, A.H. 1954c, *Publ. astr. Soc. Pacific*, **57**, 171.
Joy, A.H. 1956, *Astrophys. J.*, **124**, 317.
Joy, A.H. 1960, in *Stellar Atmospheres*, ed. J.L. Greenstein, Univ. Chicago Press, Chicago p. 653.
Jurcevic, J.S., Honeycutt, R.K., Schlegel, E.M. & Webbink, R. 1994, *Publ. astr. Soc. Pacific*, **106**, 481.
Kahabka, P. & Pietsch, W. 1993, *Lect. Notes Phys.*, **416**, 71.
Kahane, C., Audinos, P., Barnbaum, C. & Morris, M. 1993, preprint.
Kahoutek, L. & Pauls, R. 1980, *Astr. Astrophys.*, **92**, 200.
Kaisig, M. 1989a, *Astr. Astrophys.*, **218**, 89.
Kaisig, M. 1989b, *Astr. Astrophys.*, **218**, 102.
Kaitchuck, R. 1989, *Publ. astr. Soc. Pacific*, **101**, 1129.
Kaitchuck, R.H., Honeycutt, R.K. & Schlegel, E.M. 1983, *Astrophys. J.*, **267**, 239.
Kaitchuck, R.H., Mansperger, C.S. & Hantzios, P.A. 1988, *Astrophys. J.*, **330**, 305.
Kaitchuck, R.H., Hantzios, P.A., Kakaletris, P., Honeycutt, R.K. & Schlegel, E.M. 1987, *Astrophys. J.*, **317**, 765.
Kaitchuck, R.H., Schlegel, E.M., Honeycutt, R.K., Horne, K., Marsh, T.R., White, J.C. & Mansperger, C.S. 1994, *Astrophys. J. Suppl.*, **93**, 519.

Kallman, T.R. 1983, *Astrophys. J.*, **272**, 238.
Kallman, T. 1987, in *The Physics of Accretion onto Compact Objects*, eds. K.O. Mason, M.G. Watson & N.E. White, Dordrecht, Reidel, p. 269.
Kallman, T.R. & Jensen, K.A. 1985, *Astrophys. J.*, **299**, 277.
Kallman, T.R., Schlegel, E.M., Serlemitsos, P.J., Petre, R., Marshall, F., Jahoda, K., Boldt, E.A., Holt, S.S., Mushotzky, R.F., Swank, J., Szymkowiak, A.E., Kelley, R.L., Smale, A., Arnaud, K. & Weaver, K. 1993, *Astrophys. J.*, **411**, 869.
Kaluzny, J. 1989, *Acta Astr.*, **39**, 235.
Kaluzny, J. 1990, *Mon. Not. R. astr. Soc.*, **245**, 547.
Kaluzny, J. & Chlebowski, T. 1988, *Astrophys. J.*, **332**, 287.
Kaluzny, J. & Chlebowski, T. 1989, *Acta Astr.*, **39**, 35.
Kaluzny, J. & Semeniuk, I. 1987, *Acta Astr.*, **37**, 349.
Kaluzny, J. & Semeniuk, I. 1988, *Inf. Bull. Var. Stars*, No. 3145.
Kamata, Y. & Koyama, K. 1993, *Astrophys. J.*, **405**, 307.
Kamata, Y., Tawara, Y. & Koyama, K. 1991, *Astrophys. J.*, **379**, L65.
Kaplan, S.A. & Pikel'ner, S.B. 1970, *The Interstellar Medium*, Harvard Univ. Press, Cambridge.
Kapteyn, J.C. 1901, *Astr. Nach.*, **157**, 202.
Kato, S. 1978, *Mon. Not. R. astr. Soc.*, **185**, 629.
Kato, S. 1984, *Publ. astr. Soc. Japan*, **36**, 55.
Kato, T. 1990a, *Inf. Bull. Var. Stars*, No. 3522.
Kato, M. 1990b, *Astrophys. J.*, **355**, 277.
Kato, M. 1990c, *Astrophys. J.*, **362**, L17.
Kato, T. 1991a, *Inf. Bull. Var. Stars*, No. 3671.
Kato, T. 1991b, *Int. Astr. Union Circ.*, No. 5379.
Kato, M. 1991c, *Astrophys. J.*, **366**, 369 and 471.
Kato, M. 1991d, *Astrophys. J.*, **369**, 471.
Kato, T. 1993, *Pub. astr. Soc. Japan*, **45**, L67.
Kato, M. 1994a, *Astr. Astrophys.*, **281**, L49.
Kato, T. 1994b, *Proc. Workshop Cataclys. Var.*, Padova, in press.
Kato, T. & Fujino, S. 1987, *Var. Star Bull. Japan.*, **3**, 10.
Kato, M. & Hachisu, I. 1988, *Astrophys. J.*, **329**, 808.
Kato, M. & Hachisu, I. 1989, *Astrophys. J.*, **346**, 424.
Kato, M. & Hachisu, I. 1991, *Astrophys. J.*, **373**, 620.
Kato, M. & Hachisu, I. 1994, *Astrophys. J.*, **437**, 802.
Kato, T. & Hirata, R. 1990, *Inf. Bull. Var. Stars*, No. 3489.
Kato, S. & Horiuchi, T. 1985, *Publ. astr. Soc. Japan*, **37**, 399.
Kato, S. & Horiuchi, T. 1986, *Publ. astr. Soc. Japan*, **38**, 313.
Kato, M. & Iben, I. 1992, *Astrophys. J.*, **394**, L47.
Kato, T., Fujino, S. & Iida, M. 1989, *Var. Star Bull. Japan*, **9**, 33.
Kato, T., Hirata, R. & Mineshige, S. 1992, *Pub. Astr. Soc. Japan*, **44**, L215.
Kato, T., Fujino, S., Iida, M., Makiguchi, N. & Kashino, M. 1988, *Var. Star Bull. Japan*, **5**, 18.
Katz, J.I. 1973, *Nature Phys. Sci.*, **246**, 87.
Katz, J.I. 1975, *Astrophys. J.*, **200**, 298.
Katz, J.I. 1989, *Mon. Not. R. astr. Soc.*, **239**, 751.
Katz, J.I. 1991, *Astrophys. J.*, **374**, L59.
Kaul, C.L., Kaul, R.K. & Bhat, C.L. 1993, *Astr. Astrophys.*, **272**, 501.
Kawaler, S.D. 1988, *Astrophys. J.*, **333**, 236.
Keenan, P.C. & McNeil, R.C. 1976, *An Atlas of Spectra of the Cooler Stars*, Ohio State Univ. Press, Columbus, Ohio.
Kelly, B.D., Kilkenny, D. & Cooke, J.A. 1981, *Mon. Not. R. astr. Soc.*, **196**, 91P.
Kemic, S.B. 1974, *JILA Report*, No. 113.
Kemp, J.C., Swedlund, J.B. & Wolstencroft, R.D. 1974, *Astrophys. J.*, **193**, L15.
Kenyon, S. 1986, *The Symbiotic Stars*, Cambridge University Press, Cambridge
Kenyon, S.J. 1988a, *Int. Astr. Union Colloq. No. 103*, p. 11.
Kenyon, S. 1988b, *Int. Astr. Union Colloq. No. 103*, p. 161.
Kenyon, S.J. & Berriman, G. 1988, *Astr. J.*, **95**, 526.
Kenyon, S.J. & Gallagher, J.S. 1983, *Astr. J.*, **88**, 666.

Kenyon, S.J. & Garcia, M.R. 1986, *Astr. J.*, **91**, 125.
Kenyon, S.J. & Truran, J.W. 1983, *Astrophys. J.*, **273**, 280.
Kenyon, S.J. & Webbink, R.F. 1984, *Astrophys. J.*, **279**, 252.
Kenyon, S.J., Mikolajewski, J., Mikolajewski, M., Polidan, R.S. & Slovak, M. 1993, *Astr. J.*, **106**, 1573.
Kepler, S.O., 1990, *Rev. Mex. Astr. Astrophys.*, **21**, 335.
Kepler, S.O., Steiner, J.E. & Jablonski, F. 1989, *Int. Astr. Union Colloq. No. 114*, p. 443.
Kern, J.R. & Bookmyer, B.B. 1986, *Publ. astr. Soc. Pacific*, **98**, 1336.
Kholopov, P.N. & Efremov, Yu. N. 1976, *Var. Stars*, **20**, 277.
Kilkenny, D., Marang, F. & Menzies, J.W. 1994, *Mon. Not. R. astr. Soc.*, **267**, 535.
Kim, S.-W., Wheeler, J.C. & Mineshige, S. 1992, *Astrophys. J.*, **384**, 269.
King, A.R. 1985a, *Nature*, **313**, 291.
King, A.R. 1985b, *Mon. Not. R. astr. Soc.*, **217**, 23P.
King, A.R. 1986, *Ann. N.Y. Acad. Sci.*, **470**, 320.
King, A.R. 1988, *Quart. J. Roy. astr. Soc.*, **29**, 1.
King, A.R. 1989a, *Mon. Not. R. astr. Soc.*, **241**, 365.
King, A.R. 1989b, in *Classical Novae*, eds. M.F. Bode & A. Evans, Wiley, Chichester, p. 17.
King, A.R. 1993, *Mon. Not. R. astr. Soc.*, **261**, 144.
King, A.R. & Lasota, J.-P. 1984, *Astr. Astrophys.*, **140**, L16.
King, A.R. & Lasota, J.-P. 1990, *Mon. Not. R. astr. Soc.*, **247**, 214.
King, A.R. & Lasota, J.-P. 1991, *Astrophys. J.*, **378**, 674.
King, A.R. & Shaviv, G. 1984a, *Nature*, **308**, 519.
King, A.R. & Shaviv, G. 1984b, *Mon. Not. R. astr. Soc.*, **211**, 883.
King, A.R. & Watson, M.G. 1987, *Mon. Not. R. astr. Soc.*, **227**, 205.
King, A.R. & Whitehurst, R. 1991, *Mon. Not. R. astr. Soc.*, **250**, 152.
King, A.R. & Williams, G.A. 1983, *Mon. Not. R. astr. Soc.*, **205**, 57P.
King, A.R. & Williams, G.A. 1985, *Mon. Not. R. astr. Soc.*, **215**, 1P.
King, A.R. & Wynn, G.A. 1993, in *White Dwarfs: Advances in Observation and Theory*, ed. M.A. Barstow, Kluwer, Dordrecht p. 371.
King, A.R., Frank, J. & Ritter, H. 1985, *Mon. Not. R. astr. Soc.*, **213**, 181.
King, A.R., Frank, J. & Whitehurst, R. 1990, *Mon. Not. R. astr. Soc.*, **244**, 731.
King, A.R., Raine, D.J. & Jameson, R.F. 1978, *Astr. Astrophys.*, **70**, 327.
King, A.R., Regev, O. & Wynn, G.A. 1991, *Mon. Not. R. astr. Soc.*, **251**, 30P.
King, A.R., Watson, M.G. & Heise, J. 1985, *Nature*, **313**, 290.
King, A.R., Frank, J., Jameson, R.F. & Sherrington, M.R. 1983, *Mon. Not. R. astr. Soc.*, **203**, 677.
Kiplinger, A.L. 1979a, *Astrophys. J.*, **234**, 997.
Kiplinger, A.L. 1979b, *Astr. J.*, **84**, 655.
Kiplinger, A.L. 1980, *Astrophys. J.*, **236**, 839.
Kiplinger, A.L. 1988, private communication.
Kiplinger, A.L. & Nather, R.E. 1975, *Nature*, **255**, 125.
Kiplinger, A.L., Sion, E.M. & Szkody, P. 1991, *Astrophys. J.*, **366**, 569.
Kippenhahn, R. 1981, *Astr. Astrophys.*, **102**, 293.
Kippenhahn, R. & Meyer-Hofmeister, F. 1977, *Astr. Astrophys.*, **54**, 539.
Kippenhahn, R. & Weigert, A. 1990, *Stellar Structure & Evolution*, Springer, Berlin.
Kirbiyik, H., 1982, *Mon. Not. R. astr. Soc.*, **200**, 907.
Klare, G., Krautter, J., Wolf, B., Stahl, O., Vogt, N., Wargau, W. & Rahe, J. 1982, *Astr. Astrophys.*, **113**, 76.
Kley, W. 1989a, *Astr. Astrophys.*, **208**, 98.
Kley, W. 1989b, *Astr. Astrophys.*, **222**, 141.
Kley, W. 1989c, in *Theory of Accretion Discs*, eds. F. Meyer, W.J. Duschl, J. Frank & E. Meyer-Hofmeister, Kluwer, Dordrecht, p. 289.
Kley, W., 1990, in *Accretion-Powered Compact Binaries*, ed. C.W. Mauche, Cambridge University Press, Cambridge, p. 301.
Kley, W. 1991, *Astr. Astrophys.*, **247**, 95.
Kley, W. & Hensler, G. 1987, *Astr. Astrophys.*, **172**, 124.
Kley, W. & Lin, D.N.C. 1992, *Astrophys. J.*, **397**, 600.
Knapp, G.R., Rauch, K.P. & Wilcots, E.M. 1990, *Astr. Soc. Pacific Conf. Ser.*, **12**, 151.

Knobloch, E. 1992, *Mon. Not. R. astr. Soc.*, **255**, 25P.
Knott, G. 1882, *Observatory,* **5**, 110.
Knott, G. 1896, *Mem. R. astr. Soc.*, **52**, 1.
Kodaira, K. 1983, *Int. Astr. Union Colloq. No. 71*, p. 561.
Koen, C. 1976. M.Sc. Thesis, Univ. Cape Town (published in *Pub. Dept. Astr. Univ. Cape Town*, No. 2, 1981.)
Koen, C. 1986, *Mon. Not. R. astr. Soc.*, **223**, 529.
Koen, C. 1988, *Astrophys. Sp. Sci.*, **141**, 347.
Koen, C. 1992, *Astr. Soc. Pacific Conf. Ser.*, **30**, 127.
Koester, D. & Schoenberner, D. 1986, *Astr. Astrophys.*, **154**, 125.
Kolb, U. 1993, *Astr. Astrophys.*, **271**, 149.
Kolb, U. & de Kool, M. 1994, *Astr. Astrophys.* in press.
Kolb, U. & Ritter, H. 1990, *Astr. Astrophys.*, **236**, 385.
Kolb, U. & Ritter, H. 1992, *Astr. Astrophys.*, **254**, 213.
Kondo, Y., van Flandern, T.C. & Wolff, C.L. 1983, *Astrophys. J.*, **273**, 716.
Königl, A. 1989, *Astrophys. J.*, **342**, 208.
Kopal, Z. 1959, *Close Binary Systems*, Chapman & Hall, London.
Kopal, Z. 1978, *Dynamics of Close Binary Systems*, Reidel, Dordrecht.
Kopylov, I.M. 1954, *Dok. Akad. Nauk SSR*, **99**, 515.
Kopylov, I.M., Somov, N.N. & Somova, T.A. 1991, *Astrofiz. Issled. Izv. Spets. Astrofiz. Obs.*, **31**, 15.
Kopylov, I.M., Lipovetskij, V.A., Somov, N.N., Somova, T.A. & Stepanyan, D.A. 1988, *Astrophys.*, **28**, 168.
Korth, S. 1990, *J. Amer. Assoc. Var. Star Obs.*, **19**, 135.
Korvacs, G. 1980, *Acta Astr.*, **31**, 207.
Kovetz, A. & Prialnik, D. 1985, *Astrophys. J.*, **291**, 812.
Kovetz, A. & Prialnik, D. 1990, *Lect. Notes Phys.*, **369**, 394.
Kovetz, A., Prialnik, D. & Shara, M.M. 1988, *Astrophys. J.*, **325**, 828.
Koyama, K., Takano, S., Tawara, Y., Matsumoto, T., Noguchi, K., Fukui, Y., Iwata, T., Ohashi, N., Tatematsu, K., Takahashi, N., Umemoto, T., Hodapp, K.W., Rayner, J. & Makishima, K. 1991, *Astrophys. J.*, **377**, 240.
Kozlowska, A. 1988, *Acta Astr.*, **38**, 13.
Kraft, R.P. 1956, *Carnegie Inst. Yearbook*, Carnegie Inst., Washington, p. 56.
Kraft, R.P. 1958, *Astrophys. J.*, **127**, 625.
Kraft, R.P. 1959, *Astrophys. J.*, **130**, 110.
Kraft, R.P. 1961a, *Science*, **134**, 1433.
Kraft, R.P. 1961b, *Astrophys. J.*, **134**, 171.
Kraft, R.P. 1962, *Astrophys. J.*, **135**, 408.
Kraft, R.P. 1963, *Adv. Astr. Astrophys.*, **2**, 43.
Kraft, R.P. 1964a, *Astrophys. J.*, **139**, 457.
Kraft, R.P. 1964b, in *First Conference on Faint Blue Stars*, ed. W.J. Luyten, Univ. Minn. Press, Minneapolis, p. 77.
Kraft, R.P. 1967, *Astrophys. J.*, **150**, 551.
Kraft, R.P. & Luyten, W.J. 1965, *Astrophys. J.*, **142**, 1041.
Kraft, R.P., Krzeminski, W. & Mumford, G.S. 1969, *Astrophys. J.*, **158**, 589.
Kraft, R.P., Matthews, J. & Greenstein, J.L. 1962, *Astrophys. J.*, **136**, 312.
Kraicheva, Z. & Genkov, V. 1992, *Inf. Bull. Var. Stars*, No. 3697.
Krautter, J. & Williams, R.E. 1989, *Astrophys. J.*, **341**, 968.
Krautter, J., Klaas, U. & Radons, G. 1987, *Astr. Astrophys.*, **181**, 373.
Krautter, J., Klare, G., Wolf, B., Wargau, W., Drechsel, H., Rahe, J. & Vogt, N. 1981a, *Astr. Astrophys.*, **98**, 27.
Krautter, J., Klare, G., Wolf, B., Wargau, W., Drechsel, H., Rahe, J. & Vogt, N. 1981b, *Astr. Astrophys.*, **102**, 337.
Krautter, J., Beuermann, K., Leitherer, C., Oliva, E., Moorwood, A.F.M., Deul, E., Wargau, W., Klare, G., Kohoutek, L., van Paradijs, J. & Wolf, B. 1984, *Astr. Astrophys.*, **137**, 307.
Krautter, J., Beuermann, K., Finkenzeller, U., Heske, A., Kollatschny, W., Neckel, T., Ögelman, H., Pakull, M., Schulte-Ladbeck, R. & Strupat, W. 1986, in *RS Ophiuchi (1985) and the Recurrent Nova Phenomenon*, ed. M.F. Bode, VNU Sci. Press, Utrecht, p. 93.

Kriz, S. & Hubeney, I. 1986, *Bull. Astr. Inst. Czech.*, **37**, 129.
Kruszewski, A. 1966, *Adv. Astr. Astrophys.*, **4**, 233.
Kruszewski, A. 1967, *Acta Astr.*, **17**, 297.
Kruszewski, A. & Semeniuk, I. 1992, *Acta Astr.*, **42**, 311.
Kruszewski, A. & Semeniuk, I. 1993, *Acta Astr.*, **43**, 127.
Kruszewski, A., Semeniuk, I. & Duerbeck, H.W. 1983, *Acta Astr.*, **33**, 339.
Kruszewski, A., Merwe, R., Heise, J., Chlebowski, J., van Diik, W. & Bakker, R. 1981, *Sp. Sci. Rev.*, **30**, 221.
Kruytbosch, W.E. 1928, *Bull. Astr. Inst. Neth.*, **144**, 145.
Krzeminski, W. 1965, *Astrophys. J.*, **142**, 1051.
Krzeminski, W. 1972, *Acta Astr.*, **22**, 387.
Krzeminski, W. & Kraft, R.P. 1964, *Astrophys. J.*, **140**, 921.
Krzeminski, W. & Serkowski, K. 1977, *Astrophys. J.*, **216**, L45.
Krzeminski, W. & Smak, J. 1971, *Acta Astr.*, **21**, 133.
Krzeminski, W. & Vogt, N. 1985, *Astr. Astrophys.* **144**, 124.
Kubiak, M. 1984a, *Acta Astr.*, **34**, 331.
Kubiak, M. 1984b, *Acta Astr.*, **34**, 397.
Kubiak, M. & Krzeminski, W. 1989, *Publ. astr. Soc. Pacific*, **101**, 669.
Kubiak, M. & Krzeminski, W. 1992, *Acta Astr.*, **42**, 177.
Kudritzki, R.P. & Reimers, D. 1978, *Astr. Astrophys.*, **70**, 227.
Kudritzki, R.P. & Simon, K.P. 1978, *Astr. Astrophys.*, **70**, 653.
Kudritzki, R.P., Simon, K.P., Lynas-Grey, A.E., Kilkenny, D. & Hill, P.W. 1982, *Astr. Astrophys.*, **106**, 254.
Kuerster, M. & Barwig, H. 1988, *Astr. Astrophys.*, **199**, 201.
Kuijpers, J. 1985, *Radio Stars*, eds. R.M. Hjellming & D.M. Gibson, Reidel, Dordrecht, p. 3.
Kuijpers, J. & Pringle, J.E. 1982, *Astr. Astrophys.*, **114**, L4.
Kuiper, G.P. 1941, *Astrophys. J.*, **93**, 133.
Kukarkin, B.V. 1977, *Mon. Not. R. astr. Soc.*, **180**, 5P.
Kukarkin. B. & Kholopov, P.N. 1975, *Astr. Tsirk.*, No. 889.
Kukarkin, B.V. & Parenago, P.P. 1934, *Var. Star Bull.*, **4**, 44.
Kukarkin, B.V. & Parenago, P.P. 1948, *General Catalogue of Variable Stars*, USSR Acad. Sci., Moscow.
Kukarkin, B.V., Parenago, P.P., Efremov, Yu I. & Kholopov, P.N. 1958, *General Catalogue of Variable Stars*, USSR Acad. Sci., Moscow, 2nd edition.
Kulkarni, S. & Narayan, R. 1988, *Astrophys. J.*, **335**, 755.
Kumar, S. 1986, *Mon. Not. R. astr. Soc.*, **223**, 225.
Kumar, S. 1989, in *Theory of Accretion Disks*, eds. F. Meyer, W.J. Duschl, J. Frank & E. Meyer-Hofmeister, Kluwer, Dordrecht, p. 297.
Kuperus, M. & Ionson, J. 1985, *Astr. Astrophys.*, **148**, 309.
Kurochka, L.N. & Maslennikova, L.B. 1970, *Sol. Phys.*, **11**, 33.
Kurochkin, N.E. & Shugarov, S. Yu. 1980, *Astr. Tsrik.*, No. 1114.
Kurucz, R. 1979, *Astrophys. J. Suppl.*, **40**, 1.
Kuulkers, E. 1990, M.Sc. Thesis, Univ. Amsterdam.
Kuulkers, E., van Amerongen, S., van Paradijs, J. & Röttgering, H. 1991a, *Astr. Astrophys.*, **252**, 605.
Kuulkers, E., Hollander, A., Oosterbroek, T. & van Paradijs, J. 1991b, *Astr. Astrophys.*, **242**, 401.
Kuulkers, E. & Rutten, R.G.M. 1994. In preparation.
Kwok, S. 1983, *Mon. Not. R. astr. Soc.*, **202**, 1149.
Kylafis, N.D. & Lamb, D.Q. 1979, *Astrophys. J.*, **228**, L105.
Kylafis, N.D. & Lamb, D.Q. 1982, *Astrophys. J. Suppl.*, **48**, 239.
Lacy, C.H. 1977, *Astrophys. J. Suppl.*, **34**, 479.
la Dous, C. 1989a, *Astr. Astrophys.*, **211**, 131.
la Dous, C. 1989b, *Mon. Not. R. astr. Soc.*, **238**, 935.
la Dous, C. 1989c, *IUE-ULDA Access Guide No. 1*, ESA SP-1114, Paris.
la Dous, C. 1990, *Sp. Sci. Rev.*, **52**, 203.
la Dous, C. 1991, *Astr. Astrophys.*, **252**, 100.
la Dous, C., Verbunt, F., Schoembs, R., Argyle, R.W., Jones, D.H.P., Schwarzenberg-Czerny, A., Hassall, B.J.M., Pringle, J.E. & Wade, R.A. 1985, *Mon. Not. R. astr. Soc.*, **212**, 231.

Lamb, D.Q. 1974, *Astrophys. J.*, **192**, L129.
Lamb, D.Q. 1985, in *Cataclysmic Variables and Low-Mass X-Ray Binaries*, eds. D.Q. Lamb and J. Patterson, Reidel, Dordrecht, 179.
Lamb, D.Q. 1988, in *Polarized Radiation of Circumstellar Origin*, eds. G.V. Coyne *et al.*, Vatican Obs., Vatican, p. 151.
Lamb, F.K. 1989, in *Timing Neutron Stars*, eds. H. Ögelman & E.P.J. van den Heuvel, Kluwer, p. 649.
Lamb, F.K. & Ghosh, P. 1991, in *Particle Acceleration near Accreting Compact Objects*, eds. J. van Paradijs, M. van der Klis & A. Achterberg, North Holland, Amsterdam, p. 37.
Lamb, D.Q. & Masters, A.R. 1979, *Astrophys. J.*, **234**, L117.
Lamb, D.Q. & Melia, F. 1986, in *The Physics of Accretion onto Compact Objects*, eds. K.O. Mason *et al.*, Springer-Verlag, Berlin, p. 113.
Lamb, D.Q. & Melia, F. 1987, *Astrophys. Sp. Sci.*, **131**, 511.
Lamb, D.Q. & Melia, F. 1988, in *Polarized Radiation of Circumstellar Origin*, eds. G.V. Coyne *et al.*, Obs., Vatican, Vatican, p. 45.
Lamb, D.Q. & Patterson, J. 1983, *Int. Astr. Union Colloq. No. 72*, p. 229.
Lamb, F.K., Aly, J.-J., Cook, M.C. & Lamb, D.Q. 1983, *Astrophys. J.*, **274**, L71.
Lamb, F.K., Shibazaki, N., Alpar, M.A. & Shaham, J. 1985, *Nature*, **317**, 681.
Lambert, D.L. & Slovak, M.H. 1981, *Publ. astr. Soc. Pacific*, **93**, 477.
Lambert, D.L., Slovak, M.H., Shields, G.A. & Ferland, G.J. 1981, in *The Universe at Ultraviolet Wavelengths*, ed. R.D. Chapman, NASA CP 2171, Washington, p. 461.
Lance, C.M., McCall, M.L. & Uomoto, A.K. 1988, *Astrophys. J. Suppl.*, **66**, 151.
Landau, L. & Lifschitz, E. 1958, *The Classical Theory of Fields*, Pergamon, Oxford.
Landolt, A.U. 1970, *Publ. astr. Soc. Pacific*, **82**, 86.
Landolt, A.U. & Drilling, J.S. 1986, *Astr. J.*, **91**, 1372.
Lang, K.R. 1974, *Astrophysical Formulae*, Springer-Verlag, New York.
Langer, S.H., Chanmugam, G. & Shaviv, G. 1981, *Astrophys. J.*, **245**, L23.
Langer, S.H., Chanmugam, G. & Shaviv, G. 1982, *Astrophys. J.*, **258**, 289.
Lanning, H.H. & Semeniuk, I. 1981, *Acta Astr.*, **31**, 175.
Lanzafame, G., Belvedere, G. & Molteni, D. 1992, *Mon. Not. R. astr. Soc.*, **258**, 152.
Lanzafame, G., Belvedere, G. & Molteni, D. 1993, *Mon. Not. R. astr. Soc.*, **263**, 839.
Larson, R.B. 1978, *J. Comp. Phys.*, **27**, 397.
Larsson, S. 1985, *Astr. Astrophys.*, **145**, L1.
Larsson, S. 1987a, *Astr. Astrophys.*, **181**, L15.
Larsson, S. 1987b, *Astrophys. Sp. Sci.*, **130**, 187.
Larsson, S. 1988, in *The Physics of Compact Objects*, eds. N.E. White & L.G. Filipov, Pergamon, Oxford, p. 305.
Larsson, S. 1989a, *Astr. Astrophys.*, **217**, 146.
Larsson, S. 1989b, in *Accretion Powered Compact Binaries*, ed. C.W. Mauche, Cambridge University Press, Cambridge, p. 279.
Larsson, S. 1992, *Astr. Astrophys.*, **265**, 133.
Latham, D.W., Liebert, J. & Steiner, J.E. 1981, *Astrophys. J.*, **246**, 919.
Lázaro, C., Solheim, J.-E. & Arévalo, M.J. 1989, *Int. Astr. Union Colloq. No. 114*, p. 458.
Lázaro, C., Martinez-Pais, G., Arévalo, M.J. & Solheim, J.-E. 1991, *Astr. J.*, **101**, 196.
Lecar, M., Wheeler, J.C. & McKee, C.F. 1976, *Astrophys. J.*, **205**, 556.
Leibowitz, E.M., Mendelson, H., Mashal, E., Prialnik, D. & Seitter, W.C. 1992, *Astrophys. J.*, **385**, L49.
Leibowitz, E.M., Mendelson, H., Bruch, A., Duerbeck, H.W., Seitter, W.C. & Richter, G.A. 1994, *Astrophys. J.*, **421**, 771.
Leising, M.D. & Clayton, D.D. 1987, *Astrophys. J.*, **323**, 159.
Lemm, K., Patterson, J., Thomas, G. & Skillman, D.R. 1993, *Publ. astr. Soc. Pacific*, **105**, 1120.
Lenouvel, F. & Daguillon, J. 1954, *Ann. d'Astrophys.*, **17**, 416.
Li, J., Wu, K. & Wickramasinghe, D.T. 1994, *Mon. Not. R. astr. Soc.*, **268**, 61.
Li, Y., Jiang, Z., Chen, J. & Wei, M. 1990, *Chin. Astr. Astrophys.*, **14**, 359.
Liang, E.D.T. & Price, R.H. 1977, *Astrophys. J.*, **218**, 247.
Libbrecht, K.G. & Woodard, M.F. 1990, *Nature*, **345**, 779.
Liebert, J. 1976, *Publ. astr. Soc. Pacific.* **88**, 490.

Liebert, J. & Probst, R.G. 1987, *Ann. Rev. Astr. Astrophys.*, **25**, 473.
Liebert, J. & Stockman, H.S. 1979, *Astrophys. J.*, **229**, 652.
Liebert, J. & Stockman, H.S. 1985, in *Cataclysmic Variables and Low-Mass X-Ray Binaries*, eds. D.Q. Lamb and J. Patterson, Reidel, Dordrecht, p. 151.
Liebert, J., Angel, J.R.P., Stockman, H.S. & Beaver, E.A. 1978a, *Astrophys. J.*, **225**, 181.
Liebert, J., Stockman, H.S., Angel, J.R.P., Woolf, N.J., Hege, K. & Margon, B. 1978b, *Astrophys. J.*, **225**, 201.
Liebert, J., Stockman, H.S., Williams, R.E., Tapia, S., Green, R.F., Rautenkranz, D., Ferguson, D.H. & Szkody, P. 1982a, *Astrophys. J.*, **256**, 594.
Liebert, J., Tapia, S., Bond, H.E. & Grauer, A.D. 1982b, *Astrophys. J.*, **254**, 232.
Liebert, J., Tweedy, R.W., Napiwotzi, R. & Fulbright, S. 1994, *Astrophys. J.*, **441**, 424.
Lightman, A.P. 1974, *Astrophys. J.*, **194**, 429.
Lightman, A.P. & Eardley, D.M. 1974, *Astrophys. J.*, **187**, L1.
Liller, M.H. 1980a, *Inf. Bull. Var. Stars*, No. 1743.
Liller, M.H. 1980b, *Astr. J.*, **85**, 1092.
Liller, W. & Mayer, B. 1987, *Publ. astr. Soc. Pacific*, **99**, 606.
Liller, W., Shao, C.Y., Mayer, B., Garnavich, P. & Harbrecht, R.P. 1975, *Int. Astr. Union Circ.*, No. 2848.
Lin, D.N.C. 1975, *Mon. Not. R. astr. Soc.*, **170**, 379.
Lin, D.N.C. 1981, *Astrophys. J.*, **246**, 972.
Lin, D.N.C. 1989, in *Theory of Accretion Discs*, eds. F. Meyer, W.J. Duschl, J. Frank & E. Meyer-Hofmeister, Kluwer, Dordrecht, p. 89.
Lin, D.N.C. & Papaloizou, J. 1980, *Mon. Not. R. astr. Soc.*, **191**, 37.
Lin, D.N.C. & Papaloizou, J. 1988, in *Critical Observations Versus Physical Models for Close Binary Systems*, ed. K.C. Leung, Gordon & Breach, New York, p. 317.
Lin, D.N.C. & Pringle, J.E. 1976, *Int. Astr. Union Symp. No. 73*, p. 237.
Lin, D.N.C. & Shields, G.A. 1986, *Astrophys. J.*, **305**, 28.
Lin, D.N.C., Papaloizou, J. & Faulkner, J. 1985, *Mon. Not. R. astr. Soc.*, **212**, 105.
Lin, D.N.C., Williams, R.E. & Stover, R.J. 1988, *Astrophys. J.*, **327**, 234.
Lines, H.C., Lines, R.D. & McFaul, T.G. 1988, *Astr. J.*, **95**, 1505.
Linnell, A.P. 1949, *Sky & Tel.*, **8**, 166.
Linnell, A.P. 1950, *Harv. Circ.* No. 455.
Linsky, J.L., Bornmann, P.L., Carpenter, K.G., Wing, R.F., Giampapa, M.S., Worden, S.P. & Hege, E.K. 1982, *Astrophys. J.*, **260**, 670.
Lipunova, N.A. & Shugarov, S. Yu. 1990, *J. Amer. Assoc. Var. Star Obs.*, **19**, 40.
Lipunova, N.A. & Shugarov, S. Yu. 1991, *Inf. Bull. Var. Stars*, No. 3580.
Livio, M. 1982, *Astr. Astrophys.*, **112**, 190.
Livio, M. 1989, *Sp. Sci. Rev.*, **50**, 299.
Livio, M. 1992a, *Astr. Soc. Pacific Conf. Ser.*, **29**, 269.
Livio, M. 1992b, *Astr. Soc. Pacific Conf. Ser.*, **22**, 316.
Livio, M. 1992c, *Astrophys. J.*, **393**, 516.
Livio, M. 1993a, in *Accretion Disks in Compact Stellar Systems*, ed. J.C. Wheeler, World Sci. Publ., Singapore, p. 243.
Livio, M. 1993b, *22nd SAAS FEE Advance Course., Interacting Binaries*, eds. H. Nussbaumer & A. Orr, Springer-Verlag, Berlin, p. 135.
Livio, M. 1993c, *Int. Astr. Union Symp. No. 155*, in press.
Livio, M. & Pringle, J.E. 1992, *Mon. Not. R. astr. Soc.*, **259**, 23P.
Livio, M. & Shara, M. 1987, *Astrophys. J.*, **319**, 819.
Livio, M. & Shaviv, G. 1977, *Astr. Astrophys.*, **55**, 95.
Livio, M. & Soker, N. 1988, *Astrophys. J.*, **329**, 764.
Livio, M. & Spruit, H.C. 1991, *Astr. Astrophys.*, **252**, 189.
Livio, M. & Truran, J.W. 1991, *Ann. N. Y. Acad. Sci.*, **617**, 126.
Livio, M. & Truran, J.W. 1992, *Astrophys. J.*, **389**, 695.
Livio, M. & Truran, J.W. 1994, *Astrophys. J.*, **425**, 797.
Livio, M. & Verbunt, F. 1988, *Mon. Not. R. astr. Soc.*, **232**, 1P.
Livio, M. & Warner, B. 1984, *Observatory*, **104**, 152.
Livio, M., Govarie, A. & Ritter, H. 1991, *Astr. Astrophys.*, **246**, 84.

Livio, M., Prialnik, D. & Regev, O. 1989, *Astrophys. J.*, **341**, 299.
Livio, M., Shankar, A. & Truran, J.W. 1988, *Astrophys. J.*, **325**, 282.
Livio, M., Soker, N. & Dgani, R. 1986, *Astrophys. J.*, **305**, 267.
Livio, M., Shankar, A., Burkert, A. & Truran, J.W. 1990, *Astrophys. J.*, **356**, 250.
Lloyd, H.M., O'Brien, T.J., Bode, M.F., Predehl, P., Schmitt, J.H.M.M., Trümper, J., Watson, M.G. & Pounds, K.A. 1992, *Nature*, **356**, 222.
Lockyer, J.N. 1914, *Pub. Solar Phys. Committee*, 'Phenomena of New Stars', p. 5.
Lohsen, E. 1977, *Inf. Bull. Var. Stars*, No. 1264.
Lombardi, R., Giovannelli, F. & Gaudenzi, S. 1987, *Astrophys. Sp. Sci.*, **130**, 275.
Long, K.S. 1993, in *Cataclysmic Variables and Related Physics*, eds. O. Regev & G. Shaviv, Inst. Phys. Publ., Bristol, p. 24.
Long, K.S., Blair, W.D., Davidsen, A.F., Bowyers, C.W., van Dyke, W.D., Durrance, S.T., Feldman, P.D., Henry, R.C., Kriss, G.A., Kruk, J.W., Moos, H.W., Vancura, O., Ferguson, H.C. & Kimble, R.A. 1991, *Astrophys. J.*, **381**, L25.
Long, K.S., Blair, W.P., Bowers, C.W., Sion, E.M. & Hubeny, I. 1993, *Astrophys. J.*, **405**, 327.
Long, K.S., Wade, R.A., Blair, W.P., Davidsen, A.F. & Hubeny, I. 1994a, *Astrophys. J.*, **426**, 704.
Long, K.S., Sion, E.M., Huang, M. & Szkody, P. 1994b, *Astrophys. J.*, **424**, L49.
Longmore, A.J., Lee, T.J., Allen, D.A. & Adams, D.J. 1981, *Mon. Not. R. astr. Soc.*, **195**, 825.
Lortet-Zuckermann, M.C. 1967, *Compt. Rend.*, **B265**, 826.
Lu, W. & Hutchings, J.B. 1985, *Publ. astr. Soc. Pacific*, **97**, 990.
Lubow, S. 1989, *Astrophys. J.*, **340**, 1064.
Lubow, S.H. 1991a, *Astrophys. J.*, **381**, 259.
Lubow, S.H. 1991b, *Astrophys. J.*, **381**, 268.
Lubow, S.H. 1992a, *Astrophys. J.*, **398**, 525.
Lubow, S.H. 1992b, *Astrophys. J.*, **401**, 317.
Lubow, S.H. & Shu, F.H. 1975, *Astrophys. J.*, **198**, 383.
Luthardt, R. 1992a, *Astr. Soc. Pacific Conf. Ser.* **29**, p. 375.
Luthardt, R. 1992b, *Rev. Mod. Astr.*, **5**, 38.
Luyten, W.J. & Haro, G. 1959, *Publ. astr. Soc. Pacific*, **71**, 469.
Luyten, W.J. & Hughes, H.S. 1965, *Pub. Univ. Minnesota*, No. 36.
Lynden-Bell, D. 1969, *Nature*, **223**, 690.
Lynden-Bell, D. & Pringle, J.E. 1974, *Mon. Not. R. astr. Soc.*, **168**, 603.
MacDonald, J. 1980, *Mon. Not. R. astr. Soc.*, **191**, 933.
MacDonald, J. 1983, *Astrophys. J.*, **267**, 732.
MacDonald, J. 1984, *Astrophys. J.*, **283**, 241.
MacDonald, J. 1986, *Astrophys. J.*, **305**, 251.
MacDonald, J., Fujimoto, M.Y. & Truran, J.W. 1985, *Astrophys. J.*, **294**, 263.
Maceroni, C. 1993, *Int. Astr. Union Symp. No.137*, in press.
Maceroni, C., Bianchini, A., Van't Veer, F., Rodonò, M. & Vio, R. 1990, *Astr. Astrophys.*, **237**, 395.
Machin, G., Allington-Smith, J., Callanan, P.J., Charles, P.A., Hassall, B.J.M., Mason, K.O., Mukai, K., Naylor, T., Smale, A.P. & van Paradijs, J. 1991, *Mon. Not. R. astr. Soc.*, **250**, 602.
MacRae, D.A. 1952, *Astrophys. J.*, **116**, 592.
Makishima, K. 1986, in *Physics of Accretion onto Compact Objects*, eds. K.O. Mason, M.G. Watson & N.E. White, Springer-Verlag, Berlin, p. 249.
Malakpur, I. 1973, *Astr. Astrophys.*, **24**, 125.
Mandel, O. 1965, *Per. Zvesdy*, **15**, 474.
Mansperger, C.S. 1990, Ph.D. Thesis, Univ. Ohio.
Mansperger, C.S. & Kaitchuck, R.H. 1990, *Astrophys. J.*, **358**, 268.
Mansperger, C.S., Kaitchuck, R.H., Garnavich, P.M., Dinshaw, N. & Zamkoff. E. 1994, *Publ. astr. Soc. Pacific*, **106**, 858.
Mantel, K.H., Marschhaeusser, H., Schoembs, R., Haefner, R. & la Dous, C. 1988, *Astr. Astrophys.*, **193**, 101.
Mantle, V.J. & Bath, G.T. 1983, *Mon. Not. R. astr. Soc.*, **202**, 151.
Maraschi, L., Treves, A., Tanzi, E.G., Mouchet, M., Lauberts, A., Motch, C., Bonnet-Bidaud, J.M. & Philips, M.M. 1985, *Astrophys. J.*, **285**, 214.
Marcy, G.W. & Moore, D. 1989, *Astrophys. J.*, **341**, 961.
Mardirossian, F., Mezzeti, M., Pucillo, M., Santin, P., Sedmak, P., Sedman, G. & Giurcin, G. 1980,

Astr. Astrophys., **85**, 29.
Margon, B. & Bolte, M. 1987, *Astrophys. J.*, **321**, L61.
Margon, B., Downes, R.A. & Gunn, J.E. 1981, *Astrophys. J.*, **247**, L89.
Margon, B., Szkody, P., Bowyer, S., Lampton, M. & Paresce, F. 1978, *Astrophys. J.*, **224**, 167.
Margon, B., Anderson, S.F., Downes, R.A., Bohlin, R.C. & Jakobsen, P. 1991, *Astrophys. J.*, **369**, L71.
Marino, B.F. & Walker, W.S.G. 1974, *Inf. Bull. Var. Stars*, No. 864.
Marino, B.F. & Walker, W.S.G. 1979, *Int. Astr. Union Colloq. No. 46*, p. 29.
Marino, B.F. & Walker, W.S.G. 1984, *South. Stars*, **30**, 389.
Marino, B.F., Walker, W.S.G., Herdman, G.C.D. & Allen, W.H. 1984, *South. Stars*, **31**, 61.
Marsh, T.R. 1987, *Mon. Not. R. astr. Soc.*, **228**, 779.
Marsh, T.R. 1988, *Mon. Not. R. astr. Soc.*, **231**, 1117.
Marsh, T.R. 1990, *Astrophys. J.*, **357**, 621.
Marsh, T.R. 1992, *Mon. Not. R. astr. Soc.*, **259**, 695.
Marsh, T.R. 1993, in *Cataclysmic Variables and Related Physics*, eds. O. Regev & G. Shaviv, Inst. Phys. Publ., Bristol, p. 7.
Marsh, T.R. & Horne, K. 1988, *Mon. Not. R. astr. Soc.*, **235**, 269.
Marsh, T.R. & Horne, K. 1990, *Astrophys. J.*, **349**, 593.
Marsh, T.R. & Pringle, J.E. 1990, *Astrophys. J.*, **365**, 677.
Marsh, T.R., Horne, K. & Rosen, S. 1991, *Astrophys. J.*, **366**, 535.
Marsh, T.R., Horne, K. & Shipman, H.L. 1987, *Mon. Not. R. astr. Soc.*, **225**, 551.
Marsh, T.R., Wade, R.A. & Oke, J.B. 1983, *Mon. Not. R. astr. Soc.*, **205**, 33P.
Marsh, T.R., Horne, K., Schlegel, E.M., Honeycutt, R.K. & Kaitchuck, R.H. 1990, *Astrophys. J.*, **364**, 637.
Martel, L. 1961, *Ann. d'Astrophys.*, **24**, 267.
Martell, P.J. & Kaitchuck, R.H. 1991, *Astrophys. J.*, **366**, 286.
Martin, P.G. 1989a, in *Classical Novae*, eds. M.F. Bode & A. Evans, Wiley & Sons, Chichester, p. 73.
Martin, P.G. 1989b, in *Classical Novae*, eds. M.F. Bode & A. Evans, Wiley & Sons, Chichester, p. 93.
Martin, P.G. 1989c, in *Classical Novae*, eds. M.F. Bode & A. Evans, Wiley & Sons, Chichester, p. 113.
Martin, J.S., Jones, D.H.P. & Smith, R.C. 1987, *Mon. Not. R. astr. Soc.*, **224**, 1031.
Martin, J.S., Friend, M.T., Smith, R.C. & Jones, D.H.P. 1989, *Mon. Not. R. astr. Soc.*, **240**, 519.
Mason, K.O. 1985, *Sp. Sci. Rev.*, **40**, 99.
Mason, K.O. 1986, *Lect. Not. Phys.*, **266**, 29.
Mason, P.A., Liebert, J. & Schmidt, G.D. 1989, *Astrophys. J.*, **346**, 941.
Mason, K.O., Rosen, S.R. & Hellier, C. 1988, *Adv. Sp. Res.*, **8**, 293.
Mason, K.O., Lampton, M., Charles, P. & Bowyer, S. 1978, *Astrophys. J.*, **226**, L129.
Mason, K.O., Reichert, G.A., Bowyer, S. & Thorstensen, J.R. 1982, *Publ. astr. Soc. Pacific*, **94**, 521.
Mason, K.O., Middleditch, J.H., Cordova, F.A., Jensen, K.A., Reichert, G., Murdin, P.G., Clark, D. & Bowyer, S. 1983a, *Astrophys. J.*, **264**, 575.
Mason, K., Cordova, F., Middleditch, J., Reichert, J., Bowyer, S., Murdin, P. & Clark, D. 1983b, *Publ. astr. Soc. Pacific*, **95**, 370.
Mason, K.O., Cordova, F.A., Watson, M.G. & King, A.R. 1988, *Mon. Not. R. astr. Soc.*, **232**, 779.
Mason, K.O., Branduardi-Raymont, G., Bromage, G.E., Buckley, D., Charles, P.A., Hassall, B.J.M., Hawkins, M.R.S., Hodgkin, S., Pike, C.D., Jomaron, C.M., Jones, D.H.P., McHardy, I., Naylor, T.T., Ponman, T.J., & Watson, M.G. 1991, *Vistas*, **34**, 343.
Mason, K.O., Watson, M.G., Ponman, T.J., Charles, P.A., Duck, S.R., Hassall, B.J.M., Howell, S.B., Ishida, M., Jones, D.H.P. & Mittaz, J.P.D. 1992, *Mon. Not. R. astr. Soc.*, **258**, 749.
Mason, K.O., Drew, J.E., Cordova, F.A., Horne, K., Hilitch, R., Knigge, C., Lanz, T. & Maylan, T. 1995, *Mon. Not. R. astr. Soc.*, in press.
Masters, A.R., Fabian, A.C., Pringle, J.E. & Rees, M.J. 1977, *Mon. Not. R. astr. Soc.*, **178**, 501.
Mateo, M. & Bolte, M. 1985, *Publ. astr. Soc. Pacific*, **97**, 45.
Mateo, M. & Szkody, P. 1984, *Astr. J.*, **89**, 863.
Mateo, M., Szkody, P. & Bolte, M. 1985, *Publ. astr. Soc. Pacific*, **97**, 45.
Mateo, M., Szkody, P. & Garnavich, P. 1991. *Astrophys. J.*, **370**, 370.
Mateo, M., Szkody, P. & Hutchings, J. 1985, *Astrophys. J.*, **288**, 292.
Matese, J.J. & Whitmire, D.P. 1983, *Astr. Astrophys.*, **117**, L7.
Matusuda, T., Sekino, N., Shima, E., Sawada, K. & Spruit, H. 1990, *Astr. Astrophys.*, **235**, 211.

Matvienko, A.N., Cherepashchuk, A.M. & Yagola, A.G. 1988, *Sov. Astr. J.*, **65**, 526.
Mauche, C.W. 1991, *Astrophys. J.*, **373**, 624.
Mauche, C.W. & Raymond, J.C. 1987, *Astrophys. J.*, **323**, 690.
Mauche, C.W., Raymond, J.C. & Cordova, F.A. 1988, *Astrophys. J.*, **335**, 829.
Mauche, C.W., Miller, G.S., Raymond, J.C. & Lamb, F.K. 1990, in *Accretion-Powered Compact Binaries*, ed. C.W. Mauche, Cambridge University Press, Cambridge, p. 195.
Mauche, C.W., Wade, R.A., Polidan, R.S. van der Woerd, H. & Paerels, F.B.S. 1991, *Astrophys. J.*, **372**, 659.
Mauche, C.W., Raymond, J.C., Buckley, D.A.H., Mouchet, M., Bonnell, J., Sullivan, D.J., Bonnet-Bidaud, J.-M. & Bunk, W.H. 1993, *Astrophys. J.*, **424**, 347.
Mayo, S., Wickramasinghe, D.T. & Whelan, J.A.J. 1980, *Mon. Not. R. astr. Soc.*, **193**, 793.
Mazeh, T., Kieboom, K. & Heise, J. 1986, *Mon. Not. R. astr. Soc.*, **221**, 513.
Mazeh, T., Tal, Y., Shaviv, G., Bruch, A. & Budell, R. 1985, *Astr. Astrophys.*, **149**, 470.
Mazeh, T., Latham, D.W., Mathieu, R.D. & Carney, B.W. 1990, in *Active Close Binaries*, ed. C. Ibanoglu, Kluwer, Dordrecht, p. 145.
McCarthy, P., Bowyer, S. & Clarke, J.T. 1986, *Astrophys. J.*, **311**, 873.
McClintock, J.E., Canizares, C.R. & Tarter, C.B. 1975, *Astrophys. J.*, **198**, 641.
McClintock, J.E., Petro, C.D., Remillard, R.A. & Ricker, G.R. 1983, *Astrophys. J.*, **266**, L27.
McCook, C.P. & Sion, E.M. 1984, *Villanova Univ. Obs. Contr.*, No. 3.
McDermott, P.N. & Taam, R.E. 1989, *Astrophys. J.*, **342**, 1019.
McDermott, P.N., Taam, R.E. & Ringwald, F.A. 1988, *Astrophys. J.*, **328**, 617.
McHardy, I.M., Pye, J.P., Fairall, A.P., Warner, B., Cropper, M. & Allen, S. 1984, *Mon. Not. R. astr. Soc.*, **210**, 663.
McHardy, I.M., Pye, J.P., Fairall, A.P. & Menzies, J.W. 1987, *Mon. Not. R. astr. Soc.*, **225**, 355.
McKee, C.F. 1990, *Astr. Soc. Pacific Conf. Ser.* **12**, 3.
McLaughlin, D.B. 1935, *Publ. astr. Soc. Pacific*, **8**, 145.
McLaughlin, D.B. 1937, *Pub. Astr. Obs. Univ. Michigan*, **6**, 107.
McLaughlin, D.B. 1938, *Pop. Astr.*, **46**, 373.
McLaughlin, D.B. 1939, *Pop. Astr.*, **47**, 410, 481, 538.
McLaughlin, D.B. 1941, *Pop. Astr.*, **49**, 292.
McLaughlin, D.B. 1942a, *Pop. Astr.*, **50**, 233.
McLaughlin, D.B. 1942b, *Pop. Astr.*, **52**, 109.
McLaughlin, D.B. 1943, *Pub. Astr. Obs. Univ. Michigan*, **8**, 149.
McLaughlin, D.B. 1944, *Astr. J.*, **51**, 20.
McLaughlin, D.B. 1945, *Publ. astr. Soc. Pacific*, **57**, 69.
McLaughlin, D.B. 1946, *Publ. astr. Soc. Pacific*, **58**, 46.
McLaughlin, D.B. 1953, *Astrophys. J.*, **117**, 279.
McLaughlin, D.B. 1960a, in *Stellar Atmospheres*, ed. J.L. Greenstein, Univ. Chicago Press, Chicago, p. 585.
McLaughlin, D.B. 1960b, *Astrophys. J.*, **131**, 739.
McMahan, R.K. 1989, *Astrophys. J.*, **336**, 409.
McNaught, R.H. 1986, *Inf. Bull. Var. Stars*, No. 2926.
Meggitt, S.M.A. & Wickramasinghe, D.T. 1982, *Mon. Not. R. astr. Soc.*, **198**, 71.
Meggitt, S.M.A. & Wickramasinghe, D.T. 1984, *Mon. Not. R. astr. Soc.*, **207**, 1.
Meggitt, S.M.A. & Wickramasinghe, D.T. 1989, *Mon. Not. R. astr. Soc.*, **236**, 31.
Meglicki, Z., Wickramasinghe, D. & Bicknell, G.V. 1993, *Mon. Not. R. astr. Soc.*, **264**, 691.
Meinel, A.B., Aveni, A.F. & Stockton, M.W. 1975, *Catalogue of Emission Lines in Astrophysical Objects*, Opt. Sci. Center, Tech. Report No. 27, Univ. Arizona, Tucson.
Meintjes, P.J. 1990, M.Sc. Thesis, Univ. Potchefstroom.
Meintjes, P.J., Raubenheimer, B.C., de Jager, O.C., Brink, C., Nel, H.I., North, A.R., van Urk, G. & Visser, B. 1992, *Astrophys. J.*, **401**, 325.
Meintjes, P.J., de Jager, O.C., Raubenheimer, B.C., Buckley, D.A.H., Koen, C., Nel, H.I. & North, A.R. 1994, *Astrophys. J.*, **434**, 292.
Meinunger, L. 1976, *Inf. Bull. Var. Stars*, No. 1168.
Meinunger, L. 1979, *Inf. Bull. Var. Stars*, No. 1677.
Melia, F. & Lamb, D.Q. 1987, *Astrophys. J.*, **321**, L139.

Melrose, D.B. & Dulk, G.A. 1982, *Astrophys. J.*, **259**, 844.
Mendelson, H., Leibowitz, E.M., Brosch, N. & Almoznino, E. 1992, *Int. Astr. Union Circ.*, No. 5509.
Mendez, R.H., Marino, R.F., Claria, J.J. & van Driel, W. 1985, *Rev. Mex. Astr. Astrophys.*, **10**, 187.
Mennickent, R. 1994, *Astr. Astrophys.*, **285**, 979.
Menzies, J.W., O'Donoghue, D. & Warner, B. 1986, *Astrophys. Sp. Sci.*, **122**, 73.
Mestel, L. 1952, *Mon. Not. R. astr. Soc.*, **112**, 598.
Mestel, L. 1965, in *Stellar Structure*, eds. L.H. Aller & D.B. McLaughlin, Univ. Chicago, Chicago, p. 297.
Mestel, L. 1968, *Mon. Not. R. astr. Soc.*, **138**, 359.
Mestel, L. 1984, *Lect. Notes Phys.*, **193**, 49.
Mestel, L. & Spruit, H.C. 1987, *Mon. Not. R. astr. Soc.*, **226**, 57.
Metz, K. 1982, *Inf. Bull. Var. Stars*, No. 2201.
Meyer, F. 1984, *Astr. Astrophys.*, **131**, 303.
Meyer, F. 1985, in *Recent Results on Cataclysmic Variables*, ed. W.R. Burke, ESA-SP 236, Paris, p. 83.
Meyer, F. & Meyer-Hofmeister, E. 1979, *Astr. Astrophys.*, **8**, 167.
Meyer, F. & Meyer-Hofmeister, E. 1981, *Astr. Astrophys.*, **104**, L10.
Meyer, F. & Meyer-Hofmeister, E. 1982, *Astr. Astrophys.*, **106**, 34.
Meyer, F. & Meyer-Hofmeister, E. 1983a, *Astr. Astrophys.*, **121**, 29.
Meyer, F. & Meyer-Hofmeister, E. 1983b, *Astr. Astrophys.*, **128**, 420.
Meyer, F. & Meyer-Hofmeister, E. 1984, *Astr. Astrophys.*, **132**, 143.
Meyer, F. & Meyer-Hofmeister, E. 1989, *Astr. Astrophys.*, **221**, 36.
Meyer, F. & Meyer-Hofmeister, E. 1994, *Astr. Astrophys.*, **288**, 175.
Meyer, F., Duschl, W.J., Frank, J. & Meyer-Hofmeister, E. (eds.) 1989, *Theory of Accretion Disks*, Kluwer, Dordrecht.
Meyer-Hofmeister, E. 1987, *Astr. Astrophys.*, **175**, 113.
Meyer-Hofmeister, E. & Meyer, F. 1988, *Astr. Astrophys.*, **194**, 135.
Meyer-Hoffmeister, E. & Ritter, H. 1993, in *The Realm of Interacting Binary Stars*, eds. J. Sahade, G. McCluskey & Y. Kondo, Dordrecht, Kluwer, p. 143.
Middleditch, J. 1982, *Astrophys. J.*, **257**, L71.
Middleditch, J. & Cordova, F.A. 1982, *Astrophys. J.*, **255**, 585.
Middleditch, J. & Nelson, J. 1979, *Bull. Amer. astr. Soc.*, **11**, 664.
Middleditch, J. & Nelson, J. 1980, *Bull. Amer. astr. Soc.*, **12**, 848.
Middleditch, J., Nelson, J.E. & Chanan, G.A. 1978, *Bull. Amer. astr. Soc.*, **9**, 557.
Middleditch, J., Imamura, J.N., Wolff, M.T. & Steiman-Cameron, T.Y. 1991, *Astrophys. J.*, **382**, 315.
Mihalas, D. 1978, *Stellar Atmospheres*, 2nd ed., Freeman, San Francisco.
Mikolajewska, J., Friedjung, M., Kenyon, S.J. & Viotti, R. (eds.) 1988, *The Symbiotic Phenomenon, Int. Astr. Union Colloq.* No. 103, Kluwer, Dordrecht.
Miles, H.G. 1976, *J. Brit. astr. Assoc.*, **86**, 335.
Milgrom, M. & Salpeter, E.E. 1975, *Astrophys. J.*, **196**, 583.
Miller, H.R. 1982, *Mon. Not. R. astr. Soc.*, **201**, 21P.
Mineshige, S. 1986, *Publ. astr. Soc. Japan*, **38**, 831.
Mineshige, S. 1987, *Astrophys. Sp. Sci.*, **130**, 331.
Mineshige, S. 1988a, *Astr. Astrophys.*, **190**, 72.
Mineshige, S. 1988b, *Astrophys. J.*, **335**, 881.
Mineshige, S. 1991, *Mon. Not. R. astr. Soc.*, **250**, 253.
Mineshige, S. & Osaki, Y. 1983, *Publ. astr. Soc. Japan*, **35**, 377.
Mineshige, S. & Osaki, Y. 1985, *Publ. astr. Soc. Japan*, **37**, 1.
Mineshige, S. &. Shields, G.A. 1990, *Astrophys. J.*, **351**, 47.
Mineshige, S. & Wood, J. 1989, *Mon. Not. R. astr. Soc.*, **241**, 259.
Mineshige, S. & Wood, J.H. 1990, *Mon. Not. R. astr. Soc.*, **247**, 43.
Minkowski, R. 1939, *Publ. astr. Soc. Pacific*, **51**, 54.
Mittaz, J.P.D., Rosen, S.R., Mason, K.O. & Howell, S.B. 1992, *Mon. Not. R. astr. Soc.*, **258**, 277.
Moffat, A.F.J. & Shara, M.M. 1984, *Publ. astr. Soc. Pacific*, **96**, 552.
Moffat, A.F.J. & Vogt, N. 1975, *Astr. Astrophys. Suppl.*, **20**, 85.
Moffet, T. & Barnes, T.G. 1974, *Astrophys. J.*, **194**, 141.
Molnar, L.A. & Kobulnicky, H.A. 1992, *Astrophys. J.*, **392**, 678.

Molteni, D., Belvedere, G. & Lanzafame, G. 1991, *Mon. Not. R. astr. Soc.*, **249**, 748.
Morris, S.C., Schmidt, G.D., Liebert, J., Stocke, J., Gioia, I. & Maccacaro, T. 1987, *Astrophys. J.*, **314**, 641.
Motch, C. 1981, *Astr. Astrophys.*, **100**, 277.
Motch, C. & Pakull, M.W. 1981, *Astr. Astrophys.*, **101**, L29.
Motch, C., Hassinger, G. & Pietsch, W. 1994, *Astr. Astrophys.*, **284**, 827.
Motch, C., van Paradijs, J. Pedersen, H., Ilovaisky, S.A. & Chevalier, C. 1982, *Astr. Astrophys.*, **110**, 316.
Motch, C., Belloni, T., Buckley, D., Gottwald, M., Hasinger, G., Ilovaisky, S.A., Pakull, M.W., Pietsch, W., Reinsch, K., Remillard, R.A., Schmitt, J.H.M.M. & Trümper, J. 1991. *Astr. Astrophys.*, **246**, L24.
Mouchet, M. 1983, *Int. Astr. Union Colloq. No. 72*, p. 173.
Mouchet, M. 1993, in *White Dwarfs: Advances in Observations and Theory*, ed. M.A. Barstow, Kluwer, Dordrecht, p. 411.
Mouchet, M. & Bonnet-Bidaud, J.M. 1984, *Fourth European IUE Conference*, ESA SP-218, Paris, p. 431.
Mouchet, M., Bonnet-Bidaud, J.M., Ilovaisky, S.A. & Chevalier, C. 1981, *Astr. Astrophys.*, **102**, 31.
Mouchet, M., Van Amerongen, S.F., Bonnet-Bidaud, J.M. & Osborne, J.P. 1987, *Astrophys. Sp. Sci.*, **131**, 613.
Mouchet, M., Bonnet-Bidaud, J.M., Buckley, D.A.H. & Tuohy, I.R. 1991, *Astr. Astrophys.*, **250**, 99.
Mukaï, K. 1988, *Mon. Not. R. astr. Soc.*, **232**, 175.
Mukai, K. & Charles, P.A. 1985, *Mon. Not. R. astr. Soc.*, **212**, 609.
Mukai, K. & Charles, P.A. 1986, *Mon. Not. R. astr. Soc.*, **222**, 1P.
Mukai, K. & Charles, P.A. 1987, *Mon. Not. R. astr. Soc.*, **226**, 209.
Mukai, K. & Corbet, R.H.D. 1987, *Publ. astr. Soc. Pacific*, **99**, 149.
Mukai, K. & Corbet, R.H.D. 1991, *Astrophys. J.*, **378**, 701.
Mukai, K. & Hellier, C. 1992, *Astrophys. J.*, **391**, 295.
Mukai, K. & Shiokawa, K. 1993, *Astrophys. J.*, **418**, 863.
Mukai, K., Charles, P.A. & Smale, A.P. 1988, *Astr. Astrophys.*, **194**, 153.
Mukai, K., Bonnet-Bidaud, J.M., Bowyer, S., Charles, P.A., Chiapetti, L., Clarke, J.T., Corbet, R.H.D., Henry, J.P., Hill, G.J., Kahn, S.M., Marashi, L., Osborne, J., Treves, A., van der Klis, M., van Paradijs, J. & Vrtilek, S.D. 1985, *Sp. Sci. Rev.*, **40**, 151.
Mukai, K., Bonnet-Bidaud, J.M., Charles, P.A., Corbet, R.H.D., Maraschi, L., Osborne, J., Smale, A.P., Treves, A., van der Klis, M. & van Paradjis, J. 1986, *Mon. Not. R. astr. Soc.*, **221**, 839.
Mukai, K., Mason, K.O., Howell, S.B., Allington-Smith, J., Callanan, P.J., Charles, P.A., Hassall, B.J.M., Machin, G., Taylor, N., Smale, A.P. & van Paradijs, J. 1990, *Mon. Not. R. astr. Soc.* **245**, 385.
Müller, G. & Hartwig, E. 1918, *Geschichte und Literatur des Lichtwechsels der bis Ende 1915 als sicher veranderlich erkannten Sterne*, Poeschel & Trepte, Leipzig.
Mumford, G.S. 1964a, *Astrophys. J.*, **139**, 476.
Mumford, G.S. 1964b, *Publ. astr. Soc. Pacific*, **76**, 57.
Mumford, G.S. 1966, *Astrophys. J.*, **146**, 411.
Mumford, G.S. 1967a, *Publ. astr. Soc. Pacific*, **79**, 283.
Mumford, G.S. 1967b, *Astr. Astrophys. Suppl.*, **15**, 1.
Mumford, G.S. 1969, *Astrophys. J.*, **156**, 125.
Mumford, G.S. 1971, *Astrophys. J.*, **165**, 369.
Mumford, G.S. 1980, *Astr. J.*, **85**, 748.
Mumford, G.S. & Krzeminski, W. 1969, *Astrophys. J. Suppl.*, **18**, 429.
Munari, U. 1992, *Astr. Astrophys.*, **257**, 163.
Munari, U. & Renzini, A. 1992, *Astr. Soc. Pacific Conf. Ser.*, **30**, 339.
Munari, U. & Whitelock, P.A. 1989, *Mon. Not. R. astr. Soc.*, **239**, 273.
Munari, U., Whitelock, P.A., Gilmore, A.C., Blanco, C., Massone, G. & Schmeer, P. 1992, *Astr. J.*, **104**, 262.
Mustel, E.R. & Baranova, L.I. 1965, *Sov. Astr. J.*, **9**, 31.
Mustel, E.R. & Boyarchuk, A.A. 1970, *Astrophys. Sp. Sci.*, **6**, 183.
Narayan, R. & Popham, R. 1993, *Nature*, **362**, 820.

Nather, R.E. 1978, *Publ. astr. Soc. Pacific*, **90**, 477.
Nather, R.E. 1985, in *Interacting Binaries*, eds. P.P. Eggleton & J.E. Pringle, Reidel, Dordrecht, p. 349.
Nather, R.E. 1989, *Int. Astr. Union Colloq. No. 114*, p. 109.
Nather, R.E. & Robinson, E.L. 1974, *Astrophys. J.*, **190**, 637.
Nather, R.E. & Stover, R.J. 1978, *Int. Astr. Union Circ.*, No. 3311.
Nather, R.E. & Warner, B. 1969, *Mon. Not. R. astr. Soc.*, **143**, 145.
Nather, R.E., Robinson, E.L. & Stover, R.J. 1981, *Astrophys. J.*, **244**, 269.
Nauenberg, M. 1972, *Astrophys. J.*, **175**, 417.
Naylor, T. 1989, *Mon. Not. R. astr. Soc.*, **238**, 587.
Naylor, T., Smale, A.P. & van Paradijs, J. 1990, *Mon. Not. R. astr. Soc.*, **245**, 385.
Naylor, T., Charles, P.A., Hassall, B.J.M., Bath, G.T., Berriman, G., Warner, B., Bailey, J. & Reinsch, K. 1987, *Mon. Not. R. astr. Soc.*, **229**, 183.
Naylor, T., Bath, G.T., Charles, P.A., Hassall, B.J.M., Sonneborn, G., van der Woerd, H. & van Paradijs, J. 1988, *Mon. Not. R. astr. Soc.*, **231**, 237.
Naylor, T., Allington-Smith, J., Callanan, P.J., Charles, P.A., Hassall, B.J.M., Machin, G., Mason, K.O., Smale, A.P. & van Paradijs, J. 1989, *Mon. Not. R. astr. Soc.*, **241**, 25P.
Naylor, T., Charles, P.A., Mukai, K. & Evans, A. 1992, *Mon. Not. R. astr. Soc.*, **258**, 449.
Neece, G.D. 1984, *Astrophys. J.*, **277**, 738.
Neill, A.E. 1992, private communication to O.C. de Jager.
Nelson, M.R. 1975, *Astrophys. J.*, **196**, L113.
Nelson, M.R. 1976, *Astrophys. J.*, **209**, 168.
Nelson, M.R. & Olson, E.C. 1976, *Astrophys. J.*, **207**, 195.
Nelson, R.F. & Spencer, R.E. 1988, *Mon. Not. R. astr. Soc.*, **234**, 1105.
Nelson, L.A., Chau, W.Y. & Rosenblum, A. 1985, *Astrophys. J.*, **299**, 658.
Nelson, L.A., Rappaport, S. & Joss, P.C. 1986, *Astrophys. J.*, **304**, 231.
Nevo, I. & Sadeh, D. 1976, *Mon. Not. R. astr. Soc.*, **177**, 167.
Nevo, I. & Sadeh, D. 1978, *Mon. Not. R. astr. Soc.*, **182**, 595.
Nofar, I., Shaviv, G. & Starrfield, S. 1991, *Astrophys. J.*, **369**, 440.
Norton, A.J. 1993, *Mon. Not. R. astr. Soc.*, **265**, 316.
Norton, A.J. & Watson, M.G. 1989a, *Mon. Not. R. astr. Soc.*, **237**, 715.
Norton, A.J. & Watson, M.G. 1989b, *Mon. Not. R. astr. Soc.*, **237**, 853.
Norton, A.J., Watson, M.G. & King, A.R. 1988, *Mon. Not. R. astr. Soc.*, **231**, 783.
Norton, A.J., Watson, M.G. & King, A.R. 1991, *Lect. Notes Phys.*, **385**, 155.
Norton, A.J., Watson, M.G., King, A.R., McHardy, I.M. & Lehto, H. 1990, in *Accretion-Powered Compact Binaries*, ed. C.W. Mauche, Cambridge University Press, Cambridge, p. 209.
Norton, A.J., Watson, M.G., King, A.R., Lehto, H.J. & McHardy, I.M. 1992a, *Mon. Not. R. astr. Soc.*, **254**, 705.
Norton, A.J., McHardy, I.M., Lehto, H.J. & Watson, M.G. 1992b, *Mon. Not. R. astr. Soc.*, **258**, 697.
Nousek, J.A. & Pravdo, S.H. 1983, *Astrophys. J.*, **266**, L39.
Nousek, J.A., Takalo, L.O., Schmidt, G.D., Tapia, S., Hill, G.J., Bond, H.E., Grauer, A.D., Stern, R.A. & Agrawal, P.C. 1984, *Astrophys. J.*, **277**, 682.
Nussbaumer, H., Schild, H., Schmid, H.M. & Vogel, M. 1988, *Astr. Astrophys.*, **198**, 179.
O'Brien, T.J., Bode, M.F. & Kahn, F.D. 1992, *Mon. Not. R. astr. Soc.*, **255**, 683.
O'Donoghue, D. 1985, in *Proc. Ninth N. American Workshop on Cataclysmic Var.*, ed. P. Szkody, Univ. Washington, Seattle, p. 98.
O'Donoghue, D. 1986, *Mon. Not. R. astr. Soc.*, **220**, 23P.
O'Donoghue, D. 1987, *Astrophys. Sp. Sci.*, **136**, 247.
O'Donoghue, D. 1990, *Mon. Not. R. astr. Soc.*, **246**, 29.
O'Donoghue, D. & Kilkenny, D. 1989, *Mon. Not. R. astr. Soc.*, **236**, 319.
O'Donoghue, D. & Soltynski, M.G. 1992, *Mon. Not. R. astr. Soc.*, **254**, 9.
O'Donoghue, D., Fairall, A.P. & Warner, B. 1987, *Mon. Not. R. astr. Soc.*, **225**, 43.
O'Donoghue, D., Menzies, J. & Hill, P. 1987a, *Int. Astr. Union Colloq. No. 95*, p. 693.
O'Donoghue, D., Menzies, J. & Hill, P.W. 1987b, *Mon. Not. R. astr. Soc.*, **227**, 295.
O'Donoghue, D., Warner, B., Wargau, W. & Grauer, A.D. 1989, *Mon. Not. R. astr. Soc.*, **240**, 41.
O'Donoghue, D., Wargau, W., Warner, B., Kilkenny, D., Martinez, P., Kanaan, A., Kepler, S.O.,

Henry, G., Winget, D.E., Clemens, J.C. & Grauer, A.D. 1990, *Mon. Not. R. astr. Soc.*, **245**, 140.
O'Donoghue, D., Chen, A., Marang, F., Mittaz, J.P.D., Winkler, H. & Warner, B. 1991, *Mon. Not. R. astr. Soc.*, **250**, 363.
O'Donoghue, D., Mason, K.O., Chen, A., Hassall, B.J.M. & Watson, M.G. 1993, *Mon. Not. R. astr. Soc.*, **265**, 545.
O'Donoghue, D., Kilkenny, D., Chen, A.-L., Stobie, R.S., Koen, C., Warner, B. & Lawson, W.A. 1994, *Mon. Not. R. astr. Soc.*, **271**, 910.
Oestreicher, R. & Seifert, W. 1988, *Astr. Astrophys.*, **190**, L29.
Oestreicher, R., Seifert, W., Wunner, G. & Ruder, H. 1990, *Astrophys. J.*, **250**, 324.
Ögelman, H. 1987, *Astr. Astrophys.*, **172**, 79.
Ögelman, H. 1990, *Int. Astr. Union Colloq. No. 122*, p. 148.
Ögelman, H., Krautter, J. & Beuermann, K. 1987, *Astr. Astrophys.*, **177**, 110.
Ögelman, H., Orio, M., Krautter, J. & Starrfield, S. 1993, *Nature*, **361**, 331.
Okazaki, A., Kitamura, M. & Yamasaki, A. 1982, *Publ. astr. Soc. Pacific*, **94**, 162.
Oke, J.B. & Wade, R.A. 1982, *Astr. J.*, **87**, 670.
Okuda, T., Ono, K., Tabata, M. & Mineshige, S. 1992, *Mon. Not. R. astr. Soc.*, **254**, 427.
Oort, J.H. 1946, *Mon. Not. R. astr. Soc.*, **106**, 159.
Oosterhoff, P.Th. 1941, *Ann. Sterrew. Leiden,* **17**, part 4.
Oppenheimer, B. 1994, *J. Amer. Assoc. Var. Star Obs.*, in press.
Orio, M. 1993, *Astr. Astrophys.*, **274**, L41.
Orio, M., Della Valle, M., Massone, G. & Ögelman, H. 1994, *Astr. Astrophys.*, **289**, L11.
Orszay, S.A. & Kells, L.C. 1980, *J. Fluid Mech.*, **96**, 159.
Osaki, Y. 1970, *Astrophys. J.*, **162**, 621.
Osaki, Y. 1974, *Pub. astr. Soc. Japan,* **26**, 429.
Osaki, Y. 1985, *Astr. Astrophys.*, **144**, 369.
Osaki, Y. 1989a, *Pub. astr. Soc. Japan,* **41**, 1005.
Osaki, Y. 1989b, in *Theory of Accretion Disks*, eds. F. Meyer, W.J. Duschl, J. Frank & E. Meyer-Hofmeister, Kluwer, Dordrecht, p. 183.
Osaki, Y. 1994, in *Theory of Accretion Discs II*, eds. W.J. Duschl *et al.*, Kluwer, Dordrecht, p. 93.
Osaki, Y., Hirose, M. & Ichikawa, S. 1993 in *Accretion Disks in Compact Stellar Systems*, ed. J.C. Wheeler, World Sci. Publ., Singapore, p. 272.
Osborne, J.P. 1986, *Int. Astr. Union Colloq. No. 93*, p. 207.
Osborne, J.P. 1988, *Mem. astr. Soc. Italy*, **59**, 117.
Osborne, J.P. 1990, in *Accretion-Powered Compact Binaries*, ed. C.W. Mauche, Cambridge University Press, Cambridge, p. 215.
Osborne, J.P. & Mukai, K. 1989, *Mon. Not. R. astr. Soc.*, **238**, 1233.
Osborne, J.P., Cropper, M.S. & Cristiani, S. 1986, *Astrophys. Sp. Sci.*, **130**, 643.
Osborne, J.P., Maraschi, L., Beuermann, K., Bonnet-Bidaud, J.-M., Charles, P.A., Chiapetti, L., Motch, C., Mouchet, M., Tanzi, E.G., Treves, A. & Mason, K.O. 1984, in *X-Ray Astronomy '84*, eds. R. Giaconi & M. Oda, Inst. Sp. & Astronautical Sci., Japan, p. 59.
Osborne, J., Rosen, R., Mason, K.O., & Beuermann, K. 1985, *Sp. Sci. Rev.*, **40**, 143.
Osborne, J.P., Bonnet-Bidaud, J.-M., Bowyer, S., Charles, P.A., Chiapetti, L., Clarke, J.T., Henry, J.P., Hill, G.J., Kahn, S., Maraschi, L., Mukai, K., Treves, A. & Vrtilek, S. 1986, *Mon. Not. R. astr. Soc.*, **221**, 823.
Osborne, J.P., Beuermann, K., Charles, P., Maraschi, L., Mukai, K. & Treves, A. 1987, *Astrophys. J.*, **315**, L123.
Osborne, J.P., Giommi, P., Angelni, L., Tagliaferri, G. & Stella, L. 1988, *Astrophys. J.*, **328**, L45.
Osborne, J.P., Beardmore, A.P., Wheatley, P.J., Hakala, P., Mason, K.O., Hassall, B.J.M. & King, A.R. 1993, in *Cataclysmic Variables and Related Physics*, eds. O. Regev & G. Shaviv, Inst. Phys. Publ., Bristol, p. 303.
Osborne, J.P., Beardmore, A.P., Wheatley, P.J., Hakala, P., Watson, M.G., Mason, K.O., Hassall, B.J.M. & King, A.R. 1994, *Mon. Not. R. astr. Soc.*, **270**, 650.
Oskanyan, A.V. 1983, *Inf. Bull. Var. Stars*, No. 2349.
Osterbrock, D.E. 1989, *Astrophysics of Gaseous Nebulae and Active Galactic Nuclei*, University Science Books, California.
Ostriker, J.P. 1971, *Ann. Rev. Astr. Astrophys.*, **9**, 353.

Ostriker, J.P. 1976, *Int. Astr. Union Symp. No. 73*, p. 206.
Ostriker, J.P. & Hesser, J.E. 1968, *Astrophys. J.*, **153**, L151.
Pacharintanakul, P. & Katz, J.I. 1980, *Astrophys. J.*, **238**, 985.
Paczynski, B. 1963, *Publ. astr. Soc. Pacific*, **75**, 278.
Paczynski, B. 1965a, *Acta Astr.*, **15**, 89.
Paczynski, B. 1965b, *Acta Astr.*, **15**, 305.
Paczynski, B. 1965c, *Acta Astr.*, **15**, 197.
Paczynski, B. 1967, *Acta Astr.*, **17**, 287.
Paczynski, B. 1971a, *Acta Astr.*, **21**, 417.
Paczynski, B. 1971b, *Ann. Rev. Astr. Astrophys.*, **9**, 183.
Paczynski, B. 1976, *Int. Astr. Union Symp. No. 73*, p. 75.
Paczynski, B. 1977, *Astrophys. J.*, **216**, 822.
Paczynski, B. 1978, in *Nonstationary Evolution of Close Binaries*, ed. A. Zytkow, Polish Scientific Publ., Warsaw, p. 89.
Paczynski, B. 1991, *Astrophys. J.*, **370**, 597.
Paczynski, B. & Schwarzenberg-Czerny, A. 1980, *Acta Astr.*, **30**, 127.
Paczynski, B. & Sienkiewicz, R. 1972, *Acta Astr.*, **22**, 73.
Paczynski, B. & Sienkiewicz, R. 1981, *Astrophys. J.*, **248**, L27.
Paczynski, B. & Sienkiewicz, R. 1983, *Astrophys. J.*, **268**, 825.
Paczynski, B., Ziolkowski, J. & Zytkow, A. 1969, in *Mass Loss from Stars*, ed. M. Hack, Reidel, Dordrecht, p. 237.
Pajdos, G. & Zola, S. 1992, *Int. Astr. Union Symp. No. 151*, p. 441.
Pakull, J. & Beuermann, K. 1987, *Astrophys. Sp. Sci.*, **131**, 641.
Pakull, M.W., Beuermann, K., van der Klis, M. & van Paradijs, J. 1988, *Astr. Astrophys.*, **203**, L27.
Pallavicini, R., Golub, L., Rosner, R., Vaiana, G.S., Ayres, T. & Linsky, J.L. 1981, *Astrophys. J.*, **248**, 279.
Panek, R.J. 1980, *Astrophys. J.*, **241**, 1077.
Panek, R.J. & Eaton, J.A. 1982, *Astrophys. J.*, **258**, 572.
Panek, R. & Holm, A.V. 1984, *Astrophys. J.*, **277**, 700.
Panek, R.J. & Howell, S.B. 1980, *Astr. J.*, **85**, 560.
Papaloizou, J.C.B. & Bath, G.T. 1975, *Mon. Not. R. astr. Soc.*, **172**, 339.
Papaloizou, J. & Pringle, J.E. 1977, *Mon. Not. R. astr. Soc.*, **181**, 441.
Papaloizou, J. & Pringle, J.E. 1978a, *Astr. Astrophys.*, **70**, L65.
Papaloizou, J. & Pringle, J. 1978b, *Mon. Not. R. astr. Soc.*, **182**, 423.
Papaloizou, J. & Pringle, J.E. 1979, *Mon. Not. R. astr. Soc.*, **189**, 293.
Papaloizou, J.C.B. & Pringle, J.E. 1983, *Mon. Not. R. astr. Soc.*, **202**, 1181.
Papaloizou, J.C.B. & Pringle, J. 1984, *Mon. Not. R. astr. Soc.*, **208**, 721.
Papaloizou, J.C.B. & Pringle, J. 1985, *Mon. Not. R. astr. Soc.*, **213**, 799.
Papaloizou, J. & Stanley, C.Q.G. 1986, *Mon. Not. R. astr. Soc.*, **220**, 593.
Papaloizou, J., Faulkner, J. & Lin, D.N.C. 1983, *Mon. Not. R. astr. Soc.*, **205**, 487.
Papaloizou, J., Pringle, J.E. & MacDonald, J. 1982, *Mon. Not. R. astr. Soc.*, **108**, 215.
Paresce, F. 1993, *Sp. Tel. Si. Inst. Newsletter*, **10**, 6.
Paresce, F., De Marchi, G. & Ferraro, F.R. 1992, *Nature*, **360**, 46.
Parker, E.N. 1975, *Astrophys. J.*, **198**, 205.
Parker, E.N. 1979, *Cosmical Fluid Dynamics*, Oxford Univ. Press, Oxford.
Parkhurst, J.A. 1897, *Pop. Astr.*, **5**, 164.
Parmar, A.N., White, N.E., Stella, L., Izzo, C. & Ferri, P. 1989, *Astrophys. J.*, **338**, 359.
Patterson, J. 1978, *Astrophys. J.*, **225**, 954.
Patterson, J. 1979a, *Astrophys. J.*, **231**, 789.
Patterson, J. 1979b, *Astrophys. J.*, **234**, 978.
Patterson, J. 1979c, *Astrophys. J.*, **233**, L13.
Patterson, J. 1979d, *Astr. J.*, **84**, 804.
Patterson, J. 1979e, Ph.D. Thesis, Univ. Texas.
Patterson, J. 1979f, *Publ. astr. Soc. Pacific*, **91**, 487.
Patterson, J. 1980, *Astrophys. J.*, **241**, 235.
Patterson, J. 1981, *Astrophys. J. Suppl.*, **45**, 517.

Patterson, J. 1984. *Astrophys. J. Suppl.*, **54**, 443.
Patterson, J. 1990a, in *Accretion-Powered Compact Binaries*, ed. C.W. Mauche, Cambridge University Press, Cambridge, p. 203.
Patterson, J. 1990b, *Imaging X-Ray Astronomy*, ed. M. Elvis, Cambridge University Press, Cambridge, p. 89.
Patterson, J. 1991, *Publ. astr. Soc. Pacific*, **103**, 1149.
Patterson, J. 1994, *Publ. astr. Soc. Pacific*, **106**, 209.
Patterson, J. & Halpern, J.P. 1990, *Astrophys. J.*, **361**, 173.
Patterson, J. & Moulden, M. 1993, *Publ. astr. Soc. Pacific*, **105**, 779.
Patterson, J. & Price, C. 1980, *Int. Astr. Union Circ.*, No. 3511.
Patterson, J. & Price, C. 1981a, *Publ. astr. Soc. Pacific*, **93**, 71.
Patterson, J. & Price, C. 1981b, *Astrophys. J.*, **243**, L83.
Patterson, J. & Raymond, J.C. 1985a, *Astrophys. J.*, **292**, 535.
Patterson, J. & Raymond, J.C. 1985b, *Astrophys. J.*, **292**, 550.
Patterson, J. & Richman, H. 1991, *Publ. astr. Soc. Pacific*, **103**, 735.
Patterson, J. & Skillman, D.R. 1994. In preparation.
Patterson, J. & Steiner, J.E. 1983, *Astrophys. J.*, **264**, L61.
Patterson, J. & Szkody, P. 1993, *Publ. astr. Soc. Pacific*, **105**, 1116.
Patterson, J. & Thomas, G. 1993, *Publ. astr. Soc. Pacific*, **105**, 59.
Patterson, J., Beuermann, K. & Africano, J. 1988, *Bull. Amer. astr. Soc.*, **20**, 1099.
Patterson, J., Halpern, J. & Shambrook, A. 1993, *Astrophys. J.*, **419**, 83.
Patterson, J., Jablonski, F.J., Koen, M.C. & O'Donoghue, D. 1994, in preparation.
Patterson, J., Robinson, E.L. & Kiplinger, A.L. 1978, *Astrophys. J.*, **226**, L137.
Patterson, J., Robinson, E.L. & Nather, R.E. 1977, *Astrophys. J.*, **214**, 144.
Patterson, J., Robinson, E.L. & Nather, R.E. 1978, *Astrophys. J.*, **224**, 570.
Patterson, J., Williams, G. & Hiltner, W.A. 1981, *Astrophys. J.*, **245**, 618.
Patterson, J., Nather, R.E., Robinson, E.L. & Handler, F. 1979, *Astrophys. J.*, **232**, 819.
Patterson, J., Branch, D., Chincarini, G. & Robinson, E.L. 1980, *Astrophys. J.*, **240**, L133.
Patterson, J., McGraw, J.T., Coleman, L. & Africano, J.L. 1981, *Astrophys. J.*, **248**, 1067.
Patterson, J., Schwartz, D.A., Bradt, H., Remillard, R., McHardy, I., Pye, J.P., Williams, G., Fesen, R.A. & Szkody, P. 1982, *Bull. Amer. astr. Soc.*, **14**, 618.
Patterson, J., Beuermann, K., Lamb, D.Q., Fabbiano, G., Raymond, J.C., Swank, J. & White, N.E. 1984, *Astrophys. J.*, **279**, 785.
Patterson, J. *et al.* 1985, see Sion (1987).
Patterson, J., Sterner, E., Halpern, J.P. & Raymond, J.C. 1992a, *Astrophys. J.*, **384**, 234.
Patterson, J., Schwartz, D.A., Pye, J.P., Blair, W.P., Williams, G.A. & Caillault, J.-P. 1992b, *Astrophys. J.*, **392**, 233.
Patterson, J., Bond, H.E., Grauer, A.D., Shafter, A.W. & Mattei, J.A. 1993a, *Publ. astr. Soc. Pacific*, **105**, 69.
Patterson, J., Thomas, G., Skillman, D.R. & Diaz, M. 1993b, *Astrophys. J. Suppl.*, **86**, 235.
Pavelin, P.E., Spencer, R.E. & Davis, R.J. 1994, *Mon. Not. R. astr. Soc.*, **269**, 779.
Pavlenko, E.P. & Pelt, J. 1991, *Astrofiz.*, **34**, 169.
Payne-Gaposchkin, C. 1941, in *XIII Colloq. Int. Astrophys: Novae & White Dwarfs*, ed. A.J. Shaler, Hermann, Paris, p. 69.
Payne-Gaposchkin, C. 1957, *The Galactic Novae*, North-Holland, Amsterdam.
Payne-Gaposchkin, C. 1977, *Astr. J.*, **82**, 665.
Payne-Gaposchkin, C. & Gaposchkin, S. 1938, *Variable Stars*, Harv. Obs. Mono. No. 5, Cambridge, Mass.
Peacock, T., Cropper, M., Bailey, J., Hough, J.H. & Wickramasinghe, D.T. 1992, *Mon. Not. R. astr. Soc.*, **259**, 583.
Peel, M. 1985, *J. Amer. Assoc. Var. Star Obs.*, **14**, 8.
Peimbert, M. & Sarmiento, A. 1984, *Astr. Express*, **1**, 97.
Penning, W.R. 1985, *Astrophys. J.*, **289**, 300.
Penning, W.R., Schmidt, G.D. & Liebert, J. 1986, *Astrophys. J.*, **301**, 881.
Penning, R.W., Ferguson, D.H., McGraw, J.T., Liebert, J. & Green, R.F. 1984, *Astrophys. J.*, **276**, 233.

Pennington, R. 1985, in *Interacting Binary Stars*, eds. J.E. Pringle & R.A. Wade, Cambridge University Press, Cambridge p. 197.
Persson, S.E. 1988, *Publ. astr. Soc. Pacific*, **100**, 710.
Peters, C.H.F. 1865, *Astr. Nach.*, **65**, 55.
Petersen, A.C. 1848, *Mon. Not. R. astr. Soc.*, **8**, 156.
Petit, E. 1946, *Publ. astr. Soc. Pacific*, **58**, 152, 213, 255.
Petit, M. 1961, Asiago Contr. **119**, 31.
Petit, M. 1987, *Variable Stars*, Wiley, Chichester.
Petitjean, P., Boisson, C. & Péquignot, D. 1990, *Astr. Astrophys.*, **240**, 433.
Petrochenko, L.N. & Shugarov, S. Yu. 1982, *Astr. Tsirk.*, No. 1230, p. 3.
Petterson, J.A. 1977a, *Astrophys. J.*, **214**, 550.
Petterson, J.A. 1977b, *Astrophys. J.*, **216**, 827.
Petterson, J. 1980, *Astrophys. J.*, **241**, 247.
Pfau, W. 1976, *Astr. Astrophys.*, **50**, 113.
Piccioni, A., Guarnieri, A., Bartolini, A. & Giovannelli, F. 1984, *Acta Astr.*, **34**, 473.
Piché, F. & Szkody, P. 1989, *Astr. J.*, **98**, 2225.
Pickles, A.J. & Visvanathan, N. 1983, *Mon. Not. R. astr. Soc.*, **204**, 463.
Pietsch, W., Voges, W., Kendziorra, E. & Pakull, M. 1987, *Astrophys. Sp. Sci.*, **130**, 281.
Piirola, V. 1988, in *Polarized Radiation of Circumstellar Origin*, eds. G.V. Coyne *et al.*, Vatican Obs., Vatican, p. 261.
Piirola, V., Coyne, G.V. & Reiz, A. 1990, *Astr. Astrophys.*, **235**, 245.
Piirola, V., Hakala, P. & Coyne, G.V. 1993, *Astrophys. J.*, **410**, L107.
Piirola, V., Reiz, A. & Coyne, G.V. 1987a, *Astr. Astrophys.*, **186**, 120.
Piirola, V., Reiz, A. & Coyne, G.V. 1987b, *Astrophys. Sp. Sci.*, **130**, 203.
Piirola, V., Reiz, A. & Coyne, G.V. 1987c, *Astr. Astrophys.*, **185**, 189.
Piirola, V., Vilhu, O., Kyröläinen, J., Shakhovskoy, N.M. & Efimov, Y.S. 1985, *Proc. ESA Workshop Recent Results on Cataclysmic Variables*, Bamberg, ESA SP-236, Paris, p. 245.
Piirola, V., Coyne, G.V., Larsson, S., Takalo, L. & Vilhu, O. 1993, *Ann. Israel Phys. Soc.*, **10**, 306.
Piirola, V., Coyne, G.V., Takalo, L., Larsson, S. & Vilhu, O. 1994, *Astr. Astrophys.*, **283**, 163.
Pinto, G. & Rosino, L. 1959, *Contr. Astrof. Asiago*, No. 106.
Piotrowski, S.L. 1975, *Acta Astr.*, **25**, 21.
Pismis, P. 1972, *Bol. Obs. Tonant. Tac.*, **6**, 197.
Plaut, L. 1965, in *Galactic Structure*, eds. A. Blaauw & M. Schmidt, Univ. Chicago Press, Chicago, p. 311.
Plavec, M. 1968, *Adv. Astr. Astrophys.*, **6**, 201.
Plavec, M. & Kratochvil, P. 1964, *Bull. Astr. Czech.*, **15**, 165.
Podsiadlowski, Ph. 1991, *Nature*, **350**, 136.
Pogson, N. 1856, *Mon. Not. R. astr. Soc.*, **16**, 185.
Pogson, N. 1857, *Mon. Not. R. astr. Soc.*, **17**, 200.
Pogson, N. 1860, *Mon. Not. R. astr. Soc.*, **21**, 32.
Pojmanski, G. 1986, *Acta Astr.*, **36**, 69.
Polidan, R.S. & Carone, T.E. 1987, *Astrophys. Sp. Sci.*, **130**, 235.
Polidan, R.S. & Holberg, J.B. 1984, *Nature*, **309**, 528.
Polidan, R.S. & Holberg, J.B. 1987, *Mon. Not. R. astr. Soc.*, **225**, 131.
Polidan, R.S., Mauche, C.W. & Wade, R.A. 1990, *Astrophys. J.*, **356**, 211.
Politano, M. & Webbink, R.F. 1989, *Int. Astr. Union Colloq. No. 114*, p. 440.
Politano, M. & Webbink, R.F. 1990, *Int. Astr. Union Colloq. No. 122*, p. 392.
Politano, M., Livio, M., Truran, J.W. & Webbink, R.F. 1989, *Lect. Notes Phys.*, **369**, 386.
Pollaco, D.L. & Bell, S.A. 1993a, *Mon. Not. R. astr. Soc.*, **262**, 377.
Pollaco, D.L. & Bell, S.A. 1993b, *Mon. Not. R. astr. Soc.*, **267**, 452.
Popham, R. & Narayan, R. 1991, *Astrophys. J.*, **370**, 604.
Popham, R. & Narayan, R. 1992, *Astrophys. J.*, **394**, 255.
Popova, M.D. & Vitrichenko, E.A. 1978, *Sov. Astr.*, **22**, 438.
Popper, D.M. 1980, *Ann. Rev. Astr. Astrophys.*, **18**, 115.
Pottasch, S. 1959, *Ann. d'Astrophys.*, **22**, 412.
Pottasch, S. 1967, *Bull. Astr. Inst. Neth.*, **19**, 227.

Pounds, K.A. *et al.*, 1993, *Mon. Not. R. astr. Soc.*, **260**, 77.
Prager, R. 1934, *Geschichte und Lichtwechsel der veranderlichen Sterne,* Verof. Univ. Sternwarte Berlin Babelesberg.
Prantzos, N., Doom, C., Arnould, M. & de Loore, C. 1986, *Astr. Astrophys.*, **304**, 695.
Prialnik, D. 1986, *Astrophys. J.*, **310**, 222.
Prialnik, D. 1990, *Lect. Notes Phys.*, **369**, 351.
Prialnik, D. & Shara, M.M. 1986, *Astrophys. J.*, **311**, 172.
Prialnik, D., Kovetz, A. & Shara, M.M. 1989, *Astrophys. J.*, **339**, 1013.
Prialnik, D., Livio, M., Shaviv, G. & Kovetz, A. 1982, *Astrophys. J.*, **257**, 312.
Priedhorsky, W.C. 1977, *Astrophys. J.*, **212**, L117.
Priedhorsky, W.C. & Holt, S.S. 1987, *Sp. Sci. Rev.*, **45**, 291.
Priedhorsky, W.C. & Krzeminski, W. 1978, *Astrophys. J.*, **219**, 597.
Priedhorsky, W.C., Krzeminski, W. & Tapia, S. 1978, *Astrophys. J.*, **225**, 542.
Pringle, J.E. 1975, *Mon. Not. R. astr. Soc.*, **170**, 633.
Pringle, J.E. 1977, *Mon. Not. R. astr. Soc.*, **178**, 195.
Pringle, J.E. 1981, *Ann. Rev. Astr. Astrophys.*, **19**, 137.
Pringle, J.E. 1985, in *Interacting Binaries*, eds. J.E. Pringle & R.A. Wade, Cambridge University Press, Cambridge, p. 1.
Pringle, J.E. 1988, *Mon. Not. R. astr. Soc.*, **230**, 587.
Pringle, J.E. 1989, *Mon. Not. R. astr. Soc.*, **236**, 107.
Pringle, J.E. 1992a, *Mon. Not. R. astr. Soc.*, **258**, 811.
Pringle, J.E. 1992b, *Rev. Mod. Astr.*, **5**, 97.
Pringle, J.E. & Rees, M.J. 1972, *Astr. Astrophys.*, **21**, 1.
Pringle, J.E. & Savonije, G.J. 1979. *Mon. Not. R. astr. Soc.*, **187**, 777.
Pringle, J.E. & Verbunt, F. 1984, in *Proc. Fourth IUE Conf.*, ESA SP-218, Paris, p. 377.
Pringle, Verbunt, F. & Wade, R.A. 1986, *Mon. Not. R. astr. Soc.*, **221**, 169.
Pringle, J.E., Bateson, F.M., Hassall, B.J.M., Heise, J., Holberg, J.B., Polidan, R.A., van Amerongen, S., van der Woerd, H., van Paradjis, J. & Verbunt, F. 1987, *Mon. Not. R. astr. Soc.*, **225**, 73.
Prinja, R.K. & Rosen, S.R. 1993, *Mon. Not. R. astr. Soc.*, **262**, L37.
Prinja, R.K., Drew, J.E. & Rosen, S.R. 1992, *Mon. Not. R. astr. Soc.*, **256**, 219.
Prinja, R.K., Rosen, S.R. & Supelli, K. 1991, *Mon. Not. R. astr. Soc.*, **248**, 40.
Protitch, M. 1958, *Perem. Zv.*, **11**, 312.
Provencal, J. 1993, Ph.D. Thesis, Univ. Texas.
Provencal, J.L., Clemens, J.C., Henry, G., Hine, B.P., Nather, R.E., Winget, D.E., Wood, M.A., Kepler, S.O., Vauclair, G., Chevreton, M., O'Donoghue, D., Warner, B., Grauer, A.D. & Ferrario, L. 1989, *Lect. Notes Phys.*, **328**, 296.
Provencal, J.L., Winget, D.E., Nather, R.E., Clemens, J.C., Hine, B.P., Henry, G., Kepler, S.O., Vauclair, G., Chevreton, M., O'Donoghue, D., Warner, B., Grauer, A.D. & Ferrario, L. 1991, in *White Dwarfs*, eds. G. Vauclair & E. Sion, Kluwer, Dordrecht, p. 449.
Pskovski, Y.P. 1972, *Sov. Astr.*, **16**, 23.
Pskovski, Y.P. 1979, *Sov. Astr. Letts.*, **5**, 209.
Puchnarewicz, E.M., Mason, K.O., Murdin, P.G. & Wickramasinghe, D.T. 1990, *Mon. Not. R. astr. Soc.*, **244**, 20P.
Pudritz, R.E. 1981a, *Mon. Not. R. astr. Soc.*, **195**, 881.
Pudritz, R.E. 1981b, *Mon. Not. R. astr. Soc.*, **195**, 897.
Pylyser, E. & Savonije, G.J. 1988a, *Astr. Astrophys.*, **191**, 57.
Pylyser, E. & Savonije, G.J. 1988b, *Astr. Astrophys.*, **208**, 52.
Quigley, R. & Africano, J. 1978, *Publ. astr. Soc. Pacific*, **90**, 445.
Quirrenbach, A., Elias, N.M., Mozurkewich, D., Armstrong, J.T., Buscher, D.F. & Hummel, C.A. 1993, *Astr. J.*, **106**, 1118.
Rahe, J., Boggess, A., Drechsel, H., Holm, A. & Krautter, J. 1980, *Astr. Astrophys.*, **88**, L9.
Raikova, D. 1990, *Lect. Notes Phys.*, **369**, 163.
Ramana Murthy, P.V. & Wolfendale, A.W. 1986, *Gamma-Ray Astronomy*, Cambridge University Press, Cambridge
Ramsay, G., Rosen, S.R., Mason, K.O., Cropper, M.S. & Watson, M.G. 1993, *Mon. Not. R. astr. Soc.*, **262**, 993.

Ramsay, G., Mason, K.O., Cropper, M., Watson, M.G. & Clayton, K.L. 1994, *Mon. Not. R. astr. Soc.*, **270**, 692.
Ramseyer, T.F., Robinson, E.L., Zhang, E., Wood, J.H. & Stiening, R.F. 1993a, *Mon. Not. R. astr. Soc.*, **260**, 209.
Ramseyer, T.,F., Dinerstein, H.L., Lester, D.F. & Provencal, J. 1993b, *Astr. J.*, **106**, 1191.
Rappaport, S. & Joss, P.C. 1984, *Astrophys. J.*, **283**, 232.
Rappaport, S., Joss, P.C. & Webbink, R.F. 1982, *Astrophys. J.*, **254**, 616.
Rappaport, S., Verbunt, F. & Joss, P. 1983, *Astrophys. J.*, **275**, 713.
Rappaport, S., Cash, W., Doxsey, R., McClintock, J. & Moore, G. 1974, *Astrophys. J.*, **187**, L5.
Ratering, C., Bruch, A. & Diaz, M. 1993, *Astr. Astrophys.*, **268**, 694.
Ratering, C., Bruch, A. & Schimpke, T. 1991, *Astr. Gesell. Abstr.*, **6**, 70.
Raubenheimer, B.C., North, A.R., de Jager, O.C., Meintjes, P.J., Brink, C., Nel, H.I., van Urk, G., Visser, B. & O'Donoghue, D. 1991, *High-energy Gamma-Ray Astronomy*, ed. J. Matthews, Amer. Inst. Phys. **220**, 70.
Raymond, J.C. & Smith, B.W. 1977, *Astrophys. J. Suppl.*, **35**, 419.
Raymond, J.C., Black, J.H., Davies, R.J., Dupree, A.K., Gursky, H., Hartmann, L. & Matilsky, T.A. 1979, *Astrophys. J.*, **230**, L95.
Rayne, M. & Whelan, J.A.J. 1981, *Mon. Not. R. astr. Soc.*, **196**, 73.
Refsdal, S. & Weigert, A. 1970, *Astr. Astrophys.*, **6**, 426.
Regev, O. 1983, *Astr. Astrophys.*, **126**, 146.
Regev, O. 1989, *Lect. Notes Phys.*, **328**, 519.
Regev, O. & Hougerat, A.A. 1988, *Astr. Astrophys.*, **232**, 81.
Regev, O. & Shara, M.M. 1989, *Astrophys. J.*, **340**, 1006.
Reid, N., Saffer, R.A. & Liebert, J. 1993, *White Dwarfs: Advances in Observation & Theory*, ed. M.A. Barstow, Kluwer, Dordrecht, p. 441.
Reimers, D. 1981, in *Physical Processes in Red Giants*, eds. I. Iben & A. Renzini, Reidel, Dordrecht, p. 269
Reimers, D., Griffin, R. & Brown, A. 1988, *Astr. Astrophys.*, **193**, 180.
Reinsch, K. 1994, *Astr. Astrophys.*, **281**, 108.
Reinsch, K. & Beuermann, K. 1990, *Astr. Astrophys.*, **240**, 360.
Reinsch, K. & Beuermann, K. 1994, *Astr. Astrophys.*, **282**, 493.
Remillard, R., Bradt, H.V., Buckley, D., Roberts, W., Schwartz, D.A., Tuohy, I.R. & Wood, K. 1986a, *Astrophys. J.*, **301**, 742.
Remillard, R.A., Bradt, H.V., McClintock, J.E., Patterson, J., Roberts, W., Schwartz, D.A. & Tapia, S. 1986b, *Astrophys. J.*, **302**, L11.
Remillard, R.A., Bradt, H., Silber, A., Tuohy, I., Brissendon, R., Buckley, D., Stroozas, B. & Schwartz, D. 1989, *Bull. Amer. astr. Soc.*, **21**, 1080.
Remillard, R.A., Stroozas, B.A., Tapia, S. & Silber, A. 1991, *Astrophys. J.*, **379**, 715.
Remillard, R.A., Silber, A.D., Schachter, J.F. & Slave, P. 1993, *Bull. Amer. astr. Soc.*, **25**, 910.
Remillard, R.A, Bradt, H.V., Bissenden, R.J.V., Buckley, D.A.H., Schwartz, D.A., Silber, A., Stroozas, B.A. & Tuohy, I.R. 1994a, *Astrophys. J*, **428**, 785.
Remillard, R.A., Schacter, J.F., Silber, A. & Slane, P. 1994b, *Astrophys. J.*, **426**, 288.
Richer, H.B. & Fahlman, G.G. 1988, *Astrophys. J.*, **325**, 218.
Richer, H.B., Auman, J.R., Isherwood, B.C., Steele, J.P. & Ulrych, T.J. 1973, *Astrophys. J.*, **180**, 107.
Richman, H.R. 1991, *J. Amer. Assoc. Var. Star. Obs.*, **20**, 173.
Richter, G.A. 1989, *Astr. Nach.*, **310**, 143.
Richter, G.A. 1992, *Astr. Soc. Pacific Conf. Ser.*, **29**, 12.
Ricketts, M.J., King, A.R. & Raine, D.J. 1979, *Mon. Not. R. astr. Soc.*, **186**, 233.
Rieutord, M. & Bonazzola, S. 1987, *Mon. Not. R. astr. Soc.*, **227**, 295.
Ringwald, F. A. 1992. Ph.D. Thesis, Dartmouth College.
Ringwald, F. A. 1994, *Mon. Not. R. astr. Soc.*, **270**, 804.
Ringwald, F.A. & Thorstensen, J.R. 1990, *Bull. Amer. astr. Soc.*, **22**, 1291.
Ritter, H. 1976, *Mon. Not. R. astr. Soc.*, **175**, 279.
Ritter, H. 1978, *Astr. Astrophys.*, **68**, 455.
Ritter, H. 1980a, *Astr. Astrophys.*, **91**, 161.
Ritter, H. 1980b, *The Messenger*, No. 21, 16.

Ritter, H. 1980c, *Astr. Astrophys.*, **85**, 362.
Ritter, H. 1980d, *Astr. Astrophys.*, **86**, 204.
Ritter, H. 1983, *Int. Astr. Union Colloq. No. 72*, p. 257.
Ritter, H. 1984, *Astr. Astrophys. Suppl.*, **57**, 385.
Ritter, H. 1985a, *Astr. Astrophys.*, **145**, 227.
Ritter, H. 1985b, *Astr. Astrophys.*, **148**, 207.
Ritter, H. 1986a, *Astr. Astrophys.*, **169**, 139.
Ritter, H. 1986b, in *The Evolution of Galactic X-Ray Binaries,* eds. J. Truemper *et al.*, Reidel, Dordrecht, p. 271.
Ritter, H. 1986c, *Astr. Astrophys.*, **168**, 105.
Ritter, H. 1987, *Astr. Astrophys. Suppl.*, **70**, 335.
Ritter, H. 1988, *Astr. Astrophys.*, **202**, 93.
Ritter, H. 1990, *Astr. Astrophys. Suppl.*, **85**, 1179.
Ritter, H. 1992, personal communication (running update of his CV catalogue).
Ritter, H. & Burkert, A. 1986, *Astr. Astrophys.*, **158**, 161.
Ritter, H. & Özkan, M.T. 1986, *Astr. Astrophys.*, **167**, 260.
Ritter, H. & Schroeder, R. 1979, *Astr. Astrophys.*, **76**, 168.
Ritter, H., Politano, M., Livio, M. & Webbink, R.F. 1991, *Astrophys. J.*, **376**, 177.
Roberts, W.J. 1974, *Astrophys. J.*, **187**, 575.
Robertson, J.A. & Frank, J. 1986, *Mon. Not. R. astr. Soc.*, **221**, 279.
Robertson, J.W., Honeycutt, R.K. & Turner, G.W. 1994, *Astr. Soc. Pacific Conf. Ser.*, **56**, 298.
Robinson, E.L. 1973a, *Astrophys. J.*, **180**, 121.
Robinson, E.L. 1973b, *Astrophys. J.*, **183**, 193.
Robinson, E.L. 1973c, *Astrophys. J.*, **186**, 347.
Robinson, E.L. 1973d, *Astrophys. J.*, **181**, 531.
Robinson, E.L. 1974, *Astrophys. J.*, **193**, 191.
Robinson, E.L. 1975, *Astr. J.*, **80**, 515.
Robinson, E.L. 1976a, *Ann. Rev. Astr. Astrophys.*, **14**, 119.
Robinson, E.L. 1976b, *Astrophys. J.*, **203**, 485.
Robinson, E.L. 1983, *Int. Astr. Union Colloq. No. 72*, p. 1.
Robinson, E.L. 1990, private communication.
Robinson, C.R. & Cordova, F.A. 1994, *Astr. Soc. Pacific Conf. Ser.*, in press.
Robinson, C.R. & Cordova, F.A. 1995, *Astrophys. J.*, **437**, 436.
Robinson, E.L. & Faulkner, J. 1975, *Astrophys. J.*, **200**, L33.
Robinson, E.L. & Nather, R.E. 1977, *Publ. astr. Soc. Pacific*, **89**, 572.
Robinson, E.L. & Nather, R.E. 1979, *Astrophys. J. Suppl.*, **39**, 461.
Robinson, E.L. & Nather, R.E. 1983, *Astrophys. J.*, **273**, 255.
Robinson, E.L. & Warner, B. 1984, *Astrophys. J.*, **277**, 250.
Robinson, E.L., Nather, R.E. & Kepler, S.O. 1982, *Astrophys. J.*, **254**, 646.
Robinson, E.L., Nather, R.E. & Kiplinger, A.L. 1974, *Publ. astr. Soc. Pacific*, **86**, 401.
Robinson, E.L., Nather, R.E. & Patterson, J. 1978, *Astrophys. J.*, **219**, 168.
Robinson, E.L., Shafter, A.W. & Balachandran, S. 1991, *Astrophys. J.*, **374**, 298.
Robinson, E.L., Shetrone, M.D. & Africano, J.L. 1991, *Astr. J.*, **103**, 1176.
Robinson, E.L., Zhang, E.-H. & Stover, R.J. 1986, *Astrophys. J.*, **305**, 732.
Robinson, E.L., Barker, E.S., Cochran, A.L., Cochran, W.D. & Nather, R.E. 1981, *Astrophys. J.*, **251**, 611.
Robinson, E.L., Shafter, A.W., Hill, J.A., Wood, M.A. & Mattei, J. 1987, *Astrophys. J.*, **313**, 772.
Rogers, F.J. & Iglesias, C.A. 1992, *Astrophys. J. Suppl.*, **79**, 507.
Romano, G. 1983, *Inf. Bull. Var. Stars*, No. 2265.
Rose, W.K. & Smith, R.L. 1972, *Astrophys. J.*, **172**, 699.
Rosen, S.R. 1987, Ph.D. Thesis, Univ. London.
Rosen, S.R. 1992, *Mon. Not. R. astr. Soc.*, **254**, 493.
Rosen, S., Done, C. & Watson, M.G. 1993, *Int. Astr. Union Circ.*, No. 5850.
Rosen, S.R., Mason, K.O. & Cordova, F.A. 1987, *Mon. Not. R. astr. Soc.*, **224**, 987.
Rosen, S.R., Mason, K.O. & Cordova, F.A. 1988, *Mon. Not. R. astr. Soc.*, **231**, 549.
Rosen, S.R., Mittaz, J.P.D. & Hakala, P.J. 1993, *Mon. Not. R. astr. Soc.*, **264**, 171.

Rosen, S.R., Branduardi-Raymont, G., Mason, K.O. & Murdin, P.G. 1989, *Mon. Not. R. astr. Soc.*, **237**, 1037.
Rosen, S.R., Mason, K.O., Mukai, K. & Williams, O.R. 1991, *Mon. Not. R. astr. Soc.*, **249**, 417.
Rosino, L. 1973, *Astr. Astrophys.*, **9**, 347.
Rosino, L. 1987, in *RS Ophiuchi (1985) and the Recurrent Nova Phenomenon*, ed. M.F. Bode, VNU Sci. Press, Utrecht, p. 1.
Rosino, L., Bianchini, A. & Rafarelli, P. 1982, *Astr. Astrophys.*, **108**, 243.
Rosino, L., Ciatti, F. & Della Valle, M. 1986, *Astr. Astrophys.*, **158**, 34.
Rosino, L., Romano, G. & Marziani, P. 1993, *Publ. astr. Soc. Pacific*, **105**, 51.
Rosner, R. 1980, in *Cool Stars, Stellar Systems and the Sun*, ed. A.K. Dupree, Springer, Berlin, p. 79.
Rosner, R., Golub, L. & Vaiana, G.S. 1985, *Ann. Rev. Astr. Astrophys.*, **23**, 413.
Rothschild, R.E., Gruber, D.E., Knight, F.K., Matteson, J.L. Nolan, P.L., Swank, J.H., Holt, S.S., Serlemitsos, P.J., Mason, K.O. & Tuohy, I.R. 1981, *Astrophys. J.*, **250**, 723.
Rozyczka, M. 1985, *Astr. Astrophys.*, **143**, 59.
Rozyczka, M. 1988, *Acta Astr.*, **38**, 175.
Rozyczka, M. & Schwarzenberg-Czerny, A. 1987, *Acta Astr.*, **37**, 141.
Rozyczka, M. & Spruit, H.C. 1993, *Astrophys. J.*, **417**, 677.
Rubenstein, E.P., Patterson, J. & Africano, J.L. 1991, *Publ. astr. Soc. Pacific*, **103**, 1258.
Rucinski, S.M. 1969, *Acta Astr.*, **19**, 245.
Rucinski, S.M. 1983, *Observatory*, **103**, 280.
Rucinski, S.M. 1984, *Observatory*, **104**, 259.
Ruden, S.P. & Pollack, J.B. 1991, *Astrophys. J.*, **375**, 740.
Ruderman, M.A. 1987, in *High-Energy Phenomena around Collapsed Stars*, ed. F. Pacini, Reidel, Dordrecht, p. 145.
Rupprecht, G. & Bues, I. 1983, *Mitt. Astr. Gesell.*, **60**, 337.
Russell, H.N. 1945, *Astrophys. J.*, **102**, 1.
Rutten, R.G.M. 1987, *Astr. Astrophys.*, **177**, 131.
Rutten, R.G.M. & Dhillon, V.S. 1992, *Astr. Astrophys.*, **253**, 139.
Rutten, R.G.M. & Dhillon, V.S. 1994, *Astr. Astrophys.*, **288**, 773.
Rutten, R.G.M., van Paradijs, J. & Tinbergen, J. 1992, *Astr. Astrophys.*, **260**, 213.
Rutten, R.G.M., Kuulkers, E., Vogt, N. & van Paradijs, J. 1992, *Astr. Astrophys.*, **265**, 159.
Rutten, R.G.M., Dhillon, V.S., Horne, K., Kullkers, E. & van Paradijs, J. 1993, *Nature*, **362**, 518.
Rybicki, G.B. & Hummer, D.G. 1983, *Astrophys. J.*, **274**, 380.
Ryu, D. & Goodman, J. 1992, *Astrophys. J.*, **388**, 438.
Saar, S.H. 1987, in *Cool Stars, Stellar Systems & the Sun*, eds. J.L. Linsky & R.E. Stencel, Springer, Berlin, p. 10.
Saar, S.H. 1991, *Int. Astr. Union Colloq. No. 130*, p. 389.
Sabbadin, F. & Bianchini, A. 1983, *Astr. Astrophys. Suppl.*, **54**, 393.
Saffer, R.A., Wade, R.A., Liebert, J., Green, R.F., Sion, E.M., Bechtold, J., Foss, D. & Kidder, K. 1993, *Astr. J.*, **105**, 1945.
Saizar, P., Starrfield, S., Ferland, G.J., Wagner, R.M., Truran, J.W., Kenyon, S.J., Sparks, W.M., Williams, R.E. & Stryker, L.L. 1991, *Astrophys. J.*, **367**, 310.
Saizar, P., Starrfield, S., Ferland, G.J., Wagner, R.M., Truran, J.W., Kenyon, S.J., Sparks, W.M., Williams, R.E. & Stryker, L.L. 1992, *Astrophys. J.*, **398**, 651.
Sambruna, R.M., Chiappetti, L., Treves, A., Bonnet-Bidaud, J.M., Bouchet, P., Maraschi, L., Motch, C. & Mouchet, M. 1991, *Astrophys. J.*, **374**, 744.
Sambruna, R.M., Chiappetti, L., Treves, A., Bonnet-Bidaud, J.M., Bouchet, P., Maraschi, L., Motch, C., Mouchet, M. & van Amerongen, S., 1992, *Astrophys. J.*, **391**, 750.
Sanford, R.F. 1943, *Publ. astr. Soc. Pacific*, **55**, 284.
Sanford, R.F. 1949, *Astrophys. J.*, **109**, 81.
Sarma, M.B.K. 1991, *Astrophys. J.*, **380**, 208.
Sarna, M.J. 1990., *Astr. Astrophys.*, **239**, 163.
Saslaw, W.C. 1968, *Mon. Not. R. astr. Soc.*, **138**, 337.
Savonije, G.J. & Papaloizou, J.C.B. 1985, in *Interacting Binaries*, eds. P.P. Eggleton & J.E. Pringle, Reidel, Dordrecht, p. 83.
Savonije, G.J., de Kool, M. & van den Heuvel, E.P.J. 1986, *Astr. Astrophys.*, **155**, 51.

Saw, D.R.B. 1982a, *J. Brit. astr. Assoc.*, **92**, 127.
Saw, D.R.B. 1982b, *J. Brit. astr. Assoc.*, **92**, 220.
Saw, D.R.B. 1983, *J. Brit. astr. Assoc.*, **93**, 2.
Sawada, K., Matsuda, T. & Hachisu, I. 1986, *Mon. Not. R. astr. Soc.*, **219**, 75.
Sawada, K., Matsuda, T., Inoue, M. & Hachisu, I. 1987, *Mon. Not. R. astr. Soc.*, **224**, 307.
Sazanov, A.V. & Shugarov, S.Yu. 1992, *Inf. Bull. Var. Stars*, No. 3744.
Schaaf, R., Pietsch, W. & Biermann, P. 1987, *Astr. Astrophys.*, **174**, 357.
Schachter, J., Filippenko, A.V., Kahn, S.M. & Paerels, F.B.S. 1991, *Astrophys. J.*, **373**, 633.
Schaefer, B. 1983, *Astrophys. J.*, **268**, 710.
Schaefer, B.E. 1987, *Astrophys. J.*, **323**, L47.
Schaefer, B.E. 1988, *Astrophys. J.*, **327**, 347.
Schaefer, B. 1990, *Astrophys. J.*, **355**, L39.
Schaefer, B.E. & Patterson, J. 1983, *Astrophys. J.*, **268**, 710.
Schaefer, B.E. & Patterson, J. 1987, *Inf. Bull. Var. Stars*, No. 3025.
Schaefer, B .E., Landolt, A.U., Vogt, N., Buckley, D., Warner, B., Walker, A.R. & Bond, H.E. 1992, *Astrophys. J. Suppl.*, **81**, 321.
Schaeidt, S., Hasinger, G. & Trümper, J. 1993, *Astr. Astrophys.*, **270**, L9.
Schaich, M., Wolf, D., Ostreicher, R. & Ruder, H. 1992, *Astr. Astrophys.*, **264**, 529.
Scharlemann, E.T. 1978, *Astrophys. J.*, **219**, 617.
Schatzman, E. 1951, Ann. d'Astrophys., **14**, 305.
Schatzman, E. 1958, Ann. d'Astrophys., **21**, 1.
Schatzman, E. 1959, Ann. d'Astrophys., **22**, 436.
Schatzman, E. 1965, in *Stellar Structure*, eds. L.H. Aller & D.B. McLaughlin, Univ. Chicago Press, Chicago, p. 327.
Schild, R.E. 1969, *Astrophys. J.*, **157**, 709.
Schlegel, E.M., Honeycutt, R.K. & Kaitchuck, R.H. 1983, *Astrophys. J. Suppl.*, **53**, 397.
Schlegel, E.M., Honeycutt, R.K. & Kaitchuck, R.H. 1986, *Astrophys. J.*, **307**, 760.
Schlegel, E.M., Kaitchuck, R.H. & Honeycutt, R.K. 1984, *Astrophys. J.*, **280**, 235.
Schmidt, T. 1957, *Z. Astrophys.*, **41**, 182.
Schmidt, G.D. 1987, *Mem. Soc. Astr. Ital.*, **58**, 77.
Schmidt, G.D. 1988, in *Polarized Radiation of Circumstellar Origin*, eds. G.V. Coyne *et al.*, Vatican Obs., Vatican, Vatican, p. 85.
Schmidt, G.D. 1990, in *Accretion-Powered Compact Binaries*, ed. C.W. Mauche, Cambridge University Press, Cambridge, p. 295.
Schmidt, G.D. & Liebert, J. 1987, *Astrophys. Sp. Sci.*, **131**, 549.
Schmidt, G. &. Norsworthy, J. 1989, *Int. Astr. Union Circ.*, No. 4866.
Schmidt, G.D. & Stockman, H.S. 1991, *Astrophys. J.*, **371**, 749.
Schmidt, G.D., Liebert, J. & Stockman, H.S. 1995, *Astrophys. J.*, **441**, 414.
Schmidt, G.D., Stockman, H.S. & Grandi, S.A. 1983, *Astrophys. J.*, **271**, 735.
Schmidt, G.D., Stockman, H.S. & Grandi, S.A. 1986, *Astrophys. J.*, **300**, 804.
Schmidt, G.D., Stockman, H.S. & Margon, B. 1981, *Astrophys. J.*, **243**, L157.
Schmidt, G.D., Bergeron, P., Liebert, J. & Saffer, R.A. 1992, *Astrophys. J.*, **394**, 603.
Schmidt, G.D., Liebert, J., Stockman, H.S. & Holberg, J. 1993, in *Cataclysmic Variables and Related Physics*, eds. O. Regev & G. Shaviv, Inst. Phys. Publ., Bristol, p. 13.
Schmidt-Kaler, Th. 1957, *Z. Astrophys.*, **41**, 182.
Schmidt-Kaler, Th. 1962, *Klein Verof. Remeis Obs.*, **34**, 109.
Schneider, D.P. & Greenstein, J.L. 1979, *Astrophys. J.*, **233**, 935.
Schneider, D.P. & Young, P. 1980a, *Astrophys. J.*, **238**, 946.
Schneider, D.P. & Young, P. 1980b, *Astrophys. J.*, **240**, 871.
Schneider, D.P., Young, P. & Schectman, S.A. 1981, *Astrophys. J.*, **245**, 644.
Schoembs, R. 1982, *Astr. Astrophys.*, **115**, 190.
Schoembs, R. 1986, *Astr. Astrophys.*, **158**, 233.
Schoembs, R. & Hartmann, K. 1983, *Astr. Astrophys.*, **128**, 37.
Schoembs, R. & Rebhan, H. 1989, *Astr. Astrophys.*, **224**, 42.
Schoembs, R. & Stolz, B. 1981, *Inf. Bull. Var. Stars*, No. 1986.
Schoembs, R. & Vogt, N. 1980, *Astr. Astrophys.*, **91**, 25.
Schoembs, R. & Vogt, N. 1981, *Astr. Astrophys.*, **97**, 185.

Schoembs, R., Dreier, H. & Barwig, H. 1987, *Astr. Astrophys.*, **181**, 50.
Schoenberner, D. 1978, *Astr. Astrophys.*, **70**, 451.
Schönberner, D. 1983, *Astrophys. J.*, **272**, 708.
Schrijver, C.J. & Zwaan, C. 1991, *Astr. Astrophys.*, **251**, 183.
Schrijver, C.J. & Zwaan, C. 1992, *Astr. Soc. Pacific Conf. Ser.*, **26**, 370.
Schrijver, J., Brinkman, A.C. & van der Woerd, H. 1987, *Astrophys. Sp. Sci.*, **130**, 261.
Schrijver, J., Brinkman, A.C., van der Woerd, H., Watson, M.G., King, A.R., van Paradijs, J. & van der Klis, M. 1985, *Sp. Sci. Rev.*, **40**, 121.
Schussler, M. 1983, *Int. Astr. Union Colloq. No. 102*, p. 213.
Schwarz, D.A., Bradt, H., Briel, U., Doxey, R.E., Fabbiano, G., Griffiths, R.E., Johnston, M.D. & Margon, B. 1979, *Astr. J.*, **84**, 1560.
Schwarz, H.E., van Amerongen, S.F., Heemskerk, M.H.M. & van Paradijs, J. 1988, *Astr. Astrophys.*, **202**, L16.
Schwarzenberg-Czerny, A. 1981, *Acta Astr.*, **31**, 241.
Schwarzenberg-Czerny, A. 1984a, *Observatory*, **104**, 27.
Schwarzenberg-Czerny, A. 1984b, *Mon. Not. R. astr. Soc.*, **208**, 57.
Schwarzenberg-Czerny, A. 1987, *Acta Astr.*, **37**, 213.
Schwarzenberg-Czerny, A. 1992, *Astr. Astrophys.*, **260**, 268.
Schwarzenberg-Czerny, A. & Rozyczka, M. 1977, *Acta Astr.*, **27**, 429.
Schwarzenberg-Czerny, A. & Rozyczka, M. 1988, *Acta Astr.*, **38**, 189.
Schwarzenberg-Czerny, A., Udalski, A. & Monier, R. 1992, *Astrophys. J.*, **401**, L19.
Schwarzenberg-Czerny, A., Ward, M., Hanes, D.A., Jones, D.H.P., Pringle, J.E., Verbunt, F. & Wade, R.A. 1985, *Mon. Not. R. astr. Soc.*, **212**, 645.
Schwope, A.D. 1990, *Rev. Mod. Astr.*, **3**, 44.
Schwope, A. & Beuermann, K. 1987, *Astrophys. Sp. Sci.*, **131**, 637.
Schwope, A.D. & Beuermann, K. 1989, *Astr. Astrophys.*, **222**, 132.
Schwope, A.D. & Beuermann, K. 1990, *Astr. Astrophys.*, **238**, 173.
Schwope, A.D. & Beuermann, K. 1993, in *White Dwarfs: Advances in Observation and Theory*, ed. M.A. Barstow, Kluwer, Dordrecht, p. 381.
Schwope, A.D., Beuermann, K. & Thomas, H.-C. 1990, *Astr. Astrophys.*, **230**, 120.
Schwope, A.D., Jordan, S. & Beuermann, K. 1993, in *Cataclysmic Variables and Related Physics*, eds. O. Regev & G. Shaviv, Inst. Phys. Publ., Bristol, p. 312.
Schwope, A.D., Thomas, H.-C. & Beuermann, K. 1993, *Astr. Astrophys.*, **271**, L25.
Schwope, A.D., Thomas, H.-C., Beuermann, K. & Naundorf, C.E. 1991, *Astr. Astrophys.*, **244**, 373.
Schwope, A.D., Thomas, H.-C., Beuermann, K. & Reinsch, K. 1993a, *Astr. Astrophys.*, **267**, 103.
Schwope, A.D., Beuermann, K., Jordan, S. & Thomas, H.-C. 1993b, *Astr. Astrophys.*, **278**, 487.
Scott, B.D. 1990, *Astrophys. J.*, **357**, L53.
Scott, A.D., Rawlings, J.M.C. & Evans, A. 1994, *Mon. Not. R. astr. Soc.*, **269**, 707.
Seaquist, E.R. 1989, in Classical Novae, eds. M.F. Bode & A. Evans, Wiley, Chichester, p. 143.
Seaquist, E.R. & Palimaka, J. 1977, *Astrophys. J.*, **217**, 781.
Seaquist, E.R., Bode, M.E., Frail, D.A., Roberts, J.A., Evans, A. & Albinson, J.S. 1989, *Astrophys. J.*, **344**, 805.
Seifert, W., Oestreicher, R., Wunner, G. & Ruder, H. 1987, *Astr. Astrophys.*, **183**, L1.
Seitter, W.C. 1969, in *Non-periodic Phenomena in Variable Stars*, ed. L. Detre, Academic Press, Budapest, p. 277.
Seitter, W.C. 1984, *Astrophys. Sp. Sci.*, **99**, 95.
Seitter, W.C. 1990, *Int. Astr. Union Colloq. No. 122*, p. 79.
Seitter, W.C. & Duerbeck, H.W. 1987, in *RS Ophiuchi (1985) and the Recurrent Nova Phenomenon*, ed. M.F. Bode, VNU Sci. Press, Utrecht, p. 71.
Sekiguchi, K. 1992a, *Nature*, **358**, 563.
Sekiguchi, K. 1992b, *Astr. Soc. Pacific Conf. Ser.* **30**, 345.
Sekiguchi, K., Nakada, Y. & Bassett, B. 1994. *Mon. Not. R. astr. Soc.*, **266**, L51.
Sekiguchi, K., Feast, M.W., Whitelock, P.A., Overbeek, M.D., Wargau, W. & Spencer-Jones, J. 1988, *Mon. Not. R. astr. Soc.*, **234**, 281.
Sekiguchi, K., Catchpole, R.M., Fairall, A.P., Feast, M.W., Kilkenny, D., Laney, C.D., Lloyd Evans, T., Marang, F. & Parker, Q.A. 1989, *Mon. Not. R. astr. Soc.*, **236**, 611.
Sekiguchi, K., Whitelock, P.A., Feast, M.W., Barrett, P.E., Caldwell, J.A.R., Carter, B.S., Catchpole,

R.M., Laing, J.D., Laney, C.D., Marang, F. & van Wyk, F. 1990a, *Mon. Not. R. astr. Soc.*, **246**, 78.
Sekiguchi, K., Stobie, R.S., Buckley, D.A.H. & Caldwell, J.A.R. 1990b, *Mon. Not. R. astr. Soc.*, **245**, 28P.
Selvelli, P.L. 1982, in *Third European IUE Conference*, ESA SP-176, Paris, p. 197.
Selvelli, P.L. & Hack, M. 1983, *Mem. Soc. Astr. Ital.*, **54**, 467.
Selvelli, P.L., Cassatella, A. & Gilmozzi, R. 1992, *Astrophys. J.*, **393**, 289.
Selvelli, P.L., Cassatella, A., Bianchini, A., Friedjung, M. & Gilmozzi, R. 1990, *Int. Astr. Union Colloq. No. 122*, p. 65.
Semeniuk, I. 1980, *Astr. Astrophys. Suppl.*, **39**, 29.
Semeniuk, I. & Kaluzny, J. 1988, *Acta Astr.*, **38**, 49.
Semeniuk, I., Schwarzenberg-Czerny, A., Duerbeck, H., Hoffman, M., Smak, J., Stepien, K. & Tremko, J. 1987, *Acta Astr.*, **37**, 197.
Serkowski, K. 1971, *Int. Astr. Union Colloq. No. 15*, p. 11.
Shafter, A.W. 1983, *Astrophys. J.*, **267**, 222.
Shafter, A.W. 1984a, Ph.D. Thesis, Univ. Calif., Los Angeles.
Shafter, A.W. 1984b, *Astr. J.*, **89**, 1555.
Shafter, A.W. 1985, *Astr. J.*, **90**, 643.
Shafter, A.W. 1992a, *Proc. 12th N. Amer. Workshop on Cataclysmic Variables & Low Mass X-ray Binaries*, San Diego State Univ., California, p. 39.
Shafter, A.W. 1992b, *Astrophys. J.*, **394**, 268.
Shafter, A.W. & Abbott, T.M.C. 1989, *Astrophys. J.*, **339**, L75.
Shafter, A.W. & Harkness, R.P. 1986, *Astr. J.*, **92**, 658.
Shafter, A.W. & Hessman, F.V. 1988, *Astr. J.*, **95**, 178.
Shafter, A.W. & Macry, J.D. 1987, *Mon. Not. R. astr. Soc.*, **228**, 193.
Shafter, A.W. & Szkody, P. 1984, *Astrophys. J.*, **276**, 305.
Shafter, A.W. & Targan, D.M. 1982, *Astr. J.* **87**, 655.
Shafter, A.W., Hessman, F.V. & Zhang, E.H. 1988, *Astrophys. J.*, **327**, 248.
Shafter, A.W., Lanning, H.H. & Ulrich, R.K. 1983, *Publ. astr. Soc. Pacific*, **95**, 206.
Shafter, A.W., Misselt, K.A. & Veal, J.M. 1993, *Publ. astr. Soc. Pacific*, **105**, 853.
Shafter, A.W., Szkody, P. & Thorstensen, J. 1986, *Astrophys. J.*, **308**, 765.
Shafter, A.W., Wheeler, J.C. & Cannizzo, J.K. 1986, *Astrophys. J.*, **305**, 261.
Shafter, A.W., Szkody, P., Liebert, J., Penning, W.R., Bond, H.E. & Grauer, A.D. 1985, *Astrophys. J.*, **290**, 707.
Shafter, A.W., Robinson, E.L., Crampton, D., Warner, B. & Prestage, R.M. 1990, *Astrophys. J.*, **354**, 708.
Shafter, A.W., Reinsch, K., Beuermann, K., Misselt, K.A., Buckley, D.A.H., Burwitz, V. & Schwope, A.D. 1995, *Astrophys. J.*, in press.
Shakhovskoy, N.M., Alexeev, I. Yu., Andronov, I.L. & Kolesnikov, S.V. 1993, in *Cataclysmic Variables and Related Physics*, eds. O. Regev & G. Shaviv, Inst. Phys. Publ., Bristol, p. 237.
Shakun, L.I. 1987, *Astr. Tsirk.*, No. 1491, 7.
Shakura, N.I. & Sunyaev, R.A. 1973, *Astr. Astrophys.*, **24**, 337.
Shankar, A., Arnett, D. & Fryxell, B.A. 1992, *Astrophys. J.*, **394**, L13.
Shara, M. 1982, *Astrophys. J.*, **261**, 649.
Shara, M.M. 1989, *Publ. astr. Soc. Pacific*, **101**, 5.
Shara, M.M. 1994a, *Astr. J.*, **107**, 1546.
Shara, M.M. 1994b, *Astrophys. J.*, in press.
Shara, M.M. & Moffat, A.F.J. 1983, *Astrophys. J.*, **264**, 560.
Shara, M.M. & Prialnik, D. 1994, *Astr. J.*, **107**, 1542.
Shara, M.M., Moffat, A.F.J. & Potter, M. 1987, *Astr. J.*, **94**, 357.
Shara, M.M., Moffat, A.F.J. & Potter, M. 1990a, *Astr. J.*, **99**, 1858.
Shara, M.M., Moffat, A.F.J. & Potter, M. 1990b, *Astr. J.*, **100**, 540.
Shara, M.M., Moffat, A.F.J. & Webbink, R.F. 1985, *Astrophys. J.*, **294**, 271.
Shara, M.M., Potter, M. & Shara, D.J. 1989, *Publ. astr. Soc. Pacific*, **101**, 985.
Shara, M.M., Prialnik, D. & Kovetz, A. 1993, *Astrophys. J.*, **406**, 220.
Shara, M.M., Moffat, A.F.J., McGraw, J.T., Dearborn, D.S., Bond, H.E., Kemper, E. & Lamontagne, R. 1984, *Astrophys. J.*, **282**, 763.
Shara, M.M., Livio, M., Moffat, A.F.J. & Orio, M. 1986a, *Astrophys. J.*, **311**, 163.

Shara, M.M., Moffat, A.F.J., Potter, M., Hogg, H.S. & Wehlau, A. 1986b, *Astrophys. J.*, **311**, 796.
Shara, M.M., Kaluzny, J., Potter, M. & Moffat, A.F.J. 1988, *Astrophys. J.*, **328**, 594.
Shara, M.M., Moffat, A.F.J., Williams, R.E. & Cohen, J.G. 1989, *Astrophys. J.*, **337**, 720.
Shara, M.M., Potter, M., Moffat, A.F.J., Bode, M. & Stephenson, F.R. 1990, *Int. Astr. Union Colloq. No. 122*, p. 57.
Shara, M.M., Moffat, A.F.J., Potter, M., Bode, M. & Stephenson, F.R. 1993, in *Cataclysmic Variables and Related Physics*, eds. O. Regev & G. Shaviv, Inst. Phys. Publ., Bristol, p. 84.
Sharov, A.S. 1972, *Sov. Astr.*, **16**, 41.
Sharov, A.S. 1993, *Astr. Lett.*, **19**, 7.
Sharov, A.S. & Alksnis, A. 1992, *Astrophys. Sp. Sci.*, **190**, 119.
Shaviv, G. 1987, *Astrophys. Sp. Sci.*, **130**, 303.
Shaviv, G. & Starrfield, S. 1987, *Astrophys. J.*, **321**, L51.
Shaviv, G. & Wehrse, R. 1986, *Astr. Astrophys.*, **159**, L5.
Shaviv, G. & Wehrse, R. 1991, *Astr. Astrophys.*, **251**, 117.
Shaviv, G. & Wehrse, R. 1993, in *Accretion Disks in Compact Stellar Systems*, ed. J.C. Wheeler, World Sci. Publ. Co., Singapore, p. 148.
Sherrington, M.R. & Jameson, R.F. 1983, *Mon. Not. R. astr. Soc.*, **205**, 265.
Sherrington, M.R., Bailey, J. & Jameson, R.F. 1984, *Mon. Not. R. astr. Soc.*, **206**, 859.
Sherrington, M.R., Jameson, R.F. & Bailey, J. 1984, *Mon. Not. R. astr. Soc.*, **210**, 1P.
Sherrington, M.R., Lawson, P.A., King, A.R. & Jameson, R.F. 1980, *Mon. Not. R. astr. Soc.*, **191**, 185.
Sherrington, M.R., Jameson, R.F., Bailey, J. & Giles, A.B. 1982, *Mon. Not. R. astr. Soc.*, **200**, 861.
Shlosman, I. & Vitello, P. 1993, *Astrophys. J.*, **409**, 372.
Shore, S.N., Foltz, C.B., Wasilewsky, A.J., Byard, P.C. & Wagner, R.M. 1982, *Publ. astr. Soc. Pacific*, **94**, 682.
Shore, S.N., Sonneborn, G., Starrfield, S., Hamuy, M., Williams, R.E., Cassatella, A. & Drechsel, H. 1991, *Astrophys. J.*, **370**, 193.
Shu, F.H. 1976, *Int. Astr. Union Symp. No. 73*, p. 253.
Shugarov, S. Yu. 1980, *Astr. Tsirk*, No. 1119.
Shugarov, S. Yu. 1982, *Sov. Astr.*, **25**, 332.
Shugarov, S. Yu. 1983a, *Sov. Astr. Letts.*, **9**, 15.
Shugarov, S. Yu. 1983b, *Var. Stars*, No. 6, p. 807.
Shylaya, B.S. 1987, *Astrophys. Sp. Sci.*, **130**, 181.
Siegel, N., Reinsch, K., Beuermann, K., van der Woerd, H. & Wolff, E. 1989, *Astr. Astrophys.*, **225**, 97.
Sienkiewicz, R. 1984, *Acta Astr.*, **34**, 325.
Silber, A.D. 1992, Ph.D. Thesis, M.I.T.
Silber, A. & Remillard, R. 1993, in *Cataclysmic Variables and Related Physics*, eds. O. Regev & G. Shaviv, Inst. Phys. Publ., Bristol, p. 317.
Silber, A., Vrtilek, S.D. & Raymond, J.C. 1994, *Astrophys. J.*, **425**, 829.
Silber, A., Bradt, H.V., Ishida, M., Ohashi, T. & Remillard, R.A. 1992, *Astrophys. J.*, **389**, 704.
Silber, A., Remillard, R.A., Horne, K. & Bradt, H.V. 1994, *Astrophys. J.*, in press.
Singh, J. & Swank, J. 1993, *Mon. Not. R. astr. Soc.*, **262**, 1000.
Singh, K.P., Agrawal, P.C. & Riegler, G.R. 1984, *Mon. Not. R. astr. Soc.*, **208**, 679.
Singh, J., Agrawal, P.C., Apparao, M.V., Rao, P.V. & Sarma, M.B.K. 1991, *Astrophys. J.*, **380**, 208.
Singh, J., Vivekananda Rao, P., Agrawal, P.C., Apparao, K.M.V., Manchanda, R.K., Sanwal, B.B. & Sarma, M.B.K. 1993, *Astrophys. J.*, **419**, 337.
Sion, E.M. 1984, *Astrophys. J.*, **282**, 612.
Sion, E. 1985a, *Astrophys. J.*, **292**, 601.
Sion, E.M. 1985b, *Astrophys. J.*, **297**, 597.
Sion, E.M. 1986, *Publ. astr. Soc. Pacific*, **98**, 821.
Sion, E.M. 1987a, *Int. Astr. Union Colloq. No. 95*, p. 413.
Sion, E.M. 1987b, *Astrophys. Sp. Sci.*, **130**, 47.
Sion, E.M. 1991, *Astr. J.*, **102**, 295.
Sion, E.M. 1993a, *Astr. J.*, **106**, 298.
Sion, E.M. 1993b, in *Cataclysmic Variables and Related Physics*, eds. O. Regev & G. Shaviv, Inst. Phys. Publ., Bristol, p. 86.

Sion, E.M. & Guinan, E.F. 1982, *Advances in Ultraviolet Astronomy*, NASA Conf. Pub. 2238, Washington, p. 460.
Sion, E.M. & Ready, C.J. 1992, *Publ. astr. Soc. Pacific*, **104**, 87.
Sion, E.M. & Szkody, P. 1990, *Int. Astr. Union Colloq. No. 122*, p. 59.
Sion, E.M., Leckenby, H.J. & Szkody, P. 1990, *Astrophys. J.*, **364**, L41.
Sion, E.M., Starrfield, S.G., Van Steenberg, M.E., Sparks, W., Truran, J.W. & Williams, R.E. 1986, *Astr. J.*, **92**, 1145.
Sion, E.M., Bruhweiler, F.C., Mullan, D. & Carpenter, K. 1989, *Astrophys. J.*, **341**, L17.
Sion, E.M., Wagner, R.M., Starrfield, S.G. & Liebert, J. 1991, in *Objective-Prism and Other Surveys*, eds. A.G.D. Philip & A.R. Upgren, L. Davis Press, N.Y., p. 163.
Sion, E.M., Long, K., Szkody, P. & Huang, M. 1994, *Astrophys. J.*, **430**, L53.
Skillman, D.R. & Patterson, J. 1988, *Astr. J.*, **96**, 976.
Skillman, D.R. & Patterson, J. 1993, *Astrophys. J.*, **417**, 298.
Skumanich, A. 1972, *Astrophys. J.*, **171**, 565.
Slovak, M.H. 1981, *Astrophys. J.*, **248**, 1059.
Slovak, M.H., Nelson, M.J. & Bless, R.C. 1988, *Bull. Amer. astr. Assoc.*, **20**, 1021.
Smak, J. 1967, *Acta Astr.*, **17**, 255.
Smak, J. 1969, *Acta Astr.*, **19**, 155.
Smak, J. 1970, *Acta Astr.*, **20**, 311.
Smak, J. 1971a, *Acta Astr.*, **21**, 15.
Smak, J. 1971b, *Acta Astr.*, **21**, 467.
Smak, J. 1971c, *Int. Astr. Union Colloq. No. 15*, p. 248.
Smak, J. 1972, *Acta Astr.*, **22**, 1.
Smak, J. 1975a, *Acta Astr.*, **25**, 371.
Smak, J. 1975b, *Acta Astr.*, **25**, 227.
Smak, J. 1976, *Acta Astr.*, **26**, 277.
Smak, J. 1979a, *Acta Astr.*, **29**, 309.
Smak, J. 1979b, *Acta Astr.*, **29**, 325.
Smak, J. 1979c, *Acta Astr.*, **29**, 469.
Smak, J. 1980, *Acta Astr.*, **30**, 267.
Smak, J. 1981, *Acta Astr.*, **31**, 395.
Smak, J. 1982a, *Acta Astr.*, **32**, 199.
Smak, J. 1982b, *Acta Astr.*, **32**, 213.
Smak, J. 1983, *Acta Astr.*, **33**, 333.
Smak, J. 1984a, *Acta Astr.*, **34**, 161.
Smak, J. 1984b, *Acta Astr.*, **34**, 317.
Smak, J. 1984c, *Acta Astr.*, **34**, 93.
Smak, J. 1984d, *Publ. astr. Soc. Pacific*, **96**, 5.
Smak, J. 1985a, *Acta Astr.*, **35**, 1.
Smak, J. 1985b, *Acta Astr.*, **35**, 357.
Smak, J. 1986, *Acta Astr.*, **36**, 211.
Smak, J. 1987, *Astrophys. Sp. Sci.*, **131**, 497.
Smak, J. 1989a, *Acta Astr.*, **39**, 317.
Smak, J. 1989b, *Acta Astr.*, **39**, 201.
Smak, J. 1989c, *Acta Astr.*, **39**, 41.
Smak, J. 1991a, in *Structure and Emission Properties of Accretion Disks*, eds. C. Bertout *et al.*, Ed. Frontières, Paris, p. 247.
Smak, J. 1991b, *Acta Astr.*, **41**, 269.
Smak, J. 1992, *Acta Astr.*, **42**, 323.
Smak, J. 1993a, *Int. Astr. Union Symp. No. 151*, p. 83.
Smak, J. 1993b, *Acta Astr.*, **43**, 121.
Smak, J. 1993c, *Acta Astr.*, **43**, 101.
Smak, J. 1994a, *Acta Astr.*, **44**, 45, 59.
Smak, J. 1994b, *Acta Astr.*, in press.
Smak, J. & Stepien, K. 1975, *Acta Astr.*, **25**, 379.
Smale, A.P., Corbet, R.H.D., Charles, P.A., Ilovaisky, S.A., Mason, K.O., Motch, C., Mukai, K.,

Naylor, T., Parmar, A.N., van der Klis, M. & van Paradijs, J. 1988, *Mon. Not. R. astr. Soc.*, **233**, 51.

Smart, W.M. 1953, *Celestial Mechanics*, Longmans, Green & Co., London.

Smith, M.A. 1979, *Publ. astr. Soc. Pacific*, **91**, 737.

Smith, R.C. 1990, in *Accretion-powered Compact Binaries*, ed. C.W. Mauche, Cambridge University Press, Cambridge, p. 89.

Smith, R.C., Fiddick, R.J., Hawkins, N.A. & Catalán, M.S. 1993, *Mon. Not. R. astr. Soc.*, **264**, 619.

Smits, D.P. 1991a, *Mon. Not. R. astr. Soc.*, **248**, 20.

Smits, D.P. 1991b, *Mon. Not. R. astr. Soc.*, **248**, 193.

Sneden, C. & Lambert, D.L. 1975, *Mon. Not. R. astr. Soc.*, **170**, 533.

Snijders, M.A.J. 1987, in *RS Ophiuchi (1987) and the Recurrent Nova Phenomenon*, ed. M.F. Bode, VNU Sci. Press, Utrecht, p. 51.

Snijders, M.A.J. 1990, *Lect. Notes Phys.*, **369**, 188.

Snijders, M.A.J., Batt, T.J., Roche, P.F., Seaton, M.J., Morton, D.C., Spoelstra, T.A.T. & Blades, J.C. 1987, *Mon. Not. R. astr. Soc.*, **228**, 329.

Soderblom, D.R. 1976, *Publ. astr. Soc. Pacific*, **88**, 517.

Soderblom, D.R. 1991, *NATO ASI Ser. C.*, **340**, 151.

Soker, N., Horpaz, A. & Livio, M. 1984, *Mon. Not. R. astr. Soc.*, **210**, 189.

Solf, J. 1983, *Astrophys. J.*, **273**, 647.

Solheim, J.-E. 1989, *Int. Astr. Union Colloq. No. 114*, p. 446.

Solheim, J.-E. 1992, *Int. Astr. Union Symp. No. 151*, p. 461.

Solheim, J.-E. 1993, in *White Dwarfs: Advances in Observation and Theory*, ed. M.A. Barstow, Kluwer, Dordrecht, p. 387.

Solheim, J.-E. & Kjeldseth-Moe, O. 1987, *Astrophys. Sp. Sci.*, **131**, 785.

Solheim, J.-E. & Sion, E.M. 1994, *Astr. Astrophys.*, **287**, 503.

Solheim, J.-E., Robinson, E.L., Nather, R.E. & Kepler, S.O. 1984, *Astr. Astrophys.*, **135**, 1.

Solheim, J.-E., Emanuelsen, P.-I., Vauclair, G., Dolez, N., Chevreton, M., Barstow, M., Sansom, A.E., Tweedy, R.W., Kepler, S.O., Kanaan, A., Fontaine, G., Bergeron, P., Grauer, A.D., Provencal, J.L., Winget, D.E., Nather, R.E., Bradley, P.A., Claver, C.E., Clemens, J.C., Kleinman, S.J., Hine, B.P., Marar, T.M.K., Seetha, S., Ashoka, B.N., Leibowitz, E.M. & Mazeh, T. 1991, in *White Dwarfs*, eds. G. Vauclair & E. Sion, Kluwer, Dordrecht, p. 431.

Sparks, W.M. 1969, *Astrophys. J.*, **156**, 569.

Sparks, W.M., Starrfield, S. & Truran, J.W. 1978, *Astrophys. J.*, **220**, 1063.

Sparks, W.M., Truran, J.W. & Starrfield, S. 1976, *Astrophys. J.*, **208**, 819.

Sparks, W.M., Sion, E.M., Starrfield, S.G. & Austin, S. 1993, in *Cataclysmic Variables and Related Physics*, eds. O. Regev & G. Shaviv, Inst. Phys. Publ., Bristol, p. 96.

Spencer Jones, H. 1931, *Cape Obs. Ann.*, **10**, part 9.

Spruit, H.C. 1987, *Astr. Astrophys.*, **184**, 173.

Spruit, H.C. 1989, in *Theory of Accretion Disks*, eds. F. Meyer, W.J. Duschl, J. Frank & E. Meyer-Hofmeister, Kluwer, Dordrecht, p. 325.

Spruit, H.C. 1991, *Rev. Mod. Astr.*, **4**, 197.

Spruit, H.C. & Ritter, H. 1983, *Astr. Astrophys.*, **124**, 267.

Spruit, H.C. & Taam, R.E. 1990, *Astr. Astrophys.*, **229**, 475.

Spruit, H.C. & Taam, R.E. 1993a, *Astr. Astrophys.*, in press.

Spruit, H.C. & Taam, R.E. 1993b, *Astrophys. J.*, **402**, 593.

Starrfield, S. 1986, in *Radiation Hydrodynamics*, eds. D. Mihalas & K.-H. Winkler, Reidel, Dordrecht, p. 225.

Starrfield, S. 1988, in *Multiwavelength Astrophysics*, ed. F.A. Cordova, Cambridge University Press, Cambridge, p. 159.

Starrfield, S. 1989, in *Classical Novae*, eds. M.F. Bode & A. Evans, Wiley, Chichester, p. 39.

Starrfield, S. 1990, *Int. Astr. Union Colloq. No. 122*, p. 127.

Starrfield, S. 1992, *Rev. Mod. Astr.*, **5**, 73.

Starrfield, S. & Snijders, M.A.J. 1987, *Astrophys. Sp. Sci. Libr.*, **129**, 377.

Starrfield, S., Sparks, W.M. & Shaviv, G. 1988, *Astrophys. J.*, **325**, L35.

Starrfield, S., Sparks, W.M. & Truran, J.W. 1974, *Astrophys. J.*, **192**, 647.

Starrfield, S., Sparks, W.M. & Truran, J.W. 1976, *Int. Astr. Union Symp. No. 73, p. 155.*

Starrfield, S., Sparks, W.M. & Truran, J.W. 1985, *Astrophys. J.*, **291**, 136.

Starrfield, S. Wagner, R.M., Liebert, J. & Sion, E.M. 1991, *Bull. Amer. astr. Soc.*, **23**, 20.
Staubert, R., König, M., Friedrich, S., Lamer, G., Sood, R.K., James, S.D. & Sharma, D.P. 1994, *Astr. Astrophys.*, **288**, 513.
Stauffer, J.R. 1987, *Astr. J.*, **94**, 996.
Stauffer, J., Spinrad, H. & Thorstensen, J. 1979, *Publ. astr. Soc. Pacific*, **91**, 59.
Stauffer, J.R., Giampapa, M.S., Herbst, W., Vincent, J.M., Hartmann, L.W. & Stern, R.A. 1991, *Astrophys. J.*, **374**, 142.
Stefano, R. Di. & Rappaport, S. 1994, *Astrophys. J.*, in press.
Steiman-Cameron, T.Y. & Imamura, J.N. 1988, *Bull. Amer. astr. Soc.*, **20**, 647.
Steiman-Cameron, T.Y., Imamura, J.N. & Steiman-Cameron, D.V. 1989, *Astrophys. J.*, **339**, 434.
Steiner, J.E., Schwartz, D.A., Jablonski, F.J., Busko, I.C., Watson, M.G., Pye, J.P. & McHardy, I.M. 1981, *Astrophys. J.*, **249**, L21.
Stella, L. & Rosner, R. 1984, *Astrophys. J.*, **277**, 312.
Stella, L., Beuermann, K. & Patterson, J. 1986, *Astrophys. J.*, **306**, 225.
Stephenson, F.R. 1976, *Quart. J. R. astr. Soc.*, **17**, 121.
Stephenson, F.R. 1986, in *RS Ophiuchi and the Recurrent Nova Phenomenon*, ed. M.F. Bode, VNU Sci. Press, Utrecht, p. 105.
Stephenson, C.B., Sanduleak, N. & Schild, R.E. 1968, *Astrophys. Letts.*, **2**, 247.
Stepién, K. 1991, *Acta Astr.*, **41**, 1.
Stepinski, T.F. 1991, *Publ. astr. Soc. Pacific*, **103**, 777.
Stepinski, T.F. & Levy, E.H. 1988, *Astrophys. J.*, **331**, 416.
Sterken, C., Vogt, N., Freeth, R., Kennedy, H.D., Marino, B.F., Page, A.A. & Walker, W.S.G. 1983, *Astr. Astrophys.*, **118**, 325.
Sterne, T.E. & Campbell, L. 1934, *Ann. Harv. Coll. Obs.*, **90**, 201.
Stickland, D.J., Penn, C.J., Seaton, M.J., Snijders, M.A.J. & Storey, P.J. 1981, *Mon. Not. R. astr. Soc.*, **197**, 107.
Stickland, D.J., Kelly, B.D., Cooke, J.A., Coulson, I., Engelbrecht, C., Kilkenny, D. & Spencer-Jones, J. 1984, *Mon. Not. R. astr. Soc.*, **206**, 819.
Stiening, R.F., Hildebrand, R.H. & Spillar, E.J. 1979, *Publ. astr. Soc. Pacific*, **91**, 384.
Stiening, R.F., Dragovan, M. & Hildebrand, R.H. 1982, *Publ. astr. Soc. Pacific*, **94**, 672.
Stienon, F.M. 1971, *Publ. astr. Soc. Pacific*, **83**, 363.
Still, M.D., Dhillon, V.S. & Jones, D.H.P. 1995, *Mon. Not. R. astr. Soc.*, **273**, 863.
Still, M.D., Dhillon, V.S. & Marsh, T.R. 1993, in *Cataclysmic Variables and Related Physics*, eds. O. Regev & G. Shaviv, Inst. Phys. Publ., Bristol, p. 321.
Still, M.D., Marsh, T.R., Dhillon, V.S. & Horne, K. 1994, *Mon. Not. R. astr. Soc.*, **267**, 957.
Stockman, H.S. 1988, in *Polarized Radiation of Circumstellar Origin*, eds. G.V. Coyne *et al.*, Vatican Obs., Vatican, p. 237.
Stockman, H.S. & Lubenow, A.F. 1987, *Astrophys. Sp. Sci.*, **131**, 607.
Stockman, H.S. & Sargent, T.A. 1979, *Astrophys. J.*, **227**, 197.
Stockman, H.S., Liebert, J. & Bond, H.E. 1979, *Int. Astr. Union Colloq. No. 53*, p. 334.
Stockman, H.S., Schmidt, G.D. & Lamb, D.Q. 1988, *Astrophys. J.*, **332**, 282.
Stockman, H.S., Schmidt, G.D., Angel, J.R.P., Liebert, J., Tapia, S. & Beaver, E.A. 1977, *Astrophys. J.*, **217**, 815.
Stockman, H.S., Foltz, C.B., Schmidt, G.D. & Tapia, S. 1983, *Astrophys. J.*, **271**, 725.
Stockman, H.S., Schmidt, G.D., Berriman, G., Liebert, J., Moore, R.L. & Wickramasinghe, D.T. 1992, *Astrophys. J.*, **401**, 628.
Stockman, H.S., Schmidt, G.D., Liebert, J. & Holberg, J.B. 1994, *Astrophys. J.*, **430**, *323*.
Stolz, B. & Schoembs, R. 1984, *Astr. Astrophys.*, **132**, 187.
Stover, R.J. 1981a, *Astrophys. J.*, **248**, 684.
Stover, R.J. 1981b, *Astrophys. J.*, **249**, 673.
Stover, R.J. 1983, *Publ. astr. Soc. Pacific*, **95**, 18.
Stover, R.J., Robinson, E.L. & Nather, R.E. 1981, *Astrophys. J.*, **248**, 696.
Stover, R.J., Robinson, E.L., Nather, R.E. & Montemayor, T.J. 1980, *Astrophys. J.*, **240**, 597.
Stratton, F.J.M. 1920, *Pub. Solar Phys. Obs. Camb.*, **4**, part 1.
Stratton, F.J.M. 1928, *Handb. der Astrophys.*, **6**, 254.
Stratton, F.J.M. 1936, *Pub. Solar Phys. Obs. Camb.*, **4**, 133.
Stratton, F.J.M. 1950, *Trans. Int. Astr. Union*, **7**, 305.

Stratton, F.J.M. & Manning, W.H., 1939, *An Atlas of Nova Herculis 1934*, Pub. Solar Phys. Obs. Camb., Cambridge, England.
Strittmater, P.A., Woolf, N.J., Thompson, R.I., Wilkerson, S., Angel, J.R.P., Stockman, H.S., Gilbert, G., Grandi, S.A., Larson, H. & Fink, U. 1977, *Astrophys. J.*, **216**, 23.
Struve, O. 1948, *Astrophys. J.*, **108**, 153.
Sulkanen, M.E., Brasure, L.W. & Patterson, J. 1981, *Astrophys. J.*, **244**, 579.
Sutherland, R.S. & Dopita, M.A. 1993, *Astrophys. J. Suppl.*, **88**, 253.
Swank, J.H. 1979, *Int. Astr. Union Colloq. No. 53*, p. 135.
Swank, J.H., Fabian, A.C. & Ross, R.R. 1984, *Astrophys. J.*, **280**, 734.
Swank, J.H., Lampton, M., Boldt, E., Holt, S. & Serlemitsos, P. 1977, *Astrophys. J.*, **216**, L71.
Swank, J.H., Boldt, E.A., Holt, S.S., Rothschild, R.E. & Serlemitsos, P.J. 1978, *Astrophys. J.*, **226**, L133.
Swedlund, J.B., Kemp, J.C. & Wolstencroft, R.D. 1974, *Astrophys. J.*, **193**, L11.
Swings, P. & Jose, P.D. 1949, *Astrophys. J.*, **110**, 475.
Swings, P. & Jose, P.D. 1952, *Astrophys. J.*, **116**, 229.
Swope, H.H. 1940, *Harvard Obs. Bull.*, No. 913, p. 11.
Szkody, P. 1976a, *Astrophys. J.*, **207**, 190.
Szkody, P. 1976b, *Astrophys. J.*, **210**, 168.
Szkody, P. 1976c, *Astrophys. J.*, **207**, 824.
Szkody, P. 1977, *Astrophys. J.*, **217**, 140.
Szkody, P. 1981a, *Astrophys. J.*, **247**, 577.
Szkody, P. 1981b, *Publ. astr. Soc. Pacific*, **93**, 456.
Szkody, P. 1982, *Astrophys. J.*, **261**, 200.
Szkody, P. 1985a, in *Cataclysmic Variables and Low-Mass X-Ray Binaries*, eds. D.Q. Lamb & J. Patterson, Reidel, Dordrecht, p. 385.
Szkody, P. 1985b, *Recent Results on Cataclysmic Variables*, ESA SP-236, Paris, p. 39.
Szkody, P. 1985c, *Astr. J.*, **90**, 1837.
Szkody, P. 1987a, *Astrophys. J. Suppl.*, **63**, 685.
Szkody, P. 1987b, *Astr. J.*, **94**, 1055.
Szkody, P. 1987c, *Astrophys. Sp. Sci.*, **130**, 69.
Szkody, P. 1988, *Publ. astr. Soc. Pacific*, **100**, 791.
Szkody, P. 1992, *Astr. Soc. Pacific Conf. Ser.* **29**, 326.
Szkody, P. 1993, in *Cataclysmic Variables and Related Physics*, eds. O. Regev & G. Shaviv, Inst. Phys. Publ., Bristol, p. 148.
Szkody, P. & Brownlee, D.E. 1977, *Astrophys. J.*, **212**, L113.
Szkody, P. & Capps, R.W. 1980, *Astr. J.*, **85**, 882.
Szkody, P. & Cropper, M. 1988, in *Multiwavelength Astrophysics*, ed. F. Cordova, Cambridge University Press, Cambridge, p. 109.
Szkody, P. & Crosa, L. 1981, *Astrophys. J.*, **251**, 620.
Szkody, P. & Downes, R.A. 1982, *Publ. astr. Soc. Pacific*, **94**, 328.
Szkody, P. & Feinswog, L. 1988, *Astrophys. J.*, **334**, 422.
Szkody, P. & Howell, S.B. 1989, *Astr. J.*, **97**, 1176.
Szkody, P. & Howell, S.B. 1992a, *Astrophys. J.*, **403**, 743.
Szkody, P. & Howell, S.B. 1992b, *Astrophys. J. Suppl.*, **78**, 537.
Szkody, P. & Ingram, D. 1992, *Int. Astr. Union Circ.*, No. 5516.
Szkody, P. & Ingram, D. 1994, *Astrophys. J.*, **420**, 830
Szkody, P. & Kiplinger, A.L. 1985, *Astrophys. J.*, **366**, 569.
Szkody, P. & Margon, B. 1980, *Astrophys. J.*, **236**, 862.
Szkody, P. & Mateo, M. 1984, *Astrophys. J.*, **280**, 729.
Szkody, P. & Mateo, M. 1986a, *Astrophys. J.*, **301**, 286.
Szkody, P. & Mateo, M. 1986b, *Astr. J.*, **92**, 483.
Szkody, P. & Mateo, M. 1988, *Publ. astr. Soc. Pacific*, **100**, 1111.
Szkody, P. & Mattei, J.A. 1984, *Publ. astr. Soc. Pacific*, **96**, 988.
Szkody, P. & Piché, F. 1990, *Astrophys. J.*, **361**, 235.
Szkody, P. & Shafter, A.W. 1983, *Publ. astr. Soc. Pacific*, **95**, 509.
Szkody, P. & Sion, E.M. 1988, *Int. Astr. Union Colloq. No. 114*, p. 92.
Szkody, P. & Sion, E.M. 1989, *Lect. Notes Phys.*, **328**, 92.

Szkody, P. & Wade, R.A. 1980, *Publ. astr. Soc. Pacific*, **92**, 806.
Szkody, P. & Wade, R.A. 1981, *Astrophys. J.*, **251**, 201.
Szkody, P., Bailey, J.A. & Hough, J.H. 1983, *Mon. Not. R. astr. Soc.*, **203**, 749.
Szkody, P., Downes, R. & Mateo, M. 1988, *Publ. astr. Soc. Pacific*, **100**, 362.
Szkody, P., Downes, R. & Mateo, M. 1990, *Publ. astr. Soc. Pacific*, **102**, 1310.
Szkody, P., Howell, S.B. & Kennicutt, R. 1986, *Publ. astr. Soc. Pacific*, **98**, 1151.
Szkody, P., Kii, T. & Osaki, Y. 1990, *Astr. J.*, **100**, 546.
Szkody, P., Liebert, J. & Panek, R.J. 1985, *Astrophys. J.*, **293**, 321.
Szkody, P., Mattei, J.A. & Mateo, M. 1985, *Publ. astr. Soc. Pacific*, **97**, 264.
Szkody, P., Michalsky, J.J. & Stokes, G.M. 1982, *Publ. astr. Soc. Pacific*, **94**, 137.
Szkody, P., Osborne, J. & Hassall, B.J.M. 1988, *Astrophys. J.*, **328**, 243.
Szkody, P., Piché, F. & Feinswog, L. 1990, *Astrophys. J. Suppl.*, **73**, 441.
Szkody, P., Raymond, J.C. & Capps, R.W. 1982, *Astrophys. J.*, **257**, 686.
Szkody, P., Shafter, A.W. & Cowley, A.P. 1984, *Astrophys. J.*, **282**, 236.
Szkody, P., Cordova, F.A., Tuohy, I.R., Stockman, H.S., Angel, R.P.J. & Wisniewski, W. 1980, *Astrophys. J.*, **241**, 1070.
Szkody, P., Schmidt, E., Crosa, L. & Schommer, R. 1981a, *Astrophys. J.*, **246**, 233.
Szkody, P., Crosa, L., Bothun, G.D., Downes, R.A. & Schommer, R.A. 1981b, *Astrophys. J.*, **249**, L61.
Szkody, P., Howell, S.B., Mateo, M. & Kreidl, T.J.N. 1989, *Publ. astr. Soc. Pacific*, **101**, 899.
Szkody, P., Mattei, J.A., Waagen, E.O. & Stablein, C. 1991, *Astrophys. J. Suppl.*, **76**, 359.
Szkody, P., Williams, R.E., Margon, B., Howell, S.B. & Mateo, M. 1992, *Astrophys. J.*, **387**, 357.
Szkody, P., Hoard, D.W., Patterson, J., Moulden, M., Howell, S. & Garnavich, P. 1994, *Astr. Soc. Pacific Conf. Ser.*, **56**, 350.
Sztanjo, M. 1978, *Inf. Bull. Var. Stars*, No. 1710.
Taam, R.E. 1983, *Astrophys. J.*, **268**, 361.
Taam, R.E. 1988, in *Critical Observations versus Physical Models for Close Binary Systems*, ed. K.C. Leung, Gordon & Breach, New York, p. 365.
Taam, R.E. 1989, *Highlights Astr.*, **8**, 155.
Taam, R.E. & Bodenheimer, P. 1989, *Astrophys. J.*, **337**, 849.
Taam, R.E. & Bodenheimer, P. 1991, *Astrophys. J.*, **373**, 246.
Taam, R.E. & McDermott, P.N. 1987, *Astrophys. J.*, **319**, L83.
Taam, R.E. & Spruit, H.C. 1989, *Astrophys. J.*, **345**, 972.
Taam, R.E., Flannery, B.P. & Faulkner, J. 1980, *Astrophys. J.*, **239**, 1017.
Tagger, M., Pellat, R. & Coroniti, F.V., 1992, *Astrophys. J.*, **393**, 708.
Takalo, L.O. & Nousek, J.A. 1988, *Astrophys. J.*, **327**, 328.
Tanzi, E.G., Tarenghi, M., Treves, A., Howarth, I.D., Willis, A.J. & Wilson, R. 1980, *Astr. Astrophys.*, **83**, 270.
Tapia, S. 1977a, *Astrophys. J.*, **212**, L125.
Tapia, S. 1977b, *Int. Astr. Union Circ.*, No. 3054.
Tassoul, J.-L. 1978, *Theory of Rotating Stars*, Princeton Univ. Press, Princeton p. 248.
Tassoul, J.-L. 1988, *Astrophys. J.*, **324**, L71.
Tassoul, J.-P. & Tassoul, M. 1992, *Astrophys. J.*, **395**, 259.
Tayler, R.J. 1980, *Mon. Not. R. astr. Soc.*, **191**, 135.
Taylor, A.R., Seaquist, E.R., Hollis, J.M. & Pottasch, S.R. 1987, *Astr. Astrophys.*, **183**, 38.
Taylor, A.R., Hjellming, R.,M., Seaquist, E.R. & Gehrz, R.D. 1988, *Nature*, **335**, 235.
Taylor, A.R., Davis, R.J., Porcas, R.W. & Bode, M.F. 1989, *Mon. Not. R. astr. Soc.*, **237**, 81.
Tempesti, P. 1975, *Inf. Bull. Var. Stars*, No. 974.
Thackeray, A.D., Wesselink, A.J. & Oosterhoff, P. Th. 1950, *Bull. astr. Inst. Nethl.*, **11**, 193.
Thomas, J.H. 1979, *Nature*, **280**, 662.
Thorstensen, J.R. 1986, *Astr. J.*, **91**, 940.
Thorstensen, J.R. & Freed, I.W. 1985, *Astr. J.*, **90**, 2082.
Thorstensen, J.R. & Ringwald, F.A. 1994. Preprint.
Thorstensen, J.R. & Wade, R.A. 1986, *Astrophys. J.*, **309**, 721.
Thorstensen, J.R., Davis, M.K. & Ringwald, F.A. 1991, *Astr. J.*, **102**, 683.
Thorstensen, J.R., Patterson, J. & Thomas, G. 1993, in preparation.
Thorstensen, J.R., Schommer, R.A. & Charles, P.A. 1983, *Publ. astr. Soc. Pacific*, **95**, 564.

Thorstensen, J.R., Smak, J. & Hessman, F.V. 1985, *Publ. astr. Soc. Pacific*, **97**, 437.
Thorstensen, J.R., Wade, R.A. & Oke, J.B. 1986, *Astrophys. J.*, **309**, 721.
Thorstensen, J.R., Ringwald, F.A., Wade, R.A., Schmidt, G.D. & Nosworthy, J.E. 1991, *Astr. J.*, **102**, 272.
Tomaney, A.B. & Shafter, A.W. 1993, *Astrophys. J.*, **411**, 640.
Torbett, M.V. & Campbell, B. 1987, *Astrophys. J.*, **318**, L29.
Tout, C.A. & Pringle, J.E. 1992a, *Mon. Not. R. astr. Soc.*, **256**, 269.
Tout, C.A. & Pringle, J.E. 1992b, *Mon. Not. R. astr. Soc.*, **259**, 604.
Tout, C.A., Pringle, J.E. & la Dous, C. 1993, *Mon. Not. R. astr. Soc.*, **265**, L5.
Treves, A. 1978, *Astr. Astrophys.*, **67**, 441.
Treves, A., Maraschi, L. & Abramowicz, M. 1988, *Publ. astr. Soc. Pacific*, **100**, 427.
Trümper, J., Hasinger, G., Aschenbach, B., Bräuninger, H., Briel, U.G., Burkert, W., Fink, H., Pfefferman, E., Pietsch, W., Predehl, P., Schmitt, J.H.M.M., Voges, W., Zimmerman, U. & Beuermann, K. 1991, *Nature*, **349**, 579.
Truran, J.W. 1981, *Prog. Part. Nucl. Phys.*, **6**, 177.
Truran, J.W. 1982, in *Essays in Nuclear Astrophysics*, eds. C.A. Barnes, D.D. Clayton, J.W. Truran & D. Schramm, Cambridge University Press, Cambridge, p. 467.
Truran, J.W. 1985, in *Production and Distribution of CNO Elements*, ed. I.J. Danziger, ESO, Garching, p. 211.
Truran, J.W. 1990, in *Physics of Classical Novae*, eds. A. Cassatella & R. Viotti, Springer-Verlag, Berlin, p. 373.
Truran, J.W. & Livio, M. 1986, *Astrophys. J.*, **308**, 721.
Truran, J.W. & Livio, M. 1989, *Lect. Notes Phys.*, **328**, 498.
Truran, J.W., Livio, M., Hayes, J., Starrfield, S. & Sparks, W.M. 1988, *Astrophys. J.*, **324**, 345.
Tucker, W.H. 1975, *Radiation Processes in Astrophysics*, M.I.T. Press, Cambridge, Mass.
Tukey, J.W. 1971, *Exploratory Data Analysis*, Addison-Wesley, Reading, Massachusetts.
Tuohy, I.R., Visvanathan, N. & Wickramasinghe, D.T. 1985, *Astrophys. J.*, **289**, 721.
Tuohy, I.R., Lamb, F.K., Garmire, G.P. & Mason, K.O. 1978, *Astrophys. J.*, **226**, L17.
Tuohy, I.R., Mason, K.O., Garmire, G.P. & Lamb, F.K. 1981, *Astrophys. J.*, **245**, 183.
Tuohy, I.R., Buckley, D.A.H., Remillard, R.A., Bradt, H.V. & Schwartz, D.A. 1986, *Astrophys. J.*, **311**, 275.
Tuohy, I.R., Ferrario, L., Wickramasinghe, D.T. & Hawkins, M.R.S. 1988, *Astrophys. J.*, **328**, L59.
Tuohy, I.R., Remillard, R.A., Brissenden, R.J.V. & Bradt, H.V. 1990, *Astrophys. J.*, **359**, 204.
Turner, H.H. 1906, *Mon. Not. R. astr. Soc.*, **67**, 119.
Turner, H.H. 1907, *Mon. Not. R. astr. Soc.*, **67**, 316.
Turner, K.C. 1985, in *Radio Stars*, eds. R.M. Hjellming & D.M. Gibson, Reidel, Dordrecht, p. 283.
Tutukov, A.V. & Fedorova, A.V. 1989, *Sov. Astr.*, **33**, 606.
Tutukov, A.V., Fedorova, A.V. & Yungelson, L.R. 1982, *Pis'ma Astr. Zh.*, **8**, 365.
Tylenda, R. 1977, *Acta Astr.*, **27**, 235.
Tylenda, R. 1978, *Acta Astr.*, **28**, 333.
Tylenda, R. 1981a, *Acta Astr.*, **31**, 127.
Tylenda, R. 1981b, *Acta Astr.*, **31**, 267.
Udalski, A. 1988a, *Acta Astr.*, **38**, 315.
Udalski, A. 1988b, *Inf. Bull. Var. Stars*, No. 3239.
Udalski, A. 1990a, *Inf. Bull. Var. Stars*, No. 3425.
Udalski, A. 1990b, *Astr. J.*, **100**, 226.
Udalski, A. & Pych, W. 1992, *Acta Astr.*, **42**, 285.
Udalski, A. & Schwarzenberg-Czerny, A. 1989, *Acta Astr.*, **39**, 125.
Udalski, A. & Szymanski, M. 1988, *Acta Astr.*, **38**, 215.
Ulla, A.M. & Solheim, J.-E. 1990, *Astrophys. Sp. Sci.*, **169**, 189.
Ulla, A.M. & Solheim, J.-E. 1991, *NATO A.S.I. Ser. C.*, **336**, 441.
Ulrich, R.K. & Burger, H.L. 1976, *Astrophys. J.*, **206**, 509.
Vahia, M.N. & Rao, A.R. 1988, *Astr. Astrophys.*, **207**, 55.
Vahia, M.N., Rao, A.R. & Singh, R.K. 1991, *Astr. Astrophys.*, **250**, 424.
Vaidya, J., Agrawal, P.C., Apparao, K.M.V., Manchanda, R.K., Vivekand Rao, P. & Sarma, M.B.K. 1988, *Astr. Astrophys. Suppl.*, **75**, 43.
Valentijn, E.A. 1990, *Nature*, **346**, 153.

van Amerongen, S. & van Paradijs, J. 1989, *Astr. Astrophys.*, **219**, 195.
van Amerongen, S., Augusteijn, T. & van Paradijs, J. 1987, *Mon. Not. R. astr. Soc.*, **228**, 377.
van Amerongen, S., Bovenschen, H. & van Paradijs, J. 1987, *Mon. Not. R. astr. Soc.*, **229**, 245.
van Amerongen, S., Kuulkers, E. & van Paradijs, J. 1990, *Mon. Not. R. astr. Soc.*, **242**, 522.
van Amerongen, S., Kraakman, H., Damen, E., Tjemkes, S. & van Paradijs, J. 1985, *Mon. Not. R. astr. Soc.*, **215**, 45P.
van Amerongen, S., Damen, E., Groot, M., Kraakman, H. & van Paradijs, J. 1987, *Mon. Not. R. astr. Soc.*, **225**, 93.
van Ballegooijen, A.A. 1989, in *Accretion Disks & Magnetic Fields in Astrophysics*, ed. G. Belvedere, Kluwer, Dordrecht, p. 99.
van Biesbroek, D. 1904, *Ann. Obs. Belg.*, **13**, 21.
Vandenberg, D.A., Hartwick, F.D.A., Dawson, P. & Alexander, D.R. 1983, *Astrophys. J.*, **266**, 747.
van den Bergh, S. 1977, *Publ. astr. Soc. Pacific*, **89**, 637.
van den Bergh, S. 1988, *Publ. astr. Soc. Pacific*, **100**, 8.
van den Bergh, S. 1991, *Publ. astr. Soc. Pacific*, **103**, 609.
van den Bergh, S. & Pritchet, C.J. 1986, *Publ. astr. Soc. Pacific*, **98**, 110.
van den Bergh, S. & Younger, P.F. 1987, *Astr. Astrophys. Suppl.*, **70**, 125.
van den Heuvel, E.P.J., Bhattacharya, D., Nomoto, K. & Rappaport, S.A. 1992, *Astr. Astrophys.*, **262**, 97.
van der Bilt, J. 1908, *Rech. Astr. Utrecht*, Vol. 3.
van der Hucht, K.A., Jurriens, T.A., Olnon, F.M., The, P.S., Wesselius, P.R. & Williams, P.M. 1985, *Astr. Astrophys.*, **145**, L13.
van der Woerd, H. 1987, *Astrophys. Sp. Sci.*, **130**, 225.
van der Woerd, H. 1988, *Adv. Sp. Res.*, **8**, 265.
van der Woerd, H. & Heise, J. 1987, *Mon. Not. R. astr. Soc.*, **225**, 141.
van der Woerd, H. & van Paradijs, J. 1987, *Mon. Not. R. astr. Soc.*, **224**, 271.
van der Woerd, H., de Kool, M. & van Paradijs, J. 1984, *Astr. Astrophys.*, **131**, 137.
van der Woerd, H., Heise, J. & Bateson, F. 1986, *Astr. Astrophys.*, **156**, 252.
van der Woerd, H., Heise, J., Paerels, F., Beuermann, K., van der Klis, M., Motch, C. & van Paradijs, J. 1987, *Astr. Astrophys.*, **182**, 219.
van der Woerd, H, van der Klis, M., van Paradijs, J., Beuermann, K. & Motch, C. 1988, *Astrophys. J.*, **330**, 911.
van Horn, H.M., Wesemael, F. & Winget, D.E. 1980, *Astrophys. J.*, **235**, L143.
van Paradijs, J. 1983, *Astr. Astrophys.*, **125**, L16.
van Paradijs, J. 1985, *Astr. Astrophys.*, **144**, 199.
van Paradijs, J. 1986, *Mon. Not. R. astr. Soc.*, **218**, 31P.
van Paradijs, J., Kraakman, H. & van Amerongen, S. 1989, *Astr. Astrophys. Suppl.*, **79**, 205.
van Paradijs, J., van der Klis, M. & Pedersen, H. 1989, *Astr. Astrophys.*, **225**, L5.
van Paradijs, J., Verbunt, F., van den Heuvel, E.P.J., van der Linden, Th.J., Brand, J. & van Leeuwen, F. 1981, *Astr. Astrophys. Suppl.*, **46**, 89.
van Paradijs, J., Charles, P.A., Harlaftis, E.T., Arevalo, M.J., Baruch, J.E.F., Callanan, P.J., Casares, J., Dhillon, V.S., Gimenez, A., Gonzalez, R., Martinez-Pais, I.G., Jones, D.H.P., Hassall, B.J.M., Hellier, C., Kidger, M.R., Lazaro, C., Marsh, T.R., Mason, K.O., Mukai, K., Naylor, T., Reglero, V., Rutten, R.G.M. & Smith, R.C. 1994, *Mon. Not. R. astr. Soc.*, **267**, 465.
van Teeseling, A., Verbunt, F. & Heise, J. 1993, *Astr. Astrophys.*, **270**, 159.
van't Veer, F. & Maceroni, C. 1988, *Astr. Astrophys.*, **199**, 183.
Veeder, G.J. 1974, *Astr. J.*, **79**, 1056.
Vennes, S., Thorstensen, J.R., Thejll, P. & Shipman, H.L. 1991, *Astrophys. J.*, **372**, L37.
Verbunt, F. 1982, *Sp. Sci. Rev.*, **32**, 379.
Verbunt, F. 1984, *Mon. Not. R. astr. Soc.*, **209**, 227.
Verbunt, F. 1987, *Astr. Astrophys. Suppl.*, **71**, 339.
Verbunt, F. 1991, *Adv. Sp. Res.*, **11**, 57.
Verbunt, F. & Hut, P. 1983, *Astr. Astrophys.*, **127**, 161.
Verbunt, F. & Meylan, G. 1988, *Astr. Astrophys.*, **203**, 297.
Verbunt, F. & Rappaport, S. 1988, *Astrophys. J.*, **332**, 193.
Verbunt, F. & Zwaan, C. 1981, *Astr. Astrophys.*, **100**, L7.
Verbunt, F., van den Heuvel, E.P.J., van der Linden, Th.J., Brand, J., van Leeuwen, F. & van

Paradijs, J. 1980, *Astr. Astrophys.*, **86**, L10.
Verbunt, F., Pringle, J.E., Wade, R.A., Echevarria, J., Jones, D.H.P., Argyle, R.W., Schwarzenberg-Czerny, A., la Dous, C. & Schoembs, R. 1984, *Mon. Not. R. astr. Soc.*, **210**, 197.
Verbunt, F., Hassall, B.J.M., Pringle, J.E., Warner, B. & Marang, F. 1987, *Mon. Not. R. astr. Soc.*, **225**, 113.
Vidal, N.V. & Wickramasinghe, D.T. 1974, *Astr. Astrophys.*, **36**, 309.
Vila, S.C. 1978, *Astrophys. J.*, **223**, 979.
Vilhu, O. & Moss, D. 1986, *Astr. J.*, **92**, 1178.
Vilhu, O. & Walter, F.M. 1987, *Astrophys. J.*, **321**, 958.
Villata, M. 1992, *Mon. Not. R. astr. Soc.*, **257**, 450.
Viotti, R. 1988, *Int. Astr. Union Colloq. No. 103*, p. 269.
Viotti, R. 1990, *Int. Astr. Union Colloq. No. 122*, p. 416.
Vishniac, E.T. 1994, *Astr. Soc. Pacific Conf. Ser.*, **56**, 25.
Vishniac, E.T. & Diamond, P. 1989, *Astrophys. J.*, **347**, 435.
Vishniac, E.T. & Diamond, P. 1992, *Astrophys. J.*, **398**, 561.
Vishniac, E.T., Jin, L. & Diamond, P.H. 1990, *Astrophys. J.*, **365**, 648.
Vishniac, E.T., Jin, L. & Diamond, P.H. 1991, in *Proc. Twelfth Int. Symp. Nonlinear Acoustics*, in press.
Visvanathan, N. & Tuohy, I. 1983, *Astrophys. J.*, **275**, 709.
Visvanathan, N. & Wickramasinghe, D.T. 1979, *Nature*, **281**, 47.
Visvanathan, N. & Wickramasinghe, D.T. 1981, *Mon. Not. R. astr. Soc.*, **196**, 275.
Vitello, P.A.J. & Shlosman, I. 1993, *Astrophys. J.*, **410**, 815.
Vogel, H.C. 1874, *Astr. Nach.*, **84**, 113.
Vogel, M. & Nussbaumer, H. 1992, *Astr. Astrophys.*, **259**, 525.
Vogt, N. 1974, *Astr. Astrophys.*, **36**, 369.
Vogt, N. 1975, *Astr. Astrophys.*, **41**, 15.
Vogt, N. 1976, *Int. Astr. Union Symp. No. 73*, p. 147.
Vogt, N. 1979, *Mitt. Astr. Gesell.*, **45**, 158.
Vogt, N. 1980, *Astr. Astrophys.*, **88**, 66.
Vogt, N. 1981, *Mitt. Astr. Gesell.*, **57**, 79.
Vogt, N. 1982, *Astrophys. J.*, **252**, 563.
Vogt, N. 1983a, *Astr. Astrophys.*, **118**, 95.
Vogt, N. 1983b, *Astr. Astrophys.*, **128**, 29.
Vogt, N. 1983c, *Astr. Astrophys. Suppl.*, **53**, 21.
Vogt, N. 1989, in *Classical Novae*, eds. M.F. Bode & A. Evans, Wiley, Chichester, p. 225.
Vogt, N. 1990, *Astrophys. J.*, **356**, 609.
Vogt, N. & Bateson, F.M. 1982, *Astr. Astrophys. Suppl.*, **48**, 383.
Vogt, N. & Breysacher, J. 1980, *Astrophys. J.*, **235**, 945.
Vogt, N. & Semeniuk, I. 1980, *Astr. Astrophys.*, **89**, 223.
Vogt, N., Krzeminski, W. & Sterken, C. 1980, *Astr. Astrophys.*, **85**, 106.
Vogt, N., Schoembs, R., Krzeminski, W. & Pedersen, H. 1981, *Astr. Astrophys.*, **94**, L29.
Vogt, N., Barrera, L.H., Barwig, H. & Mantel, K.H. 1990, in *Accretion-Powered Compact Binaries*, ed. C.W. Mauche, Cambridge University Press, Cambridge, p. 391.
Vojkhanskaya, N.F. 1973, *Astrofiz. Issled., Isv. Spets. Astrofiz. Obs.*, **5**, 89.
Vojkhanskaya, N.F. 1980, *Sov. Astr. Zh.*, **57**, 520.
Vojkhanskaya, N.F. 1983, Sov. *Astr. J.*, **26**, 558.
Vojkhanskaya, N.F. 1984, *Sov. Astr.*, **26**, 665.
Vojkhanskaya, N.F. 1985a, *Pis'ma Astr. Zh.*, **11**, 916.
Vojkhanskaya, N.F. 1985b, *Sov. Astr. Letts.*, **11**, 385.
Vojkhanskaya, N.F. 1986a, *Astr. Zh.*, **63**, 516.
Vojkhanskaya, N.F. 1986b, *Pis'ma Astr. Zh.*, **12**, 468.
Vojkhanskaya, N.F. 1988, *Pis'ma Ast. Zh.*, **65**, 797.
Vojkhanskaya, N.F. 1989, *Close Binary Systems of AM Her Type*, Publ. of USSR Special Astrophys. Obs. Crimea.
Vojkhanskaya, N.F., Gnedin, Yu. N., Borisov, N.V., Natsvlishvili, N.F. & Fabrika, S.N. 1987, *Pis'ma Ast. Zh.*, **13**, 495.
Volkov, I.M., Shugarov, S.Yu. & Seregina, T.M. 1986, *Astr. Tsirk.*, No. 1418.

Voloshina, I.B. 1986, *Sov. Astr. Letts.*, **12**, 89.
Voloshina, I.B. & Lyutyi, V.M. 1984, *Sov. Astr. Letts.*, **10**, 319.
Voloshina, I.B. & Shugarov. S. Yu. 1989, *Sov. Astr. Letts.*, **15**, 312.
Volpi, A., Natali, G. & D'Antona, F. 1988, *Astr. Astrophys.*, **193**, 87.
Vrielmann, S. & Bruch, A. 1993, *Astr. Gesell. Abstr. Ser. 9*, p. 154.
Vrtilek, S.D., Silber, A., Raymond, J.C. & Patterson, J. 1994, *Astrophys. J.*, **425**, 787.
Wachmann, A.A. 1935, *Astr. Nach.*, **255**, 341.
Wade, R.A. 1979, *Astr. J.*, **84**, 562.
Wade, R.A. 1981, *Astrophys. J.*, **246**, 215.
Wade, R.A. 1982, *Astr. J.*, **87**, 1558.
Wade, R.A. 1984, *Mon. Not. R. astr. Soc.*, **208**, 381.
Wade, R.A. 1985, in *Interacting Binaries*, eds. P.P. Eggleton & J.E. Pringle, Reidel, Dordrecht, p. 289.
Wade, R.A. 1988, *Astrophys. J.*, **335**, 394.
Wade, R.A. 1990, *Int. Astr. Union Colloq. No. 122*, p. 179.
Wade, R.A. & Horne, K. 1988, *Astrophys. J.*, **324**, 411.
Wade, R.A. & Ward, M.J. 1985, in *Interacting Binary Stars*, eds. J.E. Pringle & R.A. Wade, Cambridge University Press, Cambridge, p. 129.
Wagner, R.M., Sion, E.M., Liebert, J. & Starrfield, S. 1988, *Astrophys. J.*, **328**, 213.
Walker, M.F. 1954a, *Publ. astr. Soc. Pacific*, **66**, 71.
Walker, M.F. 1954b, *Publ. astr. Soc. Pacific*, **66**, 230.
Walker, M.F. 1956, *Astrophys. J.*, **123**, 68.
Walker, M.F. 1957, *Int. Astr. Union Symp. No. 3*, p. 46.
Walker, M.F. 1958, *Astrophys. J.*, **127**, 319.
Walker, M.F. 1961, *Astrophys. J.*, **134**, 171.
Walker, M.F. 1963a, *Astrophys. J.*, **137**, 485.
Walker, M.F. 1963b, *Astrophys. J.*, **138**, 313.
Walker, M.F. 1965a, *Mitt. Sternw. Budapest*, No. 57.
Walker, M.F. 1965b, *Sky & Tel.*, **29**, 23.
Walker, A.R. 1977, *Mon. Not. R. astr. Soc.*, **178**, 248.
Walker, A.R. 1979, *Publ. astr. Soc. Pacific*, **91**, 46.
Walker, M.F. 1981, *Astrophys. J.*, **248**, 256.
Walker, M.F. & Bell, M. 1980, *Astrophys. J.*, **237**, 89.
Walker, M.F. & Chincarini, G. 1968, *Astrophys. J.*, **154**, 157.
Walker, M.F. & Herbig, G.H. 1954, *Astrophys. J.*, **120**, 278.
Walker, W.S.G. & Marino, B. 1978, *Pub. Var. Star Sect. R. astr. Soc. New Zealand*, **6**, 73.
Wallerstein, G. 1963, *Publ. astr. Soc. Pacific*, **75**, 26.
Wallerstein, G. & Cassinelli, J.P. 1968, *Publ. astr. Soc. Pacific*, **95**, 135.
Walter, F.M. & Bowyer, S. 1981, *Astrophys. J.*, **245**, 671.
Wampler, E.J. 1967, *Astrophys. J.*, **149**, L101.
Wang, Y.-M. 1987, *Astr. Astrophys.*, **183**, 257.
Wargau, W., Rahe, J. & Vogt, N. 1983, *Astr. Astrophys.*, **117**, 283.
Wargau, W., Drechsel, H., Rahe, J. & Bruch, A. 1983, *Mon. Not. R. astr. Soc.*, **204**, 35P.
Wargau, W., Bruch, A., Drechsel, H., Rahe, J. & Schoembs, R. 1984, *Astrophys. Sp. Sci.*, **99**, 145.
Warner, B. 1971, *Publ. astr. Soc. Pacific*, **83**, 817.
Warner, B. 1972a, *Mon. Not. R. astr. Soc.*, **158**, 425.
Warner, B. 1972b, *Mon. Not. R. astr. Soc.*, **159**, 315.
Warner, B. 1973a, *Mon. Not. astr. Soc. S. Af.*, **32**, 120.
Warner, B. 1973b, *Mon. Not. R. astr. Soc.*, **163**, 25P.
Warner, B. 1973c, *Mon. Not. R. astr. Soc.*, **162**, 189.
Warner, B. 1974a, *Mon. Not. R. astr. Soc.*, **168**, 235.
Warner, B. 1974b, *Mon. Not. astr. Soc. S. Af.*, **33**, 21.
Warner, B. 1975a, *Mon. Not. R. astr. Soc.*, **173**, 37P.
Warner, B. 1975b, *Mon. Not. R. astr. Soc.*, **170**, 219.
Warner, B. 1976a, *Int. Astr. Union Symp. No. 73*, p. 85.
Warner, B. 1976b, *Observatory*, **96**, 49.
Warner, B. 1978, *Acta Astr.*, **28**, 303.
Warner, B. 1979, in *White Dwarfs and Variable Degenerate Stars*, eds. H. van Horne & V.

Weidemann, Rochester, N.Y., p. 417.
Warner, B. 1980, *Mon. Not. R. astr. Soc.*, **190**, 69P.
Warner, B. 1981, *Mon. Not. R. astr. Soc.*, **195**, 101.
Warner, B. 1983a, in *Cataclysmic Variables and Related Objects*, eds. M. Livio & G. Shaviv, Reidel, Dordrecht, p. 155.
Warner, B. 1983b, *Inf. Bull. Var. Stars*, No. 2397.
Warner, B. 1983c, *Inf. Bull. Var. Stars*, No. 2295.
Warner, B. 1985a, in *Cataclysmic Variables & Low-Mass X-Ray Binaries*, eds. D.Q. Lamb & J. Patterson, Reidel, Dordrecht, p. 269.
Warner, B. 1985b, in *Interacting Binaries*, eds. P.P. Eggleton & J.E. Pringle, Reidel, Dordrecht, p. 367.
Warner, B. 1985c, in *Proc. 9th N. Amer. Workshop on Cataclys. Var.*, ed. P. Szkody, Univ. Washington, Seattle, p. 142.
Warner, B. 1985d, *Mon. Not. R. astr. Soc.*, **217**, 1P.
Warner, B. 1986a, *Mon. Not. R. astr. Soc.*, **219**, 347.
Warner, B. 1986b, *Astrophys. Sp. Sci.*, **118**, 271.
Warner, B. 1986c, *Mon. Not. astr. Soc. S. Af.*, **45**, 117.
Warner, B. 1986d, *Mon. Not. R. astr. Soc.*, **219**, 751.
Warner, B. 1986e, *Mon. Not. R. astr. Soc.*, **222**, 11.
Warner, B. 1986f, *J. Amer. Assoc. Var. Star Obs.*, **15**, 163.
Warner, B. 1987a, *Mon. Not. R. astr. Soc.*, **227**, 23.
Warner, B. 1987b, *Lect. Notes Phys.*, **274**, 384.
Warner, B. 1988a, *Nature*, **336**, 129.
Warner, B. 1988b, *High Speed Astronomical Photometry*, Cambridge University Press, Cambridge.
Warner, B. 1990, *Astrophys. Sp. Sci.*, **164**, 79.
Warner, B. & Brickhill, A.J. 1974, *Mon. Not. R. astr. Soc.*, **166**, 673.
Warner, B. & Brickhill, A.J. 1978, *Mon. Not. R. astr. Soc.*, **182**, 777.
Warner, B. & Cropper, M.S. 1984, *Mon. Not. R. astr. Soc.*, **206**, 261.
Warner, B. & Harwood, J. 1973, *Inf. Bull. Var. Stars*, No. 756.
Warner, B. & Livio, M. 1987, *Astrophys. J.*, **322**, L95.
Warner, B. & McGraw, J.T. 1981, *Mon. Not. R. astr. Soc.*, **196**, 59P.
Warner, B. & Nather, R.E. 1971, *Mon. Not. R. astr. Soc.*, **152**, 219.
Warner, B. & Nather, R.E. 1972a, *Mon. Not. R. astr. Soc.*, **156**, 305.
Warner, B. & Nather, R.E. 1972b, *Mon. Not. R. astr. Soc.*, **159**, 429.
Warner, B. & Nather, R.E. 1972c, *Mon. Not. R. astr. Soc.*, **156**, 297.
Warner, B. & Nather, R.E., 1988, *Inf. Bull. Var. Stars*, No. 3140.
Warner, B. & O'Donoghue, D. 1987, *Mon. Not. R. astr. Soc.*, **224**, 733.
Warner, B. & O'Donoghue, D. 1988, *Mon. Not. R. astr. Soc.*, **233**, 705.
Warner, B. & Robinson, E.L. 1972a, *Nature Phys. Sci.*, **239**, 2.
Warner, B. & Robinson, E.L. 1972b, *Mon. Not. R. astr. Soc.*, **159**, 101.
Warner, B. & Thackeray, A.D. 1975, *Mon. Not. R. astr. Soc.*, **172**, 433.
Warner, B. & van Citters, G.W. 1974, *Observatory*, **94**, 116.
Warner, B. & Wickramasinghe, D.T. 1991, *Mon. Not. R. astr. Soc.*, **248**, 370.
Warner, B., O'Donoghue, D. & Allen, S. 1985, *Mon. Not. R. astr. Soc.*, **212**, 9P.
Warner, B., O'Donoghue, D. & Fairall, A.P. 1981, *Mon. Not. R. astr. Soc.*, **196**, 705.
Warner, B., O'Donoghue, D. & Wargau, W. 1989, *Mon. Not. R. astr. Soc.*, **238**, 73.
Warner, B., Peters, W.L., Hubbard, W.B. & Nather, R.E. 1972, *Mon. Not. R. astr. Soc.*, **159**, 321.
Warren, J.K., Vallerga, J.V., Mauche, C.W., Mukai, K. & Siegmund, O.H.W. 1993, *Astrophys. J.*, **414**, L69.
Warwick, R.S., Turner, M.J.L., Watson, M.G. & Willingale, R. 1985, *Nature*, **317**, 218.
Watanabe, M., Hirosawa, K., Kato, T. & Narumi, H. 1989, *Var. Star. Bull. Japan*, **10**, 40.
Watson, M.G. 1986, in *Physics of Accretion onto Compact Objects*, eds. K.O. Mason, M.G. Watson & N.E. White, Springer-Verlag, Berlin, p. 97.
Watson, M.G., King, A.R. & Heise, J. 1985, *Sp. Sci. Rev.*, **40**, 127.
Watson, M.G., King, A.R. & Osborne, J. 1985, *Mon. Not. R. astr. Soc.*, **212**, 917.
Watson, M.G., King, A.R. & Williams, G.A. 1987, *Mon. Not. R. astr. Soc.*, **226**, 867.
Watson, M.G., Mayo, S.K. & King, A.R. 1980, *Mon. Not. R. astr. Soc.*, **192**, 689.
Watson, M.G., Sherrington, M.R. & Jameson, R.F. 1978, *Mon. Not. R. astr. Soc.*, **187**, 79P.

Watson, M.G., King, A.R., Jones, M.H. & Motch, C. 1989, *Mon. Not. R. astr. Soc.*, **237**, 299.
Watson, M.G., Rosen, S.R., O'Donoghue, D., Buckley, D., Warner, B., Hellier, C., Ramseyer, T., Done, C. & Madejski, G. 1995, *Mon. Not. R. astr. Soc.*, **273**, 681.
Watts, D.J., Greenhill, J.G., Hill, P.W. & Thomas, D.M. 1982, *Mon. Not. R. astr. Soc.*, **200**, 1039.
Watts, D.J., Giles, A.B., Greenhill, J.G., Hill, K. & Bailey, J. 1985, *Mon. Not. R. astr. Soc.*, **215**, 83.
Watts, D.J., Bailey, J., Hill, P.W., Greenhill, J.G., McCowage, C. & Carty, T. 1986, *Astr. Astrophys.*, **154**, 197.
Weaver, H.F. 1951, *Astrophys. J.*, **113**, 320.
Weaver, H. 1974, *Highlights of Astr.*, **3**, 509.
Webbink, R.F. 1976, *Nature*, **262**, 271.
Webbink, R.F. 1979a, *Int. Astr. Union Colloq. No. 46*, p. 102.
Webbink, R.F. 1979b, *Int. Astr. Union Colloq. No. 53,* p. 426.
Webbink, R.F. 1985, in *Interacting Binaries*, eds. J.E. Pringle & R.A. Wade, Cambridge University Press, Cambridge, p. 39.
Webbink, R.F. 1990a, in *Accretion-Powered Compact Binaries*, ed. C.W. Mauche, Cambridge University Press, Cambridge p. 177.
Webbink, R.F. 1990b, *Int. Astr. Union Colloq. No. 122*, p. 405.
Webbink, R.F. 1991, private communication.
Webbink, R.F. & Iben, I. 1987, *Int. Astr. Union Colloq. No. 95*, p. 445.
Webbink, R.F., Rappaport, S. & Savonije, G.J. 1983, *Astrophys. J.*, **270**, 678.
Webbink, R.F., Livio, M., Truran, J.W. & Orio, M. 1987, *Astrophys. J.*, **314**, 653.
Weber, E. & Davis, L. 1967, *Astrophys. J.*, **148**, 217.
Weekes, T.C. 1988, *Phys. Reports*, **160**, 1.
Wegner, G. 1972, *Astrophys. Letts.*, **12**, 219.
Weight, A., Evans, A., Naylor, T., Wood, J.H. & Bode, M.F. 1994, *Mon. Not. R. astr. Soc.*, **266**, 761.
Weiss, A. & Truran, J.W. 1990, *Astr. Astrophys.*, **238**, 178.
Wellmann, P. 1952, *Z. Astrophys.*, **31**, 123.
Wells, L.D. 1896, *Harv. Coll. Obs. Circ.*, No. 12.
Welsh, W.F., Horne, K. & Gomer, R. 1993, *Astrophys. J.*, **410**, L39.
Welsh, W.F., Horne, K. & Oke, J.B. 1993, *Astrophys. J.*, **406**, 229.
Wenzel, W. 1963, *Mitt. Ver. Sterne*, Nos. 754 and 755.
Wenzel, W.W. 1979, *Inf. Bull. Var. Stars*, No.1720.
Wenzel, W. 1983a, *Inf. Bull. Var. Stars*, No. 2262.
Wenzel, W. 1983b, *Mitt. Verand. Sterne*, **9**, 141.
Wenzel, W. 1987a, *Astr. Nach.* **308**, 75.
Wenzel, W. 1987b, *Inf. Bull. Var. Stars*, No. 3086.
Wenzel, W. 1993a, *Inf. Bull. Var. Stars*, No. 3874.
Wenzel, W. 1993b, *Inf. Bull. Var. Stars*, No. 3890.
Wenzel, W. & Fuhrmann, B. 1983, *Mitt. Verand. Sterne.*, 9, 175.
Wenzel, W. & Fuhrmann, B. 1989, *Inf. Bull. Var. Stars*, No. 3391.
Wenzel, W., Banny, M.I. & Andronov, I.L. 1988, *Mitt. Sonnenberg*, **11**, 141.
Wenzel, W., Hudec, R., Schult, R. & Tremko, J. 1993, *Contrib. Astr. Obs. Skalnaté Pleso,* **22**, 69.
Wesselink, A.J. 1939, *Astrophys. J.*, **89**, 659.
Wesselink, A.J. 1969, *Mon. Not. R. astr. Soc.*, **144**, 297.
West, S.C., Berriman, G. & Schmidt, G.D. 1987, *Astrophys. J.*, **322**, L35.
Westin, B.A.M. 1980, *Astr. Astrophys.*, **81**, 74.
Whelan, J.A.J., Rayne, M.W. & Brunt, C.C. 1979, *Int. Astr. Union Colloq. No. 46*, p. 39.
Wheeler, J.C. 1991, *Astr. Soc. Pacific Conf. Ser.*, **20**, 483.
White, N. 1981, *Astrophys. J.*, **244**, L85.
White, N.E. & Marshall, F.E. 1980, *Int. Astr. Union Circ.*, No. 3514.
White, N.E. & Marshall, F.E. 1981, *Astrophys. J.*, **249**, L25.
White, N.E. & Stella, L. 1987, *Mon. Not. R. astr. Soc.*, **231**, 325.
White, J.C., Honeycutt, R.K. & Horne, K. 1993, *Astrophys. J.*, **412**, 278.
Whitehurst, R. 1988a, *Mon. Not. R. astr. Soc.*, **233**, 529.
Whitehurst, R. 1988b, *Mon. Not. R. astr. Soc.*, **232**, 35.
Whitehurst, R. 1994, *Mon. Not. R. astr. Soc.*, **266**, 35.
Whitehurst, R. & King, A. 1991, *Mon. Not. R. astr. Soc.*, **249**, 25.

Whitehurst, R., Bath, G.T. & Charles, P.A. 1984, *Nature*, **309**, 768.
Whitelock, P.A. 1987, *Publ. astr. Soc. Pacific*, **99**, 573.
Whittaker, E.T. 1911, *Mon. Not. R. astr. Soc.*, **71**, 686.
Whyte, C.A. & Eggleton, P.P. 1980, *Mon. Not. R. astr. Soc.*, **190**, 801.
Wickramasinghe, D.T. 1988a, in *Polarized Radiation of Circumstellar Origin*, eds. G.V. Coyne *et al.*, Vatican Obs., Vatican, p. 3.
Wickramasinghe, D.T. 1988b, in *Polarized Radiation of Circumstellar Origin*, eds. G.V. Coyne *et al.*, Vatican Obs., Vatican, p. 199.
Wickramasinghe, D.T. 1990, in *Spectral Line Shapes*, eds. L. Frommhold & J. Keto, A.I.P. New York, **6**, 574.
Wickramasinghe, D.T. 1993, in *Cataclysmic Variables and Related Physics*, eds. O. Regev & G. Shaviv, Inst. Phys. Publ., Bristol, p. 213.
Wickramasinghe, D.T. & Ferrario, L. 1988, *Astrophys. J.*, **334**, 412.
Wickramasinghe, D.T. & Martin, B. 1979, *Mon. Not. R. astr. Soc.*, **188**, 165.
Wickramasinghe, D.T. & Martin, B. 1985, *Mon. Not. R. astr. Soc.*, **212**, 353.
Wickramasinghe, D.T. & Meggitt, S.M.A. 1982, *Mon. Not. R. astr. Soc.*, **198**, 975.
Wickramasinghe, D.T. & Meggitt, S.M.A. 1985a, *Mon. Not. R. astr. Soc.*, **214**, 605.
Wickramasinghe, D.T. & Meggitt, S.M.A. 1985b, *Mon. Not. R. astr. Soc.*, **216**, 857.
Wickramasinghe, D.T. & Visvanathan, N. 1980, *Mon. Not. R. astr. Soc.*, **191**, 589.
Wickramasinghe, D.T. & Wu, K. 1991, *Mon. Not. R. astr. Soc.*, **253**, 11P.
Wickramasinghe, D.T. & Wu, K. 1993, *Mon. Not. R. astr. Soc.*, **266**, L1.
Wickramasinghe, D.T., Ferrario, L. & Bailey, J. 1989, *Astrophys. J.*, **342**, L35.
Wickramasinghe, D.T., Reid, I.N. & Bessell, M.S. 1984, *Mon. Not. R. astr. Soc.*, **211**, 37P.
Wickramasinghe, D.T., Stobie, R.S. & Bessell, M.S. 1982, *Mon. Not. R. astr. Soc.*, **200**, 605.
Wickramasinghe, D.T., Tuohy, I.R. & Visvanathan, N. 1987, *Astrophys. J.*, **318**, 326.
Wickramasinghe, D.T., Visvanathan, N. & Tuohy, I.R. 1984, *Astrophys. J.*, **286**, 328.
Wickramasinghe, D.T., Wu, K. & Ferrario, L. 1991, *Mon. Not. R. astr. Soc.*, **249**, 460.
Wickramasinghe, D.T., Achilleos, N., Wu. K. & Boyle, B.J. 1990, *Int. Astr. Union Circ.*, No. 4962.
Wickramasinghe, D.T., Ferrario, L., Cropper, M.S. & Bailey, J. 1991a, *Mon. Not. R. astr. Soc.*, **251**, 137.
Wickramasinghe, D.T., Bailey, J., Meggitt, S.M.A., Ferrario, L., Hough, J. & Tuohy, I.R. 1991b, *Mon. Not. R. astr. Soc.*, **251**, 28.
Wickramasinghe, D.T., Cropper, M.S., Mason, K.O. & Garlick, M. 1991c, *Mon. Not. R. astr. Soc.*, **250**, 692.
Wickramasinghe, D.T., Ferrario, L., Bailey, J.A., Drissen, L., Dopita, M.A., Shara, M. & Hough, J.H. 1993. *Mon. Not. R. astr. Soc.*, **265**, L29.
Williams, R.E. 1977, *Int. Astr. Union Colloq. No. 42*, p. 242.
Williams, R.E. 1980, *Astrophys. J.*, **235**, 939.
Williams, R.E. 1982, *Astrophys. J.*, **261**, 170.
Williams, G.A. 1983, *Astrophys. J. Suppl.*, **53**, 523.
Williams, R.E. 1985, in *Production and Distribution of CNO Elements*, ed. I.J. Danziger, ESO, Garching, p. 225.
Williams, R.E. 1989, *Astr. J.*, **97**, 1752.
Williams, R.E. 1990, *Int. Astr. Union Colloq. No. 122*, p. 215.
Williams, G.A. 1991, *Astr. J.*, **101**, 1929.
Williams, R.E. 1992, *Astr. J.*, **104**, 725.
Williams, R.E. & Ferguson, D.H. 1982, *Astrophys. J.*, **257**, 672.
Williams, R.E. & Ferguson, D.H. 1983, *Int. Astr. Union Colloq. No. 72*, p. 97.
Williams, R.E. & Gallagher, J.S. 1979, *Astrophys. J.*, **228**, 482.
Williams, G. & Hiltner, W.A. 1980, *Publ. astr. Soc. Pacific*, **92**, 178.
Williams, G. & Hiltner, W.A. 1982, *Astrophys. J.*, **252**, 277.
Williams, G.A. & Hiltner, W.A. 1984, *Mon. Not. R. astr. Soc.*, **211**, 629.
Williams, G.A. & Shipman, H.L. 1988, *Astrophys. J.*, **326**, 738.
Williams, G.A., King, A.R. & Brooker, J.R.E. 1987, *Mon. Not. R. astr. Soc.*, **226**, 725.
Williams, G.A., King, A.R. & Watson, M.G. 1984, *Mon. Not. R. astr. Soc.*, **207**, 17P.
Williams, R.E., Phillips, M.M. & Hamuy, M. 1994, *Astrophys. J. Suppl.* **90**, 297.
Williams, R.E., Woolf, N.J., Hege, E.K., Moore, R.L. & Kopriva, D.A. 1978, *Astrophys. J.*, **224**, 171.

Williams, G., Johns, M., Price, C., Hiltner, A., Boley, F., Maker, S. & Mook, D. 1979, *Nature*, **281**, 48.
Williams, R.,E., Sparks, W.M., Gallagher, J.S., Ney, E.P., Starrfield, S. & Truran, J.W. 1981, *Astrophys. J.*, **251**, 221.
Williams, R.E, Ney, E.P., Sparks, W.M., Starrfield, S. & Truran, J.W. 1985, *Mon. Not. R. astr. Soc.*, **212**, 753.
Williams, R.E., Hamuy, M., Phillips, M.M., Heathcote, S.R., Wells, L. & Navarette, M. 1991, *Astrophys. J.*, **376**, 721.
Williger, G., Berriman, G., Wade, R.A. & Hassall, B.J.M. 1988, *Astrophys. J.*, **333**, 277.
Wilson, O.C. 1966a, *Astrophys. J.*, **144**, 695.
Wilson, O.C. 1966b, *Science*, **151**, 1487.
Wilson, J.W., Miller, R.H., Africano, J.L., Goodrich, B.D., Mahaffey, C.T. & Quigley, R.J. 1986, *Astr. Astrophys. Suppl.*, **66**, 323.
Winget, D.E. *et al.* 1994, *Astrophys. J.*, **430**, 839.
Winkler, L. 1977, *Astr. J.*, **82**, 1008.
Wlodarczyk, K. 1986, *Acta Astr.*, **36**, 395.
Wlodarczyk, K. 1988, in *Critical Observations versus Physical Models for Close Binary Systems*, ed. K.C. Leung, Gordon & Breach, New York, p. 371.
Wolf, B. 1977, *Int. Astr. Union Colloq. No. 42*, p. 151.
Wolf, S., Mantel, K.H., Horne, K., Barwig, H., Schoembs, R. & Baernbanter, O. 1993, *Astr. Astrophys.*, **273**, 160.
Wolff, M.T., Gardener, J. & Wood, K.S. 1989, *Astrophys. J.*, **346**, 833.
Wolff, M.T., Wood, K.S. & Imamura, J.N. 1991, *Astrophys. J.*, **375**, L31.
Wood, P.R. 1977, *Astrophys. J.*, **217**, 530.
Wood, J.H. 1987, *Mon. Not. R. astr. Soc.*, **228**, 797.
Wood, J.H. 1990a, private communication.
Wood, J.H. 1990b, *Mon. Not. R. astr. Soc.*, **243**, 219.
Wood, J.H. 1992, *Publ. astr. Soc. Pacific*, **104**, 780.
Wood, J.H. & Crawford, C.S. 1986, *Mon. Not. R. astr. Soc.*, **222**, 645.
Wood, J.H. & Horne, K. 1990, *Mon. Not. R. astr. Soc.*, **242**, 606.
Wood, J.H. & Marsh, T.R. 1991, *Astrophys. J.*, **381**, 551.
Wood, J.H. Abbott, T.M.C. & Shafter, A.W. 1992, *Astrophys. J.*, **393**, 729.
Wood, J.H., Horne, K. & Vennes, S. 1992, *Astrophys. J.*, **385**, 294.
Wood, K.S., Imamura, J.N. & Wolff, M.T. 1992, *Astrophys. J.*, **398**, 593.
Wood, J.H., Irwin, M.J. & Pringle, J.E. 1985, *Mon. Not. R. astr. Soc.*, **214**, 475.
Wood, J.H., Zhang, E.H. & Robinson, E.L. 1993, *Mon. Not. R. astr. Soc.*, **261**, 103.
Wood, J., Horne, K. Berriman, G., Wade, R., O'Donoghue, D. & Warner, B. 1986, *Mon. Not. R. astr. Soc.*, **219**, 629.
Wood, M.A., Winget, D.E., Nather, R.E., Hessman, F.V., Liebert, J., Kurtz, D.W., Wesemael, F. & Wegner, G. 1987, *Astrophys. J.*, **313**, 757.
Wood, J.H., Marsh, T.R., Robinson, E.L., Stiening, R.F., Horne, K., Stover, R.J., Schoembs, R., Allen, S.L., Bond, H.E., Jones, D.H.P., Grauer, A.D. & Ciardullo, R. 1989a, *Mon. Not. R. astr. Soc.*, **239**, 809.
Wood, J.H., Horne, K., Berriman, G. & Wade, R.A. 1989b, *Astrophys. J.*, **341**, 974.
Wood, J.H., Robinson, E.L., Bless, R.C., Dolan, J.F., Elliot, J.L. & van Citters, G.W. 1993, in *Cataclysmic Variables and Related Physics*, eds. O. Regev & G. Shaviv, Inst. Phys. Publ., Bristol, p. 19.
Woods, J.A., Drew, J.E. & Verbunt, F. 1990, *Mon. Not. R. astr. Soc.*, **245**, 323.
Woods, J.A., Verbunt, F., Collier Cameron, A., Drew, J.E. & Piters, A. 1992, *Mon. Not. R. astr. Soc.*, **255**, 237.
Woodward, C.E., Gehrz, R.D. Jones, T.J. & Lawrence, G.F. 1992, *Astrophys. J.*, **384**, L41.
Wright, A.E., Cropper, M.S., Stewart, R.T., Nelson, G.J. & Slee, O.B. 1988, *Mon. Not. R. astr. Soc.*, **231**, 319.
Wu, K. & Chanmugam, G. 1988, *Astrophys. J.*, **331**, 861.
Wu, K. & Chanmugam, G. 1989, *Astrophys. J.*, **344**, 889.
Wu, K. & Chanmugam, G. 1990, *Astrophys. J.*, **354**, 625.
Wu, C.-C. & Kester, D. 1977, *Astr. Astrophys.*, **58**, 331.

Wu, C.-C. & Panek, R.J. 1982, *Astrophys. J.*, **262**, 244.
Wu, C.-C. & Panek, R.J. 1983, *Astrophys. J.*, **271**, 754.
Wu, K. & Wickramasinghe, D.T. 1990, *Mon. Not. R. astr. Soc.*, **246**, 686.
Wu, K. & Wickramasinghe, D.T. 1991, *Mon. Not. R. astr. Soc.*, **252**, 386.
Wu, K. & Wickramasinghe, D.T. 1992, *Mon. Not. R. astr. Soc.*, **256**, 329.
Wu, K. & Wickramasinghe, D.T. 1993a, *Mon. Not. R. astr. Soc.*, **260**, 141.
Wu, K. & Wickramasinghe, D.T. 1993b, *Proc. Astr. Soc. Australia,* **10**, 325.
Wu, K. & Wickramasinghe, D.T. 1993c, *Mon. Not. R. astr. Soc.*, **265**, 115.
Wu, K., Chanmugam, G. & Shaviv, G. 1992, *Astrophys. J.*, **397**, 232.
Wu, K., Wickramasinghe, D.T. & Warner, B. 1994. *Mon. Not. R. astr. Soc.*, submitted.
Wu, C.-C., Panek, R.J., Holm, A.V., Raymond, J.C., Hartmann, L.W. & Swank, J.H. 1989, *Astrophys. J.*, **339**, 443.
Wyckoff, S. & Wehinger, P.A. 1977, *Int. Astr. Union Colloq. No. 42*, p. 201.
Wynn, G.A. & King, A.R. 1992, *Mon. Not. R. astr. Soc.*, **255**, 83.
Yamasaki, A., Okazaki, A. & Kitamura, M. 1983, *Pub. astr. Soc. Japan*, **35**, 423.
Yi, I., Soon-Wook, K., Vishniac, E.T. & Wheeler, J.C. 1992, *Astrophys. J.*, **391**, L25.
Yoshida, K., Inoue, H. & Osaki, Y. 1992, *Pub. astr. Soc. Japan,* **44**, 537.
Young, P. & Schneider, D.P. 1979, *Astrophys. J.*, **230**, 502.
Young, P. & Schneider, D.P. 1980, *Astrophys. J.*, **238**, 955.
Young, P. & Schneider, D.P. 1981, *Astrophys. J.*, **247**, 960.
Young, A., Ajir, F. & Thurman, G. 1989, *Publ. astr. Soc. Pacific*, **101**, 1017.
Young, P., Schneider, D.P. & Schectman, S.A. 1981a, *Astrophys. J.*, **245**, 1043.
Young, P., Schneider, D.P. & Schectman, S.A. 1981b, *Astrophys. J.*, **245**, 1035.
Young, P., Schneider, D.P. & Schectman, S.A. 1981c, *Astrophys. J.*, **244**, 259.
Young, P., Schneider, D.P., Sargent, W.L.W. & Boksenberg, A. 1982, *Astrophys. J.*, **252**, 269.
Yudin, B. & Munari, U. 1993, *Astr. Astrophys.*, **270**, 165.
Zahn, J.-P. 1966, *Ann. d'Astrophys.*, **29**, 489.
Zahn, J.-P. 1977, *Astr. Astrophys.*, **57**, 383.
Zahn, J.-P. 1989, *Astr. Astrophys.*, **220**, 112.
Zahn, J.-P. & Bouchet, L. 1989, *Astr. Astrophys.*, **223**, 112.
Zanstra, H. 1931, *Publ. Dom. Astrophys. Obs.*, **4**, 209.
Zapolsky, H.S. & Salpeter, E.E. 1969, *Astrophys. J.*, **158**, 809.
Zhang, E. 1989, *Chin. Astr. Astrophys.*, **13**, 2.
Zhang, Z.-Y. & Chen, J.-S. 1992, *Astr. Astrophys.*, **261**, 493.
Zhang, E.-H. & Robinson, E.L. 1987, *Astrophys. J.*, **321**, 813.
Zhang, E.-H. & Robinson, E.L. 1994, *Astr. Soc. Pacific Conf. Ser.* **56**, 358.
Zhang, E.-H., Robinson, E.L. & Nather, R.E. 1986, *Astrophys. J.*, **305**, 740.
Zhang, E., Robinson, E.L., Ramseyer, T.R., Shetrone, M.D. & Stiening, R.F. 1991, *Astrophys. J.*, **381**, 534.
Zinner, E. 1938, *Astr. Nach.*, **265**, 345.
Zola, S. 1989, *Acta Astr.*, **39**, 45.
Zuckerman, B. & Aller, L.H. 1986, *Astrophys. J.*, **301**, 772.
Zuckerman, M.-C. 1954, *Ann. d'Astrophys.*, **17**, 243.
Zuckerman, M.-C. 1961, *Ann. d'Astrophys.*, **24**, 431.
Zuckerman, B., Becklin, E.E., McLean, I.S. & Patterson, J. 1992, *Astrophys. J.*, **400**, 465.
Zwaan, C. 1986, in *Cool Stars, Stellar Systems and the Sun*, eds. M. Zelik & D. Gibson, Springer, Berlin, p. 19.
Zylstra, G. 1988, Ph.D. Thesis, Univ. Illinois.
Zylstra, G.J. 1989, *Bull. Amer. astr. Assoc.*, **21**, 1079.

Object index*

*The ordering follows that of *Astronomy and Astrophysics Abstracts*

Subject index

Printed in the United States
By Bookmasters